MATHEMATICAL METHODS
FOR PHYSICS
Using MATLAB and Maple

MATHEMATICAL METHODS

FOR PHYSICS

Using MATLAB and Maple

James R. Claycomb, PhD

MERCURY LEARNING AND INFORMATION

Dulles, Virginia
Boston, Massachusetts
New Delhi

Publisher: David Pallai
MERCURY LEARNING AND INFORMATION

22841 Quicksilver Drive
Dulles, VA 20166
info@merclearning.com
www.merclearning.com
(800) 232-0223

James R. Claycomb. *Mathematical Methods for Physics Using MATLAB and Maple*
ISBN: 978-1-68392-098-4

The publisher recognizes and respects all marks used by companies, manufacturers, and developers as a means to distinguish their products. All brand names and product names mentioned in this book are trademarks or service marks of their respective companies. Any omission or misuse (of any kind) of service marks or trademarks, etc. is not an attempt to infringe on the property of others.

Library of Congress Control Number: 2017960715

181920321   This book is printed on acid-free paper in the United States of America.

Our titles are available for adoption, license, or bulk purchase by institutions, corporations, etc. For additional information, please contact the Customer Service Dept. at 800-232-0223(toll free).

Dedicated to the memory of my Father,
Jackson Reynolds Claycomb (1928–2017),
Engineer, pilot, and inventor who flew out on
the wings of Hurricane Harvey

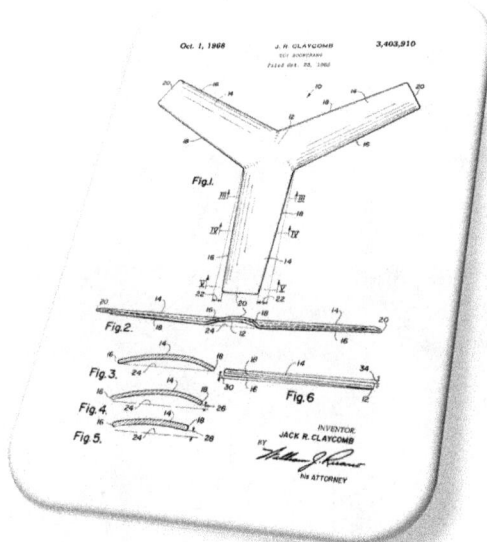

CONTENTS

PREFACE

The first part of this textbook (Chapters 1–9) includes a review of mathematical methods in physics that have direct applicability to problems encountered in physics courses at the undergraduate and beginning graduate level. Topics in the first nine chapters include basic mathematical operations (algebra, trigonometry, and complex numbers), vectors and matrices, calculus, vector calculus, ordinary differential equations, special functions, Fourier series and integral transformations, partial differential equations, and complex analysis. The second part of the textbook (Chapters 10–16) includes examples from classical mechanics, electrodynamics, quantum mechanics, statistical mechanics, special relativity, general relativity, and relativistic quantum mechanics. The independent learner will also benefit from the textbook in learning to apply Maple and MATLAB to his or her own research. The textbook is suitable for advanced undergraduates and beginning graduate students taking introductory courses on theoretical physics and mathematical methods in physics. The text may be used as a supplement to core physics classes and to supply reinforcement in mathematical techniques used in physics.

Some key features of the textbook include:
- Many simple examples that target specific skill sets
- End-of-chapter exercises that are specifically designed for skill building
- Examples demonstrating the use of Maple software at the end of most sections
- Examples using MATLAB software at the end of chapters
- Key examples using the Maple Physics package and the MATLAB PDE toolbox
- Mathematical analogies in table form help bridge the gap between topics such as electricity and magnetism
- The Lagrangian approach is demonstrated in chapters on mechanics, electromagnetics, special relativity, general relativity and relativistic quantum mechanics

It is left to the reader or the professor to decide if a given exercise should be solved using Maple, MATLAB, by hand, or by using a combination of approaches. One often finds that computer software only provides a partial result and that further analysis and reasoning is necessary to forward the problem and to check the validity of results.

1 **F**UNDAMENTALS

Chapter

Chapter Outline

1.1 Algebra
1.2 Trigonometry
1.3 Complex Numbers
1.4 Elements of Calculus

1.1 ALGEBRA

In this section, common algebraic equations in physics and astronomy are reviewed, including systems of equations and inverse, exponential and logarithmic functions. Techniques of completing the square, obtaining common denominators and partial fraction decomposition are covered. The numerical solution of transcendental equations is then discussed.

1.1.1 Systems of Equations

Many problems in introductory physics require the simultaneous solution of N equations for N unknowns. Systems of linear equations are solved by substitution, addition, and matrix methods.

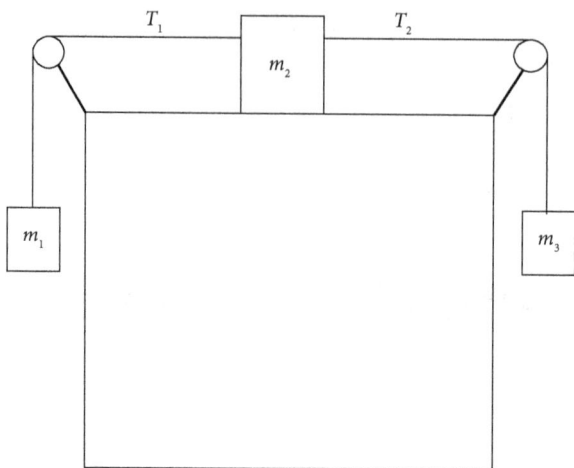

Figure 1.1.1: Three masses are connected by two cables that travel over frictionless pulleys. Gravity causes the system to accelerate with the top mass m_2 moving to the right. The coefficient of friction between m_2 and the table is μ_k.

Example 1.1.1

From the illustration in Figure 1.1.1, consider a system of three force equations with unknown tensions T_1, T_2 and acceleration a

$$T_1 - m_1 g = m_1 a$$

$$T_2 - \mu_k m_2 g - T_1 = m_2 a \qquad (1.1.1)$$

$$m_3 g - T_2 = m_3 a$$

Solve for the acceleration a for $m_3 > m_1$

Solution: We may use the addition method adding the left-hand and right-hand sides

$$m_3 g - m_1 g - \mu_k m_2 g = \left(m_1 + m_2 + m_3 \right) a \qquad (1.1.2)$$

thus eliminating the unknown tensions T_1 and T_2 with the resulting acceleration

$$a = \frac{m_3 g - m_1 g - \mu_k m_2 g}{\left(m_1 + m_2 + m_3 \right)}. \qquad (1.1.3)$$

We can verify this solution is consistent with Newton's second law with the acceleration inversely proportional to the total mass and directly proportional to the net force acting on the system. We could have also used the substitution method by solving the first and third equations for T_1 and T_2 and substituting those expressions into the second equation to obtain the acceleration. Once the acceleration is known, we can then find the unknown tensions.

1.1.2 Completing the Square

It is often desirable to express a quadratic form such as

$$y = ax^2 + bx \qquad (1.1.4)$$

as a squared term plus a constant. This is done by factoring the coefficient of x^2 and adding and subtracting half the coefficient of the linear term squared

$$y = a\left(x^2 + \frac{b}{a}x + \frac{b^2}{4a^2} \right) - \frac{b^2}{4a} \qquad (1.1.5)$$

thus

$$y = a\left(x + \frac{b}{2a} \right)^2 - \frac{b^2}{4a}. \qquad (1.1.6)$$

Graphically this equation corresponds to a parabola with vertex $(-b/2a, -b^2/4a)$. The technique of completing the square may be used to express exponential functions in a form that can be easily integrated. For example, after completing the square in the argument of the exponential function

$$e^{\left(ax^2 + bx\right)} = e^{-\frac{b^2}{4a}} e^{a\left(x + \frac{b}{2a}\right)^2}, \qquad (1.1.7)$$

a substitution method could then be applied in evaluating its integral.

1.1.3 Common Denominator

Rational expressions such as

$$\frac{1}{x} + \frac{1}{x+1} \qquad (1.1.8)$$

may be combined by obtaining a common denominator

$$\frac{1}{x}\left(\frac{x+1}{x+1} \right) + \frac{1}{x+1}\left(\frac{x}{x} \right) = \frac{2x+1}{x^2+x}. \qquad (1.1.9)$$

Example 1.1.2

Find the equivalent resistance R_{eff} of a parallel combination of resistors R_1 and R_2 where

$$\frac{1}{R_{eff}} = \frac{1}{R_1} + \frac{1}{R_2} \qquad (1.1.10)$$

Solution: Obtaining a common denominator

$$\frac{1}{R_{eff}} = \frac{1}{R_1}\frac{R_2}{R_2} + \frac{1}{R_2}\frac{R_1}{R_1} = \frac{R_2 + R_1}{R_1 R_2} \qquad (1.1.11)$$

we have

$$R_{eff} = \frac{R_1 R_2}{R_2 + R_1}. \qquad (1.1.12)$$

1.1.4 Partial Fractions Decomposition

Rational expressions may be separated using the method of partial fractions.

Example 1.1.3

Separate

$$\frac{1}{x^2 + x} = \frac{1}{x(x+1)} \qquad (1.1.13)$$

Solution: We write

$$\frac{1}{x(x+1)} = \frac{A}{x} + \frac{B}{x+1} \qquad (1.1.14)$$

where A and B are to be determined. Multiplying both sides by $x(x + 1)$

$$A(x + 1) + Bx = 1 \qquad (1.1.15)$$

and equating coefficients of powers of x on both sides gives $A = 1$, $A + B = 0$ and $B = -1$ so that

$$\frac{1}{x^2 + x} = \frac{1}{x} + \frac{-1}{x+1}. \qquad (1.1.16)$$

Partial fraction decompositions are useful for evaluating integrals with factorable polynomials in the denominator.

1.1.5 Inverse Functions

The functions $f(x)$ and $g(x)$ are inverse functions if

$$f(g(x)) = g(f(x)) = x \qquad (1.1.17)$$

The inverse of $f(x)$ is written as $f^{-1}(x)$ so that

$$f(f^{-1}(x)) = f^{-1}(f(x)) = x. \qquad (1.1.18)$$

The graphs of a function and its inverse have reflection symmetry about the line $y = x$. To compute the inverse of a function $f(x)$ we write it as $y = f(x)$, interchange $x \leftrightarrow y$ and solve for

$$y = f^{-1}(x) \qquad (1.1.19)$$

Example 1.1.4

In computing the inverse of

$$f(x) = \frac{1}{x-1} \text{ we write } y = \frac{1}{x-1} \qquad (1.1.20)$$

Solution: Interchanging $x \leftrightarrow y$ gives $x = \dfrac{1}{y-1}$ and $y = \dfrac{1}{x}+1$ thus

$$f^{-1}(x) = \frac{1}{x}+1. \qquad (1.1.21)$$

To verify that this is the inverse we check

$$f^{-1}(f(x)) = \frac{1}{\dfrac{1}{x-1}}+1 = x \text{ and } f\left(f^{-1}(x)\right) = \frac{1}{\dfrac{1}{x}+1-1} = x \qquad (1.1.22)$$

1.1.6 Exponential and Logarithmic Equations

The logarithm and exponential functions are inverse functions so that if

$$x = a^y \qquad (1.1.23)$$

we may solve for the exponent

$$y = \log_a x \qquad (1.1.24)$$

where a is the base of the logarithm. We frequently encounter base 10

$$\log_{10} x = \log x \qquad (1.1.25)$$

and base e, or natural logarithms

$$\log_e x = \ln x. \qquad (1.1.26)$$

Note that $\log_a a = \log 10 = \ln e = 1$. Also, $\log_a 1 = \log 1 = \ln 1 = 0$. To summarize the inverse properties of logarithms

$$\ln(e^x) = e^{\ln x} = x$$
$$\log(10^x) = 10^{\log x} = x \qquad (1.1.27)$$
$$\log_a(a^x) = a^{\log_a x} = x$$

1.1.7 Logarithms of Powers, Products and Ratios

An important power rule of logarithms is

$$\ln x^n = n\ln x. \tag{1.1.28}$$

Thus, $\ln 1000 = \ln 10^3 = 3\ln 10$.

The log of a product is the sum of logs

$$\ln(a \cdot b) = \ln a + \ln b \tag{1.1.29}$$

It is important to remember that $\ln(a + b) \neq \ln a + \ln b$.

The product of multiple terms such as

$$\ln\left(\prod_i a_i\right) = \sum_i \ln a_i \tag{1.1.30}$$

is often encountered in statistical mechanics.

The log of a ratio is given by a difference of logs

$$\ln\frac{b}{a} = \ln b - \ln a \tag{1.1.31}$$

so that

$$\ln\frac{1}{a} = \ln 1 - \ln a = -\ln a \tag{1.1.32}$$

or equivalently,

$$\ln\frac{1}{a} = \ln a^{-1} = -\ln a \tag{1.1.33}$$

by the power rule.

1.1.8 Radioactive Decay

Example 1.1.5

Calculate the half-life of radioactive nuclei. The number of radioactive nuclei N is treated as a continuous variable

$$N(t) = N_0 e^{-\lambda t} \tag{1.1.34}$$

where N_0 is the number of nuclei at $t = 0$ and λ is a decay constant that depends on the type of nuclei.

Solution: Setting $N(t) = \frac{1}{2}N_0$ we find the time t such that

$$e^{-\lambda t} = \frac{1}{2} \tag{1.1.35}$$

Taking the natural logarithm of both sides of this equation gives

$-\lambda t = \ln\frac{1}{2}$ and we obtain the half-life $t = \dfrac{\ln 2}{\lambda}$.

1.1.9 Transcendental Equations

Transcendental equations often involve algebraic or trigonometric equations that must be solved by graphical or numerical means. For example, the equation

$$5 = (5 - x)e^x. \tag{1.1.36}$$

is encountered in the theory of black-body radiation where $x = hc/\lambda k_B T$. This equation cannot be solved analytically. In quantum mechanics, the solutions to transcendental equations often give the energies of bound state systems.

1.1.10 Even and Odd Functions

The graphs of even functions have reflection symmetry about the y-axis. A function $f(x)$ is even if $f(-x) = f(x)$. Examples of even functions include

$$f(x) = \frac{1}{x^2 + 9}, \ f(x) = e^{-3|x|} \text{ and } f(x) = e^{-3x^2}. \tag{1.1.37}$$

The total area between $f(x)$ and the x-axis is the same to the left and to the right of the y-axis.

The graphs of odd functions are flipped when reflected about the y-axis. A function $f(x)$ is odd if $f(-x) = f(x)$. Examples of odd functions include

$$f(x) = \frac{x}{x^2 + 9}, \ f(x) = \frac{1}{x} \text{ and } f(x) = xe^{-3x^2} \tag{1.1.38}$$

An important property of odd functions is that the areas above and below the x-axis are equal. The product of two odd functions and the product of two even functions are even. The product of an even and an odd function is odd. Many functions such as $f(x) = x^2 + x^3$ are neither even nor odd.

1.1.11 Examples in Maple

The following worksheet illustrates basic algebraic operations in Maple including factoring, collecting and combining terms, expanding expressions, solving equations, and selecting parts of an expression. The numerical evaluation of fundamental constants is then demonstrated. Maple examples in this textbook are presented in "document mode" where a semicolon (;) following most Maple commands is not required. The reader is encouraged to view the Quick Reference Card from the Maple Help Menu to compare document mode vs. worksheet mode, common operations, and important syntax.

Key Maple operations: *assume, collect, combine, convert, denom, evalf, expand, factor, GetConstants, lhs, normal, numer, rhs, solve*

Maple packages: *with(ScientificConstants)*

restart

Factoring

$factor(x^3 - 1)$

$$(x - 1)(x^2 + x + 1)$$

$factor((x - 1)y + (x - 1)y^2 + (x - 1) \cdot y^3)$

$$(x - 1) \, y \, (y^2 + y + 1)$$

Collecting Terms

$poly := a \cdot x^2 - b \cdot x^2$

$$a \, x^2 - b \, x^2$$

$collect(poly, x)$

$$(a - b) \, x^2$$

Combining Terms

$combine(\ln(a) - \ln(b), \ln)$

$$\ln(a) - \ln(b)$$

$assume(a > 0, b > 0)$
$combine(\ln(a) - \ln(b), \ln)$

$$\ln\left(\frac{a \sim}{b \sim}\right)$$

$combine(\exp(x) \cdot \exp(y))$

$$e^{x+y}$$

$combine(\text{sqrt}(a + b) \cdot \text{sqrt}(a - b), radical)$

$$\sqrt{(a \sim + b \sim)(a \sim - b \sim)}$$

Expanding Expressions

$expand((x + y) \cdot (x - y^2))$

$$-xy^2 - y^3 + x^2 + xy$$

$expand\left(\dfrac{x^2 + 2x + 1}{x - 2}\right)$

$$\dfrac{x^2}{x-2} + \dfrac{2x}{x-2} + \dfrac{1}{x-2}$$

$expand((x - y) \cdot (q + r + s))$

$$qx + rx + sx - q - r - s$$

Partial Expansions

$expand((x - 1) \cdot (q + r + s), x - 1)$

$$(x - 1)q + (x - 1)r + (x - 1)s$$

Clear Variable

$a := \text{'}a\text{'}$

$$a$$

Solving Equations

$solve\left(\left\{\dfrac{1}{2} \cdot k \cdot x^2 = \dfrac{1}{2} m \cdot v^2 + \dfrac{1}{2} M \cdot V^2, M \cdot V + m \cdot v = 0\right\}, \{v, V\}\right);$

$$\left\{V = -\dfrac{m \, RootOf\left((Mm + m^2)\_Z^2 - kM\right)x}{M}, v = RootOf\left((Mm + m^2)\_Z^2 - kM\right)x\right\}$$

$convert(\%, radical)$

$$\left\{V = -\dfrac{m\sqrt{\dfrac{kM}{Mm + m^2}}\,x}{M}, v = \sqrt{\dfrac{kM}{Mm + m^2}}\,x\right\}$$

$eq1 := T1 - m1 \cdot g = m1 \cdot a$

$$-gm1 + T1 = m1a$$

$$eq2 := T2 - T1 - \mu_k \cdot m2 \cdot g = m2 \cdot a$$

$$-gm2\mu_k - T1 + T2 = m2a$$

$$eq3 := m3 \cdot g - T2 = m3 \cdot a$$

$$gm3 - T2 = m3a$$

$$solve(\{eq1, eq2, eq3\}, \{a, T1, T2\});$$

$$\left\{ T1 = -\frac{m1g(m2\mu_k - m2 - 2m3)}{m1 + m2 + m3}, T2 = -\frac{m3g(m2\mu_k + 2m1 + m2)}{m1 + m2 + m3}, \right.$$
$$\left. a = -\frac{g(m2\mu_k + m1 - m3)}{m1 + m2 + m3} \right\}$$

$$restart$$

Selecting Parts of an Expression

$$eq1 := m \cdot g \cdot h = \frac{1}{2} m \cdot v^2 + m \cdot g \cdot 2 \cdot R$$

$$mgh = \frac{1}{2} mv^2 + 2mgR$$

$$rhs(eq1)$$

$$\frac{1}{2} mv^2 + 2mgR$$

$$lhs(eq1)$$

$$mgh$$

$$term := \frac{m_1 - m_2}{m_1 + m_2} \cdot v_{i1}$$

$$\frac{(m_1 - m_2)v_{i1}}{m_1 + m_2}$$

$$numer(term)$$

$$(m_1 - m_2) v_{i1}$$

$$denom(term)$$

$$m_1 + m_2$$

$v_{i1} + term$

$$v_{i1} + \frac{(m_1 - m_2)v_{i1}}{m_1 + m_2}$$

$normal(v_{i1} + term)$

$$\frac{2v_{i1}m_1}{m_1 + m_2}$$

Numerical Evaluation

$convert(3.1416, rational)$

$$\frac{3927}{1250}$$

$evalf(\text{Pi}, 11)$

$$3.1415926536$$

$convert(\%, rational)$

$$\frac{103993}{33102}$$

$evalf(\%, 11)$

$$3.1415926530$$

$restart$

Fundamental Constants

$with(ScientificConstants)$

[AddConstant, AddElement, AddProperty, Constant, Element, GetConstant, GetConstants, GetElement, GetElements, GetError, GetIsotopes, GetProperties, GetProperty, GetUnit, GetValue, HasConstant, HasElement, HasProperty, ModifyConstant, ModifyElement]

$GetConstants(\ )$

$A[r](alpha), A[r](d), A[r](e), A[r](h), A[r](n), A[r](p), E_h, F, G, G_0, K_J, M_{Earth}, M_{Sun}, M_u, N_A, \Phi_0, R, R_{Earth}, R_K, R_\infty, V_m, Z_0, a_0, a_e, a_\mu, a, b, c, c_{1,L}, c_1, c_2, e, \epsilon_0, g, g_e, g_\mu, g_n, g_p, \gamma_e, \gamma_n, \gamma_p, gamma\_prime_h, gamma\_prime_p, h, \hbar, k, l_p, \lambda_{C,\mu}, \lambda_{C,n},$

$\lambda_{C,p}, \lambda_{C,\tau}, \lambda_C, m_p, m_\alpha, m_d, m_e, m[e]/m[mu], m_h, m_\mu, m_n, m_p, m_\tau, m[tau]c\wedge2,$
$m_u, \mu_0, \mu_B, \mu_N, \mu_d, mu[d]/mu[e], \mu_e, mu[e]/mu[p], mu[e]/mu\_prime[p],$
$\mu_\mu, \mu_n, mu[n]/mu\_prime[p], \mu_p, mu\_prime_h, mu\_prime[h]/mu\_prime[p],$
$mu\_prime_p, n_0, r_e, \sigma_p, \sigma_e, sigma\_prime_p, t_p$

$c := Constant(c, units)$

$$Constant_{SI}(c)\, \frac{\text{m}}{\text{s}}$$

$evalf(c)$

$$2.99792458\ 10^8\, \frac{\text{m}}{\text{s}}$$

$hbar := Constant(hbar, units)$

$$Constant_{SI}(\hbar)\, \frac{\text{m}^2\ \text{kg}}{\text{s}}$$

$G := Constant(G, units)$

$$Constant_{SI}(G)\, \frac{\text{m}^3}{\text{kg s}^2}$$

$Plancktime: =evalf\left(sqrt\left(\dfrac{hbar \cdot G}{c^3}\right)\right)$

$$1.616097442\ 10^{-35}\, \frac{\sqrt{\dfrac{\dfrac{\text{m}^2\ \text{kg}}{\text{s}}\ \dfrac{\text{m}^3}{\text{kg s}^2}}{\dfrac{\text{m}}{\text{s}}}}}{\dfrac{\text{m}}{\text{s}}}$$

$simplify(\%)$

$$1.616097442\ 10^{-35}\ \text{m}$$

$Planckmass: =simplify\left(evalf\left(sqrt\left(\dfrac{hbar \cdot c}{G}\right)\right)\right)$

$$2.17664105 \ 10^{-8} \ \text{kg}$$

$$Plancktime: =\text{simplify}\left(evalf\left(sqrt\left(\frac{hbar \cdot G}{c^5}\right)\right)\right)$$

$$5.390720810 \ 10^{-44} \ \text{s}$$

$$\epsilon_0: = Constant(\epsilon_0, units)$$

$$Constant_{SI}(\epsilon_0) \, \frac{A^2 s^4}{\text{kg m}^2}$$

$$e: = Constant(e, units)$$

$$Constant_{SI}(e) \ C$$

$$finestructure: =\text{simplify}\left(evalf\left(\frac{1}{4 \cdot Pi \cdot \epsilon_0} \cdot \frac{e^2}{hbar \cdot c}\right)\right)$$

$$0.007297352532$$

$$\frac{1}{finestructure}$$

$$137.0359998$$

1.2 TRIGONOMETRY

Trigonometry commonly encountered in physics is covered, including polar coordinate transformations and trigonometric identities; systems of equations involving trig functions are also reviewed below. The law of cosines and transcendental equations involving trigonometric equations are also discussed.

1.2.1 Polar Coordinates

In two dimensions, the Cartesian coordinates (x, y) are related to the polar coordinates (r, θ)

$$x = r \cos \theta \tag{1.2.1}$$

$$y = r \sin \theta \tag{1.2.2}$$

By dividing these equations, we obtain

$$\tan \theta = \frac{y}{x} \text{ so that } \theta = \tan^{-1}\left(\frac{y}{x}\right) \tag{1.2.3}$$

Squaring and adding x and y gives

$$x^2 + y^2 = r^2 \left(\cos^2 \theta + \sin^2 \theta\right) \tag{1.2.4}$$

or
$$x^2 + y^2 = r^2 \ \text{ since } \ \cos^2 \theta + \sin^2 \theta = 1 \tag{1.2.5}$$

1.2.2 Common Identities

The identities

$$\cos^2 \theta = \frac{1}{2}(1 + \cos 2\theta) \tag{1.2.6}$$

$$\sin^2 \theta = \frac{1}{2}(1 - \cos 2\theta) \tag{1.2.7}$$

frequently occur in electromagnetics and quantum mechanics. Adding these we verify

$$\sin^2 \theta + \cos^2 \theta = 1. \tag{1.2.8}$$

The double angle identity

$$\sin(2\theta) = 2\sin \theta \cos \theta \tag{1.2.9}$$

is encountered in introductory physics problems involving projectile motion. This identity can be verified by substitution of $\theta = \phi$ in the relation

$$\sin(\theta + \phi) = \sin \theta \cos \phi + \sin \phi \cos \theta . \tag{1.2.10}$$

Identities involving the addition of sine and cosine functions

$$\sin a + \sin b = 2 \sin\left(\frac{a+b}{2}\right)\cos\left(\frac{a-b}{2}\right) \tag{1.2.11}$$

$$\cos a + \cos b = 2 \cos\left(\frac{a+b}{2}\right)\cos\left(\frac{a-b}{2}\right) \tag{1.2.12}$$

are encountered in problems involving the superposition of waves resulting in constructive and destructive interference.

Example 1.2.1

Add the two waves

$$\psi_1 = A \cos(k_1 x + \omega_1 t) \qquad \psi_2 = A \cos(k_2 x + \omega_2 t) \tag{1.2.13}$$

in a region of space where the position of linear superposition holds.

Solution: Using identity (1.2.12) the superposition of the waves is

$$\psi_1 + \psi_2 = 2A\cos(\bar{k}x + \bar{\omega}t)\cos(\Delta kx + \Delta\omega t) \tag{1.2.14}$$

where $\bar{k} = (k_1 + k_2)/2$, $\bar{\omega} = (\omega_1 + \omega_2)/2$, $\Delta k = (k_1 - k_2)$ and $\Delta\omega = (\omega_1 - \omega_2)$.
$$\tag{1.2.15}$$

1.2.3 Law of Cosines

For triangles with sides a, b and c

$$a^2 = b^2 + c^2 - 2bc\cos\alpha \tag{1.2.16}$$

where a is the side opposite a. For right triangles, a is the hypotenuse and $a^2 = b^2 + c^2$. If $\alpha = \pi$ then $a^2 = (b + c)^2$ and if $\alpha = 0$ then $a^2 = (b - c)^2$.

The law of sines gives

$$\frac{\sin\alpha}{a} = \frac{\sin\beta}{b} = \frac{\sin\gamma}{c} \tag{1.2.17}$$

where the angles β and γ are opposite sides b and c, respectively.

1.2.4 Systems of Equations

Systems of equations involving trig functions such as

$$a\cos(\theta) + b\sin(\phi) = 0 \tag{1.2.18}$$
$$a\sin(\theta) + b\cos(\phi) = c \tag{1.2.19}$$

occur in mechanical equilibrium problems involving forces and torques. Many such systems can be solved by squaring and adding equations, dividing equations, and by substitution.

1.2.5 Transcendental Equations

Transcendental equations frequently involve trigonometric functions such as

$$\tan x = -\left(\frac{a^2}{x^2} - 1\right)^{-1/2} \tag{1.2.20}$$

encountered in quantum mechanics. These require a numerical solution or root finding.

Maple Examples

Basic trig operations are demonstrated in the following Maple worksheet, including expanding, combining and converting trig expressions, solving trigonometric equations, and numerically solving transcendental equations.

Key Maple operations: *combine, convert, expand, fsolve, plot, simplify, solve*

restart

Expanding Trig Expressions

expand$(\sin(q + r + s))$

$$\sin(q) \cos(r) \cos(s) - \sin(q) \sin(r) \sin(s) + \cos(q) \sin(r) \cos(s)$$
$$+ \cos(q) \cos(r) \sin(s)$$

expand$(2 \cdot x)$

$2\sin(x) \cos(x)$

Combining Trig Expressions

combine$(\sin(q) \cos(r) \cos(s) - \sin(q) \sin(r) \sin(s) + \cos(q) \sin(r) \cos(s)$
$+ \cos(q) \cos(r) \sin(s))$

$$(\sin(q + r + s)$$

Converting Trig Expressions

convert$(\text{sech}(x), \exp)$

$$\frac{2}{e^x + e^{-x}}$$

convert$(\sin(x)^3, \exp)$

$$\frac{1}{8} I \left(e^{Ix} - e^{-Ix} \right)^3$$

convert$\left(\frac{2}{e^x + e^{-x}}, trig \right)$

$$\frac{1}{\cosh(x)}$$

convert$\left(\frac{1}{8} I \left(e^{Ix} - e^{-Ix} \right)^3, trig \right)$

$$\sin(x)^3$$

Simplifying Trig Expressions

$Range: = \dfrac{v_i^2 \cdot 2\sin(theta) \cdot \cos(theta)}{g}$

$$\dfrac{2v_i^2 \sin(\theta)\cos(\theta)}{g}$$

$simplify(Range, \{2 \cdot \sin(theta) \cdot \cos(theta) = \sin(2 \cdot theta)\})$

$$\dfrac{v_i^2 \sin(2\theta)}{g}$$

Trig Equations

$eq1 := Ry \cdot \sin(theta) = a + b$

$$Ry \cdot \sin(\theta) = a + b$$

$eq\,2 := Rx \cdot \cos(theta) = a$

$$Rx \cdot \cos(\theta) = a$$

$eq\,3 := \dfrac{rhs(eq1)}{rhs(eq2)} = \dfrac{lhs(eq1)}{lhs(eq2)}$

$$\dfrac{a+b}{a} = \dfrac{Ry\sin(\theta)}{Rx\cos(\theta)}$$

$simplify(eq3)$

$$\dfrac{a+b}{a} = \dfrac{Ry\sin(\theta)}{Rx\cos(\theta)}$$

$simplify\left(eq3, \left\{\dfrac{\sin(theta)}{\cos(theta)} = \tan(theta)\right\}\right)$

$$\dfrac{a+b}{a} = \dfrac{Ry\tan(\theta)}{Rx}$$

$solve(eq3, theta)$

$$\arctan\left(\dfrac{Rx(a+b)}{Ry\,a}\right)$$

*eq 4: = A·*cos(delta) = 1

$$A \cos(\delta) = 1$$

*eq5 : = A·*sin(delta) = −1

$$A \sin(\delta) = -1$$

solve({*eq4, eq5*}, {*A·*delta})

$\{A = -2\ RootOf(2\_Z^2 - 1), \delta = \arctan(\ RootOf(2\_Z^2 - 1), -\ RootOf(2\_Z^2 - 1))\}$

convert(%, *radical*)

$$A = -\sqrt{2}, \delta = \frac{3}{4}\pi$$

restart

Transcendental Equations

$$plot\left(\{\tan(x), 3\cdot x\}, x = -\frac{Pi}{2}\dots\frac{Pi}{2}\right)$$

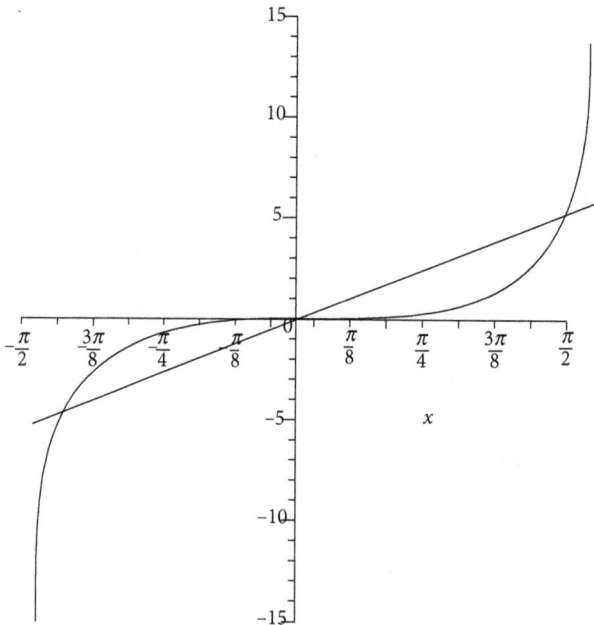

Figure 1.2.1: Plot of the functions tan(x) and 3x.

$$fsolve\left(\tan(x) = 3\cdot x, x = \frac{Pi}{4}\dots\frac{Pi}{2}\right)$$

$$1.324194450$$

$$fsolve\left(\tan(x)=3\cdot x, x=-\frac{Pi}{2}...-\frac{Pi}{4}\right)$$

$$-1.324194450$$

$$fsolve\left(\tan(x)=3\cdot x, x=-\frac{Pi}{8}...\frac{Pi}{8}\right)$$

$$0.$$

$$TransEqn: = \tan(x)=\frac{1}{x}$$

$$\tan(x)=\frac{1}{x}$$

$$plot\left(\{rhs(TransEqn),\ lhs(TransEqn)\}, x=-\frac{Pi}{2}...\frac{Pi}{2}\right)$$

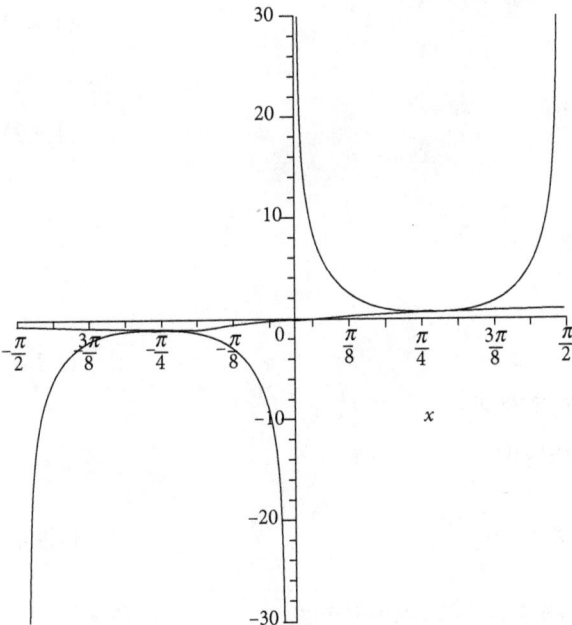

Figure 1.2.2: Plot of the functions tan(x) and 1/x.

$$fsolve\left(TransEqn, x=\frac{Pi}{4}...\frac{Pi}{2}\right)$$

$$0.8603335890$$

$$fsolve\left(TransEqn, x = -\frac{Pi}{2}...-\frac{Pi}{4}\right)$$

$$-0.8603335890$$

1.3 COMPLEX NUMBERS

Complex numbers have applicability in most areas of physics, including both classical and quantum mechanics as well as electricity and magnetism. In this section, we review some of the most basic and essential properties of complex numbers required for physics and differential equations.

1.3.1 Complex Roots

The solution to the quadratic equation

$$az^2 + bz + c = 0 \text{ is } z = \frac{-b \pm \sqrt{b^2 - 4ac}}{2a}. \tag{1.3.1}$$

If $b^2 < 4ac$ then $z = \frac{-b \pm i\sqrt{4ac - b^2}}{2a}$ \hfill (1.3.2)

where $i = \sqrt{-1}$.

Example 1.3.1

Solve

$$z^2 + z + 1 = 0 \tag{1.3.3}$$

and find the real and imaginary parts of z.

Solution: The quadratic equation gives

$$z = \frac{-1 \pm \sqrt{3}i}{2} \tag{1.3.4}$$

with $\text{Re}(z) = -1/2$ and $\text{Im}(z) = \pm\sqrt{3}/2$, respectively.

1.3.2 Complex Arithmetic

Any complex number can be written as a sum of real and imaginary parts. Given two complex numbers

$z_1 = a + ib$ and $z_2 = c + id$ if $z_1 = z_2$ then $a = c$ and $b = d$.

Addition and subtraction of complex numbers:

$$z_1 \pm z_2 = (a \pm c) + i(b \pm d) \tag{1.3.5}$$

Powers of i:

$$
\begin{aligned}
i^0 &= 1 \\
i^1 &= i \\
i^2 &= -1 \\
i^3 &= i^2\, i = -i \\
i^4 &= i^2\, i^2 = 1 \\
i^5 &= i^4\, i = i
\end{aligned}
\tag{1.3.6}
$$

Example 1.3.2

Calculate i^{19}

Solution: Since $i^4 = 1$ we can write

$$i^{19} = (i^4)^4 i^3 = -i \tag{1.3.7}$$

1.3.3 Complex Conjugate

Given the complex number

$$z = x + iy \tag{1.3.8}$$

we may form the complex conjugate of z (denoted as z^*) by replacing i by $-i$

$$z^* = x - iy. \tag{1.3.9}$$

Note that the product of a number and its complex conjugate will always be a real number

$$z^* z = (x - iy)(x + iy) = x^2 + y^2. \tag{1.3.10}$$

This is the same as zz^*.

Adding a complex number to its complex conjugate will also give a real number

$$z + z^* = x + iy + x - iy = 2x. \tag{1.3.11}$$

Subtracting z and z^* gives a pure imaginary number

$$z - z^* = x + iy - (x - iy) = 2iy. \tag{1.3.12}$$

Any function of a complex number can be written as a sum of real and imaginary parts

$$f(z) = \text{Re } f(z) + i \text{ Im } f(z). \tag{1.3.13}$$

1.3.4 Euler's Formula

Euler's formula relates the circular sine and cosine functions to a complex exponential function

$$e^{i\theta} = \cos(\theta) + i\sin(\theta). \tag{1.3.14}$$

Since $\cos(\theta)$ is an even function and $\sin(\theta)$ is an odd function

$$e^{-i\theta} = \cos(\theta) - i\sin(\theta). \tag{1.3.15}$$

Adding equations (1.3.14) and (1.3.15), we obtain

$$\cos(\theta) = \frac{e^{i\theta} + e^{-i\theta}}{2}. \tag{1.3.16}$$

Subtracting these equations, we find

$$\sin(\theta) = \frac{e^{i\theta} - e^{-i\theta}}{2i}. \tag{1.3.17}$$

1.3.5 Complex Plane

A complex number $z = x + iy$ may be graphed on the complex plane where

$$\text{Re}(z) = x \text{ and } \text{Im}(z) = y. \tag{1.3.18}$$

We define the magnitude of a complex number as

$$|z| = \sqrt{z^* z} = \sqrt{x^2 + y^2} = r \tag{1.3.19}$$

and

$$\theta = \tan^{-1}\frac{\text{Im}(z)}{\text{Re}(z)}. \tag{1.3.20}$$

1.3.6 Polar Form of Complex Numbers

Euler's relation enables us to represent a complex number z in polar coordinates $x = r\cos\theta$ and $y = r\sin\theta$

$$z = x + iy = r(\cos\theta + i\sin\theta) \tag{1.3.21}$$

or

$$z = re^{i\theta}. \tag{1.3.22}$$

Example 1.3.3

Express $z = 1 + i$ in polar form

Solution: $|z| = \sqrt{2}$ and $\theta = \tan^{-1}(1) = \pi / 4$ thus

$$z = \sqrt{2} e^{i\pi/4}. \tag{1.3.23}$$

1.3.7 Powers of Complex Numbers

To compute powers of a complex number we use polar form

$$z^n = r^n e^{in\theta}. \tag{1.3.24}$$

Roots are computed as

$$\sqrt[n]{z} = r^{1/n} e^{i\theta/n}. \tag{1.3.25}$$

Example 1.3.4

Write $\sqrt{1+i}$ in polar form

Solution: $z = 1 + i = \sqrt{2} e^{i\pi/4}$ so that $z^{1/2} = 2^{1/4} e^{i\pi/8}$. \hfill (1.3.26)

1.3.8 Hyperbolic Functions

Hyperbolic functions involve real exponentials where circular functions are expressed as exponentials with imaginary arguments. Table 1.3.1 compares circular and hyperbolic functions.

Table 1.3.1: Circular and hyperbolic functions.

$\sin(\theta) = \dfrac{e^{i\theta} - e^{-i\theta}}{2i}$	$\sinh(\theta) = \dfrac{e^{\theta} - e^{-\theta}}{2}$
$\cos(\theta) = \dfrac{e^{i\theta} + e^{-i\theta}}{2}$	$\cosh(\theta) = \dfrac{e^{\theta} + e^{-\theta}}{2}$
$\tan(\theta) = \dfrac{\sin(\theta)}{\cos(\theta)}$	$\tanh(\theta) = \dfrac{\sinh(\theta)}{\cosh(\theta)}$
$\csc(\theta) = \dfrac{1}{\sin(\theta)}$	$\operatorname{csch}(\theta) = \dfrac{1}{\sinh(\theta)}$
$\sec(\theta) = \dfrac{1}{\cos(\theta)}$	$\operatorname{sech}(\theta) = \dfrac{1}{\cosh(\theta)}$

(Contd.)

$\cot(\theta) = \dfrac{\cos(\theta)}{\sin(\theta)}$	$\coth(\theta) = \dfrac{\cosh(\theta)}{\sinh(\theta)}$
$\cos^2(\theta) + \sin^2(\theta) = 1$	$\cosh^2(\theta) - \sinh^2(\theta) = 1$

From Table 1.3.1 we see that the hyperbolic cosine is an even function where $\cosh(-x) = \cosh(x)$. The hyperbolic sine is an odd function $\sinh(-x) = -\sinh(x)$. The hyperbolic tangent and cotangent are also odd functions.

Maple Examples

Operations involving complex numbers are demonstrated in the following Maple worksheet. Examples include factoring, floating point evaluation, finding the magnitude, real and imaginary parts of complex numbers, defining a complex function, and conversion of complex numbers to trigonometric and polar forms.

Key Maple commands: *abs, Complex, convert, evalf, Im, Re, solve*

Programming: Functional operators

restart

Complex Numbers

solve($x^2 + x + 1 = 0, x$)

$$-\frac{1}{2} + \frac{1}{2}I\sqrt{3}, -\frac{1}{2} - \frac{1}{2}I\sqrt{3}$$

evalf(I^I)

$$0.2078795764 + 0.I$$

evalf($\exp(I \cdot Pi)$)

$$-1.$$

factor($x^2 + x + 1 = 0, complex$)

$$(x + 0.5000000000 + 0.8660254038I)\,(x + 0.5000000000 - 0.8660254038I)$$

psi: $= 2 + 3 \cdot I$

$$2 + 3\,I$$

psi: $= Complex(2, 3)$

$$2 + 3\,I$$

abs(psi)

$$\sqrt{13}$$

sqrt(psi·*conjugate*(psi))

$$\sqrt{13}$$

Re(psi)

2

Im(psi)

3

Functions of Complex Numbers

$f := (x, y) \rightarrow \exp(I \cdot x)^{\exp(I \cdot y)}$

$$(x, y) \rightarrow \left(e^{Ix}\right)^{e^{Iy}}$$

Converting Complex Numbers

convert(exp(I·x + y), *trig*)

$$(\cosh(y) + \sinh(y))(\cos(x) + I \sin(x))$$

$z := Complex(1,1)$

$$1 + I$$

convert(z, polar)

$$polar\left(\sqrt{2}, \frac{1}{4}\pi\right)$$

1.4 ELEMENTS OF CALCULUS

Basic techniques of integration and differentiation are reviewed in this section with examples.

1.4.1 Derivatives

The derivative of a function $f(x)$ with respect to x can be thought of as the instantaneous rate of change of the function as x is varied. The derivative is defined as

$$\frac{df}{dx} = \lim_{\Delta x \to 0} \frac{f(x + \Delta x) - f(x)}{\Delta x} \tag{1.4.1}$$

and is numerically equal to the slope of a line tangent to the graph of $f(x)$ vs. x.

Example 1.4.1

Use the definition of the derivative to find df/dx where $f(x) = x^2$.

Solution:

$$\frac{df}{dx} = \lim_{\Delta x \to 0} \frac{(x + \Delta x)^2 - (x)^2}{\Delta x}$$

$$= \lim_{\Delta x \to 0} \frac{2x\Delta x + \Delta x^2}{\Delta x} = 2x \qquad (1.4.2)$$

1.4.2 Prime and Dot Notation

Prime and dot notation provides a shorthand for writing derivatives with respect to x and t, respectively. The derivative of $f(x)$ with respect to x is written compactly as

$$\frac{df(x)}{dx} = f'(x). \qquad (1.4.3)$$

A dot denotes the time derivative of a function

$$\frac{df(t)}{dt} = \dot{f}(t). \qquad (1.4.4)$$

1.4.3 Chain Rule for Derivatives

The derivative of a function composition $f(g(x))$ is

$$\frac{d}{dx} f(g(x)) = g'(x)f'(g(x)). \qquad (1.4.5)$$

Thus

$$\frac{d}{dx} f(ax) = (ax)' f'(ax) = af'(ax). \qquad (1.4.5)$$

1.4.4 Product Rule for Derivatives

The derivative of a product of functions $f(x)\,g(x)$ is

$$\frac{d}{dx}\left[f(x)g(x)\right] = f'(x)g(x) + f(x)g'(x). \qquad (1.4.6)$$

1.4.5 Quotient Rule for Derivatives

The derivative of a ratio $f(x)/g(x)$ is

$$\frac{d}{dx}\frac{f(x)}{g(x)} = \frac{f'(x)g(x) - g'(x)f(x)}{g(x)^2}. \qquad (1.4.7)$$

1.4.6 Indefinite Integrals

Given that the function $g(x)$ is the derivative of $f(x)$

$$g(x) = \frac{df(x)}{dx} \tag{1.4.8}$$

The function $f(x)$ is said to be the antiderivative of $g(x)$

$$f(x) = \int g(x)dx + C \tag{1.4.9}$$

where C is a constant.

1.4.7 Definite Integrals

If $g(x)$ is continuous over an interval $[a, b]$ and

$$f(x) = \int g(x)dx + C \tag{1.4.10}$$

the first fundamental theorem of calculus states that

$$\int_a^b g(x)dx = f(b) - f(a). \tag{1.4.11}$$

Geometrically the definite integral above corresponds to the area under the curve $g(x)$ between $x = a$ and $x = b$.

1.4.8 Common Integrals and Derivatives

Table 1.4.1 gives some common derivatives and integrals routinely encountered in introductory physics.

Table 1.4.1: Common integrals and derivatives.

$\dfrac{d}{dx}x^n = nx^{n-1}$	$\displaystyle\int x^n dx = \frac{1}{n+1}x^{n+1} + C$
$\dfrac{d}{dx}\sin(ax) = a\cos(ax)$	$\displaystyle\int \sin(ax)d\theta = -\frac{1}{a}\cos(ax) + C$
$\dfrac{d}{dx}\cos(ax) = -a\sin(ax)$	$\displaystyle\int \cos(ax)d\theta = \frac{1}{a}\sin(ax) + C$
$\dfrac{d}{dx}e^{ax} = ae^{ax}$	$\displaystyle\int e^{ax}dx = \frac{1}{a}e^{ax} + C$
$\dfrac{d}{dx}\ln(ax) = \dfrac{a}{x}$	$\displaystyle\int \frac{1}{x}dx = \ln x + C$

1.4.9 Derivatives of Trigonometric and Hyperbolic Functions

Table 1.4.2 compares derivatives of trigonometric and hyperbolic functions. This table may also be used to determine antiderivatives such as

$$\int \frac{1}{\sqrt{1-x^2}}dx = \sin^{-1}(x)+C \quad \text{and} \quad \int \frac{1}{\sqrt{1+x^2}}dx = \sinh^{-1}(x)+C. \quad (1.4.12)$$

Table 1.4.2: Derivatives of trigonometric and hyperbolic functions.

$\dfrac{d}{dx}\sin(x) = \cos(x)$	$\dfrac{d}{dx}\sinh(x) = \cosh(x)$
$\dfrac{d}{dx}\cos(x) = -\sin(x)$	$\dfrac{d}{dx}\cosh(x) = \sinh(x)$
$\dfrac{d}{dx}\tan(x) = \sec^2(x)$	$\dfrac{d}{dx}\tanh(x) = \operatorname{sech}^2(x)$
$\dfrac{d}{dx}\cot(x) = -\csc^2(x)$	$\dfrac{d}{dx}\coth(x) = -\operatorname{csch}^2(x)$
$\dfrac{d}{dx}\sec(x) = \sec(x)\tan(x)$	$\dfrac{d}{dx}\operatorname{sech}(x) = -\operatorname{sech}(x)\tanh(x)$
$\dfrac{d}{dx}\csc(x) = -\csc(x)\cot(x)$	$\dfrac{d}{dx}\operatorname{csch}(x) = -\operatorname{csch}(x)\coth(x)$
$\dfrac{d}{dx}\sin^{-1}(x) = \dfrac{1}{\sqrt{1-x^2}}$	$\dfrac{d}{dx}\sinh^{-1}(x) = \dfrac{1}{\sqrt{1+x^2}}$
$\dfrac{d}{dx}\cos^{-1}(x) = -\dfrac{1}{\sqrt{1-x^2}}$	$\dfrac{d}{dx}\cosh^{-1}(x) = \dfrac{1}{\sqrt{x^2-1}}$
$\dfrac{d}{dx}\tan^{-1}(x) = \dfrac{1}{1+x^2}$	$\dfrac{d}{dx}\tanh^{-1}(x) = \dfrac{1}{1-x^2}$

1.4.10 Euler's Formula

In polar coordinates, $(x, y) = (r \cos \theta, r \sin \theta)$ we can express z

$$z = x + iy = r(\cos \theta + i\sin \theta). \quad (1.4.13)$$

Taking the derivative of z with respect to θ

$$\frac{dz}{d\theta} = r\left(-\sin\theta + i\cos\theta\right) \quad (1.4.14)$$

or

$$\frac{dz}{d\theta} = iz. \quad (1.4.15)$$

This equation may be integrated

$$\int \frac{dz}{z} = i \int d\theta \qquad (1.4.16)$$

giving

$$\ln(z) = i\theta + \text{const.} \qquad (1.4.17)$$

Exponentiating both sides

$$z = \text{const.} \times e^{i\theta} \qquad (1.4.18)$$

with

$$z = re^{i\theta} \qquad (1.4.19)$$

we have Euler's formula

$$e^{i\theta} = \cos(\theta) + i\sin(\theta). \qquad (1.4.20)$$

1.4.11 Integrals of Trigonometric and Hyperbolic Functions

Table 1.4.3 compares integrals of trigonometric and hyperbolic functions. This table may also be used to determine derivatives such as

$$\frac{d}{dx}\ln\sin(x) = \cot(x) \quad \text{and} \quad \frac{d}{dx}\ln\sinh(x) = \coth(x) \qquad (1.4.21)$$

Table 1.4.3: Integrals of trigonometric and hyperbolic functions.

$\int \sin(x)dx = -\cos(x) + C$	$\int \sinh(x)dx = \cosh(x) + C$
$\int \cos(x)dx = \sin(x) + C$	$\int \cosh(x)dx = \sinh(x) + C$
$\int \tan(x)dx = -\ln\cos(x) + C$	$\int \tanh(x)dx = \ln\cosh(x) + C$
$\int \cot(x)dx = \ln\sin(x) + C$	$\int \coth(x)dx = \ln\sinh(x) + C$
$\int \sec^2(x)dx = \tan(x) + C$	$\int \operatorname{sech}^2(x)dx = \tanh(x) + C$
$\int \csc^2(x)dx = -\cot(x) + C$	$\int \operatorname{csch}^2(x)dx = -\coth(x) + C$

1.4.12 Improper Integrals

Integrals over infinite intervals are defined as limits such as

$$\int_0^\infty f(x)dx = \lim_{a \to \infty} \int_0^a f(x)dx \cdot \qquad (1.4.22)$$

The integral converges if the limit is finite; otherwise the integral is divergent. One requirement for the integral above to converge is

$$\lim_{x \to \infty} f(x) = 0 \qquad (1.4.23)$$

although the integral may still diverge even if the integrand goes to zero at infinity. For example,

$$\int_1^\infty \frac{1}{x} dx = \lim_{a \to \infty} \ln(a) = \infty . \qquad (1.4.24)$$

The limit notation is frequently omitted in physics textbooks when dealing with common improper integrals such as Gaussian integrals. Improper integrals also occur when the integrand has a vertical asymptote.

1.4.13 Integrals of Even and Odd Functions

Because even functions $f_{even}(-x) = f_{even}(x)$ are symmetric about the y-axis, we can write integrals with symmetric limits

$$\int_{-L}^L f_{even}(x)dx = 2\int_0^L f_{even}(x)dx \qquad (1.4.25)$$

Thus if $\int_0^\infty e^{-x^2} dx = \frac{1}{2}\sqrt{\pi}$ we can evaluate $\int_{-\infty}^\infty e^{-x^2} dx = \sqrt{\pi}$ $\qquad (1.4.26)$

Odd functions $f_{odd}(-x) = -f_{odd}(x)$ are antisymmetric about the y-axis, with just as much area above the x-axis as below it over symmetric intervals so that

$$\int_{-L}^L f_{odd}(x)dx = 0 \qquad (1.4.27)$$

Thus, integrals such as

$$\int_{-\infty}^\infty xe^{-x^2} dx, \quad \int_{-\pi}^\pi \sin(x)dx \quad \text{and} \quad \int_{-\pi}^\pi x\cos(x)dx \qquad (1.4.28)$$

are immediately evaluated as zero.

Maple Examples

Basic calculus operations demonstrated in the Maple worksheet below include the evaluation of limits and the definition of a derivative, first derivatives, higher derivatives, evaluation of derivatives at a point, determining maxima and minima of functions, indefinite integrals, checking integrals by differentiation, definite integrals, and the numerical evaluation of integrals.

Key Maple commands: *assume, combine, D, diff, evalf, int, limit, RiemannSum, simplify*

Maple packages: *with(Student[Calculus1])*

Programming: Functional operators

restart

Limits

$$Limit\left(\frac{\sin(x)}{x}, x=0\right) = limit\left(\frac{\sin(x)}{x}, x=0\right)$$

$$\lim_{x\to0}\frac{\sin(x)}{x} = 1$$

$$Limit\left(\frac{\sin(x)^2}{x}, x=0\right) = limit\left(\frac{\sin(x)^2}{x}, x=0\right)$$

$$\lim_{x\to0}\frac{\sin(x)^2}{x} = 0$$

Directional Limits

$$Limit\left(\frac{1}{x}, x=0, right\right) = limit\left(\frac{1}{x}, x=0, right\right)$$

$$\lim_{x\to0^+}\frac{1}{x} = \infty$$

$$Limit\left(\frac{1}{x}, x=0, left\right) = limit\left(\frac{1}{x}, x=0, left\right)$$

$$\lim_{x\to0^-}\frac{1}{x} = -\infty$$

Definition of Derivative

$$f := x \rightarrow x^3 + 2 \cdot x + 10 \cdot \sin(x) \cdot \cos(x)$$

$$x \rightarrow x^3 + 2x + 10 \sin(x) \cos(x)$$

$$Limit\left(\frac{(f(x+dx)-f(x))}{dx}, dx = 0\right)$$

$$3x^2 - 10\sin(x)^2 + 10\cos(x)^2 + 2$$

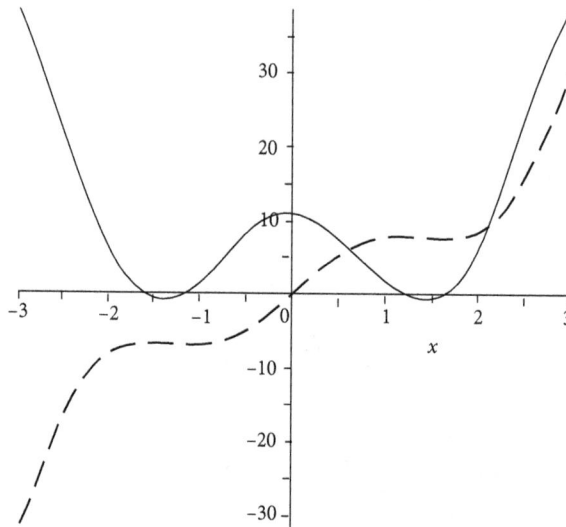

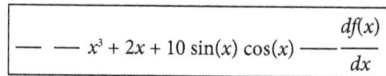

| | $x^3 + 2x + 10 \sin(x) \cos(x)$ | $\dfrac{df(x)}{dx}$ |

Figure 1.4.1: Plot of a function and its derivative.

restart

First Derivatives

$$diff(\exp(a \cdot x), x)$$

$$a\, e^{ax}$$

$$Diff(\exp(\sin(\exp(a \cdot x))), x) = diff(\exp(\sin(\exp(a \cdot x))), x)$$

$$\frac{\partial}{\partial x} e^{\sin(e^{ax})} = a e^{ax} \cos\left(e^{ax}\right) e^{\sin(e^{ax})}$$

$diff(\log(\cosh(x)), x)$

$$\frac{\sinh(x)}{\cosh(x)}$$

Higher Derivatives

$diff(\exp(a \cdot x), x, x)$

$$a^2\, e^{a\,x}$$

$diff(\exp(a \cdot x), x\$2)$

$$a^2\, e^{a\,x}$$

$diff(\exp(a \cdot x), x, x, x, x, x)$

$$a^5\, e^{a\,x}$$

$diff(\exp(a \cdot x), x\$5)$

$$a^5\, e^{a\,x}$$

Derivatives at a Point

$f := (x) \rightarrow \sin(x) \cdot x^2$

$$x \rightarrow \sin(x)\, x^2$$

$diff(f(x), x)$

$$\cos(x)\, x^2 + 2\, \sin(x)\, x$$

$subs(x = a, diff(f(x), x))$

$$\cos(a)\, a^2 + 2\, \sin(a)\, a$$

$D(f)\, (a)$

$$\cos(a)\, a^2 + 2\, \sin(a)\, a$$

restart

Determining Maxima and Minima

$f := (x) \rightarrow x^4 - 2 \cdot x^2$

$$x \rightarrow x^4 - 2\, x^2$$

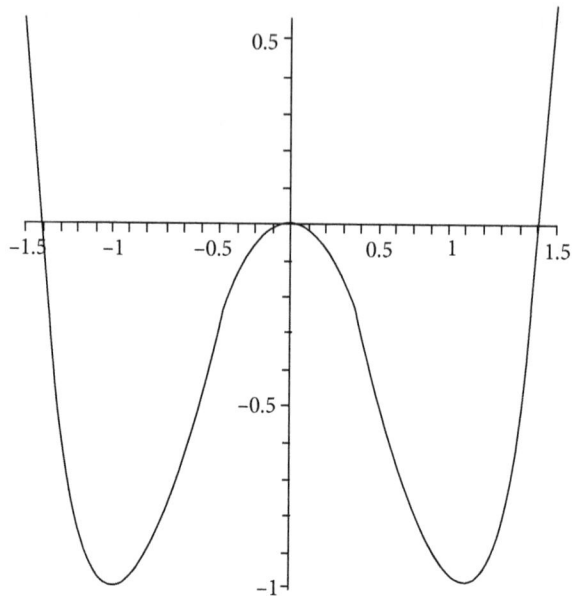

Figure 1.4.2: Plot of a function with one maximum and two minima.

$diff(f(x), x)$

$$4\,x^3 - 4\,x$$

$solve(\% = 0, x)$

$$0, 1, -1$$

Second Derivative Test

$D(D(f))\,(-1)$

$$8$$

$D(D(f))\,(0)$

$$-4$$

$D(D(f))\,(1)$

$$8$$

Thus $x = $ -1, 0, 1 corresponds to min, max and min values, respectively.

Indefinite Integrals

$int(\sin(x), x)$

$$- \cos(x)$$

$$f := (x) \to x \exp(x^2)$$

$$x \to xe^{x^2}$$

$$Int(f(x), x) = int(f(x), x)$$

$$\int xe^{x^2}\, dx = \frac{1}{2}e^{x^2}$$

Checking Integrals by Differentiation

$$g := int\left(\frac{1}{\sin(x)}, x\right)$$

$$\ln(\csc(x) - \cot(x))$$

$$h := diff(g,x)$$

$$\frac{-\csc(x)\cot(x)+1+\cot(x)^2}{\csc(x)-\cot(x)}$$

simplify(h)

$$\frac{1}{\sin(x)}$$

$$w := int(\ln(x),x)$$

$$x\ln(x) - x$$

diff(w, x)

$$\ln(x)$$

Definite Integrals

$$Int(\exp(-x), x = 0...infinity) = int(\exp(-x), x = 0...infinity)$$

$$\int_0^\infty e^{-x}dx = 1$$

$$Int\left(\frac{\sin(x)}{x}, x = -infinity...infinity\right) = int\left(\frac{\sin(x)}{x}, x = -infinity...infinity\right)$$

$$\int_{-\infty}^\infty \frac{\sin(x)}{x}dx = \pi$$

$$int\left(\frac{1}{x}, x = a...b\right)$$

Warning, unable to determine if 0 is between a and b; try to use
assumptions or use the AllSolutions option

$$\int_a^b \frac{1}{x}\, dx$$

assume(a, 'real', b, 'real'): assume(a > 0, b > 0):

$$-\ln(a\sim) + \ln(b\sim)$$

$$\text{int}\left(\frac{1}{x}, x = a\ldots b\right)$$

$$-\ln(a\sim) + \ln(b\sim)$$

combine(%)

$$\ln\left(\frac{b\sim}{a\sim}\right)$$

restart

1.4.26 Numerical Evaluations of Integrals

with(Student[Calculus1]):
RiemannSum(x.exp(-x), x = 0\ldots5.0, method = lower, output = plot, partition = 20)

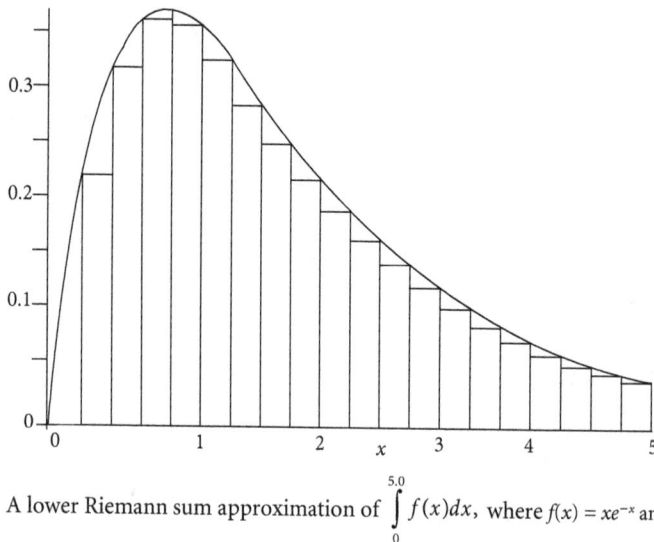

A lower Riemann sum approximation of $\int_0^{5.0} f(x)\,dx$, where $f(x) = xe^{-x}$ and the partition is uniform.

The approximate value of the integral is 0.8664812764. Number of subintervals used 20.

Figure 1.4.3: Integral calculated using a Riemann sum (method–lower).

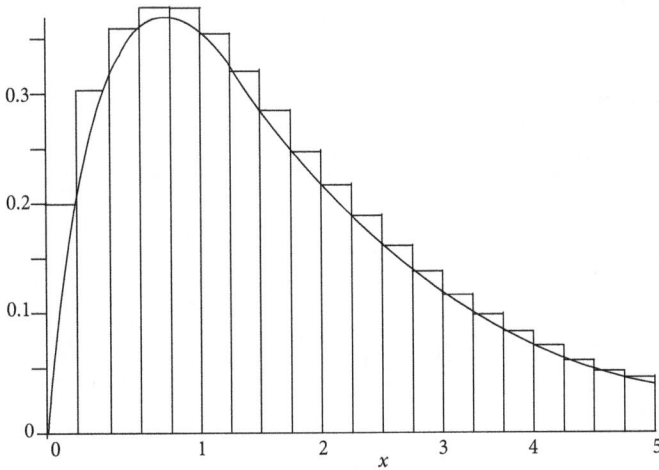

A lower Riemann sum approximation of $\int_0^{5.0} f(x)dx$, where $f(x) = xe^{-x}$ and the partition is uniform.

The approximate value of the integral is 1.041998563. Number of subintervals used 20.

Figure 1.4.4: Integral calculated using a Riemann sum (method–upper).

$$Int(\exp(-x^2 \cdot \sin(x))\ x = -1\ ...\ 1) = \text{evalf}(int(\exp(-x^2 \cdot \sin(x))\ x = -1\ ...\ 1)$$

$$\int_{-1}^{1} e^{-x^2 \sin(x)}dx = 2.113228922$$

$$f := (x, y, z) \rightarrow \cosh(x \cdot y \cdot z)$$
$$(x, y, z) \rightarrow \cosh(x \cdot y \cdot z)$$
$$evalf(Int(f(x, y, z), x = -1\ ...\ 1, y = -1\ ...\ 1, z = -1\ ...\ 1)$$
$$8.150847483$$

1.5 MATLAB Examples

MATLAB's Symbolic Math Toolbox is demonstrated performing algebraic and trigonometric operations as well as calculations involving complex numbers below. The differentiation and integration of basic functions in MATLAB are then shown.

Key MATLAB commands: *conj, diff, factor, fzero, imag, int, limit, real, simplify, solve, subs, syms*

Section 1.1 Algebra

Solving One Equation and One Unknown

```
>> clear all
>> syms x y a b
>> solve('x^2+y^2')

ans =

  y*i
 -y*i

>> solve('x^3=3')

ans =

                        3^(1/3)
  3^(1/3)*(- 1/2 + (3^(1/2)*i)/2)
   -3^(1/3)*(1/2 + (3^(1/2)*i)/2)

>> solve('x^3=3',x)

ans =

                        3^(1/3)
  3^(1/3)*(- 1/2 + (3^(1/2)*i)/2)
   -3^(1/3)*(1/2 + (3^(1/2)*i)/2)

>> solve('x^2+y^2',y)

ans =

  x*i
 -x*i

>> solve('x^2+y^2=9',y)

ans =

  (3 - x)^(1/2)*(x + 3)^(1/2)
 -(3 - x)^(1/2)*(x + 3)^(1/2)
```

Solving Systems of Algebraic Equations

```
>> S=solve('x+a*y=3','b*x-a*y=5',x,y)

S =

    x: [1x1 sym]
    y: [1x1 sym]

>> S.x
```

```
ans =

8/(b + 1)

>> S.y

ans =

(3*b - 5)/(a + a*b)
```

Factoring Expressions

```
>> factor(x^5-y^5)

ans =

(x - y)*(x^4 + x^3*y + x^2*y^2 + x*y^3 + y^4)
```

Simplifying Expressions

```
>> syms x y
>> simplify((x^2-y^2)/(x+y))

ans =

x - y
```

Substitution

```
>> syms x y

>> subs(x^2+y^2,{x,y},{3,sym('r')})

ans =

r^2 + 9
```

Section 1.2 Trigonometry

```
>> simplify(cos(x)^2+sin(x)^2)

ans =

1

>> x = fzero(@(x)tan(x)-x,10)

x =

    7.8540

>> fzero('cot(x)-x',1)
```

```
ans =

    0.8603

>> F=@(x)cot(x)-x;
>> fzero(F,1)

ans =

    0.8603
```

Section 1.3 Complex Numbers

```
>> syms a b real
>> psi = a+1i*b

psi =

a + b*i
>> real(psi)

ans =

a

>> imag(psi)

ans =

b

>> conj(psi)

ans =

a - b*i

>> simplify(conj(psi)*psi)

ans =

a^2 + b^2
```

Section 1.4 Elements of Calculus

```
>> clear all
>> syms x

>> limit(sin(x)/x,x,0)

ans =
```

```
1

>> f=x*cos(x)

f =

x*cos(x)

>> diff(f,x)

ans =

cos(x) - x*sin(x)
```

Higher derivatives

```
>> diff(f,x,2)

ans =

- 2*sin(x) - x*cos(x)

>> diff(f,x,3)

ans =

x*sin(x) - 3*cos(x)
```

Integration

```
>> int(x*exp(-x))

ans =

-(x + 1)/exp(x)
```

1.5.1 Functional Calculator

The functional calculator *funtool* featured in MATLAB's Symbolic Math Toolbox can be used to plot and perform operations on functions. Operations in funtool include algebraic operations, finding the inverse function, function composition as well as differentiation and integration of functions of one variable. The calculator is launched by typing 'funtool' at the Command line.

1.6 EXERCISES

Section 1.1 Algebra

1. Given

$$P = n\alpha \left(E + \frac{P}{3\varepsilon_0} \right) \text{ and } P = (\varepsilon_r - 1)\varepsilon_0 E$$

show that

$$\frac{n\alpha}{3\varepsilon_0} = \left(\frac{\varepsilon_r - 1}{\varepsilon_r + 2} \right).$$

This result is known as the Clausius-Mossotti relation.

2. For a block of mass m attached to a spring with force constant k obeying Hooke's law
$F = -kx = ma$
Conservation of energy gives

$$\frac{1}{2}mv^2 + \frac{1}{2}kx^2 = \frac{1}{2}kA^2, \text{ where } -A \leq x \leq A .$$

(a) Calculate the velocity v and acceleration a of the mass at $x = 0$ and at $x = A$. Note that the velocity of the mass is zero where the acceleration is maximal and vice versa.
(b) Calculate the velocity and acceleration of the mass at $x = A/2$.

3. Consider a two-particle collision in one dimension. Momentum conservation gives

$$m_1 v_{1i} + m_2 v_{2i} = m_1 v_{1f} + m_2 v_{2f} .$$

Energy conservation gives

$$\frac{1}{2}m_1 v_{1i}^2 + \frac{1}{2}m_2 v_{2i}^2 = \frac{1}{2}m_1 v_{1f}^2 + \frac{1}{2}m_2 v_{2f}^2 .$$

Show that if m_2 is initially at rest so that $v_{2i} = 0$ the final velocities are

$$v_{1f} = \frac{m_1 - m_2}{m_1 + m_2} v_{1i}$$

$$v_{2f} = \frac{2m_1}{m_1 + m_2} v_{1i} .$$

4. Kepler's third law may be obtained from Newton's second law for circular orbits where the centripetal acceleration is supplied by the gravitational force

$$m\frac{v^2}{r} = G\frac{mM}{r^2} \text{ where the velocity } v = \frac{2\pi r}{T} .$$

Show that the square of the orbital period T^2 is proportional to the cube of the orbital radius r^3.

5. The escape speed from a body of mass M is

$$v_{esc} = \sqrt{\frac{2GM}{r}}$$

Show that the escape speed is $\sqrt{2}$ times the orbital speed for circular orbits.

6. Bernoulli's equation is a statement of conservation of energy density for ideal streamline fluid flow. For any two points in the flow

$$\frac{1}{2}\rho v_1^2 + \rho g h_1 + P_1 = \frac{1}{2}\rho v_2^2 + \rho g h_2 + P_2$$

Use the continuity equation $A_1 v_1 = A_2 v_2$ to eliminate v_1 in Bernoulli's equation and solve for v_2 in special cases where
(a) both points in the streamline flow are at atmospheric pressure $P_1 = P_2 = P_0$
(b) both points are at the same height $h_1 = h_2$
(c) one point is at vacuum $P_1 = 0$ and the other is at atmospheric $P_2 = P_0$
Show that $P_1 - P_2 = \rho g(h_2 - h_1)$ if the fluid speed is everywhere zero.

7. The energy levels of hydrogen may be obtained from Newton's second law for circular orbits where the centripetal acceleration is supplied by the electric force between the electron and proton

$$m\frac{v^2}{r} = \frac{1}{4\pi\varepsilon_0}\frac{q_e^2}{r^2}$$

Using the total energy of the bound system

$$E = \frac{1}{2}mv^2 - \frac{1}{4\pi\varepsilon_0}\frac{q_e^2}{r}$$

and Bohr's angular momentum quantization hypothesis $L = mvr = n\dfrac{h}{2\pi}$ where $n = 1, 2, 3, \ldots$

show that

$$E_n = -\frac{E_1}{n^2} \text{ and } r_n = n^2 a_0$$

where $E_1 = \dfrac{m_e q_e^4}{8\varepsilon_0 h^2} = 13.6 \text{ eV}$ and $a_0 = \dfrac{\varepsilon_0 h^2}{\pi m_e q_e^2} = 0.0529 \text{ nm}$

8. Solve for the effective capacitance C_{eff} of a series combination of three capacitors

$$\frac{1}{C_{\text{eff}}} = \frac{1}{C_1} + \frac{1}{C_2} + \frac{1}{C_3} \, .$$

9. Perform a partial fractions decomposition of

$$\frac{1}{x^2 + 2x}$$

10. Identify if the following functions are even, odd, or neither

$$f(x) = \frac{1}{x - 1}$$

$$f(x) = \frac{1}{x^2 + 1}$$

$$f(x) = \frac{x^3}{x^2 + 1}$$

$$f(x) = \left| \frac{x}{|x| + 1} \right|$$

11. Solve Shockley's diode equation

$$I \simeq I_0 \left(e^{\frac{qV}{k_B T}} - 1 \right)$$

for the Boltzmann constant k_B

12. Simplify rational expression by factoring the numerator and denominator

$$\frac{x^2 + x}{x^3 + x^2}$$

Factor the expressions

$$x^3 + 4x^2 - x - 4$$
$$x^4 - x^3 + x - 1$$

13. Find the energy E of a photon with a wavelength λ equal to the Planck length λ_P where $E = hf$, $\lambda f = c$ and $\ell_P = \sqrt{\dfrac{\hbar G}{c^3}}$. Convert your answer to tons of TNT where one ton is equal to 4.184×10^9 joules.

14. Obtain a common denominator to simplify the expression

$$\frac{\dfrac{v^2}{c^2}}{\left(1 - \dfrac{v^2}{c^2}\right)^{3/2}} - \frac{1}{\left(1 - \dfrac{v^2}{c^2}\right)^{1/2}}$$

Section 1.2 Trigonometry

15. The horizontal and vertical components of the tension F_T in a conical pendulum of length L as shown in Figure 1.6.1 are given by

$$F_T \sin\theta = m\frac{v^2}{r} \text{ and } F_T \cos\theta = mg.$$

Show that the angle

$$\theta = \tan^{-1}\left(\frac{v^2}{rg}\right)$$

Solve for the pendulum period T where $v = \dfrac{2\pi r}{T}$ and $r = L\sin\theta$

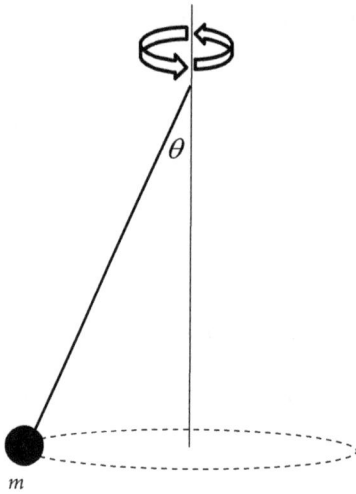

Figure 1.6.1: Conical pendulum of length L.

16. Numerically find at least one solution to the following transcendental equations

(a) $e^{-x} = \dfrac{x}{2}$.

(b) $\tan(x) = x$

(c) $\sin(x) = \dfrac{x}{2}$

17. Plot the following function
$$y(t) = 3\sin(2t) + 3\sin(3t)$$

18. Plot the following functions on the same graph
$$y(t) = e^{-3t}\sin(2t) \qquad y(t) = e^{-3t}$$

19. Evaluate the expression
$$(1 - \cos n\pi)(1 - \cos m\pi)$$
for even and odd values of m and n.

Section 1.3 Complex Numbers

20. Show that

$$\sqrt{i} = \frac{1+i}{\sqrt{2}}$$

21. Calculate the real and imaginary parts of $z = \sqrt{3+2i}$

22. Given
$$\psi = (1+i)e^{-3i}$$
calculate $\psi^*\psi$, $\text{Re}(\psi)$, and $\text{Im}(\psi)$

23. Given
$$\psi(t) = e^{-3it} + e^{-2it}$$
calculate $\psi^*\psi$, $\text{Re}(\psi)$, and $\text{Im}(\psi)$

24. Using the relations
$$\cos(\theta) = \frac{e^{i\theta} + e^{-i\theta}}{2} \quad \text{and} \quad \sin(\theta) = \frac{e^{i\theta} - e^{-i\theta}}{2i}$$

show that $\cos^2(\theta) + \sin^2(\theta) = 1$

25. Using the relations
$$\cosh(x) = \frac{e^x + e^{-x}}{2} \quad \text{and} \quad \sinh(x) = \frac{e^x - e^{-x}}{2}$$

show that $\cosh^2(x) - \sinh^2(x) = 1$

26. Show that
$$\cosh(ix) = \cos(x)$$
$$\sinh(ix) = \sin(x)$$

$$\tanh(x) = \frac{e^{2x} - 1}{e^{2x} + 1}$$

27. Plot the Langevin function
$$L(x) = \coth(x) - \frac{1}{x}$$

Section 1.4 Elements of Calculus

28. Given that the position (in meters) of a mass attached to a spring is given by
$x(t) = 3e^{-3t}\cos(2t)$
Calculate the speed $v(t)$ and acceleration $a(t)$ where

$$v = \frac{dx}{dt} \quad \text{and} \quad a = \frac{dv}{dt} \quad \text{are in m/s and m/s}^2\text{, respectively}$$

29. Given that the position (in meters) of a particle is
$$x(t) = 3t^3 - 2t^2 + 1$$
Find $x(t)$ and $a(t)$ when $v(t) = 0$ m/s

30. The position (in meters) of an object falling in a resistive medium is described by $y(t) = 1 - e^{-t}$

Plot $y(t)$, $\dot{y}(t)$ and $\ddot{y}(t)$ on the same graph. What is the initial acceleration $\ddot{y}(0)$?

31. Given that the velocity of a particle is $v(t) = e^{-t}$ calculate the distance x that the particle travels between $t_i = 0$ and $t_f = \infty$ seconds where

$$x = \int_{t_i}^{t_f} v(t)dt$$

What is the initial acceleration $\dot{v}(0)$?

32. The Gaussian function $e^{-\alpha x^2}$ is encountered in finding solutions to the quantum harmonic oscillator. Calculate the following derivatives

$$\frac{d^2}{dx^2}e^{-\alpha x^2}$$

$$\frac{d}{d\alpha}e^{-\alpha x^2}$$

33. Verify the following integrals by differentiating the result

$$\int \tanh(ax)dx = \frac{1}{a}\ln\cosh(ax) + \text{const.}$$

$$\int \tan(ax)dx = \frac{1}{a}\ln\cos(ax) + \text{const.}$$

$$\int e^x \sin(x)dx = \frac{e^x}{2}(\sin x - \cos x) + \text{const.}$$

$$\int x \sin(x)dx = \sin x - x\cos x + \text{const.}$$

34. Given $f(x) = e^{3x}$
calculate $f'(0), f''(0)$ and $f'''(0)$

35. Given $f(t) = e^{3t}\cos(t)$

calculate $\dot{f}(0)$, $\ddot{f}(0)$ and $\dddot{f}(0)$

2 VECTORS AND MATRICES

Chapter

Chapter Outline

2.1 VECTORS AND SCALARS IN PHYSICS

Physical quantities in nature can be scalars (such as mass, temperature, or density) or vectors (such as force or electric and magnetic fields). Both scalars and vectors are described by a numerical factor depending on the system of units used. Common units in physics are the MKS (meter, kilogram, second) and the CGS (centimeter, gram, second) systems. Vector quantities also have directionality. Two scalar quantities with the same units may be added or subtracted directly. Scalars with differing units may be multiplied or divided. For example, the density of an object is its mass divided by its volume. The directionality of vectors must be taken into account during addition, subtraction, and multiplication, however.

In this textbook, vectors are designated with boldfaced symbols such as **A**. Vectors with unit magnitude are usually boldfaced with a hat such as $\hat{\mathbf{k}}$.

2.1.1 Vector Addition and Unit Vectors

Scalars are tensors of rank zero while vectors are tensors of rank one. A vector may be represented graphically by an arrow with a specified length (or magnitude) and a direction. Vectors with the same magnitude, length, and units are said to be equal. A vector translated in space remains unchanged unless it is rotated or stretched. Vectors may be added graphically by arranging them tip-to-tail in any order. The vector sum, known as the resultant, is found by constructing a vector from the tail of the first vector to the tip of the last vector. A vector **A** may be represented in Cartesian coordinates as

$$\mathbf{A} = A_x\hat{\mathbf{i}} + A_y\hat{\mathbf{j}} + A_z\hat{\mathbf{k}} \tag{2.1.1}$$

where $\hat{\mathbf{i}}$, $\hat{\mathbf{j}}$, and $\hat{\mathbf{k}}$ are unit vectors in the respective x-, y-, and z-directions. The components of **A** along each coordinate axis are A_x, A_y, and A_z. The length, or magnitude, of **A** can be determined from the Pythagorean theorem as $|\mathbf{A}| = \sqrt{A_x^2 + A_y^2 + A_z^2}$.

Given the vector $\mathbf{B} = B_x\hat{\mathbf{i}} + B_y\hat{\mathbf{j}} + B_z\hat{\mathbf{k}}$ $\tag{2.1.2}$

the sum **A** + **B** is

$$\mathbf{A} + \mathbf{B} = \left(A_x + B_x\right)\hat{\mathbf{i}} + \left(A_y + B_y\right)\hat{\mathbf{j}} + \left(A_z + B_z\right)\hat{\mathbf{k}} \tag{2.1.3}$$

Vector addition is commutative where **A** + **B** = **B** + **A**. Subtraction is similarly performed

$$\mathbf{A} - \mathbf{B} = \left(A_x - B_x\right)\hat{\mathbf{i}} + \left(A_y - B_y\right)\hat{\mathbf{j}} + \left(A_z - B_z\right)\hat{\mathbf{k}} \tag{2.1.4}$$

2.1.2 Scalar Product of Vectors

The scalar product, also known as the dot product, between vectors **A** and **B** is defined as

$$\mathbf{A} \cdot \mathbf{B} = |\mathbf{A}||\mathbf{B}|\cos(\theta) = AB\cos(\theta) \tag{2.1.5}$$

where $A = |\mathbf{A}|$ and $B = |\mathbf{B}|$. The angle θ between **A** and **B** can be determined by

$$\theta = \cos^{-1}\left(\frac{\mathbf{A} \cdot \mathbf{B}}{AB}\right) \tag{2.1.6}$$

The dot product between like unit vectors is unity $\hat{i}\cdot\hat{i}=\hat{j}\cdot\hat{j}=\hat{k}\cdot\hat{k}=1$ and zero between unlike unit vectors $\hat{i}\cdot\hat{j}=\hat{j}\cdot\hat{k}=\hat{k}\cdot\hat{i}=0$. To calculate the dot product between two vectors expressed in terms of the unit vectors we simply multiply their respective components and add

$$\mathbf{A}\cdot\mathbf{B}=A_x B_x + A_y B_y + A_z B_z \tag{2.1.7}$$

The dot product is commutative where $\mathbf{A}\cdot\mathbf{B}=\mathbf{B}\cdot\mathbf{A}$. The magnitude of a vector is also expressed as a dot product where $\mathbf{A}\cdot\mathbf{A}=A^2$ and

$$|\mathbf{A}|=\sqrt{\mathbf{A}\cdot\mathbf{A}}=\sqrt{A_x^{\,2}+A_y^{\,2}+A_z^{\,2}} \tag{2.1.8}$$

2.1.3 Vector Cross Product

The cross product (or vector product) between two vectors $\mathbf{A}$ and $\mathbf{B}$ denoted as $\mathbf{A}\times\mathbf{B}$ forms a third vector with magnitude

$$|\mathbf{A}\times\mathbf{B}|=AB\sin(\theta) \tag{2.1.9}$$

where θ is the angle between $\mathbf{A}$ and $\mathbf{B}$. If $\mathbf{A}$ and $\mathbf{B}$ are parallel then $\mathbf{A}\times\mathbf{B}=0$.

If $\mathbf{A}\times\mathbf{B}=\mathbf{C}$ then $\mathbf{A}$ and $\mathbf{B}$ are both perpendicular to $\mathbf{C}$ so that $\mathbf{A}\cdot\mathbf{C}=\mathbf{B}\cdot\mathbf{C}=0$. Also $\mathbf{A}\times\mathbf{B}=-\mathbf{B}\times\mathbf{A}$. The cross products of the unit vectors are

$$\begin{aligned} \hat{i}\times\hat{j}=\hat{k} \quad & \hat{j}\times\hat{i}=-\hat{k} \\ \hat{j}\times\hat{k}=\hat{i} \quad & \hat{k}\times\hat{j}=-\hat{i} \\ \hat{k}\times\hat{i}=\hat{j} \quad & \hat{i}\times\hat{k}=-\hat{j} \end{aligned} \tag{2.1.10}$$

The cross products of like unit vectors are zero $\hat{i}\times\hat{i}=\hat{j}\times\hat{j}=\hat{k}\times\hat{k}=0$.

Example 2.1.1

Calculate the cross product between $\mathbf{A}=3\hat{i}-2\hat{j}$ and $\mathbf{B}=\hat{j}-\hat{k}$

Solution:

$$\mathbf{A}\times\mathbf{B}=\left(3\hat{i}-2\hat{j}\right)\times\left(\hat{j}-\hat{k}\right)=3\underbrace{\left(\hat{i}\times\hat{j}\right)}_{\hat{k}}-3\underbrace{\left(\hat{i}\times\hat{k}\right)}_{-\hat{j}}-2\underbrace{\left(\hat{j}\times\hat{j}\right)}_{0}+2\underbrace{\left(\hat{j}\times\hat{k}\right)}_{\hat{i}} \tag{2.1.11}$$

$$=2\hat{i}+3\hat{j}+3\hat{k}$$

Now we can check if this vector is perpendicular to $\mathbf{A}$ and $\mathbf{B}$

$$\mathbf{A}\cdot\left(\mathbf{A}\times\mathbf{B}\right)=\left(3\hat{i}-2\hat{j}\right)\cdot\left(2\hat{i}+3\hat{j}+3\hat{k}\right)=6-6=0$$

$$\mathbf{B}\cdot\left(\mathbf{A}\times\mathbf{B}\right)=\left(\hat{j}-\hat{k}\right)\cdot\left(2\hat{i}+3\hat{j}+3\hat{k}\right)=3-3=0 \tag{2.1.12}$$

The cross product may also be calculated using a determinant (expanding across the top row)

$$\mathbf{A} \times \mathbf{B} = \begin{vmatrix} \hat{\mathbf{i}} & \hat{\mathbf{j}} & \hat{\mathbf{k}} \\ A_x & A_y & A_z \\ B_x & B_y & B_z \end{vmatrix} = \hat{\mathbf{i}}\left(A_y B_z - B_y A_z\right) - \hat{\mathbf{j}}\left(A_x B_z - B_x A_z\right) + \hat{\mathbf{k}}\left(A_x B_y - B_x A_y\right)$$

(2.1.13)

Determinants are discussed in Section 2.3 of this chapter.

Example 2.1.2

Using a determinant to calculate the cross product in the previous example

Solution

$$\mathbf{A} \times \mathbf{B} = \begin{vmatrix} \hat{\mathbf{i}} & \hat{\mathbf{j}} & \hat{\mathbf{k}} \\ 3 & -2 & 0 \\ 0 & 1 & -1 \end{vmatrix} = \hat{\mathbf{i}}(2-0) - \hat{\mathbf{j}}(-3-0) + \hat{\mathbf{k}}(3-0) = 2\hat{\mathbf{i}} + 3\hat{\mathbf{j}} + 3\hat{\mathbf{k}} \quad (2.1.14)$$

Determinants may also be used to calculate the cross product or the curl in other coordinate systems as discussed in Chapter 4.

2.1.4 Triple Vector Products

The triple cross product is frequently encountered where

$$\mathbf{A} \times \mathbf{B} \times \mathbf{C} = \mathbf{B}(\mathbf{A} \cdot \mathbf{C}) - \mathbf{C}(\mathbf{A} \cdot \mathbf{B}) \qquad (2.1.15)$$

This rule is known as the "BAC – CAB" rule.

A determinant may represent the dot product of a vector with a cross product

$$\mathbf{A} \cdot (\mathbf{B} \times \mathbf{C}) = \begin{vmatrix} A_x & A_y & A_z \\ B_x & B_y & B_z \\ C_x & C_y & C_z \end{vmatrix} \qquad (2.1.16)$$

This determinant gives the volume of a parallelepiped spanned by the three vectors. The determinant is the same upon the interchange of any row or column so that

$$\mathbf{A} \cdot (\mathbf{B} \times \mathbf{C}) = \mathbf{B} \cdot (\mathbf{C} \times \mathbf{A}) = \mathbf{C} \cdot (\mathbf{A} \times \mathbf{B}) \qquad (2.1.17)$$

Observe that

$$\mathbf{A}\cdot(\mathbf{A}\times\mathbf{C}) = \begin{vmatrix} A_x & A_y & A_z \\ A_x & A_y & A_z \\ C_x & C_y & C_z \end{vmatrix} = 0 \qquad (2.1.18)$$

since $\mathbf{A}\times\mathbf{C}$ is perpendicular to $\mathbf{A}$. Also $\mathbf{A}\times(\mathbf{A}\times\mathbf{B}) = \mathbf{B}\times(\mathbf{A}\times\mathbf{B}) = 0$.

2.1.5 The Position Vector

Consider the vector $\mathbf{C} = \mathbf{A} - \mathbf{B}$. If we dot $\mathbf{C}$ with itself

$$\mathbf{C}\cdot\mathbf{C} = (\mathbf{A}-\mathbf{B})\cdot(\mathbf{A}-\mathbf{B}) \qquad (2.1.19)$$

we obtain the law of cosines

$$C^2 = A^2 + B^2 - 2AB\cos\theta \qquad (2.1.20)$$

In physics, we frequently represent vector positions from an origin of coordinates as $\mathbf{r}$ and $\mathbf{r}'$ for source and field quantities, respectively. The quantity $\mathbf{r} - \mathbf{r}'$ denotes the difference from the source to field point as shown in Figure 2.1.1. The position vector is independent of the origin of coordinates. In terms of the Cartesian unit vectors

$$\mathbf{r}-\mathbf{r}' = (x-x')\,\hat{\mathbf{i}} + (y-y')\hat{\mathbf{j}} + (z-z')\hat{\mathbf{k}} \qquad (2.1.21)$$

and

$$|\mathbf{r}-\mathbf{r}'| = \sqrt{(x-x')^2 + (y-y')^2 + (z-z')^2} \qquad (2.1.22)$$

Note that the vector

$$\hat{\mathbf{n}} = \frac{\mathbf{r}-\mathbf{r}'}{|\mathbf{r}-\mathbf{r}'|} \qquad (2.1.23)$$

represents a unit vector pointing in the direction $\mathbf{r} - \mathbf{r}'$.

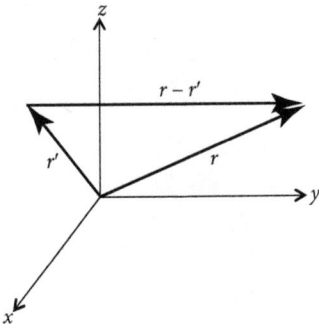

Figure 2.1.1: The position vector $r - r'$.

2.1.6 Expressing Vectors in Different Coordinate Systems

Unit vectors may be expressed in different coordinate systems. For example, in cylindrical coordinates (r, ϕ, z)

$$\hat{i} = \cos(\phi)\hat{r} - \sin(\phi)\hat{\phi}$$
$$\hat{j} = \sin(\phi)\hat{r} + \cos(\phi)\hat{\phi} \qquad (2.1.24)$$
$$\hat{k} = \hat{z}$$

Maple Examples

Vector operations are demonstrated in the Maple worksheet below. A vector **v** is specified by the notation $\langle v_x, v_y, v_z \rangle$ with Cartesian unit vectors denoted as e_x, e_y, e_z in Maple. Equivalent syntaxes for vector magnitude, dot and cross products are also demonstrated in the following examples.

Key Maple commands: *ChangeBasis, DotProduct, CrossProduct, Norm, PlotVector, VectorSpace*

Maple packages: *with(VectorCalculus):*   *with(Student[VectorCalculus]):* *with(Student[LinearAlgebra]):*

restart

Define Vectors

With(VectorCalculus) :
V1 := ⟨1,2,3⟩

$$V1 := e_x + 2e_y + 3e_z$$

V2 := ⟨−1,5,9⟩

$$V2 := -e_x + 5e_y + 9e_z$$

Dot Product

DotProduct(V1, V2)

$$36$$

V1.V2

$$36$$

Vector Magnitude

Norm (V2)

$$\sqrt{107}$$

$sqrt(V1 \cdot V2)$

$$\sqrt{107}$$

Cross Product

$V3 := CrossProduct(V1, V2)$

$$V3 := 3e_x - 12e_y + 7e_z$$

$V1 \&x V2$

$$3e_x - 12e_y + 7e_z$$

Show That V3 Is Orthogonal to V1 and V2

$DotProduct(V1, V3)$

$$0$$

$DotProduct(V2, V3)$

$$0$$

Position Vector

$r := \langle x, y, z \rangle$

$$r := (x)e_x + (y)e_y + (z)e_z$$

$rp := \langle xp, yp, zp \rangle$

$$rp := (xp)e_x + (yp)e_y + (zp)e_z$$

$r - rp$

$$(x - xp)e_x + (y - yp)e_y + (z - zp)e_z$$

Unit Vector in the Direction of (r – rp)

$$n\_hat := \frac{r - rp}{((r - rp) \cdot (r - rp))^{\frac{1}{2}}}$$

$$n\_hat := \left(\frac{x - xp}{\sqrt{(x - xp)^2 + (y - yp)^2 + (z - zp)^2}} \right) e_x$$

$$+ \left(\frac{y - yp}{\sqrt{(x - xp)^2 + (y - yp)^2 + (z - zp)^2}} \right) e_y$$

$$+ \left(\frac{z - zp}{\sqrt{(x - xp)^2 + (y - yp)^2 + (z - zp)^2}} \right) e_z$$

n_hat.n_hat

$$\frac{(x-xp)^2}{(x-xp)^2+(y-yp)^2+(z-zp)^2}+\frac{(y-yp)^2}{(x-xp)^2+(y-yp)^2+(z-zp)^2}$$
$$+\frac{(z-zp)^2}{(x-xp)^2+(y-yp)^2+(z-zp)^2}$$

simplify(%)

$$1$$

restart

Vectors in Spherical and Cylindrical Coordinates

With(Physics[Vectors]):
Setup(mathematicalnotation = true)

$$[mathematicalnotation = true]$$

$$R := x\cdot\_i + y\cdot\_j + z\cdot\_k$$

$$R := x\hat{i} + y\hat{j} + z\hat{k}$$

ChangeBasis(R, spherical)

$$\left(x\sin(\theta)\cos(\phi)+y\sin(\theta)\sin(\phi)+z\cos(\theta)\right)\hat{r}$$
$$+\left(x\cos(\phi)\cos(\theta)+y\cos(\theta)\sin(\phi)-z\sin(\theta)\right)\hat{\theta}$$
$$+\left(\cos(\phi)y-\sin(\phi)x\right)\hat{\phi}$$

ChangeBasis(R, cylindrical)

$$\left(x\cos(\phi)+y\sin(\phi)\right)\hat{\rho}+\left(\cos(\phi)y-\sin(\phi)x\right)\hat{\phi}+z\hat{k}$$

$$S := rho\cdot\_rho + z\cdot\_k$$

$$S := z\hat{k} + \rho\hat{\rho}$$

ChangeBasis(S, cartesian)

$$\rho\cos(\phi)\hat{i} + \rho\sin(\phi)\hat{j} + z\hat{k}$$

restart

Plot Vectors

With(VectorCalculus):
$$r := \langle 1, 0, 1 \rangle$$

$$r := e_x + e_z$$

$rp := \langle -2, 1, 1 \rangle$

$$rp := -2e_x + e_y + e_z$$

$PlotVector([r, -rp])$

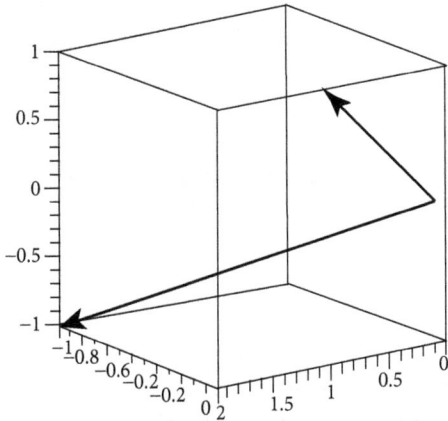

Figure 2.1.2: Plotting the vectors r and −rp in 3D.

$With(Student[LinearAlgebra]) :$
$VectorSumPlot(r, -rp)$

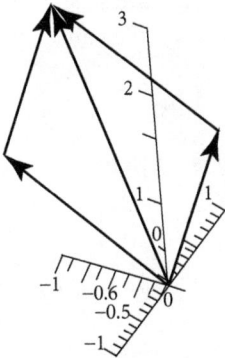

The sum of 2 vectors, showing the resultant in black and the parallelogram(s) of addition
Figure 2.1.3: Maple output plotting the sum of vectors r and −rp in 3D.

$With(Student[VectorCalculus]) :$
$vs1 := VectorSpace(`cartesian`[x, y, z], [0,0,0]) :$
$vs2 := VectorSpace(`cartesian`[x, y, z], [rp[1], rp[21], rp[31]]) :$
$PlotVector([vs1 :- Vector([r[1], r[2], r[3]]), vs1 :- Vector([rp[1], rp[2], rp[3]]), vs2 :- Vector([r[1]-rp[1], r[2]-rp[2], r[3]-rp[3]])], scaling = constrained)$

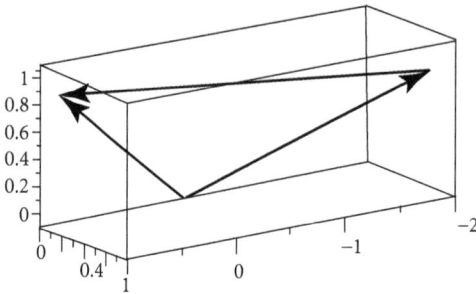

Figure 2.1.4: Plotting the sum of vectors *r* and *–rp* in 3D using the package *Student[VectorCalculus]*.

2.2 MATRICES IN PHYSICS

Matrices are rectangular arrays of elements consisting of M rows and N columns. A matrix can be represented by a bold faced symbol $\mathbf{A}$ or $[a_{ij}]$ with elements denoted by a_{ij} where the indices i and j correspond to the row and column numbers. A matrix may also be written as a symbol with indices that may appear raised or lowered, such as $T^{\mu\nu}$ or $g_{\mu\nu}$. Matrices are tensors of rank two that have widespread applications in math and science. Important uses of matrices in physics can be found in classical mechanics, electromagnetism, quantum mechanics, special relativity, general relativity, and relativistic quantum mechanics.

2.2.1 Matrix Dimension

A matrix $\mathbf{A}$ with dimension $[M \times N]$, (pronounced "M by N") is written

$$\mathbf{A} = \begin{pmatrix} a_{11} & a_{12} & \cdots & a_{1N} \\ a_{21} & a_{22} & \cdots & a_{2N} \\ \vdots & \vdots & \cdots & \vdots \\ a_{M1} & a_{M2} & \cdots & a_{MN} \end{pmatrix} \tag{2.2.1}$$

The most commonly encountered matrices in physics are square, single column (column vectors) and single row (row vectors). If $M = N$ then $\mathbf{A}$ is a square matrix

$$\mathbf{A} = \begin{pmatrix} a_{11} & a_{12} & \cdots & a_{1N} \\ a_{21} & a_{22} & \cdots & a_{2N} \\ \vdots & \vdots & \cdots & \vdots \\ a_{N1} & a_{N2} & \cdots & a_{NN} \end{pmatrix} \tag{2.2.2}$$

A matrix $\mathbf{B}$ with M rows and one column is known as a column vector

$$\mathbf{B} = \begin{pmatrix} a_1 \\ a_2 \\ \vdots \\ a_M \end{pmatrix} \tag{2.2.3}$$

A row vector is a matrix with one row and N columns

$$\mathbf{A} = \begin{pmatrix} a_1 & a_2 & \cdots & a_N \end{pmatrix} \tag{2.2.4}$$

Square and column matrices will also be useful in representing systems of linear equations and systems of differential equations. Matrix elements may be real numbers, complex numbers, functions, or operators.

2.2.2 Matrix Addition and Subtraction

Matrices of the same dimension may be added or subtracted.

Example 2.2.1

Given the 2×2 matrices

$$\mathbf{A} = \begin{pmatrix} a & b \\ c & d \end{pmatrix} \text{ and } \mathbf{B} = \begin{pmatrix} 1 & 2 \\ 3 & 4 \end{pmatrix} \tag{2.2.5}$$

$$\mathbf{A} \pm \mathbf{B} = \begin{pmatrix} a & b \\ c & d \end{pmatrix} \pm \begin{pmatrix} 1 & 2 \\ 3 & 4 \end{pmatrix} = \begin{pmatrix} a \pm 1 & b \pm 2 \\ c \pm 3 & d \pm 4 \end{pmatrix} \tag{2.2.6}$$

Constant multipliers multiply each term in the matrix

$$3\mathbf{A} = 3 \begin{pmatrix} a & b \\ c & d \end{pmatrix} = \begin{pmatrix} 3a & 3b \\ 3c & 3d \end{pmatrix} \tag{2.2.7}$$

2.2.3 Matrix Integration and Differentiation

Operations of differentiation and integration map to each element of a matrix.

Example 2.2.2

Given

$$\mathbf{C}(x) = \begin{pmatrix} 1 & x \\ x^2 & x^3 \end{pmatrix} \tag{2.2.8}$$

the integral of $\mathbf{C}(x)$ is

$$\int \mathbf{C}(x)\,dx = \begin{pmatrix} x & \dfrac{x^2}{2} \\ \dfrac{x^3}{3} & \dfrac{x^4}{4} \end{pmatrix} + \text{const. matrix} \tag{2.2.9}$$

The derivative of $\mathbf{C}(x)$ is

$$\frac{d}{dx}\mathbf{C}(x) = \begin{pmatrix} 0 & 1 \\ 2x & 3x^2 \end{pmatrix} \tag{2.2.10}$$

2.2.4 Matrix Multiplication and Commutation

A matrix product is performed by multiplying row by column.

Example 2.2.3

Given the matrices

$$\mathbf{A} = \begin{pmatrix} a & b \\ c & d \end{pmatrix} \text{ and } \mathbf{B} = \begin{pmatrix} 1 & 2 \\ 3 & 4 \end{pmatrix} \tag{2.2.11}$$

the first element of the matrix product $\mathbf{AB}$ is obtained by multiplying the first row of matrix $\mathbf{A}$ times the first column of matrix $\mathbf{B}$, and so on. The number of columns of $\mathbf{A}$ must equal the number of rows of $\mathbf{B}$. For example,

$$\mathbf{AB} = \begin{pmatrix} a & b \\ c & d \end{pmatrix}\begin{pmatrix} 1 & 2 \\ 3 & 4 \end{pmatrix} = \begin{pmatrix} a+3b & 2a+4b \\ c+3d & 2c+4d \end{pmatrix} \tag{2.2.12}$$

Note that the order of matrix multiplication is important. Here $\mathbf{AB} \neq \mathbf{BA}$ since

$$\mathbf{BA} = \begin{pmatrix} 1 & 2 \\ 3 & 4 \end{pmatrix}\begin{pmatrix} a & b \\ c & d \end{pmatrix} = \begin{pmatrix} a+2c & b+2d \\ 3a+4c & 3b+4d \end{pmatrix} \tag{2.2.13}$$

Two matrices whose product is independent of the order of multiplication are said to commute. The commutator between matrices $\mathbf{A}$ and $\mathbf{B}$, denoted by square brackets, is defined as

$$\begin{bmatrix} \mathbf{A}, \mathbf{B} \end{bmatrix} = \mathbf{AB} - \mathbf{BA} \tag{2.2.14}$$

Example 2.2.4

For the matrices above

$$
\begin{aligned}
\left[\mathbf{A},\mathbf{B}\right] &= \begin{pmatrix} a+3b & 2a+4b \\ c+3d & 2c+4d \end{pmatrix} - \begin{pmatrix} a+2c & b+2d \\ 3a+4c & 3b+4d \end{pmatrix} \\
&= \begin{pmatrix} 3b-2c & 2a+3b-2d \\ -3a-3c+3d & 2c-3b \end{pmatrix} \neq \begin{pmatrix} 0 & 0 \\ 0 & 0 \end{pmatrix}
\end{aligned}
\tag{2.2.15}
$$

Thus matrices **A** and **B** do not commute. The anticommutator between two matrices **A** and **B**, denoted by curly brackets, is defined as

$$
\{\mathbf{A},\mathbf{B}\} = \mathbf{AB} + \mathbf{BA}
\tag{2.2.16}
$$

If $\{\mathbf{A}, \mathbf{B}\} = 0$ then $\mathbf{AB} = -\mathbf{BA}$ and the matrices are said to anticommute.

Matrices with M rows and N columns can only be multiplied by matrices with N rows and P columns to yield a matrix with M rows and P columns, or $[M \times N]$ $[N \times P] = [M \times P]$.

Example 2.2.5

A $[1 \times 3][3 \times 1] = [1 \times 1]$ is given by

$$
\begin{pmatrix} 1 & 2 & 3 \end{pmatrix} \begin{pmatrix} a \\ b \\ c \end{pmatrix} = a + 2b + 3c
\tag{2.2.17}
$$

Example 2.2.6

A $[3 \times 1][1 \times 3] = [3 \times 3]$ is given by

$$
\begin{pmatrix} a \\ b \\ c \end{pmatrix} \begin{pmatrix} 1 & 2 & 3 \end{pmatrix} = \begin{pmatrix} a & 2a & 3a \\ b & 2b & 3b \\ c & 2c & 3c \end{pmatrix}
\tag{2.2.18}
$$

2.2.5 Direct Product

The direct product, or Kronecker product, between two matrices **A** and **B** is denoted $\mathbf{A} \otimes \mathbf{B}$.

Example 2.2.7

Given the matrices

$$
\mathbf{A} = \begin{pmatrix} a & b \\ c & d \end{pmatrix} \text{ and } \mathbf{B} = \begin{pmatrix} 1 & 2 \\ 3 & 4 \end{pmatrix}
\tag{2.2.19}
$$

the direct product is constructed

$$\mathbf{A} \otimes \mathbf{B} = \begin{pmatrix} a\mathbf{B} & b\mathbf{B} \\ c\mathbf{B} & d\mathbf{B} \end{pmatrix} = \begin{pmatrix} 1a & 2a & 1b & 2b \\ 3a & 4a & 3b & 4b \\ 1c & 2c & 1d & 2d \\ 3c & 4c & 3d & 4d \end{pmatrix} \tag{2.2.20}$$

2.2.6 Identity Matrix

The identity matrix $\mathbf{I}$ is a square matrix with ones along the diagonal and zeros elsewhere. The $[3 \times 3]$ identity matrix is written as

$$\mathbf{I} = \begin{pmatrix} 1 & 0 & 0 \\ 0 & 1 & 0 \\ 0 & 0 & 1 \end{pmatrix} \tag{2.2.21}$$

The identity matrix can be expressed as a Kronecker delta symbol d_{ij} where

$$\delta_{ij} = \begin{cases} 1 & i = j \\ 0 & i \neq j \end{cases} \tag{2.2.22}$$

An $[N \times N]$ identity multiplying any $[N \times M]$ matrix returns the same $[N \times M]$ matrix so that $\mathbf{IA} = \mathbf{A}$.

Example 2.2.8

The $[3 \times 3]$ identity acting on a column vector

$$\begin{pmatrix} 1 & 0 & 0 \\ 0 & 1 & 0 \\ 0 & 0 & 1 \end{pmatrix} \begin{pmatrix} a \\ b \\ c \end{pmatrix} = \begin{pmatrix} a \\ b \\ c \end{pmatrix} \tag{2.2.23}$$

A square matrix commutes with the identity matrix of the same dimension $\mathbf{IA} = \mathbf{AI}$.

2.2.7 Transpose of a Matrix

The transpose of a matrix $\mathbf{A}$, written as $\mathbf{A}^T$ (or sometimes as $\tilde{\mathbf{A}}$), is obtained by interchanging the rows and columns of $\mathbf{A}$. Given $\mathbf{A} = [a_{ij}]$ then $\mathbf{A}^T = [a_{ji}]$.

Example 2.2.9

Given

$$\mathbf{A} = \begin{pmatrix} 1 & 2 & 3 \\ 4 & 5 & 6 \\ 7 & 8 & 9 \end{pmatrix} \qquad (2.2.24)$$

then

$$\mathbf{A}^T = \begin{pmatrix} 1 & 4 & 7 \\ 2 & 5 & 8 \\ 3 & 6 & 9 \end{pmatrix} \qquad (2.2.25)$$

The transpose operation is equivalent to flipping a matrix about its main diagonal. The transpose of a column vector will give a row vector

$$\begin{pmatrix} a \\ b \\ c \end{pmatrix}^T = \begin{pmatrix} a & b & c \end{pmatrix} \qquad (2.2.26)$$

and vice versa

$$\begin{pmatrix} a & b & c \end{pmatrix}^T = \begin{pmatrix} a \\ b \\ c \end{pmatrix} \qquad (2.2.27)$$

Additional properties of the transpose include

$$\left(\mathbf{A}^T \right)^T = \mathbf{A}, \qquad (2.2.28)$$

$$\left(\mathbf{A} + \mathbf{B} \right)^T = \mathbf{A}^T + \mathbf{B}^T \qquad (2.2.29)$$

and

$$\left(\mathbf{AB} \right)^T = \mathbf{B}^T \mathbf{A}^T \qquad (2.2.30)$$

2.2.8 Symmetric and Antisymmetric Matrices

A symmetric matrix is equal to its transpose $\mathbf{A} = \mathbf{A}^T$.

Example 2.2.10

$$\text{If } \mathbf{A} = \begin{pmatrix} 1 & 0 & 3 \\ 0 & 1 & 0 \\ 3 & 0 & 9 \end{pmatrix} \text{ then } \mathbf{A}^T = \begin{pmatrix} 1 & 0 & 3 \\ 0 & 1 & 0 \\ 3 & 0 & 9 \end{pmatrix} \tag{2.2.31}$$

An antisymmetric is equal to the negative of its transpose $\mathbf{A} = -\mathbf{A}^T$.

Example 2.2.11

$$\text{If } \mathbf{A} = \begin{pmatrix} 0 & 1 \\ -1 & 0 \end{pmatrix} \text{ then } \mathbf{A}^T = \begin{pmatrix} 0 & -1 \\ 1 & 0 \end{pmatrix} = -\mathbf{A} \tag{2.2.32}$$

2.2.9 Diagonal Matrix

Diagonal matrices such as

$$\mathbf{A} = \begin{pmatrix} 1 & 0 & 0 \\ 0 & 1 & 0 \\ 0 & 0 & 9 \end{pmatrix} \tag{2.2.33}$$

have $a_{ij} = \delta_{ij} = 0$ for $i \neq j$ and are also symmetric.

2.2.10 Tridiagonal Matrix

A tridiagonal matrix

$$\mathbf{A} = \begin{pmatrix} a_1 & b_1 & 0 & \cdots & & 0 \\ c_1 & a_2 & b_2 & \ddots & & \vdots \\ 0 & c_2 & \ddots & \ddots & & 0 \\ \vdots & \ddots & \ddots & & a_{N-1} & b_{N-1} \\ 0 & \cdots & 0 & & c_{N-1} & a_N \end{pmatrix} \tag{2.2.34}$$

may have nonzero elements along the main diagonal $a_1, a_2, \ldots a_N$ as well as upper adjacent $b_1, b_2, \ldots b_{N-1}$ and lower adjacent diagonals $c_1, c_2, \ldots c_{N-1}$. Tridiagonal matrices are encountered in numerical solutions to diffusion equations such as the Schrödinger equation using the Crank-Nicolson method.

2.2.11 Orthogonal Matrices

An orthogonal matrix multiplied by its transpose is equal to the identity matrix

$$\mathbf{A}\mathbf{A}^T = \mathbf{I} \tag{2.2.35}$$

Example 2.2.12

The matrix

$$\mathbf{A} = \begin{pmatrix} 0 & 1 \\ -1 & 0 \end{pmatrix} \qquad (2.2.36)$$

is orthogonal since

$$\mathbf{A}\mathbf{A}^T = \begin{pmatrix} 0 & 1 \\ -1 & 0 \end{pmatrix}\begin{pmatrix} 0 & -1 \\ 1 & 0 \end{pmatrix} = \begin{pmatrix} 1 & 0 \\ 0 & 1 \end{pmatrix} \qquad (2.2.37)$$

Orthogonal matrices also have the properties $\mathbf{A}^T = \mathbf{A}^{-1}$ and $\mathbf{A}\mathbf{A}^T = \mathbf{A}^T\mathbf{A}$.

2.2.12 Complex Conjugate of a Matrix

The complex conjugate of a matrix A, written as A^*, is formed by taking the complex conjugate of each element of A. Given $A = a_{ij}$ then $A^* = a_{ij}^*$. For example, if

$$A = \begin{pmatrix} 1+i & 0 & i \\ 0 & 0 & 0 \\ -i & 0 & 1-i \end{pmatrix} \text{ then } A^* = \begin{pmatrix} 1-i & 0 & -i \\ 0 & 0 & 0 \\ i & 0 & 1+i \end{pmatrix} \qquad (2.2.38)$$

2.2.13 Matrix Adjoint (Hermitian Conjugate)

To form the Hermitian conjugate (or adjoint) of a matrix $\mathbf{A}$, written as $\mathbf{A}^\dagger$, the complex conjugate of each element of A is taken followed by the transpose. Given $\mathbf{A} = a_{ij}$ then $\mathbf{A}^\dagger = a_{ij}^*$. For example, if

$$\mathbf{A} = \begin{pmatrix} 1+i & 0 & i \\ 0 & 0 & 0 \\ -i & 0 & 1-i \end{pmatrix} \text{ then } \mathbf{A}^\dagger = \begin{pmatrix} 1-i & 0 & i \\ 0 & 0 & 0 \\ -i & 0 & 1+i \end{pmatrix} \qquad (2.2.39)$$

A Hermitian matrix is equal to its Hermitian conjugate. In quantum mechanics, Hermitian matrices correspond to physical observables.

Example 2.2.13

Given the matrix

$$\mathbf{A} = \begin{pmatrix} 1 & 1+i & i \\ 1-i & 2 & 0 \\ -i & 0 & 1 \end{pmatrix} \qquad (2.2.40)$$

we see that $\mathbf{A}^\dagger = \mathbf{A}$, thus $\mathbf{A}$ is Hermitian.

The diagonal elements of a Hermitian matrix must be real. A Hermitian matrix can be constructed from a non-Hermitian matrix $\mathbf{B}$ by adding its Hermitian conjugate $\mathbf{B}^\dagger$.

2.2.14 Unitary Matrix

A matrix is said to be unitary if the identity matrix is obtained when multiplied by its Hermitian conjugate.

Example 2.2.14

Given

$$\mathbf{A} = \begin{pmatrix} i & 0 \\ 0 & -i \end{pmatrix} \text{ then } \mathbf{A}^\dagger = \begin{pmatrix} -i & 0 \\ 0 & i \end{pmatrix} \tag{2.2.41}$$

and

$$\mathbf{A}\mathbf{A}^\dagger = \begin{pmatrix} i & 0 \\ 0 & -i \end{pmatrix} \begin{pmatrix} -i & 0 \\ 0 & i \end{pmatrix} = \begin{pmatrix} 1 & 0 \\ 0 & 1 \end{pmatrix} = \mathbf{I} \tag{2.2.42}$$

Thus, $\mathbf{A}^\dagger = \mathbf{A}^{-1}$ and $\mathbf{A}^\dagger \mathbf{A} = \mathbf{I}$.

2.2.15 Partitioned Matrix

A matrix may be partitioned into submatrices.

Example 2.2.15

The gamma matrix

$$\gamma^0 = \begin{pmatrix} 1 & 0 & 0 & 0 \\ 0 & 1 & 0 & 0 \\ 0 & 0 & -1 & 0 \\ 0 & 0 & 0 & -1 \end{pmatrix} \tag{2.2.43}$$

may be written in a more compact way

$$\gamma^0 = \begin{pmatrix} \mathbf{I} & 0 \\ 0 & -\mathbf{I} \end{pmatrix} \text{ where } \mathbf{I} = \begin{pmatrix} 1 & 0 \\ 0 & 1 \end{pmatrix} \tag{2.2.44}$$

2.2.16 Matrix Trace

The trace of a matrix $\mathbf{A}$ is the sum of its diagonal elements

$$\mathrm{Tr}\left(\mathbf{A}\right)=\sum_{i}a_{ii}=a_{11}+a_{22}+\cdots \tag{2.2.45}$$

Example 2.2.16

Given the matrix

$$\mathbf{A}=\begin{pmatrix}1 & 0 & 1\\ 0 & 0 & 0\\ 1 & 0 & 1\end{pmatrix} \tag{2.2.46}$$

we have

$$\mathrm{Tr}\left(\mathbf{A}\right)=1+0+1=2 \tag{2.2.47}$$

2.2.17 Matrix Exponentiation

The exponential function of an $N\times N$ square matrix $\mathbf{A}$ is defined as

$$e^{\mathbf{A}}=\sum_{n=0}^{\infty}\frac{1}{n!}\mathbf{A}^{n}=\mathbf{I}+\mathbf{A}+\frac{1}{2}\mathbf{AA}+\frac{1}{6}\mathbf{AAA}+\cdots, \tag{2.2.48}$$

where $\mathbf{I}$ is the $N\times N$ identity matrix. For $a_{ij}\ll 1$

$$e^{\mathbf{A}}\approx\mathbf{I}+\mathbf{A} \tag{2.2.49}$$

Maple Examples

Matrix operations demonstrated in the Maple worksheet below include defining matrices, specifying individual elements, addition and subtraction, multiplication, mapping operations of integration and differentiation on matrices, commutation and anticommutation, direct products, transpose, Hermitian conjugate, trace and matrix exponential. Equivalent syntaxes for defining matrices are shown.

Key Maple commands: *diff, HermitianTranspose, int, KroneckerProduct, Matrix, MatrixExponential, Trace, Transpose*

Maple packages: *with(LinearAlgebra)*

restart

Define Matrices

with(LinearAlgebra) :
A := Matrix([[1, 2], [3, 4]])

$$\begin{bmatrix} 1 & 2 \\ 3 & 4 \end{bmatrix}$$

$$B := \langle\langle 1, 3 \rangle \mid \langle 2, 4 \rangle\rangle$$

$$\begin{bmatrix} 1 & 2 \\ 3 & 4 \end{bmatrix}$$

Matrix Elements

$B(1, 1)$

$$1$$

$B(1, 2)$

$$2$$

$B(2, 1)$

$$3$$

$B(2, 2)$

$$4$$

Matrix Addition and Subtraction

$A + B$

$$\begin{bmatrix} 2 & 4 \\ 6 & 8 \end{bmatrix}$$

$A - B$

$$\begin{bmatrix} 0 & 0 \\ 0 & 0 \end{bmatrix}$$

Matrix Multiplication

$A.B$

$$\begin{bmatrix} 7 & 10 \\ 15 & 22 \end{bmatrix}$$

$$V1 := \langle 1, 2, 3 \rangle$$

$$\begin{bmatrix} 1 \\ 2 \\ 3 \end{bmatrix}$$

$V2 := \langle 1|2|3 \rangle$

$V1.V2$

$$\begin{bmatrix} 1 & 2 & 3 \\ 2 & 4 & 6 \\ 3 & 6 & 9 \end{bmatrix}$$

$V2.V1$

$$14$$

Matrix Integration and Differentiation

$U := \langle\langle x, x^3 \rangle \mid \langle x^2, x^4 \rangle\rangle$

$$\begin{bmatrix} x & x^2 \\ x^3 & x^4 \end{bmatrix}$$

$Diff(U, x) = map(diff, U, x)$

$$\frac{\partial}{\partial x}\begin{bmatrix} x & x^2 \\ x^3 & x^4 \end{bmatrix} = \begin{bmatrix} 1 & 2x \\ 3x^2 & 4x^3 \end{bmatrix}$$

$Int(U, x) = map(int, U, x)$

$$\int \begin{bmatrix} x & x^2 \\ x^3 & x^4 \end{bmatrix} dx = \begin{bmatrix} \dfrac{1}{2}x^2 & \dfrac{1}{3}x^3 \\ \dfrac{1}{4}x^4 & \dfrac{1}{5}x^5 \end{bmatrix}$$

Matrix Commutation and Anticommutation

$A := \langle\langle 1, 3 \rangle \mid \langle 2, 4 \rangle\rangle$

$$\begin{bmatrix} 1 & 2 \\ 3 & 4 \end{bmatrix}$$

$B := \langle\langle a, c \rangle \mid \langle b, d \rangle\rangle$

$$\begin{bmatrix} a & b \\ c & d \end{bmatrix}$$

$A.B - B.A$

$$\begin{bmatrix} 2c - 3b & -3b + 2d - 2a \\ 3a + 3c - 3d & 3b - 2c \end{bmatrix}$$

$A.B + B.A$

$$\begin{bmatrix} 2a + 2c + 3b & 5b + 2d + 2a \\ 3a + 5c + 3d & 3b + 8d + 2c \end{bmatrix}$$

Kronecker (Direct) Product

$KroneckerProduct(A, B)$

$$\begin{bmatrix} a & b & 2a & 2b \\ c & d & 2c & 2d \\ 3a & 3b & 4a & 4b \\ 3c & 3c & 4c & 4d \end{bmatrix}$$

$KroneckerProduct(B, A)$

$$\begin{bmatrix} a & 2a & b & 2b \\ 3a & 4a & 3b & 4b \\ c & 2c & d & 2d \\ 3c & 4c & 3d & 4d \end{bmatrix}$$

Matrix Transpose

$V1 := \langle 1, 2, 3 \rangle$

$$\begin{bmatrix} 1 \\ 2 \\ 3 \end{bmatrix}$$

$Transpose(V1)$

$$\begin{bmatrix} 1 & 2 & 3 \end{bmatrix}$$

$V1^{\%T}$

$$\begin{bmatrix} 1 & 2 & 3 \end{bmatrix}$$

Hermitian Conjugate (Adjoint)

$V1 := \langle I, 0, 2 - I \rangle$

$$\begin{bmatrix} I \\ 0 \\ 2 - I \end{bmatrix}$$

HermitianTranspose(V1)

$$\begin{bmatrix} -I & 0 & 2+I \end{bmatrix}$$

V1$^{\%H}$

$$\begin{bmatrix} -I & 0 & 2+I \end{bmatrix}$$

$$\psi := \frac{1}{sqrt(2)} \big\langle \exp(I \cdot x), 0, \exp(-I \cdot x) \big\rangle$$

$$\begin{bmatrix} \dfrac{1}{2}\sqrt{2}e^{Ix} \\[2mm] 0 \\[2mm] \dfrac{1}{2}\sqrt{2}e^{-Ix} \end{bmatrix}$$

$\psi^{\%H} . \psi$

$$\frac{1}{2} e^{-\overline{Ix}} e^{Ix} + \frac{1}{2} e^{\overline{Ix}} e^{-Ix}$$

assume(x, 'real')
$\psi^{\%H} . \psi$

$$1$$

$A := \big\langle \langle 1+I, I \rangle \,|\, \langle 1-I, 4 \cdot I \rangle \big\rangle$

$$\begin{bmatrix} 1+I & 1-I \\ I & 4I \end{bmatrix}$$

$A^{\%H}$

$$\begin{bmatrix} 1-I & -I \\ 1+I & -4I \end{bmatrix}$$

Re(%)

$$\begin{bmatrix} 1 & 0 \\ 1 & 0 \end{bmatrix}$$

Im(%%)

$$\begin{bmatrix} -1 & -1 \\ 1 & -4 \end{bmatrix}$$

Matrix Trace

$$R := \langle\langle 1,5,9 \rangle \,|\, \langle 2,3,6 \rangle \,|\, \langle 0,1,8 \rangle\rangle$$

$$\begin{bmatrix} 1 & 2 & 0 \\ 5 & 3 & 1 \\ 9 & 6 & 8 \end{bmatrix}$$

$Trace(R)$

$$12$$

Matrix Exponential

$$Jy := \langle\langle 0|0|\text{-}I \rangle, \langle 0|0|0 \rangle, \langle I|0|0 \rangle\rangle$$

$$\begin{bmatrix} 0 & 0 & -I \\ 0 & 0 & 0 \\ I & 0 & 0 \end{bmatrix}$$

$MatrixExponential(I \cdot Jy,\, phi)$

$$\begin{bmatrix} \cos(\phi) & 0 & \sin(\phi) \\ 0 & 1 & 0 \\ -\sin(\phi) & 0 & \cos(\phi) \end{bmatrix}$$

2.3 MATRIX DETERMINANT AND INVERSE

In this section, matrix determinants and matrix inverses are discussed. Determinants and inverses are encountered in solving systems of equations common in physics, finding eigenvalues and eigenvectors, and matrix diagonalization. The determinant of a matrix $\mathbf{A}$ is indicated by $|\mathbf{A}|$ or as $\det(\mathbf{A})$. Given the 2×2 matrix

$$\mathbf{A} = \begin{pmatrix} a & b \\ c & d \end{pmatrix} \tag{2.3.1}$$

the determinant is simply $|\mathbf{A}| = ad - cb$.

In general, any row or column may be expanded about to calculate the determinant of an $N \times N$ matrix. The determinant is the sum of a_{ij} multiplied by determinants of the $(N-1) \times (N-1)$ submatrices formed by striking out the ith row and the jth column.

Example 2.3.1

The determinant of the 3×3 matrix

$$\mathbf{A} = \begin{pmatrix} a_{11} & a_{12} & a_{13} \\ a_{21} & a_{22} & a_{23} \\ a_{31} & a_{32} & a_{33} \end{pmatrix} \tag{2.3.2}$$

may be obtained by expanding across the top row

$$|\mathbf{A}| = a_{11}(-1)^{1+1}\begin{vmatrix} a_{22} & a_{23} \\ a_{32} & a_{33} \end{vmatrix} + a_{12}(-1)^{1+2}\begin{vmatrix} a_{21} & a_{23} \\ a_{31} & a_{33} \end{vmatrix} + a_{13}(-1)^{1+3}\begin{vmatrix} a_{21} & a_{22} \\ a_{31} & a_{32} \end{vmatrix} \tag{2.3.3}$$

giving

$$|\mathbf{A}| = a_{11}(a_{22}a_{33} - a_{32}a_{23}) - a_{12}(a_{21}a_{33} - a_{31}a_{23}) + a_{13}(a_{21}a_{32} - a_{31}a_{22}) \tag{2.3.4}$$

It is easier to expand about the row or column containing more zeros.

Example 2.3.2

We may calculate the determinant of

$$\mathbf{A} = \begin{pmatrix} 1 & 3 & 9 \\ 2 & 6 & 0 \\ 5 & 7 & 0 \end{pmatrix} \tag{2.3.5}$$

by expanding about the third column

$$|\mathbf{A}| = 9(-1)^{(1+3)}\begin{vmatrix} 2 & 6 \\ 5 & 7 \end{vmatrix} + 0(-1)^{(2+3)}\begin{vmatrix} 1 & 3 \\ 5 & 7 \end{vmatrix} + 0(-1)^{(3+3)}\begin{vmatrix} 1 & 3 \\ 2 & 6 \end{vmatrix} = -144 \tag{2.3.6}$$

2.3.1 Matrix Inverse

The inverse of the matrix $\mathbf{A}$ is denoted $\mathbf{A}^{-1}$. The identity matrix results from multiplying a matrix and its inverse

$$\mathbf{A}^{-1}\mathbf{A} = \mathbf{A}\mathbf{A}^{-1} = \mathbf{I} \tag{2.3.7}$$

The matrix inverse may be computed as

$$\mathbf{A}^{-1} = \frac{1}{|\mathbf{A}|}\mathbf{A}^{C\dagger} \tag{2.3.8}$$

where $\mathbf{A}^C$ is a matrix of cofactors

$$\mathbf{A}^C = \begin{pmatrix} a_{11}^c & \cdots & a_{1N}^c \\ \vdots & \ddots & \vdots \\ a_{N1}^c & \cdots & a_{NN}^c \end{pmatrix} \tag{2.3.9}$$

For a 3 × 3 matrix

$$\mathbf{A}^C = \begin{pmatrix} a_{11}^c & a_{12}^c & a_{13}^c \\ a_{21}^c & a_{22}^c & a_{23}^c \\ a_{31}^c & a_{32}^c & a_{33}^c \end{pmatrix} \tag{2.3.10}$$

the cofactors are

$$a_{11}^c = (-1)^{(1+1)} \begin{vmatrix} a_{22} & a_{23} \\ a_{32} & a_{33} \end{vmatrix} \quad a_{12}^c = (-1)^{(1+2)} \begin{vmatrix} a_{21} & a_{23} \\ a_{31} & a_{33} \end{vmatrix} \quad a_{13}^c = (-1)^{(1+3)} \begin{vmatrix} a_{21} & a_{22} \\ a_{31} & a_{32} \end{vmatrix}$$

$$a_{21}^c = (-1)^{(2+1)} \begin{vmatrix} a_{12} & a_{13} \\ a_{32} & a_{33} \end{vmatrix} \quad a_{22}^c = (-1)^{(2+2)} \begin{vmatrix} a_{11} & a_{13} \\ a_{31} & a_{33} \end{vmatrix} \quad a_{23}^c = (-1)^{(2+3)} \begin{vmatrix} a_{11} & a_{12} \\ a_{31} & a_{32} \end{vmatrix}$$

$$a_{31}^c = (-1)^{(3+1)} \begin{vmatrix} a_{12} & a_{13} \\ a_{22} & a_{23} \end{vmatrix} \quad a_{32}^c = (-1)^{(3+2)} \begin{vmatrix} a_{11} & a_{13} \\ a_{21} & a_{23} \end{vmatrix} \quad a_{33}^c = (-1)^{(3+3)} \begin{vmatrix} a_{11} & a_{12} \\ a_{21} & a_{22} \end{vmatrix}$$

$$\tag{2.3.11}$$

The inverse of a 2 × 2 matrix is simply

$$\begin{pmatrix} a & b \\ c & d \end{pmatrix}^{-1} = \frac{1}{ad - bc} \begin{pmatrix} d & -b \\ -c & a \end{pmatrix} \tag{2.3.12}$$

2.3.2 Singular Matrices

A singular matrix is a matrix that has a zero determinant as well as an undefined inverse.

Example 2.3.3

The singular matrix

$$\mathbf{A} = \begin{pmatrix} 1 & 0 & 1 \\ 0 & 0 & 0 \\ 1 & 0 & 1 \end{pmatrix} \tag{2.3.13}$$

has $|\mathbf{A}| = 0$ and is thus noninvertible. Matrices with linearly dependent rows or columns are singular.

2.3.3 Systems of Equations

The system of linear equations

$$a_1 x + b_1 y + c_1 z = d_1$$
$$a_2 x + b_2 y + c_2 z = d_2 \qquad (2.3.14)$$
$$a_3 x + b_3 y + c_3 z = d_3$$

is written in matrix form

$$\begin{pmatrix} a_1 & b_1 & c_1 \\ a_2 & b_2 & c_2 \\ a_3 & b_3 & c_3 \end{pmatrix} \begin{pmatrix} x \\ y \\ z \end{pmatrix} = \begin{pmatrix} d_1 \\ d_2 \\ d_3 \end{pmatrix} \qquad (2.3.15)$$

Multiplying both sides by the inverse of the coefficient matrix

$$\begin{pmatrix} x \\ y \\ z \end{pmatrix} = \begin{pmatrix} a_1 & b_1 & c_1 \\ a_2 & b_2 & c_2 \\ a_3 & b_3 & c_3 \end{pmatrix}^{-1} \begin{pmatrix} d_1 \\ d_2 \\ d_3 \end{pmatrix} \qquad (2.3.16)$$

Example 2.3.4

The system

$$2x + y - z = 2$$
$$2x + 2y - z = 4 \qquad (2.3.17)$$
$$-x - y + z = -1$$

is written in matrix form

$$\begin{pmatrix} 2 & 1 & -1 \\ 2 & 2 & -1 \\ -1 & -1 & 1 \end{pmatrix} \begin{pmatrix} x \\ y \\ z \end{pmatrix} = \begin{pmatrix} 2 \\ 4 \\ -1 \end{pmatrix} \qquad (2.3.18)$$

and the coefficient matrix is inverted

$$\begin{pmatrix} x \\ y \\ z \end{pmatrix} = \begin{pmatrix} 2 & 1 & -1 \\ 2 & 2 & -1 \\ -1 & -1 & 1 \end{pmatrix}^{-1} \begin{pmatrix} 2 \\ 4 \\ -1 \end{pmatrix} \qquad (2.3.19)$$

with solution $(x, y, z) = (1, 2, 2)$.

Geometrically the solution to the three linear equations corresponds to finding the intersection of three planes at the point (x, y, z). There is no solution if the three planes are parallel. There are infinitely many solutions if the planes intersect along one, two or three lines and the inverse of the coefficient matrix is undefined.

Cramer's rule may also be used to solve a system of equations where the solution (x, y, z) is determined from ratios of determinants. For three equations and three unknowns

$$x = \frac{\begin{vmatrix} d_1 & b_1 & c_1 \\ d_2 & b_2 & c_2 \\ d_3 & b_3 & c_3 \end{vmatrix}}{\begin{vmatrix} a_1 & b_1 & c_1 \\ a_2 & b_2 & c_2 \\ a_3 & b_3 & c_3 \end{vmatrix}} \quad y = \frac{\begin{vmatrix} a_1 & d_1 & c_1 \\ a_2 & d_2 & c_2 \\ a_3 & d_3 & c_3 \end{vmatrix}}{\begin{vmatrix} a_1 & b_1 & c_1 \\ a_2 & b_2 & c_2 \\ a_3 & b_3 & c_3 \end{vmatrix}} \quad z = \frac{\begin{vmatrix} a_1 & b_1 & d_1 \\ a_2 & b_2 & d_2 \\ a_3 & b_3 & d_3 \end{vmatrix}}{\begin{vmatrix} a_1 & b_1 & c_1 \\ a_2 & b_2 & c_2 \\ a_3 & b_3 & c_3 \end{vmatrix}} \quad (2.3.20)$$

where the constants d_1, d_2, and d_3 replace the first, second and third columns in the top coefficient matrix determinants in calculating x, y, and z, respectively. For a system of two equations

$$a_1 x + b_1 y = d_1$$
$$a_2 x + b_2 y = d_2 \quad (2.3.21)$$

Cramer's rule gives the unknowns

$$x = \frac{\begin{vmatrix} d_1 & b_1 \\ d_2 & b_2 \end{vmatrix}}{\begin{vmatrix} a_1 & b_1 \\ a_2 & b_2 \end{vmatrix}} \quad y = \frac{\begin{vmatrix} a_1 & d_1 \\ a_2 & d_2 \end{vmatrix}}{\begin{vmatrix} a_1 & b_1 \\ a_2 & b_2 \end{vmatrix}} \quad (2.3.22)$$

Maple Examples

Matrix examples in the Maple worksheet below include matrix determinant, matrix inverse, and the use of matrices in solving systems of linear equations.

Key Maple commands: *Determinant, implicitplot3d, MatrixInverse, solve*

Maple packages: *with(LinearAlgebra): with(plots):*

restart

Matrix Determinant

$with(LinearAlgebra):$
$C := \langle\langle 1|1|0\rangle, \langle -1|0|1\rangle, \langle 1|0|-2\rangle\rangle$

$$\begin{bmatrix} 1 & 1 & 0 \\ -1 & 0 & 1 \\ 1 & 0 & -2 \end{bmatrix}$$

$Determinant(C)$

$$-1$$

Matrix Inverse

$MatrixInverse(C)$

$$\begin{bmatrix} 0 & -2 & -1 \\ 1 & 2 & 1 \\ 0 & -1 & -1 \end{bmatrix}$$

$C.MatrixInverse(C)$

$$\begin{bmatrix} 1 & 0 & 0 \\ 0 & 1 & 0 \\ 0 & 0 & 1 \end{bmatrix}$$

Systems of Equations

$with(plots):$

$eqns := [x + 2 \cdot y = -3, 3 \cdot x + 2 \cdot y = 3, x + z = 1,]$

$$[x + 2y = -3, 3x + 2y = 3, x + z = 1,]$$

$Implicitplot3d(eqns, x = -5 \ldots 5, y = -5 \ldots 5, z = -5 \ldots 5, color = [blue, green, red],$
$scaling = constrained, axes = boxed)$

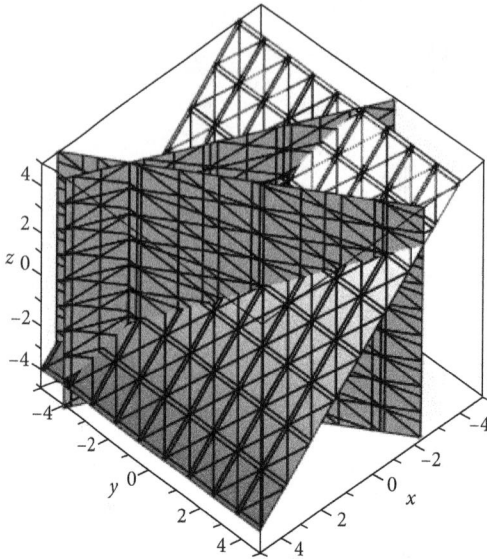

Figure 2.3.1: Maple output plotting the intersection of three planes.

solve(eqns, {x, y, z}

$$\{x = 3, y = -3, z = -2\}$$

2.4 EIGENVALUES AND EIGENVECTORS

In physics, the eigenvalue problem arises in the analysis of rotating bodies, small oscillations and the calculation of energies in quantum systems. In quantum mechanics states corresponding to specific energy values can be represented by column vectors. The eigenvalue equation is written

$$\mathbf{AX} = \lambda\mathbf{X} \qquad (2.4.1)$$

where $\mathbf{A}$ is an $N \times N$ matrix. The $\mathbf{X}$ are $N \times 1$ eigenvectors to be determined and l are the eigenvalues. The eigenvalue equation can be expressed as

$$\mathbf{AX} = \lambda\mathbf{IX} \qquad (2.4.2)$$

or as

$$\left(\mathbf{A} - \lambda\mathbf{I}\right)X = 0 \qquad (2.4.3)$$

where $\mathbf{I}$ is the $N \times N$ identity matrix. The polynomial equation formed by setting the determinant equal to zero

$$\det\left(\mathbf{A} - \lambda\mathbf{I}\right) = 0 \qquad (2.4.4)$$

is called the characteristic equation. Those values of l that satisfy the characteristic equation are the eigenvalues. Once the eigenvalues are found the eigenvector $\mathbf{X}$ corresponding to each may then be determined.

Example 2.4.1

Find the eigenvalues and eigenvectors corresponding to the matrix

$$\mathbf{A} = \begin{pmatrix} 1 & 0 \\ -1 & 2 \end{pmatrix} \tag{2.4.5}$$

Solution: Subtracting $\lambda \mathbf{I}$ from $\mathbf{A}$

$$\mathbf{A} - \lambda \mathbf{I} = \begin{pmatrix} 1-\lambda & 0 \\ -1 & 2-\lambda \end{pmatrix} \tag{2.4.6}$$

and setting the determinant equal to zero we obtain the characteristic equation

$$\det(\mathbf{A} - \lambda \mathbf{I}) = (1-\lambda)(2-\lambda) = 0 \tag{2.4.7}$$

The resulting eigenvalues are

$$\lambda_1 = 1, \lambda_2 = 2 \tag{2.4.8}$$

To find the eigenvector $\mathbf{X}_1$ corresponding to $\lambda_1 = 1$ we solve

$$\mathbf{A}\mathbf{X}_1 = \lambda_1 \mathbf{X}_1 \tag{2.4.9}$$

This gives

$$\begin{pmatrix} 1 & 0 \\ -1 & 2 \end{pmatrix} \begin{pmatrix} x_1 \\ x_2 \end{pmatrix} = 1 \cdot \begin{pmatrix} x_1 \\ x_2 \end{pmatrix} \tag{2.4.10}$$

or

$$\begin{aligned} x_1 &= x_1 \\ -x_1 + 2x_2 &= x_2 \end{aligned} \tag{2.4.11}$$

Choosing $x_1 = 1$ gives $x_2 = 1$ so that

$$\mathbf{X}_1 = \begin{pmatrix} 1 \\ 1 \end{pmatrix} \tag{2.4.12}$$

To find the eigenvector $\mathbf{X}_2$ corresponding to $\lambda_2 = 2$ we solve

$$\mathbf{A}\mathbf{X}_2 = \lambda_2 \mathbf{X}_1. \tag{2.4.13}$$

This gives

$$\begin{pmatrix} 1 & 0 \\ -1 & 2 \end{pmatrix} \begin{pmatrix} x_1 \\ x_2 \end{pmatrix} = 2 \cdot \begin{pmatrix} x_1 \\ x_2 \end{pmatrix} \tag{2.4.14}$$

or

$$x_1 = 2x_1$$
$$-x_1 + 2x_2 = 2x_2 \tag{2.4.15}$$

The first equation gives $x_1 = 0$. For the second equation we take $x_2 = 1$ so that

$$\mathbf{X}_2 = \begin{pmatrix} 0 \\ 1 \end{pmatrix} \tag{2.4.16}$$

The normalized eigenvectors are

$$\mathbf{X}_1 = \frac{1}{\sqrt{2}} \begin{pmatrix} 1 \\ 1 \end{pmatrix} \text{ and } \mathbf{X}_2 = \begin{pmatrix} 0 \\ 1 \end{pmatrix} \tag{2.4.17}$$

Example 2.4.2

Find the eigenvalues and eigenvectors corresponding to the matrix

$$\mathbf{A} = \begin{pmatrix} 1 & 0 & 1 \\ 0 & 1 & 0 \\ 1 & 1 & 0 \end{pmatrix} \tag{2.4.18}$$

Solution: Subtracting $\lambda \mathbf{I}$ from $\mathbf{A}$

$$\mathbf{A} - \lambda \mathbf{I} = \begin{pmatrix} 1-\lambda & 0 & 1 \\ 0 & 1-\lambda & 0 \\ 1 & 1 & -\lambda \end{pmatrix} \tag{2.4.19}$$

Taking the determinant by expanding across the top row

$$\det(\mathbf{A} - \lambda \mathbf{I}) = (1-\lambda) \begin{vmatrix} 1-\lambda & 0 \\ 1 & -\lambda \end{vmatrix} - 0 + 1 \begin{vmatrix} 0 & 1-\lambda \\ 1 & 1 \end{vmatrix} = 0 \tag{2.4.20}$$

and we have the characteristic equation

$$(1-\lambda)(\lambda^2 - \lambda) - (1-\lambda) = 0 \tag{2.4.21}$$

This is a cubic equation that we can solve by factoring $(1-\lambda)$

$$(1-\lambda)(\lambda^2 - \lambda - 1) = 0 \tag{2.4.22}$$

The resulting eigenvalues are

$$\lambda_1 = 1, \lambda_2 = \frac{1+\sqrt{5}}{2}, \lambda_3 = \frac{1-\sqrt{5}}{2} \tag{2.4.23}$$

To find the eigenvector $\mathbf{X}_1$ corresponding to $\lambda_1 = 1$ we solve

$$\mathbf{AX}_1 = \lambda_1\mathbf{X}_1 \tag{2.4.24}$$

This gives

$$\begin{pmatrix} 1 & 0 & 1 \\ 0 & 1 & 0 \\ 1 & 1 & 0 \end{pmatrix}\begin{pmatrix} x_1 \\ x_2 \\ x_3 \end{pmatrix} = \begin{pmatrix} x_1 \\ x_2 \\ x_3 \end{pmatrix} \tag{2.4.25}$$

and we have three equations

$$\begin{aligned} x_1 + x_3 &= x_1 \\ x_2 &= x_2 \\ x_1 + x_2 &= x_3 \end{aligned} \tag{2.4.26}$$

with solution $x_1 = -1$, $x_2 = 1$ and $x_3 = 0$ so that

$$\mathbf{X}_1 = \begin{pmatrix} -1 \\ 1 \\ 0 \end{pmatrix} \tag{2.4.27}$$

For the second eigenvector $\mathbf{X}_2$ we must solve $\mathbf{AX}_2 = \lambda_2\mathbf{X}_2$

$$\begin{pmatrix} 1 & 0 & 1 \\ 0 & 1 & 0 \\ 1 & 1 & 0 \end{pmatrix}\begin{pmatrix} x_1 \\ x_2 \\ x_3 \end{pmatrix} = \frac{1+\sqrt{5}}{2}\begin{pmatrix} x_1 \\ x_2 \\ x_3 \end{pmatrix} \tag{2.4.28}$$

giving

$$\begin{aligned} x_1 + x_3 &= \tfrac{1}{2}\left(1+\sqrt{5}\right)x_1 \\ x_2 &= \tfrac{1}{2}\left(1+\sqrt{5}\right)x_2 \\ x_1 + x_2 &= \tfrac{1}{2}\left(1+\sqrt{5}\right)x_3 \end{aligned} \tag{2.4.29}$$

The second equation gives $x_2 = 0$

$$\begin{aligned} x_3 &= \tfrac{1}{2}\left(-1+\sqrt{5}\right)x_1 \\ x_1 &= \tfrac{1}{2}\left(1+\sqrt{5}\right)x_3 \end{aligned} \tag{2.4.30}$$

One choice is

$$\begin{aligned} x_3 &= 2 \\ x_1 &= \left(1+\sqrt{5}\right) \end{aligned} \tag{2.4.31}$$

and

$$\mathbf{X}_2 = \begin{pmatrix} 1+\sqrt{5} \\ 0 \\ 2 \end{pmatrix} \tag{2.4.32}$$

A similar analysis solving $\mathbf{AX}_3 = \lambda_3\mathbf{X}_3$ gives

$$x_3 = \tfrac{1}{2}\left(-1-\sqrt{5}\right)x_1$$
$$x_1 = \tfrac{1}{2}\left(1-\sqrt{5}\right)x_3 \tag{2.4.33}$$

and

$$\mathbf{X}_3 = \begin{pmatrix} 1-\sqrt{5} \\ 0 \\ 2 \end{pmatrix} \tag{2.4.34}$$

Thus we have the normalized eigenvectors

$$\mathbf{X}_1 = \frac{1}{\sqrt{2}}\begin{pmatrix} -1 \\ 1 \\ 0 \end{pmatrix}$$

$$\mathbf{X}_2 = \frac{1}{\sqrt{10+2\sqrt{5}}}\begin{pmatrix} 1+\sqrt{5} \\ 0 \\ 2 \end{pmatrix}$$

$$\mathbf{X}_3 = \frac{1}{\sqrt{10-2\sqrt{5}}}\begin{pmatrix} 1-\sqrt{5} \\ 0 \\ 2 \end{pmatrix} \tag{2.4.35}$$

Example 2.4.3

Find the imaginary eigenvalues and eigenvectors of the matrix

$$\mathbf{A} = \begin{pmatrix} 0 & -1 \\ 1 & 0 \end{pmatrix} \tag{2.4.36}$$

Solution: Subtracting $\lambda\mathbf{I}$ from $\mathbf{A}$

$$\mathbf{A} - \lambda\mathbf{I} = \begin{pmatrix} -\lambda & -1 \\ 1 & -\lambda \end{pmatrix} \tag{2.4.37}$$

Taking the determinant

$$\det\left(\mathbf{A} - \lambda\mathbf{I}\right) = \lambda^2 + 1 = 0 \tag{2.4.38}$$

the resulting eigenvalues are

$$\lambda_1 = i, \lambda_2 = -i \tag{2.4.39}$$

To find the eigenvector $\mathbf{X}_1$ corresponding to $\lambda_1 = i$ we solve

$$\mathbf{A}\mathbf{X}_1 = \lambda_1\mathbf{X}_1 \tag{2.4.40}$$

This gives

$$\begin{pmatrix} 0 & -1 \\ 1 & 0 \end{pmatrix}\begin{pmatrix} x_1 \\ x_2 \end{pmatrix} = i \cdot \begin{pmatrix} x_1 \\ x_2 \end{pmatrix} \tag{2.4.41}$$

or

$$\begin{aligned} -x_2 &= ix_1 \\ x_1 &= ix_2 \end{aligned} \tag{2.4.42}$$

Choosing $x_1 = i$ gives $x_2 = 1$ so that

$$\mathbf{X}_1 = \begin{pmatrix} i \\ 1 \end{pmatrix} \tag{2.4.43}$$

To find the eigenvector $\mathbf{X}_2$ corresponding to $\lambda_2 = -i$ we solve

$$\mathbf{A}\mathbf{X}_2 = \lambda_2\mathbf{X}_1 \tag{2.4.44}$$

This gives

$$\begin{pmatrix} 0 & -1 \\ 1 & 0 \end{pmatrix}\begin{pmatrix} x_1 \\ x_2 \end{pmatrix} = -i \cdot \begin{pmatrix} x_1 \\ x_2 \end{pmatrix} \tag{2.4.45}$$

or

$$\begin{aligned} -x_2 &= -ix_1 \\ x_1 &= -ix_2 \end{aligned} \tag{2.4.46}$$

Choosing $x_1 = -i$ gives $x_2 = 1$ so that

$$\mathbf{X}_2 = \begin{pmatrix} -i \\ 1 \end{pmatrix} \tag{2.4.47}$$

The normalized eigenvectors are

$$\mathbf{X}_1 = \frac{1}{\sqrt{2}}\begin{pmatrix} i \\ 1 \end{pmatrix} \text{ and } \mathbf{X}_2 = \frac{1}{\sqrt{2}}\begin{pmatrix} -i \\ 1 \end{pmatrix} \tag{2.4.48}$$

2.4.1 Matrix Diagonalization

The technique of matrix diagonalization has application in the dynamics of rigid bodies and quantum mechanics. Consider the matrix

$$\mathbf{A} = \begin{pmatrix} 1 & 0 \\ -1 & 2 \end{pmatrix} \tag{2.4.49}$$

with normalized eigenvectors

$$\mathbf{X}_1 = \frac{1}{\sqrt{2}} \begin{pmatrix} 1 \\ 1 \end{pmatrix} \text{ and } \mathbf{X}_2 = \begin{pmatrix} 0 \\ 1 \end{pmatrix} \tag{2.4.50}$$

corresponding to eigenvalues $\lambda_1 = 1$ and $\lambda_2 = 2$. We may calculate a diagonal matrix

$$\mathbf{D} = \begin{pmatrix} \lambda_1 & 0 \\ 0 & \lambda_2 \end{pmatrix} \tag{2.4.51}$$

from the operation

$$\mathbf{D} = \mathbf{P}^{-1}\mathbf{A}\mathbf{P} \tag{2.4.52}$$

where $\mathbf{P}$ is formed from the eigenvectors

$$\mathbf{P} = \begin{pmatrix} \dfrac{1}{\sqrt{2}} & 0 \\ \dfrac{1}{\sqrt{2}} & 1 \end{pmatrix} \tag{2.4.53}$$

The inverse is computed as

$$\mathbf{P}^{-1} = \sqrt{2} \begin{pmatrix} 1 & 0 \\ -\dfrac{1}{\sqrt{2}} & \dfrac{1}{\sqrt{2}} \end{pmatrix} \tag{2.4.54}$$

As a check, we verify

$$\mathbf{P}^{-1}\mathbf{P} = \sqrt{2} \begin{pmatrix} 1 & 0 \\ -\dfrac{1}{\sqrt{2}} & \dfrac{1}{\sqrt{2}} \end{pmatrix} \begin{pmatrix} \dfrac{1}{\sqrt{2}} & 0 \\ \dfrac{1}{\sqrt{2}} & 1 \end{pmatrix}$$

$$= \sqrt{2} \begin{pmatrix} \dfrac{1}{\sqrt{2}} & 0 \\ 0 & \dfrac{1}{\sqrt{2}} \end{pmatrix} = \begin{pmatrix} 1 & 0 \\ 0 & 1 \end{pmatrix} \tag{2.4.55}$$

Now to diagonalize **A**

$$\mathbf{P}^{-1}\mathbf{AP} = \sqrt{2}\begin{pmatrix} 1 & 0 \\ -\dfrac{1}{\sqrt{2}} & \dfrac{1}{\sqrt{2}} \end{pmatrix}\begin{pmatrix} 1 & 0 \\ -1 & 2 \end{pmatrix}\begin{pmatrix} \dfrac{1}{\sqrt{2}} & 0 \\ \dfrac{1}{\sqrt{2}} & 1 \end{pmatrix}$$

$$= \sqrt{2}\begin{pmatrix} 1 & 0 \\ -\dfrac{1}{\sqrt{2}} & \dfrac{1}{\sqrt{2}} \end{pmatrix}\begin{pmatrix} \dfrac{1}{\sqrt{2}} & 0 \\ -\dfrac{1}{\sqrt{2}}+\dfrac{2}{\sqrt{2}} & 2 \end{pmatrix} \qquad (2.4.56)$$

$$= \sqrt{2}\begin{pmatrix} \dfrac{1}{\sqrt{2}} & 0 \\ 0 & \dfrac{2}{\sqrt{2}} \end{pmatrix} = \begin{pmatrix} 1 & 0 \\ 0 & 2 \end{pmatrix} = \mathbf{D}$$

Thus, we have constructed a diagonal matrix whose elements are the eigenvalues of **A**.

Maple Examples

Matrix eigenvalue, eigenvector, and diagonalization calculations are demonstrated in the Maple worksheet below.

Key Maple commands: Eigenvectors, evalf, MatrixInverse, Simplify

Maple packages: *with* (*LinearAlgebra*)

restart

Eigenvalues and Eigenvectors

with(*LinearAlgebra*) :
$C := \langle\langle 1|2|3\rangle, \langle 4|5|6\rangle, \langle 7|8|9\rangle\rangle$

$$\begin{bmatrix} 1 & 2 & 3 \\ 4 & 5 & 6 \\ 7 & 8 & 9 \end{bmatrix}$$

$(evals, evecs) := Eigenvectors(C)$

$$
\begin{bmatrix} \dfrac{15}{2} + \dfrac{3}{2}\sqrt{33} \\[2ex] \dfrac{15}{2} - \dfrac{3}{2}\sqrt{33} \\[2ex] 0 \end{bmatrix}, \quad
\begin{bmatrix} \dfrac{4}{\dfrac{11}{2} + \dfrac{3}{2}\sqrt{33}} & \dfrac{4}{\dfrac{11}{2} - \dfrac{3}{2}\sqrt{33}} & 1 \\[4ex] \dfrac{1}{2}\dfrac{\dfrac{19}{2} + \dfrac{3}{2}\sqrt{33}}{\dfrac{11}{2} + \dfrac{3}{2}\sqrt{33}} & \dfrac{1}{2}\dfrac{\dfrac{19}{2} - \dfrac{3}{2}\sqrt{33}}{\dfrac{11}{2} - \dfrac{3}{2}\sqrt{33}} & -2 \\[4ex] 1 & 1 & 1 \end{bmatrix}
$$

Matrix of Eigenvectors

$evecs$

$$
\begin{bmatrix} \dfrac{4}{\dfrac{11}{2} + \dfrac{3}{2}\sqrt{33}} & \dfrac{4}{\dfrac{11}{2} - \dfrac{3}{2}\sqrt{33}} & 1 \\[4ex] \dfrac{1}{2}\dfrac{\dfrac{19}{2} + \dfrac{3}{2}\sqrt{33}}{\dfrac{11}{2} + \dfrac{3}{2}\sqrt{33}} & \dfrac{1}{2}\dfrac{\dfrac{19}{2} - \dfrac{3}{2}\sqrt{33}}{\dfrac{11}{2} - \dfrac{3}{2}\sqrt{33}} & -2 \\[4ex] 1 & 1 & 1 \end{bmatrix}
$$

Specific Eigenvalue with Eigenvector

$evals[1], evecs[1 \dots 3, 1]$

$$
\dfrac{15}{2} + \dfrac{3}{2}\sqrt{33}, \quad
\begin{bmatrix} \dfrac{4}{\dfrac{11}{2} + \dfrac{3}{2}\sqrt{33}} \\[4ex] \dfrac{1}{2}\dfrac{\dfrac{19}{2} + \dfrac{3}{2}\sqrt{33}}{\dfrac{11}{2} + \dfrac{3}{2}\sqrt{33}} \\[4ex] 1 \end{bmatrix}
$$

$evals[2], evecs[1 \ldots 3, 2]$

$$\frac{15}{2} - \frac{3}{2}\sqrt{33}, \begin{bmatrix} \dfrac{4}{\dfrac{11}{2} - \dfrac{3}{2}\sqrt{33}} \\[3ex] \dfrac{1 \dfrac{19}{2} - \dfrac{3}{2}\sqrt{33}}{2 \dfrac{11}{2} - \dfrac{3}{2}\sqrt{33}} \\[3ex] 1 \end{bmatrix}$$

$evals[3], evecs[1 \ldots 3, 3]$

$$0, \begin{bmatrix} 1 \\ -2 \\ 1 \end{bmatrix}$$

Matrix Diagonalization

$simplify(MatrixInverse(evecs).C.evecs)$

$$\begin{bmatrix} \dfrac{15}{2} + \dfrac{3}{2}\sqrt{33} & 0 & 0 \\[2ex] 0 & \dfrac{15}{2} - \dfrac{3}{2}\sqrt{33} & 0 \\[2ex] 0 & 0 & 0 \end{bmatrix}$$

$evalf(\%)$

$$\begin{bmatrix} 16.11684397 & 0 & 0 \\ 0 & -1.116843970 & 0 \\ 0 & 0 & 0 \end{bmatrix}$$

2.5 ROTATION MATRICES

In this section, we review matrices that transform the components of a vector under rotation of the coordinate system. The transformation of 2D vector components by coordinate rotation is first discussed. This result is extended to consider the transformation of 3D vector components by rotations about all three coordinate axes.

2.5.1 Rotations in Two Dimensions

Consider the vector

$$\mathbf{A} = A_x\hat{\mathbf{i}} + A_y\hat{\mathbf{j}} \tag{2.5.1}$$

If the vector makes an angle q with respect to the x-axis, then its components are expressed as

$$A_x = A\cos\theta$$
$$A_y = A\sin\theta \tag{2.5.2}$$

We will now look at components of $\mathbf{A}$ in a coordinate system that has been rotated counterclockwise about the z-axis by an angle f as shown in Figure 2.5.1.

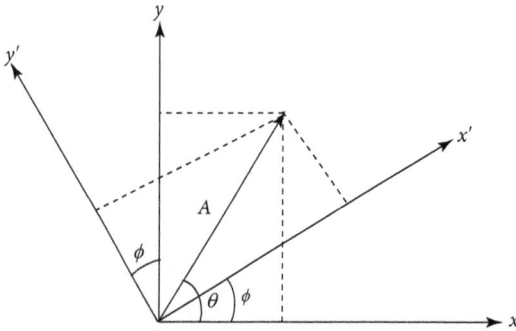

Figure 2.5.1: Rotation of the coordinate system by f.

In the transformed coordinate system $\theta' = \theta - \phi$ so the components

$$A_{x'} = A\cos\theta'$$
$$A_{y'} = A\sin\theta' \tag{2.5.3}$$

To develop a relation between the components of $\mathbf{A}$ in the primed and unprimed coordinate systems we use the identities

$$\cos(\theta') = \cos(\theta - \phi) = \cos\theta\cos\phi + \sin\theta\sin\phi$$
$$\sin(\theta') = \sin(\theta - \phi) = \sin\theta\cos\phi - \cos\theta\sin\phi \tag{2.5.4}$$

and

$$A_{x'} = \underbrace{A\cos\theta}_{A_x}\cos\phi + \underbrace{A\sin\theta}_{A_y}\sin\phi$$
$$A_{y'} = \underbrace{A\sin\theta}_{A_y}\cos\phi - \underbrace{A\cos\theta}_{A_x}\sin\phi \tag{2.5.5}$$

Thus we have that

$$A_{x'} = A_x \cos\phi + A_y \sin\phi$$
$$A_{y'} = -A_x \sin\phi + A_y \cos\phi \tag{2.5.6}$$

Writing the vector $\mathbf{A} = A_x\hat{\mathbf{i}} + A_y\hat{\mathbf{j}}$ as $\begin{pmatrix} A_x \\ A_y \end{pmatrix}$ we express the rotation about the z-axis in matrix form

$$\begin{pmatrix} A_{x'} \\ A_{y'} \end{pmatrix} = \begin{pmatrix} \cos\phi & \sin\phi \\ -\sin\phi & \cos\phi \end{pmatrix}\begin{pmatrix} A_x \\ A_y \end{pmatrix} \tag{2.5.7}$$

or as

$$\mathbf{A}' = R(\phi)\mathbf{A} \tag{2.5.8}$$

The matrix $R(\phi) = \begin{pmatrix} \cos\phi & \sin\phi \\ -\sin\phi & \cos\phi \end{pmatrix}$ is referred to as a rotation matrix. Operation by $R_z(\phi)$ may either be interpreted as a counterclockwise rotation of the coordinate system or a clockwise rotation of the vector about the z-axis. A clockwise rotation of the coordinate system is performed with $R(-\phi) = \begin{pmatrix} \cos\phi & -\sin\phi \\ \sin\phi & \cos\phi \end{pmatrix}$. If we first rotate by ϕ and then by $-\phi$ according to

$$\mathbf{A}' = R(-\phi)R(\phi)\mathbf{A} \tag{2.5.9}$$

the vector components should be unchanged

$$\begin{pmatrix} A_{x'} \\ A_{y'} \end{pmatrix} = \begin{pmatrix} \cos\phi & -\sin\phi \\ \sin\phi & \cos\phi \end{pmatrix}\begin{pmatrix} \cos\phi & \sin\phi \\ -\sin\phi & \cos\phi \end{pmatrix}\begin{pmatrix} A_x \\ A_y \end{pmatrix} \tag{2.5.10}$$

and

$$\begin{pmatrix} A_{x'} \\ A_{y'} \end{pmatrix} = \begin{pmatrix} \cos^2\phi + \sin^2\phi & 0 \\ 0 & \cos^2\phi + \sin^2\phi \end{pmatrix}\begin{pmatrix} A_x \\ A_y \end{pmatrix} \tag{2.5.11}$$

Using the relation $\cos^2\phi + \sin^2\phi = 1$

$$\begin{pmatrix} A_{x'} \\ A_{y'} \end{pmatrix} = \begin{pmatrix} 1 & 0 \\ 0 & 1 \end{pmatrix}\begin{pmatrix} A_x \\ A_y \end{pmatrix} \tag{2.5.12}$$

and we see that the components of $\mathbf{A}$ are unchanged

$$\begin{pmatrix} A_{x'} \\ A_{y'} \end{pmatrix} = \begin{pmatrix} A_x \\ A_y \end{pmatrix} \tag{2.5.13}$$

Observe that $R(-\phi)R(\phi) = \mathbf{I}$ and that $R(-\phi) = R^T(\phi)$ so that $RR^T = \mathbf{I}$. Since $RR^{-1} = \mathbf{I}$ we have the orthogonal property of rotation matrices $R^T = R^{-1}$.

2.5.2 Rotations in Three Dimensions

We can extend this analysis to three dimensions where

$$\mathbf{A} = A_x \hat{\mathbf{i}} + A_y \hat{\mathbf{j}} + A_z \hat{\mathbf{k}} \tag{2.5.14}$$

and consider the rotation of vectors about each coordinate axis. Rotation of the coordinate axes about the z-axis by the rotation matrix

$$R_z = \begin{pmatrix} \cos(\phi) & \sin(\phi) & 0 \\ -\sin(\phi) & \cos(\phi) & 0 \\ 0 & 0 & 1 \end{pmatrix} \tag{2.5.15}$$

will leave z-component A_z unaffected as can be seen from the operation

$$\begin{pmatrix} A_{x'} \\ A_{y'} \\ A_{z'} \end{pmatrix} = \begin{pmatrix} \cos(\phi) & \sin(\phi) & 0 \\ -\sin(\phi) & \cos(\phi) & 0 \\ 0 & 0 & 1 \end{pmatrix} \begin{pmatrix} A_x \\ A_y \\ A_z \end{pmatrix} \tag{2.5.16}$$

where $A_{z'} = A_z$. Similarly, the rotation matrix

$$R_x = \begin{pmatrix} 1 & 0 & 0 \\ 0 & \cos(\theta) & \sin(\theta) \\ 0 & -\sin(\theta) & \cos(\theta) \end{pmatrix} \tag{2.5.17}$$

produces a counterclockwise rotation of the coordinate axes by an angle θ about the x-axis leaving A_x unaffected

$$\begin{pmatrix} A_{x'} \\ A_{y'} \\ A_{z'} \end{pmatrix} = \begin{pmatrix} 1 & 0 & 0 \\ 0 & \cos(\theta) & \sin(\theta) \\ 0 & -\sin(\theta) & \cos(\theta) \end{pmatrix} \begin{pmatrix} A_x \\ A_y \\ A_z \end{pmatrix} \tag{2.5.18}$$

where $A_{x'} = A_x$. As well

$$R_y = \begin{pmatrix} \cos(\psi) & 0 & \sin(\psi) \\ 0 & 1 & 0 \\ -\sin(\psi) & 0 & \cos(\psi) \end{pmatrix} \tag{2.5.19}$$

generates a rotation about the y-axis by an angle ψ

$$
\begin{pmatrix} A_{x'} \\ A_{y'} \\ A_{z'} \end{pmatrix} = \begin{pmatrix} \cos(\psi) & 0 & \sin(\psi) \\ 0 & 1 & 0 \\ -\sin(\psi) & 0 & \cos(\psi) \end{pmatrix} \begin{pmatrix} A_x \\ A_y \\ A_z \end{pmatrix}
$$

(2.5.20)

where $A_{y'} = A_y$. One may perform subsequent rotations about different axes simply by multiplying the rotation matrices. Rotation about the x-axis followed by a rotation about the y-axis is achieved by

$$
\mathbf{A'} = R_y R_x \mathbf{A}
$$

(2.5.21)

$$
\begin{pmatrix} A_{x'} \\ A_{y'} \\ A_{z'} \end{pmatrix} = \begin{pmatrix} \cos(\psi) & 0 & \sin(\psi) \\ 0 & 1 & 0 \\ -\sin(\psi) & 0 & \cos(\psi) \end{pmatrix} \begin{pmatrix} 1 & 0 & 0 \\ 0 & \cos(\theta) & \sin(\theta) \\ 0 & -\sin(\theta) & \cos(\theta) \end{pmatrix} \begin{pmatrix} A_x \\ A_y \\ A_z \end{pmatrix}
$$

(2.5.22)

$$
\begin{pmatrix} A_{x'} \\ A_{y'} \\ A_{z'} \end{pmatrix} = \begin{pmatrix} \cos(\psi) & -\sin(\psi)\sin(\theta) & \sin(\psi)\cos(\theta) \\ 0 & \cos(\theta) & \sin(\theta) \\ -\sin(\psi) & -\cos(\psi)\sin(\theta) & \cos(\psi)\cos(\theta) \end{pmatrix} \begin{pmatrix} A_x \\ A_y \\ A_z \end{pmatrix}
$$

(2.5.23)

If we first rotate about the y-axis and then rotate about the x-axis

$$
\begin{pmatrix} A_{x'} \\ A_{y'} \\ A_{z'} \end{pmatrix} = \begin{pmatrix} 1 & 0 & 0 \\ 0 & \cos(\theta) & \sin(\theta) \\ 0 & -\sin(\theta) & \cos(\theta) \end{pmatrix} \begin{pmatrix} \cos(\psi) & 0 & \sin(\psi) \\ 0 & 1 & 0 \\ -\sin(\psi) & 0 & \cos(\psi) \end{pmatrix} \begin{pmatrix} A_x \\ A_y \\ A_z \end{pmatrix}
$$

(2.5.24)

and we obtain a different result

$$
\begin{pmatrix} A_{x'} \\ A_{y'} \\ A_{z'} \end{pmatrix} = \begin{pmatrix} \cos(\psi) & 0 & \sin(\psi) \\ 0 & \cos(\theta) & \sin(\theta)\cos(\psi) \\ -\cos(\theta)\sin(\psi) & -\sin(\theta) & \cos(\theta)\cos(\psi) \end{pmatrix} \begin{pmatrix} A_x \\ A_y \\ A_z \end{pmatrix}
$$

(2.5.25)

We see that $R_y R_x \neq R_x R_y$ so that the order of rotation is important and the rotation matrices do not commute— $\left[R_x, R_y \right] \neq 0$.

2.5.3 Infinitesimal Rotations

Consider the transformation of $\mathbf{A}$ by rotation of the coordinate system about the z-axis by a small angle ε by the rotation matrix

$$R_z\left(\varepsilon\right)=\begin{pmatrix}1 & \varepsilon & 0 \\ -\varepsilon & 1 & 0 \\ 0 & 0 & 1\end{pmatrix}=\begin{pmatrix}1 & 0 & 0 \\ 0 & 1 & 0 \\ 0 & 0 & 1\end{pmatrix}+\begin{pmatrix}0 & \varepsilon & 0 \\ -\varepsilon & 0 & 0 \\ 0 & 0 & 0\end{pmatrix} \qquad (2.5.26)$$

The components of **A** in the new coordinate system are

$$\begin{pmatrix}A_{x'} \\ A_{y'} \\ A_{z'}\end{pmatrix}=\begin{pmatrix}A_x \\ A_y \\ A_z\end{pmatrix}+\begin{pmatrix}0 & \varepsilon & 0 \\ -\varepsilon & 0 & 0 \\ 0 & 0 & 0\end{pmatrix}\begin{pmatrix}A_x \\ A_y \\ A_z\end{pmatrix} \qquad (2.5.27)$$

Similarly, matrices generating infinitesimal rotations of the coordinate system about the x- and y- axes are

$$R_x\left(\varepsilon\right)=\begin{pmatrix}1 & 0 & 0 \\ 0 & 1 & -\varepsilon \\ 0 & \varepsilon & 1\end{pmatrix}$$

$$R_y\left(\varepsilon\right)=\begin{pmatrix}1 & 0 & -\varepsilon \\ 0 & 1 & 0 \\ \varepsilon & 0 & 1\end{pmatrix} \qquad (2.5.28)$$

The rotation matrices are related to the angular momentum matrices.

Maple Examples

Rotation matrices are demonstrated in the Maple worksheet below corresponding to successive rotations about three coordinate axes. The transpose and inverse of rotation matrices are compared.

Key Maple commands: *determinant, evalf, simplify, transpose, unapply*

Maple packages: *with(LinearAlgebra)*:

Programming: Function operation using '*unapply*'

restart

Rotation Matrices

with(LinearAlgebra) :
$Rx := \langle\langle 1, 0, 0\rangle \,|\, \langle 0, \cos(\text{theta}), -\sin(\text{theta})\rangle \,|\, \langle 0, \sin(\text{theta}), \cos(\text{theta})\rangle\rangle$

$$\begin{bmatrix}1 & 0 & 0 \\ 0 & \cos(\theta) & \sin(\theta) \\ 0 & -\sin(\theta) & \cos(\theta)\end{bmatrix}$$

$Ry := \langle\langle\cos(\text{psi}), 0, -\sin(\text{psi})\rangle \mid \langle 0, 1, 0\rangle \mid \langle\sin(\text{psi}), 0, \cos(\text{psi})\rangle\rangle$

$$\begin{bmatrix} \cos(\psi) & 0 & \sin(\psi) \\ 0 & 1 & 0 \\ -\sin(\psi) & 0 & \cos(\psi) \end{bmatrix}$$

$Rz := \langle\langle\cos(\text{phi}), -\sin(\text{psi}), 0\rangle \mid \langle\sin(\text{phi}), \cos(\text{phi}), 0\rangle \mid \langle 0, 0, 1\rangle\rangle$

$$\begin{bmatrix} \cos(\phi) & \sin(\phi) & 0 \\ -\sin(\phi) & \cos(\phi) & 0 \\ 0 & 0 & 1 \end{bmatrix}$$

$T := Rx.Ry.Rz$

$[[\cos(\psi)\cos(\phi), \cos(\psi)\sin(\phi), \sin(\psi)],$
 $[-\sin(\theta)\sin(\psi)\cos(\phi) - \cos(\theta)\sin(\phi), -\sin(\theta)\sin(\psi)\sin(\phi) + \cos(\theta)\cos(\phi),$
 $\sin(\theta)\cos(\psi)],$
 $[-\cos(\theta)\sin(\psi)\cos(\phi) + \sin(\theta)\sin(\phi), -\cos(\theta)\sin(\psi)\sin(\phi) - \sin(\theta)\cos(\phi),$
 $\cos(\theta)\cos(\psi)]$

$Transpose(T)$

$[[\cos(\psi)\cos(\phi), -\sin(\theta)\sin(\psi)\cos(\phi) - \cos(\theta)\sin(\phi), -\cos(\theta)\sin(\psi)\cos(\phi)$
 $+ \sin(\theta)\sin(\phi)],$
 $[\cos(\psi)\sin(\phi), -\sin(\theta)\sin(\psi)\sin(\phi) + \cos(\theta)\cos(\phi), -\cos(\theta)\sin(\psi)\sin(\phi)$
 $-\sin(\theta)\cos(\phi)],$
 $[\sin(\psi), \sin(\theta)\cos(\psi), \cos(\theta)\cos(\psi)]$

$simplify(T.Transpose(T))$

$$\begin{bmatrix} 1 & 0 & 0 \\ 0 & 1 & 0 \\ 0 & 0 & 1 \end{bmatrix}$$

$simplify(Determinant(T))$

$$1$$

Create a matrix function of theta, psi and phi

$U := unapply(T, \text{theta}, \text{psi}, \text{phi}) :$

$U(\text{theta}, 0, 0)$

$$\begin{bmatrix} 1 & 0 & 0 \\ 0 & \cos(\theta) & \sin(\theta) \\ 0 & -\sin(\theta) & \cos(\theta) \end{bmatrix}$$

$U(0, \text{psi}, 0)$

$$\begin{bmatrix} \cos(\psi) & 0 & \sin(\psi) \\ 0 & 1 & 0 \\ -\sin(\psi) & 0 & \cos(\psi) \end{bmatrix}$$

$U(0, 0, \text{phi})$

$$\begin{bmatrix} \cos(\phi) & \sin(\phi) & 0 \\ -\sin(\phi) & \cos(\phi) & 0 \\ 0 & 0 & 1 \end{bmatrix}$$

$U\left(\dfrac{Pi}{2}, \dfrac{Pi}{3}, \dfrac{Pi}{4}\right)$

$$\begin{bmatrix} \dfrac{1}{4}\sqrt{2} & \dfrac{1}{4}\sqrt{2} & \dfrac{1}{2}\sqrt{3} \\ -\dfrac{1}{4}\sqrt{3}\sqrt{2} & -\dfrac{1}{4}\sqrt{3}\sqrt{2} & \dfrac{1}{2} \\ \dfrac{1}{2}\sqrt{2} & -\dfrac{1}{2}\sqrt{2} & 0 \end{bmatrix}$$

evalf(%)

$$\begin{bmatrix} 0.3535533905 & 0.3535533905 & 0.8660254040 \\ -0.6123724358 & -0.6123724358 & 0.5000000000 \\ 0.7071067810 & -0.7071067810 & 0 \end{bmatrix}$$

$V1 := \langle 1, 2, 3 \rangle$

$$\begin{bmatrix} 1 \\ 2 \\ 3 \end{bmatrix}$$

$U(.01, .01, .01).V1$

$$\begin{bmatrix} 1.04989816999400 \\ 2.01969668162800 \\ 2.96960252316400 \end{bmatrix}$$

$U(\text{theta}, 0, 0).U(-\text{theta}, 0, 0)$

$$\begin{bmatrix} 1 & 0 & 0 \\ 0 & \cos(\theta)^2 + \sin(\theta)^2 & 0 \\ 0 & 0 & \cos(\theta)^2 + \sin(\theta)^2 \end{bmatrix}$$

simplify(%)

$$\begin{bmatrix} 1 & 0 & 0 \\ 0 & 1 & 0 \\ 0 & 0 & 1 \end{bmatrix}$$

2.6 MATLAB EXAMPLES

As an abbreviation for "Matrix Laboratory," MATLAB is the software of choice when handling large arrays for numerical computation. Examples of numeric and symbolic operations on vectors and matrices in MATLAB are shown below.

Key MATLAB commands: *conj, cross, det, dot, inv, kron, min, rand, simple, sym, syms, transpose*

Section 2.1 Vectors and Scalars in Physics

Specifying Vectors

```
>> % row vector
>> v1 = [ 1 2 3]
v1 =
     1     2     3
>> % column vector formed as the transpose of v1
>> v1'
ans =
     1
     2
     3
```

Vector Addition

```
>> v1 = [ 1 2 3];
>> v2 = [-1 -1 -3];
>> v1+v2
ans =
     0     1     0
```

Scalar Product of Vectors

```
>> v1 = [1 0 1];
>> v2 = [2 1 2];
>> dot(v1,v2)
ans =
     4
>> v1*v2'  % alternate form of the dot product
ans =
     4
>> sqrt(v1*v1') % magnitude of a vector
ans =
    1.4142
>> v1/sqrt(v1*v1') % normalized vector
ans =
    0.7071         0    0.7071
```

Vector Cross Product

```
>> v1 = [-1 0 1];
>> v2 = [0 1 0];
>> cross(v1,v2)
ans =
    -1     0    -1
```

Triple Vector Products

```
>> v1=[1 0 0];
>> v2=[-1 0 1];
>> v3 = [0 1 0];
>> cross(v1,cross(v2,v3))
ans =
     0     1     0
```

Section 2.2 Matrices in Physics

Specifying Matrices and Matrix Elements

```
>> A = zeros(5) % 5 x 5 zero matrix
A =
     0     0     0     0     0
     0     0     0     0     0
     0     0     0     0     0
     0     0     0     0     0
     0     0     0     0     0
>> % A = zeros(5,5) produces the same output
>> A = rand (5) % 5 x 5 random matrix
A =
    0.8147    0.0975    0.1576    0.1419    0.6557
    0.9058    0.2785    0.9706    0.4218    0.0357
    0.1270    0.5469    0.9572    0.9157    0.8491
    0.9134    0.9575    0.4854    0.7922    0.9340
    0.6324    0.9649    0.8003    0.9595    0.6787
  >> A(1,1) % element at first row and first column
```

```
ans =
    0.8147
>> A(5,1) % element at fifth row and first column
ans =
    0.6324
>> max(A) % returns max values in each column
ans =
    0.9134    0.9649    0.9706    0.9595    0.9340
>> max(max(A)) % returns max element in A
ans =
    0.9706
>> min(min(A)) % returns min element in A
ans =
    0.0357
```

Adding and Subtracting Matrices

```
>> A = [1 0 -1 ; 0 1 0; -1 0 1]
A =
    1    0   -1
    0    1    0
   -1    0    1
>> B = ones(3) % 3 x 3 ones matrix
B =
    1    1    1
    1    1    1
    1    1    1
>> A + B
ans =
    2    1    0
    1    2    1
    0    1    2
>> A- B
ans =
    0   -1   -2
   -1    0   -1
   -2   -1    0
```

Matrix Multiplication

```
>> A = [1 2 ; 3 4];
>> B = [0 -1 ; -1 0];
>> A*B
ans =
   -2   -1
   -4   -3
>> B*A
ans =
   -3   -4
   -1   -2
```

Matrix Commutation and Anticommutation

```
>> A*B-B*A % commutator [A,B]
ans =
```

```
      1     3
     -3    -1
>> A*B+B*A % anti-commutator {A,B}
ans =
     -5    -5
     -5    -5
```

Kronecker (Direct) Product

```
>> A=[1 0; 0 1]
A =
     1     0
     0     1
>> B=[2 -2 ; 1 5]
B =
     2    -2
     1     5
>> kron(A,B) % compute the Kronecker product
ans =
     2    -2     0     0
     1     5     0     0
     0     0     2    -2
     0     0     1     5
>> kron(B,A)
ans =
     2     0    -2     0
     0     2     0    -2
     1     0     5     0
     0     1     0     5
```

Hermitian Conjugate (Adjoint)

```
>> A = [ i 0 ; 1+i 2*i]
A =
        0 + 1.0000i          0
   1.0000 + 1.0000i          0 + 2.0000i
>> A' % compute the adjoint of A
ans =
        0 - 1.0000i    1.0000 - 1.0000i
        0              0 - 2.0000i
>> conj(transpose(A)) % equivalent method to compute adjoint
ans =
        0 - 1.0000i    1.0000 - 1.0000i
        0              0 - 2.0000i
```

Matrix Exponential

```
>> syms q % define a symbolic matrix
>> A = sym([0 1 0; -1 0 0; 0 0 0]);
>> expm(q*A) % compute matrix exponential

ans =
[  cos(q),  sin(q),      0]
[ -sin(q),  cos(q),      0]
[      0,       0,      1]
```

Section 2.3 Matrix Determinant and Inverse

```
>> A = [ 1 0 1 ; 2 -1 2 ; 1 1 0 ]
A =
     1     0     1
     2    -1     2
     1     1     0
>> det(A) % compute the determinate of A
ans =
     1
>> inv(A) % compute the inverse of A
ans =
    -2     1     1
     2    -1     0
     3    -1    -1
>> A*inv(A) % obtain the identity matrix
ans =
     1     0     0
     0     1     0
     0     0     1
>> inv(A)*A
ans =
     1     0     0
     0     1     0
     0     0     1
```

Section 2.4 Eigenvalues and Eigenvectors

```
>> A= [ 1 0 1; 0 1 1; -1 0 1]
A =
     1     0     1
     0     1     1
    -1     0     1
>> eig(A) % find the eigenvalues of the matrix A
ans =
   1.0000
   1.0000 + 1.0000i
   1.0000 - 1.0000i
>> [T,E]=eig(A) % returns a matrix T with columns corresponding
     to the eigenvectors of A and a matrix E with diagonal elements
     corresponding to the eigenvalues of A
T =
        0             0.5774             0.5774
   1.0000             0.5774             0.5774
        0                  0 + 0.5774i        0 - 0.5774i
E =
   1.0000                  0                  0
        0             1.0000 + 1.0000i        0
        0                  0             1.0000 - 1.0000i
>> % the matrix E may also be constructed by pre- and post-multiplying
     A by inv(T) and T, respectively
>> inv(T)*A*T   % diagonalize the matrix A
ans =
   1.0000                  0                  0
```

```
                  0            1.0000 + 1.0000i         0
                  0                  0            1.0000 - 1.0000i
>> % symbolically evaluate the eigenvalues and eigenvectors of the
   matrix B
>> syms a
>> B = sym ([ a 0 -a; 0 a 0 ; -a 0 a])
B =
[  a,  0,  -a]
[  0,  a,   0]
[ -a,  0,   a]
>> [T, E]=eig(B)
T =
[ 1, -1, 0]
[ 0,  0, 1]
[ 1,  1, 0]
E =
[ 0,   0, 0]
[ 0, 2*a, 0]
[ 0,   0, a]
>> inv(T)*B*T   % equivalent diagonalization of B
ans =
[ 0,   0, 0]
[ 0, 2*a, 0]
[ 0,   0, a]
```

Section 2.5 Rotation Matrices

```
>> syms q r s;
>> % rotation about the z-axis
>> Rz = [cos(q) sin(q) 0 ; -sin(q) cos(q) 0 ; 0 0 1]
Rz =
[  cos(q),   sin(q),        0]
[ -sin(q),   cos(q),        0]
[       0,        0,        1]
>> % rotation about the y-axis
>> Ry = [cos(r) 0 sin(r); 0 1 0; -sin(r) 0 cos(r)]
Ry =
[  cos(r),        0,   sin(r)]
[       0,        1,        0]
[ -sin(r),        0,   cos(r)]
>> % rotation about the x-axis
>> Rx = [1 0 0 ; 0 cos(s) sin(s) ; 0 -sin(s) cos(s)]
Rx =
[       1,        0,        0]
[       0,   cos(s),   sin(s)]
[       0,  -sin(s),   cos(s)]
>> % successive rotation about the y- and x-axes
>> Rx*Ry
ans =
[        cos(r),               0,          sin(r)]
[ -sin(s)*sin(r),         cos(s),   sin(s)*cos(r)]
[ -cos(s)*sin(r),        -sin(s),   cos(s)*cos(r)]
>> % rotation matrices do not commute
>> Rx*Ry-Ry*Rx
```

```
ans =
[                    0,          sin(s)*sin(r), -cos(s)*sin(r)+sin(r)]
[        -sin(s)*sin(r),                     0, -sin(s)+sin(s)*cos(r)]
[ -cos(s)*sin(r)+sin(r), -sin(s)+sin(s)*cos(r),                     0]
>> % successive rotations about the x-axis
>> Rx*Rx
ans =
[                1,              0,              0]
[                0, cos(s)^2-sin(s)^2,   2*cos(s)*sin(s)]
[                0,  -2*cos(s)*sin(s), cos(s)^2-sin(s)^2]
>> simple(Rx*Rx)
% MATLAB outputs several simplifications and we choose
combine(trig):
[        1,          0,          0]
[        0,    cos(2*s),    sin(2*s)]
[        0,   -sin(2*s),    cos(2*s)]
>> det(Rx) % rotation matrices have unit determinate
ans =
cos(s)^2+sin(s)^2
>> simple(det(Rx))
simplify:
1
>> transpose(Rx)*Rx % transpose is equal to the inverse
ans =
[                1,              0,              0]
[                0, cos(s)^2+sin(s)^2,              0]
[                0,              0, cos(s)^2+sin(s)^2]
```

2.7 EXERCISES

Section 2.1 Vectors and Scalars in Physics

1. Given the vectors $\mathbf{A} = \hat{\mathbf{i}} - 2\hat{\mathbf{j}} + \hat{\mathbf{k}}$, $\mathbf{B} = 2\hat{\mathbf{i}} - 2\hat{\mathbf{j}} + 3\hat{\mathbf{k}}$ and $\mathbf{C} = \hat{\mathbf{i}} + 2\hat{\mathbf{j}} + \hat{\mathbf{k}}$ calculate
(a) $\mathbf{A} - \mathbf{B} + \mathbf{C}$
(b) the angle between $\mathbf{A}$ and $\mathbf{B}$
(c) the magnitude $|\mathbf{A} + \mathbf{B}|$
(d) the dot product $\mathbf{A} \times (\mathbf{B} + \mathbf{C})$
(e) the cross product $\mathbf{A} \times (\mathbf{B} + \mathbf{C})$
(f) the vector product $\mathbf{A} \times (\mathbf{B} \times \mathbf{C})$
(g) the triple cross product $\mathbf{A} \times (\mathbf{B} \times \mathbf{C})$

2. Given the vectors $\mathbf{r} = 2\hat{\mathbf{i}} - 3\hat{\mathbf{j}}$, $\mathbf{r}' = \hat{\mathbf{i}} - 2\hat{\mathbf{j}} + \hat{\mathbf{k}}$ calculate
(a) $\mathbf{r} - \mathbf{r}'$

(b) $\left| \mathbf{r} - \mathbf{r}' \right|$

(c) $\dfrac{\mathbf{r} - \mathbf{r}'}{\left| \mathbf{r} - \mathbf{r}' \right|^3}$

(d) Show that $\hat{\mathbf{n}} = \dfrac{\mathbf{r}-\mathbf{r}'}{|\mathbf{r}-\mathbf{r}'|}$ has unit magnitude (calculate $\hat{\mathbf{n}} \cdot \hat{\mathbf{n}}$)

3. Given the vectors $\mathbf{A} = 2\hat{\mathbf{i}} - \hat{\mathbf{j}}$ and $\mathbf{B} = \hat{\mathbf{i}} - \hat{\mathbf{j}} + \hat{\mathbf{k}}$ find a vector with unit magnitude that is perpendicular to both $\mathbf{A}$ and $\mathbf{B}$.

4. Express the vector $\mathbf{A} = 2x\hat{\mathbf{i}} - y\hat{\mathbf{j}}$ in spherical and cylindrical coordinates.

Section 2.2 Matrices in Physics

5. Show that the Pauli matrices $\sigma_1, \sigma_2,$ and σ_3 are Hermitian and unitary

$$\sigma_1 = \begin{pmatrix} 0 & 1 \\ 1 & 0 \end{pmatrix} \quad \sigma_2 = \begin{pmatrix} 0 & -i \\ i & 0 \end{pmatrix} \quad \sigma_3 = \begin{pmatrix} 1 & 0 \\ 0 & -1 \end{pmatrix}$$

6. Show that the Pauli matrices σ_1 and σ_2 satisfy

$$\left(\sigma_1 + \sigma_2\right)^T = \sigma_1^{\ T} + \sigma_2^{\ T}$$
$$\left(\sigma_1\sigma_2\right)^T = \sigma_2^{\ T}\sigma_1^{\ T}$$

7. Show that the Pauli matrices $\sigma_1, \sigma_2,$ and σ_3 obey the commutation relations

$$\left[\sigma_1, \sigma_2\right] = i\sigma_3$$
$$\left[\sigma_2, \sigma_3\right] = i\sigma_1$$
$$\left[\sigma_3, \sigma_1\right] = i\sigma_2$$

8. Calculate the transpose of $(\sigma_1\sigma_2)^\dagger$

9. Calculate the direct product $\sigma_2 \otimes I_2$ where

$$I_2 = \begin{pmatrix} 1 & 0 \\ 0 & 1 \end{pmatrix}$$

10. Indicate if the following matrices are symmetric, antisymmetric, or neither
    (a) The electromagnetic field tensor

$$F^{\mu\nu} = \begin{pmatrix} 0 & E_x/c & E_y/c & E_z/c \\ -E_x/c & 0 & B_z & -B_y \\ -E_y/c & -B_z & 0 & B_x \\ -E_z/c & B_y & -B_x & 0 \end{pmatrix}$$

(b) The Lorentz transformation matrix

$$\Lambda = \begin{pmatrix} \gamma & -\gamma\dfrac{v}{c} & 0 & 0 \\ -\gamma\dfrac{v}{c} & \gamma & 0 & 0 \\ 0 & 0 & 1 & 0 \\ 0 & 0 & 0 & 1 \end{pmatrix}$$

11. Show that $\mathrm{Tr}(\mathbf{A}\otimes\mathbf{B}) = \mathrm{Tr}(\mathbf{A})\mathrm{Tr}(\mathbf{B})$

12. Given the matrix

$$\mathbf{A} = \begin{pmatrix} 1+i & 0 & i \\ 0 & 0 & 0 \\ -i & 0 & 1-i \end{pmatrix}$$

show that $\mathbf{A}+\mathbf{A}^{\dagger}$ is Hermitian

13. Given the matrices

$$J_x = \begin{pmatrix} 0 & 0 & 0 \\ 0 & 0 & -i \\ 0 & i & 0 \end{pmatrix} \quad J_y = \begin{pmatrix} 0 & 0 & -i \\ 0 & 0 & 0 \\ i & 0 & 0 \end{pmatrix} \quad J_z = \begin{pmatrix} 0 & -i & 0 \\ i & 0 & 0 \\ 0 & 0 & 0 \end{pmatrix}$$

show that

$$i\left[J_x,J_y\right] = J_z \quad i\left[J_z,J_x\right] = J_y \quad i\left[J_y,J_z\right] = J_x$$

14. Calculate $e^{iJ_x\theta}$, $e^{iJ_y\psi}$, and $e^{iJ_z\phi}$ given the matrices above.

15. Cube the following matrix

$$J_x = \begin{pmatrix} 0 & 0 & 0 \\ 0 & 0 & -i \\ 0 & i & 0 \end{pmatrix}$$

Section 2.3 Matrix Determinant and Inverse

16. Calculate the determinant of the Lorentz transformation matrix

$$\Lambda = \begin{pmatrix} \gamma & -\gamma\dfrac{v}{c} & 0 & 0 \\ -\gamma\dfrac{v}{c} & \gamma & 0 & 0 \\ 0 & 0 & 1 & 0 \\ 0 & 0 & 0 & 1 \end{pmatrix} \qquad \text{where} \quad \gamma = \left(1 - \dfrac{v^2}{c^2}\right)^{-1/2}$$

17. Calculate the determinants of the Pauli spin matrices

$$\sigma_1 = \begin{pmatrix} 0 & 1 \\ 1 & 0 \end{pmatrix} \qquad \sigma_2 = \begin{pmatrix} 0 & -i \\ i & 0 \end{pmatrix} \qquad \sigma_3 = \begin{pmatrix} 1 & 0 \\ 0 & -1 \end{pmatrix}$$

18. Determine if the following matrices are invertible

(a) $\begin{pmatrix} 1 & 0 & 1 \\ 1 & -2 & 1 \\ 0 & 0 & 1 \end{pmatrix}$

(b) $\begin{pmatrix} 1 & 2 & 1 \\ -1 & -2 & 0 \\ 1 & 2 & 1 \end{pmatrix}$

(c) $\begin{pmatrix} 1 & 0 & 1 \\ 0 & -2 & 0 \\ -1 & 0 & 1 \end{pmatrix}$

19. Find the value of x such that the matrix

$$\begin{pmatrix} 1 & 2 & 3 \\ x & 5 & 6 \\ 7 & 8 & 9 \end{pmatrix}$$

is singular.

20. Find the inverse of the following matrix

$$\begin{pmatrix} 2 & 0 \\ 0 & -1 \end{pmatrix}$$

21. Calculate the inverse of the following matrix

$$\begin{pmatrix} 1 & 0 & 1 \\ 0 & -2 & 0 \\ 0 & 0 & 1 \end{pmatrix}$$

22. Given two square matrices $\mathbf{A}$ and $\mathbf{B}$ show that $|\mathbf{AB}| = |\mathbf{A}||\mathbf{B}|$.

23. Given $\mathbf{A} = \begin{pmatrix} 1 & 0 & 1 \\ 0 & -2 & 0 \\ 0 & 0 & 1 \end{pmatrix}$ show that $|\mathbf{A}^{-1}| = \dfrac{1}{|\mathbf{A}|}$

24. Use inverse matrices to solve the following systems of equations

$x + 2y = 0$
$x - y = 1$

$2x - y = 0$
$-x + y = 1$

25. Use a matrix inverse to verify the solution to

$$x - y + z = 1$$
$$x - y - z = 2$$
$$x + y - z = 1$$
$$(x, y, z) = (1, -\frac{1}{2}, -\frac{1}{2})$$

26. Use a matrix inverse to verify the solution to

$$x + y + z = 2$$
$$x - y - z = 2$$
$$x + y - z = 0$$
$$(x, y, z) = (2, -1, 1)$$

27. Use a Cramer's rule to verify the solutions to

$$x - y - z = 0 \qquad\qquad x + y + z = 0$$
$$x + y + z = 0 \qquad\qquad x - y - z = 1$$
$$x + y - z = 0 \qquad \text{and } x + y - z = 0$$
$$(x, y, z) = (0, 0, 0) \qquad\qquad (x, y, z) = \left(\frac{1}{2}, -\frac{1}{2}, 0\right)$$

28. In general, the Jacobian determinant can be computed for N functions of N variables. Given the functions $u(x, y)$ and $v(x, y)$ the Jacobian determinant is

$$J\left(\frac{u, v}{x, y}\right) = \begin{vmatrix} \dfrac{\partial u}{\partial x} & \dfrac{\partial u}{\partial y} \\ \dfrac{\partial v}{\partial x} & \dfrac{\partial v}{\partial y} \end{vmatrix}$$

Calculate J for $u = e^{xy}$ and $v = x \sin y$.

Section 2.4 Eigenvalues and Eigenvectors

29. Calculate the characteristic equations whose roots determine the eigenvalues of the following matrices

(a) $\begin{pmatrix} 1 & 3 \\ 4 & 2 \end{pmatrix}$

(b) $\begin{pmatrix} 1 & 0 & 0 \\ 0 & 2 & 0 \\ 0 & 0 & 3 \end{pmatrix}$

(c) $\begin{pmatrix} \cos\theta & \sin\theta & 0 \\ -\sin\theta & \cos\theta & 0 \\ 0 & 0 & 1 \end{pmatrix}$

(d) $\begin{pmatrix} I_{xx} & I_{xy} & I_{xz} \\ I_{yx} & I_{yy} & I_{yz} \\ I_{zx} & I_{zy} & I_{zz} \end{pmatrix}$

30. Show that the characteristic equation of a 2×2 matrix $\mathbf{A} = \begin{pmatrix} a_{11} & a_{12} \\ a_{21} & a_{22} \end{pmatrix}$ is given by

$$\lambda^2 - \text{Tr}(\mathbf{A})\lambda + |\mathbf{A}| = 0$$

31. Calculate the eigenvalues and normalized eigenvectors of the following matrices

(a) $\begin{pmatrix} 0 & 1 \\ -1 & 1 \end{pmatrix}$

(b) $\begin{pmatrix} 1 & 0 & 0 \\ 0 & 2 & 0 \\ 0 & 0 & 3 \end{pmatrix}$

32. Show that the trace of $A = \begin{pmatrix} 0 & 1 & 0 \\ 1 & 0 & 0 \\ 0 & 0 & 2 \end{pmatrix}$ is equal to the sum of its eigenvalues and the determinant of A is equal to the product of its eigenvalues.

33. Diagonalize the following matrices

(a) $\begin{pmatrix} 1 & 3 \\ 2 & -1 \end{pmatrix}$

(b) $\begin{pmatrix} 1 & 1 & -1 \\ 2 & 0 & 2 \\ -1 & 0 & -2 \end{pmatrix}$

Section 2.5 Rotation Matrices

34. Verify the following trigonometric identities for $\theta = \phi$
$$\cos(\theta') = \cos(\theta - \phi) = \cos\theta\cos\phi + \sin\theta\sin\phi$$
$$\sin(\theta') = \sin(\theta - \phi) = \sin\theta\cos\phi - \cos\theta\sin\phi$$

35. Calculate the determinants $|R_x|$, $|R_y|$ and $|R_z|$

36. Calculate

(a) $R_x R_y X$ where $X = \begin{pmatrix} 1 \\ 0 \\ -1 \end{pmatrix}$ for $\theta = \psi = \dfrac{\pi}{2}$

(b) $R_y R_x X$ where $X = \begin{pmatrix} 1 \\ 0 \\ -1 \end{pmatrix}$ for $\theta = \psi = \dfrac{\pi}{2}$

37. Calculate the commutator $\left[R_x, R_y \right]$ for $\theta = \psi = \dfrac{\pi}{2}$

38. Show that $\left(R_x R_y \right)^T = \left(R_x R_y \right)^{-1}$ for $\theta = \psi = \dfrac{\pi}{2}$

39. Calculate $R_x R_y R_z$ for infinitesimal rotations with $\theta = \psi = \phi = \varepsilon$. Neglect terms of $O(\varepsilon^2)$ and higher.

40. Show that the rotation matrices

$$R_x(\theta) = e^{iJ_x\theta} \quad R_y(\psi) = e^{iJ_y\psi} \quad R_z(\phi) = e^{iJ_z\phi}$$

are generated by the matrices

$$J_x = \begin{pmatrix} 0 & 0 & 0 \\ 0 & 0 & -i \\ 0 & i & 0 \end{pmatrix} \quad J_y = \begin{pmatrix} 0 & 0 & -i \\ 0 & 0 & 0 \\ i & 0 & 0 \end{pmatrix} \quad J_z = \begin{pmatrix} 0 & -i & 0 \\ i & 0 & 0 \\ 0 & 0 & 0 \end{pmatrix}$$

Chapter **3** **CALCULUS**

Chapter Outline

3.1 SINGLE-VARIABLE CALCULUS

Differential and integral calculus techniques are reviewed in this section. Examples of common integrals encountered in physics are given.

3.1.1 Critical Points

A function $f(x)$ has a critical point where the slope is zero at x_0 if

$$\left. \frac{df(x)}{dx} \right|_{x_0} = 0 \tag{3.1.1}$$

The critical point (or extremum) may be a local maximum, local minimum or an inflection point. To determine the type of critical point a second derivative test is used

Local maximum:
$$\left.\frac{d^2 f(x)}{dx^2}\right|_{x_0} < 0 \qquad (3.1.2)$$

Local minimum:
$$\left.\frac{d^2 f(x)}{dx^2}\right|_{x_0} > 0 \qquad (3.1.3)$$

Inflection point:
$$\left.\frac{d^2 f(x)}{dx^2}\right|_{x_0} = 0 \qquad (3.1.4)$$

Example 3.1.1

Find the extremum of $f(x) = x^2$ and determine the type of critical point.

Solution: Solving $f'(x) = 2x = 0$ we find $x = 0$ is an extremum. Performing the second derivative test $f''(x)\big|_{x=0} = 2 > 0$ so that $x = 0$ is a minimum.

3.1.2 Integration with Substitution

Integrals of the form
$$\int f(g(x))g'(x)dx \qquad (3.1.5)$$
may be transformed
$$\int f(u)du \qquad (3.1.6)$$
by making the substitutions $u = g(x)$ and $du = g'(x)dx$. Once integrated, $g(x)$ is substituted back in for u. The limits of definite integrals
$$\int_a^b f(g(x))g'(x)dx \qquad (3.1.7)$$
may be transformed as
$$\int_{u(a)}^{u(b)} f(u)du \qquad (3.1.8)$$

Alternatively, the original limits may be plugged back in after $g(x)$ is substituted for u.

Example 3.1.2

Evaluate

$$\int \sin\theta \cos\theta d\theta \tag{3.1.9}$$

Solution: Making the substitution $u = \sin\theta$ and $du = \cos\theta \, d\theta$

$$\int u\,du = \frac{u^2}{2} = \frac{1}{2}\sin^2\theta \tag{3.1.10}$$

Example 3.1.3

Evaluate

$$\int_0^\pi \frac{\sin\theta d\theta}{\sqrt{R^2 + z^2 - 2Rz\cos\theta}} \tag{3.1.11}$$

Solution: Substituting $u = R^2 + z^2 - 2\,Rz\cos\theta$ and $du = 2\,Rz\sin\theta\,d\theta$. The limits are now from $u = R^2 + z^2 - 2\,Rz\cos(0)$ to $u = R^2 + z^2 - 2\,Rz\cos(\pi)$

$$\frac{1}{2Rz}\int_{R^2+z^2-2Rz}^{R^2+z^2+2Rz} \frac{du}{\sqrt{u}} = \frac{1}{Rz}\sqrt{u}\Big|_{R^2+z^2-2Rz}^{R^2+z^2+2Rz} = \frac{\sqrt{R^2+z^2+2Rz} - \sqrt{R^2+z^2-2Rz}}{Rz} = \frac{2}{R} \tag{3.1.12}$$

Instead of transforming the limits we can back-substitute for u and plug in the original limits

$$\frac{1}{Rz}\sqrt{R^2+z^2-2Rz\cos\theta}\,\Big|_0^\pi \tag{3.1.13}$$

3.1.3 Work-Energy Theorem

A famous u-substitution integral is encountered when calculating the kinetic energy of a mass under an applied force. The relativistic form of the work-kinetic energy theorem is given by the integral

$$W = \int_{x_i}^{x_f} \gamma^3 m \frac{dv}{dt} dx = m \int_{v_i}^{v_f} \frac{v\,dv}{\left(1 - \dfrac{v^2}{c^2}\right)^{3/2}} \tag{3.1.14}$$

This integral can be evaluated by making the substitution $u = 1 - \dfrac{v^2}{c^2}$ so $du = -\dfrac{2v}{c^2} dv$

$$W = -\frac{mc^2}{2} \int_{1}^{1-\frac{v^2}{c^2}} u^{-3/2} du \qquad (3.1.15)$$

where the limits $u_i = 1$ and $u_f = 1 - \dfrac{v^2}{c^2}$ taking $v_i = 0$ and $v_f = v$. From the work-kinetic energy theorem $W = KE$

$$KE = \frac{mc^2}{\sqrt{1 - \dfrac{v^2}{c^2}}} - mc^2 = (\gamma - 1) mc^2 \qquad (3.1.16)$$

where the initial kinetic energy is zero.

3.1.4 Integration by Parts

The derivative of a product of functions $u(x)$ $v(x)$ is obtained using the product rule

$$\frac{d}{dx}(uv) = u\frac{dv}{dx} + v\frac{du}{dx} \qquad (3.1.17)$$

Integrating both sides of this expression we obtain

$$uv = \int u\,dv + \int v\,du \qquad (3.1.18)$$

within a constant. Thus,

$$\int u\,dv = uv - \int v\,du \qquad (3.1.19)$$

Evaluating the definite integrals by parts

$$\int_{a}^{b} u\,dv = \left[uv\right]_{a}^{b} - \int_{a}^{b} v\,du \qquad (3.1.20)$$

This technique is often useful for integrating the product of a polynomial function and an exponential, logarithmic or trigonometric function.

Example 3.1.4

Evaluate

$$\int_0^\infty x e^{-\beta x}\,dx \tag{3.1.21}$$

Solution: Making the substitutions

$$u = x \qquad\qquad dv = e^{-\beta x}\,dx \tag{3.1.22}$$

$$du = dx \qquad\qquad v = -\frac{1}{\beta} e^{-\beta x}$$

$$\int_0^\infty x e^{-\beta x}\,dx = \left[-\frac{1}{\beta} x e^{-\beta x} \right]_0^\infty + \frac{1}{\beta} \int_0^\infty e^{-\beta x}\,dx = \frac{1}{\beta^2} \tag{3.1.23}$$

Example 3.1.5

Evaluate once again

$$\int \sin\theta \cos\theta\,d\theta \tag{3.1.24}$$

Solution: Making the substitutions

$$u = \sin\theta \quad dv = \cos\theta\,d\theta \tag{3.1.25}$$
$$du = \cos\theta\,d\theta \quad v = \sin\theta$$

or, sometimes written in table form, we "multiply along the diagonal" and "subtract the integral along the bottom"

$$\sin\theta \qquad\qquad \cos\theta\,d\theta$$
$$\searrow \tag{3.1.26}$$
$$\cos\theta\,d\theta \xleftarrow{}_{-\int} \sin\theta$$

to obtain

$$\int \sin\theta \cos\theta\,d\theta = \sin^2\theta - \int \sin\theta \cos\theta\,d\theta \tag{3.1.27}$$

Notice that the same integral appears on the left- and right-hand sides. We can then solve for the integral (suppressing the constant).

$$\int \sin\theta \cos\theta\,d\theta = \frac{1}{2}\sin^2\theta \tag{3.1.28}$$

3.1.5 Integration with Partial Fractions

Partial fractions can sometimes expand integrals with polynomials in the denominator.

Example 3.1.6

Evaluate the integral

$$\int \frac{dx}{x^3 - x} \tag{3.1.29}$$

Solution: Factoring $x^3 - x = x(x^2 - 1) = x(x + 1)(x - 1)$ (3.1.30)

$$\frac{1}{x(x+1)(x-1)} = \frac{A}{x} + \frac{B}{(x+1)} + \frac{C}{(x-1)} \tag{3.1.31}$$

so we must have

$$1 = A(x+1)(x-1) + Bx(x-1) + Cx(x+1) \tag{3.1.32}$$

or

$$1 = A(x^2 - 1) + B(x^2 - x) + C(x^2 + x) \tag{3.1.33}$$

Equating coefficients of like powers of x on both sides gives

$$\begin{aligned} 1 &= -A \\ 0 &= A + B + C \\ 0 &= -B + C \end{aligned} \tag{3.1.34}$$

so that $A = -1$, $B = C = 1/2$ and

$$\begin{aligned} \int \frac{dx}{x^3 - x} &= -\ln(x) + \frac{1}{2}\ln(x+1) + \frac{1}{2}\ln(x-1) + \text{const.} \\ &= \frac{1}{2}\ln\left(\frac{x^2 - 1}{x^2}\right) + \text{const.} \end{aligned} \tag{3.1.35}$$

after combining logarithm terms.

Example 3.1.7

Evaluate the integral

$$\int \frac{dx}{x^3 + x} \tag{3.1.36}$$

Solution: Factoring $x^3 + x = x(x^2 + 1)$ $\qquad$ (3.1.37)

$$\frac{1}{x(x^2+1)} = \frac{A}{x} + \frac{Bx}{(x^2+1)} \qquad (3.1.38)$$

we must have

$$1 = A(x^2+1) + Bx^2 \qquad (3.1.39)$$

Equating coefficients of like powers of x on both sides gives

$$\begin{aligned} 1 &= A \\ 0 &= A + B \end{aligned} \qquad (3.1.40)$$

so that $A = 1, B = -1$

$$\int \frac{dx}{x^3+x} = \int \frac{dx}{x} - \int \frac{xdx}{x^2+1} \qquad (3.1.41)$$

$$= \ln(x) - \frac{1}{2}\ln(x^2+1) = \ln\left(\frac{x}{\sqrt{x^2+1}}\right) + \text{const.}$$

after combining logarithm terms.

3.1.6 Integration by Trig Substitution

Integrals involving radical expression can often by solved by trigonometric substitution.

Example 3.1.8

Evaluate the integral

$$\int_{-L}^{L} \frac{dz}{\sqrt{r^2+z^2}} \qquad (3.1.42)$$

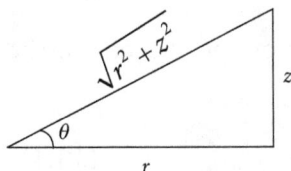

Figure 3.1.1: Triangle for trig substitution.

Solution: Referring to the right triangle in Figure 3.1.1 we identify

$$\tan\theta = \frac{z}{r}$$

$$\cos\theta = \frac{r}{\sqrt{r^2 + z^2}}$$

$$\sin\theta = \frac{z}{\sqrt{r^2 + z^2}} \tag{3.1.43}$$

Since $z = r \tan \theta$ we have $dz = r \sec^2 \theta \, d\theta$. Writing the integrand in terms of $\cos \theta$ we have that

$$\int \frac{dz}{\sqrt{r^2 + z^2}} = \int \frac{d\theta}{\cos\theta} \tag{3.1.44}$$

Multiplying top and bottom by $\cos \theta$ and breaking the integrand by partial fractions gives

$$\int \frac{d\theta}{\cos\theta} = \int \frac{\cos\theta}{\cos^2\theta} d\theta = \int \frac{\cos\theta}{1 - \sin^2\theta} d\theta = \int \frac{du}{1 - u^2} = \frac{1}{2} \int \left(\frac{1}{1+u} + \frac{1}{1-u} \right) du \tag{3.1.45}$$

The integral

$$\int \left(\frac{1}{1+u} + \frac{1}{1-u} \right) du = \ln(1+u) - \ln(1-u) \tag{3.1.46}$$

becomes

$$\int \left(\frac{1}{1+u} + \frac{1}{1-u} \right) du = \ln\left(1 + \frac{z}{\sqrt{r^2 + z^2}} \right) - \ln\left(1 - \frac{z}{\sqrt{r^2 + z^2}} \right) \tag{3.1.47}$$

Plugging in the original limits

$$\int_{-L}^{L} \frac{dz}{\sqrt{r^2 + z^2}} = \ln\left(\frac{\sqrt{r^2 + L^2} + L}{\sqrt{r^2 + L^2} - L} \right) \tag{3.1.48}$$

3.1.7 Differentiating Across the Integral Sign

It is often useful to interchange operations of differentiation and integration. The Leibnitz integral rule

$$\frac{d}{dy} \int f(x, y) \, dx = \int \frac{d}{dy} f(x, y) \, dx \tag{3.1.49}$$

can be used where $f(x, y)$ is continuous and differentiable. If the limits of a definite integral are functions of the differentiating variable

$$\frac{d}{dy}\int_{a(y)}^{b(y)} f(x,y)\,dx = \int_{a(y)}^{b(y)} \frac{d}{dy} f(x,y)\,dx + f(x,b(y))\frac{db}{dy} - f(x,a(y))\frac{da}{dy} \quad (3.1.50)$$

We may differentiate across the integral sign to evaluate integrals of the form

$$\int_0^\infty x^n e^{-\beta x}\,dx \quad (3.1.51)$$

We encounter such integrals when normalizing the radial wave function of the hydrogen atom in quantum mechanics and computing expectation values of position and momentum.

Example 3.1.9

Use the simple integral

$$\int_0^\infty e^{-\beta x}\,dx = \frac{1}{\beta} \quad (3.1.52)$$

to evaluate the more complicated integral

$$\int_0^\infty x e^{-\beta x}\,dx \quad (3.1.53)$$

Solution: Using the Leibnitz integral rule treating β as a continuous variable where

$$\int_0^\infty x e^{-\beta x}\,dx = -\frac{\partial}{\partial\beta}\int_0^\infty e^{-\beta x}\,dx = -\frac{\partial}{\partial\beta}\frac{1}{\beta} = \frac{1}{\beta^2} \quad (3.1.54)$$

Increasing powers of x are evaluated by successive differentiation

$$\int_0^\infty x^n e^{-\beta x}\,dx = \left(-\frac{\partial}{\partial\beta}\right)^n \int_0^\infty e^{-\beta x}\,dx = \left(-\frac{\partial}{\partial\beta}\right)^n \frac{1}{\beta} = \frac{n!}{\beta^{n+1}} \quad (3.1.55)$$

3.1.8 Integrals of Logarithmic Functions

The derivative of the natural logarithm

$$\frac{d}{dx}\ln x = \frac{1}{x} \quad (3.1.56)$$

for $x > 0$ so that

$$\int \frac{dx}{x} = \ln x + \text{const.} \quad (3.1.57)$$

Example 3.1.10

Show that

$$\int \ln x\, dx = x \ln x - x + \text{const.} \tag{3.1.58}$$

Solution: By taking the derivative

$$\frac{d}{dx}(x \ln x - x + \text{const.}) = \ln x + \frac{x}{x} - 1 = \ln x \tag{3.1.59}$$

Also, from the chain rule for $f(x) > 0$

$$\frac{d}{dx}\ln f(x) = \frac{f'(x)}{f(x)} \tag{3.1.60}$$

so that

$$\int \frac{f'(x)}{f(x)}\, dx = \ln f(x) + \text{const.} \tag{3.1.61}$$

Example 3.1.11

Show that

$$\int \frac{y+h}{x^2+(y+h)^2}\, dy = \ln \sqrt{x^2+(y+h)^2} + \text{const.} \tag{3.1.62}$$

Solution:

$$\frac{d}{dy}\ln \sqrt{x^2+(y+h)^2} = \frac{\frac{1}{2}\left(x^2+(y+h)^2\right)^{-1/2} 2(y+h)}{\sqrt{x^2+(y+h)^2}} = \frac{y+h}{x^2+(y+h)^2} \tag{3.1.63}$$

Example 3.1.12

Show that

$$\int \frac{1}{\sin \theta}\, d\theta = \ln\left(\tan \frac{\theta}{2}\right) \tag{3.1.64}$$

Solution:

$$\frac{d}{d\theta}\ln\left(\tan \frac{\theta}{2}\right) = \frac{1}{\tan \frac{\theta}{2}}\frac{d}{d\theta}\tan \frac{\theta}{2} = \frac{1}{2}\frac{\sec^2 \frac{\theta}{2}}{\tan \frac{\theta}{2}} = \frac{1}{2\sin \frac{\theta}{2}\cos \frac{\theta}{2}} = \frac{1}{\sin \theta} \tag{3.1.65}$$

Maple Examples

Integrals involving radical, polynomial, exponential, and trig functions are demonstrated on the Maple worksheet below.

Key Maple commands: *assume, int*

restart

Integrals Involving Radicals

$$int\left(\frac{1}{\text{sqrt}\left(r^2+z^2\right)}, z\right)$$

$$\ln\left(z+\sqrt{r^2+z^2}\right)$$

$$int\left(\frac{1}{\text{sqrt}\left(r^2+z^2\right)}, z=-L\ldots L\right)$$

$$-\ln\left(-L+\sqrt{L^2+r^2}\right)+\ln\left(L+\sqrt{L^2+r^2}\right)$$

$$int\left(\frac{1}{\left(r^2+z^2\right)^{\frac{3}{2}}}, z\right)$$

$$\frac{z}{\sqrt{r^2+z^2}\,r^2}$$

$$int\left(\frac{z}{\left(r^2+z^2\right)^{\frac{3}{2}}}, z\right)$$

$$-\frac{1}{\sqrt{r^2+z^2}}$$

Integrals Involving Polynomials

$$int\left(\frac{1}{a-u}, u\right)$$

$$-\ln(a-u)$$

$$int\left(\frac{1}{a^2-u^2}, u\right)$$

$$-\frac{1}{2}\frac{\ln(u-a)}{a}+\frac{1}{2}\frac{\ln(u+a)}{a}$$

$$int\left(\frac{u}{a-u}, u\right)$$

$$-u - a\ln(u-a)$$

$$int\left(\frac{u}{a^2 - u^2}, u\right)$$

$$-\frac{1}{2}\ln\left(-a^2 + u^2\right)$$

$$int\left(\frac{u^2}{a^2 - u^2}, u\right)$$

$$-u - \frac{1}{2}a\ln(u-a) + \frac{1}{2}a\ln(u+a)$$

Integrals Involving Trig Functions

$int(\cos(\text{theta})^2, \text{theta})$

$$\frac{1}{2}\cos(\theta)\sin(\theta) + \frac{1}{2}\theta$$

$int(\sin(\text{theta})^2, \text{theta})$

$$-\frac{1}{2}\cos(\theta)\sin(\theta) + \frac{1}{2}\theta$$

$int(\sin(\text{theta})^2 \cdot \cos(\text{theta})^2, \text{theta})$

$$-\frac{1}{4}\sin(\theta)\cos(\theta)^3 + \frac{1}{8}\cos(\theta)\sin(\theta) + \frac{1}{8}\theta$$

$int(\cos(\text{theta})^2, \text{theta} = 0...2 \cdot \text{Pi})$

$$\pi$$

$int(\sin(\text{theta})^2, \text{theta} = 0...2 \cdot \text{Pi})$

$$\pi$$

$int(\sin(\text{theta})^2 \cdot \cos(\text{theta})^2, \text{theta} = 0...2 \cdot \text{Pi})$

$$\frac{1}{4}\pi$$

$$\mathrm{int}\left(\frac{1}{\cos(\text{theta})},\text{theta}\right)$$

$$\ln(\sec(\theta) + \tan(\theta))$$

$$\mathrm{int}\left(\frac{1}{\sin(\text{theta})},\text{theta}\right)$$

$$\ln(\csc(\theta) - \cot(\theta))$$

$$\mathrm{int}\left(\frac{\sin(\text{theta})}{\mathrm{sqrt}\left(r^2 + z^2 - 2 \cdot r \cdot z \cdot \cos(\text{theta})\right)},\text{theta}\right)$$

$$\frac{\sqrt{r^2 + z^2 - 2rz\cos(\theta)}}{rz}$$

$$\mathrm{int}\left(\frac{\sin(\text{theta})}{\mathrm{sqrt}\left(r^2 + z^2 - 2 \cdot r \cdot z \cdot \cos(\text{theta})\right)^{\frac{3}{2}}},\text{theta}\right)$$

$$\frac{2\left(r^2 + z^2 - 2rz\cos(\theta)\right)^{1/4}}{rz}$$

Integrals Involving Exponential Functions

$$int(\exp(-a \cdot x), x)$$

$$-\frac{e^{-ax}}{a}$$

$$int(x{\cdot}\exp(-a \cdot x), x)$$

$$-\frac{(ax+1)e^{-ax}}{a^2}$$

$$int(x^2{\cdot}\exp(-a \cdot x), x)$$

$$-\frac{\left(a^2 x^2 + 2ax + 2\right)e^{-ax}}{a^3}$$

$$assume(a > 0)$$
$$int(\exp(-a \cdot x), x = 0\ldots\text{infinity})$$

$$\frac{1}{a\,\sim}$$

$int(\exp(-a \cdot x), x = 0\ldots\text{infinity})$

$$\frac{1}{a\,\sim^2}$$

$int(x^2{\cdot}\exp(-a \cdot x), x = 0\ldots\text{infinity})$

$$\frac{2}{a\,\sim^3}$$

$assume(n, \text{'integer'})$
$int(\sin(n{\cdot}x)\exp(-a \cdot x), x = 0\ldots\text{infinity})$

$$\frac{n\,\sim}{a\,\sim^2 + n\,\sim^2}$$

$int(\cos(n{\cdot}x)\exp(-a \cdot x), x = 0\ldots\text{infinity})$

$$\frac{a\,\sim}{a\,\sim^2 + n\,\sim^2}$$

$int(x^n{\cdot}\exp(-a \cdot x), x = 0\ldots\text{infinity})$

$$a\,\sim^{-1\,n} \ \Gamma(n\,\sim +1)$$

$int(\sin(n{\cdot}theta), theta = 0\ldots2 \cdot \text{Pi})$

$$0$$

$int(\sin(n{\cdot}theta), theta = 0\ldots\text{Pi})$

$$\frac{-1 + (-1)^{n\sim}}{n\,\sim}$$

3.2 MULTIVARIABLE CALCULUS

Partial derivatives, multiple integrals, and orthogonal coordinate systems are reviewed in this section. Examples of finding and characterizing extrema of function and performing volume and surface integrals in Cartesian, cylindrical, and spherical coordinates are discussed.

3.2.1 Partial Derivatives

Partial derivatives are used when differentiating a multivariable function with respect to a single variable. The partial derivative is defined like the derivative of a function of one variable. The partial derivative of $f(x, y, z, t)$ with respect to x is given by the following limit

$$\frac{\partial f}{\partial x} = \lim_{\Delta x \to 0} \frac{f(x+\Delta x, y, z, t) - f(x, y, z, t)}{\Delta x} \tag{3.2.1}$$

The symbol $\partial f / \partial x$ is read as "the partial of f with respect to x." Similarly, the partial of f with respect to y is

$$\frac{\partial f}{\partial y} = \lim_{\Delta x \to 0} \frac{f(x, y+\Delta y, z, t) - f(x, y, z, t)}{\Delta y} \tag{3.2.2}$$

Example 3.2.1

Given

$$f(x, t) = e^{-\alpha x} \cos(\omega t) \tag{3.2.3}$$

we may calculate the first partial derivatives

$$\frac{\partial f}{\partial x} = -\alpha e^{-\alpha x} \cos(\omega t)$$

$$\frac{\partial f}{\partial t} = -\omega e^{-\alpha x} \sin(\omega t) \tag{3.2.4}$$

the second partial derivatives

$$\frac{\partial^2 f}{\partial x^2} = \alpha^2 e^{-\alpha x} \cos(\omega t)$$

$$\frac{\partial^2 f}{\partial t^2} = -\omega^2 e^{-\alpha x} \cos(\omega t) \tag{3.2.5}$$

as well as mixed and higher partial derivatives

$$\frac{\partial^3 f}{\partial t \partial x^2} = -\omega \alpha^2 e^{-\alpha x} \sin(\omega t)$$

$$\frac{\partial^3 f}{\partial t^3} = \omega^3 e^{-\alpha x} \sin(\omega t) \tag{3.2.6}$$

3.2.2 Critical Points

A function of two variables $f(x, y)$ has a critical point at a point (x_0, y_0) where

$$\left.\frac{\partial f(x,y)}{\partial x}\right|_{(x_0,y_0)} = 0 \tag{3.2.7}$$

$$\left.\frac{\partial f(x,y)}{\partial y}\right|_{(x_0,y_0)} = 0 \tag{3.2.8}$$

The critical point may be a local maximum, local minimum, or a saddle point. To determine the type of critical point, a second derivative test is used

$$D(x_0, y_0) = \left.\left(\frac{\partial^2 f}{\partial^2 x} + \frac{\partial^2 f}{\partial^2 y} - \left(\frac{\partial^2 f}{\partial x \partial y}\right)^2\right)\right|_{(x_0,y_0)} \tag{3.2.9}$$

Local maximum: $D(x_0, y_0) > 0$ and $\left.\frac{\partial^2 f(x,y)}{\partial x^2}\right|_{(x_0,y_0)} < 0$ $\tag{3.2.10}$

Local minimum: $D(x_0, y_0) > 0$ and $\left.\frac{\partial^2 f(x,y)}{\partial x^2}\right|_{(x_0,y_0)} > 0$ $\tag{3.2.11}$

Saddle point: $D(x_0, y_0) < 0$ $\tag{3.2.12}$

3.2.3 Double Integrals

A double integral of a scalar function $f(x, y)$ over a region R is defined as a double Riemann sum

$$\int_R f(x,y)\,da = \lim_{\substack{N,M\to\infty \\ \Delta a\to 0}} \sum_{i=1}^{N}\sum_{j=1}^{M} f(x_i, y_j)\,\Delta a \tag{3.2.13}$$

where $\Delta a = \Delta x \Delta y$.

Example 3.2.2

Integrate $f(x, y) = x^2 y$ over the unit square with $da = dxdy$

$$\int_R f(x,y)\,da = \int_0^1\int_0^1 x^2 y\,dxdy = \left.\frac{x^3}{3}\right|_0^1 \left.\frac{y^2}{2}\right|_0^1 = \frac{1}{6} \tag{3.2.14}$$

The order of integration is inward to outward.

Example 3.2.3

Integrate the same function over half of the unit square consisting of the area between the line $y = x$ and the x-axis. We first integrate over y from $y = 0$ to $y = x$ and then integrate over x between 0 and 1

$$\int_0^1 \int_0^x (x^2 y) dy dx = \int_0^1 \left(x^2 \frac{y^2}{2} \Big|_0^x \right) dx = \int_0^1 \left(\frac{x^4}{2} \right) dx = \frac{x^5}{10} \Big|_0^1 = \frac{1}{10} \qquad (3.2.15)$$

Example 3.2.4

Obtain the area of half of the unit square

$$\int_0^1 \int_0^x dy dx = \int_0^1 y \Big|_0^x dx = \int_0^1 x dx = \frac{1}{2} \qquad (3.2.16)$$

To find the area A bound between two functions $y = f(x)$ and $y = g(x)$ over an interval $[a, b]$ we first integrate over y and then x

$$A = \int_a^b \int_{f(x)}^{g(x)} dy dx \qquad (3.2.17)$$

Example 3.2.5

Find the area between $y = x^2$ and $y = x$ over the interval $[0, 1]$

$$A = \int_0^1 \int_{x^2}^x dy dx = \int_0^1 (x - x^2) dx = \frac{1}{2} - \frac{1}{3} = \frac{1}{6} \qquad (3.2.18)$$

Example 3.2.6

Find the area of the unit circle. In polar coordinates $da = r dr d\theta$ and

$$A = \int_0^{2\pi} \int_0^R r dr d\theta = \pi R^2 \qquad (3.2.19)$$

Example 3.2.7

Integrate $f(x, y) = x^2 y$ over the upper half of the unit circle. We must first perform a polar coordinate transformation where $x = r \cos \theta$ and $y = r \sin \theta$

$$\int_R f(x, y) da = \int_0^\pi \int_0^1 (r \cos \theta)^2 (r \sin \theta) r dr d\theta$$

$$= \int_0^\pi \int_0^1 r^4 dr \cos^2 \theta \sin \theta d\theta = -\frac{1}{5} \int_1^{-1} u^2 du = \frac{2}{15} \qquad (3.2.20)$$

3.2.4 Triple Integrals

The triple integral of a scalar function $f(x, y, z)$ over a volume is defined as a triple Riemann sum

$$\int_{vol} f(x, y, z)dv = \lim_{\substack{N,M,O \to \infty \\ \Delta v \to 0}} \sum_{i=1}^{N} \sum_{j=1}^{M} \sum_{k=1}^{O} f(x_i, y_j, z_k)\Delta v \qquad (3.2.21)$$

where $\Delta v = \Delta x \Delta y \Delta z$.

Example 3.2.8

Integrate $f(x, y, z) = xy^2z^3$ over the unit cube where $dv = dxdydz$

$$\int_{R} f(x, y, z)dv = \int_0^1\int_0^1\int_0^1 xy^2z^3 dxdydz = \left.\frac{x^2}{2}\right|_0^1 \left.\frac{y^3}{3}\right|_0^1 \left.\frac{z^4}{4}\right|_0^1 = \frac{1}{24} \qquad (3.2.22)$$

The order of integration is inward to outward as it is for double integrals. Triple integrals in physics are used to calculate quantities such as mass, charge, and probability.

3.2.5 Orthogonal Coordinate Systems

Common coordinate systems used in physics are the Cartesian, cylindrical and spherical coordinate systems. These are orthogonal systems with coordinate surfaces that intersect at right angles.

3.2.6 Cartesian Coordinates

Orthogonal surfaces in the Cartesian coordinate system (x, y, z) are planes described by

$$x = \text{const.}$$
$$y = \text{const.} \qquad (3.2.23)$$
$$z = \text{const.}$$

with coordinate ranges $-\infty < x < \infty, \ -\infty < y < \infty, \ -\infty < z < \infty$.

3.2.7 Cylindrical Coordinates

The cylindrical coordinates (r, ϕ, z) are related to the Cartesian coordinates

$$x = r\cos\phi$$
$$y = r\sin\phi \qquad (3.2.24)$$
$$z = \text{const.}$$

The coordinate surfaces are circular cylinders of constant r

$$x^2 + y^2 = r^2,$$ (3.2.25)

half planes parallel to the z-axis described by

$$\tan\phi = \frac{y}{x}$$ (3.2.26)

and planes perpendicular to the z-axis $z = z$. The coordinate ranges of the cylindrical coordinates are $0 \le r < \infty$, $-\infty < z < \infty$, $0 \le \phi < 2\pi$. Note that θ is frequently used for the angular coordinate in cylindrical polar coordinates.

3.2.8 Spherical Coordinates

The spherical coordinates (r, θ, ϕ) are related to the Cartesian coordinates

$$
\begin{aligned}
x &= r\sin\theta\cos\phi \\
y &= r\sin\theta\sin\phi \\
z &= r\cos\theta
\end{aligned}
$$ (3.2.27)

The coordinate surfaces are spheres of constant r

$$x^2 + y^2 + z^2 = r^2$$ (3.2.28)

cones of constant θ described by

$$\theta = \tan^{-1}\frac{\sqrt{x^2+y^2}}{z}$$ (3.2.29)

and half planes of constant ϕ parallel to the z-axis

$$\phi = \tan^{-1}\frac{y}{x}$$ (3.2.30)

Coordinate ranges of the spherical coordinates are $0 \le r < \infty$, $0 \le \theta \le \pi$, $0 \le \phi < 2\pi$.

3.2.9 Line, Volume, and Surface Elements

Surface elements da and volume elements dv may be determined from components of the line element $d\hat{\ell}$ in a given coordinate system as shown in Table 3.2.1. There will be three possible surface elements corresponding to faces of the volume element. In Cartesian coordinates, the volume element is a cube with square surface elements. Integration over volume elements in Cartesian, cylindrical, and spherical coordinates will give the volume of a cube, cylinder and sphere, respectively.

$$\int_{cube} dv = \int_0^L \int_0^L \int_0^L dxdydz = L^3$$

(3.2.31)

$$\int_{cylinder} dv = \int_0^{2\pi} \int_0^H \int_0^R rdrdzd\phi = \pi R^2 H$$

(3.2.32)

$$\int_{sphere} dv = \int_0^{2\pi} \int_0^\pi \int_0^R r^2 \sin\theta drd\theta d\phi = \frac{4}{3}\pi R^3$$

(3.2.33)

Table 3.2.1: Line, surface, and volume elements in Cartesian, cylindrical, and spherical coordinates.

Coordinate System	Line Element	Possible Surface Elements	Volume Element
Cartesian	$d\hat{\ell} = dx\hat{\mathbf{i}} + dy\hat{\mathbf{j}} + dz\hat{\mathbf{k}}$	$d\mathbf{a} = \begin{cases} dxdy\hat{\mathbf{k}} \\ dydz\hat{\mathbf{i}} \\ dzdx\hat{\mathbf{j}} \end{cases}$	$dv = dxdydz$
Cylindrical	$d\hat{\ell} = dr\hat{\mathbf{r}} + rd\phi\hat{\varphi} + dz\hat{\mathbf{z}}$	$d\mathbf{a} = \begin{cases} rd\phi dz\hat{\mathbf{r}} \\ rd\phi dr\hat{\mathbf{z}} \\ drdz\hat{\varphi} \end{cases}$	$dv = rdrdzd\phi$
Spherical	$d\hat{\ell} = dr\hat{\mathbf{r}} + rd\theta\hat{\theta} + r\sin\theta d\phi\hat{\varphi}$	$d\mathbf{a} = \begin{cases} r^2 \sin\theta d\theta d\phi\hat{\mathbf{r}} \\ r\sin\theta d\theta dr\hat{\varphi} \\ rd\phi dr\hat{\theta} \end{cases}$	$dv = r^2\sin\theta drd\theta d\phi$

Maple Examples

Partial derivatives of multivariate functions are demonstrated in the Maple worksheet below as well as multiple integrals in Cartesian, cylindrical, and spherical coordinates. Extrema including maxima, minima, and saddle points of surface plots are evaluated. Various coordinate systems are plotted in 2D and 3D.

Possible coordinate systems: bipolarcylindrical, bispherical, cardioidal, cardioidcylindrical, casscylindrical, confocalellip, confocalparab, conical, cylindrical, ellcylindrical, ellipsoidal, hypercylindrical, invcasscylindrical, invellcylindrical, invoblspheroidal, invprospheroidal, logcoshcylindrical, logcylindrical, maxwellcylindrical, oblatespheroidal, paraboloidal, paraboloidal2, paracylindrical, prolatespheroidal, rectangular, rosecylindrical, sixsphere, spherical, tangentcylindrical, tangentsphere, and toroidal.

Key Maple commands: *coordplot, coordplot3d, evalf, plot3d, simplify, subs*

Maple packages: *with(plots)*:

Programming: Function operation using '→'

restart

Partial Derivatives

$$f := (x, y, z) \rightarrow \sin(y^x) \cdot \cosh(z)$$

$$f := (x, y, z) \mapsto \sin(y^x) \cosh(z)$$

$$\frac{\partial}{\partial x} f(x, y, z)$$

$$y^x \ln(y) \cos(y^x) \cosh(z)$$

$$\frac{\partial^2}{\partial x \partial y} f(x, y, z)$$

$$\frac{y^x x \cos(y^x) \sinh(z)}{y}$$

$$\frac{\partial^2}{\partial z \partial y} f(x, y, z)$$

$$\frac{y^x x \cos(y^x) \sinh(z)}{y}$$

restart

Determining Maxima and Minima

$$f := (x, y) \rightarrow (x^2 - 2 \cdot y^2) \cdot \exp(-(x^2 + y^2))$$

$$f := (x, y) \mapsto (x^2 - 2y^2)e^{-x^2 - y^2}$$

$$plot3d(f(x, y), x = -2...2, y = -2...2)$$

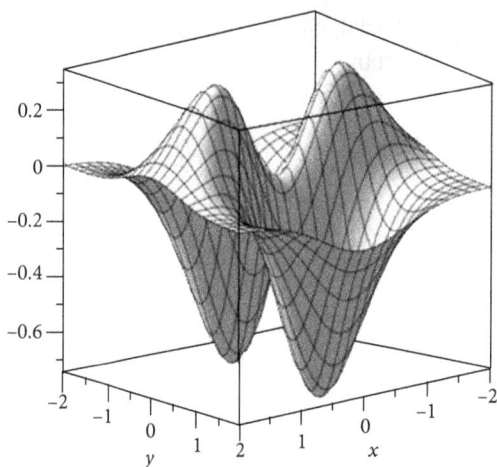

Figure 3.2.1: Surface plot of a function with two maxima, two minima, and a saddle point.

$$CritPt := solve(\{diff(f(x, y), x) = 0, diff(f(x, y), y) = 0\}, \{x, y\})$$

$$CritPt := \{x = 0, y = 0\}, \{x = 1, y = 0\}, \{x = -1, y = 0\}, \{x = 0, y = 1\}, \{x = 0, y = -1\}$$

Second Derivative Test

$$Extrema := simplify(diff(f(x, y), x, x) \cdot diff(f(x, y), y, y) - (diff(f(x, y), x, y))^2)$$

$$Extrema := -8\left(x^6 + \left(-3y^2 - \frac{1}{2}\right)x^4 + \left(23y^2 - \frac{9}{2}\right)x^2 + 4y^6 - 8y^4 - 3y^2 + 1\right)e^{-2x^2 - 2y^2}$$

$$evalf(subs(\{x = 0, y = 0\}, Extrema))$$
$$-8$$

$$evalf(subs(\{x = 1, y = 1\}, Extrema))$$
$$3.248046797$$

$$evalf(subs(\{x = -1, y = 0\}, Extrema))$$
$$3.248046797$$

$$evalf(subs(\{x = 0, y = 1\}, Extrema))$$
$$6.496093594$$

$$evalf(subs(\{x = 0, y = -1\}, Extrema))$$
$$6.496093594$$

Thus, (x = 0, y=0) is a saddle point. The other extrema are maxima or minima.

Double Integrals

$$g := (x, y) \rightarrow (x \cdot y)$$

$$g := (x, y) \mapsto (yx)$$

$$int(g(x, y), x = 0...1)$$

$$\frac{y}{2}$$

$$Int(Int(g(x, y), x = 0...1), y = 0 ... 1) = int(int(g(x, y), x = 0...1), y = 0 ... 1)$$

$$\int_0^1 \int_0^1 yx\,dx\,dy = \frac{1}{4}$$

Double Integrals in Polar Coordinates

$$Int(Int(r, r = 0...R), theta = 0 ... 2\cdot Pi) = int(int(r, r = 0...R), theta = 0 ... 2\cdot Pi)$$

$$\int_0^{2\pi} \int_0^R r\,dr\,d\theta = R^2 \pi$$

Triple Integrals

$$h := (x, y, z) \to (x \cdot y \cdot z)$$

$$h := (x, y, z) \mapsto y\,x\,z$$

$$Int(Int(Int(h(x, y, z), x = 0...1), y = 0 ... 1), z = 0 ... 1) = int(int(int(h(x, y, z),$$
$$x = 0...1), y = 0 ... 1), z = 0 ... 1)$$

$$\int_0^1 \int_0^1 \int_0^1 yxz\,dx\,dy\,dz = \frac{1}{8}$$

Triple Integrals in Cylindrical Coordinates

$$Int(Int(Int(r, r = 0...R), theta = 0 ... 2\cdot Pi), z = 0 ... H) = int(int(int(r, r = 0...R),$$
$$theta = 0 ... 2\cdot Pi), z = 0 ... H)$$

$$\int_0^H \int_0^{2\pi} \int_0^R r\,dr\,d\theta\,dz = R^2 \pi H$$

Triple Integrals in Spherical Coordinates

$$Int(Int(Int(r^2 \cdot sin(theta), r = 0...R), theta = 0 ... Pi), phi = 0 ... 2\cdot Pi) =$$
$$int(int(int(r^2 \cdot sin(theta), r = 0...R), theta = 0 ... Pi), phi = 0 ... 2\cdot Pi)$$

$$\int_0^{2\pi} \int_0^{\pi} \int_0^R r^2 \sin(\theta)\,dr\,d\theta\,d\phi = \frac{4R^3 \pi}{3}$$

Integrals Using the Palette

$$evalf\left(\int_0^1 \int_0^{z^2} \int_0^y \frac{\exp(x \cdot y \cdot z)}{sqrt\left(x^2 + y^2 + z^2\right) + 1}\,dx\,dy\,dz \right)$$

$$0.05697298256$$

Coordinate Systems

$with(plots):$
$coordplot(polar, labeling = true, color = "Black")$

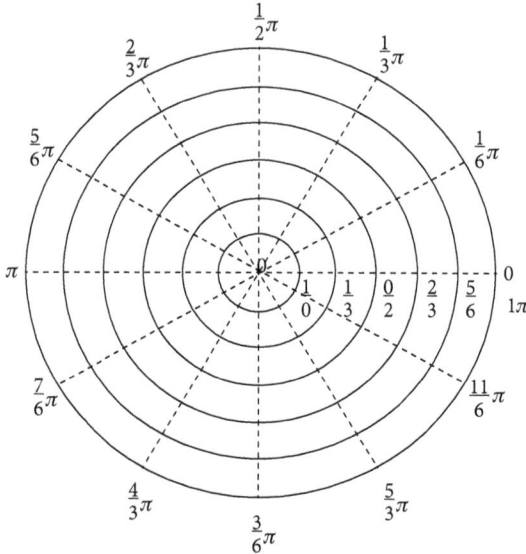

Figure 3.2.2: Plot of polar coordinates.

$coordplot(bipolar, color = "Black")$

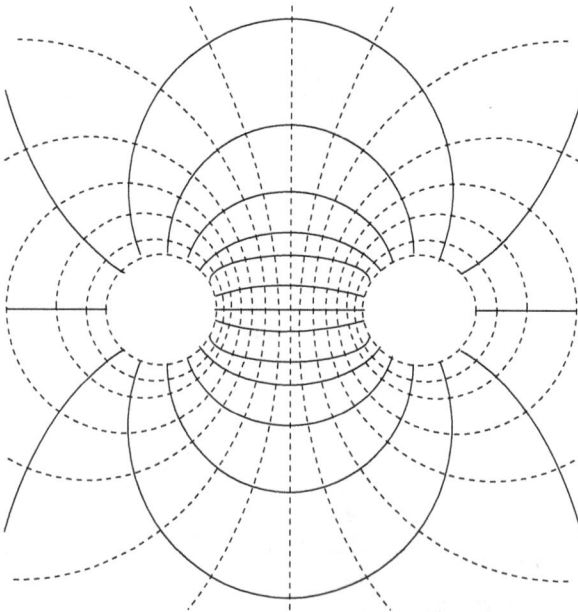

Figure 3.2.3: Plot of bipolar coordinates.

coordplot3d(rectangular, color = "Black")

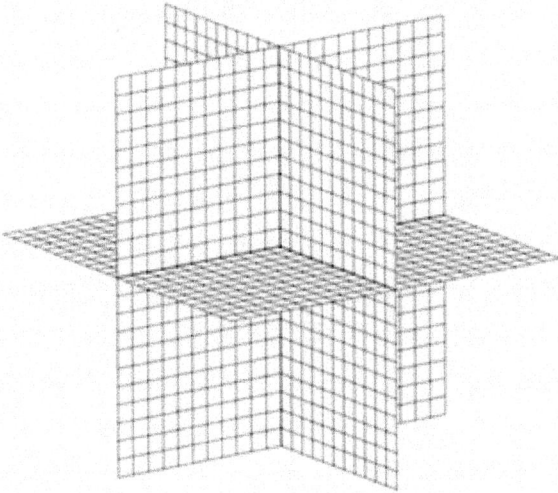

Figure 3.2.4: Plot of Cartesian coordinates. Orthogonal surfaces are planes.

coordplot3d(cylindrical, color = "Black")

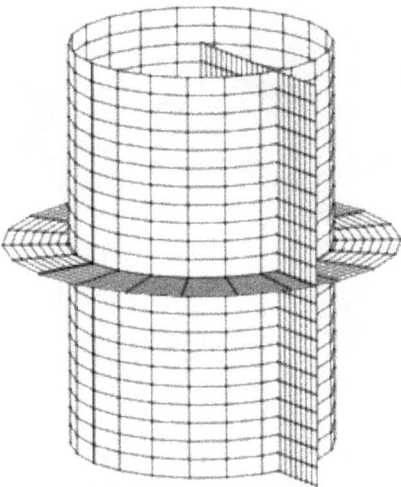

Figure 3.2.5: Plot of cylindrical coordinate surfaces. Orthogonal surfaces are cylinders, planes and half planes.

coordplot3d(spherical, color = "Black")

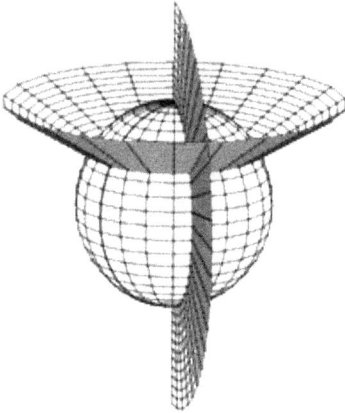

Figure 3.2.6: Plot of spherical coordinate surfaces.
Orthogonal surfaces are spheres, cones (one cone shown), and half planes.

$$coordplot3d(oblatespheroidal, color = \text{``Black''})$$

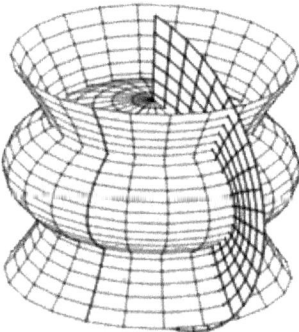

Figure 3.2.7: Plot of oblate spheroidal coordinate surfaces.
Orthogonal surfaces are oblate spheroids, one-sheet hyperboloids, and half planes.

$$coordplot3d(prolatespheroidal, color = \text{``Black''})$$

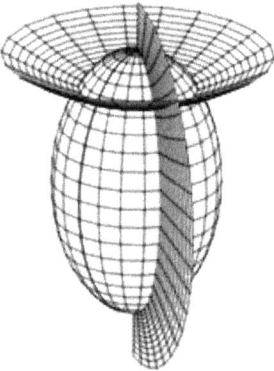

Figure 3.2.8: Plot of prolate spheroidal coordinate surfaces.
Orthogonal surfaces are prolate spheroids, two-sheet hyperboloids (one sheet shown), and half planes

coordplot3d(toroidal, color = "Black")

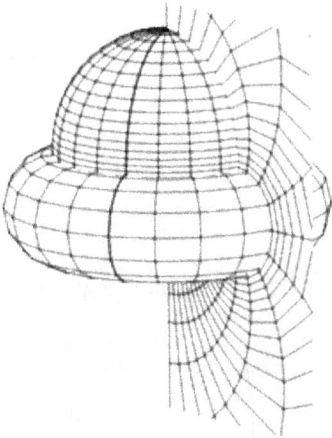

Figure 3.2.9: Plot of toroidal coordinate surfaces. Orthogonal surfaces are spheres, toroids, and half planes.

3.3 GAUSSIAN INTEGRALS

Gaussian integrals of the form

$$\int_{-\infty}^{\infty} x^n e^{-\alpha x^2} dx \tag{3.3.1}$$

occur in the quantum harmonic oscillator, Maxwell-Boltzmann velocity distribution, and path integrals in quantum field theory.

Example 3.3.1

To solve the integral

$$I = \int_{-\infty}^{\infty} e^{-\alpha x^2} dx \tag{3.3.2}$$

we first perform a polar coordinate transformation

$$I^2 = \int_{-\infty}^{\infty} e^{-\alpha x^2} dx \int_{-\infty}^{\infty} e^{-\alpha y^2} dy = \int_{-\infty}^{\infty} \int_{-\infty}^{\infty} e^{-\alpha(x^2 + y^2)} dx dy \tag{3.3.3}$$

Substituting $x^2 + y^2 \to r^2$ and $dx dy \to r dr d\theta$ gives

$$I^2 = \int_{0}^{2\pi} \int_{0}^{\infty} e^{-\alpha r^2} r dr d\theta \tag{3.3.4}$$

and substituting $u = \alpha r^2$ and $du = 2\alpha r dr$

$$I^2 = 2\pi \frac{1}{2\alpha} \rightarrow I = \sqrt{\frac{\pi}{\alpha}} \qquad (3.3.5)$$

Because the integral is symmetric about $x = 0$

$$\int_0^\infty e^{-\alpha x^2} dx = \frac{1}{2}\sqrt{\frac{\pi}{\alpha}} \qquad (3.3.6)$$

Now we can use the Leibnitz integral rule differentiating across the integral sign and treating a as a continuous variable

$$\int_{-\infty}^\infty x^2 e^{-\alpha x^2} dx = -\frac{\partial}{\partial \alpha} \int_{-\infty}^\infty e^{-\alpha x^2} dx = -\frac{\partial}{\partial \alpha}\sqrt{\frac{\pi}{\alpha}} = \frac{1}{2}\sqrt{\frac{\pi}{\alpha^3}} \qquad (3.3.7)$$

Next

$$\int_{-\infty}^\infty x^4 e^{-\alpha x^2} dx = \left(-\frac{\partial}{\partial \alpha}\right)^2 \int_{-\infty}^\infty e^{-\alpha x^2} dx = \left(-\frac{\partial}{\partial \alpha}\right)^2 \sqrt{\frac{\pi}{\alpha}} = \frac{3}{4}\sqrt{\frac{\pi}{\alpha^5}} \qquad (3.3.8)$$

and in general

$$\int_{-\infty}^\infty x^{2n} e^{-\alpha x^2} dx = \left(-\frac{\partial}{\partial \alpha}\right)^n \sqrt{\frac{\pi}{\alpha}} \qquad n = 1,\ 2,\ 3,\ldots \qquad (3.3.9)$$

For odd powers of x

$$\int_{-\infty}^\infty x^{2n+1} e^{-\alpha x^2} dx = 0 \qquad (3.3.10)$$

since the integrand is antisymmetric about $x = 0$. However

$$\int_0^\infty x e^{-\alpha x^2} dx = \frac{1}{2\alpha} \qquad (3.3.11)$$

so that the Leibnitz integral rule can be used to evaluate higher odd powers

$$\int_0^\infty x^{2n+1} e^{-\alpha x^2} dx = \left(-\frac{\partial}{\partial \alpha}\right)^n \frac{1}{2\alpha} \qquad n = 1,\ 2,\ 3,\ldots \qquad (3.3.12)$$

3.3.1 Error Functions

A related Gaussian integral with a finite upper limit is defined as a function of the upper limit x

$$\text{erf}(x) = \frac{2}{\sqrt{\pi}} \int_0^x \exp(-\xi^2) d\xi \qquad (3.3.13)$$

where erf(x) is known as the error function. A Gaussian integral with finite lower limit x defines the complementary error function erfc(x) where

$$\text{erfc}(x) = \frac{2}{\sqrt{\pi}} \int_x^\infty \exp(-\xi^2) d\xi \qquad (3.3.14)$$

The prefactors $2/\sqrt{\pi}$ are chosen so that

$$\text{erf}(x) + \text{erfc}(x) = \frac{2}{\sqrt{\pi}} \int_0^\infty \exp(-\xi^2) d\xi = 1 \qquad (3.3.15)$$

If the numerical value erf(x) is known, we can immediately find erfc(x) = 1 − erf(x). For small values of $x \ll 1$ we have $\text{erf}(x) \approx 2x/\sqrt{\pi}$. Also, the error function is an odd function since $erf(-x) = -erf(x)$.

Maple Examples

The computation of even and odd Gaussian integrals by differentiation of a kernel is shown in the Maple worksheet below.

Key Maple commands: *diff, int, simplify*

Programming: For loops

restart

Gaussian Integrals

assume(alpha > 0)
for *n* **from** 0 **to** 10 **do**
Int(x^n·exp(−alpha*x^2), x = 0..infinity) = simplify(int(x^n·exp(−alpha*x^2), x = 0..infinity))
end

$$\int_0^\infty e^{-\alpha \sim x^2} dx = \frac{1}{2} \frac{\sqrt{\pi}}{\sqrt{\alpha \sim}}$$

$$\int_0^\infty e^{-\alpha\sim x^2}\,dx = \frac{1}{2\alpha\sim}$$

$$\int_0^\infty x^2 e^{-\alpha\sim x^2}\,dx = \frac{1}{4}\frac{\sqrt{\pi}}{\alpha\sim^{3/2}}$$

$$\int_0^\infty x^3 e^{-\alpha\sim x^2}\,dx = \frac{1}{2\alpha\sim^2}$$

$$\int_0^\infty x^4 e^{-\alpha\sim x^2}\,dx = \frac{3}{8}\frac{\sqrt{\pi}}{\alpha\sim^{5/2}}$$

$$\int_0^\infty x^5 e^{-\alpha\sim x^2}\,dx = \frac{1}{\alpha\sim^3}$$

$$\int_0^\infty x^6 e^{-\alpha\sim x^2}\,dx = \frac{15}{16}\frac{\sqrt{\pi}}{\alpha\sim^{7/2}}$$

$$\int_0^\infty x^7 e^{-\alpha\sim x^2}\,dx = \frac{3}{\alpha\sim^4}$$

$$\int_0^\infty x^8 e^{-\alpha\sim x^2}\,dx = \frac{105}{32}\frac{\sqrt{\pi}}{\alpha\sim^{9/2}}$$

$$\int_0^\infty x^9 e^{-\alpha\sim x^2}\,dx = \frac{12}{\alpha\sim^5}$$

$$\int_0^\infty x^{10} e^{-\alpha\sim x^2}\,dx = \frac{945}{64}\frac{\sqrt{\pi}}{\alpha\sim^{11/2}}$$

Even Gaussian Integrals by Differentiation of the Kernel I_0

$$I_0 := \frac{1}{2}\sqrt{\frac{\pi}{\text{alpha}}}$$

$$\frac{1}{2}\sqrt{\frac{\pi}{\alpha\sim}}$$

for n **from** 2 **to** 10 **by** 2 **do** $simplify\left((-1)^{\frac{n}{2}}\cdot diff\left(I_0, alpha\$\frac{n}{2}\right)\right)$**end**

$$\frac{1}{4}\frac{\sqrt{\pi}}{\alpha\sim^{3/2}}$$

$$\frac{3}{8}\frac{\sqrt{\pi}}{\alpha\sim^{5/2}}$$

$$\frac{15}{16}\frac{\sqrt{\pi}}{\alpha\sim^{7/2}}$$

$$\frac{105}{32}\frac{\sqrt{\pi}}{\alpha\sim^{9/2}}$$

$$\frac{945}{64}\frac{\sqrt{\pi}}{\alpha\sim^{11/2}}$$

Odd Gaussian Integrals by Differentiation of the Kernel I$_1$

$$I_1 := \frac{1}{2\cdot\text{alpha}}$$

$$\frac{1}{2\alpha\sim}$$

for *n* **from** 3 **to** 9 **by** 2 **do** *simplify*$\left((-1)^{\frac{(n-1)}{2}} \cdot \textit{diff}\left(I_1, alpha\$\frac{(n-1)}{2} \right) \right)$**end**

$$\frac{1}{2\alpha\sim^2}$$

$$\frac{1}{\alpha\sim^3}$$

$$\frac{3}{\alpha\sim^4}$$

$$\frac{12}{\alpha\sim^5}$$

3.4 SERIES AND APPROXIMATIONS

In this section geometric, Taylor and Maclaurin series are discussed. Topics include index shifting, convergence of series, and the binomial approximation. Approximations based on the binomial theorem are also covered.

3.4.1 Geometric Series

Consider the series

$$S = 1 + x + x^2 + x^3 + \cdots + x^n \tag{3.4.1}$$

Euclid noticed that multiplying both sides by x and then adding one to both sides results in

$$1 + xS = 1 + x + x^2 + x^3 + \cdots + x^{n+1}$$
$$= S + x^{n+1} \tag{3.4.2}$$

Solving for S gives

$$S = \frac{1 - x^{n+1}}{1 - x} \tag{3.4.3}$$

Now if $|x| < 1$ then $x^{n+1} \to 0$ as $n \to \infty$ so that

$$\frac{1}{1-x} = \sum_{n=0}^{\infty} x^n \tag{3.4.4}$$

Additional series may be derived from the geometric series such as

$$\frac{1}{1-x^2} = 1 + x^2 + x^4 + x^6 + \cdots \tag{3.4.5}$$

$$\frac{1}{1-x^2} = \sum_{n=0}^{\infty} x^{2n} \tag{3.4.6}$$

Differentiating the geometric series

$$\frac{d}{dx}\frac{1}{1-x} = \frac{d}{dx}\sum_{n=0}^{\infty} x^n \tag{3.4.7}$$

reveals that

$$\frac{1}{(1-x)^2} = \sum_{n=0}^{\infty} n x^{n-1} \qquad \text{for } |x| < 1 \tag{3.4.8}$$

Subsequent differentiation gives

$$\frac{2}{(1-x)^3} = \sum_{n=0}^{\infty} n(n-1) x^{n-2} \tag{3.4.9}$$

The sum may be started from $n = 2$ since the first two terms are zero.

3.4.2 Taylor Series

A Taylor series relates the value of a function $f(x)$ to the function evaluated at a nearby point $f(a)$ where $f(x)$ is continuous and differentiable over some region containing x and a. The expansion of $f(x)$ about $x = a$ is

$$f(x) = f(a) + (x-a)\frac{df}{dx}\bigg|_{x=a} + \frac{1}{2!}(x-a)^2\frac{d^2 f}{dx^2}\bigg|_{x=a} + \frac{1}{3!}(x-a)^3\frac{d^3 f}{dx^3}\bigg|_{x=a} + \cdots$$

(3.4.10)

that can be written compactly as

$$f(x) = \sum_{n=0}^{\infty}\frac{1}{n!}(x-a)^n\frac{d^n f(x)}{dx^n}\bigg|_{x=a}$$

(3.4.11)

3.4.3 Maclaurin Series

The expansion of $f(x)$ about $x = 0$ is known as a Maclaurin series

$$f(x) = \sum_{n=0}^{\infty}\frac{1}{n!}x^n\frac{d^n f(x)}{dx^n}\bigg|_{x=0}$$

(3.4.12)

Expanding $\sin(\theta)$ and $\cos(\theta)$ in a Taylor series about $\theta = 0$

$$\sin(\theta) = \theta - \frac{\theta^3}{3!} + \frac{\theta^5}{5!} + \cdots = \sum_{n=0}^{\infty}(-1)^n\frac{\theta^{2n+1}}{(2n+1)!}$$

(3.4.13)

$$\cos(\theta) = 1 - \frac{\theta^2}{2!} + \frac{\theta^4}{4!} + \cdots = \sum_{n=0}^{\infty}(-1)^n\frac{\theta^{2n}}{(2n)!}$$

(3.4.14)

Example 3.4.1

Expand $e^{i\theta}$ in a Taylor series about $\theta = 0$ to show that $e^{i\theta} = \cos(\theta) + i\sin(\theta)$
Solution: Using

$$e^z = \sum_{n=0}^{\infty}\frac{z^n}{n!}$$

(3.4.15)

$$e^{i\theta} = \sum_{n=0}^{\infty}\frac{(i\theta)^n}{n!} = 1 + i\theta - \frac{\theta^2}{2!} - i\frac{\theta^3}{3!} + \frac{\theta^4}{4!} + i\frac{\theta^5}{5!} - \frac{\theta^6}{6!}$$

(3.4.16)

and separating real and imaginary terms

$$e^{i\theta} = \left(1 - \frac{\theta^2}{2!} + \frac{\theta^4}{4!} - \frac{\theta^6}{6!} + \cdots\right) + i\left(\theta - \frac{\theta^3}{3!} + \frac{\theta^5}{5!} - \cdots\right)$$

(3.4.17)

we find that $e^{i\theta} = \cos(\theta) + i\sin(\theta)$.

3.4.4 Index Labels

Any symbol may be used for indexing summations and series. For example

$$\sum_{m=0}^{\infty} a_m \left(x - x_0\right)^m = \sum_{n=0}^{\infty} a_n \left(x - x_0\right)^n \qquad (3.4.18)$$

The index of summation is often referred to a "dummy" index. It is sometimes convenient to shift indices when combining summations. We can begin the summation at $m = 1$ by shifting our indices to $m - 1$

$$\sum_{m=0}^{\infty} a_m \left(x - x_0\right)^m = \sum_{m=1}^{\infty} a_{m-1} \left(x - x_0\right)^{m-1} \qquad (3.4.19)$$

3.4.5 Convergence of Series

The power series

$$\sum_{n=0}^{\infty} a_n \left(x - x_0\right)^n \qquad (3.4.20)$$

is said to converge if

$$\lim_{M \to \infty} \sum_{n=0}^{M} a_n \left(x - x_0\right)^n \qquad (3.4.21)$$

converges. The power series converges absolutely if

$$\sum_{n=0}^{\infty} \left| a_n \left(x - x_0\right)^n \right| \qquad (3.4.22)$$

converges. Note that a power series converges if it converges absolutely. However, the converse is not necessarily true.

3.4.6 Ratio Test

The ratio text compares subsequent terms in a series

$$\lim_{n \to \infty} \left| \frac{a_{n+1} \left(x - x_0\right)^{n+1}}{a_n \left(x - x_0\right)^n} \right| = \left| x - x_0 \right| \lim_{n \to \infty} \left| \frac{a_{n+1}}{a_n} \right| = L \qquad (3.4.23)$$

if $L < 1$ then the power series converges. If $L = 1$ the test is inconclusive. Several other convergence tests may also be used such as the integral test.

3.4.7 Integral Test

The integral test for convergence of the series $\sum_{n=k}^{\infty} a_n$ is performed by replacing the series by an improper integral

$$\sum_{n=k}^{\infty} a_n > \int_{k}^{\infty} f(n)dn \tag{3.4.24}$$

If the improper integral converges then the series converges. If the integral diverges then the sum also diverges.

Example 3.4.2

Applying the integral test for convergence of the harmonic series

$$\sum_{n=1}^{\infty} \frac{1}{n} = 1 + \frac{1}{2} + \frac{1}{3} + \frac{1}{4} + \cdots \tag{3.4.25}$$

Solution: The integral test gives

$$\lim_{M \to \infty} \int_{1}^{M} \frac{1}{n} dn = \lim_{M \to \infty} \ln(M) \to \infty \tag{3.4.26}$$

Example 3.4.3

Test the Riemann zeta function $\zeta(1 + \varepsilon)$ for convergence

$$\zeta(1+\varepsilon) = \sum_{n=1}^{\infty} \frac{1}{n^{(1+\varepsilon)}} = 1 + \frac{1}{2^{1+\varepsilon}} + \frac{1}{3^{1+\varepsilon}} + \frac{1}{4^{1+\varepsilon}} + \cdots \tag{3.4.27}$$

Solution: Applying the integral test

$$\lim_{M \to \infty} \int_{1}^{M} \frac{1}{n^{(1+\varepsilon)}} dn = \lim_{M \to \infty} \left(-\frac{n^{-\varepsilon}}{\varepsilon} \Big|_{1}^{M} \right) = \frac{1}{\varepsilon} \lim_{M \to \infty} \left(1 - \frac{1}{M^{\varepsilon}} \right) < \frac{1}{\varepsilon} \tag{3.4.28}$$

Thus $\zeta(1 + \varepsilon)$ converges for any $\varepsilon > 0$. The divergent harmonic series is obtained for $\varepsilon = 0$.

3.4.8 Binomial Theorem

The binomial theorem gives an expansion of a binomial raised to a power n

$$(a+x)^n = a^n + \binom{n}{1} a^{n-1} x + \binom{n}{2} a^{n-2} x^2 + \cdots + x^n \tag{3.4.29}$$

where the binomial coefficients are defined as

$$\binom{n}{k} = \frac{n!}{k!(n-k)!} \tag{3.4.30}$$

Example 3.4.4

Expand $(a + x)^4$ using the binomial theorem

Solution: $(a+x)^4 = a^4 + \binom{4}{1}a^3 x + \binom{4}{2}a^2 x^2 + \binom{4}{3}ax^3 + x^4 \tag{3.4.31}$

$$(a+x)^4 = a^4 + \frac{4!}{1!(4-1)!}a^3 x + \frac{4!}{2!(4-2)!}a^2 x^2 + \frac{4!}{3!(4-3)!}ax^3 + x^4 \tag{3.4.32}$$

$$= a^4 + 4a^3 x + 6a^2 x^2 + 4ax^3 + x^4$$

3.4.9 Binomial Approximations

A very useful approximation based on the binomial theorem for $a = 1$ is obtained from

$$(1+x)^n = 1 + \binom{n}{1}x + \binom{n}{2}x^2 + \cdots + x^n \tag{3.4.33}$$

For values of $x \ll 1$ terms of order x^2 and higher may be neglected so that

$$(1+x)^n \approx 1 + nx \tag{3.4.34}$$

Example 3.4.5

Find an approximation of the velocity

$$v = c\sqrt{1 - \left(1 - \frac{R_s}{R}\right)^2} \tag{3.4.35}$$

Solution: The binomial approximation gives for $R_s \ll R$

$$\left(1 - \frac{R_s}{R}\right)^2 \approx 1 - 2\frac{R_s}{R} \tag{3.4.36}$$

Thus, the velocity $v \approx c\sqrt{2\frac{R_s}{R}}$

Maple Examples

Examples in the Maple worksheet below present finite and infinite sums as well as series expansions including Taylor, Maclaurin, and multivariate expansions. The conversion of functions to formal power series is shown. Pascal's triangle is constructed with the use of binomial coefficients.

Key Maple commands: *add, convert, expand, evalf, FormalPowerSeries, mtaylor, sum*

Maple packages: *with(plots)*:

Programming: For loops

Special functions: binomial

restart

Finite Sums

$$sum\left(\frac{1}{k!}, k=0...5\right)$$

$$\frac{163}{60}$$

$$add\left(\frac{1}{k!}, k=0...5\right)$$

$$\frac{163}{60}$$

$$sum(f(k), k)$$

$$\sum_k f(k)$$

Infinite Sums

$$sum\left(\frac{1}{3^n}, n=0...\text{infinity}\right)$$

$$\frac{3}{2}$$

$$sum\left(\left(\frac{999}{1000}\right)^n, n=0...\text{infinity}\right)$$

$$sum\left(\left(\frac{998}{1000}\right)^n, n = 0\ldots\text{infinity}\right)$$
$$1000$$

$$sum\left(\left(\frac{997}{1000}\right)^n, n = 0\ldots\text{infinity}\right)$$
$$500$$

$$\frac{1000}{3}$$

Series

$$series\left(\frac{1}{1-x}, x\right)$$

$$1 + x + x^2 + x^3 + x^4 + x^5 + O(x^6)$$

$$series\left(\frac{1}{1-x^2}, x\right)$$

$$1 + x^2 + x^4 + O(x^6)$$

$$series\left(\frac{1}{1-x^3}, x\right)$$

$$1 + x^3 + O(x^6)$$

$$series\left(\frac{1}{1+x^2}, x\right)$$

$$1 - x^2 + x^4 + O(x^6)$$

$$series(x^{-x}, x = 0, 5)$$

$$1 - \ln(x)x + \frac{1}{2}\ln(x)^2 x^2 - \frac{1}{6}\ln(x)^3 x^3 + \frac{1}{24}\ln(x)^4 x^4 + O(x^5)$$

$$series(\text{sqrt}(\cos(\cos(x))), x = 0, 10):$$
$$evalf(\%)$$

$$0.7350525872 + 0.2861941445\, x^2 - 0.1255054201\, x^4 + 0.04539280320\, x^6 - 0.02503855750\, x^8 + O(x^{10})$$

$series(\cos(x \cos(theta))), theta = 0, 10)$

$$\cos(x) + \frac{1}{2}\sin(x)x\theta^2 + \left(-\frac{1}{8}\cos(x)x^2 - \frac{1}{24}\sin(x)x\right)\theta^4$$
$$+\left(\frac{1}{48}\cos(x)x^2 - \sin(x)\left(-\frac{1}{720}x + \frac{1}{48}x^3\right)\right)\theta^6$$
$$+\left(\cos(x)\left(\frac{1}{384}x^4 - \frac{1}{640}x^2\right) - \sin(x)\left(-\frac{1}{192}x^3 + \frac{1}{40320}x\right)\right)\theta^8 + O(\theta^{10})$$

$type(\%, series)$

true

Taylor Series

$taylor(\exp(-x)\cdot\sin(x)\cdot x, x = 1, 4);$

$$e^{-1}\sin(1) + e^{-1}\cos(1)(x-1) - e^{-1}\sin(1)(x-1)^2$$
$$+\left(-\frac{2}{3}e^{-1}\cos(1) + \frac{1}{3}e^{-1}\sin(1)\right)(x-1)^3 + O\left((x-1)^4\right)$$

$type(\%, \text{'}taylor\text{'})$

true

Maclaurin Series

$taylor(\exp(-x)\cdot x, x = 0);$

$$x - x^2 + \frac{1}{2}x^3 - \frac{1}{6}x^4 + \frac{1}{24}x^5 + O(x^6)$$

$taylor\left(\dfrac{\sin(x)}{x}, x = 0\right);$

$$1 - \frac{1}{6}x^2 + \frac{1}{120}x^4 + O(x^6)$$

Multivariate Series

$mtaylor(sqrt(\cos(x^2 + y^2)), [x, y]);$

$$1 - \frac{1}{4}x^4 - \frac{1}{2}y^2x^2 - \frac{1}{4}y^4$$

Converting to Power Series

$$convert\left(\frac{\sin(x)}{x}, FormalPowerSeries, x\right)$$

$$\sum_{k=0}^{\infty} \frac{(-1)^k x^{2k}}{(2k+1)!}$$

$convert(exp(-x), FormalPowerSeries, x)$

$$\sum_{k=0}^{\infty} \frac{(-1)^k x^k}{k!}$$

$convert(arctan(x), FormalPowerSeries, x)$

$$\sum_{k=0}^{\infty} \frac{(-1)^k x^{2k+1}}{2k+1}$$

Binomial Coefficients

$binomial(5,3);$

$$10$$

$expand((1+x)^5)$

$$x^5 + 5x^4 + 10x^3 + 10x^2 + 5x + 1$$

Pascal's Triangle

for n **from** 0 **to** 6 **do** $expand((a+b)^n)$ **end**

$$1$$
$$a + b$$
$$a^2 + 2ab + b^2$$
$$a^3 + 3a^2b + 3ab^2 + b^3$$
$$a^4 + 4a^3b + 6a^2b^2 + 4ab^3 + b^4$$
$$a^5 + 5a^4b + 10a^3b^2 + 10a^2b^3 + 5ab^4 + b^5$$
$$a^6 + 6a^5b + 15a^4b^2 + 20a^3b^3 + 15a^2b^4 + 6ab^5 + b^6$$

3.5 SPECIAL INTEGRALS

3.5.1 Integral Functions

Many important integrals that are commonly encountered in physics have no analytical solution. Several special functions are defined in terms of integrals such as the error function erf(x)

$$\text{erf}(x) = \frac{2}{\sqrt{\pi}} \int_0^x \exp\left(-\xi^2\right) d\xi \tag{3.5.1}$$

and the complementary error function erfc(x)

$$\text{erfc}(x) = \frac{2}{\sqrt{\pi}} \int_x^\infty \exp\left(-\xi^2\right) d\xi \tag{3.5.2}$$

related to the Gaussian integrals in Section 3.3. Other special functions have integral representation such as the Bessel functions in Section 6.6

$$J_0(x) = \frac{1}{\pi} \int_0^\pi \cos\left(x\sin\theta\right) d\theta \tag{3.5.3}$$

$$J_n(x) = \frac{1}{\pi} \int_0^\pi \cos\left(n\theta - x\sin\theta\right) d\theta \quad n = 1,2,3\ldots \tag{3.5.4}$$

3.5.2 Elliptic Integrals

Elliptic integrals may be expressed in Jacobi or the Legendre forms shown below. Incomplete elliptic integrals have two arguments (k, ϕ) while complete elliptic integrals have one argument $(k, \phi = \pi/2)$

Incomplete elliptic integral of the first kind:

$$F(k,\phi) = \int_0^\phi \frac{d\phi}{\left(1 - k^2 \sin\phi\right)^{1/2}} \tag{3.5.5}$$

Incomplete elliptic integral of the second kind:

$$E(k,\phi) = \int_0^\phi \left(1 - k^2 \sin\phi\right)^{1/2} d\phi \tag{3.5.6}$$

Complete elliptic integral of the first kind:

$$F\left(k,\frac{\pi}{2}\right)=\int_0^{\pi/2}\frac{d\phi}{\left(1-k^2\sin\phi\right)^{1/2}} \tag{3.5.7}$$

Complete elliptic integral of the second kind:

$$E\left(k,\frac{\pi}{2}\right)=\int_0^{\pi/2}\left(1-k^2\sin\phi\right)^{1/2}d\phi \tag{3.5.8}$$

The period T of a simple pendulum of length L and amplitude θ_0 may be expressed as a complete elliptic integral of the first kind

$$T=4\sqrt{\frac{L}{g}}F\left(k,\frac{\pi}{2}\right)\text{ with }k=\sin\frac{\theta_0}{2} \tag{3.5.9}$$

The perimeter of an ellipse P is given by a complete elliptic integral of the second kind $P=4aE\left(k,\dfrac{\pi}{2}\right)$ with $k=\sqrt{1-\dfrac{b^2}{a^2}}$ equal to the eccentricity of the ellipse with semi-major axis a and semi-minor axis b. Complete elliptic integrals are usually written with the $\pi/2$ suppressed $F(k)$ and $E(k)$.

3.5.3 Gamma Functions

The integral representation of the gamma function $\Gamma(s)$ is

$$\Gamma(s)=\int_0^\infty e^{-x}x^{s-1}dx \tag{3.5.10}$$

If n is a positive integer

$$\Gamma(n)=(n-1)! \tag{3.5.11}$$

For positive real numbers s

$$\Gamma(s+1)=s\Gamma(s) \tag{3.5.12}$$

Using the integral form of the gamma function we may verify that $\Gamma(2) = 1!$ by integrating by parts

$$\Gamma(2)=\int_0^\infty e^{-x}xdx=\left(-e^{-x}x\right)_0^\infty+\int_0^\infty e^{-x}dx=1 \tag{3.5.13}$$

For negative real values of s

$$\Gamma(s) = \frac{1}{s}\Gamma(s+1) \tag{3.5.14}$$

so that for negative integers $\Gamma(-1,-2...) \to \infty$.

3.5.4 Riemann Zeta Function

The Riemann zeta function $\zeta(z)$ is defined as

$$\zeta(z) = \sum_{n=1}^{\infty} n^{-z} \text{ for } |z| \geq 1 \tag{3.5.15}$$

Convergence of $\zeta(1 + \varepsilon)$ for any $\varepsilon > 0$ was investigated in Section 3.4 using the integral test. The zeta function is related to the gamma function

$$\zeta(n) = \frac{1}{\Gamma(n)} \int_0^{\infty} \frac{x^{n-1}}{e^x - 1} dx \tag{3.5.16}$$

The zeta function is related to the prime numbers p

$$\zeta(z) = \prod_p \frac{1}{1 - p^{-z}} \tag{3.5.17}$$

For example $\zeta(2)$ is the product

$$\zeta(2) = \left(\frac{1}{1-2^{-2}}\right)\left(\frac{1}{1-3^{-2}}\right)\left(\frac{1}{1-5^{-2}}\right)\left(\frac{1}{1-7^{-2}}\right)\cdots \tag{3.5.18}$$

In quantum field theory the zeta function is useful in the renormalization of divergent quantities such as the zero-point energy of vacuum modes ω_n

$$\underbrace{\frac{1}{2}\sum_n \hbar|\omega_n|}_{\text{diverges}} \xrightarrow{\text{renormalization}} \underbrace{\frac{1}{2}\sum_n \hbar|\omega_n||\omega_n|^{-s}}_{\text{finite}} \tag{3.5.19}$$

In renormalization procedures involving the zeta function, the $s \to 0$ limit is found for the calculation of physical quantities.

3.5.5 Writing Integrals in Dimensionless Form

An integral may contain variables and parameters that have units. By making suitable substitutions we may essentially factor the units outside of the integral and obtain an expression that is a product of physical constants with dimensions and an integral that is just a numerical factor. Writing integrals in dimensionless form is useful for dimensional analysis as well as for simplifying the integration. Consider the integral

$$\int_0^\infty \frac{e^{-kr}}{r^2+a^2}\,dr$$

(3.5.20)

where r and a have units of length and k has units of inverse length. If we define the dimensionless variable $x = kr$ then $dx = kdr$. Substitution gives

$$\int_0^\infty \frac{e^{-x}}{\left(\dfrac{x}{k}\right)^2+a^2}\frac{dx}{k} = k\underbrace{\int_0^\infty \frac{e^{-x}}{x^2+(ka)^2}\,dx}_{\text{numerical factor}}$$

(3.5.21)

where the limits of the integral are now from $x = 0$ to $x = \infty$. The product ka is a dimensionless number so the original integral is expressed as a numerical factor times the parameter k.

3.5.6 Black-Body Radiation

The integral

$$\int_0^\infty \frac{8\pi f^2}{c^3}\frac{hf}{e^{\frac{hf}{k_BT}}-1}\,df$$

(3.5.22)

is encountered in statistical mechanics in calculating the total power radiated by a body at temperature T. The integral may be cast in dimensionless form by making the substitution

$$x = \frac{hf}{k_BT} \text{ so that } df = \frac{k_BT}{h}dx \text{ and } f^3df = \left(\frac{k_BT}{h}\right)^4 x^3 dx$$

(3.5.23)

Both hf and k_BT have units of energy so that x is a dimensionless variable. The temperature dependence and physical constants are now factored out

$$\underbrace{\frac{8\pi h}{c^3}\left(\frac{k_BT}{h}\right)^4}_{\substack{\text{physical}\\\text{constants}}}\underbrace{\int_0^\infty \frac{x^3}{e^x-1}\,dx}_{\substack{\text{numerical}\\\text{factor}}}$$

(3.5.24)

The integral is evaluated by multiplying the top and bottom of the integrand by e^{-x}

$$\int_0^\infty x^3 \frac{e^{-x}}{1-e^{-x}} dx \tag{3.5.25}$$

using the result that

$$\frac{1}{1-e^{-x}} = \sum_{n=0}^\infty e^{-nx} \tag{3.5.26}$$

Factoring the sum outside of the integral

$$\int_0^\infty x^3 e^{-x} \sum_{n=0}^\infty e^{-nx} dx = \sum_{n=0}^\infty \int_0^\infty x^3 e^{-(n+1)x} dx \tag{3.5.27}$$

letting $u = (n+1)x$ and we have that

$$= \underbrace{\sum_0^\infty \frac{1}{(n+1)^4}}_{\frac{\pi^4}{90}} \underbrace{\int_0^\infty u^3 e^{-u} du}_{6} = \frac{\pi^4}{15} \tag{3.5.28}$$

and the integral is a numerical factor equal to $\pi^4/15$

$$\frac{8\pi h}{c^3} \left(\frac{k_B T}{h}\right)^4 \frac{\pi^4}{15} \tag{3.5.29}$$

Factoring out the temperature dependence shows that the power radiated by a black body is proportional to the fourth power of temperature T

$$\left(\frac{8\pi^4 k_B^4}{15 h^3 c^3}\right) T^4 \tag{3.5.30}$$

The Stephan-Boltzmann constant

$$\sigma = \frac{c}{4}\left(\frac{8\pi^4 k_B^4}{15 h^3 c^3}\right) = \frac{2\pi^4 k_B^4}{15 h^3 c^2} \tag{3.5.31}$$

is equal to $\sigma = 5.67 \times 10^{-8} \frac{\text{Watts}}{\text{m}^2 \text{K}^4}$ and the radiated flux F in units of power per area Watts/m^2 is

$$F = \sigma T^4 \tag{3.5.32}$$

The black-body radiation integral can be written as a product of Riemann zeta and gamma functions

$$\int_0^\infty \frac{x^3}{e^x - 1} dx = \Gamma(4)\zeta(4) \tag{3.5.33}$$

where in general

$$\int_0^\infty \frac{x^{n-1}}{e^x - 1} dx = \Gamma(n)\zeta(n) \tag{3.5.34}$$

Maple Examples

The black-body integral is expressed in terms of gamma and Riemann zeta functions in the Maple worksheet below. The gamma function is plotted in 1D and 3D. Zeros of the gamma function are shown by plotting 1/gamma.
Key Maple commands: *complexplot3d, plot, int*

Maple packages: *with(plots)*:

Special functions: GAMMA, Zeta

restart

Black-Body Integral

$$Int\left(\frac{x^3}{\exp(x)-1}, x = 0 \ldots \text{infinity}\right) = int\left(\frac{x^3}{\exp(x)-1}, x = 0 \ldots \text{infinity}\right)$$

$$\int_0^\infty \frac{x^3}{e^x - 1} dx = \frac{1}{15}\pi^4$$

GAMMA(4)·Zeta(4)

$$\frac{1}{15}\pi^4$$

$$Int\left(\frac{x^4}{\exp(x)-1}, x = 0 \ldots \text{infinity}\right) = int\left(\frac{x^4}{\exp(x)-1}, x = 0 \ldots \text{infinity}\right)$$

$$\int_0^\infty \frac{x^4}{e^x - 1} dx = 24\zeta(5)$$

GAMMA(5)·Zeta(5)

$$24\zeta(5)$$

Plot of the Gamma Function

restart
plot(GAMMA, -5 ... 4, 0 ...3)

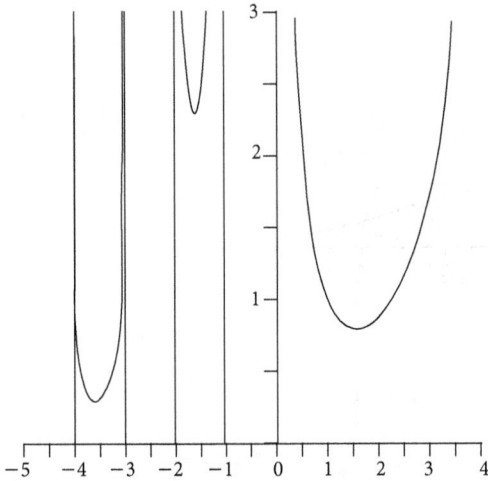

Figure 3.5.1: Plot of the gamma function.

with(plots) :
complexplot3d(abs(GAMMA(z)), z = -4 − 4 * I ... 4 + 4 * I, view = [-4 ... 4, -4 ... 4, 0 ... 6], orientation = [-50, 70, 0]):

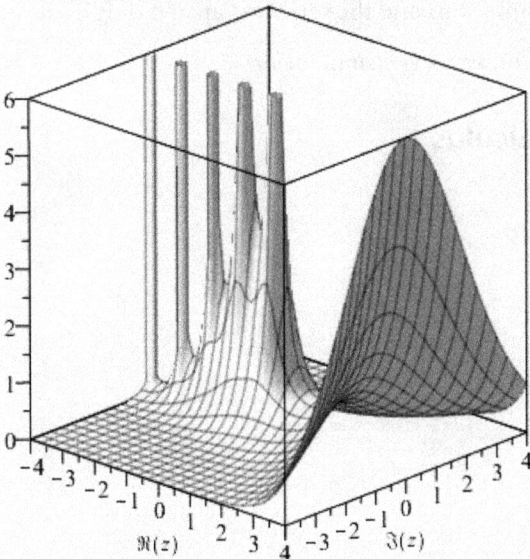

Figure 3.5.2: 3D Plot of the gamma function in the complex plane.

Plot of the 1/Gamma Function

$$plot\left(\frac{1}{\text{GAMMA}},-4\ldots4,-7\ldots7\right)$$

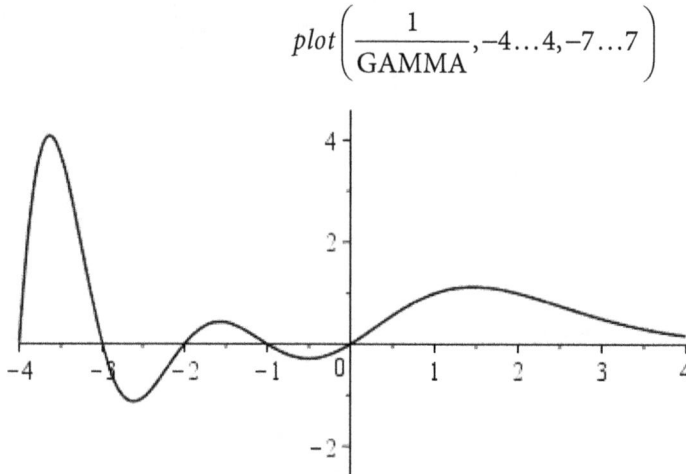

Figure 3.5.3: Plot of the 1/gamma function showing zeros at -3, -2, and 0.

3.6 MATLAB EXAMPLES

The use of MATLAB's Symbolic Math Toolbox is demonstrated in this section in solving single- and multivariable integration and differentiation. Topics include the calculation of maxima and minima of functions, multiple integrals, Gaussian and special integrals, symbolic summations and the series expansion of functions.

Key MATLAB commands: *ezplot3, int, syms, symsum, taylor*

Section 3.1 Single Variable Calculus

Symbolic Integration and Differentiation

```
>> int(log(x),x)
ans =
x*(log(x) - 1)
>> diff(ans,x)
ans =
log(x)
>> F=x^2+x*exp(x)
F =
x*exp(x) + x^2
>> diff(F,x,2)
ans =
2*exp(x) + x*exp(x) + 2
>> int(F,x,0,2)
ans =
exp(2) + 11/3
```

Maxima and Minima

```
>> syms x real
>> f=x^4-9*x^2
f =
x^4 - 9*x^2
>> diff(f,x)
ans =
4*x^3 - 18*x
>> solve(ans)
ans =
             0
  (3*2^(1/2))/2
 -(3*2^(1/2))/2
```

Section 3.2 Multivariable Calculus

```
3D Parametric Curve
>> syms t
>> ezplot3(sin(t)/t,cos(t),t,[0,20])
>> xlabel('x')
>> ylabel('y')
>> zlabel('t')
```

Double Integral

```
>> syms x y
>> int(int(x*y,y,0,x),x,0,1)
ans =
1/8
```

Triple Integral

```
>> syms x y z
>> int(int(int(x*y*z/sqrt(x^2+y^2+z^2),x),y),z)
ans =
(x^2 + y^2 + z^2)^(1/2)*(z^2*((2*x^2)/15 + (2*y^2)/15) + (2*x^2*y^2)/15
    + x^4/15 + y^4/15 + z^4/15)
```

Section 3.3 Gaussian Integrals

```
>> syms x a
>> int(exp(-a*x^2),x,0,inf)
Warning: Explicit integral could not be found.

ans =

piecewise([a = 1, pi^(1/2)/2], [a < 0, Inf], [Re(a) < 0 and pi/2 <
    abs(arg(a)) and not a < 0, int(1/exp(x^2*a), x = 0..Inf)], [0 <=
    Re(a) or abs(arg(a)) <= pi/2 and a <> 0, pi^(1/2)/(2*a^(1/2))])
```

```
>> syms a positive
>> int(exp(-a*x^2),x,0,inf)

ans =

pi^(1/2)/(2*a^(1/2))
>> int(x^2*exp(-a*x^2),x,0,inf)

ans =

pi^(1/2)/(4*a^(3/2))
```

Section 3.4 Series and Approximations

Symbolic Summation

```
>> syms x k
>> symsum(exp(-k*x),k,-1,1)

ans =

1/exp(x) + exp(x) + 1
>> syms n
>> symsum(1/n^2,1,inf)

ans =

pi^2/6
>> symsum((-1)^n/2^n,0,inf)

ans =

2/3
```

Taylor Expansion

```
>> syms x
>> taylor(sin(x)/x,x,5)
ans =
x^4/120 - x^2/6 + 1
>> syms m v c
>> E= m*c^2/sqrt(1-v^2/c^2)
E =
(c^2*m)/(1 - v^2/c^2)^(1/2)
>> taylor(E,v,3)
ans =
m*c^2 + (m*v^2)/2
```

Section 3.5 Special Integrals

```
>> syms x y
>> int(exp(-y^2),y,0,x)
ans =
(pi^(1/2)*erf(x))/2
```

```
>> int(exp(-y^2),y,x,inf)
ans =
-(pi^(1/2)*(erf(x) - 1))/2
>> int(cos(x*sin(y)),y,0,pi)
ans =
pi*besselj(0, x)
```

3.7 EXERCISES

Section 3.1 Single-Variable Calculus

1. The displacement vector of a particle moving with constant speed in a unit circle is given by

$$\mathbf{r}(t) = \cos(\omega t)\hat{\mathbf{i}} + \sin(\omega t)\hat{\mathbf{j}}$$

Calculate (a) $\dot{\mathbf{r}}$ (b) $\ddot{\mathbf{r}}$ (c) $|\mathbf{r}|$ (d) $|\dot{\mathbf{r}}|$ (e) $|\ddot{\mathbf{r}}|$

2. A particle moves with time-dependent speed

$$v(t) = v_0 \lambda t \exp(-\lambda t)$$

Calculate the total distance x traveled from $t_i = 0$ to $t_f = \infty$

$$x = \int_{t_i}^{t_f} v(t)\,dt$$

Calculate the acceleration of the particle $a(t) = \dot{v}(t)$ evaluated at time $t = 0$.

3. The Lennard-Jones formula models the potential energy between two neutral atoms

$$U(r) = 4\varepsilon \left(\left(\frac{r_0}{r}\right)^{12} - 2\left(\frac{r_0}{r}\right)^6 \right)$$

Calculate the equilibrium distance where the force is zero

$$F = -\frac{dU}{dr} = 0$$

Calculate the distance where the force is maximal

$$\frac{dF}{dr} = 0$$

What is the value of the force at its maximum value?

4. Evaluate the u-substitution integral

$$\int \sin^5(\theta)d\theta$$

by writing $\sin^5(\theta) = \sin(\theta)\sin^4(\theta) = \sin(\theta)(1-\cos^2\theta)^2$ and letting $u = \cos\theta$ This is the procedure for integrating odd powers of $\sin(\theta)$. A similar procedure is used for integrating odd powers of $\cos(\theta)$.

5. Evaluate the u-substitution integrals

(a) $\int \sin(\theta)\sqrt{\cos(\theta)}d\theta$

(b) $\int \sin^4(\theta)\cos(\theta)d\theta$

(c) $\int e^{\sin(\theta)}\cos\theta d\theta$

(d) $\int_0^1 \dfrac{4x^3+6x^2+1}{\left(x^4+2x^3+x+1\right)^{3/2}}dx$

6. Calculate the following integrals

(a) $\int_0^L \sin^2\left(\dfrac{3\pi x}{L}\right)dx$

(b) $\int_0^L \cos^2\left(\dfrac{4\pi x}{L}\right)dx$

7. Evaluate the following integrals by parts

(a) $\int r^2 \ln(r)dr$

(b) $\int x\ln(x)dx$

8. Show that the integral

$$\int_{-\frac{\lambda}{4}}^{\frac{\lambda}{4}} \cos\left(\dfrac{2\pi}{\lambda}z\right)\exp\left(-i\dfrac{2\pi}{\lambda}z\cos\theta\right)dz = \dfrac{\lambda}{\pi}\dfrac{\cos\left(\dfrac{\pi}{2}\cos\theta\right)}{\sin^2\theta}$$

using integration by parts.

9. Use the result that

$$\int_0^\infty e^{-\beta x}dx = \dfrac{1}{\beta}$$

to calculate the following integrals using the Leibnitz integral rule

(a) $\int_0^\infty x^3 e^{-\beta x}dx$

(b) $\int_0^\infty x^4 e^{-\beta x}dx$

(c) $\int_0^\infty x^5 e^{-\beta x}dx$

10. Given $\int_0^\infty x^n e^{-\beta x}dx = \dfrac{n!}{\beta^{n+1}}$ calculate $\int_0^\infty x^2 e^{-2x}dx$

Section 3.2 Multivariable Calculus

11. Given $f(x,y,z) = e^z \sin(x)\cos(y)$
calculate the following partial derivatives

$$\frac{\partial f}{\partial x}, \frac{\partial f}{\partial y}, \frac{\partial^2 f}{\partial y^2}, \text{ and } \frac{\partial^2 f}{\partial y \partial z}$$

12. Given

$$f(x,y) = e^{-\alpha(x^2+y^2)}$$

show that

$$\frac{\partial^2 f}{\partial x \partial y} = \frac{\partial^2 f}{\partial y \partial x}$$

13. Evaluate the double integral $\int_0^1 \int_0^1 x^2 y \, dx \, dy$

14. Evaluate the double integral $\iint (x + y^2) dx \, dy$ over the first quadrant of the unit circle using a polar coordinate transformation

15. Evaluate the triple integrals in Cartesian coordinates

$$\int_0^1 \int_0^1 \int_0^1 x^2 yz \, dx \, dy \, dz$$

$$\int_0^1 \int_0^z \int_0^y xyz^2 \, dx \, dy \, dz$$

$$\int_0^a \int_0^a \int_0^a \rho(x,y,z) \, dx \, dy \, dz \text{ where } \rho(x,y,z) = \rho_0 e^{-x/a} e^{-y/a} e^{-z/a}$$

16. Evaluate the volume integral in spherical coordinates

$$\int_0^{\frac{\pi}{2}} \int_0^{\frac{\pi}{2}} \int_0^R r \cos\theta \, dv$$

17. Evaluate the volume integral in cylindrical coordinates

$$\int_0^{2\pi} \int_{-1}^1 \int_0^R z^2 r \, dv$$

Section 3.3 Gaussian Integrals

18. Given $\int\limits_{0}^{\infty} e^{-ax^2}\,dx = \frac{1}{2}\sqrt{\frac{\pi}{\alpha}}$ calculate $\int\limits_{0}^{\infty} x^6 e^{-ax^2}\,dx$

19. Given $\int\limits_{0}^{\infty} x e^{-ax^2}\,dx = \frac{1}{2\alpha}$ calculate $\int\limits_{0}^{\infty} x^7 e^{-ax^2}\,dx$

20. Show that $\int\limits_{-\infty}^{\infty} \exp\left(-\frac{1}{2}\alpha x^2 + Jx\right)dx = \left(\frac{2\pi}{\alpha}\right)^{\frac{1}{2}} \exp\left(\frac{J^2}{2\alpha}\right)$

by completing the square in the exponent

21. Calculate the following integrals

$$\int\limits_{-\infty}^{\infty} (x+1)^2 e^{-\lambda x^2}\,dx$$

$$\int\limits_{0}^{\infty} (x-1)^3 e^{-\lambda x^2}\,dx$$

$$\int\limits_{-\infty}^{\infty} e^{-ax^2+bx}\,dx$$

22. Calculate the following integrals

$$\int\limits_{-\infty}^{\infty} x e^{-\lambda(x-a)^2}\,dx$$

$$\int\limits_{-\infty}^{\infty} x^2 e^{-\lambda(x-a)^2}\,dx$$

Section 3.4 Series and Approximations

23. Find the Taylor series for $f(x)$ centered about $x = a$ where

$$f(x) = \sqrt{1-x}$$

24. Find the Taylor series for $f(x)$ centered about $x = a$ where

$$f(x) = e^{-x^2}$$

25. Find the Maclaurin series for $f(x)$ where

$$f(x) = \frac{x^3}{e^x - 1}$$

26. Use the binomial expansion to approximation to show that

$$\frac{1}{\sqrt{1 - \left(\dfrac{v}{c}\right)^2}} \approx 1 + \frac{1}{2}\left(\frac{v}{c}\right)^2$$

27. Use the binomial expansion to approximate the following expression where $x \ll 1$

$$\sqrt{3 - x}$$

$$\frac{1}{(1-x)^{3/2}}$$

28. Perform the integral test for convergence of the series

$$\sum_{n=1}^{\infty} \frac{1}{n^2}$$

Section 3.5 Special Integrals

29. Perform a Maclaurin expansion of the complete elliptic integral of the first kind

$$F\left(k, \frac{\pi}{2}\right)$$

30. Perform a Maclaurin expansion of the complete elliptic integral of the second kind

$$E\left(k, \frac{\pi}{2}\right)$$

31. Plot $F\left(k, \dfrac{\pi}{2}\right)$ and $E\left(k, \dfrac{\pi}{2}\right)$ from $k = 0$ to $k = 1$.

32. Plot the Fresnel integrals $S(x)$ vs. $C(x)$ defined by

$$C(x) = \int_0^x \cos\left(\frac{\pi}{2}u^2\right) du$$

$$S(x) = \int_0^x \sin\left(\frac{\pi}{2}u^2\right) du$$

from $x = -4$ to $x = 4$

33. Express the following integrals in dimensionless form and evaluate the integrals

$$\int_0^\infty e^{-kx} \sin(kx)dx$$

where x has units of length and k has units of inverse length.

$$\int_0^\infty \frac{e^{-i\omega t}}{1+(\omega t)^2}dt$$

where t has units of time and ω has units of inverse time (frequency).

34. Write the triple integral

$$\int_{-L}^L \int_{-L}^L \int_{-L}^L \frac{1}{\left(x^2+y^2+z^2\right)^{-3/2}} dxdydz$$

in dimensionless form by making the substitutions $u = x/L$, $v = y/L$ and $w = z/L$. Also transform the limits of integration.

35. Use the integral representation of the gamma function to show that $\Gamma(1) = 1$

36. Show that the perimeter of an ellipse with zero eccentricity is that of a circle $P = 2\pi a$

37. Use the integral representation of the gamma function for $\Gamma(s+1)$

$$\Gamma(s+1) = \int_0^\infty e^{-x}x^s dx$$

integrating by parts to show that $\Gamma(s+1) = s\Gamma(s)$

38. Numerically evaluate the following integrals encountered in the theory of Bose-Einstein condensation

$$\int_0^\infty \frac{\sqrt{x}}{e^x-1}dx$$

$$\int_0^\infty \frac{x^{3/2}}{e^x-1}dx$$

39. Write a program to estimate the Riemann zeta function $\zeta(2)$ from the first 100 prime numbers. Hint: The Maple command *ithprime(i)* returns the ith prime number. For example, *ithprime(1)* $= 2$ and *ithprime(100)* $= 541$.

Chapter 4 VECTOR CALCULUS

Chapter Outline

4.1 VECTOR AND SCALAR FIELDS

Vector fields have both magnitude and direction while scalar fields are directionless. Both scalar and vector fields can have physical units and be spatially varying. A vector field can be described by the spatial variation of a scalar field. In this section, we seek a mathematical description of the spatial distribution of scalar and vector fields.

4.1.1 Scalar Fields

Examples of scalar fields include temperature $T(\mathbf{r})$ and pressure $P(\mathbf{r})$ with values that depend on coordinates $\mathbf{r}$. A two-dimensional scalar field such as the temperature

$$T(x,y) = T_0 e^{-\left(x^2+y^2\right)} \tag{4.1.1}$$

may describe the temperature in a plate with a hot spot at the center. A scalar field describes the electrical potential of a point charge located at $r = 0$ in spherical coordinates

$$V(r) = \frac{1}{4\pi\varepsilon_0}\frac{q}{r^2} \tag{4.1.2}$$

Scalar fields may be represented graphically using contour plots, surface plots or density plots. Contour plots show lines of constant field values. Surface plots represent scalar field values by different heights. Density plots may use variations in grayscale or color to illustrate regions with different field values.

4.1.2 Vector Fields

Examples of vector fields include the electric field $\mathbf{E}(\mathbf{r})$, magnetic field $\mathbf{B}(\mathbf{r})$ and the velocity field $\mathbf{v}(\mathbf{r})$ of a fluid. A two-dimensional vector field has x and y components that can vary over space

$$\mathbf{F}(x,y) = F_x(x,y)\hat{\mathbf{i}} + F_y(x,y)\hat{\mathbf{j}} \tag{4.1.3}$$

The magnitude of a vector field will produce a scalar field

$$\left|\mathbf{F}(x,y)\right| = \sqrt{F_x(x,y)^2 + F_y(x,y)^2} \tag{4.1.4}$$

We may plot a vector field by drawing arrows with lengths proportional to the magnitude of the field at equally spaced locations on a grid. The vector field $\mathbf{F} = \hat{\mathbf{i}}$ is plotted with arrows of the same length all pointing in the x-direction. Arrows corresponding to the vector field $\mathbf{F} = x\hat{\mathbf{i}}$ point to the right for $x > 0$ and to the left for $x < 0$ with increasing length as $|x|$ increases.

4.1.3 Field Lines

Lines that are tangent to the vector field are known as field lines and can be used to visualize a given vector field. In two dimensions, the equation describing field lines are a solution to the equation

$$\frac{dy}{dx} = \frac{F_y(x,y)}{F_x(x,y)} \tag{4.1.5}$$

where dy/dx gives the slope of the line tangent to the vector field at the point (x, y).

Example 4.1.1

Find the equation of the field lines corresponding to

$$\mathbf{F} = x\hat{\mathbf{i}} - y\hat{\mathbf{j}} \tag{4.1.6}$$

We identify $F_x = x$ and $F_y = -y$ so that

$$\frac{dy}{dx} = \frac{-y}{x} \tag{4.1.7}$$

This equation is solved by separating variables

$$\frac{dy}{y} = -\frac{dx}{x} \tag{4.1.8}$$

and integrating

$$\int \frac{dy}{y} = -\int \frac{dx}{x} \tag{4.1.9}$$

with the result

$$\ln(y) = -\ln(x) + \text{const.} \tag{4.1.10}$$

where we have combined both integration constants on the right-hand side. Exponentiating this expression

$$e^{\ln(y)} = e^{\ln\left(\frac{1}{x}\right) + \text{const.}} \tag{4.1.11}$$

gives us an equation for the field lines

$$y = \text{const.} \frac{1}{x} \tag{4.1.12}$$

where $e^{\text{const.}}$ is a constant. Positive constant values give hyperbolas in the first and third quadrants. Negative constant values give hyperbolas in the second and fourth quadrants. There are infinitely many field lines corresponding to all values of the constant. Usually, just a few lines are required to visualize the field, however. A similar analysis shows that the field lines corresponding to the vector field

$$\mathbf{F} = x\hat{\mathbf{i}} + y\hat{\mathbf{j}} \tag{4.1.13}$$

are straight lines passing through the origin according to

$$y = \text{const. } x \qquad (4.1.14)$$

with slopes of the field lines given by the constant.

Maple Examples

Examples of scalar field and vector field plots are given in the Maple worksheet below. Scalar field plots include contour, surface and combined surface and contour plots. Vector fields are plotted in 2D and 3D including combined vector field and field line plots.

Key Maple commands: *Contourplot, Ezsurfc, Fieldplot, Fieldplot3d, FlowLine, Plot3d, VectorField*

Maple packages: with(plots): with(MTM): with(Student[VectorCalculus]):

restart

Scalar Field Plots

with(plots) :
$g := exp(-(x^2+y^2))\cdot\sin(x\cdot y)$
$$g := e^{-x^2-y^2}\sin(x,y)$$

contourplot(g, x = -3 ... 3, y = -3 ... 3, color = "Black")

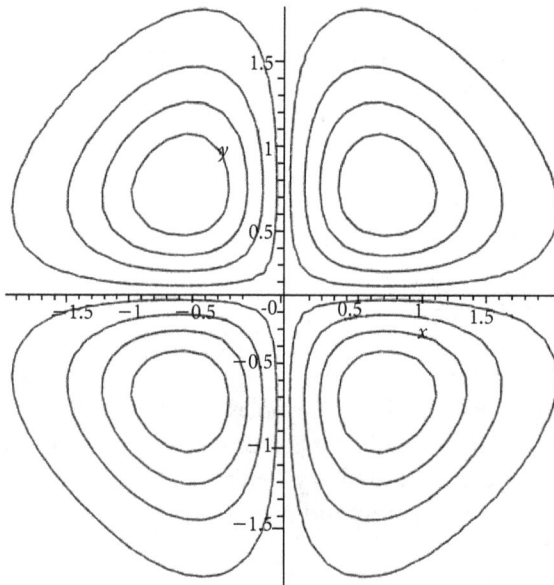

Figure 4.1.1 Contour plot of scalar field.

$plot3d(g, x = \text{-}3 \ldots 3, y = \text{-}3 \ldots 3, color = \text{“Gray”})$

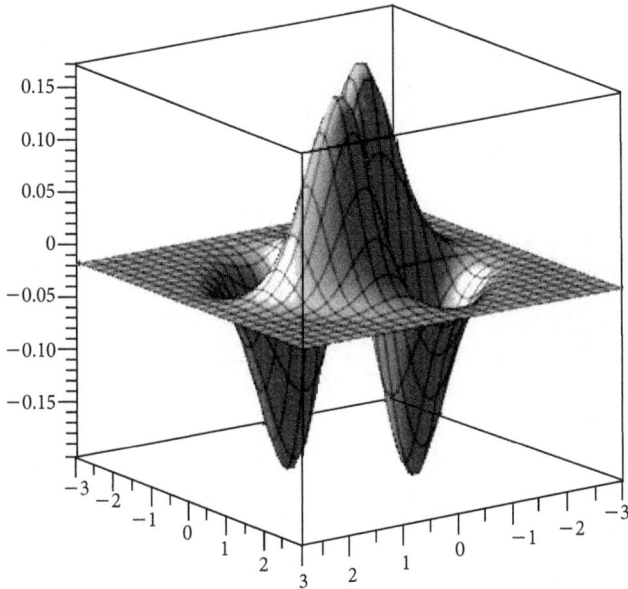

Figure 4.1.2: Surface plot of scalar field.

$with(MTM):$
$ezsurfc(g, [-3, 3, -3, 3])$

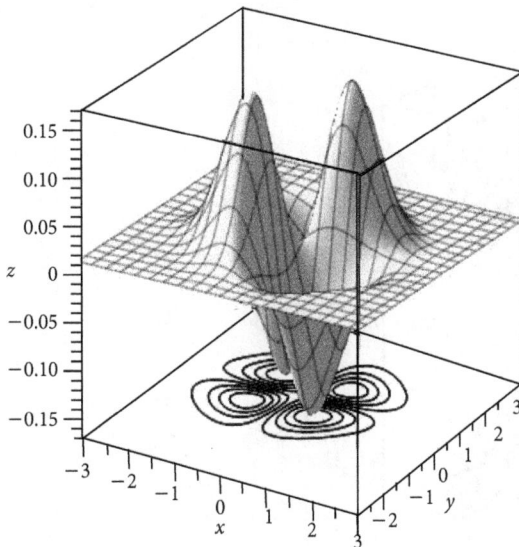

Figure 4.1.3: Surface plot of scalar field with contours.

Vector Field Plots

$$fieldplot([-y, x]\ x = -2 \ldots 2, y = -2 \ldots 2)$$

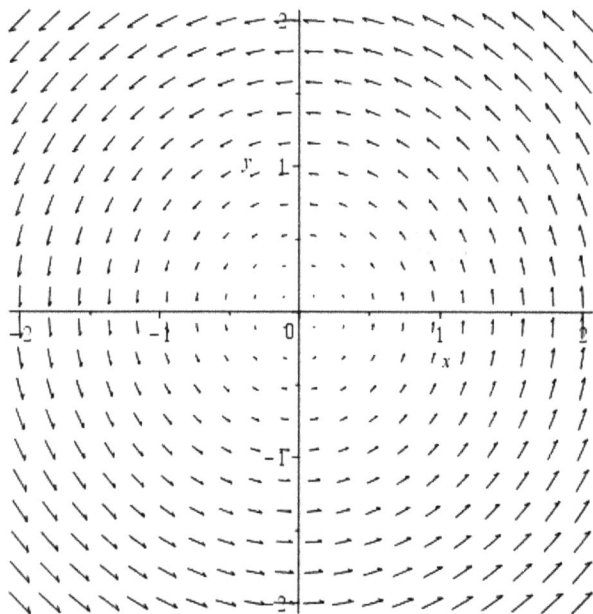

Figure 4.1.4: Plot of vector field

$$fieldplot3d([-y, x, z]\ x = -2 \ldots 2, y = -2 \ldots 2, z = -2 \ldots 2, color = \text{"Black"})$$

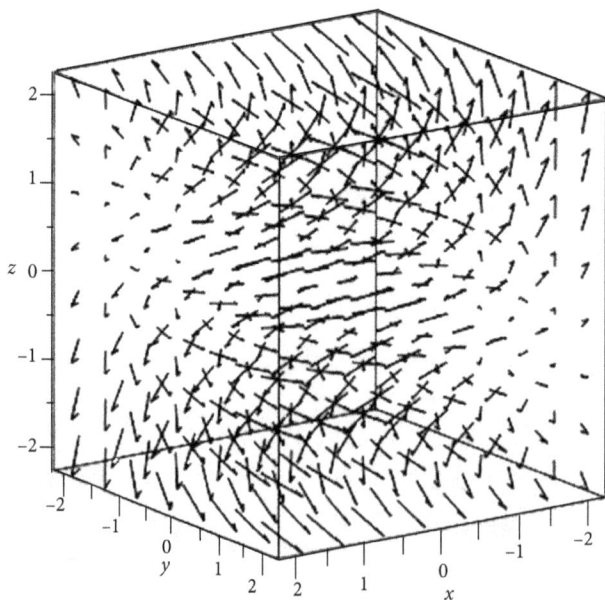

Figure 4.1.5: 3D vector field plot.

Field Line Plots

$with(Student[VectorCalculus]):$
$FlowLine(VectorField(\langle -y, x \rangle), [\langle 0, 1 \rangle, \langle 0, 2 \rangle, \langle 0, 3 \rangle, \langle 0, 4 \rangle], scaling = constrained)$

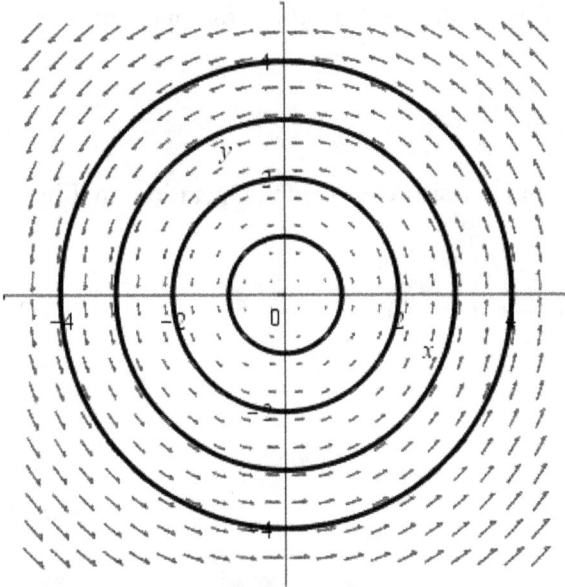

Arrows of the vector field, and the flow line(s) emanating from the given initial point(s)
Figure 4.1.6: Vector field displayed with field lines.

$FlowLine(VectorField(\langle -y, x, z \rangle), [\langle 0, 1, -2 \rangle, \langle 0, 1, -1 \rangle, \langle 0, 1, 1 \rangle, \langle 0, 1, 2 \rangle])$

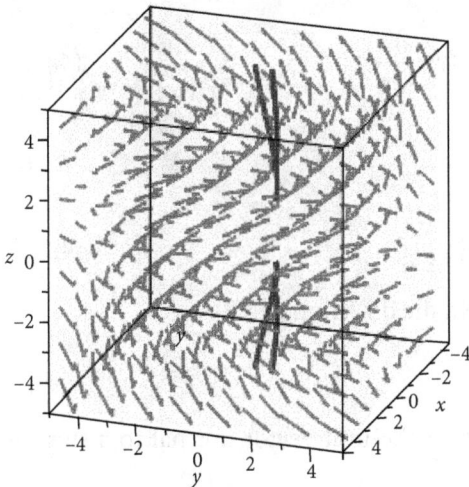

Arrows of the vector field, and the flow line(s) emanating from the given initial point(s)
Figure 4.1.7: 3D vector field plot with field lines.

4.2 GRADIENT OF SCALAR FIELDS

4.2.1 Gradient in Cartesian Coordinates

The gradient of a scalar field T is written as ∇T where the operator ∇ is a directional derivative defined as

$$\nabla \equiv \frac{\partial}{\partial x}\hat{\mathbf{i}} + \frac{\partial}{\partial y}\hat{\mathbf{j}} + \frac{\partial}{\partial z}\hat{\mathbf{k}} \tag{4.2.1}$$

Consider a scalar field $T(x, y, z)$ in Cartesian coordinates. Operation on T by ∇ gives a vector field

$$\nabla T = \left(\frac{\partial}{\partial x}\hat{\mathbf{i}} + \frac{\partial}{\partial y}\hat{\mathbf{j}} + \frac{\partial}{\partial z}\hat{\mathbf{k}}\right)T = \frac{\partial T}{\partial x}\hat{\mathbf{i}} + \frac{\partial T}{\partial y}\hat{\mathbf{j}} + \frac{\partial T}{\partial z}\hat{\mathbf{k}} \tag{4.2.2}$$

Example 4.2.1

Calculate the gradient of the scalar field

$$T = e^{-\left(x^2+y^2+z^2\right)} \tag{4.2.3}$$

Solution: In Cartesian coordinates

$$\nabla T = \left(-2x\hat{\mathbf{i}} - 2y\hat{\mathbf{j}} - 2z\hat{\mathbf{k}}\right)e^{-\left(x^2+y^2+z^2\right)} \tag{4.2.4}$$

The gradient function ∇T is related to the total differential

$$dT = \frac{\partial T}{\partial x}dx + \frac{\partial T}{\partial y}dy + \frac{\partial T}{\partial z}dz \tag{4.2.5}$$

formed by the dot product

$$dT = \nabla T \cdot d\mathbf{r} \tag{4.2.6}$$

where

$$d\mathbf{r} = dx\hat{\mathbf{i}} + dy\hat{\mathbf{j}} + dz\hat{\mathbf{k}} \tag{4.2.7}$$

Note that dT is maximal when $d\mathbf{r}$ points in the direction of ∇T.

4.2.2 Unit Normal

The gradient can also be used to construct the unit vector normal to a surface defined by

$$T(x, y, z) = \text{const.} \tag{4.2.8}$$

The gradient ∇T is a vector normal to a surface of constant T. Thus, a unit vector normal to the surface is obtained by dividing by the magnitude of the gradient

$$\hat{\mathbf{n}} = \frac{\nabla T}{|\nabla T|} \tag{4.2.9}$$

Example 4.2.2

Find a unit vector normal to the sphere described by

$$x^2 + y^2 + z^2 = 1 \tag{4.2.10}$$

Solution: Here we have

$$\hat{\mathbf{n}} = \frac{2x\hat{\mathbf{i}} + 2y\hat{\mathbf{j}} + 2z\hat{\mathbf{k}}}{\sqrt{(2x)^2 + (2y)^2 + (2z)^2}} = x\hat{\mathbf{i}} + y\hat{\mathbf{j}} + z\hat{\mathbf{k}} \tag{4.2.11}$$

We can verify that we have found a unit vector by forming the dot product $\hat{\mathbf{n}} \cdot \hat{\mathbf{n}} = x^2 + y^2 + z^2 = 1$.

Example 4.2.3

Find the unit vector normal to the paraboloid of revolution

$$z = x^2 + y^2 \tag{4.2.12}$$

Solution: We must first write the equation in the form $T(x, y, z) = $ const., or $x^2 + y^2 - z = 0$, which gives

$$\hat{\mathbf{n}} = \frac{2x\hat{\mathbf{i}} + 2y\hat{\mathbf{j}} - \hat{\mathbf{k}}}{\sqrt{(2x)^2 + (2y)^2 + 1}} \tag{4.2.13}$$

and we see that $\hat{\mathbf{n}} \cdot \hat{\mathbf{n}} = 1$ as required.

4.2.3 Gradient in Curvilinear Coordinates

The gradient of a scalar T may be written in orthogonal curvilinear coordinates (u_1, u_2, u_3) as

$$\nabla T = \frac{1}{g_{11}^{1/2}} \frac{\partial T}{\partial u_1} \hat{\mathbf{e}}_1 + \frac{1}{g_{22}^{1/2}} \frac{\partial T}{\partial u_2} \hat{\mathbf{e}}_2 + \frac{1}{g_{33}^{1/2}} \frac{\partial T}{\partial u_3} \hat{\mathbf{e}}_3 \tag{4.2.14}$$

where $(\hat{\mathbf{e}}_1, \hat{\mathbf{e}}_2, \hat{\mathbf{e}}_3)$ are unit vectors along the coordinate axes. The line element is

$$ds^2 = dx^2 + dy^2 + dz^2 = g_{11}du_1^2 + g_{22}du_2^2 + g_{33}du_3^2 \tag{4.2.15}$$

In terms of the Cartesian coordinates, the metric coefficients g_{ii} are

$$g_{ii} = \left(\frac{\partial x}{\partial u_i}\right)^2 + \left(\frac{\partial y}{\partial u_i}\right)^2 + \left(\frac{\partial z}{\partial u_i}\right)^2 \qquad (4.2.16)$$

Note that in Cartesian coordinates $g_{11} = g_{22} = g_{33} = 1$.

4.2.4 Cylindrical Coordinates

In cylindrical coordinates, we have $(u_1, u_2, u_3) = (r, \phi, z)$ and $(x, y, z) = (r \cos \phi, r \sin \phi, z)$

$$g_{11} = \left(\frac{\partial r \cos \phi}{\partial r}\right)^2 + \left(\frac{\partial r \sin \phi}{\partial r}\right)^2 + \left(\frac{\partial z}{\partial r}\right)^2 = \cos^2 \phi + \sin^2 \phi = 1 \qquad (4.2.17)$$

$$g_{22} = \left(\frac{\partial r \cos \phi}{\partial \phi}\right)^2 + \left(\frac{\partial r \sin \phi}{\partial \phi}\right)^2 + \left(\frac{\partial z}{\partial \phi}\right)^2 = r^2 \sin^2 \phi + r^2 \cos^2 \phi = r^2 \qquad (4.2.18)$$

$$g_{33} = \left(\frac{\partial r \cos \phi}{\partial z}\right)^2 + \left(\frac{\partial r \sin \phi}{\partial z}\right)^2 + \left(\frac{\partial z}{\partial z}\right)^2 = 1 \qquad (4.2.19)$$

Hence

$$\nabla T = \frac{\partial T}{\partial r}\hat{\mathbf{r}} + \frac{\partial T}{\partial z}\hat{\mathbf{z}} + \frac{1}{r}\frac{\partial T}{\partial \phi}\hat{\phi} \qquad (4.2.20)$$

Example 4.2.4

Compute the gradient of

$$T = r \cos \phi \qquad (4.2.21)$$

Solution: In cylindrical coordinates, the gradient is

$$\nabla T = \cos \phi \hat{\mathbf{r}} - \sin \phi \hat{\phi} \qquad (4.2.22)$$

4.2.5 Spherical Coordinates

In spherical coordinates, we have $(u_1, u_2, u_3) = (r, \theta, \phi)$ and
$(x, y, z) = (r \sin \theta \cos \phi, r \sin \theta \sin \phi, r \cos \theta)$

$$g_{11} = \left(\frac{\partial r \sin \theta \cos \phi}{\partial r}\right)^2 + \left(\frac{\partial r \sin \theta \sin \phi}{\partial r}\right)^2 + \left(\frac{\partial r \cos \theta}{\partial r}\right)^2$$
$$= \sin^2 \theta \cos^2 \phi + \sin^2 \theta \sin^2 \phi + \cos^2 \theta \qquad (4.2.23)$$
$$= \sin^2 \theta \left(\cos^2 \phi + \sin^2 \phi\right) + \cos^2 \theta = \sin^2 \theta + \cos^2 \theta = 1$$

$$g_{22} = \left(\frac{\partial r \sin\theta \cos\phi}{\partial\theta}\right)^2 + \left(\frac{\partial r \sin\theta \sin\phi}{\partial\theta}\right)^2 + \left(\frac{\partial r \cos\theta}{\partial\theta}\right)^2$$

$$= r^2\left(\cos^2\theta\cos^2\phi + \cos^2\theta\sin^2\phi + \sin^2\theta\right) \tag{4.2.24}$$

$$= r^2\left(\cos^2\theta\left(\cos^2\phi + \sin^2\phi\right) + \sin^2\theta\right) = r^2\left(\cos^2\theta + \sin^2\theta\right) = r^2$$

$$g_{33} = \left(\frac{\partial r \sin\theta \cos\phi}{\partial\phi}\right)^2 + \left(\frac{\partial r \sin\theta \sin\phi}{\partial\phi}\right)^2 + \left(\frac{\partial r \cos\theta}{\partial\phi}\right)^2$$

$$= r^2\sin^2\theta\sin^2\phi + r^2\sin^2\theta\cos^2\phi \tag{4.2.25}$$

$$= r^2\sin^2\theta\left(\sin^2\phi + \cos^2\phi\right) = r^2\sin^2\theta$$

hence

$$\nabla T = \frac{\partial T}{\partial r}\hat{\mathbf{r}} + \frac{1}{r}\frac{\partial T}{\partial\theta}\hat{\boldsymbol{\theta}} + \frac{1}{r\sin\theta}\frac{\partial T}{\partial\phi}\hat{\boldsymbol{\phi}} \tag{4.2.26}$$

Example 4.2.5

Calculate the gradient of

$$T = \frac{1}{r}\sin\theta \tag{4.2.27}$$

Solution: In spherical coordinates, the gradient is

$$\nabla T = -\frac{1}{r^2}\sin\theta\hat{\mathbf{r}} + \frac{1}{r^2}\cos\theta\hat{\boldsymbol{\theta}} \tag{4.2.28}$$

4.2.6 Scalar Field from the Gradient

Given a vector field **F**, one may seek to find if it could have resulted from the gradient of a scalar field Ω.

Example 4.2.6

The vector field

$$\mathbf{F} = x\hat{\mathbf{i}} + y\hat{\mathbf{j}} + z\hat{\mathbf{k}} \tag{4.2.29}$$

could have resulted from the gradient of the scalar field

$$\Omega = \frac{x^2}{2} + \frac{y^2}{2} + \frac{z^2}{2} \tag{4.2.30}$$

Example 4.2.7

The vector field in cylindrical coordinates

$$\mathbf{F} = -\frac{1}{r}\hat{\mathbf{r}} \qquad (4.2.31)$$

may be obtained from the gradient of

$$\Omega = \ln(r) \qquad (4.2.32)$$

Maple Examples

Examples of gradient field calculations are given in the Maple worksheet below. The unit normal to a surface is computed. Gradient fields are calculated and plotted in Cartesian, cylindrical, and spherical coordinates. Scalar potentials corresponding to vector fields in Cartesian, cylindrical and spherical coordinates are obtained.

Key Maple commands:

VectorField, ScalarPotential , SetCoordinates, Gradient, fieldplot3d

Maple packages: *with(plots): with(MTM): with(Student[VectorCalculus]):*

restart

Gradient in Cartesian Coordinates

with(VectorCalculus) :

with(plots) :

SetCoordinates('cartesian'[x, y, z]);

$$cartesian_{x,\,y,\,z}$$

$$G:= x^2 + y^2 + z^2$$

$$x^2 + y^2 + z^2$$

Gradient(G);

$$2x\overline{e}_x + 2y\overline{e}_y + 2z\overline{e}_z$$

Unit Normal to a Surface

$$Z:= x^2 + y^2$$

$$x^2 + y^2$$

$$\frac{Gradient(Z)}{\text{sqrt}\,(Gradient(Z)\cdot Gradient(Z))}$$

$$\left(\frac{x}{\sqrt{x^2+y^2}}\right)\overline{e}_x + \left(\frac{y}{\sqrt{x^2+y^2}}\right)\overline{e}_y$$

Gradient Plot in Cartesian Coordinates

$fieldplot3d(Gradient(G), x = -2 \ldots 2, y = -2 \ldots 2, z = -2 \ldots 2)$

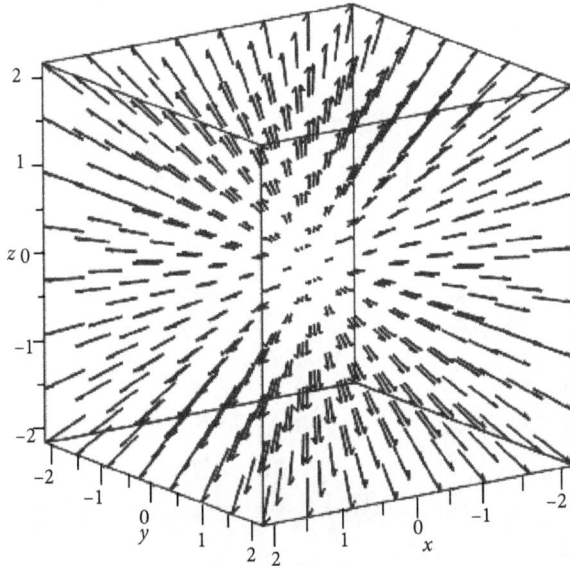

Figure 4.2.1: 3D plot of a field gradient.

Gradient in Cylindrical Coordinates

$SetCoordinates('cylindrical'[r, \text{theta}, z]);$

$$cylindrical_{r,\,\theta,\,\phi}$$

$_{H:} = r \cdot \sinh(z) \cdot \cos(\text{theta})$

$$r \cdot \sinh(z) \cdot \cos(\theta)$$

$Gradient(H);$

$$\left(\sinh(z)\cos(\theta)\right)\overline{e}_r - \sinh(z)\sin(\theta)\overline{e}_y + \left(r\cosh(z)\cos(\theta)\right)\overline{e}_z$$

Gradient Plot in Cylindrical Coordinates

$fieldplot3d(Gradient(H), r = 0 \ldots 2, theta = 0 \ldots 2 \cdot Pi, z = -2 \ldots 2)$

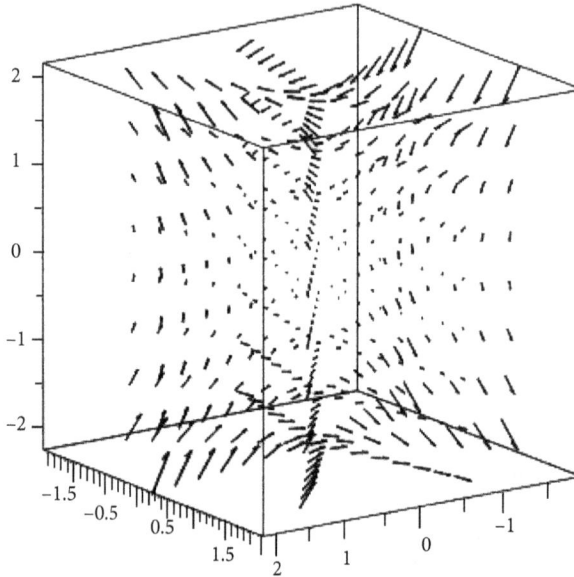

Figure 4.2.2: 3D plot of a field gradient in cylindrical coordinates.

Gradient in Spherical Coordinates

$SetCoordinates(`spherical'[r, theta, phi])$

$$spherical_{r,\theta,\phi}$$

$S: = r \cdot \sin(theta)$

$$r \cdot \sin(\theta)$$

$Gradient(S);$

$$(\sin(\theta))\overline{e}_r + (\cos(\theta))\overline{e}_\theta$$

Plot of Gradient in Spherical Coordinates

$fieldplot3d(Gradient(S), r = 0 \ldots 2, theta = 0 \ldots Pi, phi = -2 \ldots 2 \cdot Pi)$

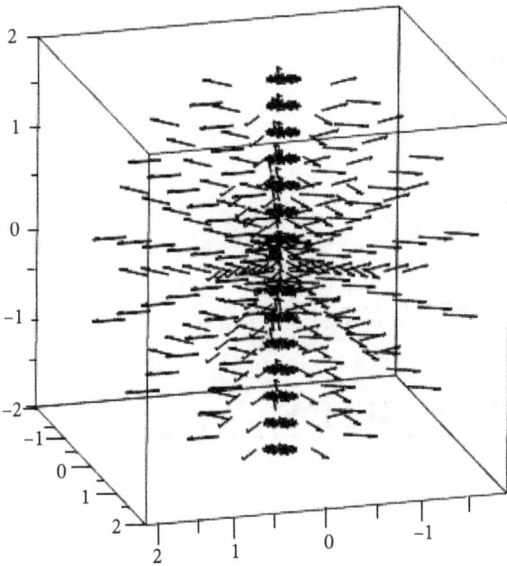

Figure 4.2.3: 3D plot of a field gradient in spherical coordinates.

Scalar Potential in Cartesian Coordinates

$with(VectorCalculus)$:
$SetCoordinates(`cartesian'[x, y, z])$;

$$cartesian_{x, y, z}$$

$v = VectorField(\langle x^2, y^3, z \rangle)$

$$(x^2)\overline{e}_x + (y^3)\overline{e}_y + (z)\overline{e}_z$$

$ScalarPotential(v)$

$$\frac{1}{3}x^3 + \frac{1}{4}y^4 + \frac{1}{2}z^2$$

Scalar Potential in Spherical Coordinates

$SetCoordinates(`spherical'[r, theta, phi])$

$$spherical_{r, \theta, f}$$

$v = VectorField(\langle r, 0, 0 \rangle)$

$$(r)\overline{e}_r$$

$ScalarPotential(v)$

$$\frac{1}{2}r^2$$

Scalar Potential in Cylindrical Coordinates

$SetCoordinates('cylindrical'\,[r, phi, z]);$

$$cylindrical_{r,\,\phi,\,z}$$

$v = VectorField(\langle r, 0, z^2\rangle)$

$$(r)\overline{e}_r+(z^2)\overline{e}_z$$

$ScalarPotential(v)$

$$\frac{1}{2}r^2\cos(\phi)^2+\frac{1}{2}r^2\sin(\phi)^2+\frac{1}{3}z^3$$

4.3 DIVERGENCE OF VECTOR FIELDS

4.3.1 Flux through a Surface

Consider a three-dimensional vector field

$$\mathbf{F}(x,y,z)=F_x\left(x,y,z\right)\hat{\mathbf{i}}+F_y\left(x,y,z\right)\hat{\mathbf{j}}+F_z\left(x,y,z\right)\hat{\mathbf{k}} \tag{4.3.1}$$

The flux of **F** through a surface is defined as surface integral

$$\text{flux}=\int\mathbf{F}\cdot d\mathbf{a} \tag{4.3.2}$$

where surface element $d\mathbf{a}$ is normal to the surface.

Example 4.3.1

The flux through the $z=0$ plane due to the vector field

$$\mathbf{F}=e^{-\left(x^2+y^2\right)}\hat{\mathbf{k}} \tag{4.3.3}$$

is given by the surface integral

$$\int\mathbf{F}\cdot d\mathbf{a}=\int_0^{2\pi}\int_0^{\infty}e^{-r^2}r\,dr\,d\theta=\pi \tag{4.3.4}$$

after making the polar coordinate transformation $d\mathbf{a}=r\,dr\,d\theta\hat{\mathbf{k}}$ and $x^2+y^2=r^2$.

4.3.2 Divergence of a Vector Field

We now consider the flux of a vector field through a closed surface. The divergence of a vector field at a point is related to the flux though a closed surface surrounding the point. The surface encloses a volume ΔV. The divergence is defined as limit as $\Delta V\rightarrow 0$ of the flux divided by ΔV or

$$\text{div }\mathbf{F}\equiv\lim_{\Delta V\rightarrow 0}\frac{1}{\Delta V}\int\mathbf{F}\cdot d\mathbf{a} \tag{4.3.5}$$

If we choose a cubical volume our surface integral has six terms. Surface elements at x and $x+\Delta x$ are $-\hat{\mathbf{i}}\Delta y\Delta z$ and $\hat{\mathbf{i}}\Delta y\Delta z$, respectively. The flux through these two surfaces is $-F_x\left(x,y,z\right)\Delta y\Delta z$ and $F_x\left(x+\Delta x,y,z\right)\Delta y\Delta z$. With similar terms for the other four faces we have

$$
\begin{aligned}
\int \mathbf{F}\cdot d\mathbf{a} =& \left[F_x\left(x+\Delta x,y,z\right)-F_x\left(x,y,z\right)\right]\Delta y\Delta z \\
&+\left[F_y\left(x,y+\Delta y,z\right)-F_y\left(x,y,z\right)\right]\Delta x\Delta z \\
&+\left[F_z\left(x,y,z+\Delta z\right)-F_z\left(x,y,z\right)\right]\Delta x\Delta y
\end{aligned}
$$

$$(4.3.6)$$

Dividing the surface integral by $\Delta V = \Delta x\Delta y\Delta z$

$$
\frac{1}{\Delta x\Delta y\Delta z}\int \mathbf{F}\cdot d\mathbf{a} = \frac{\left[F_x\left(x+\Delta x,y,z\right)-F_x\left(x,y,z\right)\right]}{\Delta x}
$$
$$
+\frac{\left[F_y\left(x,y+\Delta y,z\right)-F_y\left(x,y,z\right)\right]}{\Delta y}+\frac{\left[F_z\left(x,y,z+\Delta z\right)-F_z\left(x,y,z\right)\right]}{\Delta z}
$$

$$(4.3.7)$$

and taking the limit as the $\Delta V \to 0$

$$
\lim_{\Delta V \to 0}\frac{1}{\Delta V}\int \mathbf{F}\cdot d\mathbf{a} = \frac{\partial F_x}{\partial x}+\frac{\partial F_y}{\partial y}+\frac{\partial F_z}{\partial z}
$$

$$(4.3.8)$$

We may obtain the divergence of a vector field $\mathbf{F}$ by forming the dot product with ∇ where

$$
\nabla\cdot\mathbf{F} = \left(\frac{\partial}{\partial x}\hat{\mathbf{i}}+\frac{\partial}{\partial y}\hat{\mathbf{j}}+\frac{\partial}{\partial z}\hat{\mathbf{k}}\right)\cdot\left(F_x\hat{\mathbf{i}}+F_y\hat{\mathbf{j}}+F_z\hat{\mathbf{k}}\right)
$$

$$(4.3.9)$$

gives a scalar function

$$
\nabla\cdot\mathbf{F} = \left(\frac{\partial F_x}{\partial x}+\frac{\partial F_y}{\partial y}+\frac{\partial F_z}{\partial z}\right)
$$

$$(4.3.10)$$

Example 4.3.2

Calculate the divergence of the vector field

$$
\mathbf{F} = \sin\left(x\right)\hat{\mathbf{i}}+e^{-y}\hat{\mathbf{j}}+\frac{1}{z}\hat{\mathbf{k}}
$$

$$(4.3.11)$$

Solution: In Cartesian coordinates

$$
\nabla\cdot\mathbf{F} = \cos\left(x\right)-e^{-y}-\frac{1}{z^2}
$$

$$(4.3.12)$$

Notice that a function of the form

$$\mathbf{F} = F_x(y,z)\hat{\mathbf{i}} + F_y(x,z)\hat{\mathbf{j}} + F_z(x,y)\hat{\mathbf{k}} \qquad (4.3.13)$$

will have zero divergence since

$$\frac{\partial}{\partial x}F_x(y,z) = \frac{\partial}{\partial y}F_y(x,z) = \frac{\partial}{\partial z}F_z(x,y) = 0 \qquad (4.3.14)$$

Thus, we can see from inspection that the vector field

$$\mathbf{F} = \sinh(y)e^{-z}\hat{\mathbf{i}} + \cos(x^z)\hat{\mathbf{j}} + e^{-(x^2+y^2)}\hat{\mathbf{k}} \qquad (4.3.15)$$

has zero divergence.

4.3.3 Gradient in Curvilinear Coordinates

In curvilinear coordinates (u_1, u_2, u_3) the divergence of a vector

$$\nabla \cdot \mathbf{F} = \frac{1}{g^{1/2}}\left\{\frac{\partial}{\partial u_1}\left[\left(\frac{g}{g_{11}}\right)^{1/2}F_1\right] + \frac{\partial}{\partial u_2}\left[\left(\frac{g}{g_{22}}\right)^{1/2}F_2\right] + \frac{\partial}{\partial u_3}\left[\left(\frac{g}{g_{33}}\right)^{1/2}F_3\right]\right\} \qquad (4.3.16)$$

where

$$g = \begin{vmatrix} g_{11} & g_{12} & g_{13} \\ g_{21} & g_{22} & g_{23} \\ g_{31} & g_{32} & g_{33} \end{vmatrix} = g_{11}g_{22}g_{33} \qquad (4.3.17)$$

4.3.4 Cylindrical Coordinates

In cylindrical coordinates, we have $(u_1, u_2, u_3) = (r, \phi, z)$ and $(x, y, z) = (r\cos\phi, r\sin\phi, z)$

$$g_{11} = 1, g_{22} = r^2, g_{33} = 1, \text{ so that } g = r^2 \qquad (4.3.18)$$

$$\nabla \cdot \mathbf{F} = \frac{1}{r}\left\{\frac{\partial}{\partial r}[rF_r] + \frac{\partial}{\partial z}[rF_z] + \frac{\partial}{\partial \phi}[F_\phi]\right\} = \frac{1}{r}\frac{\partial}{\partial r}(rF_r) + \frac{\partial F_z}{\partial z} + \frac{1}{r}\frac{\partial F_\phi}{\partial \phi} \qquad (4.3.19)$$

Example 4.3.3

Find the divergence of

$$\mathbf{F} = r\cos\phi\,\hat{\mathbf{r}} - r\sin\phi\,\hat{\phi} \qquad (4.3.20)$$

Solution: In cylindrical coordinates

$$\nabla \cdot \mathbf{F} = \frac{1}{r}\frac{\partial}{\partial r}(r^2\cos\phi) - \frac{1}{r}\frac{\partial}{\partial \phi}(r\sin\phi) = 2\cos\phi - \cos\phi = \cos\phi \qquad (4.3.21)$$

4.3.5 Spherical Coordinates

In spherical coordinates, we have $(u_1, u_2, u_3) = (r, \theta, \phi)$ and

$$(x, y, z) = (r \sin\theta \cos\phi, r \sin\theta \sin\phi, r \cos\theta) \tag{4.3.22}$$

$$g_{11} = 1, g_{22} = r^2, g_{33} = r^2, \text{ so that } g = r^4 \sin^2\theta \tag{4.3.23}$$

$$\nabla \cdot \mathbf{F} = \frac{1}{r^2 \sin\theta} \left\{ \frac{\partial}{\partial r} \left[r^2 \sin\theta F_r \right] + \frac{\partial}{\partial \theta} \left[r \sin\theta F_\theta \right] + \frac{\partial}{\partial \phi} \left[r F_\phi \right] \right\}$$

$$= \frac{1}{r^2} \frac{\partial \left(r^2 F_r \right)}{\partial r} + \frac{1}{r \sin\theta} \frac{\partial \left(\sin\theta F_\theta \right)}{\partial \theta} + \frac{1}{r \sin\theta} \frac{\partial F_\phi}{\partial \phi} \tag{4.3.24}$$

Example 4.3.4

Find the divergence of

$$\mathbf{F} = r \sin\theta \hat{\mathbf{r}} - r \hat{\boldsymbol{\theta}} \tag{4.3.25}$$

Solution: In spherical coordinates

$$\nabla \cdot \mathbf{F} = \frac{1}{r^2} \frac{\partial \left(r^3 \sin\theta \right)}{\partial r} - \frac{1}{r \sin\theta} \frac{\partial \left(r \sin\theta \right)}{\partial \theta} = 3 \sin\theta - \cot\theta \tag{4.3.26}$$

Maple Examples

Examples of divergence calculations of vector fields in Cartesian, cylindrical and spherical coordinates are given in the Maple worksheet below. The flux of a vector field through closed surfaces is computed.

Key Maple commands: *Assume, Divergence, DotProduct, Flux, SetCoordinates, VectorField*

Maple packages: *with(VectorCalculus):*

restart

Divergence in Cartesian Coordinates

with(VectorCalculus) :
SetCoordinates('cartesian'[x, y, z]);

$$cartesian_{x, y, z}$$

F: = VectorField(⟨-y, x, 0⟩)

$$F := (-y)\overline{e}_x + (x)\overline{e}_y + (0)\overline{e}_z$$

Divergence(F)

$$0$$

G: = VectorField(⟨-y, x, z⟩)

$$G := (-y)\overline{e}_x + (x)\overline{e}_y + (z)\overline{e}_z$$

Divergence(G)

$$1$$

DotProduct(Del, G)

$$1$$

Divergence in Cylindrical Coordinates

SetCoordinates('cylindrical'[r, phi, z]);

$$cylindrical_{r,\phi,z}$$

$$v := VectorField\left(\left\langle \frac{1}{r^2}, 0, 0 \right\rangle\right)$$

$$v := \left(\frac{1}{r^2}\right)\overline{e}_r + (0)\overline{e}_\phi + (0)\overline{e}_z$$

Divergence(v)

$$-\frac{1}{r^3}$$

Divergence in Spherical Coordinates

SetCoordinates('spherical'[r, phi, theta])

$$spherical_{r,\phi,\theta}$$

$$w := VectorField\left(\left\langle \sin(phi), 0, \frac{1}{r^2} \right\rangle\right)$$

$$w := (\sin(\phi))\overline{e}_r + (0)\overline{e}_\phi + \left(\frac{1}{r^2}\right)\overline{e}_\theta$$

Divergence(w)

$$\frac{2\sin(\phi)}{r}$$

restart

Flux through a Surface

with(VectorCalculus) :
SetCoordinates('spherical'[r, theta, phi])

$$spherical_{r,\theta,\phi}$$

$$v := VectorField\left(\left\langle \frac{1}{r^2}, 0, 0 \right\rangle\right)$$

$$v := \left(\frac{1}{r^2}\right)\overline{e}_r + (0)\overline{e}_\theta + (0)\overline{e}_\phi$$

assume(R > 0)
Flux(v, Sphere(⟨0, 0, 0⟩, R))

$$4\pi$$

Divergence(v)

$$0$$

Here, $\nabla \cdot v$ is a delta function that is zero everywhere except the origin.

SetCoordinates('cartesian'[x, y, z]);

$$cartesian_{x, y, z}$$

w := *VectorField*(⟨x, y, z⟩)

$$w := (x)\overline{e}_x + (y)\overline{e}_y + (z)\overline{e}_z$$

Flux(w, Box(0 … 1, 0 … 1, 0 … 1))

$$3$$

Flux(w, Sphere(⟨0, 0, 0⟩, 2))

$$32\pi$$

with(Student[VectorCalculus]) :
Flux(w, Box(0 … 1, 0 … 1, 0 … 1), output = plot, scaling = constrained)

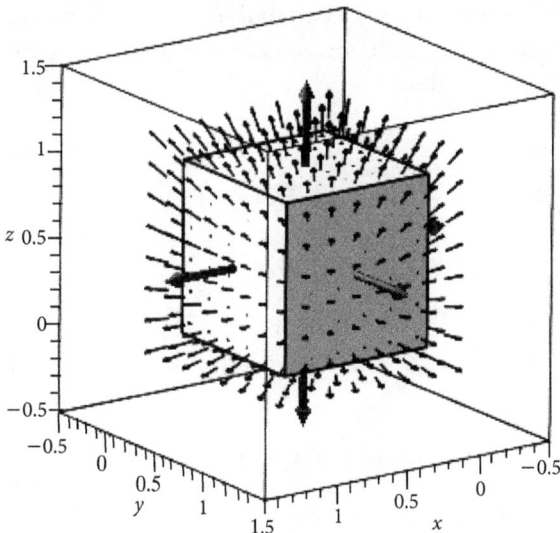

The vector field arrows, the surface through which the field passes, and vectors normal to the surface.
Figure 4.3.1: Flux through a cube showing surface normal vectors.

Flux(w, Sphere(⟨1, 1, 1⟩, 1))

$$4\pi$$

Flux(w, Sphere(⟨1, 1, 1⟩, 1), output = plot, scaling = constrained)

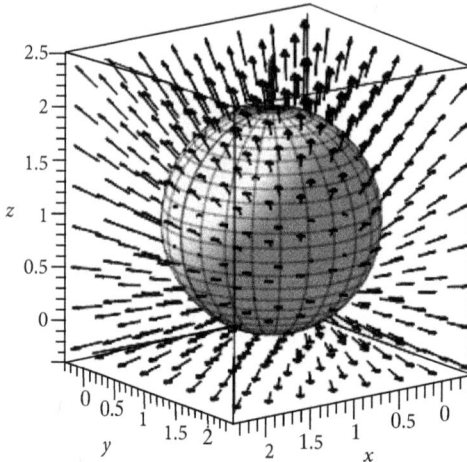

The vector field arrows, the surface through which the field passes, and vectors normal to the surface.
Figure 4.3.2: Flux through a sphere showing a surface normal vector at the top of the sphere.

4.4 CURL OF VECTOR FIELDS

The line integral and its relation to the curl are discussed in this section. Examples of the curl are given in Cartesian and curvilinear coordinates including cylindrical and spherical. The vector potential is introduced.

4.4.1 Line Integral

Given a three-dimensional vector field

$$\mathbf{F}(x, y, z) = F_x\left(x, y, z\right)\hat{\mathbf{i}} + F_y\left(x, y, z\right)\hat{\mathbf{j}} + F_z\left(x, y, z\right)\hat{\mathbf{k}} \tag{4.4.1}$$

we may form the line integral of **F** along a path Γ as

$$\int_\Gamma \mathbf{F} \cdot d\hat{\ell} \tag{4.4.2}$$

The line element $d\hat{\ell} = dx\hat{\mathbf{i}} + dy\hat{\mathbf{j}} + dz\hat{\mathbf{k}}$ is tangential to the path and

$$\mathbf{F} \cdot d\hat{\ell} = F_x\left(x, y, z\right)dx + F_y\left(x, y, z\right)dy + F_z\left(x, y, z\right)dz \tag{4.4.3}$$

so our line integral is

$$\int_\Gamma \mathbf{F} \cdot d\hat{\ell} = F_x dx + F_y dy + F_z dz \tag{4.4.4}$$

If $\mathbf{F}$ is a force vector, then our line integral would equal the work done by the force along the path Γ.

Example 4.4.1

Calculate the line integral of the vector field

$$\mathbf{F} = x^2\hat{\mathbf{i}} + y^2\hat{\mathbf{j}} + z^2\hat{\mathbf{k}} \tag{4.4.5}$$

along a path consisting of two segments from $(x, y, z) = (0, 0, 0) \rightarrow (1, 0, 0) \rightarrow (1, 0, 1)$.

Solution: Along the first segment $\Gamma_1 = (0, 0, 0) \rightarrow (1, 0, 0)$ there is no change in y or z so $dy = dz = 0$ and $\mathbf{F} \cdot d\hat{\ell} = x^2 dx$

$$\int_{\Gamma_1} \mathbf{F} \cdot d\hat{\ell} = \int_0^1 x^2 dx = \frac{1}{3} \tag{4.4.6}$$

Along the second segment $\Gamma_2 = (1, 0, 0) \rightarrow (1, 0, 1)$ where $dx = dy = 0$ and $\mathbf{F} \cdot d\hat{\ell} = z^2 dz$

$$\int_{\Gamma_2} \mathbf{F} \cdot d\hat{\ell} = \int_0^1 z^2 dz = \frac{1}{3} \tag{4.4.7}$$

thus

$$\int_{\Gamma_1} \mathbf{F} \cdot d\hat{\ell} + \int_{\Gamma_2} \mathbf{F} \cdot d\hat{\ell} = \frac{2}{3} \tag{4.4.8}$$

Example 4.4.2

Integrate the vector field

$$\mathbf{F} = y\hat{\mathbf{i}} - x\hat{\mathbf{j}} + z\hat{\mathbf{k}} \tag{4.4.9}$$

around the unit circle in the x-y plane.

Solution: Since $dz = 0$

$$\oint_\Gamma \mathbf{F} \cdot d\hat{\ell} = \oint_\Gamma F_x dx + F_y dy = \oint_\Gamma y dx - x dy \tag{4.4.10}$$

Performing a polar coordinate transformation along the unit circle with $r = 1$, $x = \cos\theta$, $y = \sin\theta$, $dx = -\sin\theta d\theta$ and $dy = \cos\theta d\theta$

$$\oint_{\Gamma} ydx - xdy = \int_{0}^{2\pi} \left(\sin^2 \theta + \cos^2 \theta \right) d\theta = 2\pi \qquad (4.4.11)$$

4.4.2 Curl of a Vector Field

The curl of a vector field at a point is related to the closed line integral around a path Γ enclosing an area Δa. The component of the curl normal to Δa is defined as the limit as $\Delta a \to 0$ of the closed line integral divided by Δa

$$\left(\text{curl } \mathbf{F} \right) \cdot \hat{\mathbf{n}} \equiv \lim_{\Delta a \to 0} \frac{1}{\Delta a} \oint_{\Gamma} \mathbf{F} \cdot d\hat{\ell} \qquad (4.4.12)$$

where $\hat{\mathbf{n}}$ is the unit vector normal to Δa. By choosing rectangular contours where $\hat{\mathbf{n}} = \hat{\mathbf{i}}, \hat{\mathbf{j}}, \hat{\mathbf{k}}$

$$\left(\text{curl } \mathbf{F} \right) \cdot \hat{\mathbf{i}} = \frac{\partial F_z}{\partial y} - \frac{\partial F_y}{\partial z} \qquad (4.4.13)$$

$$\left(\text{curl } \mathbf{F} \right) \cdot \hat{\mathbf{j}} = \frac{\partial F_x}{\partial z} - \frac{\partial F_z}{\partial x} \qquad (4.4.14)$$

$$\left(\text{curl } \mathbf{F} \right) \cdot \hat{\mathbf{k}} = \frac{\partial F_y}{\partial x} - \frac{\partial F_x}{\partial y} \qquad (4.4.15)$$

and we find that curl $\mathbf{F} = \nabla \times \mathbf{F}$.

4.4.3 Curl in Cartesian Coordinates

The curl of a vector field gives another vector field. In Cartesian coordinates the curl of $\mathbf{F}$ is given by

$$\nabla \times \mathbf{F} = \left(\frac{\partial}{\partial x} \hat{\mathbf{i}} + \frac{\partial}{\partial y} \hat{\mathbf{j}} + \frac{\partial}{\partial z} \hat{\mathbf{k}} \right) \times \left(F_x \hat{\mathbf{i}} + F_y \hat{\mathbf{j}} + F_z \hat{\mathbf{k}} \right) \qquad (4.4.16)$$

The curl may be expressed as a determinant

$$\nabla \times \mathbf{F} = \begin{vmatrix} \hat{\mathbf{i}} & \hat{\mathbf{j}} & \hat{\mathbf{k}} \\ \dfrac{\partial}{\partial x} & \dfrac{\partial}{\partial y} & \dfrac{\partial}{\partial z} \\ F_x & F_y & F_z \end{vmatrix} \qquad (4.4.17)$$

Expanding across the top row gives

$$\nabla \times \mathbf{F} = \left(\frac{\partial F_z}{\partial y} - \frac{\partial F_y}{\partial z} \right) \hat{\mathbf{i}} - \left(\frac{\partial F_z}{\partial x} - \frac{\partial F_x}{\partial z} \right) \hat{\mathbf{j}} + \left(\frac{\partial F_y}{\partial x} - \frac{\partial F_x}{\partial y} \right) \hat{\mathbf{k}} \qquad (4.4.18)$$

Example 4.4.3

Calculate the curl of $\mathbf{F} = y\hat{\mathbf{i}} - x\hat{\mathbf{j}} + z\hat{\mathbf{k}}$
Solution:

$$\nabla \times \mathbf{F} = \begin{vmatrix} \hat{\mathbf{i}} & \hat{\mathbf{j}} & \hat{\mathbf{k}} \\ \dfrac{\partial}{\partial x} & \dfrac{\partial}{\partial y} & \dfrac{\partial}{\partial z} \\ y & -x & z \end{vmatrix} = \left(\frac{\partial z}{\partial y} + \frac{\partial x}{\partial z} \right) \hat{\mathbf{i}} - \left(\frac{\partial z}{\partial x} - \frac{\partial y}{\partial z} \right) \hat{\mathbf{j}} + \left(-\frac{\partial x}{\partial x} - \frac{\partial y}{\partial y} \right) \hat{\mathbf{k}} = -2\hat{\mathbf{k}}$$

$$(4.4.19)$$

Notice that the curl of a vector field involves derivatives of each component with respect to the other two variables so that

$$\mathbf{F}(x, y, z) = F_x \left(x \right) \hat{\mathbf{i}} + F_y \left(y \right) \hat{\mathbf{j}} + F_z \left(z \right) \hat{\mathbf{k}} \qquad (4.4.20)$$

would have zero curl.

4.4.4 Curl in Curvilinear Coordinates

In curvilinear coordinates, the curl may be expressed in determinant form

$$\nabla \times \mathbf{F} = \begin{vmatrix} \hat{\mathbf{e}}_1 \left(g_{11}/g \right)^{1/2} & \hat{\mathbf{e}}_2 \left(g_{22}/g \right)^{1/2} & \hat{\mathbf{e}}_3 \left(g_{33}/g \right)^{1/2} \\ \dfrac{\partial}{\partial u_1} & \dfrac{\partial}{\partial u_2} & \dfrac{\partial}{\partial u_3} \\ g_{11}^{1/2} F_1 & g_{22}^{1/2} F_2 & g_{33}^{1/2} F_3 \end{vmatrix} \qquad (4.4.21)$$

4.4.5 Cylindrical Coordinates

In cylindrical coordinates, we have $(u_1, u_2, u_3) = (r, \phi, z)$ and $(x, y, z) = (r\cos\phi, r\sin\phi, z)$

$$g_{11} = 1, \; g_{22} = r^2, \; g_{33} = 1, \text{ so that } g = r^2 \qquad (4.4.22)$$

$$\nabla \times \mathbf{F} = \begin{vmatrix} \dfrac{1}{r}\hat{\mathbf{r}} & \hat{\boldsymbol{\phi}} & \dfrac{1}{r}\hat{\mathbf{z}} \\ \dfrac{\partial}{\partial r} & \dfrac{\partial}{\partial \phi} & \dfrac{\partial}{\partial z} \\ F_r & rF_\phi & F_z \end{vmatrix} \qquad (4.4.23)$$

$$\nabla \times \mathbf{F} = \frac{1}{r}\left(\frac{\partial}{\partial \phi}\left(F_z\right) - \frac{\partial}{\partial z}\left(rF_\phi\right)\right)\hat{\mathbf{r}}$$

$$-\left(\frac{\partial F_z}{\partial r} - \frac{\partial F_r}{\partial z}\right)\hat{\phi}$$

$$+\frac{1}{r}\left(\frac{\partial}{\partial r}\left(rF_\phi\right) - \frac{\partial}{\partial \phi}\left(F_r\right)\right)\hat{\mathbf{z}}$$

(4.4.24)

Example 4.4.4

Find the curl of

$$\mathbf{F} = z\hat{\mathbf{r}} - r\hat{\phi}$$

(4.4.25)

Solution:

In cylindrical coordinates

$$\nabla \times \mathbf{F} = \frac{1}{r}\frac{\partial}{\partial z}\left(r^2\right)\hat{\mathbf{r}} + \left(\frac{\partial z}{\partial z}\right)\hat{\phi} + \frac{1}{r}\left(\frac{\partial}{\partial r}\left(-r^2\right) - \frac{\partial}{\partial \phi}\left(z\right)\right)\hat{\mathbf{z}}$$

(4.4.26)

$$= -2\hat{\mathbf{z}} + \hat{\phi}$$

4.4.6 Spherical Coordinates

In spherical coordinates, we have $(u_1, u_2, u_3) = (r, \theta, \phi)$ and

$$(x, y, z) = (r\sin\theta\cos\phi, r\sin\theta\sin\phi, r\cos\theta)$$

$$g_{11} = 1, g_{22} = r^2, g_{33} = r^2\sin^2\theta, \text{ so that } g = r^4\sin^2\theta$$

(4.4.27)

$$\nabla \times \mathbf{F} = \begin{vmatrix} \hat{\mathbf{r}}\dfrac{1}{r^2\sin\theta} & \hat{\boldsymbol{\theta}}\dfrac{1}{r\sin\theta} & \hat{\phi}\dfrac{1}{r} \\[2mm] \dfrac{\partial}{\partial r} & \dfrac{\partial}{\partial \theta} & \dfrac{\partial}{\partial \phi} \\[2mm] F_r & rF_\theta & r\sin\theta F_\phi \end{vmatrix}$$

(4.4.28)

$$\nabla \times \mathbf{F} = \hat{\mathbf{r}}\frac{1}{r\sin\theta}\left(\frac{\partial}{\partial \theta}\left(\sin\theta F_\phi\right) - \frac{\partial}{\partial \phi}\left(F_\theta\right)\right)$$

$$-\hat{\boldsymbol{\theta}}\frac{1}{r\sin\theta}\left(\frac{\partial}{\partial r}\left(r\sin\theta F_\phi\right) - \frac{\partial}{\partial \phi}\left(F_r\right)\right) + \hat{\phi}\frac{1}{r}\left(\frac{\partial}{\partial r}\left(rF_\theta\right) - \frac{\partial}{\partial \theta}\left(F_r\right)\right)$$

(4.4.29)

Example 4.4.5

Calculate the curl of

$$\mathbf{F} = \phi\hat{\mathbf{r}} - r\hat{\boldsymbol{\theta}} \tag{4.4.30}$$

Solution: In spherical coordinates it is given by

$$\nabla \times \mathbf{F} = \hat{\mathbf{r}} \frac{1}{r\sin\theta}\left[\frac{\partial}{\partial\phi}(r)\right] - \hat{\boldsymbol{\theta}}\frac{1}{r\sin\theta}\left(-\frac{\partial}{\partial\phi}(\phi)\right)$$
$$+ \hat{\boldsymbol{\phi}}\frac{1}{r}\left(\frac{\partial}{\partial r}(-r^2) - \frac{\partial}{\partial\theta}(\phi)\right) = \hat{\boldsymbol{\theta}}\frac{1}{r\sin\theta} - 2\hat{\phi} \tag{4.4.31}$$

4.4.7 Vector Potential

Given a vector **B** field one may seek to find if it could have resulted from the curl of a vector field **A**.

Example 4.4.6

The vector field

$$\mathbf{B} = -2\hat{\mathbf{k}} \tag{4.4.32}$$

could have resulted from the curl of the vector field

$$\mathbf{A} = y\hat{\mathbf{i}} - x\hat{\mathbf{j}} \tag{4.4.33}$$

In magnetostatics the magnetic field **B** is given by the curl of a vector potential $\mathbf{B} = \nabla \times \mathbf{A}$.

Maple Examples

Examples of path integrals of scalar fields and line integrals of vector fields are shown in the Maple worksheet below. The curl of vector fields is evaluated in Cartesian, cylindrical, and spherical coordinates. Vector potential calculations are demonstrated in Cartesian and cylindrical coordinates.

Key Maple commands: *Circle, CrossProduct, Curl, Ellipse, LineInt, LineSegments, Path, PathInt, SetCoordinates, simplify, VectorField, VectorPotential*

Maple packages: *with(VectorCalculus)*:

restart

Path Integrals of Scalar Fields

with(VectorCalculus) :
g: = *x·y*

$$xy$$

$$PathInt(g, [x, y] = LineSegments(\langle 0, 0 \rangle, \langle 1, 0 \rangle, \langle 1, 1 \rangle))$$

$$\frac{1}{2}$$

$$PathInt(g, [x, y] = Circle(\langle 0, 0 \rangle, 1))$$

$$0$$

$$PathInt(1, [x, y] = Ellipse(3 \cdot x^2 + 4 \cdot y^2 - 1))$$

$$\frac{4}{3} \sqrt{3} \text{EllipticE}\left(\frac{1}{2}\right)$$

Line Integrals of Vector Fields

$$with(VectorCalculus):$$
$$SetCoordinates('cartesian'[x, y])$$

$$cartesian_{x, y}$$

$$F: = VectorField(\langle -y, x \rangle)$$

$$-y\overline{e}_x + (x)\overline{e}_y$$

$$LineInt(F, LineSegments(\langle 1, 0 \rangle, \langle 1, 1 \rangle \langle 0, 1 \rangle, \langle 0, 0 \rangle, \langle 1, 0 \rangle))$$

$$2$$

$$LineInt(F, Circle(\langle 0, 0 \rangle, r))$$

$$2\pi r^2$$

$$LineInt(F, Path(\langle \cos(t), \sin(t) \rangle, t = 0 \dots 2 \, pi))$$

$$2\pi$$

$$F: = VectorField(\langle x, y \rangle)$$

$$(x)\overline{e}_x + (y)\overline{e}_y$$

$$LineInt(F, Line(\langle 0, 0 \rangle, \langle 1, 1 \rangle))$$

$$1$$

$$LineInt(F, Circle(\langle 0, 0 \rangle, r))$$

$$0$$

$$LineInt(F, LineSegments(\langle 1, 0 \rangle, \langle 1, 1 \rangle \langle 0, 1 \rangle, \langle 0, 0 \rangle, \langle 1, 0 \rangle))$$

$$0$$

Curl in Cartesian Coordinates

$$SetCoordinates('cartesian'[x, y, z])$$

$$cartesian_{x, y, z}$$

$F := VectorField(\langle -y, x, 0 \rangle)$

$$-y\overline{e}_x + (x)\overline{e}_y$$

$Curl(F)$

$$2\overline{e}_z$$

$CrossProduct(Del, F)$

$$2\overline{e}_z$$

$F := VectorField(\langle x, y, 0 \rangle)$

$$(x)\overline{e}_x + (y)\overline{e}_y$$

$Curl(F)$

$$0\overline{e}_z$$

Curl in Cylindrical Coordinates

$with(VectorCalculus)$:
$SetCoordinates(`cylindrical`[r, phi, z])$;

$$cylindrical_{r,\phi,z}$$

$F := VectorField(\langle r, z, phi \rangle)$

$$F := (r)\overline{e}_r + (z)\overline{e}_\phi + (\phi)\overline{e}_z$$

$Curl(F)$

$$\left(\frac{1-r}{r}\right)\overline{e}_r + (0)\overline{e}_\phi + \left(\frac{z}{r}\right)\overline{e}_z$$

$CrossProduct(Del, F)$

$$\left(\frac{1-r}{r}\right)\overline{e}_r + (0)\overline{e}_\phi + \left(\frac{z}{r}\right)\overline{e}_z$$

Curl in Spherical Coordinates

$with(VectorCalculus)$:
$SetCoordinates(`spherical`[r, theta, phi])$

$$spherical_{r,\theta,\phi}$$

$v := VectorField\left(\left\langle \frac{1}{r^2}, 0, 0 \right\rangle\right)$

$$v := \left(\frac{1}{r^2}\right)\overline{e}_r + (0)\overline{e}_\theta + (0)\overline{e}_\phi$$

$Curl(v)$

$$(0)\overline{e}_r + (0)\overline{e}_\theta + (0)\overline{e}_\phi$$

$$w := VectorField\left(\left\langle \sin(phi), 0, \frac{1}{r^2}\right\rangle\right)$$

$$w := \left(\sin(\phi)\right)\overline{e}_r + (0)\overline{e}_\theta + \left(\frac{1}{r^2}\right)\overline{e}_\phi$$

$Curl(w)$

$$\left(\frac{\cos(\phi)}{r^3\sin(\phi)}\right)\overline{e}_r + \left(\frac{\cos(\phi)+\dfrac{\sin(\theta)}{r^2}}{r\sin(\theta)}\right)\overline{e}_\theta + (0)\overline{e}_\phi$$

$restart$

Vector Potential

$with(VectorCalculus):$
$SetCoordinates(`cartesian`[x, y, z])$

$$cartesian_{x,y,z}$$

$v := VectorField(\langle z, -x, y\rangle)$

$$(z)\overline{e}_x - x\overline{e}_y + (y)\overline{e}_z$$

$VectorPotential(v)$

$$\left(-xz - \frac{1}{2}y^2\right)\overline{e}_x - \frac{1}{2}z^2\overline{e}_y$$

$Curl(\%)$

$$(z)\overline{e}_x - x\overline{e}_y + (y)\overline{e}_z$$

Vector Potential of a Uniform Magnetic Field

$SetCoordinates(`cylindrical`[r, phi, z]);$

$$cylindrical_{r,\phi,z}$$

$v := VectorField(\langle 0, 0, B\rangle)$

$$(B)\overline{e}_z$$

$VectorPotential(v)$

$$-Br\sin(\phi)\cos(\phi)\overline{e}_r + \left(Br\sin(\phi)^2\right)\overline{e}_\phi$$

$SimplifyCurl(\%)$

$$(B)\overline{e}_z$$

The vector potential is not unique. For example,

$$v := VectorField\left(\left\langle 0, \frac{1}{2}r \cdot B, 0\right\rangle\right)$$

$$Curl(v) \qquad \begin{array}{c} \dfrac{1}{2}rB\overline{e}_\phi \\[2mm] (B)\overline{e}_z \end{array}$$

4.5 LAPLACIAN OF SCALAR AND VECTOR FIELDS

The divergence of a gradient field $\nabla \cdot \nabla T(x, y, z)$ results in a scalar field

$$\nabla^2 T = \left(\frac{\partial}{\partial x}\hat{\mathbf{i}} + \frac{\partial}{\partial y}\hat{\mathbf{j}} + \frac{\partial}{\partial z}\hat{\mathbf{k}} \right) \cdot \left(\frac{\partial T}{\partial x}\hat{\mathbf{i}} + \frac{\partial T}{\partial y}\hat{\mathbf{j}} + \frac{\partial T}{\partial z}\hat{\mathbf{k}} \right) = \frac{\partial^2 T}{\partial x^2} + \frac{\partial^2 T}{\partial y^2} + \frac{\partial^2 T}{\partial z^2} \qquad (4.5.1)$$

The Laplacian operator in Cartesian coordinates is defined as

$$\nabla^2 = \frac{\partial^2}{\partial x^2} + \frac{\partial^2}{\partial y^2} + \frac{\partial^2}{\partial z^2} \qquad (4.5.2)$$

Example 4.5.1

Calculate the Laplacian of the scalar field

$$T = \sin(kx)\cos(ky)e^{-\alpha z} \qquad (4.5.3)$$

Solution: In Cartesian coordinates

$$\nabla^2 T = \left(\alpha^2 - 2k^2 \right)\sin(kx)\cos(ky)e^{-\alpha z} \qquad (4.5.4)$$

4.5.1 Laplacian in Curvilinear Coordinates

In curvilinear coordinates (u_1, u_2, u_3) the scalar Laplacian may be written

$$\nabla^2 T = \frac{1}{g^{1/2}} \left\{ \frac{\partial}{\partial u_1}\left(\frac{g^{1/2}}{g_{11}}\frac{\partial T}{\partial u_1} \right) + \frac{\partial}{\partial u_2}\left(\frac{g^{1/2}}{g_{22}}\frac{\partial T}{\partial u_2} \right) + \frac{\partial}{\partial u_3}\left(\frac{g^{1/2}}{g_{33}}\frac{\partial T}{\partial u_3} \right) \right\} \qquad (4.5.5)$$

4.5.2 Cylindrical Coordinates

In cylindrical coordinates, we have $(u_1, u_2, u_3) = (r, \phi, z)$ and $(x, y, z) = (r\cos\phi, r\sin\phi, z)$

$$g_{11} = 1, g_{22} = r^2, g_{33} = r^2 \sin^2\theta, \text{ so that } g = r^2 \qquad (4.5.6)$$

$$\nabla^2 T = \frac{1}{r}\left\{ \frac{\partial}{\partial r}\left(r\frac{\partial T}{\partial r} \right) + \frac{\partial}{\partial z}\left(r\frac{\partial T}{\partial z} \right) + \frac{\partial}{\partial \phi}\left(\frac{1}{r}\frac{\partial T}{\partial \phi} \right) \right\}$$

$$= \frac{1}{r}\frac{\partial}{\partial r}\left(r\frac{\partial T}{\partial r} \right) + \frac{\partial^2 T}{\partial z^2} + \frac{1}{r^2}\frac{\partial^2 T}{\partial \phi^2} \qquad (4.5.7)$$

Example 4.5.2

Find the Laplacian of

$$T = r^2 \cosh(z)\cos\phi \tag{4.5.8}$$

Solution: In cylindrical coordinates

$$\nabla^2 T = \frac{1}{r}\frac{\partial}{\partial r}\left(r\frac{\partial r^2}{\partial r}\right)\cosh(z)\cos\phi + \frac{\partial^2 \cosh(z)}{\partial z^2}r^2\cos\phi + \frac{1}{r^2}\frac{\partial^2 \cos\phi}{\partial\phi^2}r^2\cosh(z)$$

$$= 4\cosh(z)\cos\phi + r^2\cosh(z)\cos\phi - \cos\phi\cosh(z)$$

$$= (3+r^2)\cosh(z)\cos\phi$$

$$\tag{4.5.9}$$

4.5.3 Spherical Coordinates

In spherical coordinates, we have $(u_1, u_2, u_3) = (r, \theta, \phi)$ and $(x, y, z) = (r\sin\theta\cos\phi, r\sin\theta\sin\phi, r\cos\theta)$

$$g_{11} = 1,\ g_{22} = r^2,\ g_{33} = r^2\sin^2\theta,\ \text{so that } g = r^4\sin^2\theta \tag{4.5.10}$$

$$\nabla^2 T = \frac{1}{r^2\sin\theta}\left\{\frac{\partial}{\partial r}\left(r^2\sin\theta\frac{\partial T}{\partial r}\right) + \frac{\partial}{\partial\theta}\left(\sin\theta\frac{\partial T}{\partial\theta}\right) + \frac{\partial}{\partial\phi}\left(\frac{1}{\sin\theta}\frac{\partial T}{\partial\phi}\right)\right\} \tag{4.5.11}$$

Example 4.5.3

Find the Laplacian of

$$T = \frac{1}{r^2}\sin\theta \tag{4.5.12}$$

Solution: In spherical coordinates

$$\nabla^2 T = \frac{1}{r^2\sin\theta}\left\{\frac{\partial}{\partial r}\left(r^2\sin\theta\frac{\partial}{\partial r}\left(\frac{1}{r^2}\sin\theta\right)\right) + \frac{\partial}{\partial\theta}\left(\sin\theta\frac{\partial}{\partial\theta}\left(\frac{1}{r^2}\sin\theta\right)\right)\right\}$$

$$= \frac{2\sin\theta}{r^4} + \frac{1}{r^2}\left(\cos^2\theta - \sin^2\theta\right)$$

$$\tag{4.5.13}$$

4.5.4 The Vector Laplacian

In Cartesian coordinates the Laplacian of a vector field may be expressed as

$$\nabla^2 \mathbf{F} = \frac{\partial^2}{\partial x^2}F_x(x,y,z)\hat{\mathbf{i}} + \frac{\partial^2}{\partial y^2}F_y(x,y,z)\hat{\mathbf{j}} + \frac{\partial^2}{\partial z^2}F_z(x,y,z)\hat{\mathbf{k}} \tag{4.5.14}$$

In curvilinear coordinates, the form of the vector Laplacian is complicated and will be discussed in Section 4.6.

Maple Examples

The Laplacian of scalar fields is calculated in Cartesian, cylindrical, and spherical coordinates in the Maple worksheet below.

Key Maple commands: *Divergence, expand, Gradient, Laplacian, SetCoordinates*

Maple packages: *with(VectorCalculus)*:

restart

Laplacian in Cartesian Coordinates

with(VectorCalculus) :
SetCoordinates('cartesian'[x, y, z])

$$cartesian_{x, y, z}$$

Laplacian(f(x, y, z))

$$\frac{\partial^2}{\partial x^2} f\left(x,y,z\right) + \frac{\partial^2}{\partial y^2} f\left(x,y,z\right) + \frac{\partial^2}{\partial z^2} f\left(x,y,z\right)$$

Divergence(Gradient(f(x, y, z)))

$$\frac{\partial^2}{\partial x^2} f\left(x,y,z\right) + \frac{\partial^2}{\partial y^2} f\left(x,y,z\right) + \frac{\partial^2}{\partial z^2} f\left(x,y,z\right)$$

$g := x^2 + y^2 + z^2$

$$x^2 + y^2 + z^2$$

Laplacian(g)

$$6$$

restart

Laplacian in Cylindrical Coordinates

with(VectorCalculus) :
SetCoordinates('cylindrical'[r, phi, z]);

$$cylindrical_{r, \phi, z}$$

expand(Laplacian(f(r, z)))

$$\frac{\frac{\partial}{\partial r} f\left(r,z\right)}{r} + \frac{\partial^2}{\partial r^2} f\left(r,z\right) + \frac{\partial^2}{\partial z^2} f\left(r,z\right)$$

$$g := \frac{1}{r} \cdot \cos(\text{phi}) \cdot \sinh(z)$$

$$\frac{\cos(\phi)\sinh(z)}{r}$$

Laplacian(g)

$$\frac{\cos(\phi)\sinh(z)}{r}$$

Divergence(Gradient(g))

$$\frac{\cos(\phi)\sinh(z)}{r}$$

restart

Laplacian in Spherical Coordinates

with(VectorCalculus) :
SetCoordinates('spherical'[r, theta, phi])

$$spherical_{r,\,\theta,\,\phi}$$

expand(Laplacian(f(r, phi)))

$$\frac{2\left(\frac{\partial}{\partial r} f(r,\phi)\right)}{r} + \frac{\partial^2}{\partial r^2} f(r,\phi) + \frac{\frac{\partial^2}{\partial \phi^2} f(r,\phi)}{r^2 \sin(\theta)^2}$$

$$g := r \cdot \cos(\text{theta})\sin(\text{phi})$$

$$r \cdot \cos(\theta)\sin(\phi)$$

Laplacian(g)

$$-\frac{\cos(\theta)\sinh(\phi)}{r\sin(\theta)^2}$$

Divergence(Gradient(g))

$$-\frac{\cos(\theta)\sinh(\phi)}{r\sin(\theta)^2}$$

4.6 VECTOR IDENTITIES

In this section, first and second derivatives involving the divergence, curl, and gradient of vector and scalar functions are covered. The vector Laplacian for curvilinear coordinates is introduced.

4.6.1 First Derivatives

The gradient, divergence and curl of sums is straightforward

$$\nabla(f+g) = \nabla f + \nabla g \tag{4.6.1}$$

$$\nabla \cdot (\mathbf{F}+\mathbf{G}) = \nabla \cdot \mathbf{F} + \nabla \cdot \mathbf{G} \tag{4.6.2}$$

$$\nabla \times (\mathbf{F}+\mathbf{G}) = \nabla \times \mathbf{F} + \nabla \times \mathbf{G} \tag{4.6.3}$$

4.6.2 First Derivatives of Products

The gradient of a product of scalar functions is

$$\nabla(fg) = g\nabla f + f\nabla g \tag{4.6.4}$$

The divergence of a scalar function times a vector function is

$$\nabla \cdot (f\mathbf{F}) = \mathbf{F} \cdot \nabla f + f\nabla \cdot \mathbf{F} \tag{4.6.5}$$

Example 4.6.5

Verify the above identity for the divergence of a scalar times a vector. Writing out the terms on the right-hand side

$$\mathbf{F} \cdot \nabla f = F_x \frac{\partial f}{\partial x} + F_y \frac{\partial f}{\partial y} + F_z \frac{\partial f}{\partial z} \tag{4.6.6}$$

$$f\nabla \cdot \mathbf{F} = f\frac{\partial F_x}{\partial x} + f\frac{\partial F_y}{\partial y} + f\frac{\partial F_z}{\partial z} \tag{4.6.7}$$

On the left-hand side

$$\nabla \cdot (f\mathbf{F}) = \frac{\partial}{\partial x}(fF_x) + \frac{\partial}{\partial y}(fF_y) + \frac{\partial}{\partial z}(fF_z) \tag{4.6.8}$$

Using the chain rule, we find

$$\nabla \cdot (f\mathbf{F}) = F_x \frac{\partial f}{\partial x} + f\frac{\partial F_x}{\partial x} + F_y \frac{\partial f}{\partial y} + f\frac{\partial F_y}{\partial y} + F_z \frac{\partial f}{\partial z} + f\frac{\partial F_z}{\partial z} \tag{4.6.9}$$

is equal to the sum of terms on the right. This identity is encountered in electrostatics.

The curl of a scalar function times a vector function is

$$\nabla \times (f\mathbf{F}) = f\nabla \times \mathbf{F} - \mathbf{F} \times \nabla f \tag{4.6.10}$$

The gradient of the dot product and the curl of the cross product are

$$\nabla(\mathbf{F} \cdot \mathbf{G}) = (\mathbf{F} \cdot \nabla)\mathbf{G} + (\mathbf{G} \cdot \nabla)\mathbf{F} + \mathbf{G} \times \nabla \times \mathbf{F} + \mathbf{F} \times \nabla \times \mathbf{G} \tag{4.6.11}$$

$$\nabla \times (\mathbf{F} \times \mathbf{G}) = (\mathbf{G} \cdot \nabla)\mathbf{F} - (\mathbf{F} \cdot \nabla)\mathbf{G} + \mathbf{F}(\nabla \cdot \mathbf{G}) - \mathbf{G}(\nabla \cdot \mathbf{F}) \tag{4.6.12}$$

4.6.3 Second Derivatives

The following second derivatives are frequently encountered and should be memorized

$$\nabla \times \nabla f = 0 \tag{4.6.13}$$

$$\nabla \cdot \nabla \times \mathbf{F} = 0 \tag{4.6.14}$$

$$\nabla \times \nabla \times \mathbf{F} = \nabla(\nabla \cdot \mathbf{F}) - \nabla^2 \mathbf{F} \tag{4.6.15}$$

4.6.4 Vector Laplacian

Identity (4.6.15) allows us to calculate the Laplacian of a vector field in curvilinear coordinates

$$\nabla^2 \mathbf{F} = \nabla(\nabla \cdot \mathbf{F}) - \nabla \times \nabla \times \mathbf{F} \tag{4.6.16}$$

$$
\begin{aligned}
\nabla^2 \mathbf{F} = \hat{\mathbf{e}}_1 &\left\{ \frac{1}{g_{11}^{1/2}} \frac{\partial \nabla \cdot \mathbf{F}}{\partial u_1} + \left(\frac{g_{11}}{g}\right)^{1/2} \left[\frac{\partial \Gamma_2}{\partial u_3} - \frac{\partial \Gamma_3}{\partial u_2}\right] \right\} \\
+ \hat{\mathbf{e}}_2 &\left\{ \frac{1}{g_{22}^{1/2}} \frac{\partial \nabla \cdot \mathbf{F}}{\partial u_2} + \left(\frac{g_{22}}{g}\right)^{1/2} \left[\frac{\partial \Gamma_3}{\partial u_1} - \frac{\partial \Gamma_1}{\partial u_3}\right] \right\} \\
+ \hat{\mathbf{e}}_3 &\left\{ \frac{1}{g_{33}^{1/2}} \frac{\partial \nabla \cdot \mathbf{F}}{\partial u_3} + \left(\frac{g_{33}}{g}\right)^{1/2} \left[\frac{\partial \Gamma_1}{\partial u_2} - \frac{\partial \Gamma_2}{\partial u_1}\right] \right\}
\end{aligned} \tag{4.6.17}
$$

where the Γ symbols are

$$
\Gamma_1 = \frac{g_{11}}{g^{1/2}} \left\{ \frac{\partial}{\partial u_2} \left(g_{33}^{1/2} F_3\right) - \frac{\partial}{\partial u_3} \left(g_{22}^{1/2} F_2\right) \right\}
$$

$$
\Gamma_2 = \frac{g_{22}}{g^{1/2}} \left\{ \frac{\partial}{\partial u_3} \left(g_{11}^{1/2} F_1\right) - \frac{\partial}{\partial u_1} \left(g_{33}^{1/2} F_3\right) \right\} \tag{4.6.18}
$$

$$
\Gamma_3 = \frac{g_{33}}{g^{1/2}} \left\{ \frac{\partial}{\partial u_1} \left(g_{22}^{1/2} F_2\right) - \frac{\partial}{\partial u_2} \left(g_{11}^{1/2} F_1\right) \right\}
$$

and the divergence

$$
\nabla \cdot \mathbf{F} = \frac{1}{g^{1/2}} \left\{ \frac{\partial}{\partial u_1} \left[\left(\frac{g}{g_{11}}\right)^{1/2} F_1\right] + \frac{\partial}{\partial u_2} \left[\left(\frac{g}{g_{22}}\right)^{1/2} F_2\right] + \frac{\partial}{\partial u_3} \left[\left(\frac{g}{g_{33}}\right)^{1/2} F_3\right] \right\} \tag{4.6.19}
$$

with
$$g = \begin{vmatrix} g_{11} & g_{12} & g_{13} \\ g_{21} & g_{22} & g_{23} \\ g_{31} & g_{32} & g_{33} \end{vmatrix} = g_{11}g_{22}g_{33} \qquad (4.6.20)$$

Maple Examples

Second derivatives of vector fields are evaluated in the Maple worksheet below. The vector Laplacian is calculated in Cartesian and spherical coordinates.

Key Maple commands: *Curl, Divergence, expand, Gradient, Laplacian, SetCoordinates, VectorField*

Maple packages: *with(VectorCalculus): with(Physics): with(Vectors):*

restart

Second Derivatives

with(VectorCalculus) :
SetCoordinates('cartesian'[x, y, z])

$$cartesian_{x, y, z}$$

F: = VectorField(⟨f(x, y, z), g(x, y, z), h(x, y, z)⟩)

$$F := \left(f\left(x,y,z\right) \right)\overline{e}_x + \left(g\left(x,y,z\right) \right)\overline{e}_y + \left(h\left(x,y,z\right) \right)\overline{e}_z$$

Divergence(Curl(F))

$$0$$

Curl(Gradient(f(x, y, z)))

$$(0)\overline{e}_x + (0)\overline{e}_y + (0)\overline{e}_z$$

Vector Laplacian in Cartesian Coordinates

F: = VectorField(⟨fx(x, y), fy(x, y), 0⟩)

$$F := \left(fx\left(x,y\right) \right)\overline{e}_x + \left(fy\left(x,y\right) \right)\overline{e}_y + (0)\overline{e}_z$$

Laplacian(F)

$$\left(\frac{\partial^2}{\partial x^2} fx\left(x,y\right) + \frac{\partial^2}{\partial y^2} fx\left(x,y\right) \right)\overline{e}_x + \left(\frac{\partial^2}{\partial x^2} fx\left(x,y\right) + \frac{\partial^2}{\partial y^2} fx\left(x,y\right) \right)\overline{e}_y + (0)\overline{e}_z$$

Vector Laplacian in Spherical Coordinates

SetCoordinates('spherical'[r, theta, phi])

$$spherical_{r,\,\theta,\,\phi}$$

$$w := VectorField\left(\left\langle \sin(phi), 0, \frac{1}{r^2}\right\rangle\right)$$

$$w := \left(\sin(\phi)\right)\overline{e}_r + (0)\overline{e}_\theta + \left(\frac{1}{r^2}\right)\overline{e}_\phi$$

Laplacian(w)

$$\left(-\frac{2\sin(\phi)}{r^2} - \frac{\sin(\phi)}{r^2\sin(\theta)^2}\right)\overline{e}_r + (0)\overline{e}_\theta + \left(\frac{2\cos(\phi)}{r^2\sin(\theta)} - \frac{-\dfrac{1}{r^3} + \dfrac{\cos(\theta)^2}{r^3\sin(\theta)^2}}{r}\right)\overline{e}_\phi$$

Gradient(Divergence(w)) − Curl(Curl(w))

$$\left(-\frac{2\sin(\phi)}{r^2} - \frac{\sin(\phi)^2}{r^2\sin(\theta)^2}\right)\overline{e}_r + (0)\overline{e}_\theta + \left(\frac{2\cos(\phi)}{r^2\sin(\theta)} - \frac{-\dfrac{1}{r^3} + \dfrac{\cos(\theta)^2}{r^3\sin(\theta)^2}}{r}\right)\overline{e}_\phi$$

restart
with(Physics) :
with(Vectors) :

[&x, '+', '.', *ChangeBasis, ChangeCoordinates, Component, Curl, DirectionalDiff, Divergence, Gradient, Identify, Laplacian,* ∇, *Norm, Setup, diff*]

%Curl(%Curl(A))

$$\nabla \times (\nabla \times A)$$

expand(%)

$$\nabla(\nabla \cdot A) - \nabla^2 A$$

4.7 INTEGRAL THEOREMS

In this section, the fundamental theorem of gradients is introduced. Integral theorems are discussed including Gauss's divergence theorem and Stokes's theorem. Bernoulli's equation is derived from the Navier-Stokes equation.

4.7.1 Gradient Theorem

The fundamental theorem of gradients states that the line integral of the gradient of a scalar function ∇f between points a and b depends only on the value of f at the end points

$$\int_a^b \nabla f \cdot d\hat{\ell} = f(a) - f(b) \tag{4.7.1}$$

To show this, we consider the dot product between the gradient

$$\nabla f = \frac{\partial f}{\partial x}\hat{\mathbf{i}} + \frac{\partial f}{\partial y}\hat{\mathbf{j}} + \frac{\partial f}{\partial z}\hat{\mathbf{k}} \tag{4.7.2}$$

and the line element

$$d\hat{\ell} = dx\hat{\mathbf{i}} + dy\hat{\mathbf{j}} + dz\hat{\mathbf{k}} \tag{4.7.3}$$

giving

$$\nabla f \cdot d\hat{\ell} = \frac{\partial f}{\partial x}dx + \frac{\partial f}{\partial y}dy + \frac{\partial f}{\partial z}dz = df \tag{4.7.4}$$

so that

$$\int_a^b \nabla f \cdot d\hat{\ell} = \int_a^b df = f(a) - f(b) \tag{4.7.5}$$

The line integral above is also independent of the integration path between a and b. Also

$$\oint \nabla f \cdot d\hat{\ell} = 0 \tag{4.7.6}$$

If the closed line integral of a vector field $\mathbf{F}$ is not zero

$$\oint \mathbf{F} \cdot d\hat{\ell} \neq 0 \tag{4.7.7}$$

then $\mathbf{F}$ cannot be obtained from the gradient of a scalar field.

4.7.2 Divergence Theorem

Gauss's divergence theorem states that the flux of a vector field $\mathbf{F}$ through a closed surface is equal to the divergence of $\mathbf{F}$ integrated over the enclosed volume

$$\int_{surf} \mathbf{F} \cdot \hat{\mathbf{n}} \, da = \int_{vol} \nabla \cdot \mathbf{F} \, dv \tag{4.7.8}$$

4.7.3 Cartesian Coordinates

Example 4.7.1

Verify Gauss's divergence theorem over the unit cube in the first octant given the vector field $\mathbf{F} = x\hat{\mathbf{i}}$ in Cartesian coordinates.
Solution: Computing the divergence

$$\nabla \cdot \mathbf{F} = \frac{\partial}{\partial x}(F_x) = \frac{\partial}{\partial x}(x) = 1 \tag{4.7.9}$$

and the volume integral

$$\int_{\text{vol}} \nabla \cdot \mathbf{F} \, dv = \int_0^1 \int_0^1 \int_0^1 dx \, dy \, dz = 1 \tag{4.7.10}$$

Now $\mathbf{F} \cdot \hat{\mathbf{n}} = 0$ on all sides of the cube except the face at $x=1$ where $\mathbf{F} \cdot \hat{\mathbf{n}} = x\hat{\mathbf{i}} \cdot \hat{\mathbf{i}}\big|_{x=1} = 1$. The surface integral is thus computed

$$\int_{\text{surf}} \mathbf{F} \cdot \hat{\mathbf{n}} \, da = \int_0^1 \int_0^1 dy \, dz = 1 \tag{4.7.11}$$

in agreement with the divergence theorem.

4.7.4 Cylindrical Coordinates

Example 4.7.2

Verify Gauss's divergence theorem in cylindrical coordinates given the vector field

$$\mathbf{F} = r^2 \hat{\mathbf{r}} \tag{4.7.12}$$

with the volume bounded by the cylindrical surface $r = R$ and the planes $z = 0$ and $z = H$.
Solution: Computing the divergence

$$\nabla \cdot \mathbf{F} = \frac{1}{r}\frac{\partial}{\partial r}(rF_r) = \frac{1}{r}\frac{\partial}{\partial r}(r^3) = 3r \tag{4.7.13}$$

and the volume integral

$$\int_{\text{vol}} \nabla \cdot \mathbf{F} \, dv = \int_0^H \int_0^{2\pi} \int_0^R (3r) \, r \, dr \, d\theta \, dz = 2\pi H R^3 \tag{4.7.14}$$

Now $\mathbf{F} \cdot \hat{\mathbf{n}} = 0$ on the top and bottom end caps of the cylinder and $\mathbf{F} \cdot \hat{\mathbf{n}} = r^2 \hat{\mathbf{r}} \cdot \hat{\mathbf{r}}\big|_{r=R} = R^2$ on the cylindrical surface

$$\int\limits_{\text{surf}} \mathbf{F}\cdot\hat{\mathbf{n}}\,da = \int\limits_{\text{surf}} R^2\,da = R^2 \int\limits_0^H \int\limits_0^{2\pi} R\,d\theta dz = 2\pi HR^3 \qquad (4.7.15)$$

4.7.5 Stokes's Curl Theorem

Stokes's curl theorem relates the vector field F integrated around a closed path Γ to $\nabla \times \mathbf{F}$ integrated over any capping surface bound by the contour

$$\oint\limits_\Gamma \mathbf{F}\cdot d\hat{\ell} = \int\limits_{\text{surf}} \nabla\times\mathbf{F}\cdot\hat{\mathbf{n}}\,da \qquad (4.7.16)$$

where $d\hat{\ell}$ is tangential to the path and $\hat{\mathbf{n}}$ is a unit vector normal to the surface. The direction of $\hat{\mathbf{n}}$ is determined by a right-hand rule. If fingers of the right hand are pointed in the direction of $d\hat{\ell}$ then the thumb points in the direction of $\hat{\mathbf{n}}$. Line integrals may be performed as in Section 4.4 with the line element in Cartesian coordinates

$$d\hat{\ell} = dx\hat{\mathbf{i}} + dy\hat{\mathbf{j}} + dz\hat{\mathbf{k}} \qquad (4.7.17)$$

so that

$$\oint\limits_\Gamma \mathbf{F}\cdot d\hat{\ell} = \oint\limits_\Gamma F_x dx + F_y dy + F_z dz \qquad (4.7.18)$$

Example 4.7.3

Verify Stokes's theorem given the vector field

$$\mathbf{F} = x^2 y\hat{\mathbf{i}} \qquad (4.7.19)$$

over the surface bound by the unit circle in the x-y plane.
Solution: We calculate the curl

$$\nabla\times\mathbf{F} = -x^2\hat{\mathbf{k}} \qquad (4.7.20)$$

Taking the integration contour to be counterclockwise around the circle the right-hand rule gives $\hat{\mathbf{n}} = \hat{\mathbf{k}}$ and the surface integral is

$$\int\limits_{\text{surf}} \nabla\times\mathbf{F}\cdot\hat{\mathbf{n}}\,da = -\int\limits_{\text{surf}} x^2\,da \qquad (4.7.21)$$

A polar coordinate transformation gives $da = rdrd\theta$ and $x = r\cos\theta$ so that

$$-\int\limits_{\text{surf}} x^2\,da = -\int\limits_0^1 r^3 dr \int\limits_0^{2\pi} \cos^2\theta d\theta = -\frac{1}{4}\int\limits_0^{2\pi} \frac{1}{2}(1+\cos 2\theta)\,d\theta = -\frac{\pi}{4} \qquad (4.7.22)$$

Next, we compute $\mathbf{F} \cdot d\hat{\ell} = x^2 y dx$ so our line integral

$$\oint_\Gamma \mathbf{F} \cdot d\hat{\ell} = \oint_{r=1} x^2 y dx \tag{4.7.23}$$

On the boundary of the unit circle $x = \cos\theta$, $y = \sin\theta$ and $dx = -\sin\theta d\theta$ so,

$$\oint_{r=1} x^2 y dx = -\int_0^{2\pi} \cos^2\theta \sin^2\theta d\theta \tag{4.7.24}$$

Using $\cos^2\theta = (1 + \cos2\theta)/2$ and $\sin^2\theta = (1 - \cos2\theta)/2$ we have $\cos^2\theta \sin^2\theta = (1 - \cos^2 2\theta)/4$

$$-\int_0^{2\pi} \cos^2\theta \sin^2\theta d\theta = -\int_0^{2\pi} \frac{1}{4}\left(1 - \cos^2 2\theta\right) d\theta$$

$$= -\frac{\pi}{2} + \frac{1}{4}\int_0^{2\pi} \cos^2 2\theta d\theta = -\frac{\pi}{2} + \frac{\pi}{4} + \frac{1}{4}\frac{1}{4}\sin 4\theta\Big|_0^{2\pi} = -\frac{\pi}{4} \tag{4.7.25}$$

in agreement with Stokes's theorem.

Example 4.7.4

Verify Stokes's theorem given the vector field

$$\mathbf{F} = y^2 \hat{\mathbf{i}} \tag{4.7.26}$$

over the surface bound by one-quarter of the unit circle in the first quadrant of the x-y plane.

Solution: The integration contour is taken to be counterclockwise around the figure below.

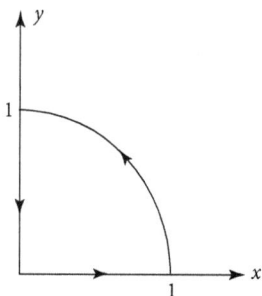

Figure 4.7.1: Counterclockwise contour in the first quadrant.

We calculate the curl

$$\nabla \times \mathbf{F} = -2y\hat{\mathbf{k}} \tag{4.7.27}$$

For a counterclockwise contour, the right-hand rule gives $\hat{\mathbf{n}} = \hat{\mathbf{k}}$ and the surface integral is computed as

$$\int_{\text{surf}} \nabla \times \mathbf{F} \cdot \hat{\mathbf{n}} \, da = -2 \int_{\text{surf}} y^2 \, da \qquad (4.7.28)$$

Using $da = r dr d\theta$ and $y = r\sin\theta$ and we have

$$-2 \int_{\text{surf}} y^2 \, da = -2 \int_0^1 r^2 dr \int_0^{\pi/2} \sin\theta d\theta = -\frac{2}{3} \qquad (4.7.29)$$

Next, we compute $\mathbf{F} \cdot d\hat{\ell} = y^2 dx$ so our line integral

$$\oint_\Gamma \mathbf{F} \cdot d\hat{\ell} = \int_{\text{bottom}} y^2 dx + \int_{\text{curve}} y^2 dx + \int_{\text{left}} y^2 dx \qquad (4.7.30)$$

On the bottom $y = 0$, on the left $dx = 0$, so that the first and third integrals are zero. On the curve we have $r = 1, x = \cos\theta, y = \sin\theta$ and $dx = -\sin\theta d\theta$ so we have

$$\int_{\text{curve}} y^2 dx = -\int_0^{\pi/2} \sin^3\theta d\theta = -\int_0^{\pi/2} \left(1 - \cos^2\theta\right)\sin\theta d\theta = \int_1^0 \left(1 - u^2\right) du = -\frac{2}{3} \quad (4.7.31)$$

in agreement with Stokes's theorem.

4.7.6 Navier-Stokes Equation

The Navier-Stokes equation describing the flow of a fluid with viscosity η and density ρ in a gravitation field is given by

$$\rho \frac{\partial \mathbf{v}}{\partial t} + \rho\left(\mathbf{v} \cdot \nabla\right)\mathbf{v} = -\nabla P + \eta \nabla^2 \mathbf{v} - \rho \mathbf{g} \qquad (4.7.32)$$

Bernoulli's equation is obtained neglecting viscous drag and acceleration of the fluid. In the approximation where $\eta \nabla^2 \mathbf{v}$ and $\rho \partial \mathbf{v} / \partial t$ are small

$$\rho\left(\mathbf{v} \cdot \nabla\right)\mathbf{v} = -\nabla P - \rho \mathbf{g} \qquad (4.7.33)$$

Integrating both sides of this equation along a vertical path

$$\rho \int \left(\mathbf{v} \cdot \nabla\right)\mathbf{v} \cdot d\hat{\ell} = -\int \nabla P \cdot d\hat{\ell} - \rho \int \mathbf{g} \cdot d\hat{\ell} \qquad (4.7.34)$$

with $\mathbf{g} = g\hat{\mathbf{j}}$ and $d\hat{\ell} = dy\hat{\mathbf{j}}$

$$\rho \int_{v_1}^{v_2} v \, dv = -\int_{P_1}^{P_2} dP - \rho g \int_{y_1}^{y_2} dy \qquad (4.7.35)$$

gives Bernoulli's equation

$$P_1 + \frac{1}{2}\rho v_1^2 + \rho g y_1 = P_2 + \frac{1}{2}\rho v_2^2 + \rho g y_2 \qquad (4.7.36)$$

4.8 MATLAB EXAMPLES

In this section, we demonstrate the visualization of scalar and vector fields in MATLAB. The numerical gradient of scalar fields as well as the divergence and curl of vector fields are calculated with graphical output. Symbolic operations of divergence, curl, gradient and Laplacian are carried out in Cartesian coordinates

Key MATLAB commands: *contour,contourf, curl, del2, divergence, ezcontour, ezcontourf, ezmesh, gradient, isosurface, laplacian, meshc, meshgrid, quiver, quiver3, subplot, syms*

Section 4.1 Vector and Scalars Fields

Contour and Surface Plots

The following example demonstrates styles of surface and contour plots of a scalar field $G(x, y)$. MATLAB code is entered at the Command Prompt.

```
>> [x,y]=meshgrid(-2:0.1:2,-2:0.1:2);
>> G=sin(y)*exp(-x.^2-y.^2);
>> subplot(2,2,1); mesh(x,y,G)
>> subplot(2,2,2); contour(x,y,G)
>> subplot(2,2,3); meshc(x,y,G)
>> subplot(2,2,4); contourf(x,y,G)
```

The Rotate 3D tool may be used to view surface plots from different orientations. The same plots above may be generated using the Symbolic Math Toolbox as shown below.

```
>> syms x y
>> colormap(bone)
>> subplot(2,2,1); ezmesh(x*exp(-x^2-y^2))
>> subplot(2,2,2); ezcontour(x*exp(-x^2-y^2))
>> subplot(2,2,3); ezmeshc(x*exp(-x^2-y^2))
>> subplot(2,2,4); ezcontourf(x*exp(-x^2-y^2))
```

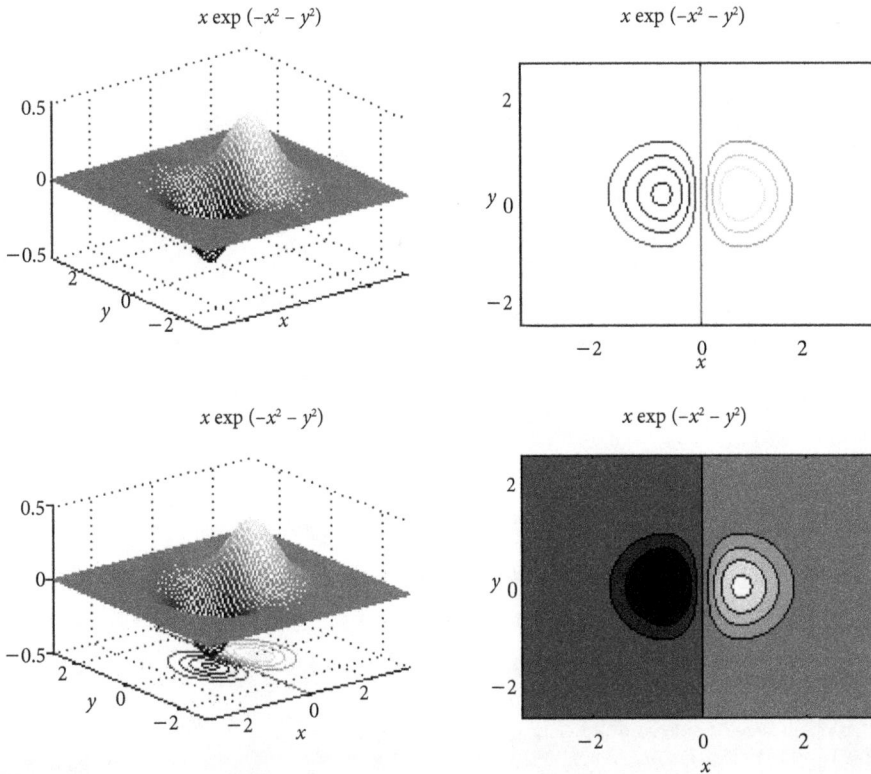

Figure 4.8.1: Subplots of surface and contour plots including surface (top left), contour (top right), surface with contours (bottom left) and filled contours (bottom right).

Section 4.2 Gradient of Scalar Fields

The gradient of a scalar field may be numerically computed and visualized in MATLAB using the following syntax entered at the Command Prompt:

```
>> [x,y]=meshgrid(-2:0.1:2,-2:0.1:2);
>> F=exp(-x.^2-y.^2);
>> quiver(x,y,Fx,Fy,'k')
>> title('Gradient Field')
>> xlabel('x')
>> ylabel('y')
```

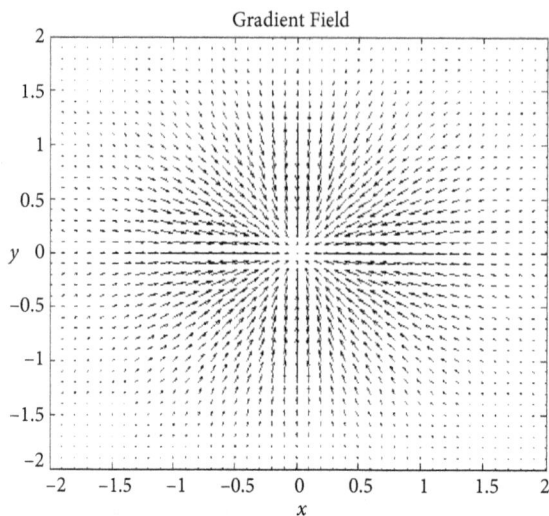

Figure 4.8.2: Gradient field plot.

In the example below, the gradient vector plot is superimposed on a contour plot of the scalar field

```
>> quiver(x,y,Fx,Fy,'k')
>> hold on
>> contour(x,y,F,'k')
>> title('Gradient and Scalar Fields')
>> xlabel('x')
>> ylabel('y')
```

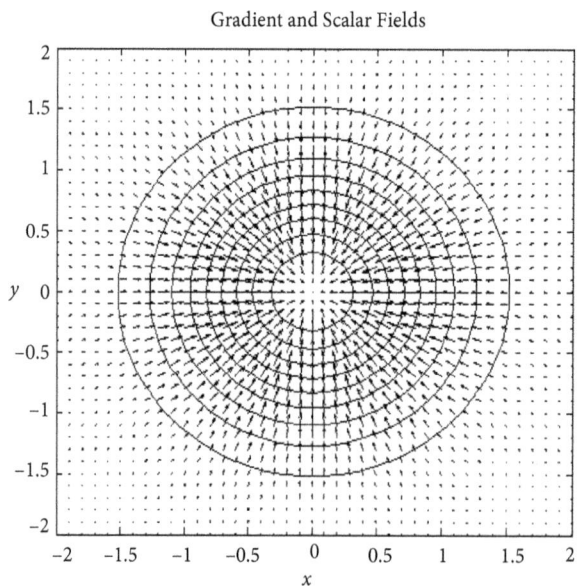

Figure 4.8.3: Gradient field plot with contours of the scalar field.

The gradient of a 3D scalar field $F(x, y, z)$ is plotted below. The isosurface $F(x, y, z) = 0.01$ is superimposed on the gradient field.

```
v=-1:0.25:1;
[x,y,z]=meshgrid(v);
F=(x.^2+y.^2-z.^2);
[Fx,Fy,Fz]=gradient(F);
quiver3(x,y,z,Fx,Fy,Fz,'k')
axis tight
hold on
isosurface(x,y,z,F,0.01)
hold off
alpha(0.7)
colormap(bone)
xlabel('x')
ylabel('y')
zlabel('z')
```

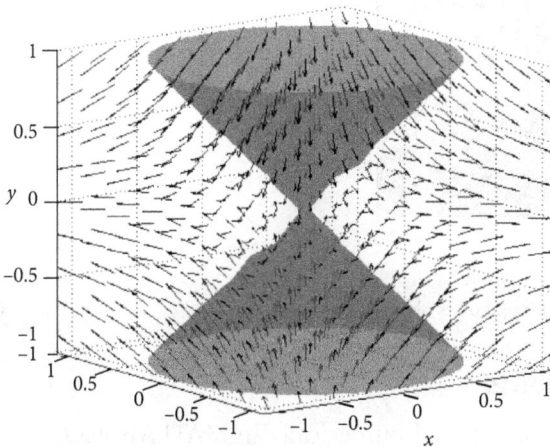

Figure 4.8.4: 3D gradient field plot with isosurface displayed.

Section 4.3 Divergence of Vector Fields

The divergence of the vector field $(\cos(3x), \sin(2y))$ may be numerically computed in MATLAB using the following syntax executed as a script file. The divergence is plotted as a filled contour plot with regions of positive divergence appearing lighter and regions of negative divergence having a darker shading. An arrow plot of the vector field is superimposed on the divergence plot.

```
[x,y]=meshgrid(-1.5:0.1:1.5,-1.5:0.1:1.5);
Fx=cos(3*x);
Fy=sin(2*y);
cav=divergence(x,y,Fx,Fy);
contourf(x,y,cav)
hold on
quiver(x,y,Fx,Fy,'k')
```

```
hold off
title('Vector Field with Divergence')
xlabel('x')

ylabel('y')

colormap bone
```

Vector Field with Divergence

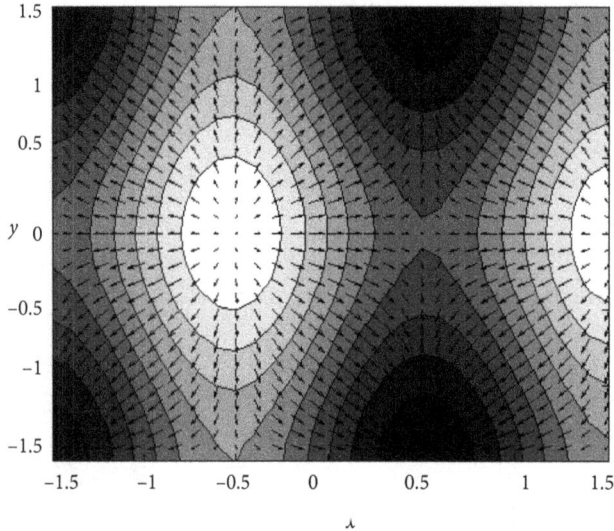

Figure 4.8.5: Vector field plotted with a filled contour plot of its divergence.

Section 4.4 Curl of Vector Fields

The curl of a vector field may be computed numerically in MATLAB using the following syntax:

```
[x,y]=meshgrid(-1.5:0.1:1.5,-1.5:0.1:1.5);
Fx=cos(3*x-y);
Fy=sin(x+3*y);
cav=curl(x,y,Fx,Fy);
contourf(x,y,cav)
hold on
quiver(x,y,Fx,Fy,'k')
hold off
title('Vector Field with Curl')
xlabel('x')
ylabel('y')
colormap bone
```

Vector Field with Curl

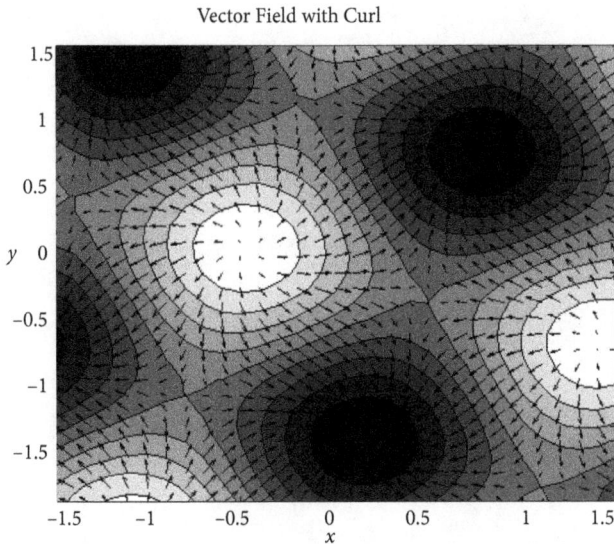

Figure 4.8.6: Vector field plotted with a filled contour plot of its curl.

Section 4.5 Laplacian of Scalar and Vector Fields

The scalar field $V(x, y)$ and its Laplacian are computed in the following MATLAB script file. The Laplacian and the field are displayed as subplots for comparison.

```
[x,y]=meshgrid(-1.5:0.1:1.5,-1.5:0.1:1.5);
V=x.*y.*exp(-x.^2-y.^2);
lap=4*del2(V);
subplot(2,1,1);
mesh(x,y,V)
axis equal
colormap bone
title('Scalar Field')
subplot(2,1,2);
mesh(x,y,lap)
axis equal
title('Laplacian of Scalar Field')
colormap bone
```

Scalar Field

Laplacian of Scalar Field

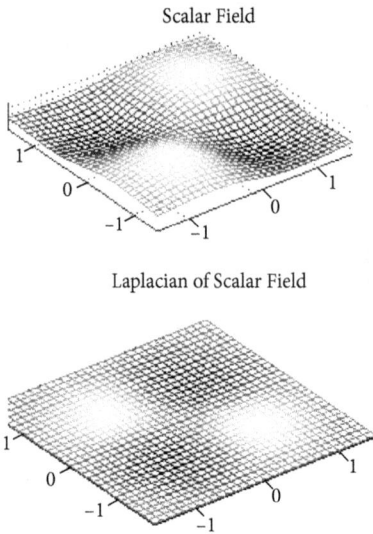

Figure 4.8.7: Subplots of a scalar field (top) and its Laplacian (bottom).

Section 4.6 Vector Identities

```
>> syms x y z
>> f=-x^2+y^3+z^4
f =
- x^2 + y^3 + z^4
>> gradient(f,[x,y,z])
ans =
  -2*x
 3*y^2
 4*z^3
>> laplacian(f,[x,y,z])
ans =
12*z^2 + 6*y - 2
>> divergence([2*x^2 , cos(y), exp(z)],[x , y, z])
ans =
4*x + exp(z) - sin(y)
>> curl([y^2 , x, x*y],[x , y, z])
ans =
       x
      -y
 1 - 2*y
>> g = x*y*cos(z)
g =
x*y*cos(z)
>> curl(gradient(g),[x,y,z])
ans =
 0
 0
 0
```

```
>> curl(curl([y*z,cos(x),exp(x*z)],[x,y,z]),[x,y,z])

ans =

 exp(x*z) + x*z*exp(x*z)
               cos(x)
         -z^2*exp(x*z)
>> divergence(curl([z^2,x*y,z*x],[x,y,z]),[x,y,z])

ans =

 0
```

4.9 EXERCISES

Section 4.1 Vector and Scalar Fields

1. Create a surface plot of the scalar function $g(x,y) = \sin(x^2 + y^2)e^{-(x^2+y^2)}$

2. Create a contour plot of the scalar function $g(x,y) = xye^{-(x^2+y^2)}$

3. Given $\mathbf{F} = y\hat{\mathbf{i}} + x\hat{\mathbf{j}}$, use the relation

   $\dfrac{dy}{dx} = \dfrac{F_y}{F_x}$ to show that the equation for the field lines is $y^2 - x^2 = \text{const.}$

4. Create an arrow plot of the vector field $\mathbf{F} = -y\hat{\mathbf{i}} + x\hat{\mathbf{j}}$. Use the relation

   $\dfrac{dy}{dx} = \dfrac{F_y}{F_x}$ to write an equation for the field lines. Plot the vector field and the field lines.

5. Create a 3D vector field plot of $\mathbf{F} = -y\hat{\mathbf{i}} + x\hat{\mathbf{j}} - z\hat{\mathbf{k}}$

Section 4.2 Gradient of Scalar Fields

6. Calculate the gradient of the following scalar functions in Cartesian coordinates

   $G(x,y,z) = xyz$
   $G(x,y,z) = e^x e^y e^z$

7. Show that

$$\nabla \ln\sinh\left(\frac{xyz}{abc}\right) = \frac{1}{abc}\coth\left(\frac{xyz}{abc}\right)\left(yz\hat{\mathbf{i}} + xz\hat{\mathbf{j}} + xy\hat{\mathbf{k}}\right)$$

8. Use the gradient operator to find the unit vector normal to the ellipsoid

$$\frac{x^2}{a^2}+\frac{y^2}{b^2}+\frac{z^2}{c^2}=1$$

9. Find the unit normal to the plane described by $2x+y-z=3$

10. Calculate the gradient of the following scalar functions in cylindrical coordinates

$$G(r,\phi,z)=r^2e^{-z}\cos\phi$$
$$G(r,z)=z\ln(r)$$

11. Calculate the gradient of the following scalar functions in spherical coordinates

$$G(r,\theta,\phi)=r\cos\theta e^{i\phi}$$

$$G(r,\theta)=\frac{\cos\theta}{r}$$

Section 4.3 Divergence of Vector Fields

12. Calculate the divergence of the following vector fields in Cartesian coordinates

$$\mathbf{F}(x,y,z)=e^x\hat{\mathbf{i}}+e^y\hat{\mathbf{j}}+e^z\hat{\mathbf{k}}$$

$$\mathbf{F}(x,y,z)=y\hat{\mathbf{i}}+z\hat{\mathbf{j}}+x\hat{\mathbf{k}}$$

13. Find a vector field F(x,y,z) with divergence in Cartesian coordinates given by

$$\nabla\cdot\mathbf{F}(x,y,z)=x+y+z$$

14. Show that the divergence of the following vector function in Cartesian coordinates is zero

$$\mathbf{F}(x,y,z)=x\hat{\mathbf{i}}-y\hat{\mathbf{j}}+xy\hat{\mathbf{k}}$$

15. Calculate the divergence of the following vector fields in cylindrical coordinates $(u_1, u_2, u_3)=(r,\phi,z)$

$$\mathbf{F}(r,\phi,z)=r^2\hat{\mathbf{r}}+z\hat{\mathbf{z}}+r\sin\phi\hat{\phi}$$

$$\mathbf{F}(r,\phi,z)=r^2\hat{\mathbf{r}}+z\hat{\mathbf{z}}+e^{i\phi}\hat{\phi}$$

16. Find a vector field F(r, ϕ, z) with divergence in cylindrical coordinates given by

$$\nabla\cdot\mathbf{F}(r,\phi,z)=r^2$$

17. Calculate the divergence of the following vector fields in spherical coordinates $(u_1, u_2, u_3) = (r, \phi, \theta)$

$$\mathbf{F}(r,\phi,\theta) = r^2\hat{\mathbf{r}} + \cos\theta\,\hat{\boldsymbol{\theta}} + r\cos\theta\,\hat{\boldsymbol{\phi}}$$

$$\mathbf{F}(r,\phi,\theta) = r^2\hat{\mathbf{r}} + r\sin\theta\,\hat{\boldsymbol{\theta}} + e^{i\phi}\,\hat{\boldsymbol{\phi}}$$

18. Find a vector field $\mathbf{F}(r,\theta,\phi)$ with divergence in spherical coordinates given by

$$\nabla\cdot\mathbf{F}(r,\theta,\phi) = r^2$$

Section 4.4 Curl of Vector Fields

19. Given $\mathbf{F} = x\hat{\mathbf{i}} + y\hat{\mathbf{j}} + z\hat{\mathbf{k}}$ calculate the line integral $\int_a^b \mathbf{F}\cdot d\hat{\ell}$

where $a = (0,0,0)$ and $b = (1,1,1)$ along the paths
(a) $(0,0,0) \to (1,0,0) \to (1,1,0) \to (1,1,1)$
(b) $(0,0,0) \to (0,0,1) \to (0,1,1) \to (1,1,1)$

20. Calculate the curl of the following vector functions in Cartesian coordinates

$$\mathbf{F}(x,y,z) = e^x\hat{\mathbf{i}} + e^y\hat{\mathbf{j}} + e^z\hat{\mathbf{k}}$$

$$\mathbf{F}(x,y,z) = y\hat{\mathbf{i}} + z\hat{\mathbf{j}} + x\hat{\mathbf{k}}$$

21. Show that the curl of the following vector function in Cartesian coordinates is zero

$$\mathbf{F}(x,y,z) = e^x\hat{\mathbf{i}} + \cos(y)\hat{\mathbf{j}} + z^2\hat{\mathbf{k}}$$

22. Create 3D vector field plots of $\mathbf{F} = y\hat{\mathbf{i}} - x\hat{\mathbf{k}}$ and $\nabla\times\mathbf{F}$ displayed together.

23. Calculate the curl of the following vector functions in cylindrical coordinates $(u_1, u_2, u_3) = (r, \phi, z)$

$$\mathbf{F}(r,\phi,z) = r^2\hat{\mathbf{r}} + z\hat{\mathbf{z}} + r\sin\phi\,\hat{\boldsymbol{\phi}}$$

$$\mathbf{F}(r,\phi,z) = r^2\hat{\mathbf{r}} + z\hat{\mathbf{z}} + e^{i\phi}\,\hat{\boldsymbol{\phi}}$$

24. Calculate the curl of the following scalar functions in spherical coordinates $(u_1, u_2, u_3) = (r, \theta, \phi)$

$$\mathbf{F}(r,\theta,\phi) = r^2\hat{\mathbf{r}} + \cos\theta\,\hat{\boldsymbol{\theta}} + r\cos\theta\,\hat{\boldsymbol{\phi}}$$

$$\mathbf{F}(r,\theta,\phi) = r^2\hat{\mathbf{r}} + r\sin\theta\,\hat{\boldsymbol{\theta}} + e^{i\phi}\,\hat{\boldsymbol{\phi}}$$

Section 4.5 Laplacian of Scalar and Vector Fields

25. Fill in the blanks with "vector" or "scalar"

(a) The divergence of a_____ field gives a _____ field
(b) The gradient of a _____ field gives a _____ field
(c) The curl of a _____ field gives a _____ field
(d) The Laplacian of a ____/ ____field gives a____/ ____ field

26. Calculate the Laplacian of the following scalar functions in Cartesian coordinates

$G(x, y, z) = xyz$

$G(x, y, z) = e^x e^y e^z$

27. Calculate the Laplacian of the following scalar functions in cylindrical coordinates

$G(r, \phi, z) = r^2 e^{-z} \cos \phi$

$G(r, z) = z \ln(r)$

28. Calculate the Laplacian of the following scalar functions in spherical coordinates

$G(r, \theta, \phi) = r \cos \theta e^{i\phi}$

$$G(r, \theta) = \frac{\cos \theta}{r}$$

29. Show that in spherical coordinates the Laplacian

$$\nabla^2 = \frac{1}{r^2}\left(\frac{\partial}{\partial r} r^2 \frac{\partial}{\partial r}\right) + \frac{1}{r^2}\left[\frac{1}{\sin\theta}\frac{\partial}{\partial\theta}\left(\sin\theta\frac{\partial}{\partial\theta}\right) + \frac{1}{\sin^2\theta}\frac{\partial^2}{\partial^2\phi}\right]$$

is equivalent to

$$\nabla^2 = \frac{\partial^2}{\partial^2 r} + \frac{2}{r}\frac{\partial}{\partial r} + \frac{1}{r^2}\left(\frac{\partial^2}{\partial^2\theta} + \cot\theta\frac{\partial}{\partial\theta} + \csc^2\theta\frac{\partial^2}{\partial^2\phi}\right)$$

30. Calculate the Laplacian of the following vector fields in Cartesian coordinates

$$\mathbf{F}(x, y, z) = e^x \hat{\mathbf{i}} + e^y \hat{\mathbf{j}} + e^z \hat{\mathbf{k}}$$

$$\mathbf{F}(x, y, z) = y\hat{\mathbf{i}} + z\hat{\mathbf{j}} + x\hat{\mathbf{k}}$$

Section 4.6 Vector Identities

31. Given $\mathbf{R} = x\hat{\mathbf{i}} + y\hat{\mathbf{j}} + z\hat{\mathbf{k}}$ show that

(a) $\nabla \cdot \mathbf{R} = 3$

(b) $\nabla \times \mathbf{R} = 0$

(c) $\nabla |\mathbf{R}| = \dfrac{\mathbf{R}}{|\mathbf{R}|}$

(d) $\nabla |\mathbf{R}|^n = n |\mathbf{R}|^{n-2} \mathbf{R}$

(e) $\nabla^2 |\mathbf{R}|^n = n(n+1) |\mathbf{R}|^{n-2}$

32. Verify that the identity

$$\nabla \times (f\mathbf{F}) = f\nabla \times \mathbf{F} - \mathbf{F} \times \nabla f$$

holds for

$$f(x, y) = ye^x \text{ and } \mathbf{F} = x\hat{\mathbf{j}}$$

33. Verify that the identity

$$\nabla \times (\mathbf{F} \times \mathbf{G}) = (\mathbf{G} \cdot \nabla)\mathbf{F} - (\mathbf{F} \cdot \nabla)\mathbf{G} + \mathbf{F}(\nabla \cdot \mathbf{G}) - \mathbf{G}(\nabla \cdot \mathbf{F})$$

holds for $\mathbf{F} = yx\hat{\mathbf{j}}$ and $\mathbf{G} = \hat{\mathbf{k}}$

34. Calculate the Laplacian of the following vector field in cylindrical coordinates

$$\mathbf{F}(r, \phi, z) = r\cos\theta\hat{\mathbf{r}} + r\sin\phi\hat{\boldsymbol{\phi}}$$

35. Calculate the Laplacian of the following vector field in spherical coordinates

$$\mathbf{F}(r, \theta, \phi) = r\cos\theta\hat{\boldsymbol{\phi}}$$

Section 4.7 Integral Theorems

36. Given $f = xyz$ calculate the line integral $\displaystyle\int_a^b \nabla f \cdot d\boldsymbol{\ell}$

where $a = (0, 0, 0)$ and $b = (1, 1, 1)$ along the paths
(a) $(0, 0, 0) \to (1, 0, 0) \to (1, 1, 0) \to (1, 1, 1)$
(b) $(0, 0, 0) \to (0, 0, 1) \to (0, 1, 1) \to (1, 1, 1)$

37. Verify Gauss's divergence theorem in Cartesian coordinates over the unit cube in the first octant $\mathbf{F} = xy\hat{\mathbf{i}}$

38. Verify Gauss's divergence theorem in cylindrical coordinates over a volume bounded by a cylindrical surface of radius R. The cylinder is coaxial with the z-axis with end caps located at $z = 0$ and $z = h$. The vector field inside the cylinder is

$$\mathbf{F}(r, \phi, z) = r^2\hat{\mathbf{r}} + z\hat{\mathbf{z}} + r\sin\phi\hat{\boldsymbol{\phi}}.$$

39. Verify Gauss's divergence theorem in spherical coordinates over a volume bounded by a spherical surface of radius R. The vector field inside the sphere is $\mathbf{F}(r,\theta,\phi)=r^2\hat{\mathbf{r}}$.

40. Verify Stokes's theorem given the vector field

$$\mathbf{F} = y\hat{\mathbf{i}} - x\hat{\mathbf{j}}$$

over the area bound by the unit square in the first quadrant of the x-y plane. Take the direction of the integration contour to be counterclockwise.

41. Verify Stokes's theorem given the vector field

$$\mathbf{F} = y^2\hat{\mathbf{i}}$$

over the area bound by the unit circle in the x-y plane. Take the direction of the integration contour to be counterclockwise around the circle.

42. Torricelli (1608–1647) gives an approximation using Bernoulli's equation and the continuity equation $A_1 v_1 = A_2 v_2$ in cases where $A_1 \gg A_2$ so that v_1^2 is negligible compared to v_2^2. If both points in the flow are at atmospheric pressure $P_1 = P_2 = P_0$, show that the efflux speed $v_2 = \sqrt{2g\left(h_1 - h_2\right)}$.

Chapter **5** **ORDINARY**
DIFFERENTIAL
EQUATIONS

Chapter Outline

5.1 CLASSIFICATION OF DIFFERENTIAL EQUATIONS

In this section, the classification of differential equations by order and degree is discussed. Examples of differential equations that are solvable by direct integration and exact equations are given. The Sturm-Liouville form is introduced.

5.1.1 Order

Ordinary differential equations (ODEs) contain derivatives with respect to a single independent variable. The order of a differential equation refers to the order of the highest derivative in the equation.

Example 5.1.1

The differential equation

$$\frac{d^2 y(t)}{dt^2} = -\frac{k}{m} y(t) \tag{5.1.1}$$

describing simple harmonic motion of a block of mass m attached to a spring with force constant k is second order.

Example 5.1.2

The differential equation

$$EI \frac{d^4 y}{dx^4} = w(x) \tag{5.1.2}$$

describing the deflection $y(x)$ of a beam with flexural rigidity EI and external load $w(x)$ is fourth order.

5.1.2 Degree

The exponent of the highest derivative is the degree of the differential equation.

Example 5.1.3

The differential equation

$$\left(\frac{d^2 y}{dx^2} \right)^3 + \frac{dy}{dx} + y = 0 \tag{5.1.3}$$

is second order with degree = 3.

5.1.3 Solution by Direct Integration

Sufficiently simple differential equations may be solved by direct integration.

Example 5.1.4

Solve the following differential equation

$$\frac{d^3 y}{dx^3} = 0 \tag{5.1.4}$$

Solution: Successive integration gives

$$\frac{d^2y}{dx^2} = c_1 \tag{5.1.5}$$

$$\frac{dy}{dx} = c_1 x + c_2 \tag{5.1.6}$$

and

$$y = c_1 \frac{x^2}{2} + c_2 x + c_3 \tag{5.1.7}$$

The constants c_1, c_2, and c_3 are determined by initial conditions. In this example $y(0) = c_3$, $y'(0) = c_2$, and $y''(0) = c_1$. In general, the number of constants to be determined by initial conditions is equal to the order of the differential equation.

5.1.4 Exact Differential Equations

A differential equation

$$N(x,y)\frac{dy}{dx} + M(x,y) = 0 \tag{5.1.8}$$

written in the form

$$M(x,y)dx + N(x,y)dy = 0 \tag{5.1.9}$$

is said to be exact if

$$\frac{\partial M(x,y)}{\partial y} = \frac{\partial N(x,y)}{\partial x} \tag{5.1.10}$$

The general solution is $f(x,y) = $ const. such that

$$f(x,y) = \int M(x,y)dx + g(y) \tag{5.1.11}$$

and

$$f(x,y) = \int N(x,y)dy + h(x) \tag{5.1.12}$$

Example 5.1.5

Verify that the differential equation

$$2xydx + x^2dy = 0 \tag{5.1.13}$$

is exact and find the general solution

Solution: Identifying $M(x, y) = 2xy$, and $N(x, y) = x^2$ we find the differential equation is exact since

$$\frac{\partial(2xy)}{\partial y} = \frac{\partial(x^2)}{\partial x} = 2x \tag{5.1.14}$$

Finding the general solution, we have that

$$f(x,y) = \int 2xy\,dx + g(y) = x^2 y + g(y) + \text{const.} \tag{5.1.15}$$

and

$$f(x,y) = \int x^2 dy + h(x) = x^2 y + h(x) + \text{const.} \tag{5.1.16}$$

so that $g(y) = h(x) = 0$ and

$$f(x,y) = x^2 y + \text{const.} \tag{5.1.17}$$

5.1.5 Sturm-Liouville Form

Many differential equations arising in physics can be written in the Sturm-Liouville form

$$\frac{d}{dx}\left(p(x)\frac{dy}{dx}\right) - q(x)y + \lambda w(x)y = 0 \tag{5.1.18}$$

where the function $y(x)$ and the eigenvalues λ are to be determined.

Example 5.1.6

Bessel's differential equation

$$x^2 \frac{d^2 y}{dx^2} + x\frac{dy}{dx} + (\lambda x^2 - n^2)y = 0 \tag{5.1.19}$$

may be written in Sturm-Liouville form dividing by x

$$x\frac{d^2 y}{dx^2} + \frac{dy}{dx} - \frac{n^2}{x}y + \lambda xy = 0 \tag{5.1.20}$$

and combining the first two terms

$$\frac{d}{dx}\left(x\frac{dy}{dx}\right) - \frac{n}{x}y + xy = \tag{5.1.21}$$

we identify

$$p(x) = x,\ q(x) = n^2/x \text{ and } w(x) = x \tag{5.1.22}$$

Maple Examples

Homogeneous and inhomogeneous first order differential equations are solved in the Maple worksheet below.

Key Maple commands: *diff, dsolve, plot, rhs*

restart

Classification of Differential Equations

With(DEtools):
Deq:= diff(y(x), x, x) + sin(x)·diff(y(x), x)² = 0

$$\frac{d^2}{dx^2}y(x)+\sin(x)\left(\frac{d^2}{dx^2}y(x)\right)^2=0$$

odeadvisor(Deq)

$$[[\_2nd\_order, \_missing\_y], [\_2nd\_order, \_reducible, \_mu\_y\_y1]]$$

dsolve(Deq)

$$y(x)=\frac{\_C2\sqrt{\_C1^2-1}+2\arctan\left(\dfrac{(\_C1+1)\tan\left(\dfrac{1}{2}x\right)}{\sqrt{\_C1^2-1}}\right)}{\sqrt{\_C1^2-1}}$$

Deq:= diff(y(x), x, x, x) + x·diff(y(x), x) + y(x) = sin(x)

$$\frac{d^3}{dx^3}y(x)+x\left(\frac{d}{dx}y(x)\right)+y(x)=\sin(x)$$

odeadvisor(Deq)

$$[[\_3rd\_order, \_exact\_linear\_nonhomogeneous]]$$

dsolve(Deq)

$$y(x)=AiryAi(-x)\_C3+AiryBi(-x)\_C2$$
$$-\pi\left(\left(\int AiryBi(-x)(\cos(x)-\_C1)dx\right)AiryAi(-x)-\right.$$
$$\left.\left(\int AiryAi(-x)(\cos(x)-\_C1)dx\right)AiryBi(-x)\right)$$

5.2 FIRST ORDER DIFFERENTIAL EQUATIONS

In this section, we review the solution of first order homogeneous and inhomogeneous differential equations. Integrating factors are introduced.

5.2.1 Homogeneous Equations

First order homogeneous equations can be integrated directly.

Example 5.2.1

The first order differential equation

$$\dot{x} + \beta x = 0 \qquad (5.2.1)$$

written as

$$\frac{dx}{dt} = -\beta x \qquad (5.2.2)$$

may be solved by separating variables

$$\frac{dx}{x} = -\beta dt \qquad (5.2.3)$$

Integrating both sides

$$\ln(x) = -\beta t + \text{const.} \qquad (5.2.4)$$

and exponentiating gives

$$x(t) = x(0)e^{-\beta t} \qquad (5.2.5)$$

5.2.2 Inhomogeneous Equations

To solve an inhomogeneous differential equation

$$\dot{x} + \beta(t)x = f(t) \qquad (5.2.6)$$

we first consider the first order differential equation with

$$\dot{x} + \beta(t)x = 0 \qquad (5.2.7)$$

Separating variables

$$\frac{dx}{x} = -\beta(t)dt \qquad (5.2.8)$$

Integrating

$$x(t) = e^{-\int \beta(t)dt} \qquad (5.2.9)$$

Now multiply both sides of the differential equation by the integrating factor $e^{\int \beta(t)dt}$

$$(\dot{x} + \beta(t)x)e^{\int \beta(t)dt} = f(t)e^{\int \beta(t)dt} \qquad (5.2.10)$$

We recognize the left-hand side to be the time derivative of the product of x and the integrating factor

$$\left(\dot{x}+\beta(t)x\right)e^{\int\beta(t)dt}=\frac{d}{dt}\left(x\,e^{\int\beta(t)dt}\right) \tag{5.2.11}$$

so that

$$\frac{d}{dt}\left(x\,e^{\int\beta(t)dt}\right)=f(t)e^{\int\beta(t)dt} \tag{5.2.12}$$

and

$$x(t)=e^{-\int\beta(t)dt}\left[\int f(t)e^{\int\beta(t)dt}\,dt+\text{const.}\right] \tag{5.2.13}$$

Example 5.2.2

Solve the inhomogeneous differential equation

$$\dot{x}+t^2x=e^{-t} \tag{5.2.14}$$

Solution: With $\beta(t)=t^2$ and $f(t)=e^{-t}$ we have

$$x(t)=e^{-\int t^2dt}\left[\int e^{-t}\,e^{\int t^2dt}\,dt+\text{const.}\right] \tag{5.2.15}$$

and

$$x(t)=\exp\left(-\frac{t^3}{3}\right)\left[\int\exp\left(\frac{t^3}{3}-t\right)dt+\text{const.}\right] \tag{5.2.16}$$

Maple Examples

Homogeneous and inhomogeneous first order equations are solved in the Maple worksheet below.

Key Maple commands: *diff, dsolve, plot*

restart

Homogeneous First Order Equations

Deq:= diff($y(t), t$) + $f(t)\cdot y(t)$ = 0

$$\frac{d}{dt}y(t)+f(t)y(t)=0$$

dsolve(Deq)

$$y(t)=\_C1e^{\int(-f(t))dt}$$

Deq:= diff($y(x), x$) + cosh(x)$y(x)$ = 0

$$\frac{d}{dt}y(x)+\cosh(x)y(x)=0$$

$dsolve(\{Deq, y(0) = 1\})$

$$y(x) = e^{-\sinh(x)}$$

Inhomogeneous First Order Equations

$Deq := diff(y(t), t) + f(t) \cdot y(t) = g(t)$

$$\frac{d}{dt} y(t) + f(t) y(t) = g(t)$$

$dsolve(Deq)$

$$y(t) = \left(\int g(t) e^{\int f(t) dt} \, dt + \_C1 \right) e^{\int (-f(t)) dt}$$

$Deq := diff(y(x), x) + x \cdot y(x) = (1 - \exp(-x))$

$$\frac{d}{dx} y(x) + xy(x) = 1 - e^{-x}$$

$dsolve(\{Deq, y(0) = 0\})$

$$y(x) = \left(\frac{1}{2} I \sqrt{\pi} e^{-\frac{1}{2}} \sqrt{2} erf\left(\frac{1}{2} I \sqrt{2} x - \frac{1}{2} I \sqrt{2} \right) - \frac{1}{2} I \sqrt{\pi} \sqrt{2} erf\left(\frac{1}{2} I \sqrt{2} x \right) \right.$$
$$\left. + \frac{1}{2} I \sqrt{\pi} e^{-\frac{1}{2}} \sqrt{2} erf\left(\frac{1}{2} I \sqrt{2} \right) \right) e^{-\frac{1}{2}x^2}$$

$plot(rhs(\%), x = 0 \ldots 6)$

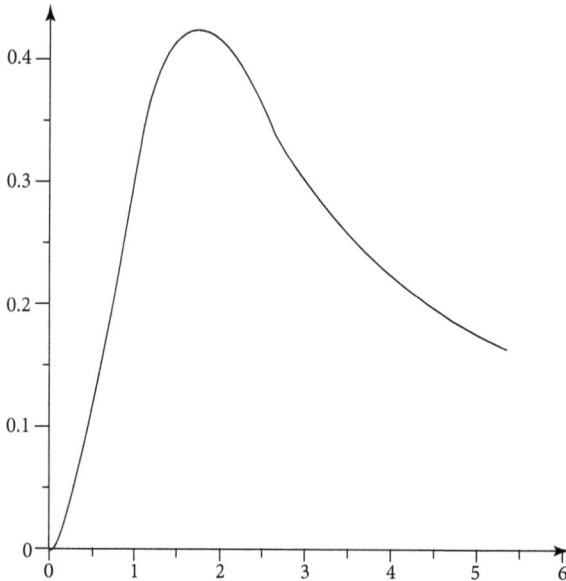

Figure 5.2.1: Plot of the solution to an inhomogeneous first order differential equation.

5.3 LINEAR, HOMOGENEOUS WITH CONSTANT COEFFICIENTS

Linear, homogeneous differential equations with constant coefficients are discussed in this section. Examples include the overdamped, underdamped, and critically damped harmonic oscillator. Solutions to higher order linear differential equations are introduced.

5.3.1 Damped Harmonic Oscillator

The second order homogeneous differential equation with constant coefficients given by

$$m\ddot{x} + b\dot{x} + kx = 0 \tag{5.3.1}$$

corresponds to a damped harmonic oscillator such as a block of mass m on a spring with force constant k and damping coefficient b. We assume a trial solution

$$x(t) = e^{rt} \tag{5.3.2}$$

where our task is to determine values of r. Substituting the trial solution into the differential equation, we obtain

$$mr^2 e^{rt} + bre^{rt} + ke^{rt} = 0 \tag{5.3.3}$$

The factor e^{rt} cancels and we obtain the characteristic equation

$$mr^2 + br + k = 0 \tag{5.3.4}$$

whose solutions determine r

$$r = \frac{-b \pm \sqrt{b^2 - 4mk}}{2m} \tag{5.3.5}$$

thus,

$$x(t) = e^{-\frac{b}{2m}t}\left(c_1 \exp\left(\frac{\sqrt{b^2 - 4mk}}{2m}t \right) + c_2 \exp\left(-\frac{\sqrt{b^2 - 4mk}}{2m}t \right) \right) \tag{5.3.6}$$

where the constants c_1 and c_2 are determined by the initial conditions $x(0)$, and $\dot{x}(0)$. If $b^2 > 4mk$ the roots are real and we say that the motion is overdamped. If $b^2 < 4mk$ the motion is underdamped and if $b^2 = 4mk$ the motion is critically damped. For undamped motion $b = 0$. Below we compare undamped, overdamped, underdamped, and critically damped motion when $x(0) = 0$.

5.3.2 Undamped Motion

Example 5.3.1

Choosing $b = 0$, $m = 1$ and $k = 1$ the square root $\sqrt{b^2 - 4mk} = 2i$ and $r = \pm 2i$ thus

$$x(t) = c_1 e^{2it} + c_2 e^{-2it} \tag{5.3.7}$$

We can also write $x(t)$ in terms of circular functions using Euler's formula

$$x(t) = c_1 \left[\cos(2t) + i\sin(2t) \right] + c_2 \left[\cos(2t) - i\sin(2t) \right] \tag{5.3.8}$$

Collecting $\cos(2t)$ and $\sin(2t)$ terms

$$x(t) = (c_1 + c_2)\cos(2t) + i(c_1 - c_2)\sin(2t) \tag{5.3.9}$$

and defining new constants $(c_1 + c_2) \to c_1$ and $i(c_1 - c_2) \to c_2$ \hfill (5.3.10)

$$x(t) = c_1 \cos(2t) + c_2 \sin(2t) \tag{5.3.11}$$

If $x(0) = 0$ then $c_1 = 0$ and

$$x(t) = c_2 \sin(2t) \tag{5.3.12}$$

where $\dot{x}(0)$ is used to determine the constant c_2. We could have chosen circular functions from the beginning as is often preferable when working with mechanical problems and bound state problems in quantum mechanics. Complex exponentials are often the functions of choice for wave motion and scattering problems.

5.3.3 Overdamped Motion

Example 5.3.2

For parameters $m = 1$, $b = 3$ and $k = 1$ we have $\sqrt{b^2 - 4mk} = \sqrt{5}$ and $r = -3/2$ $\pm\sqrt{5}/2$ thus,

$$x(t) = c_1 e^{\frac{-3+\sqrt{5}}{2}t} + c_2 e^{\frac{-3-\sqrt{5}}{2}t} \tag{5.3.13}$$

For $x(0) = 0$ we have $0 = c_1 + c_2 \to c_2 = -c_1$. This gives

$$x(t) = c_1 e^{\frac{-3}{2}t} \left(e^{\frac{\sqrt{5}}{2}t} - e^{\frac{-\sqrt{5}}{2}t} \right) = c_1 e^{\frac{-3}{2}t} \sinh\left(\frac{\sqrt{5}}{2}t \right) \tag{5.3.14}$$

where we have absorbed a factor of two into c_1.

5.3.4 Underdamped Motion

Example 5.3.3

Choosing parameters $m = 1$, $b = 1$ and $k = 1$ we have $\sqrt{b^2 - 4mk} = \sqrt{3}i$ and $r = -1/2 \pm \sqrt{3}i/2$ thus,

$$x(t) = c_1 e^{\frac{-1+\sqrt{3}i}{2}t} + c_2 e^{\frac{-1-\sqrt{3}i}{2}t} \qquad (5.3.15)$$

Taking $x(0) = 0$ as before $c_2 = -c_1$ and

$$x(t) = c_1 e^{-\frac{1}{2}t}\left(e^{\frac{\sqrt{3}i}{2}t} - e^{-\frac{\sqrt{3}i}{2}t}\right) = c_1 e^{-\frac{1}{2}t} \sin\left(\frac{\sqrt{3}}{2}t\right) \qquad (5.3.16)$$

absorbing $2i$ into c_1.

5.3.5 Critically Damped Oscillator

Example 5.3.4

For parameters $m = 1$, $b = 2$ and $k = 1$ the square root $\sqrt{b^2 - 4mk} = 0$ and we have a repeated root $r = -1$. For repeated roots the second exponential is multiplied by t to ensure linear independence

$$x(t) = c_1 e^{-t} + c_2 t e^{-t} \qquad (5.3.17)$$

Now the boundary condition $x(0) = 0$ gives $c_1 = 0$ and

$$x(t) = c_2 t e^{-t} \qquad (5.3.18)$$

5.3.6 Higher Order Differential Equations

The characteristic equation for higher order differential equations of the form

$$a_n \frac{d^n x(t)}{dt^n} + a_{n-1} \frac{d^{n-1}x(t)}{dt^{n-1}} + \cdots + a_1 \frac{dx(t)}{dt} + a_0 x(t) = 0 \qquad (5.3.19)$$

with constant coefficients an is obtained by substituting $x(t) = e^{rt}$ as before

$$a_n r^n + a_{n-1} r^{n-1} + \cdots a_1 r + a_0 = 0 \qquad (5.3.20)$$

Example 5.3.5

To solve the third order differential equation

$$\dddot{x}(t) + \ddot{x}(t) + \dot{x}(t) + x(t) = 0 \qquad (5.3.21)$$

we obtain the characteristic equation

$$r^3 + r^2 + r + 1 = 0 \tag{5.3.22}$$

that can be factored

$$r^2(r+1) + (r+1) = (r+1)(r^2+1) = 0 \tag{5.3.23}$$

so that $r = -1, i, -i$ and

$$x(t) = c_1 e^{-t} + c_2 e^{-it} + c_3 e^{it} \tag{5.3.24}$$

or using circular functions

$$x(t) = \underbrace{c_1 e^{-t}}_{\text{transient}} + \underbrace{c_2 \cos(t) + c_3 \sin(t)}_{\text{steady state}} \tag{5.3.25}$$

We can identify short duration (transient) behavior in the decaying exponential and long duration (steady state) behavior. The constants c_1, c_2 and c_3 determined from $x(0)$, $\dot{x}(0)$ and $\ddot{x}(0)$ will have values depending on the choice of complex exponentials or circular functions. In either case, both solutions will be mathematically identical. Here we have three initial conditions. In general, the number of initial conditions will be the same as the order of the differential equation.

Maple Examples

The characteristic polynomial is found for a homogeneous linear differential equation in the Maple worksheet below. Solutions to differential equations corresponding to underdamped, critically damped and overdamped oscillations are found and plotted together.

Key Maple commands: *diff, dsolve, eval, plot, rhs, simplify, solve, subs*

restart

Characteristic Polynomial

Deq:= diff(x(t), t$3) + diff(x(t), t$2) + diff(x(t), t) + x(t) = 0

$$Deq := \frac{d^3}{dt^3} x(t) + \frac{d^2}{dt^2} x(t) + \frac{d}{dt} x(t) + x(t) = 0$$

$$charPoly := simplify\left(eval\left(\frac{subs\left(x(t) = \exp(r \cdot t), Deq\right)}{\exp(r \cdot t)} \right) \right)$$

$$charPoly := r^3 + r^2 + r + 1 = 0$$

solve(charPoly, r)

$$-1, I, -1$$

dsolve(Deq)

$$x(t) = \_C1\, e^{-t} + \_C2 \sin(t) + \_C3 \cos(t)$$

Damped Oscillations

$Deq := m \cdot diff(x(t), t\$2) + b \cdot diff(x(t), t) + k \cdot x(t) = 0$

$$Deq := m\left(\frac{d^2}{dt^2}x(t)\right) + b\left(\frac{d}{dt}x(t)\right) + kx(t) = 0$$

$$charPoly := simplify\left(eval\left(\frac{subs\left(x(t) = \exp(r \cdot t), Deq\right)}{\exp(r \cdot t)}\right)\right)$$

$$charPoly := mr^2 + br + k = 0$$

$solve(charPoly, r)$

$$\frac{-b + \sqrt{b^2 - 4mk}}{2m}, -\frac{b + \sqrt{b^2 - 4mk}}{2m}$$

Underdamped

$undDamp := subs(\{m = 1, b = 1, k = 1\}, Deq)$

$$undDamp := \frac{d^2}{dt^2}x(t) + \frac{d}{dt}x(t) + x(t) = 0$$

$dsolve(\{undDamp, x(0) = 1, D(x)(0) = 0\}, x(t));$

$$x(t) = \frac{\sqrt{3}e^{-\frac{t}{2}}\sin\left(\frac{\sqrt{3}t}{2}\right)}{3} + e^{-\frac{t}{2}}\cos\left(\frac{\sqrt{3}t}{2}\right)$$

$xUnd := rhs(\%)$

$$xUnd := \frac{\sqrt{3}e^{-\frac{t}{2}}\sin\left(\frac{\sqrt{3}t}{2}\right)}{3} + e^{-\frac{t}{2}}\cos\left(\frac{\sqrt{3}t}{2}\right)$$

Critically Damped

$critDamp := subs(\{m = 1, b = 2, k = 1\}, Deq)$

$$critDamp := \frac{d^2}{dt^2}x(t) + 2\left(\frac{d}{dt}x(t)\right) + x(t) = 0$$

$dsolve(\{critDamp, x(0) = 1, D(x)(0) = 0\}, x(t));$

$$x(t) = e^{-t} + e^{-t}t$$

$xCrit := rhs(\%)$

$$xCrit = e^{-t} + e^{-t}t$$

Overdamped

$overDamp := subs(\{m = 1, b = 3, k = 1\}, Deq)$

$$overDamp := \frac{d^2}{dt^2}x(t) + 3\left(\frac{d}{dt}x(t)\right) + x(t) = 0$$

$dsolve(\{overDamp, x(0) = 1, D(x)(0) = 0\}, x(t));$

$$x(t) = \left(\frac{1}{2} + \frac{3\sqrt{5}}{10}\right)e^{\frac{(\sqrt{5}-3)t}{2}} + \left(\frac{1}{2} - \frac{3\sqrt{5}}{10}\right)e^{-\frac{(3+\sqrt{5})t}{2}}$$

$xOver := rhs(\%)$

$$xOver := \left(\frac{1}{2} + \frac{3\sqrt{5}}{10}\right)e^{\frac{(\sqrt{5}-3)t}{2}} + \left(\frac{1}{2} - \frac{3\sqrt{5}}{10}\right)e^{-\frac{(3+\sqrt{5})t}{2}}$$

$plot([xUnd, xCrit, xOver], t = 0 \ldots 10, labels = ["t", "x"], legend = ["Under damped",$
"Critically damped", "Over damped"], $linestyle = [solid, dash, dashdot], title =$
"Damped Oscillations")

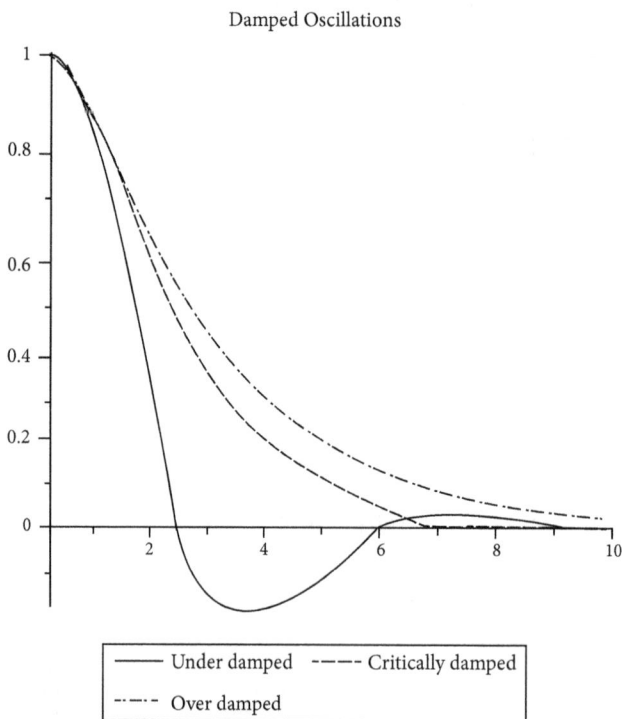

Figure 5.3.1: Plot of the position vs. time of an oscillator that is underdamped (bottom curve), critically damped (middle curve), and overdamped (top curve).

5.4 LINEAR INDEPENDENCE

The condition for linear independence of functions $y_1(x)$, $y_2(x)$, ..., $y_n(x)$ is

$$c_1 y_1(x) + c_2 y_2(x) + \cdots + c_n y_n(x) = 0 \qquad (5.4.1)$$

only if the constants $c_1 = 0, c_2 = 0$, ..., $c_n = 0$. If the equation above can be satisfied for any of the $c_1 \cdots c_n$ not equal to zero, then the functions are linearly dependent.

5.4.1 Wronskian Determinant

The Wronskian determinant

$$W = \begin{vmatrix} y_1 & y_2 & \cdots & y_n \\ y_1' & y_2' & \cdots & y_n' \\ \vdots & \vdots & \ddots & \vdots \\ y_1^{(n-1)} & y_2^{(n-1)} & y_3^{(n-1)} & y_n^{(n-1)} \end{vmatrix} \qquad (5.4.2)$$

may be used to test for the linear independence of *n functions*. If $W = 0$ then y_1... y_n are linearly dependent. If $W \neq 0$ then the functions are linearly independent.

Example 5.4.1

Determine if the functions $y_1(x) = 3\cos(x)$, $y_2(x) = 2\sin(x)$ are linearly dependent or linearly independent

Solution: We evaluate the determinant

$$W = \begin{vmatrix} y_1 & y_2 \\ y_1' & y_2' \end{vmatrix} = \begin{vmatrix} 3\cos(x) & 2\sin(x) \\ -3\sin(x) & 2\cos(x) \end{vmatrix} = 6\left(\cos^2(x) + \sin^2(x)\right) = 6 \qquad (5.4.3)$$

Thus $W \neq 0$ and y_1, y_2 are linearly independent.

Maple Examples

The Wronskian determinant is used to test for the linear independence or linear dependence of sets of functions in the Maple worksheet below.

Key Maple commands: *Wronskian*

Maple packages: *with(VectorCalculus):*

restart

Wronskian Determinant: Linearly Independent Functions

$with(VectorCalculus):$

$Wronskian([\cosh(t), \sin(t), t]),t)$

$$\begin{bmatrix} \cosh(t) & \sin(t) & t \\ \sinh(t) & \cos(t) & 1 \\ \cosh(t) & -\sin(t) & 0 \end{bmatrix}$$

$Wronskian\left(\left[2x, \dfrac{1}{x}\right], x, 'determinant'\right)$

$$\begin{bmatrix} 2x & \dfrac{1}{x} \\ 2 & -\dfrac{1}{x^2} \end{bmatrix}, -\dfrac{4}{x}$$

$Wronskian([t, t^2, t^3, t^4, t^5]),t, 'determinant')$

$$\begin{bmatrix} t & t^2 & t^3 & t^4 & t^5 \\ 1 & 2t & 3t^2 & 4t^3 & 5t^4 \\ 0 & 2 & 6t & 12t^2 & 20t^3 \\ 0 & 0 & 6 & 24t & 60t^2 \\ 0 & 0 & 0 & 24 & 120t \end{bmatrix}, 288t^5$$

Wronskian Determinant: Linearly Dependent Functions

$Wronskian([\exp(x), \exp(-x), \cosh(x)], x, 'determinant')$

$$\begin{bmatrix} e^x & e^{-x} & \cosh(x) \\ e^x & e^{-x} & \sinh(x) \\ e^x & e^{-x} & \cosh(x) \end{bmatrix}, 0$$

$Wronskian([\exp(I \cdot x), \exp(-I \cdot x), 2 \cdot \cos(x) + 3 \cdot \sin(x)], x, 'determinant')$

$$\begin{bmatrix} e^{Ix} & e^{-Ix} & 2\cos(x) + 3\sin(x) \\ Ie^x & -Ie^{-Ix} & -2\sin(x) + 3\cos(x) \\ -e^{Ix} & -e^{-Ix} & -2\cos(x) - 3\sin(x) \end{bmatrix}, 0$$

$Wronskian([t, 2t, t^3, t^4]), t, \text{'determinant'})$

$$\begin{bmatrix} t & 2t & t^3 & t^4 \\ 1 & 2 & 3t^2 & 4t^3 \\ 0 & 0 & 6t & 12t^2 \\ 0 & 0 & 6 & 24t \end{bmatrix}, 0$$

5.5 INHOMOGENEOUS WITH CONSTANT COEFFICIENTS

In this section, examples of linear, second order, inhomogeneous differential equations with constant coefficients are given. Such second order equations can model damped harmonic oscillators with driving.

Example 5.5.1

Find a solution to the inhomogeneous equation

$$\ddot{x} + 2\dot{x} + x = \sin t \tag{5.5.1}$$

Solution: The procedure is to first find a solution to the homogeneous equation

$$\ddot{x} + 2\dot{x} + x = 0 \tag{5.5.2}$$

The characteristic equation

$$r^2 + 2r + 1 = 0 \tag{5.5.3}$$

factors

$$(r + 1)(r + 1) = 0 \tag{5.5.4}$$

with repeated roots $r = -1$

$$x_c(t) = c_1 e^{-t} + c_2 t e^{-t} \tag{5.5.5}$$

where the solution to the homogeneous equation x_c is known as the complementary solution. Next, we guess the form of the particular solution x_p to the inhomogeneous equation

$$x_p(t) = A \sin t + B \cos t \tag{5.5.6}$$

and our job is to find A and B such that

$$\ddot{x}_p + 2\dot{x}_p + x_p = \sin t \tag{5.5.7}$$

Taking first and second derivatives of (5.5.6)

$$\dot{x}_p(t) = A \cos t - B \sin t \tag{5.5.8}$$

$$\ddot{x}_p(t) = -A \sin t - B \cos t \tag{5.5.9}$$

and substituting into the differential equation

$$\ddot{x}_p + 2\dot{x}_p + x_p = 2A\cos t - 2B\sin t$$

(5.5.10)

Since the right-hand side of (5.5.10) should be equal to $\sin t$ we have $A = 0$ and $B = -1/2$.

Combining our complementary and particular solutions

$$x(t) = x_c(t) + x_p(t) = c_1 e^{-t} + c_2 t e^{-t} - \frac{1}{2}\cos t$$

(5.5.11)

Example 5.5.2

Solve the inhomogeneous differential equation

$$\ddot{x} + x = 3 + 2t^2$$

(5.5.12)

Solution: To find the complementary solution we solve the homogeneous equation

$$\ddot{x} + x = 0$$

(5.5.13)

with characteristic equation

$$r^2 + 1 = 0$$

(5.5.14)

that factors

$$(r + i)(r - i) = 0$$

(5.5.15)

with roots $r = \pm i$. Thus, our complementary solution is

$$x_c(t) = c_1 e^{it} + c_2 e^{-it}$$

(5.5.16)

Next we guess the form of the particular solution

$$x_p(t) = A + Bt + Ct^2$$

(5.5.17)

Differentiating (5.5.17) twice

$$\dot{x}_p(t) = B + 2Ct$$

(5.5.18)

$$\ddot{x}_p(t) = 2C$$

(5.5.19)

and substituting we have

$$\ddot{x}_p + x_p = 2C + A + Bt + Ct^2$$

(5.5.20)

Requiring that

$$2C + A + Bt + Ct^2 = 3 + 2t^2$$

(5.5.21)

gives $C = 2, B = 0, 2C + A = 3 \rightarrow A = -1$.

Now we have that

$$x(t) = x_c(t) + x_p(t) = c_1 e^{it} + c_2 e^{-it} + 2t^2 - 1$$

(5.5.22)

or in terms of circular functions

$$x(t) = c_1 \sin t + c_2 \cos t + 2t^2 - 1$$

(5.5.23)

Maple Examples

Second order inhomogeneous linear differential equations are solved in the Maple worksheet below. Solutions to differential equations corresponding to driven underdamped, critically damped, and overdamped oscillations are found and plotted together. The oscillator response to step and impulse driving functions is calculated and plotted. Resonance curves are computed for three different damping factors.

Key Maple commands: *coeff, convert, diff, dsolve, expand, rhs, plot, solve, subs*

Special functions: *Heaviside*

restart

Second Order Inhomogeneous Differential Equations

Deq:= diff(x(t), t$2) + x(t) = 0.5*cos(0.8*t);

$$Deq := \frac{d^2}{dt^2}x(t) + x(t) = 0.5\cos(0.8t)$$

dsolve(Deq, x(0) = 0, D(x)(0) = 0}, x(t));

$$x(t) = -\frac{25\cos(t)}{18} + \frac{25\cos\left(\dfrac{4t}{5}\right)}{18}$$

Damped Driven Oscillations

Deq:= m·diff(x(t), t$2) + b·diff(x(t), t) + k·(x(t) = A·sin(2·t);

$$Deq := m\left(\frac{d^2}{dt^2}x(t)\right) + b\left(\frac{d}{dt}x(t)\right) + kx(t) = A\sin(2t)$$

Underdamped Driven Oscillations

undDamp:= subs({m = 1, b = 1, k = 1, A = 1}, Deq)

$$undDamp := \frac{d^2}{dt^2}x(t) + \frac{d}{dt}x(t) + x(t) = \sin(2t)$$

dsolve({undDamp, x(0) = 0, D(x)(0)= 0}, x(t));

$$x(t) = \frac{14e^{-\frac{t}{2}}\sqrt{3}\sin\left(\dfrac{\sqrt{3}t}{2}\right)}{39} + \frac{2e^{-\frac{t}{2}}\cos\left(\dfrac{\sqrt{3}t}{2}\right)}{13} - \frac{2\cos(2t)}{13} - \frac{3\sin(2t)}{13}$$

$xUnd := rhs(\%)$

$$xUnd := \frac{14e^{-\frac{t}{2}}\sqrt{3}\sin\left(\frac{\sqrt{3}t}{2}\right)}{39} + \frac{2e^{-\frac{t}{2}}\cos\left(\frac{\sqrt{3}t}{2}\right)}{13} - \frac{2\cos(2t)}{13} - \frac{3\sin(2t)}{13}$$

Critically Damped Driven Oscillations

$critDamp := subs(\{m = 1, b = 2, k = 1, A = 1\}, Deq)$

$$critDamp := \frac{d^2}{dt^2}x(t) + 2\left(\frac{d}{dt}x(t)\right) + x(t) = \sin(2t)$$

$dsolve(\{critDamp, x(0) = 1, D(x)(0) = 0\}, x(t));$

$$x(t) = \frac{4e^{-t}}{25} + \frac{2e^{-t}t}{5} - \frac{4\cos(2t)}{25} - \frac{3\sin(2t)}{25}$$

$xCrit := rhs(\%)$

$$xCrit = \frac{4e^{-t}}{25} + \frac{2e^{-t}t}{5} - \frac{4\cos(2t)}{25} - \frac{3\sin(2t)}{25}$$

Overdamped Driven Oscillations

$overDamp := subs(\{m = 1, b = 3, k = 1, A = 1\}, Deq)$

$$overDamp := \frac{d^2}{dt^2}x(t) + 3\left(\frac{d}{dt}x(t)\right) + x(t) = \sin(2t)$$

$dsolve(\{overDamp, x(0) = 0, D(x)(0) = 0\}, x(t));$

$$x(t) = e^{\frac{(\sqrt{5}-3)t}{2}}\left(\frac{1}{15} + \frac{\sqrt{5}}{15}\right) + e^{-\frac{(3+\sqrt{5})t}{2}}\left(\frac{1}{15} - \frac{\sqrt{5}}{15}\right) - \frac{2\cos(2t)}{15} - \frac{\sin(2t)}{15}$$

$xOver := rhs(\%)$

$$xOver := e^{\frac{(\sqrt{5}-3)t}{2}}\left(\frac{1}{15} + \frac{\sqrt{5}}{15}\right) + e^{-\frac{(3+\sqrt{5})t}{2}}\left(\frac{1}{15} - \frac{\sqrt{5}}{15}\right) - \frac{2\cos(2t)}{15} - \frac{\sin(2t)}{15}$$

$Plot([xUnd, xCrit, xOver], t = 0 \ldots 20, labels = [\text{``t''}, \text{``x(t)''}], legend =$
$[\text{``Under damped''}, \text{``Critically damped''}, \text{``Over damped''}], linestyle =$
$[solid, dash, dot], title = \text{``Damped Driven Oscillations''})$

Damped Driven Oscillations

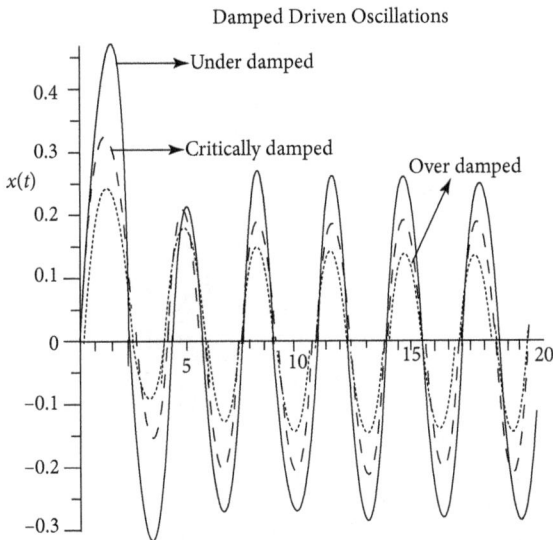

Figure 5.5.1: Plot of the position vs. time of a driven oscillator that is underdamped (large amplitude), critically damped (medium amplitude), and overdamped (small amplitude).

Response to a Step Function

$HeaviDeq := m \cdot diff(x(t), t\$2) + b \cdot diff(x(t), t) + k \cdot x(t) = A \cdot \text{Heaviside}(t);$

$$HeaviDeq := m\left(\frac{d^2}{dt^2}x(t)\right) + b\left(\frac{d}{dt}x(t)\right) + kx(t) = A\,Heaviside(t)$$

$Heq := subs(\{m = 1, b = 1, k = 1, A = 1\}, HeaviDeq)$

$$Heq := \frac{d^2}{dt^2}x(t) + \frac{d}{dt}x(t) + x(t) = Heaviside(t)$$

$dsolve(\{Heq, x(0) = 0, D(x)(0) = 0\}, x(t));$

$$x(t) = -\frac{Heaviside(t)\left(e^{-\frac{t}{2}}\sqrt{3}\sin\left(\frac{\sqrt{3}t}{2}\right) + 3e^{-\frac{t}{2}}\cos\left(\frac{\sqrt{3}t}{2}\right) - 3\right)}{3}$$

$plot([rhs(\%)], t = -2 \ldots 10, labels = [\text{``}t\text{''}, \text{``}x(t)\text{''}])$

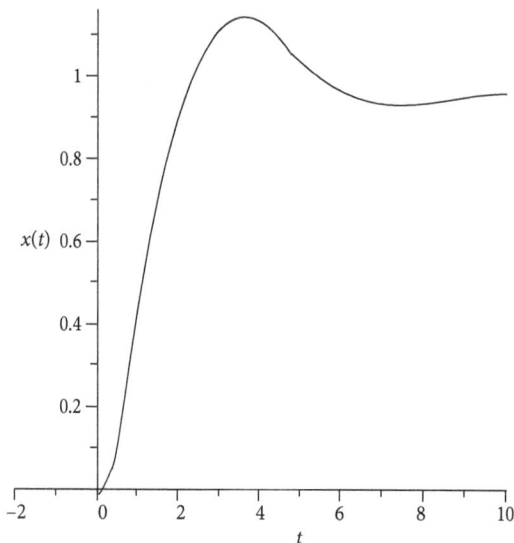

Figure 5.5.2: Plot of the position vs. time of a damped oscillator in response to a step function.

Response to an Impulse Function

ImpulseDeq:= m·diff(x(t), t\$2) + b·diff(x(t), t) + k·x(t) = A·(Heaviside(t) − Heaviside(t − 1));

$$ImpulseDeq := m\left(\frac{d^2}{dt^2}x(t)\right) + b\left(\frac{d}{dt}x(t)\right) + kx(t)$$
$$= A\left(Heaviside(t) - Heaviside(t-1)\right)$$

Ieq:= subs({m = 1, b = 1, k = 1, A = 1}, ImpulseDeq)

$$Ieq := \frac{d^2}{dt^2}x(t) + \frac{d}{dt}x(t) + x(t) = Heaviside(t) - Heaviside(t-1)$$

dsolve({Ieq, x(0) = 0, D(x)(0)= 0}, x(t));

$$x(t) = Heaviside(t) - \frac{e^{-\frac{t}{2}}\sqrt{3}\sin\left(\frac{\sqrt{3}t}{2}\right)Heaviside(t)}{3} - Heaviside(t-1)$$

$$+ \frac{\sqrt{3}Heaviside(t-1)e^{-\frac{t}{2}+\frac{1}{2}}\sin\left(\frac{\sqrt{3}(t-1)}{2}\right)}{3}$$

$$+ Heaviside(t-1)e^{-\frac{t}{2}+\frac{1}{2}}\cos\left(\frac{\sqrt{3}(t-1)}{2}\right) - e^{-\frac{t}{2}}\cos\left(\frac{\sqrt{3}t}{2}\right)Heaviside(t)$$

$plot([rhs(\%), A\cdot(\text{Heaviside}(t) - \text{Heaviside}(t - 1))], t = -2 \ldots 15, \text{labels} = [\text{"}t\text{"},$
$\text{"}x(t)\text{"}])$

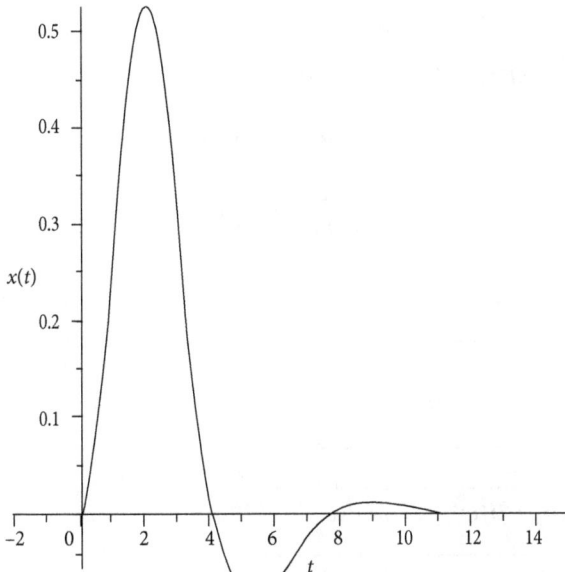

Figure 5.5.3: Plot of the position vs. time of a damped oscillator in response to an impulse function.

Resonance Curves

$Eqn:= m\cdot diff(y(t), t, t) + beta\cdot diff(y(t), t) + k\cdot y(t) = \sin(omega\cdot t)$

$$HeaviDeq := m\left(\frac{d^2}{dt^2}y(t)\right) + \beta\left(\frac{d}{dt}y(t)\right) + ky(t) = \sin(\omega t)$$

$sol:= dsolve(Eqn)$

$$sol := y(t) = e^{\frac{\left(-\beta+\sqrt{\beta^2-4km}\right)t}{2m}} \_C2 + e^{\frac{\left(\beta+\sqrt{\beta^2-4km}\right)t}{2m}} \_C1$$
$$+\frac{\left(-m\omega^2 +k\right)\sin(\omega t)-\cos(\omega t)\beta\omega}{m^2\omega^4 +\left(\beta^2 -2km\right)\omega^2 +k^2}$$

$$steadyState := \frac{\left(-m\omega^2 +k\right)\sin(\omega t)-\cos(\omega t)\beta\omega}{m^2\omega^4 +\left(\beta^2 -2km\right)\omega^2 +k^2}$$

$$steadyState := \frac{\left(-m\omega^2 +k\right)\sin(\omega t)-\cos(\omega t)\beta\omega}{m^2\omega^4 +\left(\beta^2 -2km\right)\omega^2 +k^2}$$

$sinCos: = A\cdot expand(\cos(omega\cdot t - delta))$

$$sinCos: = A(\cos(wt)\cos(\delta) + \sin(wt)\sin(\delta))$$
$$sinEqn: = coeff(sinCos, \sin(omega·t)) = coeff(steadyState, \sin(omega·t))$$

$$\sin Eqn := A\sin(\delta) = \frac{-m\omega^2 + k}{m^2\omega^4 + (\beta^2 - 2km)\omega^2 + k^2}$$

$$cosEqn: = coeff(sinCos, \cos(omega·t)) = coeff(steadyState, \cos(omega·t))$$

$$\cos Eqn := A\cos(\delta) = -\frac{\beta\omega}{m^2\omega^4 + (\beta^2 - 2km)\omega^2 + k^2}$$

$Solve(\{sinEqn, cosEqn\}, \{A, delta\}):$
$convert(\%[1], radical)$

$$A = \sqrt{\frac{1}{m^2\omega^4 + \beta^2\omega^2 - 2km\omega^2 + k^2}}$$

$$A := (m, k, beta, omega) \rightarrow \sqrt{\frac{1}{m^2\omega^4 + \beta^2\omega^2 - 2km\omega^2 + k^2}}$$

$$A := (m, k, \beta, \omega) \rightarrow \sqrt{\frac{1}{m^2\omega^4 + \beta^2\omega^2 - 2km\omega^2 + k^2}}$$

$Plot([A(1, 1, .7, omega), A(1, 1, 0.5, omega), A(1, 1, 0.2, omega)], omega = 0 \ldots 2,$
$\quad labels = ["\omega", "A(\omega)"], legend = ["beta = 1", "beta = 0.5", "beta = 0.2"],$
$\quad linestyle = [solid, dash, dashdot], title = "Resonance Curves")$

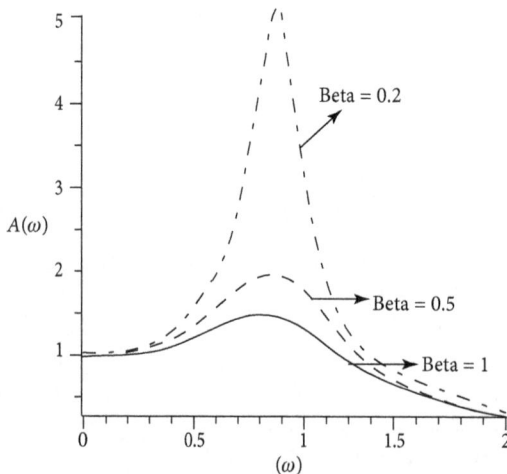

Resonance Curves

Figure 5.5.4: Plot of amplitude vs. frequency of an oscillator with large damping (bottom curve), medium damping (middle curve), and small damping (top curve).

5.6 POWER SERIES SOLUTIONS TO DIFFERENTIAL SOLUTIONS

Power series solutions to second order differential equations in standard form are covered in this section. Singular points are discussed. Examples include power series solutions to Airy's, Hermite's, Bessel's and Legendre's differential equations.

5.6.1 Standard Form

The differential equation

$$f(x)\frac{d^2y}{dx^2}+g(x)\frac{dy}{dx}+h(x)y=0 \tag{5.6.1}$$

is written in standard form by dividing by the coefficient of y''

$$\frac{d^2y}{dx^2}+P(x)\frac{dy}{dx}+Q(x)y=0 \tag{5.6.2}$$

Ordinary points are at locations where $P(x)$ and $Q(x)$ are analytic. $P(x)$ and $Q(x)$ are not analytic at singular points. Polynomial functions are analytic for all x while rational functions are analytic except where the denominator is zero.

Example 5.6.1

$x=0$ is an ordinary point of the differential equation

$$\frac{d^2y}{dx^2}+x^2\frac{dy}{dx}+xy=0 \tag{5.6.3}$$

since x^2 and x are analytic at $x=0$.

A power series solution may be found to the standard form of a differential equation centered on an ordinary point.

5.6.2 Airy's Differential Equation

Example 5.6.2

The differential equation

$$\frac{d^2y}{dx^2}+xy=0 \tag{5.6.4}$$

is in standard form with $P(x)=0$ and $Q(x)=x$. Assume a power series solution of the form

$$y(x) = \sum_{n=0}^{\infty} a_n x^n \qquad (5.6.5)$$

to find the coefficients a_n

Solution: We take derivatives of $y(x)$

$$y'(x) = \sum_{n=0}^{\infty} a_n n x^{n-1} \qquad (5.6.6)$$

$$y''(x) = \sum_{n=0}^{\infty} a_n n(n-1) x^{n-2} \qquad (5.6.7)$$

and substitute expansions of $y(x)$ and $y''(x)$ into the differential equation

$$\sum_{n=0}^{\infty} a_n n(n-1) x^{n-2} + \sum_{n=0}^{\infty} a_n x^{n+1} = 0 \qquad (5.6.8)$$

Shifting the index of the second sum

$$\sum_{n=0}^{\infty} a_n n(n-1) x^{n-2} + \sum_{n=3}^{\infty} a_{n-3} x^{n-2} = 0 \qquad (5.6.9)$$

The first and second terms of the first sum are zero. We can combine the sums beginning from $n = 3$

$$2a_2 + \sum_{n=3}^{\infty} \left(a_n n(n-1) + a_{n-3} \right) x^{n-2} = 0 \qquad (5.6.10)$$

The coefficient of each power of x must be zero so that $a_2 = 0$ and

$$a_n n(n-1) + a_{n-3} = 0 \qquad (5.6.11)$$

We thus obtain the recursion relation

$$a_n = -\frac{a_{n-3}}{n(n-1)} \qquad (5.6.12)$$

Starting from $n = 3$

$$a_3 = -\frac{a_0}{3 \cdot 2}$$

$$a_4 = -\frac{a_1}{4 \cdot 3}$$

$$a_5 = -\frac{a_2}{5 \cdot 4} = 0 \text{ since } a_2 = 0$$

$$a_6 = -\frac{a_3}{6\cdot 5} = \frac{a_0}{6\cdot 5\cdot 3\cdot 2}$$

$$a_7 = -\frac{a_4}{7\cdot 6} = \frac{a_1}{7\cdot 6\cdot 4\cdot 3}$$

$$a_8 = -\frac{a_5}{8\cdot 7} = 0 \text{ since } a_5 = 0$$

$$a_9 = -\frac{a_6}{9\cdot 8} = -\frac{a_0}{9\cdot 8\cdot 6\cdot 5\cdot 3\cdot 2}$$

$$a_{10} = -\frac{a_7}{10\cdot 9} = -\frac{a_1}{10\cdot 9\cdot 7\cdot 6\cdot 4\cdot 3}$$

Our series solution

$$y(x) = a_0 + a_1 x + 0 + a_3 x^3 + a_4 x^4 + 0 + a_6 x^6 + a_7 x^7 + 0 + a_9 x^9 + a_{10} x^{10} + \cdots \quad (5.6.13)$$

becomes

$$y(x) = a_0 + a_1 x - \frac{a_0}{3\cdot 2}x^3 - \frac{a_1}{4\cdot 3}x^4 + \frac{a_0}{6\cdot 5\cdot 3\cdot 2}x^6 + \frac{a_1}{7\cdot 6\cdot 4\cdot 3}x^7$$

$$-\frac{a_0}{9\cdot 8\cdot 6\cdot 5\cdot 3\cdot 2}x^9 - \frac{a_1}{10\cdot 9\cdot 7\cdot 6\cdot 4\cdot 3}x^{10} + \cdots \qquad (5.6.14)$$

The general solution can now be expressed as

$$y(x) = a_0 y_1(x) + a_1 y_2(x) \qquad (5.6.15)$$

with constants a_0 and a_1 to be determined from initial conditions and

$$y_1(x) = 1 - \frac{1}{3\cdot 2}x^3 + \frac{1}{6\cdot 5\cdot 3\cdot 2}x^6 - \frac{1}{9\cdot 8\cdot 6\cdot 5\cdot 3\cdot 2}x^9 + \cdots \qquad (5.6.16)$$

$$y_2(x) = x - \frac{1}{4\cdot 3}x^4 + \frac{1}{7\cdot 6\cdot 4\cdot 3}x^7 - \frac{1}{10\cdot 9\cdot 7\cdot 6\cdot 4\cdot 3}x^{10} + \cdots \qquad (5.6.17)$$

5.6.3 Hermite's Differential Equation

Example 5.6.3

The differential equation

$$\frac{d^2 y}{dx^2} - 2x\frac{dy}{dx} + 2\lambda y = 0 \qquad (5.6.18)$$

has application in the solution of the quantum harmonic oscillator. Proceed with a power series solution where

$$y = \sum_{n=0}^{\infty} a_n x^n \qquad (5.6.19)$$

to find a recursion relation for the coefficients a_n

Solution: Taking derivatives

$$y' = \sum_{n=0}^{\infty} n a_n x^{n-1} \qquad (5.6.20)$$

$$y'' = \sum_{n=0}^{\infty} n(n-1) a_n x^{n-2} \qquad (5.6.21)$$

Substituting back into the original differential equation

$$\sum_{n=0}^{\infty} n(n-1) a_n x^{n-2} - 2 \sum_{n=0}^{\infty} n a_n x^n + 2\lambda \sum_{n=0}^{\infty} a_n x^n = 0 \qquad (5.6.22)$$

Shifting the index of the first term $n \rightarrow n + 2$

$$\sum_{n=-2}^{\infty} (n+2)(n+1) a_{n+2} x^n - 2 \sum_{n=0}^{\infty} n a_n x^n + 2\lambda \sum_{n=0}^{\infty} a_n x^n = 0 \qquad (5.6.23)$$

The $n = -2$ and $n = -1$ terms of the first sum are zero so we can combine terms

$$\sum_{n=0}^{\infty} \left[(n+2)(n+1) a_{n+2} - 2n a_n + 2\lambda a_n \right] x^n = 0 \qquad (5.6.24)$$

and

$$(n+2)(n+1) a_{n+2} - 2n a_n + 2\lambda a_n = 0 \qquad (5.6.25)$$

for all x^n giving the recursion relation

$$a_{n+2} = \frac{2(n-\lambda)}{(n+2)(n+1)} a_n \qquad (5.6.26)$$

5.6.4 Singular Points

A differential equation in standard form possesses singular points where either $P(x)$ or $Q(x)$ are nonanalytic. A singular point is regular if $(x - x_0)P(x)$ and $(x - x_0)^2 Q(x)$ and are analytic. The Frobenius theorem states that at least one solution of the form

$$y = \sum_{m=0}^{\infty} a_m (x - x_0)^{m+k} \qquad (5.6.27)$$

exists if $x = x_0$ is a regular singular point.

5.6.5 Bessel's Differential Equation

Example 5.6.4

Find a recurrence relation for the coefficients in the series solution to Bessel's differential equation

$$x^2 \frac{d^2 y}{dx^2} + x \frac{dy}{dx} + (x^2 - n^2) y = 0 \qquad (5.6.28)$$

about $x = 0$.

Solution: In standard form

$$\frac{d^2 y}{dx^2} + \frac{1}{x} \frac{dy}{dx} + \left(1 - \frac{n^2}{x^2}\right) y = 0 \qquad (5.6.29)$$

we see that $x_0 = 0$ is a singular point

$$(x - x_0) P(x) = x \frac{1}{x} = 1 \ \rightarrow \ \text{analytic} \qquad (5.6.30)$$

$$(x - x_0)^2 Q(x) = x^2 - n^2 \ \rightarrow \ \text{analytic} \qquad (5.6.31)$$

so we seek a power series solution of the form

$$y = \sum_{m=0}^{\infty} a_m x^{m+k} \qquad (5.6.32)$$

Evaluating derivatives

$$y' = \sum_{m=0}^{\infty} a_m (m+k) x^{m+k-1} \qquad (5.6.33)$$

$$y'' = \sum_{m=0}^{\infty} a_m (m+k)(m+k-1) x^{m+k-2} \qquad (5.6.34)$$

and substituting into the differential equation

$$x^2 \sum_{m=0}^{\infty} a_m (m+k)(m+k-1) x^{m+k-2} + x \sum_{m=0}^{\infty} a_m (m+k) x^{m+k-1} + (x^2 - n^2) \sum_{m=0}^{\infty} a_m x^{m+k} = 0 \qquad (5.6.35)$$

$$\sum_{m=0}^{\infty} a_m (m+k)(m+k-1) x^{m+k} + \sum_{m=0}^{\infty} a_m (m+k) x^{m+k} + \sum_{m=0}^{\infty} a_m x^{m+k+2} - n^2 \sum_{m=0}^{\infty} a_m x^{m+k} = 0 \qquad (5.6.36)$$

Combining sums containing x^{m+k}

$$\sum_{m=0}^{\infty} a_m \left[(m+k)(m+k-1)+(m+k)-n^2 \right] x^{m+k} + \sum_{m=0}^{\infty} a_m x^{m+k+2} = 0 \qquad (5.6.37)$$

Simplifying the first sum and shifting the index of the second sum

$$\sum_{m=0}^{\infty} a_m \left[(m+k)^2 - n^2 \right] x^{m+k} + \sum_{m=2}^{\infty} a_{m-2} x^{m+k} = 0 \qquad (5.6.38)$$

Combining the sums

$$a_0 (k^2 - n^2) x^k + a_1 \left[(1+k)^2 - n^2 \right] x^{1+k} + \sum_{m=2}^{\infty} \left(a_m \left[(m+k)^2 - n^2 \right] + a_{m-2} \right) x^{m+k} = 0$$

$$(5.6.39)$$

The coefficient of each power of x must be zero so

$$a_0 (k^2 - n^2) = 0 \qquad (5.6.40)$$

If $k = \pm n$ then $a_1 = 0$ for

$$a_1 \left[(1+k)^2 - n^2 \right] = 0 \qquad (5.6.41)$$

and

$$a_m \left[(m+n)^2 - n^2 \right] + a_{m-2} = 0 \qquad (5.6.42)$$

Our recursion relation for $m \geq 2$ is

$$a_m = -\frac{a_{m-2}}{(2n+m)m} \qquad (5.6.43)$$

Starting from $m = 2$

$$a_2 = -\frac{a_0}{(2n+2)2}$$

$$a_3 = -\frac{a_1}{(2n+3)3} = 0 \text{ since } a_1 = 0.$$

Thus, all odd coefficients are zero. Now

$$a_4 = -\frac{a_2}{(2n+4)4} = \frac{a_0}{(2n+4)(2n+2)4 \cdot 2}$$

$$a_6 = -\frac{a_4}{(2n+6)6} = -\frac{a_0}{(2n+6)(2n+4)(2n+2)6 \cdot 4 \cdot 2} = -\frac{a_0}{2^6 (n+3)(n+2)(n+1)3 \cdot 2 \cdot 1}$$

and our even coefficients are

$$a_m = a_0 \frac{(-1)^{m/2}}{2^m \left(\dfrac{m}{2}\right)! \left[\left(n+\dfrac{m}{2}\right)\cdots(n+2)(n+1)\right]} \tag{5.6.44}$$

5.6.6 Legendre's Differential Equation

Example 5.6.5

Find a recurrence relation for the coefficients in the series solution to Legendre's differential equation

$$(1-x^2)\frac{d^2 y}{dx^2} - 2x\frac{dy}{dx} + \ell(\ell+1)y = 0 \tag{5.6.45}$$

about $x=0$ and obtain an expression for the general solution.

Solution: Writing the differential equation in standard form

$$\frac{d^2 y}{dx^2} - \frac{2x}{(1-x^2)}\frac{dy}{dx} + \frac{\ell(\ell+1)}{(1-x^2)}y = 0 \tag{5.6.46}$$

we find $x=0$ is an ordinary point so we seek a series solution of the form

$$y = \sum_{m=0}^{\infty} a_m x^m \tag{5.6.47}$$

Taking derivatives

$$y' = \sum_{m=0}^{\infty} a_m m x^{m-1} \tag{5.6.48}$$

$$y'' = \sum_{m=0}^{\infty} a_m m(m-1)x^{m-2} \tag{5.6.49}$$

and substituting into the differential equation

$$\sum_{m=-2}^{\infty} a_{m+2}(m+2)(m+1)x^m - \sum_{m=0}^{\infty} a_m m(m-1)x^m - 2\sum_{m=0}^{\infty} a_m m x^m + \ell(\ell+1)\sum_{m=0}^{\infty} a_m x^m = 0 \tag{5.6.50}$$

Since the first two terms of the first sum are zero we can combine all the sums

$$\sum_{m=0}^{\infty}\left[a_{m+2}(m+2)(m+1) - a_m\left[m(m-1) + 2m - \ell(\ell+1)\right]\right]x^m = 0 \tag{5.6.51}$$

$$a_{m+2}(m+2)(m+1) - a_m\left[m(m+1) - \ell(\ell+1)\right] = 0 \tag{5.6.52}$$

$$a_{m+2} = a_m \frac{m(m+1) - \ell(\ell+1)}{(m+2)(m+1)} \tag{5.6.53}$$

$$a_2 = -a_0 \frac{\ell(\ell+1)}{2 \cdot 1}$$

$$a_3 = a_1 \frac{1 \cdot 2 - \ell(\ell+1)}{3 \cdot 2}$$

$$a_4 = a_2 \frac{2 \cdot 3 - \ell(\ell+1)}{4 \cdot 3} = -a_0 \frac{2 \cdot 3 - \ell(\ell+1)}{4 \cdot 3} \frac{\ell(\ell+1)}{2 \cdot 1}$$

$$a_5 = a_3 \frac{4 \cdot 3 - \ell(\ell+1)}{5 \cdot 4} = a_1 \frac{4 \cdot 3 - \ell(\ell+1)}{5 \cdot 4} \frac{1 \cdot 2 - \ell(\ell+1)}{3 \cdot 2}$$

Thus,

$$y(x) = a_0 y_1(x) + a_1 y_2(x) \tag{5.6.54}$$

with constants a_0 and a_1 to be determined from initial conditions and

$$y_1(x) = 1 - \frac{\ell(\ell+1)}{2 \cdot 1} x^2 - \frac{2 \cdot 3 - \ell(\ell+1)}{4 \cdot 3} \frac{\ell(\ell+1)}{2 \cdot 1} x^4 + \cdots \tag{5.6.55}$$

$$y_2(x) = x + \frac{1 \cdot 2 - \ell(\ell+1)}{3 \cdot 2} x^3 + \frac{4 \cdot 3 - \ell(\ell+1)}{5 \cdot 4} \frac{1 \cdot 2 - \ell(\ell+1)}{3 \cdot 2} x^5 + \cdots \tag{5.6.56}$$

Maple Examples

Power series solutions to differential equations are demonstrated in the Maple worksheet below. Examples include Legendre's, Laguerre's, and Hermite's differential equations. Truncated series and exact solutions are compared.

Key Maple commands: *convert, D, diff, dsolve, rhs*

restart

Truncated Series Solutions to ODEs

Deq:= diff(y(t), t, t) − 2t·y(t) = 0

$$Deq := \frac{d^2}{dt^2} y(t) - 2ty(t) = 0$$

dsolve(Deq, y(t), 'type = series')

$$y(t) = y(0) + D(y)(0)t + \frac{1}{3} y(0)t^3 + \frac{1}{6} D(y)(0)t^4 + O(t^6)$$

Formal Series Solutions to ODEs

$$dsolve(Deq, y(t), \text{'}type = formal\_series\text{'})$$

$$y(t) = \_C1\Gamma\left(\frac{2}{3}\right)\left(\sum_{\_n=0}^{\infty}\frac{\left(\frac{2}{9}\right)^{-n} t^{3\_n}}{\Gamma(\_n+1)\Gamma\left(\_n+\frac{2}{3}\right)}\right), y(t)$$

$$= \frac{2\_C1\pi\sqrt{3}\left(\sum_{\_n=0}^{\infty}\frac{\left(\frac{2}{9}\right)^{-n} t^{3\_n+1}}{\Gamma\left(\_n+\frac{4}{3}\right)\Gamma(\_n+1)}\right)}{9\Gamma\left(\frac{2}{3}\right)}$$

Legendre's Differential Equation

$$LegEqn := (1-x^2)\cdot diff(y(x), x, x) - 2\cdot x\cdot diff(y(x), x) + (nu\cdot(nu + 1))\cdot y(x) = 0$$

$$LegEqn := (-x^2+1)\left(\frac{d^2}{dx^2}y(x)\right) - 2x\left(\frac{d}{dx}y(x)\right) + \nu(\nu+1)y(x) = 0$$

$$dsolve(LegEqn, y(x), \text{'}type = formal\_series\text{'})$$

$$y(x) = \frac{\_C1\sin(\pi(\nu+1))\left(\sum_{\_n=0}^{\infty}\frac{\left(-\frac{1}{2}\right)^{-n}\Gamma(\_n+1+\nu)\Gamma(\_n-\nu)(x-1)^{-n}}{\Gamma(\_n+1)^2}\right)}{\pi}, y(x)$$

$$= \frac{\_C1\sin(\pi(\nu+1))\left(\sum_{\_n=0}^{\infty}\frac{\left(\frac{1}{2}\right)^{-n}\Gamma(\_n+1+\nu)\Gamma(\_n-\nu)(x+1)^{-n}}{\Gamma(\_n+1)^2}\right)}{\pi}$$

Laguerre's Differential Equation

$$LagEqn := x\cdot diff(y(x), x, x) + (1-x)\cdot diff(y(x), x) + n\cdot y(x) = 0$$

$$LagEqn := \left(\frac{d^2}{dx^2}y(x)\right)x + (-x+1)\left(\frac{d}{dx}y(x)\right) + ny(x) = 0$$

$dsolve(LagEqn, y(x), \text{'}type = formal\_series\text{'})$

$$y(x) = \frac{\_C1\left(\displaystyle\sum_{\_n=0}^{\infty} \frac{\Gamma(\_n+\_n)x^{-n}}{\Gamma(\_n+1)^2}\right)}{\Gamma(-n)}$$

Hermite's Differential Equation

$HermEqn:= diff(y(x), x, x) - 2 \cdot x \cdot diff(y(x), x) + 2\ m \cdot y(x) = 0$

$$HermEqn := \frac{d^2}{dx^2}\, y(x) - 2x\left(\frac{d}{dx}\, y(x)\right) + 2my(x) = 0$$

$dsolve(HermEqn, y(x), \text{'}type = formal\_series\text{'})$

$$y(x) = \frac{\_C1\left(\displaystyle\sum_{\_n=0}^{\infty} \frac{\Gamma\left(\_n-\dfrac{m}{2}\right)4^{-n}x^{2\_n}}{\Gamma(2\_n+1)}\right)}{\Gamma\left(-\dfrac{m}{2}\right)}, y(x)$$

$$\blacksquare \quad \frac{\_C1\left(\displaystyle\sum_{\_n=0}^{\infty} \frac{\Gamma\left(\_n+\dfrac{1}{2}-\dfrac{m}{2}\right)4^{-n}x^{2\_n+1}}{\Gamma(2\_n+2)}\right)}{\Gamma\left(-\dfrac{m}{2}+\dfrac{1}{2}\right)}$$

Comparing Truncated Series

$deq1:= diff(y(t), t, t) = 2 \cdot t \cdot y(t) + \sin(t)$

$$deq1 := \frac{d^2}{dt^2}\, y(t) = 2ty(t) + \sin(t)$$

$dsolve(\{deq1, y(0) = 0, D(y)(0) = 0\})$

$$y(t) = -\frac{1}{2}\left(2^{2/3}\pi\left(\mathrm{AiryAi}\left(2^{1/3}t\right)\left(\int_0^t \mathrm{AiryBi}\left(2^{1/3}\_zl\right)\sin(\_zl)d\_zl\right)\right.\right.$$

$$\left.\left. - \mathrm{AiryBi}\left(2^{1/3}t\right)\left(\int_0^t \mathrm{AiryAi}\left(2^{1/3}\_zl\right)\sin(\_zl)d\_zl\right)\right)\right)$$

$exactSol:= rhs(\%):$
$dsolve(\{deq1, y(0) = 0, D(y)(0) = 0\}, y(t), \text{'}type = series\text{'})$

$$y(t) = \frac{1}{6}t^3 - \frac{1}{120}t^5 + O(t^6)$$

$serSol1 := convert(rhs(\%), polynom)$

$$serSol1 := \frac{1}{6}t^3 - \frac{1}{120}t^5$$

$Order: = 10:$

$dsolve(\{deq1, y(0) = 0, D(y)(0) = 0\}, y(t),\text{'}type = series\text{'})$

$$y(t) = \frac{1}{6}t^3 - \frac{1}{120}t^5 + \frac{1}{90}t^6 + \frac{1}{5040}t^7 - \frac{1}{3360}t^8 + \frac{37}{120960}t^9 + O(t^{10})$$

$serSol2 := convert(rhs(\%), polynom)$

$$serSol2 := \frac{1}{6}t^3 - \frac{1}{120}t^5 + \frac{1}{90}t^6 + \frac{1}{5040}t^7 - \frac{1}{3360}t^8 + \frac{37}{120960}t^9$$

$Plot([exactSol, serSol1, serSol2], t = 0 \ldots 3, labels = [\text{"}t\text{"}, \text{"}y(t)\text{"}], legend = [\text{"}Exact$
$Solution\text{"}, \text{"}2\ Terms\ in\ Series\text{"}, \text{"}5\ Terms\ in\ Series\text{"}], linestyle = [solid, dot,$
$dash], title = \text{"}Comparing Exact Solution and Truncated Series\text{"})$

Comparing Exact Solution and Truncated Series

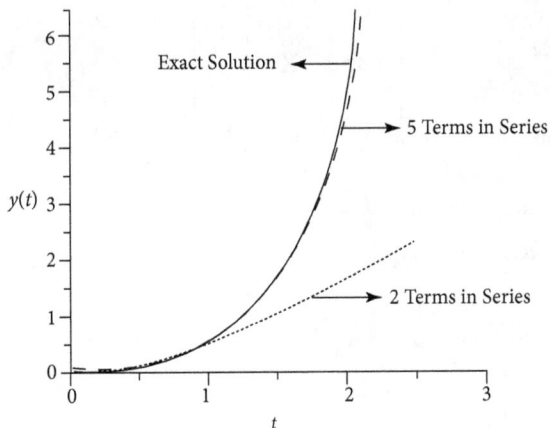

Figure 5.6.1: Comparison of exact and truncated power series solutions to a differential equation.

5.7 SYSTEMS OF DIFFERENTIAL EQUATIONS

Homogeneous and inhomogeneous systems of differential equations are discussed in this section. Topics include solution vectors, tests for linear independence, the general solution of homogeneous systems, and expressing higher order differential equations as first order systems. The differential equations describing the motion of a charged particle in crossed electric and magnetic fields are developed.

5.7.1 Homogeneous Systems

The homogeneous system of first order differential equations

$$
\begin{aligned}
\dot{x}_1 &= a_{11}x_1 + a_{12}x_2 + \cdots + a_{1n}x_n \\
\dot{x}_2 &= a_{21}x_1 + a_{22}x_2 + \cdots + a_{2n}x_n \\
&\ \vdots \qquad \vdots \qquad \vdots \\
\dot{x}_n &= a_{n1}x_1 + a_{n2}x_2 + \cdots + a_{nn}x_n
\end{aligned}
\tag{5.7.1}
$$

may be written in matrix form

$$
\dot{\mathbf{X}} = \mathbf{A}\mathbf{X}
\tag{5.7.2}
$$

where

$$
\dot{\mathbf{X}} = \begin{pmatrix} \dot{x}_1 \\ \dot{x}_2 \\ \vdots \\ \dot{x}_n \end{pmatrix}, \quad
\mathbf{A} = \begin{pmatrix} a_{11} & a_{12} & \cdots & a_{1n} \\ a_{21} & a_{22} & \cdots & a_{2n} \\ \vdots & \vdots & \ddots & \vdots \\ a_{n1} & a_{n2} & \cdots & a_{nn} \end{pmatrix}, \quad \text{and} \quad
\mathbf{X} = \begin{pmatrix} x_1 \\ x_2 \\ \vdots \\ x_n \end{pmatrix}
\tag{5.7.3}
$$

so that

$$
\begin{pmatrix} \dot{x}_1 \\ \dot{x}_2 \\ \vdots \\ \dot{x}_n \end{pmatrix} = \begin{pmatrix} a_{11} & a_{12} & \cdots & a_{1n} \\ a_{21} & a_{22} & \cdots & a_{2n} \\ \vdots & \vdots & \ddots & \vdots \\ a_{n1} & a_{n2} & \cdots & a_{nn} \end{pmatrix} \begin{pmatrix} x_1 \\ x_2 \\ \vdots \\ x_n \end{pmatrix}
\tag{5.7.4}
$$

Example 5.7.1

Write the homogeneous system below in matrix form

$$
\begin{aligned}
\dot{x} &= y \\
\dot{y} &= x - z \\
\dot{z} &= z
\end{aligned}
\tag{5.7.5}
$$

Solution: Identifying

$$\dot{\mathbf{X}} = \begin{pmatrix} \dot{x} \\ \dot{y} \\ \dot{z} \end{pmatrix}, \mathbf{A} = \begin{pmatrix} 0 & 1 & 0 \\ 1 & 0 & -1 \\ 0 & 0 & 1 \end{pmatrix} \text{ and } \mathbf{X} = \begin{pmatrix} x \\ y \\ z \end{pmatrix} \qquad (5.7.6)$$

so that

$$\begin{pmatrix} \dot{x} \\ \dot{y} \\ \dot{z} \end{pmatrix} = \begin{pmatrix} 0 & 1 & 0 \\ 1 & 0 & -1 \\ 0 & 0 & 1 \end{pmatrix} \begin{pmatrix} x \\ y \\ z \end{pmatrix} \qquad (5.7.7)$$

5.7.2 Inhomogeneous Systems

Inhomogeneous systems are of the form

$$\dot{\mathbf{X}} = \mathbf{AX} + \mathbf{F} \text{ where } \mathbf{F} = \begin{pmatrix} f_1(t) \\ f_2(t) \\ \vdots \\ x_n(f) \end{pmatrix} \qquad (5.7.8)$$

Example 5.7.2

We write the following system of inhomogeneous equations

$$\begin{aligned} \dot{x} &= y - x + 3t \\ \dot{y} &= 2x + y + t \end{aligned} \qquad (5.7.9)$$

in matrix form.

Solution:

$$\begin{pmatrix} \dot{x} \\ \dot{y} \end{pmatrix} = \begin{pmatrix} -1 & 1 \\ 2 & 1 \end{pmatrix} \begin{pmatrix} x \\ y \end{pmatrix} + \begin{pmatrix} 3t \\ t \end{pmatrix} \qquad (5.7.10)$$

5.7.3 Solution Vectors

A vector $\mathbf{X}(t) = \begin{pmatrix} x_1(t) \\ x_2(t) \\ \vdots \\ x_n(t) \end{pmatrix}$ that satisfies $\dot{\mathbf{X}} = \mathbf{AX} + \mathbf{F}$ is called a solution vector.

If the vectors $\mathbf{X}_1(t) = \begin{pmatrix} x_{11}(t) \\ x_{21}(t) \\ \vdots \\ x_{n1}(t) \end{pmatrix}$ $\mathbf{X}_2(t) = \begin{pmatrix} x_{12}(t) \\ x_{22}(t) \\ \vdots \\ x_{n2}(t) \end{pmatrix}$ $\cdots$ $\mathbf{X}_n(t) = \begin{pmatrix} x_{1n}(t) \\ x_{2n}(t) \\ \vdots \\ x_{nn}(t) \end{pmatrix}$ are

solution vectors, then $\mathbf{X}(t) = c_1 \mathbf{X}_1(t) + c_2 \mathbf{X}_2(t) + \cdots + c_n \mathbf{X}_n(t)$ is also a solution

to the system of ODEs for any constants $c_1, c_2, \ldots, c_n$ by the principle of superposition.

5.7.4 Test for Linear Independence

The set of solution vectors $\mathbf{X}_1(t), \mathbf{X}_2(t), \ldots, \mathbf{X}_n(t)$ is linearly independent if

$$c_1 \mathbf{X}_1(t) + c_2 \mathbf{X}_2(t) + \cdots + c_n \mathbf{X}_n(t) = 0 \tag{5.7.11}$$

only for all constant $c_1, c_2, \ldots, c_n$ equal to zero. If the sum is zero for any nonzero constants, then the set is linearly dependent. If the determinant of the matrix formed by the solution vectors is nonzero

$$\begin{vmatrix} x_{11} & x_{12} & \cdots & x_{1n} \\ x_{21} & x_{22} & \cdots & x_{2n} \\ \vdots & \vdots & \ddots & \vdots \\ x_{n1} & x_{n2} & \cdots & x_{nn} \end{vmatrix} \neq 0 \tag{5.7.12}$$

then the solution vectors are linearly independent. If the determinant is zero, then the solution vectors are linearly dependent.

5.7.5 General Solution of Homogeneous Systems

The general solution to the homogenous system $\dot{\mathbf{X}} = \mathbf{AX}$ is obtained by solving the characteristic equation obtained by assuming a trial solution

$$\mathbf{X}(t) = \begin{pmatrix} k_1 \\ k_2 \\ \vdots \\ k_n \end{pmatrix} e^{\lambda t} \text{ and } \dot{\mathbf{X}}(t) = \lambda \begin{pmatrix} k_1 \\ k_2 \\ \vdots \\ k_n \end{pmatrix} e^{\lambda t} \text{ so that } \mathbf{AX} = \lambda \mathbf{X} \tag{5.7.13}$$

In matrix form

$$\begin{pmatrix} a_{11} & a_{12} & \cdots & a_{1n} \\ a_{21} & a_{22} & \cdots & a_{2n} \\ \vdots & \vdots & \ddots & \vdots \\ a_{n1} & a_{n2} & \cdots & a_{nn} \end{pmatrix} \begin{pmatrix} k_1 \\ k_2 \\ \vdots \\ k_n \end{pmatrix} e^{\lambda t} = \lambda \begin{pmatrix} k_1 \\ k_2 \\ \vdots \\ k_n \end{pmatrix} e^{\lambda t} \tag{5.7.14}$$

dividing by $e^{\lambda t}$ and factoring out $\mathbf{K}$ we can write the system as $(\mathbf{A} - \lambda \mathbf{I})\mathbf{K} = 0$

$$\underbrace{\begin{pmatrix} a_{11}-\lambda & a_{12} & \cdots & a_{1n} \\ a_{21} & a_{22}-\lambda & \cdots & a_{2n} \\ \vdots & \vdots & \ddots & \vdots \\ a_{n1} & a_{n2} & \cdots & a_{nn}-\lambda \end{pmatrix}}_{(\mathbf{A}-\lambda\mathbf{I})} \underbrace{\begin{pmatrix} k_1 \\ k_2 \\ \vdots \\ k_n \end{pmatrix}}_{\mathbf{K}} = 0 \tag{5.7.15}$$

Our task is now to find the eigenvalues λ and eigenvectors $\mathbf{K}$ to determine the general solution

$$\mathbf{X}(t) = c_1 \mathbf{K}_1 e^{\lambda_1 t} + c_2 \mathbf{K}_2 e^{\lambda_2 t} + \cdots + c_n \mathbf{K}_n e^{\lambda_n t} \tag{5.7.16}$$

with constants $c_1, c_2, \ldots, c_n$ determined by the initial conditions.

Example 5.7.3

Find the general solution to the system

$$\dot{x} = x$$
$$\dot{y} = -x + 2y \tag{5.7.17}$$

Solution: Writing the system in matrix form

$$\begin{pmatrix} \dot{x} \\ \dot{y} \end{pmatrix} = \begin{pmatrix} 1 & 0 \\ -1 & 2 \end{pmatrix} \begin{pmatrix} x \\ y \end{pmatrix} \text{ with } \mathbf{A} = \begin{pmatrix} 1 & 0 \\ -1 & 2 \end{pmatrix} \tag{5.7.18}$$

we compute

$$\det(\mathbf{A} - \lambda \mathbf{I}) = \begin{vmatrix} 1-\lambda & 0 \\ -1 & 2-\lambda \end{vmatrix} = (1-\lambda)(2-\lambda) = 0 \tag{5.7.19}$$

The eigenvalues and eigenvectors are

$$\lambda_1 = 1, \lambda_1 = 2, \text{ and } \mathbf{K}_1 = \begin{pmatrix} 1 \\ 1 \end{pmatrix}, \mathbf{K}_2 = \begin{pmatrix} 0 \\ 1 \end{pmatrix} \tag{5.7.20}$$

The general solution is then

$$\mathbf{X}(t) = c_1 \begin{pmatrix} 1 \\ 1 \end{pmatrix} e^t + c_2 \begin{pmatrix} 0 \\ 1 \end{pmatrix} e^{2t}$$

(5.7.21)

Example 5.7.4

Find the general solution to the system

$$\dot{x} = x + z$$
$$\dot{y} = y$$
$$\dot{z} = x + y$$

(5.7.22)

Solution: In matrix form

$$\begin{pmatrix} \dot{x} \\ \dot{y} \\ \dot{z} \end{pmatrix} = \begin{pmatrix} 1 & 0 & 1 \\ 0 & 1 & 0 \\ 1 & 1 & 0 \end{pmatrix} \begin{pmatrix} x \\ y \\ z \end{pmatrix} \text{ with } \mathbf{A} = \begin{pmatrix} 1 & 0 & 1 \\ 0 & 1 & 0 \\ 1 & 1 & 0 \end{pmatrix}$$

(5.7.23)

we compute

$$\det(\mathbf{A} - \lambda \mathbf{I}) = \begin{vmatrix} 1-\lambda & 0 & 1 \\ 0 & 1-\lambda & 0 \\ 1 & 1 & -\lambda \end{vmatrix} = (1-\lambda)(\lambda^2 - \lambda - 1) = 0$$

(5.7.24)

The resulting eigenvalues and eigenvectors are

$$\lambda_1 = 1, \lambda_2 = \frac{1+\sqrt{5}}{2}, \lambda_3 = \frac{1-\sqrt{5}}{2} \text{ with}$$

$$\mathbf{K}_1 = \begin{pmatrix} -1 \\ 1 \\ 0 \end{pmatrix}, \mathbf{K}_2 = \begin{pmatrix} 1+\sqrt{5} \\ 0 \\ 2 \end{pmatrix} \text{ and } \mathbf{K}_3 = \begin{pmatrix} 1-\sqrt{5} \\ 0 \\ 2 \end{pmatrix}$$

(5.7.25)

and the general solution is

$$\mathbf{X}(t) = c_1 \begin{pmatrix} -1 \\ 1 \\ 0 \end{pmatrix} e^t + c_2 \begin{pmatrix} 1+\sqrt{5} \\ 0 \\ 2 \end{pmatrix} e^{\frac{1+\sqrt{5}}{2}t} + c_3 \begin{pmatrix} 1-\sqrt{5} \\ 0 \\ 2 \end{pmatrix} e^{\frac{1-\sqrt{5}}{2}t}$$

(5.7.26)

Higher Order ODEs

An nth order differential equation may be expressed as a system n first order equations with appropriate variable substitutions.

Example 5.7.5

Write the second order differential equation

$$\ddot{x} + \dot{x} + x = 0 \qquad (5.7.27)$$

as two first order equations.

Solution: We make the substitution $v = \dot{x}$ and our system is

$$\dot{x} = v \qquad (5.7.28)$$
$$\dot{v} = -x - v$$

Example 5.7.6

Write the third order equation

$$\dddot{x} + 2\ddot{x} + 3\dot{x} + 4x = \sin t \qquad (5.7.29)$$

as three first order equations.

Solution: We make the substitutions

$$x_0 = x$$
$$x_1 = \dot{x}_0 \qquad (5.7.30)$$
$$x_2 = \dot{x}_1 = \ddot{x}$$

$$\underbrace{\dddot{x}}_{\dot{x}_2} + 2\underbrace{\ddot{x}}_{x_2} + 3\underbrace{\dot{x}}_{x_1} + 4\underbrace{x}_{x_0} = \sin t \qquad (5.7.31)$$

and our system becomes

$$\dot{x}_0 = x_1$$
$$\dot{x}_1 = x_2 \qquad (5.7.32)$$
$$\dot{x}_2 = -2x_2 - 3x_1 - 4x_0 + \sin t$$

Expressing higher order differential equations as first order systems is the first step in applying numerical integration schemes such as the Runge-Kutta and Euler methods.

5.7.7 Charged Particle in Electric and Magnetic Fields

The motion of a charged particle in a region with crossed electric **E** and magnetic **B** fields

$$\mathbf{E} = E_0 \hat{\mathbf{k}}$$
$$\mathbf{B} = B_0 \hat{\mathbf{j}}$$

(5.7.33)

is obtained from the Lorentz force equation $\mathbf{F} = q\mathbf{E} + q\mathbf{v} \times \mathbf{B}$ and Newton's second law with $\mathbf{F} = m\left(\ddot{x}\hat{\mathbf{i}} + \ddot{y}\hat{\mathbf{j}} + \ddot{z}\hat{\mathbf{k}} \right)$, $\mathbf{v} = \left(\dot{x}\hat{\mathbf{i}} + \dot{y}\hat{\mathbf{j}} + \dot{z}\hat{\mathbf{k}} \right)$ and

$$\mathbf{v} \times \mathbf{B} = \begin{vmatrix} \hat{\mathbf{i}} & \hat{\mathbf{j}} & \hat{\mathbf{k}} \\ \dot{x} & \dot{y} & \dot{z} \\ 0 & B_0 & 0 \end{vmatrix} = -B_0 \dot{z}\hat{\mathbf{i}} + B_0 \dot{x}\hat{\mathbf{k}}$$

(5.7.34)

Newton's second law

$$\ddot{x}\hat{\mathbf{i}} + \ddot{y}\hat{\mathbf{j}} + \ddot{z}\hat{\mathbf{k}} = -\frac{qB_0}{m}\dot{z}\hat{\mathbf{i}} + \frac{qB_0}{m}\dot{x}\hat{\mathbf{k}} + \frac{qE_0}{m}\hat{\mathbf{k}}$$

(5.7.35)

results in a system of differential equations

$$\ddot{x} = -\frac{qB_0}{m}\dot{z}$$
$$\ddot{y} = 0$$
$$\ddot{z} = \frac{qB_0}{m}\dot{x} + \frac{qE_0}{m}$$

(5.7.36)

The first and third equations are coupled while the second equation integrates immediately to

$$y(t) = y(0) + \dot{y}(0)t$$

(5.7.37)

Making the substitution $\omega = qB_0/m$

$$\ddot{x} = -\omega\dot{z}$$
$$\ddot{z} = \omega\dot{x} + \frac{qE_0}{m}$$

(5.7.38)

and taking derivatives to decouple the x and z equations

$$\dddot{x} = -\omega\ddot{z}$$
$$\dddot{z} = \omega\ddot{x}$$

(5.7.39)

and substituting for $\ddot{z}$ gives two decoupled third order equations

$$\dddot{x} = -\omega^2\dot{x} - \frac{qE_0}{m}\omega$$

(5.7.40)

$$\dddot{z} = -\omega^2\dot{z}$$

Maple Examples

An exact solution to a system of three first order, linear differential equations is calculated in the Maple worksheet below.

Key Maple commands: *diff, dsolve*

restart

Systems of Differential Equations

$Deq1 := diff(x1(t), t) = x1(t) + x2(t) + x3(t);$

$$\frac{d}{dt}x1(t) = x1(t) + x2(t) + x3(t)$$

$Deq2 := diff(x2(t), t) = 2*x1(t) + x2(t) - x3(t);$

$$\frac{d}{dt}x2(t) = 2x1(t) + x2(t) - x3(t)$$

$Deq3 := diff(x3(t), t) = -8*x1(t) - 5*x2(t) - 3*x3(t);$

$$\frac{d}{dt}x3(t) = -8x1(t) - 5x2(t) - 3x3(t)$$

$dsolve(\{Deq1, Deq2, Deq3, x1(0) = 1, x2(0) = 0, x3(0) = 5\}, \{x1(t), x2(t), x3(t)\});$

$$\left\{ x1(t) = -7e^{-2t} + 8e^{-t}, x2(t) = \frac{23}{12}e^{2t} - \frac{32}{3}e^{-t} + \frac{35}{4}e^{-2t}, \right.$$

$$\left. x3(t) = -\frac{23}{12}e^{2t} - \frac{16}{3}e^{-t} + \frac{49}{4}e^{-2t} \right\}$$

5.8 PHASE SPACE

Phase space is an abstract mathematical space with coordinate axes corresponding to the time-dependent variables in a dynamical system. Examples of time-dependent variables include position, velocity or momentum. A point in phase space represents a possible state of the system.

5.8.1 Phase Plots

The number of coordinates in phase space is determined by degrees of freedom of the system. A particle of mass m constrained to move in the x-direction has one degree of freedom and the phase space is two-dimensional. If a force F acts on the particle so that $m\ddot{x} = F$ we may plot the resulting motion in the x - v plane

$$\dot{x} = v$$
$$\dot{v} = F / m$$

(5.8.1)

The dimensionality of the phase space is equal to the number of first order equations. Position and momentum are also frequently used as phase space coordinates where the motion may be plotted in the x - p plane

$$
\begin{aligned}
\dot{x} &= v \\
\dot{p} &= F
\end{aligned}
$$

(5.8.2)

5.8.2 Noncrossing Property

For a deterministic system, the future is determined by the initial conditions. A unique trajectory in phase space will exist for a given set of initial conditions. Two trajectories in phase space corresponding to different initial conditions will not cross. If the trajectories did cross, then past and future states of the system would become indeterminant. Trajectories of higher dimensional systems may appear to cross when plotted on a two-dimensional screen.

5.8.3 Autonomous Systems

Autonomous systems such as

$$
\begin{aligned}
\dot{x} &= x + y \\
\dot{y} &= z \\
\dot{z} &= y - z
\end{aligned}
$$

(5.8.3)

have implicit time dependence in the derivatives but time does not appear explicitly. The above system has three degrees of freedom. Nonautonomous systems such as

$$
\begin{aligned}
\dot{x} &= x + y \\
\dot{y} &= z + \cos(t) \\
\dot{z} &= y - z
\end{aligned}
$$

(5.8.4)

have explicit time dependence and time counts as a degree of freedom. Thus, the above system has four degrees of freedom.

5.8.4 Phase Space Volume

Given the autonomous system

$$
\begin{aligned}
\dot{x} &= F_x(x, y, z) \\
\dot{y} &= F_y(x, y, z) \\
\dot{z} &= F_z(x, y, z)
\end{aligned}
$$

(5.8.5)

an initial condition (x_0, y_0, z_0) will result in specific values (x, y, z) at time t. We now consider the evolution of a three-dimensional block of possible initial conditions with initial phase space volume $V_0 = \Delta x_0 \Delta y_0 \Delta z_0$ evolving to $V = \Delta x \Delta y \Delta z$. Calculating the time dependence of the volume element

$$\frac{dV}{dt} = \Delta \dot{x} \Delta y \Delta z + \Delta x \Delta \dot{y} \Delta z + \Delta x \Delta y \Delta \dot{z} \tag{5.8.6}$$

Dividing by $V = \Delta x \Delta y \Delta z$ gives

$$\frac{1}{V}\frac{dV}{dt} = \frac{\Delta \dot{x}}{\Delta x} + \frac{\Delta \dot{y}}{\Delta y} + \frac{\Delta \dot{z}}{\Delta z} \tag{5.8.7}$$

Now we have that

$$\begin{aligned} \Delta \dot{x} &= F_x(x + \Delta x, y, z) - F_x(x, y, z) \\ \Delta \dot{y} &= F_y(x, y + \Delta y, z) - F_y(x, y, z) \\ \Delta \dot{z} &= F_z(x, y, z + \Delta z) - F_z(x, y, z) \end{aligned} \tag{5.8.8}$$

Thus,

$$\frac{\Delta \dot{x}}{\Delta x} = \frac{F_x(x + \Delta x, y, z) - F_x(x, y, z)}{\Delta x} \approx \frac{\partial F_x}{\partial x} \tag{5.8.9}$$

and similarly

$$\frac{\Delta \dot{y}}{\Delta y} \approx \frac{\partial F_y}{\partial y} \text{ and } \frac{\Delta \dot{z}}{\Delta z} \approx \frac{\partial F_z}{\partial z} \tag{5.8.10}$$

so that

$$\frac{1}{V}\frac{dV}{dt} = \frac{\partial F_x}{\partial x} + \frac{\partial F_y}{\partial y} + \frac{\partial F_z}{\partial z} \tag{5.8.11}$$

This expression is known as the logarithmic divergence written compactly as

$$\frac{1}{V}\frac{dV}{dt} = \nabla \cdot \mathbf{F} \tag{5.8.12}$$

If $\nabla \cdot \mathbf{F} = 0$ the system is conservative and phase space volumes are constant. Phase space volume elements contract if $\nabla \cdot \mathbf{F} < 0$ and the system is dissipative. Phase space volumes expand where $\nabla \cdot \mathbf{F} > 0$ and the system "blows up." Integrating the logarithmic divergence

$$\ln(V) = \int \nabla \cdot \mathbf{F} dt + \text{const.} \tag{5.8.13}$$

and exponentiating gives

$$V(t) = V_0 \exp\left(\int_0^t \nabla \cdot \mathbf{F} dt'\right) \tag{5.8.14}$$

where the constant is chosen so that $V(0) = V_0$. If $\nabla \cdot \mathbf{F}$ is independent of time, then

$$V(t) = V_0 \exp(\nabla \cdot \mathbf{F}t) \tag{5.8.15}$$

Example 5.8.1

Determine if the following system is conservative, dissipative or if it blows up

$$\dot{x} = x - y + z$$
$$\dot{y} = 2x + 3z \tag{5.8.16}$$
$$\dot{z} = x + y - 3z$$

Solution: We calculate

$$\nabla \cdot \mathbf{F} = \frac{\partial \dot{x}}{\partial x} + \frac{\partial \dot{y}}{\partial y} + \frac{\partial \dot{z}}{\partial z}$$

$$= \frac{\partial}{\partial x}(x - y + z) + \frac{\partial}{\partial y}(2x + 3z) + \frac{\partial}{\partial z}(x + y - 3z) \tag{5.8.17}$$

$$= 1 + 0 - 3 = -2$$

Thus, the system is dissipative and phase volumes contract as

$$V(t) = V_0 e^{-2t} \tag{5.8.18}$$

Maple Examples

Phase portraits and the direction fields of systems of ordinary differential equations are plotted in the Maple worksheet below.

Key Maple commands: D, *dfieldplot, diff, dsolve, odeplot, phaseportrait*

Maple packages: *with(DEtools): with(plots):*

restart

Phase Portrait

with(DEtools):
with(plots):
Deq1: = diff(x(t), t) = v(t) − 0.2·x(t)

$$\frac{d}{dt}x(t) = v(t) - 0.2x(t)$$

$Deq2 := diff(v(t), t) = -0.2 \cdot x(t) + 0.2 \cdot v(t)$

$$\frac{d}{dt} v(t) = -0.2 x(t) + 0.2 v(t)$$

$sol1 := dsolve(\{Deq1, Deq2, x(0) = 3, v(0) = 0\}, \{x(t), v(t)\}, numeric)$

$$\textbf{proc}(x\_rkf45) \ldots \textbf{end proc}$$

$p2 := odeplot(sol1, [x(t), v(t)], t = 0 \ldots 40)$

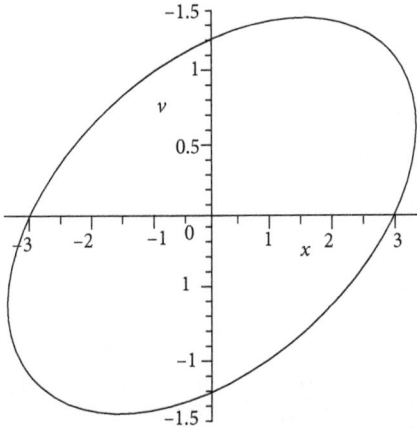

Figure 5.8.1: Phase plot of an ODE showing velocity (vertical axis) vs. position (horizontal axis).

Direction Field

$$dfieldplot\left(\frac{dy}{dx} y = \cos(x) \cdot \frac{x}{y}, y(x), x = -3 \ldots 3, y = -3 \ldots 2 \right)$$

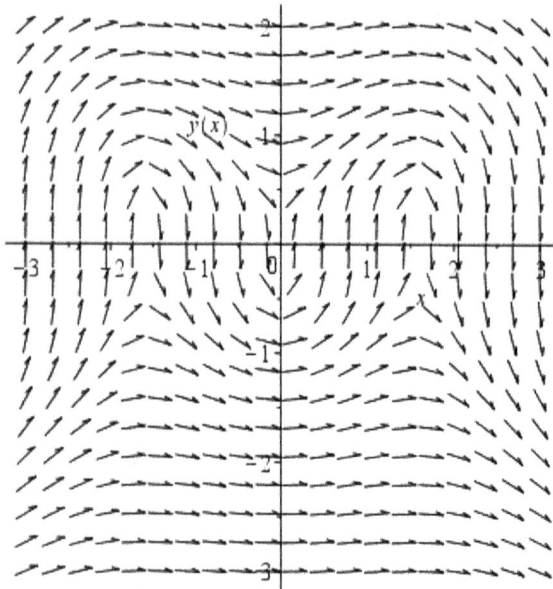

Figure 5.8.2: ODE direction field.

Direction Field and Phase Portrait

$$phaseportrait(D(y)(x) = -y(x) - x^2, y(x), x = -1 \ldots 2.5, [[y(0) = 0], [y(0) = 1], [y(0) = -1]])$$

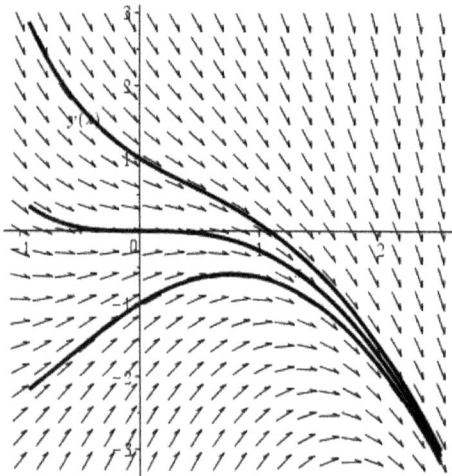

Figure 5.8.3: ODE direction field with phase plots corresponding to three different initial conditions.

5.9 NONLINEAR DIFFERENTIAL EQUATIONS

In this section, we consider nonlinear differential equations with terms that are products or nonlinear functions of dynamical variables. The predator-prey system and the nonlinear pendulum are discussed. Topics include linearization of equations near fixed points and the numerical solution of nonlinear differential equations.

5.9.1 Predator-Prey System

The predator-prey system

$$\frac{dN_1}{dt} = pN_1 - qN_1N_2$$

$$\frac{dN_2}{dt} = -rN_2 + sN_1N_2$$

(5.9.1)

describes the interaction between two species with prey population N_1 and predator population N_2 modeled as continuous variables. The nonlinear terms qN_1N_2 and sN_1N_2 represent the mortality rate of the prey and the growth rate of the predators, respectively. Mortality rate and growth rate are proportional to predator and prey populations. Interaction between the species is governed by the parameters q and s. If there is no interaction between species $q = s = 0$ and

$$\frac{dN_1}{dt} = pN_1$$

$$\frac{dN_2}{dt} = -rN_2$$

(5.9.2)

with solutions

$$N_1(t) = N_{10} \exp(pt)$$
$$N_2(t) = N_{20} \exp(-rt)$$

(5.9.3)

so that the prey population would continue to increase while the predators would die off without interaction.

5.9.2 Fixed Points

Points in the phase space where the velocities are zero are known as fixed points, or equilibria. Given the autonomous system

$$\dot{x} = F_x(x, y, z)$$
$$\dot{y} = F_y(x, y, z) \qquad (5.9.4)$$
$$\dot{z} = F_z(x, y, z)$$

the fixed points are where the derivatives are zero. The fixed points (x_0, y_0, z_0) are solutions to

$$0 = F_x(x, y, z)$$
$$0 = F_y(x, y, z)$$
$$0 = F_z(x, y, z) \qquad (5.9.5)$$

5.9.3 Linearization

An autonomous system may be linearized near a fixed point (x_0, y_0, z_0) by approximating F_x, F_y and F_z by the first terms in the Taylor expansion

$$\dot{x} = (x - x_0)\frac{\partial F_x}{\partial x}\bigg|_{(x_0, y_0, z_0)} + (y - y_0)\frac{\partial F_x}{\partial y}\bigg|_{(x_0, y_0, z_0)} + (z - z_0)\frac{\partial F_x}{\partial z}\bigg|_{(x_0, y_0, z_0)}$$

$$\dot{y} = (x - x_0)\frac{\partial F_y}{\partial x}\bigg|_{(x_0, y_0, z_0)} + (y - y_0)\frac{\partial F_y}{\partial y}\bigg|_{(x_0, y_0, z_0)} + (z - z_0)\frac{\partial F_y}{\partial z}\bigg|_{(x_0, y_0, z_0)} \qquad (5.9.6)$$

$$\dot{z} = (x - x_0)\frac{\partial F_z}{\partial x}\bigg|_{(x_0, y_0, z_0)} + (y - y_0)\frac{\partial F_z}{\partial y}\bigg|_{(x_0, y_0, z_0)} + (z - z_0)\frac{\partial F_z}{\partial z}\bigg|_{(x_0, y_0, z_0)}$$

Writing the linearized system using local coordinates $\delta x = (x - x_0)$, $\delta y = (y - y_0)$ and $\delta z = (z - z_0)$

$$\delta\dot{x} = \delta x\frac{\partial F_x}{\partial x}\bigg|_{(x_0, y_0, z_0)} + \delta y\frac{\partial F_x}{\partial y}\bigg|_{(x_0, y_0, z_0)} + \delta z\frac{\partial F_x}{\partial z}\bigg|_{(x_0, y_0, z_0)}$$

$$\delta\dot{y} = \delta x\frac{\partial F_y}{\partial x}\bigg|_{(x_0, y_0, z_0)} + \delta y\frac{\partial F_y}{\partial y}\bigg|_{(x_0, y_0, z_0)} + \delta z\frac{\partial F_y}{\partial z}\bigg|_{(x_0, y_0, z_0)} \qquad (5.9.7)$$

$$\delta\dot{z} = \delta x\frac{\partial F_z}{\partial x}\bigg|_{(x_0, y_0, z_0)} + \delta y\frac{\partial F_z}{\partial y}\bigg|_{(x_0, y_0, z_0)} + \delta z\frac{\partial F_z}{\partial z}\bigg|_{(x_0, y_0, z_0)}$$

In matrix form

$$\begin{pmatrix} \delta\dot{x} \\ \delta\dot{y} \\ \delta\dot{z} \end{pmatrix} = \begin{pmatrix} \partial F_x/\partial x & \partial F_x/\partial y & \partial F_x/\partial z \\ \partial F_y/\partial x & \partial F_y/\partial y & \partial F_y/\partial z \\ \partial F_z/\partial x & \partial F_z/\partial y & \partial F_z/\partial z \end{pmatrix}_{(x_0, y_0, z_0)} \begin{pmatrix} \delta x \\ \delta y \\ \delta z \end{pmatrix} \qquad (5.9.8)$$

The eigenvalues of the coefficient matrix will determine the type of fixed point.

5.9.4 Simple Pendulum
Example 5.9.1

The differential equation describing the motion of a simple pendulum with damping is nonlinear for large angles

$$\ddot{\theta}+\dot{\theta}+\sin(\theta)=0 \tag{5.9.9}$$

For simplicity, we set all the parameters (mass, length, damping coefficient, gravity constant) equal to one. We let $\omega=\dot{\theta}$ to obtain the system of first order equations

$$\dot{\theta}=\omega \tag{5.9.10}$$
$$\dot{\omega}=-\omega-\sin(\theta)$$

The equilibria are found from

$$0=\omega \tag{5.9.11}$$
$$0=-\omega-\sin(\theta)$$

and are $\omega_0=0$, $\theta_0=\pm n_{\text{even}}\pi$ and $\theta_0=\pm n_{\text{odd}}\pi$

Our derivative vector is

$$\mathbf{F}=\left(F_\theta,F_\omega\right)=\left(\omega,-\omega-\sin\theta\right) \tag{5.9.12}$$

Linearizing our equation near the equilibria

$$\dot{\theta}=\left(\theta-\theta_0\right)\frac{\partial F_\theta}{\partial\theta}\bigg|_{(\theta_0,\omega_0)}+\left(\omega-\omega_0\right)\frac{\partial F_\theta}{\partial\omega}\bigg|_{(\theta_0,\omega_0)} \tag{5.9.13}$$
$$\dot{\omega}=\left(\theta-\theta_0\right)\frac{\partial F_\omega}{\partial\theta}\bigg|_{(\theta_0,\omega_0)}+\left(\omega-\omega_0\right)\frac{\partial F_\omega}{\partial\omega}\bigg|_{(\theta_0,\omega_0)}$$

gives

$$\dot{\theta}=\left(\theta-\theta_0\right)\frac{\partial\omega}{\partial\theta}\bigg|_{(\theta_0,\omega_0)}+\left(\omega-\omega_0\right)\frac{\partial\omega}{\partial\omega}\bigg|_{(\theta_0,\omega_0)} \tag{5.9.14}$$
$$\dot{\omega}=\left(\theta-\theta_0\right)\frac{\partial\left(-\omega-\sin\theta\right)}{\partial\theta}\bigg|_{(\theta_0,\omega_0)}+\left(\omega-\omega_0\right)\frac{\partial\left(-\omega-\sin\theta\right)}{\partial\omega}\bigg|_{(\theta_0,\omega_0)}$$

or

$$\dot{\theta}=\left(\omega-\omega_0\right) \tag{5.9.15}$$
$$\dot{\omega}=\left(\theta-\theta_0\right)\left(-\cos\theta\right)\big|_{\theta_0}+\left(\omega-\omega_0\right)\left(-1\right)$$

We consider equilibria at the bottom and top of the motion.

Bottom: $\theta_0 = \pm n_{even}\pi$ and $\cos(\pm n_{even}\pi) = 1$

Top: $\theta_0 = \pm n_{odd}\pi$ and $\cos(\pm n_{odd}\pi) = -1$

Transforming to local coordinates we let $\delta\omega = \omega - \omega_0$ and $\delta\theta = \theta - \theta_0$. At the top $\cos\theta_0 = -1$ and

$$\dot{\delta\theta} = \delta\omega$$
$$\dot{\delta\omega} = \delta\theta - \delta\omega \qquad (5.9.16)$$

Writing the system in matrix form

$$\begin{bmatrix} \dot{\delta\theta} \\ \dot{\delta\omega} \end{bmatrix} = \underbrace{\begin{bmatrix} 0 & 1 \\ 1 & -1 \end{bmatrix}}_{A} \begin{bmatrix} \delta\theta \\ \delta\omega \end{bmatrix} \qquad (5.9.17)$$

To find the eigenvalues we take the determinant of $|A - \lambda I| = 0$

$$\det\begin{bmatrix} 0-\lambda & 1 \\ 1 & -1-\lambda \end{bmatrix} = \lambda^2 + \lambda - 1 = 0 \qquad (5.9.18)$$

This gives

$$\lambda = \frac{-1 \pm \sqrt{1+4}}{2} = \frac{-1 \pm \sqrt{5}}{2} \qquad (5.9.19)$$

and we have one positive and negative real eigenvalue corresponding to a saddle point.

At the bottom $\theta_0 = \pm n_{even}\pi$ and $\cos\theta_0 = 1$

$$\dot{\delta\theta} = \delta\omega$$
$$\dot{\delta\omega} = -\delta\theta - \delta\omega \qquad (5.9.20)$$

In matrix form

$$\begin{bmatrix} \dot{\delta\theta} \\ \dot{\delta\omega} \end{bmatrix} = \underbrace{\begin{bmatrix} 0 & 1 \\ -1 & -1 \end{bmatrix}}_{A} \begin{bmatrix} \delta\theta \\ \delta\omega \end{bmatrix} \qquad (5.9.21)$$

To find the eigenvalues take the determinant of $|A - \lambda I| = 0$

$$\det\begin{bmatrix} 0-\lambda & 1 \\ -1 & -1-\lambda \end{bmatrix} = \lambda^2 + \lambda + 1 = 0 \qquad (5.9.22)$$

$$\lambda = \frac{-1 \pm \sqrt{1-4}}{2} = \frac{-1 \pm i\sqrt{3}}{2} \qquad (5.9.23)$$

Thus, we have complex eigenvalues with negative real parts corresponding to an inward spiral.

5.9.5 Numerical Solution

The simplest scheme for numerically solving differential equations is Euler's scheme. Applied to a first order equation

$$\dot{y} = f(y,t) \qquad (5.9.24)$$

we approximate the derivative

$$\frac{y_{n+1} - y_n}{\Delta t} = f(y_n, t_n) \qquad (5.9.25)$$

For discrete time $t_n = n\Delta t$ where $n = 0,1,2\ldots$nmax and time step Δt

$$y_{n+1} = y_n + f(y_n, t_n)\Delta t \qquad (5.9.26)$$

This equation is then iteratively solved after specifying an initial condition. Higher order integration schemes, such as the Runge-Kutta method, may also be applied. The first step in applying a numerical scheme to solve a higher order differential equation is to first express the differential equation as a system of first order equations.

Maple Examples

The predator-prey model is numerically integrated in the Maple worksheet below. Euler's method is demonstrated for the numerical solution of a simple differential equation.

Key Maple commands: *display, dsolve, odeplot, plot, seq*

Maple packages: *with(plots):*

Programming: **for** loops, function statements using '$\rightarrow$'

restart

Predator-Prey Model

$p := 4$: $q := 2$: $r := 4$: $s := 5$:

$$Deq1 := \frac{d}{dt}x(t) = p \cdot x(t) - q \cdot x(t) \cdot y(t)$$

$$Deq1 := \frac{d}{dt}x(t) = 4x(t) - 2x(t)y(t)$$

$$Deq2 := \frac{d}{dt}y(t) = s \cdot x(t) \cdot y(t) - r \cdot y(t)$$

$$Deq2 := \frac{d}{dt}y(t) = 0.5x(t)y(t) - 4y(t)$$

sol1:= *dsolve*({*Deq1, Deq2, x*(0) = 5, *y*(0) = 5}, {*x*(*t*), *y*(*t*)}, *numeric*)

sol1:= **proc**(*x_rkf45*) ... **end proc**

with(*plots*):
odeplot(*sol1*, [*x*(*t*), *y*(*t*)], *t* = 0 ... 8)

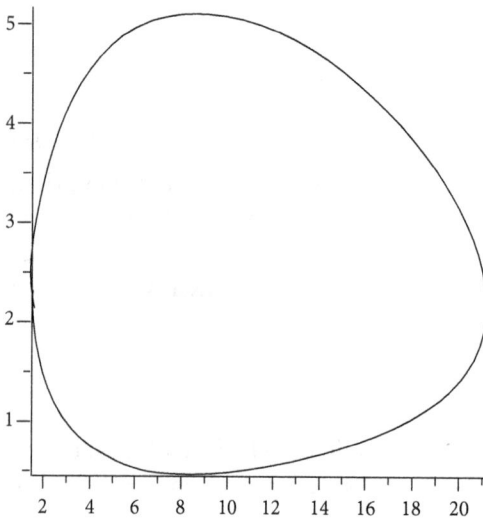

Figure 5.9.1: Phase plot of predators vs. prey.

odeplot(*sol1*, [[*t, x*(*t*)], [*t, y*(*t*)]], *t* = 0 ... 5, *legend* = [*prey, predators*], *linestyle* = [*solid, dash*])

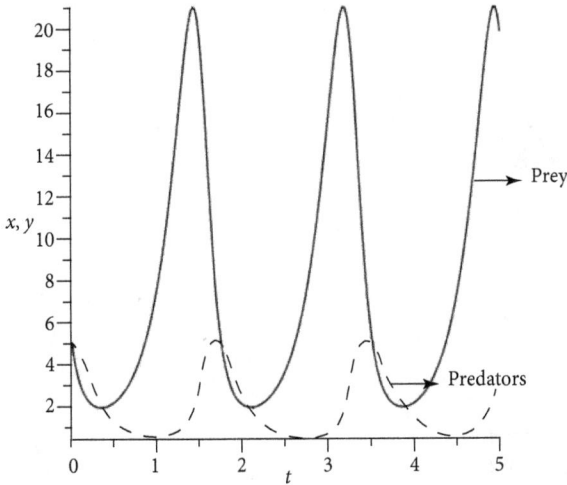

Figure 5.9.2: Time series of prey and predator populations.

Numerical Solution to Differential Equations—Euler's Method

Solve $dy/dx = y$ numerically:
Initial conditions:
$x[0] := 0 : y[0] := 1.0 :$
Step size
$h := 0.1 :$
$f := (x, y) \to y$

$$f := (x, y) \to y$$

for n **from** 1 **to** 20 **do** $x[n] := n{*}h; y[n] := y[n-1] + h{*}f(x[n-1], y[n-1])$: **od:**
$data := [seq([x[n], y[n]], n = 0 \ldots 20)]$:
$p1 := plot(data, style = point, legend = \text{“Numeric”})$:
$p2 := plot(\exp(x), x = 0 \ldots 2, legend = \text{“Exact”})$:
$display(p1, p2)$

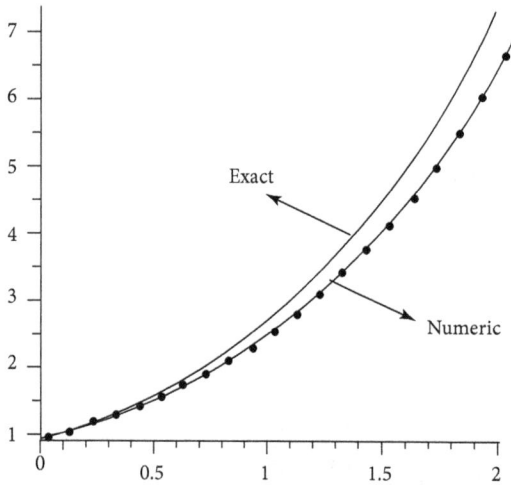

Figure 5.9.3: Comparison of Euler's method and exact solutions to a first order ODE.

5.10 MATLAB EXAMPLES

MATLAB may be used to solve single as well as systems of ordinary differential equations using the 'dsolve' command. Derivatives are represented with an uppercase D. The differential equation

$$\ddot{y} + 2\dot{y} + 3y = 0$$

would be

```
>> dsolve('D3y + 2Dy+3y=0')
```

Examples of symbolic and numerical solutions to differential equations in MATLAB are shown below.

Key MATLAB commands: *D, dsolve, global, ode45, plot3*

Section 5.2 First Order Differential Equations

```
>> clear all
>> % solve the homogeneous first order equation
>> dsolve('Dy + y=0')

ans =

C2/exp(t)
>> % first order inhomogeneous equation
>> dsolve('Dy + y = t^2')
```

```
ans =

t^2 - 2*t + C2/exp(t) + 2
```

Section 5.3 Linear, Homogeneous with Constant Coefficients

```
>> % solve the homogeneous third order differential equation
>> dsolve('D3y + D2y+D1y+y=0')
ans =
C2*cos(t) + C3*sin(t) + C4/exp(t)
>> % initial value problem
>> dsolve('D2y+D1y+y= 0','D1y(0)=0','y(0)=1')
ans =
cos((3^(1/2)*t)/2)/exp(t/2) + (3^(1/2)*sin((3^(1/2)*t)/2))/(3*exp(t/2))
```

Section 5.5 Inhomogeneous with Constant Coefficients

```
>> % solve the inhomogeneous third order differential equation
>> dsolve('D3y + D2y+D1y+y= sin(t)')

ans =

sin(t)/4 - cos(t)/4 + C2*cos(t) + C3*sin(t) + cos(t)*(cos(2*t)/8 - t/4
    + sin(2*t)/8 + 1/8) - sin(t)*(t/4 + cos(2*t)/8 - sin(2*t)/8 + 1/8)
    + C4/exp(t)
>> % initial value problem
>> dsolve('D3y + D2y+D1y+y= sin(t)','D2y(0)=0','D1y(0)=0','y(0)=1')

ans =

3/(4*exp(t)) + sin(t) + cos(t)*(cos(2*t)/8 - t/4 + sin(2*t)/8 + 1/8)
    - sin(t)*(t/4 + cos(2*t)/8 - sin(2*t)/8 + 1/8)
```

Section 5.7 Systems of Differential Equations

```
>> % solve the system of three first order equations
>> dsolve('D1y+z= 0','D1x+y=0','D1z+x=0')
ans =
    x: [1x1 sym]
    y: [1x1 sym]
    z: [1x1 sym]
>> ans.x  % returns x(t)
ans =

C3/exp(t) - (C1*exp((3*t)/2)*cos((3^(1/2)*t)/2))/(2*exp(t)) + (C2*exp((
    3*t)/2)*sin((3^(1/2)*t)/2))/(2*exp(t)) + (3^(1/2)*C2*exp((3*t)/2)*c
    os((3^(1/2)*t)/2))/(2*exp(t)) + (3^(1/2)*C1*exp((3*t)/2)*sin((3^(1/
    2)*t)/2))/(2*exp(t))
>> % initial value problem
>> dsolve('D1y+z= 0','D1x+y=0','D1z+x=0','x(0)=0','y(0)=1','z(0)=-1')
ans =
    x: [1x1 sym]
    y: [1x1 sym]
```

```
        z: [1x1 sym]
>> ans.y  % returns y(t)
ans =

(cos((3^(1/2)*t)/2)*exp(t)^(3/2))/exp(t) + (3^(1/2)*sin((3^(1/2)*t)/2)*
   exp(t)^(3/2))/(3*exp(t))
```

Section 5.9 Nonlinear Differential Equations

The following example illustrates the solution of the Rossler system consisting of three nonlinear differential equations

$$\begin{pmatrix} \dot{x} \\ \dot{y} \\ \dot{z} \end{pmatrix} = \begin{pmatrix} -y-z \\ x+ay \\ b+z(x-c) \end{pmatrix}$$

with parameters a, b and c. The M-file named 'rossler.m' is created

```
function yp=rossler(t,y)

global a b c
      yp= [-y(2)-y(3);
           y(1)+a*y(2);
           b+y(3)*(y(1)-c)];
```

The parameters, time interval $0 \leq t \leq 100$ and initial conditions $(x_0, y_0, z_0) = (0.1, 0.1, 0.1)$ are specified at the command line.

```
>> global a b c
>> a = 0.2; b=0.2; c=5.7;
>> % solve using the fourth order Runge-Kutta routine ode45
>> [t,y]=ode45(@rossler,[0, 100], [.1;.1;.1]);
>> % create a 3D plot the solution
>> plot3(y(:,1),y(:,2),y(:,3))
>> xlabel('x')
>> ylabel('y')
>> zlabel('z')
>> title('Rossler Atractor')
```

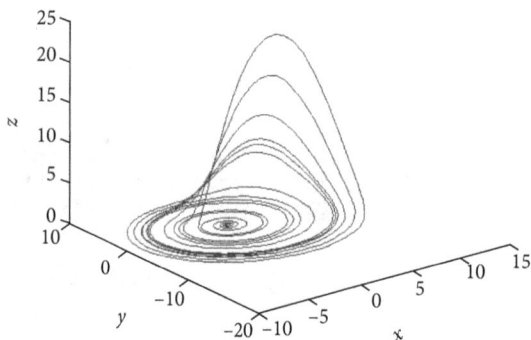

Figure 5.10.1: 3D plot of the Rossler attractor.

5.11 EXERCISES

Section 5.1 Classification of Differential Equations

1. Determine the order and degree of the following differential equations

$$\frac{d^3 y}{dx^3} - x^2 \frac{dy}{dx} = e^{-x}$$

$$\frac{d^2 r}{dt^2} - \beta \frac{dr}{dt} + \gamma r = 0$$

$$\frac{d^2 s}{dx^2} + \alpha \left(\frac{ds}{dx} \right)^2 + s \frac{ds}{dx} = 0$$

2. Solve the following differential equations by successive integration

$$\frac{d^2 y}{dt^2} = 0$$

$$\frac{d^3 y}{dx^3} = x$$

$$\frac{d^2 y}{dt^2} = e^{-t}$$

3. Solve the following differential equations by successive integration and determine the integration constants with the indicated initial conditions

$$\frac{d^4 y}{dx^4} = 0$$

where $y(0) = 1, y'(0) = 0, y''(0) = 1, y'''(0) = 1$

4. Determine if the following differential equations are exact and solve

$$y^2 dx + x dy = 0$$

$$x \frac{dy}{dx} - xy^2 = 0$$

$$\cos y dx + \sin x dy = 0$$

5. A particle of mass m is projected upward with speed v_0 from a moon with radius R and mass M

$$m \frac{d^2 r}{dt^2} = -G \frac{mM}{r^2} \text{ with } r(0) = R \text{ and } \dot{r}(0) = v_0.$$

(a) Calculate $v(r)$ making the substitution $\dfrac{d^2r}{dt^2} = \dfrac{dv}{dt} = \dfrac{dv}{dr}\dfrac{dr}{dt} = v\dfrac{dv}{dr}$ and separating variables

(b) Calculate $r(t)$

Section 5.2 First Order Differential Equations

6. The differential equation describing the variation of atmospheric pressure with height $P(y)$ is given by

$$\frac{dP}{dy} = -\frac{mg}{k_B T}P$$

Show that

$$P(y) = P_0 e^{-\frac{mgy}{k_B T}}$$

where $P(0) = P_0$ is atmospheric pressure

7. Obtain an integrating factor to solve the following differential equations

$$\frac{dy}{dx} + \frac{1}{x}y = e^{-x}$$

$$\frac{dy}{dx} + x^2 y = \sin(x)$$

$$\frac{df}{dt} + \omega f = \sin(\omega t)$$

8. Solve the following first order differential equation with the specified initial conditions

$$\frac{dy}{dx} + \frac{1}{x}y = e^{-x},\ y(0) = 0$$

$$\frac{dy}{dx} + x^2 y = \sin(x),\ y(0) = 0$$

$$\frac{df}{dt} + \omega f = \sin(\omega t),\ f(0) = 0$$

9. Solve the differential equation $m\dfrac{dv}{dt} = mg - kv^2$

10. Solve the differential equation $\dfrac{d^2 y}{dx^2} = \sqrt{1 + \left(\dfrac{dy}{dx}\right)^2}$ using the substitution

$$s = \frac{dy}{dx}$$

Section 5.3 Linear, Homogeneous with Constant Coefficients

11. Given the fourth order differential equation

$$\ddddot{x} + \dddot{x} + \ddot{x} + \dot{x} = 0$$

write the characteristic equation and the general solution to the differential equation.

12. Write the characteristic equation of the differential equation

$$\dddot{x} + 9\ddot{x} = 0$$

Solve for the roots of the characteristic equation and write the general solution (careful if repeated roots!)

13. Solve the differential equation

$$\dddot{x} - 2\ddot{x} + \dot{x} = 0$$

with initial conditions $\ddot{x}(0) = 1$, $\dot{x}(0) = 0$ and $x(0) = 0$.
Plot the solution.

14. Solve the differential equation

$$2\ddot{x} + \dot{x} - x = 0$$

with initial conditions $\dot{x}(0) = 1$ and $x(0) = 0$.
Plot the solution.

15. Find a general solution to the differential equation

$$\dddot{x} - 3\ddot{x} + 3\dot{x} - x = 0$$

16. Determine if the motion of the harmonic oscillator is underdamped, critically damped, or overdamped with parameter values below
(a) $(m, b, k) = (1, 1, 2)$
(b) $(m, b, k) = (2, 2, 1)$

Section 5.4 Linear Independence

17. Determine if $y_1(x) = e^x, y_2(x) = x$ are linearly independent or linearly dependent

18. Determine if $y_1(t) = e^t, y_2(t) = e^{-t}, y_3(t) = te^{-t}$ are linearly independent or linearly dependent

19. Determine if $y_1(x) = e^x y_2(x) = \cosh(x), y_3(x) = \sinh(x)$ are linearly independent or linearly dependent

20. Verify the linear independence of the functions t, t^2 and t^3

21. Show that the following functions are linearly dependent
$$y_1 = 3e^x, y_2 = 4e^x$$

Section 5.5 Inhomogeneous with Constant Coefficients

22. Solve the differential equation $\ddot{x} - \dot{x} + x = \cos t$

23. Solve the differential equation $\dot{x} + 2x = e^{-t}$ and plot the solution for initial conditions $\dot{x}(0) = 1$ and $x(0) = 0$

24. Solve the differential equation $\ddot{x} - x = t^2 + t + 1$

25. Solve the differential equation $\ddot{x} - \dot{x} + x = e^t$ and plot the solution for initial conditions $\dot{x}(0) = 1$ and $x(0) = 0$

26. Solve the differential equation $\ddot{x} + x = \sin t$

Section 5.6 Power Series Solutions to Differential Equations

27. Find the first four terms in the power series solution to the differential equation

$$\frac{dy}{dx} + y = x$$

28. Find a power series solution to the differential equation

$$\frac{d^2 y}{dx^2} + x^2 y = 0$$

29. Find a power series solution to the differential equation

$$\frac{dy}{dx} + x^2 y = 0$$

30. Find a power series solution to the differential equation

$$\frac{d^2 y}{dx^2} + x^2 y + \sin(x) = 0$$

31. Find a power series solution to the differential equation

$$(x^2 + 1)\frac{d^2 y}{dx^2} - x\frac{dy}{dx} + y = 0 \text{ about } x_0 = 0$$

Section 5.7 Systems of Differential Equations

32. Write the following systems of differential equation in matrix form

$$\dot{x} = 2x - y$$
$$\dot{y} = x - 2y$$

$$\dot{x} = x - y$$
$$\dot{y} = -x + 2y$$

$$\dot{x} = x - y + z$$
$$\dot{y} = x - 2y$$
$$\dot{z} = 2x + y - z$$

$$\dot{x} = 2x - y + z$$
$$\dot{y} = x - y$$
$$\dot{z} = x + y - z$$

$$\dot{x} = 2x - y + z$$
$$\dot{y} = x - y + w$$
$$\dot{z} = x + y - z$$
$$\dot{w} = -z - 2y$$

33. Find the general solutions to the systems of equations in Exercise 32.

34. Write the following third order differential equation as three first order equations

$$\dddot{x} + \ddot{x} + \dot{x} + x = 0$$

Use the substitutions $a = \ddot{x}$ and $v = \dot{x}$. Write the system in matrix form and find the general solution.

35. Write the following third order differential equation as three first order equations

$$\dddot{x} - \ddot{x} + \dot{x} - x = 0$$

Use the substitutions $a = \ddot{x}$ and $v = \dot{x}$. Write the system in matrix form and find the general solution.

36. Find the general solution to the following systems of differential equations

$$\begin{pmatrix} \dot{x} \\ \dot{y} \end{pmatrix} = \begin{pmatrix} -1 & 1 \\ 2 & 1 \end{pmatrix} \begin{pmatrix} x \\ y \end{pmatrix}$$

$$\begin{pmatrix} \dot{x} \\ \dot{y} \end{pmatrix} = \begin{pmatrix} 0 & 1 \\ 2 & 1 \end{pmatrix} \begin{pmatrix} x \\ y \end{pmatrix}$$

$$\begin{pmatrix} \dot{x} \\ \dot{y} \\ \dot{z} \end{pmatrix} = \begin{pmatrix} 0 & 1 & 0 \\ 1 & 0 & -1 \\ 0 & 0 & 1 \end{pmatrix} \begin{pmatrix} x \\ y \\ z \end{pmatrix}$$

$$\begin{pmatrix} \dot{x} \\ \dot{y} \\ \dot{z} \end{pmatrix} = \begin{pmatrix} 2 & 0 & 0 \\ 1 & 0 & -1 \\ 0 & 1 & 1 \end{pmatrix} \begin{pmatrix} x \\ y \\ z \end{pmatrix}$$

37. Determine if the solution vectors $\begin{pmatrix} 1 \\ 2 \end{pmatrix} e^{-t}$ and $\begin{pmatrix} 1 \\ -1 \end{pmatrix} e^{t}$ are linearly independent or linearly dependent.

38. Determine if the solution vectors $\begin{pmatrix} 1 \\ 0 \\ -1 \end{pmatrix} e^{-t}$, $\begin{pmatrix} 1 \\ 2 \\ 0 \end{pmatrix} e^{t}$ and $\begin{pmatrix} 1 \\ 0 \\ 1 \end{pmatrix} te^{t}$ are linearly independent or linearly dependent

39. Solve the differential equations

$$\ddot{x} = -\omega^2 \dot{x} + \frac{qE_0}{m}\omega$$

$$\ddot{z} = -\omega^2 \dot{z}$$

with initial conditions $\ddot{x}(0) = \dot{x}(0) = x(0) = 0$ and $\ddot{z}(0) = \dot{z}(0) = z(0) = 0$

Section 5.8 Phase Space

40. Create a phase plot $\dot{x}$ vs. x of the system described by the differential equation

$$\ddot{x} + \dot{x} + x = 0$$

41. Determine if the following system is conservative, dissipative or if it blows up.

$$\dot{x} = -2x + 2y$$

$$\dot{y} = 3z$$

$$\dot{z} = x + y + z$$

42. Determine if the following system is conservative, dissipative or if it blows up.

$$\dot{x} = p_x$$
$$\dot{y} = p_y$$
$$\dot{p}_x = x + y$$
$$\dot{p}_y = p_y - x$$

43. Write the following nonlinear fourth order differential equation as four first order equations

$$\dddot{x} + \ddot{x} + x^2 = 0$$

Use the substitutions

$$x_3 = \dddot{x}, \; x_2 = \ddot{x}, \; x_1 = \dot{x} \text{ and } x_0 = x$$

and calculate $\nabla \cdot \mathbf{F}$

Section 5.9 Nonlinear Differential Equations

44. Calculate the fixed points of the predator-prey system

$$\frac{dN_1}{dt} = pN_1 - qN_1N_2$$

$$\frac{dN_2}{dt} = -rN_2 + sN_1N_2$$

45. Calculate the equilibrium points of the Lorenz model

$$\dot{x} = -\sigma x + \sigma y$$
$$\dot{y} = -xz + Rx - y$$
$$\dot{z} = xy - \beta z$$

Is the system conservative or dissipative for parameter values $\sigma = 10$, $R = 28$ and $\beta = 8/3$?

46. Linearize the Lorenz model above about the equilibrium point at the origin. Determine the stability of the equilibrium point.

47. The Hénon-Heiles system models the orbits of stars in a galaxy

$$\dot{x} = p_x$$
$$\dot{y} = p_y$$
$$\dot{p}_x = -x - 2xy$$
$$\dot{p}_y = -y - x^2 + y^2$$

Calculate the equilibrium points. Is the system conservative or dissipative?

48. Linearize the Hénon-Heiles system above about all equilibrium points.

49. Write the following nonlinear fourth order differential equation as four first order equations

$$\dddot{x} + \ddot{x} + x^2 = 0$$

using the substitutions

$$x_3 = \ddot{x},\, x_2 = \ddot{x},\, x_1 = \dot{x} \text{ and } x_0 = x.$$

Integrate the system of differential equations with initial conditions $(x_{30}, x_{20}, x_{10}, x_{00}) = (0.1, 0.1, 0.1, 0.1)$ using the Euler method for 100 time steps with $\Delta t = 0.01$.

50. Integrate the Lorenz model with parameter values above with initial conditions $(x_0, y_0, z_0) = (0.1, 0.1, 0.1)$ using the Euler method for 2000 time steps with $\Delta t = 0.01$. Make a 3D plot of the solution.

Chapter **6** **SPECIAL FUNCTIONS**

Chapter Outline

6.1 DIRAC DELTA FUNCTION

The 1D Dirac delta function is defined as

$$\delta(x) = \begin{cases} 0 & x \neq 0 \\ \infty & x = 0 \end{cases} \tag{6.1.1}$$

The delta function has unit area so that

$$\int_{-\infty}^{\infty} \delta(x)\,dx = 1 \tag{6.1.2}$$

A delta function located at $x = a$ is expressed as

$$\delta(x-a) = \begin{cases} 0 & x \neq 0 \\ \infty & x = a \end{cases} \quad \text{with} \quad \int_{-\infty}^{\infty} \delta(x-a)dx = 1 \tag{6.1.3}$$

Also, we have

$$\int_{a-\varepsilon}^{a+\varepsilon} \delta(x-a)dx = 1 \tag{6.1.4}$$

where ε can be made arbitrarily small.

Because $\delta(x)$ is zero everywhere except at $x = 0$ if we multiply it by a function $f(x)$ we have that

$$f(x)\delta(x) = f(0)\delta(x) \tag{6.1.5}$$

Under the integral sign

$$\int_{-\infty}^{\infty} f(x)\delta(x)dx = f(0) \int_{-\infty}^{\infty} \delta(x)dx = f(0) \tag{6.1.6}$$

As well if we have

$$\int_{-\infty}^{\infty} f(x)\delta(x-a)dx = f(a) \int_{-\infty}^{\infty} \delta(x-a)dx = f(a) \tag{6.1.7}$$

The delta function essentially "picks out" the value $f(x)$ where the argument of the delta function is zero.

6.1.1 Representations of the Delta Function

The integral representation of the delta function

$$\delta(x-a) = \frac{1}{2\pi} \int_{-\infty}^{\infty} e^{i(x-a)t} dt \tag{6.1.8}$$

is frequently encountered as well as the analogous series form

$$\delta(x-a) = \frac{1}{2\pi} \sum_{k=-\infty}^{\infty} e^{i(x-a)k} \tag{6.1.9}$$

6.1.2 Delta Function in Higher Dimensions

The 3D delta function can be expressed as a product of 1D delta functions

$$\delta^3(\mathbf{r}) = \delta(x)\delta(y)\delta(z) \tag{6.1.10}$$

where

$$\int_{vol} \delta^3(\mathbf{r})dv = 1 \tag{6.1.11}$$

and as well

$$\int_{vol} f(\mathbf{r})\delta^3(\mathbf{r}-\mathbf{r}')dv = f(\mathbf{r}') \tag{6.1.12}$$

6.1.3 Delta Function in Spherical Coordinates

Consider a vector function F proportional to the electric field of a point charge in spherical coordinates

$$\mathbf{F}(r) = \frac{\hat{\mathbf{r}}}{r^2} \tag{6.1.13}$$

Applying Gauss's divergence theorem to this function

$$\int_{vol} \nabla \cdot \frac{\hat{\mathbf{r}}}{r^2}dv = \int_{surf} \frac{\hat{\mathbf{r}}}{r^2} \cdot \hat{\mathbf{r}}da \tag{6.1.14}$$

where the volume integral is over a sphere centered at $r=0$ and the surface integral is over the spherical surface. The right-hand side gives 4π independent of r so that the integrand on the left-hand side must be

$$\nabla \cdot \frac{\hat{\mathbf{r}}}{r^2} = 4\pi\delta(\mathbf{r}) \tag{6.1.15}$$

where

$$\int_{vol} \delta(\mathbf{r})dv = 1 \tag{6.1.16}$$

Now consider the electric field at $\mathbf{r}$ due to a point charge located at a point $\mathbf{r}'$ other than the origin. The magnitude of the displacement between $\mathbf{r}$ and $\mathbf{r}'$ is given by $|\mathbf{r}-\mathbf{r}'|$.

To get a unit vector pointing in the direction of $\mathbf{r} - \mathbf{r}'$ we replace

$$\hat{\mathbf{r}} \rightarrow \frac{\mathbf{r} - \mathbf{r}'}{|\mathbf{r} - \mathbf{r}'|} \tag{6.1.17}$$

We also replace

$$\frac{1}{r^2} \rightarrow \frac{1}{|\mathbf{r} - \mathbf{r}'|^2} \tag{6.1.18}$$

thus

$$\frac{\hat{\mathbf{r}}}{r^2} \rightarrow \frac{\mathbf{r} - \mathbf{r}'}{|\mathbf{r} - \mathbf{r}'|^3} \tag{6.1.19}$$

The divergence of this quantity is

$$\nabla \cdot \frac{\mathbf{r} - \mathbf{r}'}{|\mathbf{r} - \mathbf{r}'|^3} = 4\pi\delta(\mathbf{r} - \mathbf{r}') \tag{6.1.20}$$

To calculate the divergence of the integral

$$\nabla \cdot \int_{vol} f(\mathbf{r}') \frac{\mathbf{r} - \mathbf{r}'}{|\mathbf{r} - \mathbf{r}'|^3} dv' \tag{6.1.21}$$

we may bring the divergence inside the integral sign because the integral is over primed coordinates and ∇ is with respect to unprimed coordinates

$$\int_{vol} f(\mathbf{r}') \nabla \cdot \left(\frac{\mathbf{r} - \mathbf{r}'}{|\mathbf{r} - \mathbf{r}'|^3} \right) dv' \tag{6.1.22}$$

The divergence can now be written as a delta function and

$$\int_{vol} f(\mathbf{r}') 4\pi\delta(\mathbf{r} - \mathbf{r}') dv' = 4\pi f(\mathbf{r}) \tag{6.1.23}$$

We may take the gradient of the integral

$$\nabla \int_{vol} \frac{f(\mathbf{r}')}{|\mathbf{r} - \mathbf{r}'|} dv' = -\int_{vol} f(\mathbf{r}') \frac{\mathbf{r} - \mathbf{r}'}{|\mathbf{r} - \mathbf{r}'|^3} dv' \tag{6.1.24}$$

bringing ∇ into the integral and using

$$\nabla \frac{1}{|\mathbf{r} - \mathbf{r}'|} = -\frac{\mathbf{r} - \mathbf{r}'}{|\mathbf{r} - \mathbf{r}'|^3} \tag{6.1.25}$$

Also

$$\nabla^2 \frac{1}{|\mathbf{r}-\mathbf{r}'|} = \nabla \cdot \nabla \frac{1}{|\mathbf{r}-\mathbf{r}'|} = -\nabla \cdot \frac{\mathbf{r}-\mathbf{r}'}{|\mathbf{r}-\mathbf{r}'|^3} = -4\pi\delta(\mathbf{r}-\mathbf{r}') \qquad (6.1.26)$$

so that we may evaluate the Laplacian of the integral

$$\nabla^2 \int_{vol} \frac{f(\mathbf{r}')}{|\mathbf{r}-\mathbf{r}'|} dv' = \int_{vol} f(\mathbf{r}')\nabla^2 \left(\frac{1}{|\mathbf{r}-\mathbf{r}'|} \right) dv' = -4\pi f(\mathbf{r}) \qquad (6.1.27)$$

since ∇^2 acts on unprimed coordinates and the integral is over primed coordinates.

6.1.4 Poisson's Equation

Example 6.1.1

Obtain Poisson's equation from the integral form of the electrostatic potential $V(\mathbf{r})$ due to a charge density $\rho(\mathbf{r}')$

$$V(\mathbf{r}) = \frac{1}{4\pi\varepsilon_0} \int_{vol} \frac{\rho(\mathbf{r}')}{|\mathbf{r}-\mathbf{r}'|} dv' \qquad (6.1.28)$$

Solution: Applying the Laplacian operator to both sides

$$\nabla^2 V(\mathbf{r}) = \frac{1}{4\pi\varepsilon_0} \nabla^2 \int_{vol} \frac{\rho(\mathbf{r}')}{|\mathbf{r}-\mathbf{r}'|} dv' \qquad (6.1.29)$$

Since ∇^2 is with respect to unprimed coordinates

$$\nabla^2 V(\mathbf{r}) = \frac{1}{4\pi\varepsilon_0} \int_{vol} \rho(\mathbf{r}')\underbrace{\nabla^2 \left(\frac{1}{|\mathbf{r}-\mathbf{r}'|} \right)}_{-4\pi\delta(\mathbf{r}-\mathbf{r}')} dv' \qquad (6.1.30)$$

Thus, we obtain Poisson's equation

$$\nabla^2 V(\mathbf{r}) = -\frac{1}{\varepsilon_0}\rho(\mathbf{r}) \qquad (6.1.31)$$

6.1.5 Differential Form of Gauss's Law

Example 6.1.2

Obtain the differential form of Gauss's law from the integral expression for the electric field $E(\mathbf{r})$ due to $\rho(\mathbf{r}')$

$$\mathbf{E}(\mathbf{r}) = \frac{1}{4\pi\varepsilon_0} \int_{\text{vol}} \rho(\mathbf{r}') \frac{(\mathbf{r}-\mathbf{r}')}{|\mathbf{r}-\mathbf{r}'|^3} dv' \tag{6.1.32}$$

Solution: Applying the divergence operator to both sides

$$\nabla \cdot \mathbf{E}(r) = \frac{1}{4\pi\varepsilon_0} \nabla \cdot \int_{\text{vol}} \rho(\mathbf{r}') \frac{(\mathbf{r}-\mathbf{r}')}{|\mathbf{r}-\mathbf{r}'|^3} dv' \tag{6.1.33}$$

Since $\nabla\cdot$ acts on unprimed coordinates

$$\nabla \cdot \mathbf{E}(\mathbf{r}) = \frac{1}{4\pi\varepsilon_0} \int_{\text{vol}} \rho(\mathbf{r}') \underbrace{\nabla \cdot \frac{(\mathbf{r}-\mathbf{r}')}{|\mathbf{r}-\mathbf{r}'|^3}}_{4\pi\delta(\mathbf{r}-\mathbf{r}')} dv' \tag{6.1.34}$$

and we obtain the differential form

$$\nabla \cdot \mathbf{E}(\mathbf{r}) = \frac{1}{\varepsilon_0} \rho(\mathbf{r}) \tag{6.1.35}$$

6.1.6 Heaviside Step Function

The Heaviside step function is defined as

$$\Theta(x) = \begin{cases} 1 & x \geq 0 \\ 0 & x < 0 \end{cases} \tag{6.1.36}$$

The delta function may be expressed as the derivative of the step function

$$\delta(x) = \frac{d}{dx}\Theta(x) \tag{6.1.37}$$

Example 6.1.3

A square pulse of width $2a$ may be represented as a difference of step functions

$$\Theta(x+a) - \Theta(x-a) = \begin{cases} 1 & -a \leq x \leq a \\ 0 & |x| > a \end{cases} \tag{6.1.38}$$

Maple Examples

Integrals involving the Dirac delta function in 1D and 3D are performed in the Maple worksheet below. Examples involving the related Heaviside step function are given.

Key Maple commands: *expand, int*

Special functions: Dirac, Heaviside

restart

Dirac Delta Function

$int(\mathrm{Dirac}(x), x = \textit{-infinity} \ldots \textit{infinity})$

$$1$$

$int(\mathrm{Dirac}(x), x = \textit{-infinity} \ldots 0)$

$$\frac{1}{2}$$

$int(\mathrm{Dirac}(x), x = 1 \ldots \textit{infinity})$

$$0$$

$int(\mathrm{Dirac}(a{\cdot}x), x = \textit{-infinity} \ldots \textit{infinity})$

$$\frac{1}{|a|}$$

$Int(f(x){\cdot}\mathrm{Dirac}(x), x = \textit{-infinity} \ldots \textit{infinity}) = int(f(x){\cdot}\mathrm{Dirac}(x), x = \textit{-infinity} \ldots \textit{infinity})$

$$\int_{-\infty}^{\infty} f(x)\, Dirac(x)\, dx = f(0)$$

$Int(f(x){\cdot}\mathrm{Dirac}(a{\cdot}x), x = \textit{-infinity} \ldots \textit{infinity}) = int(f(x){\cdot}\mathrm{Dirac}(a{\cdot}x), x = \textit{-infinity} \ldots \textit{infinity})$

$$\int_{-\infty}^{\infty} f(x)\, \mathrm{Dirac}(x)\, dx = \frac{f(0)}{|a|}$$

$Int(\mathrm{Dirac}(x - x0)^{*}\exp(\textit{-}I^{*}x^{*}p), x = \textit{-infinity} \ldots \textit{infinity}) = int(\mathrm{Dirac}(x - x0)^{*}\exp(\textit{-}I^{*}x^{*}p), x = \textit{-infinity} \ldots \textit{infinity})$

$$\int_{-\infty}^{\infty} Dirac(x - x0)e^{-Ixp}\, dx = e^{-Ix0p}$$

$int(x{\cdot}\mathrm{Dirac}(2{\cdot}x - 1), x = \textit{-infinity} \ldots \textit{infinity})$

$$\frac{1}{4}$$

3D Dirac Delta Function

$\mathrm{Dirac}([x, y, z])$

$$\mathrm{Dirac}([x, y, z])$$

expand(%)

$$\text{Dirac}(x)\,\text{Dirac}(y)\,\text{Dirac}(z)$$

int(f(x, y, z)Dirac([x − a, y − b, z − c]), x = -infinity ... infinity, y = -infinity ... infinity, z = −infinity ... infinity)

$$f(a, b, c)$$

Heaviside Step Function

$$\int \text{Dirac}(x)\,dx$$

$$\text{Heaviside}(x)$$

$$\frac{d}{dx}\text{Heaviside}(x)$$

$$\text{Dirac}(x)$$

int(Heaviside(x + 1) − Heaviside(x − 1), x = -1.5 ... 1.5)

$$2$$

Int(f(x)·(Heaviside(x + 1) − Heaviside(x − 1)), x = -infinity ... infinity)
= *int*(f(x)·(Heaviside(x + 1) − Heaviside(x − 1)), x = -infinity ... infinity)

$$\int_{-\infty}^{\infty} f(x)\big(Heaviside(x+1)-Heaviside(x-1)\big)dx = \int_{-1}^{1} f(x)dx$$

6.2 ORTHOGONAL FUNCTIONS

The functions $f_n(x)$ and $f_m(x)$ are orthogonal on the interval $[a, b]$ if

$$\int_a^b w(x)f_n(x)f_m(x)dx = 0 \text{ for } m \neq n \qquad (6.2.1)$$

where $w(x)$ is a weighting function. In addition, if

$$\int_a^b w(x)f_n^2(x)dx = \int_a^b w(x)f_m^2(x)dx = 1 \qquad (6.2.2)$$

then $f_n(x)$ and $f_m(x)$ are orthonormal. For many orthogonal functions $w(x) = 1$, as we take for the examples below.

6.2.1 Expansions in Orthogonal Functions

A function $u(x)$ may be expanded as a set of functions $f_n(x)$ orthogonal on an interval $[a, b]$

$$u(x) = \sum_{n=0}^{\infty} c_n f_n(x) \tag{6.2.3}$$

To obtain the c_n we multiply both sides by $f_m(x)$ and integrate over the interval

$$\int_a^b u(x) f_m(x) dx = \int_a^b \sum_{n=0}^{\infty} c_n f_n(x) f_m(x) dx \tag{6.2.4}$$

$$\int_a^b u(x) f_m(x) dx = \sum_{n=0}^{\infty} c_n \int_a^b f_n(x) f_m(x) dx \tag{6.2.5}$$

The integral on the right is zero for all $n \neq m$ so we have $n = m$ only.

$$c_n = \int_a^b u(x) f_n(x) dx \tag{6.2.6}$$

If f_n are not normalized then there will be a constant multiplying the integral above.

6.2.2 Completeness Relation

A set of orthogonal functions $f_n(x)$ is complete on the interval $[a, b]$ if

$$\sum_{n=0}^{\infty} f_n{}^*(x') f_n(x) = \delta(x' - x) \tag{6.2.7}$$

This is also known as the closure relation. To show this we multiply both sides by $f_m{}^*(x)$ and integrate over $[a, b]$

$$\int_a^b \sum_{n=1}^{\infty} f_n{}^*(x') f_n(x) f_m{}^*(x) dx = \int_a^b f_m{}^*(x) \delta(x' - x) dx \tag{6.2.8}$$

We may factor out the sum

$$\sum_{n=1}^{\infty} f_n{}^*(x') \int_a^b f_n(x) f_m{}^*(x) dx = f_m{}^*(x') \tag{6.2.9}$$

The integral on the left may be expressed as a Kronecker delta δ_{mm}

$$\sum_{n=1}^{\infty} f_n{}^*(x')\delta_{mn} = f_m{}^*(x') \tag{6.2.10}$$

where δ_{mn} is zero if $n \neq m$.

Example 6.2.1

Show that
$P_1(x) = x$ and $P_2(x) = \dfrac{3}{2}x^2 - \dfrac{1}{2}$ are orthogonal over the interval $[-1, 1]$

Solution:
$$\int_{-1}^{1} \left(\frac{3}{2}x^3 - \frac{1}{2}x\right)dx = 0 \tag{6.2.11}$$

Note that odd functions integrated between symmetric limits are zero.

Example 6.2.2

Show that $\sin(nx)$ and $\sin(mx)$ are orthogonal over the interval $[-\pi, \pi]$ for $n \neq m$

Solution: For $n = m$ the integral $\displaystyle\int_{-\pi}^{\pi} \sin^2(nx)\,dx = \pi$ $\hspace{2cm}$ (6.2.12)

Using the identity $\sin a \sin b = \dfrac{1}{2}\left[\cos(a-b) - \cos(a+b)\right]$ for $n \neq m$

$$\int_{-\pi}^{\pi} \sin(nx)\sin(mx)\,dx = \frac{1}{2}\int_{-\pi}^{\pi}\left[\cos\left[(n-m)x\right] - \cos\left[(n+m)x\right]\right]dx$$

$$= \frac{1}{2}\left[\frac{1}{n-m}\sin\left[(n-m)x\right] - \frac{1}{n+m}\sin\left[(n+m)x\right]\right]_{-\pi}^{\pi} = 0$$

$$\tag{6.2.13}$$

6.3 LEGENDRE POLYNOMIALS

Solutions to Legendre's differential equation

$$\left(1-x^2\right)\frac{d^2 y}{dx^2} - 2x\frac{dy}{dx} + \ell(\ell+1)y = 0 \tag{6.3.1}$$

are linear combinations of Legendre polynomials of the first kind $P_\ell(x)$ and Legendre functions of the second kind $Q_\ell(x)$.

$$y = A_\ell P_\ell(x) + B_\ell Q_\ell(x) \qquad (6.3.2)$$

These polynomials and functions are encountered in potential problems in spherical coordinates with axial symmetry with argument $x = \cos\theta$. The Legendre functions of the second kind are ill behaved along the z-axis since $Q_\ell(0) \to \infty$.

The first three Legendre polynomials are given by

$$P_0(\cos\theta) = 1$$
$$P_1(\cos\theta) = \cos\theta \qquad (6.3.3)$$
$$P_2(\cos\theta) = \frac{1}{2}(3\cos^2\theta - 1)$$

6.3.1 Associated Legendre Polynomials

Solutions to the differential equation

$$(1-x^2)\frac{d^2y}{dx^2} - 2x\frac{dy}{dx} + \left[\ell(\ell+1) - \frac{m^2}{1-x^2}\right]y = 0 \qquad (6.3.4)$$

are linear combinations of associated Legendre polynomials of the first kind $P_\ell^m(x)$ and associated Legendre functions of the second kind $Q_\ell^m(x)$

$$y = A_{\ell m} P_\ell^m(x) + B_{\ell m} Q_\ell^m(x) \qquad (6.3.5)$$

where the argument $x = \cos\theta$ in spherical coordinates. They are encountered in potential problems without axial symmetry. The associated Legendre functions of the second kind become ill behaved $Q_\ell^m(0) \to \infty$ and are discarded in potential problems that include the z-axis.

6.3.2 Rodrigues' Formulas

Legendre polynomials for any $\ell = 0, 1, 2,...$ may be computed from

$$P_\ell(x) = \frac{1}{2^\ell \ell!}\frac{d^\ell}{dx^\ell}(x^2-1)^\ell \qquad (6.3.6)$$

Special values for even and odd ℓ are

$$P_{\text{odd }\ell}(0) = 0, \quad P_{\text{even }\ell}(0) = (-1)^{\ell/2}\frac{(\ell-1)!!}{\ell!!} \qquad (6.3.7)$$

$$P_\ell(1) = 1, \quad P_\ell(-x) = (-1)^\ell P_\ell(x) \qquad (6.3.8)$$

Rodrigues' formulas for the associated Legendre polynomials and functions are

$$P_\ell^m(x) = \frac{(1-x^2)^{m/2}}{2^\ell \ell!} \frac{d^{\ell+m}}{dx^{\ell+m}} (x^2-1)^\ell \tag{6.3.9}$$

$$Q_\ell^m(x) = (1-x^2)^{m/2} \frac{d^m}{dx^m} Q_\ell(x) \tag{6.3.10}$$

6.3.3 Generating Functions

The coefficients of r^ℓ in the Taylor expansion of the radical expression

$$\frac{1}{\sqrt{1-2xr+r^2}} = 1 + xr + \left(-\frac{1}{2}+\frac{3}{2}x^2\right)r^2 + \left(-\frac{3}{2}-\frac{5}{2}x^3\right)r^3$$
$$+ \left(\frac{3}{8}-\frac{15}{4}x^2+\frac{35}{8}x^4\right)r^4 + \cdots \tag{6.3.11}$$

are the Legendre polynomials $P_\ell(x)$. The radical expression is known as a generating function where

$$\frac{1}{\sqrt{1-2rx+r^2}} = \sum_{\ell=0}^{\infty} P_\ell(x) r^\ell \tag{6.3.12}$$

The generating function of the associated Legendre polynomials is

$$\frac{(2m)!\left(1-x^2\right)^{m/2} r^m}{2^m m!\left(1-2rx+r^2\right)^{m+1/2}} = \sum_{\ell=m}^{\infty} P_\ell^m(x) r^\ell \tag{6.3.13}$$

6.3.4 Orthogonality Relations

Legendre polynomials and associated Legendre polynomials are orthogonal over the interval $[-1, 1]$

$$\int_{-1}^{1} P_\ell(x) P_{\ell'}(x) dx = \frac{2}{2\ell+1} \delta_{\ell\ell'} \tag{6.3.14}$$

$$\int_{-1}^{1} P_\ell^m(x) P_{\ell'}^{m'}(x) dx = \frac{2}{2\ell+1} \frac{(\ell+m)!}{(\ell-m)!} \delta_{\ell\ell'} \delta_{mm'} \tag{6.3.15}$$

A series representation of the delta function is shown as

$$\delta(x-a) = \sum_{\ell=0}^{\infty} \left(\ell+\frac{1}{2}\right) P_\ell(x) P_\ell(a) \tag{6.3.16}$$

6.3.5 Spherical Harmonics

Associated Legendre polynomials appear in the spherical harmonics

$$Y_\ell^m(\theta,\phi) = \sqrt{\frac{2\ell+1}{4\pi}\frac{(\ell-m)!}{(\ell+m)!}}P_\ell^m(\cos\theta)e^{im\phi} \tag{6.3.17}$$

describing the angular wave function of atomic orbitals. The Y_ℓ^m are orthogonal over the surface of a sphere

$$\int_0^{2\pi}\int_0^\pi Y_\ell^m(\theta,\phi)^*Y_{\ell'}^{m'}(\theta,\phi)\sin(\theta)\,d\theta d\phi = \delta_{\ell\ell'}\delta_{mm'} \tag{6.3.18}$$

The completeness relation for spherical harmonics is given by a double sum

$$\sum_{\ell=0}^\infty\sum_{m=-\ell}^\ell Y_\ell^m(\theta',\phi')^*Y_\ell^m(\theta,\phi) = \delta(\cos\theta'-\cos\theta)\delta(\phi'-\phi) \tag{6.3.19}$$

For a given value of ℓ the sum over all possible m values is

$$\sum_{m=-\ell}^\ell Y_\ell^m(\theta,\phi)^*Y_\ell^m(\theta,\phi) = \frac{2\ell+1}{4\pi} \tag{6.3.20}$$

Maple Examples

Legendre polynomials are shown to be solutions to Legendre's differential equation and are plotted in the Maple worksheet below. Rodrigues' formula and generating functions are used to generate the Legendre and associated Legendre polynomials. Integrals involving Legendre polynomials demonstrate orthogonality. Spherical harmonics are converted into trigonometric functions and plotted in 3D. Double integrals involving spherical harmonics demonstrate orthogonality of these functions.

Key Maple commands: *Array, conjugate, diff, expand, int, plot, plot3d, subs, taylor*

Special functions: LegendreP, LegendreQ, SphericalY

Programming: for loops, function statements using '→'

restart

Legendre Polynomials

for *n* **from** 1 **to** 6 **do** *expand*(LegendreP(*n*,*x*)) **end**

$$x$$

$$-\frac{1}{2}+\frac{3x^2}{2}$$

$$\frac{5}{2}x^3-\frac{3}{2}x$$

$$\frac{3}{8}+\frac{35}{8}x^4-\frac{15}{4}x^2$$

$$\frac{63}{8}x^5-\frac{35}{4}x^3+\frac{15}{8}x$$

$$-\frac{5}{16}+\frac{231}{16}x^6-\frac{315}{16}x^4+\frac{105}{16}x^2$$

$B := Array(1 \ldots 1, 1 \ldots 5):$

Plots of Legendre Polynomials

LegendreP$(2,x)$, LegendreP$(3,x)$, LegendreP$(4,x)$, LegendreP$(5,x)$
for n **from** 1 **to** 5 **do**
$B_{1,n} := plot(\text{LegendreP}(n,x), x = -1 \ldots 1, title = [\text{LegendreP}(n,x)], color = \text{``Black''}):$
end:
B

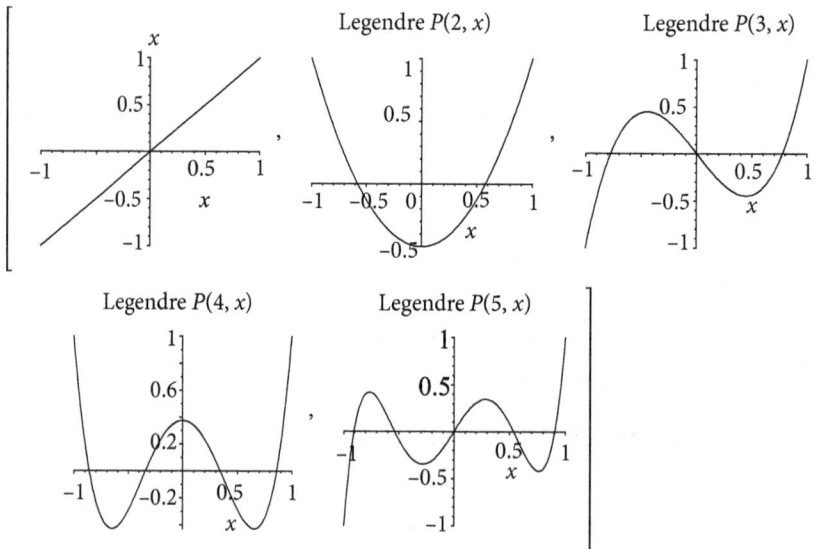

Figure 6.3.1: Plots of the first five Legendre polynomials.

for n **from** 1 **to** 3 **do** $expand(\text{LegendreQ}(n, x))$ **end**

$$\frac{x\ln(x+1)}{2}-\frac{x\ln(x-1)}{2}-1$$

$$-\frac{\ln(x+1)}{4}+\frac{\ln(x-1)}{4}+\frac{3x^2\ln(x+1)}{4}-\frac{3x^2\ln(x-1)}{4}-\frac{3x}{2}$$

$$\frac{5x^3\ln(x+1)}{4}-\frac{5x^3\ln(x-1)}{4}-\frac{3x\ln(x+1)}{4}+\frac{3x\ln(x-1)}{4}+\frac{2}{3}-\frac{5x^2}{2}$$

Legendre's Differential Equation

$LegEqn := (1-x^2)\cdot diff(y(x),x,x) - 2\cdot x\cdot diff(y(x),x) + (nu\cdot(nu+1))\cdot y(x) = 0$

$$LegEqn := (-x^2+1)\left(\frac{d^2}{dx^2}y(x)\right) - 2x\left(\frac{d}{dx}y(x)\right) + v(v+1)y(x) = 0$$

$subs(y(x) = LegendreP(nu,x) \; LegEqn)$

$$(-x^2+1)\frac{d^2}{dx^2}LegendreP(v,x) - 2x\frac{d}{dx}LegendreP(v,x)$$
$$+v(v+1)LegendreP(v,x) = 0$$

$simplify(\%)$

$$0 = 0$$

$subs(y(x) = LegendreQ(nu,x) \; LegEqn)$

$$(-x^2+1)\frac{d^2}{dx^2}LegendreQ(v,x) - 2x\frac{d}{dx}LegendreQ(v,x) + v(v+1)LegendreQ(v,x) = 0$$

$simplify(\%)$

$$0 = 0$$

$$LegEqnMod := (1-x^2)\cdot diff\left(y(x),x,x\right) - 2\cdot x\cdot diff\left(y(x),x\right)$$
$$+\left(nu\cdot(nu+1) - \frac{u^2}{1-x^2}\right)\cdot y(x) = 0$$

$$LegEqnMod := (-x^2+1)\left(\frac{d^2}{dx^2}y(x)\right) - 2x\left(\frac{d}{dx}y(x)\right) + \left(v(v+1) - \frac{u^2}{-x^2+1}\right)y(x) = 0$$

$subs(y(x) = LegendreP(nu,u,x) \; LegEqnMod)$

$$(-x^2+1)\frac{d^2}{dx^2}LegendreP(v,u,x) - 2x\frac{d}{dx}LegendreP(v,u,x)$$
$$+\left(v(v+1) - \frac{u^2}{-x^2+1}\right)LegendreP(v,u,x) = 0$$

$simplify(\%)$

$$0 = 0$$

Rodrigues' Formula: Legendre Polynomials

$$Pl := (l, x) \rightarrow \frac{1}{2^l l!} \cdot diff\left((x^2 - 1)^l, x\$l\right)$$

$$Pl := (l, x) \rightarrow \frac{\dfrac{\partial^l}{\partial x^l}\left((x^2 - 1)^l\right)}{2^l l!}$$

$Pl(3, x)$

$$x^3 + \frac{3(x^2 - 1)x}{2}$$

Rodrigues' Formula: Associated Legendre Polynomials

$$Plm := (l, m, x) \rightarrow \frac{(1 - x^2)^{\frac{m}{2}}}{2^l l!} diff\left((x^2 - 1)^l, x\$(l + m)\right)$$

$$Plm := (l, m, x) \rightarrow \frac{(-x^2 + 1)^{\frac{m}{2}}\left(\dfrac{\partial^{l+m}}{\partial x^{l+m}}\left((x^2 - 1)^l\right)\right)}{2^l l!}$$

$Plm(2, 0, x)$

$$-\frac{1}{2} + \frac{3x^2}{2}$$

$Plm(2, 1, x)$

$$3\sqrt{-x^2 + 1}\,x$$

$Plm(2, 2, x)$

$$-3x^2 + 3$$

Generating Function: Legendre Polynomials

$$taylor\left(\frac{1}{\text{sqrt}(1 - 2 \cdot r \cdot x + r^2)}, r = 0, 5\right)$$

$$1 + xr + \left(-\frac{1}{2} + \frac{3x^2}{2}\right)r^2 + \left(\frac{5}{2}x^3 - \frac{3}{2}x\right)r^3 + \left(\frac{35}{8}x^4 - \frac{15}{4}x^2 + \frac{3}{8}\right)r^4 + O(r^5)$$

Generating Function: Associated Legendre Polynomials

$$\textbf{for } m \textbf{ from } 0 \textbf{ to } 3 \textbf{ do } taylor\left(\frac{(2 \cdot m)!(1 - x^2)^{\frac{m}{2}} r^m}{2^m \cdot m!(1 - 2 \cdot r \cdot x + r^2)^{m + \frac{1}{2}}}, r = 0, 4\right) \textbf{ end}$$

$$1 + xr + \left(-\frac{1}{2} + \frac{3x^2}{2} \right) r^2 + \left(\frac{5}{2}x^3 - \frac{3}{2}x \right) r^3 + O(r^4)$$

$$\sqrt{-x^2 + 1}\,r + 3\sqrt{-x^2 + 1}\,xr^2 + \sqrt{-x^2 + 1} \left(\frac{15x^2}{2} - \frac{3}{2} \right) r^3 + O(r^4)$$

$$\left(-3x^2 + 3 \right) r^2 + 15 \left(-x^2 + 1 \right) xr^3 + O(r^4)$$

$$15 \left(-x^2 + 1 \right)^{3/2} r^3 + O(r^4)$$

Orthogonality: Legendre Polynomials

Int(LegendreP(2, *x*)·LegendreP(3, *x*), *x* = *-1* ... *1*) = *int*(LegendreP(2, *x*)·LegendreP(3, *x*), *x* = *-1* ... *1*)

$$\int_{-1}^{1} \text{LegendreP}(2, x)\, \text{LegendreP}(3, x)\,dx = 0$$

Int(LegendreP(3, *x*)·LegendreP(3, *x*), *x* = *-1* ... *1*) = *int*(LegendreP(3, *x*)·LegendreP(3, *x*), *x* = *-1* ... *1*)

$$\int_{-1}^{1} \text{LegendreP}(3, x)^2\, dx = \frac{2}{7}$$

A := *Array*(1 ... 5, 1 ... 5) :
for *i* **from** 1 **to** 5 **do**
for *j* **from** 1 **to** 5 **do**
$A_{i,j} := int(\text{LegendreP}(i, x) \cdot \text{LegendreP}(j, x), x = \text{-}1 \ldots 1)$
end
end
A

$$\begin{bmatrix} \frac{2}{3} & 0 & 0 & 0 & 0 \\ 0 & \frac{2}{5} & 0 & 0 & 0 \\ 0 & 0 & \frac{2}{7} & 0 & 0 \\ 0 & 0 & 0 & \frac{2}{9} & 0 \\ 0 & 0 & 0 & 0 & \frac{2}{11} \end{bmatrix}$$

Converting Spherical Harmonics to Trig

for *l* **from** 0 **to** 5 **do** *expand*(*convert*(SphericalY($l, 0, \theta, \phi$), LegendreP)) **end**

$$\frac{\sqrt{\dfrac{1}{\pi}}}{2}$$

$$\frac{\sqrt{3}\sqrt{\dfrac{1}{\pi}}\cos(\theta)}{2}$$

$$-\frac{\sqrt{5}\sqrt{\dfrac{1}{\pi}}}{4}+\frac{3\sqrt{5}\sqrt{\dfrac{1}{\pi}}\cos(\theta)^2}{4}$$

$$\frac{5\sqrt{7}\sqrt{\dfrac{1}{\pi}}\cos(\theta)^3}{4}-\frac{3\sqrt{7}\sqrt{\dfrac{1}{\pi}}\cos(\theta)}{4}$$

$$\frac{3\sqrt{9}\sqrt{\dfrac{1}{\pi}}}{16}+\frac{35\sqrt{9}\sqrt{\dfrac{1}{\pi}}\cos(\theta)^4}{16}-\frac{15\sqrt{9}\sqrt{\dfrac{1}{\pi}}\cos(\theta)^2}{8}$$

$$\frac{63\sqrt{11}\sqrt{\dfrac{1}{\pi}}\cos(\theta)^5}{16}-\frac{35\sqrt{11}\sqrt{\dfrac{1}{\pi}}\cos(\theta)^3}{8}+\frac{15\sqrt{11}\sqrt{\dfrac{1}{\pi}}\cos(\theta)}{16}$$

restart

Plotting Spherical Harmonics (l=2)

l: = 2;

$$l: = 2;$$

for *m* **from** 0 **to** *l* **do**
plot3d(*conjugate*(SphericalY(l, *m*, theta, phi))·SphericalY(l, *m*, theta, phi), phi = 0 ... 2·Pi, theta = 0 ... Pi, *coords* = *spherical*, *grid* = [100, 100], *title* = [*conjugate*(*Y*(*l,m*))·*Y*(*l,m*)]
end

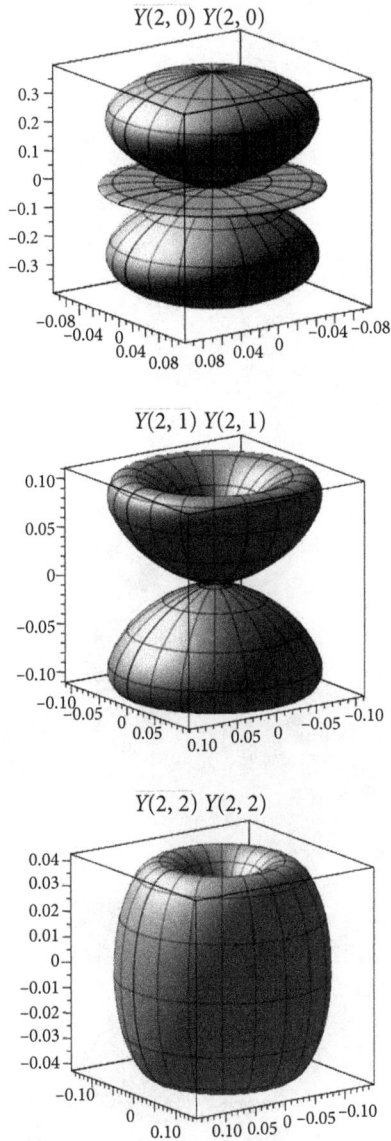

Figure 6.3.2: Plots of spherical harmonic magnitudes with l = 2.

Plotting Spherical Harmonics (m = 0)

for *l* **from** 0 **to** 3 **do**
conjugate($Y(l, 0)$)·$Y(l, 0)$;
plot3d(*conjugate*(SphericalY$(l, 0, theta, phi)$)·SphericalY$(l, 0, theta, phi)$, phi = 0
... 2·Pi, theta = 0 ... Pi, *coords = spherical, grid* = [200, 200]);
end

$\overline{Y(0,0)}\,Y(0,0)$

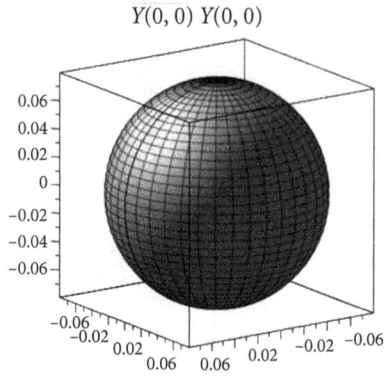

$Y(1,0)\,Y(1,0)$

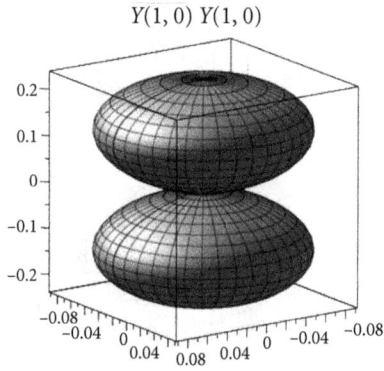

$\overline{Y(2,0)}\,Y(2,0)$

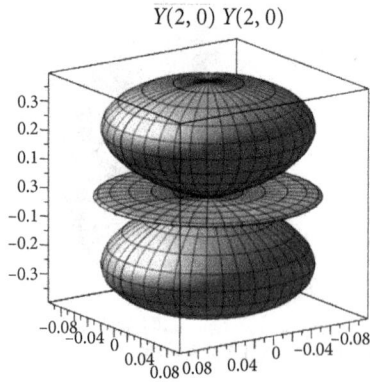

$$Y(3, 0) \ Y(3, 0)$$

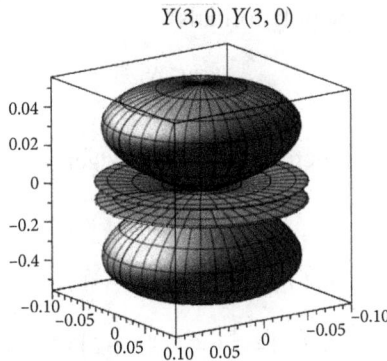

Figure 6.3.3: Plots of spherical harmonic magnitudes (m=0).

Orthogonality of Spherical Harmonics

$int(int((\text{SphericalY}(2, 2, phi, theta)) \cdot \text{SphericalY}(2, 1, phi, theta)) \cdot \sin(theta), theta = 0 \ldots \text{Pi}), phi = 0 \ldots 2 \cdot \text{Pi})$

$$0$$

$int(int((\text{SphericalY}(2, 1, phi, theta)) \cdot \text{SphericalY}(2, 1, phi, theta)) \cdot \sin(theta), theta = 0 \ldots \text{Pi}), phi = 0 \ldots 2 \cdot \text{Pi})$

$$\frac{5}{16}$$

$int(int((\text{SphericalY}(3, 1, phi, theta)) \cdot \text{SphericalY}(2, 1, phi, theta)) \cdot \sin(theta), theta = 0 \ldots \text{Pi}), phi = 0 \ldots 2 \cdot \text{Pi})$

$$0$$

6.4 LAGUERRE POLYNOMIALS

Solutions to Laguerre's differential equation

$$x\frac{d^2 y}{dx^2} + (1-x)\frac{dy}{dx} + ny = 0 \tag{6.4.1}$$

are linear combinations of Laguerre polynomials

$$y = L_n(x) \tag{6.4.2}$$

Solutions to

$$x\frac{\partial^2 y}{\partial x^2} + (m+1-x)\frac{\partial}{\partial x}y + ny = 0 \tag{6.4.3}$$

are the associated Laguerre polynomials

$$y = L_n^m(x)$$

(6.4.4)

found in the radial wavefunctions of the hydrogen atom. The first three Laguerre polynomials are

$$L_0(x) = 1$$
$$L_1(x) = 1 - x$$
$$L_2(x) = \frac{1}{2}(x^2 - 4x + 2)$$

(6.4.5)

6.4.1 Rodrigues's Formula

Laguerre polynomials for any $n = 0, 1, 2...$ may be computed from

$$L_n(x) = \frac{1}{n!} e^x \frac{d^n}{dx^n} (x^n e^{-x})$$

(6.4.6)

while the associated Laguerre polynomials are

$$L_n^m(x) = \frac{1}{n!} e^x x^{-m} \frac{d^n}{dx^n} (x^{n+m} e^{-x})$$

(6.4.7)

6.4.2 Generating Function

The coefficients of r^n in the Taylor expansion

$$\frac{1}{1-r} \exp\left(\frac{-rx}{1-r}\right) = \sum_{n=0}^{\infty} \frac{r^n}{n!} L_n(x)$$

(6.4.8)

are proportional to $L_n(x)$ while the $L_n^m(x)$ are generated by the Taylor expansion

$$\frac{1}{(1-r)^{m+1}} \exp\left(\frac{-rx}{1-r}\right) = \sum_{n=0}^{\infty} \frac{r^n}{n!} L_n^m(x)$$

(6.4.9)

6.4.3 Orthogonality Relations

$L_n(x)$ and $L_n^m(x)$ are orthogonal over the interval $[0, \infty)$

$$\int_0^{\infty} e^{-x} L_n(x) L_{n'}(x) dx = \delta_{nn'}(n!)^2$$

(6.4.10)

with weighting function $w(x) = \exp(-x)$ and

$$\int_0^\infty e^{-x} x^m L_n^m(x) L_{n'}^m(x)\,dx = \delta_{nn'} \frac{(n!)^3}{(n-m)!} \tag{6.4.11}$$

with weighting function $w(x) = \exp(-x)x^m$. An important relation is expressed as

$$\int_0^\infty e^{-x} x^{m+1} \left[L_n^m(x) \right]^2 dx = \frac{(n+m)!}{n!}(2n+m+1) \tag{6.4.12}$$

A series representation of the delta function is given by

$$\delta(x-a) = e^{-(x+a)/2} \sum_{n=0}^\infty L_n(x) L_n(a) \tag{6.4.13}$$

Maple Examples

Laguerre polynomials are shown to be solutions to Laguerre's differential equation and are plotted in the Maple worksheet below. Rodrigues's formula and generating functions are used to generate the Laguerre polynomials. Integrals involving Laguerre polynomials demonstrate orthogonality.

Key Maple commands: *Array, diff, expand, int, plot, simplify, subs, taylor*
Special functions: LaguerreL
Programming: for loops, function statements using '$\rightarrow$'

restart

Laguerre Polynomials

for *n* **from** 0 **to** 6 **do** *simplify*(LaguerreL(*n, x*)) **end**

$$1$$

$$1 - x$$

$$1 - 2x + \frac{1}{2}x^2$$

$$1 - 3x + \frac{3}{2}x^2 - \frac{1}{6}x^3$$

$$1 - 4x + 3x^2 - \frac{2}{3}x^3 + \frac{1}{24}x^4$$

$$1 - 5x + 5x^2 - \frac{5}{3}x^3 + \frac{5}{24}x^4 - \frac{1}{120}x^5$$

$$1 - 6x + \frac{15}{2}x^2 - \frac{10}{3}x^3 + \frac{5}{8}x^4 - \frac{1}{20}x^5 + \frac{1}{720}x^6$$

Plots of Laguerre Polynomials

$B := Array(1 \dots 1, 1 \dots 5):$
for m **from** 1 **to** 5 **do**
$B_{1,m} := plot(simplify(\text{LaguerreL}(m, x)), x = 0 \dots 6, title = [\text{LaguerreL}(m, x)], color =$ "Black"): **end:**
B

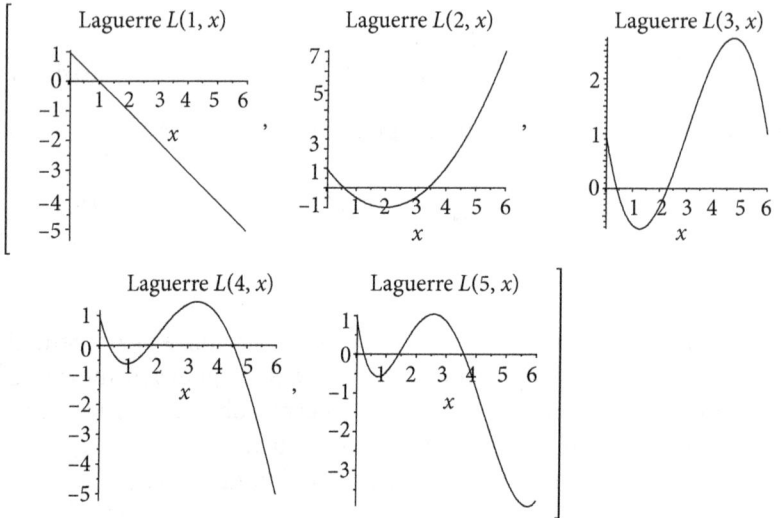

Figure 6.4.1: Plots of the first five Laguerre polynomials.

$restart$
$n := 1$

$$n := 1$$

Laguerre's Differential Equation

$LagEqn := x \cdot diff(y(x), x, x) + (1 - x) \cdot diff(y(x), x) + n \cdot y(x) = 0$

$$LagEqn := x \left(\frac{d^2}{dx^2} y(x) \right) + (1 - x) \left(\frac{d}{dx} y(x) \right) + y(x) = 0$$

$subs(y(x) = \text{LaguerreL}(n, x), LagEqn)$

$$x \frac{d^2}{dx^2} \text{LaguerreL}(1, x) + (1 - x) \frac{d}{dx} \text{LaguerreL}(1, x) + \text{LaguerreL}(1, x) = 0$$

$simplify(\%)$

$$0 = 0$$

Rodrigues' Formula

$$Ln := (n, x) \rightarrow \left(\frac{\exp(x)}{n!} \cdot diff\left(x^n \cdot \exp(-x), x\$n\right) \right)$$

$$Ln := (n, x) \rightarrow \frac{e^x \left(\dfrac{\partial^n}{\partial x^n}(x^n e^{-x}) \right)}{n!}$$

$expand(Ln(3, x))$

$$-\frac{1}{6}x^3 + \frac{3}{2}x^2 - 3x + 1$$

Generating Function

$$taylor\left(\frac{1}{1-r} \cdot \exp\left(\frac{r \cdot x}{1-r}\right), r = 0, 3 \right)$$

$$1 + (1-x)r + \left(\frac{1}{2}x^2 - 2x + 1\right)r^2 + O(r^3)$$

Orthogonality

$w := (x) \rightarrow \exp(-x)$
$Int(w(x) \cdot (\text{LaguerreL}(2, x) \cdot \text{LaguerreL}(3, x), x = 0 \ldots \text{infinity}) = int(\exp(-x) \cdot$
$(\text{LaguerreL}(2, x) \cdot \text{LaguerreL}(3, x), x = 0 \ldots \text{infinity})$

$$\int_0^\infty e^{-x} \text{LaguerreL}(2, x)\, \text{LaguerreL}(3, x) dx = 0$$

$Int(w(x) \cdot (\text{LaguerreL}(3, x) \cdot \text{LaguerreL}(3, x), x = 0 \ldots \text{infinity}) = int(\exp(-x) \cdot$
$(\text{LaguerreL}(3, x) \cdot \text{LaguerreL}(3, x), x = 0 \ldots \text{infinity})$

$$\int_0^\infty e^{-x} \text{LaguerreL}(3, x)^2\, dx = 1$$

$A := Array(1 \ldots 5, 1 \ldots 5):$
for i **from** 1 **to** 5 **do**
for j **from** 1 **to** 5 **do**
$A_{i,j} := int(w(x) \cdot \text{LaguerreL}(i, x) \cdot \text{LaguerreL}(j, x), x = 0 \ldots \text{infinity})$
end
end
A

$$\begin{bmatrix} 1 & 0 & 0 & 0 & 0 \\ 0 & 1 & 0 & 0 & 0 \\ 0 & 0 & 1 & 0 & 0 \\ 0 & 0 & 0 & 1 & 0 \\ 0 & 0 & 0 & 0 & 1 \end{bmatrix}$$

6.5 HERMITE POLYNOMIALS

The solution to Hermite's differential equation

$$\frac{d^2 y}{dx^2} - 2x\frac{dy}{dx} + 2ny = 0 \tag{6.5.1}$$

is a linear combination of Hermite polynomials,

$$y = H_n(x) \tag{6.5.2}$$

These polynomials are encountered in quantum mechanical problems such as the simple harmonic oscillator.

The first three Hermite polynomials are

$$\begin{aligned} H_0(x) &= 1 \\ H_1(x) &= 2x \\ H_2(x) &= 4x^2 - 2 \end{aligned} \tag{6.5.3}$$

6.5.1 Rodrigues' Formula

The following Rodrigues formula may be used to obtain $H_n(x)$

$$H_n(x) = (-1)^n e^{x^2} \frac{d^n}{dx^n}\left(e^{-x^2}\right) \tag{6.5.4}$$

for any $n = 0, 1, 2...$. Useful relations for even and odd n include

$$H_{\text{odd } n}(0) = 0 \tag{6.5.5}$$

$$H_{\text{even } n}(0) = (-1)^{n/2}(2)^{n/2}(n-1)!!$$

$$H_n(-x) = (-1)^n H_n(x) \tag{6.5.6}$$

6.5.2 Generating Function

Coefficients of r^n in the Taylor expansion of the exponential function

$$e^{2rx-r^2} = \sum_{n=0}^{\infty} \frac{H_n(x) r^n}{n!} \tag{6.5.7}$$

are proportional to $H_n(x)$.

6.5.4 Orthogonality

The $H_n(x)$ are orthogonal over the interval $(-\infty, \infty)$

$$\int_{-\infty}^{\infty} e^{-x^2} H_n(x) H_{n'}(x) dx = 2^n n! \sqrt{\pi} \delta_{nn'} \qquad (6.5.8)$$

with weighting function $w(x) = \exp(-x^2)$.

A series representation of the delta function is shown as

$$\delta(x-a) = \frac{e^{-(x^2+a^2)/2}}{\sqrt{\pi}} \sum_{n=0}^{\infty} \frac{1}{2^n n!} H_n(x) H_n(a) \qquad (6.5.9)$$

Maple Examples

Hermite polynomials are shown to be solutions to Hermite's differential equation and are plotted in the Maple worksheet below. Rodrigues' formula and generating functions are used to generate the Hermite polynomials. Integrals involving Hermite polynomials demonstrate orthogonality.

Key Maple commands: *Array, diff, expand, int, plot, simplify, subs, taylor*

Maple packages: *with(orthopoly):*

Special functions: HermiteH

Programming: for loops, function statements using '$\rightarrow$'

restart

Hermite Polynomials

with(orthopoly)

for *m* **from** 1 **to** 10 **do** *expand*(HermiteH(*m*, *x*)) **end**

$$2x$$
$$4x^2 - 2$$
$$8x^3 - 12x$$
$$16x^4 - 48x^2 + 12$$
$$32x^5 - 160x^3 + 120x$$
$$64x^6 - 480x^4 + 720x^2 - 120$$
$$128x^7 - 1344x^5 + 3360x^3 - 1680x$$
$$256x^8 - 3584x^6 + 13440x^4 - 13440x^2 + 1680$$
$$512x^9 - 9216x^7 + 48384x^5 - 80640x^3 + 30240x$$
$$1024x^{10} - 23040x^8 + 161280x^6 - 403200x^4 + 302400x^2 - 30240$$

Plots of Hermite Polynomials

$B := Array(1 \ldots 1, 1 \ldots 5):$

for m **from** 1 **to** 5 **do**
$B_{1,m} := plot(\text{HermiteH}(m,x), x=-2\ldots2, title=[\text{HermiteH}(m,x)], color=\text{"Black"}):$
end:
B

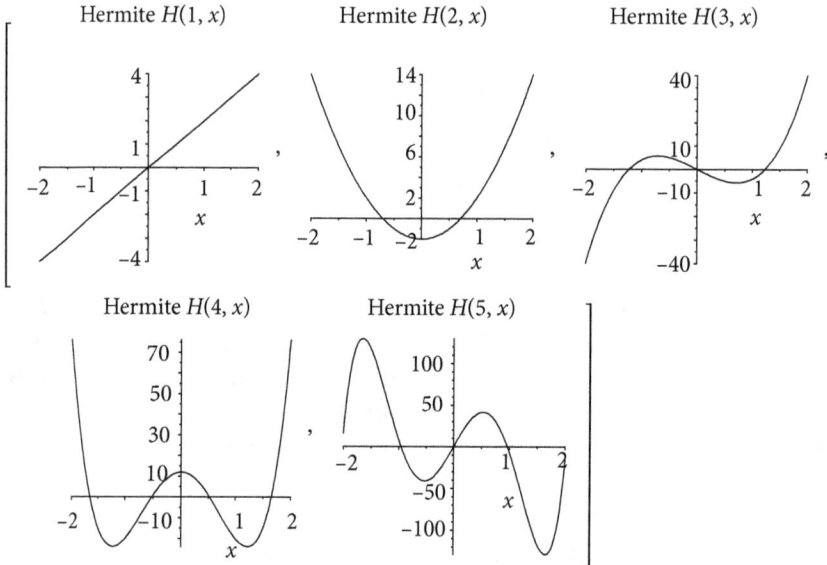

Figure 6.5.1: Plots of the first five Hermite polynomials.

Hermite's Differential Equation

$m := 3$

$$m := 3$$

$dEqn := diff(y(x), x, x) - 2 \cdot x \cdot diff(y(x), x) + 2 \cdot m \cdot y(x) = 0$

$$dEqn := \frac{d^2}{dx^2}y(x) - 2x\left(\frac{d}{dx}y(x)\right) + 6y(x) = 0$$

$subs(y(x) = \text{HermiteH}(m, x), dEqn)$

$$\frac{d^2}{dx^2}\text{HermiteH}(3,x) - 2x\frac{d}{dx}\text{HermiteH}(3,x) + 6\text{HermiteH}(3,x) = 0$$

$expand(simplify(\%))$

$$0 = 0$$

Rodrigues' Formula

$Hn := (n,x) \rightarrow ((-1)^n \exp(x^2) \cdot \text{diff}(\exp(x^2), x\$n))$

$$Hn := (n,x) \rightarrow (-1)^n e^{x^2} \left(\frac{\partial^n}{\partial x^n} e^{-x^2} \right)$$

$expand(Hn(3,x))$

$$8x^3 - 12 x$$

Generating Function

$taylor(\exp(2 \cdot r \cdot x - r^2), r = 0, 5)$

$$1 + 2xr + (2x^2 - 1)r^2 + \left(\frac{4}{3}x^3 - 2x \right)r^3 + \left(-2x^2 + \frac{1}{2} + \frac{2}{3}x^4 \right)r^4 + O(r^5)$$

Orthogonality Relations

$w := (x) \rightarrow \exp(-x^2)$

$$w := x \rightarrow e^{-x^2}$$

$Int(w(x) \cdot (\text{HermiteH}(1, \ x) \cdot \text{HermiteH}(3, \ x), \ x \ = \ -infinity \ ... \ infinity) \ = \ int(w(x) \cdot \text{HermiteH}(1, x) \cdot \text{HermiteH}(3, x), x = -infinity \ ... \ infinity)$

$$\int_{-\infty}^{\infty} e^{-x^2} \text{HermiteH}(1,x) \ \text{HermiteH}(3,x) dx = 0$$

$int(w(x) \cdot \text{HermiteH}(1, \ x) \cdot \text{HermiteH}(3, x), x = -infinity \ ... \ infinity)$

$$0$$

$A := Array(1 ... 5, 1 ... 5) :$
for i **from** 1 **to** 5 **do**
for j **from** 1 **to** 5 **do**
$A_{i,j} := int(w(x) \cdot \text{HermiteH}(i, x) \cdot \text{HermiteH}(j, x), x = -infinity \ ... \ infinity)$
end
end
A

$$\begin{bmatrix} 2\sqrt{\pi} & 0 & 0 & 0 & 0 \\ 0 & 8\sqrt{\pi} & 0 & 0 & 0 \\ 0 & 0 & 48\sqrt{\pi} & 0 & 0 \\ 0 & 0 & 0 & 384\sqrt{\pi} & 0 \\ 0 & 0 & 0 & 0 & 3840\sqrt{\pi} \end{bmatrix}$$

6.6 BESSEL FUNCTIONS

The solution to Bessel's differential equation

$$x^2 \frac{d^2 y}{dx^2} + x \frac{dy}{dx} + (x^2 - n^2)y = 0 \qquad (6.6.1)$$

is a linear combination of Bessel functions of the first kind $J_n(x)$ and the second kind $N_n(x)$

$$y = A_n J_n(x) + B_n N_n(x) \qquad (6.6.2)$$

These functions have applications in the vibrations of circular membranes, heat conduction in circular plates and electromagnetic problems in cylindrical coordinates. The Bessel functions of the second kind are ill behaved along the z-axis where $N_n(0) \to \infty$ and are discarded in regions including the z-axis.

6.6.1 Modified Bessel Functions

The solution to Bessel's modified differential equation

$$x^2 \frac{d^2 y}{dx^2} + x \frac{dy}{dx} - (x^2 + n^2)y = 0 \qquad (6.6.3)$$

is a linear combination of modified Bessel functions of the first kind $I_n(x)$ and the second kind $K_n(x)$.

$$y = A_n I_n(x) + B_n K_n(x) \qquad (6.6.4)$$

The Bessel functions of the second kind are ill behaved along the z-axis where $K_n(0) \to \infty$ and are discarded in regions including the z-axis.

6.6.2 Generating Function

The coefficients of r^n in the Taylor expansion of the generating function

$$e^{x(r-1/r)/2} = \sum_{n=-\infty}^{\infty} J_n(x)r^n \qquad (6.6.5)$$

are the Bessel functions $J_n(x)$. A similar generating function for the modified Bessel functions $I_n(x)$ is given by

$$e^{x(r+1/r)/2} = \sum_{n=-\infty}^{\infty} I_n(x)r^n \qquad (6.6.6)$$

6.6.3 Spherical Bessel Functions

Solutions to the radial part of the Helmholtz equation in spherical coordinates

$$x^2 \frac{d^2 y}{dx^2} + 2x \frac{dy}{dx} + \left[x^2 - \ell(\ell+1)\right] y = 0 \tag{6.6.7}$$

are linear combinations of spherical Bessel functions of the first kind $j_\ell(x)$ and of the second kind $n_\ell(x)$

$$y = A_\ell j_\ell(x) + B_\ell n_\ell(x) \tag{6.6.8}$$

These are related to the cylindrical Bessel functions

$$j_\ell(x) = \sqrt{\frac{\pi}{2x}} J_{\ell+1/2}(x) \tag{6.6.9}$$

$$n_\ell(x) = \sqrt{\frac{\pi}{2x}} N_{\ell+1/2}(x) \tag{6.6.10}$$

6.6.4 Rayleigh Formulas

Given the $\ell = 0$ forms

$$j_0(x) = \frac{\sin x}{x} \text{ and } n_0(x) = -\frac{\cos x}{x} \tag{6.6.11}$$

subsequent spherical Bessel functions may be obtained using the Rayleigh formulas

$$j_\ell(x) = (-x)^\ell \left(\frac{1}{x} \frac{d}{dx}\right)^\ell \left(\frac{\sin x}{x}\right) \tag{6.6.12}$$

$$n_\ell(x) = (-x)^\ell \left(\frac{1}{x} \frac{d}{dx}\right)^\ell \left(-\frac{\cos x}{x}\right) \tag{6.6.13}$$

6.6.5 Generating Functions

Spherical Bessel functions are also generated by Taylor expansions of the following functions

$$\frac{1}{x} \sin\left(\sqrt{x^2 - 2rx}\right) = \sum_{\ell=0}^{\infty} \frac{(-r)^\ell}{\ell!} j_{\ell-1}(x) \tag{6.6.14}$$

$$\frac{1}{x} \sin\left(\sqrt{x^2 + 2rx}\right) = \sum_{\ell=0}^{\infty} \frac{(-r)^\ell}{n!} n_{\ell-1}(x) \tag{6.6.15}$$

6.6.6 Useful Relations

In cylindrical coordinates, the delta function is represented by

$$\frac{1}{r}\delta(r-r') = \int_0^\infty kJ_m(kr)J_m(kr')dk \qquad (6.6.16)$$

The inverse magnitude of the position vector is

$$\frac{1}{|\mathbf{r}-\mathbf{r}'|} = \sum_{m=-\infty}^{\infty} \int_0^\infty ke^{im(\phi-\phi')}J_m(kr)J_m(kr')e^{-k|z-z'|}dk \qquad (6.6.17)$$

Integral representation of Bessel functions include

$$J_0(x) = \frac{1}{\pi}\int_0^\pi \cos(x\sin\theta)d\theta \qquad (6.6.18)$$

$$J_n(x) = \frac{1}{\pi}\int_0^\pi \cos(n\theta - x\sin\theta)d\theta \quad n=1,2,3\ldots \qquad (6.6.19)$$

Maple Examples

Bessel and Hankel functions are shown to be solutions to their respective differential equations and are plotted in the Maple worksheet below. The zeros of Bessel functions are calculated. Spherical Bessel functions are generated using Rayleigh formulas and are plotted together.

Key Maple commands: *diff, display, dsolve, evalf, expand, plot, pointplot, simplify, subs*

Maple packages: *with(orthopoly): with(plots):*

Special functions: BesselJ, BesselJZeros, BesselY, BesselI, BesselK, HankelH1, HankelH2

restart

with(orthopoly)

$$[G, H, L, P, T, U]$$

Bessel Functions

BesselEqn:= x^2diff($y(x), x, x$) + x·diff($y(x), x$) + $(x^2 - v^2)$·$y(x)$ = 0

$$BesselEqn := x^2\left(\frac{d^2}{dx^2}y(x)\right) + x\left(\frac{d}{dx}y(x)\right) + \left(-v^2 + x^2\right)y(x) = 0$$

subs($y(x)$ = BesselJ(nu, x), BesselEqn)

$$x^2 \frac{d^2}{dx^2} \mathrm{BesselJ}(v,x) + x \frac{d}{dx} \mathrm{BesselJ}(v,x) + \left(-v^2 + x^2\right) \mathrm{BesselJ}(v,x) = 0$$

simplify(%)

$$0 = 0$$

subs$(y(x) = \mathrm{BesselY}(\mathrm{nu}, x), \mathit{BesselEqn})$

$$x^2 \frac{d^2}{dx^2} \mathrm{BesselY}(v,x) + x \frac{d}{dx} \mathrm{BesselY}(v,x) + \left(-v^2 + x^2\right) \mathrm{BesselY}(v,x) = 0$$

simplify(%)

$$0 = 0$$

Modified Bessel Functions

BesselEqnMod$:= x^2 \mathrm{diff}(y(x), x, x) + x \cdot \mathrm{diff}(y(x), x) - (x^2 + y^2) \cdot y(x) = 0$

$$\mathit{BesselEqnMod} := x^2 \left(\frac{d^2}{dx^2} y(x) \right) + x \left(\frac{d}{dx} y(x) \right) - \left(v^2 + x^2\right) y(x) = 0$$

subs$(y(x) = \mathrm{BesselI}(\mathrm{nu}, x), \mathit{BesselEqnMod})$

$$x^2 \frac{d^2}{dx^2} \mathrm{BesselI}(v,x) + x \frac{d}{dx} \mathrm{BesselI}(v,x) - \left(v^2 + x^2\right) \mathrm{BesselI}(v,x) = 0$$

simplify(%)

$$0 = 0$$

subs$(y(x) = \mathrm{BesselK}(\mathrm{nu}, x), \mathit{BesselEqnMod})$

$$x^2 \frac{d^2}{dx^2} \mathrm{BesselK}(v,x) + x \frac{d}{dx} \mathrm{BesselK}(v,x) - \left(v^2 + x^2\right) \mathrm{BesselK}(v,x) = 0$$

simplify(%)

$$0 = 0$$

Hankel Functions

subs$(y(x) = \mathrm{HankelH1}(\mathrm{nu}, x), \mathit{BesselEqn})$

$$x^2 \frac{d^2}{dx^2} \mathrm{HankelH1}(v,x) + x \frac{d}{dx} \mathrm{HankelH1}(v,x) + \left(-v^2 + x^2\right) \mathrm{HankelH1}(v,x) = 0$$

simplify(%)

$$0 = 0$$

$subs(y(x) = \text{HankelH2}(\text{nu}, x), BesselEqn)$

$$x^2 \frac{d^2}{dx^2} \text{HankelH2}(v, x) + x \frac{d}{dx} \text{HankelH2}(v, x) + \left(-v^2 + x^2\right) \text{HankelH2}(v, x) = 0$$

$plot([\text{seq}(\text{BesselJ}(\text{nu}, x), \text{nu} = 0 \dots 3)], x = 0 \dots 20, legend = [J0, J1, J2, J3], linestyle = [solid, dash, dashdot, dot]):$

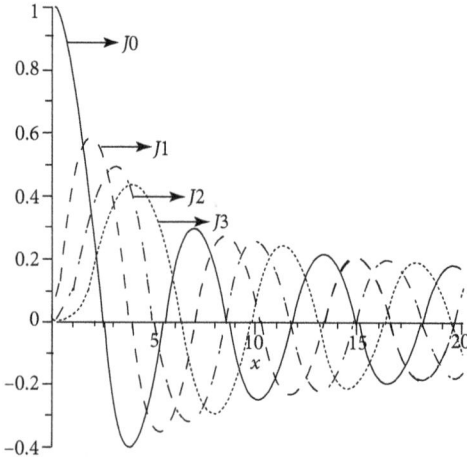

Figure 6.6.1: Plots of the first four Bessel functions of the first kind.

Zeros of Bessel Functions

$\text{BesselJZeros}(3, 1 \dots 5)$

$$\text{BesselJZeros}(3, 1 \dots 5)$$

$evalf(\%)$

6.380161896. 9.761023130, 13.01520072, 16.22346616, 19.40941523

$with(plots):$

$p1 := pointplot(\{\text{seq}([evalf(\text{BesselJZeros}(3, n)), 0], n = 0 \dots 5)\}):$

$p2 := plot(\text{BesselJ}(3, x), x = 0 \dots 20):$

$display(p1, p2)$

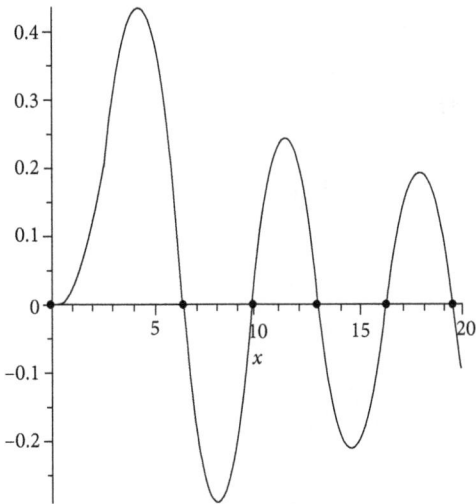

Figure 6.6.2: Plots of the Bessel functions J_3 and its zeros.

Spherical Bessel Functions

$sphBessEqn := r^2 \text{diff}(y(r), r, r) + 2 \cdot r \cdot \text{diff}(y(r), r) + (r^2 - l(l+1)) \cdot y(r) = 0$

$$sphBessEqn := r^2 \left(\frac{d^2}{dr^2} y(r) \right) + 2r \left(\frac{d}{dr} y(r) \right) + \left(r^2 - l(l+1) \right) y(r) = 0$$

$dsolve(sphBessEqn)$

$$y(r) = \frac{\_C1 \text{BesselJ}\left(\frac{\sqrt{1 + 4l(l+1)}}{2}, r \right)}{\sqrt{r}} + \frac{\_C2 \text{BesselY}\left(\frac{\sqrt{1 + 4l(l+1)}}{2}, r \right)}{\sqrt{r}}$$

Spherical Bessel Functions: Rayleigh Formulas

$$j[0] := \frac{\sin(x)}{x}$$

$$j[0] := \frac{\sin(x)}{x}$$

for *l* **from** 1 **to** 3 **do** $j[l] := simplify\left(\frac{1}{x} \cdot diff\left(j[l-1], x \right) \right)$ **end**

for *l* **from** 1 **to** 3 **do** $expand(j[l] \cdot (-x)^l)$ **end**

$$\frac{\sin(x)}{x}$$

$$-\frac{\cos(x)}{x}+\frac{\sin(x)}{x^2}$$

$$-\frac{\sin(x)}{x}-\frac{3\cos(x)}{x^2}+\frac{3\sin(x)}{x^3}$$

$$\frac{\cos(x)}{x}-\frac{6\sin(x)}{x^2}-\frac{15\cos(x)}{x^3}+\frac{15\sin(x)}{x^4}$$

$$n[0]:=-\frac{\cos(x)}{x}$$

$$n[0]:=-\frac{\cos(x)}{x}$$

for l **from** 1 **to** 3 **do** $n[l]:=simplify\left(\frac{1}{x}\cdot diff\left(n[l-1],x\right)\right)$ **end:**

for l **from** 0 **to** 3 **do** $expand(n[l]\cdot(-x)^l)$ **end**

$$-\frac{\cos(x)}{x}$$

$$-\frac{\sin(x)}{x}-\frac{\cos(x)}{x^2}$$

$$\frac{\cos(x)}{x}-\frac{3\sin(x)}{x^2}-\frac{3\cos(x)}{x^3}$$

$$\frac{\sin(x)}{x}+\frac{6\cos(x)}{x^2}-\frac{15\sin(x)}{x^3}-\frac{15\cos(x)}{x^4}$$

Spherical Bessel Function Plots

$plot([j[0],j[1],j[2],j[3]],x=0\ldots15,title ='Spherical\ Bessel\ Functions',legend = [j_0,j_1,j_2,j_3],linestyle = [solid,dash,dashdot,dot]):$

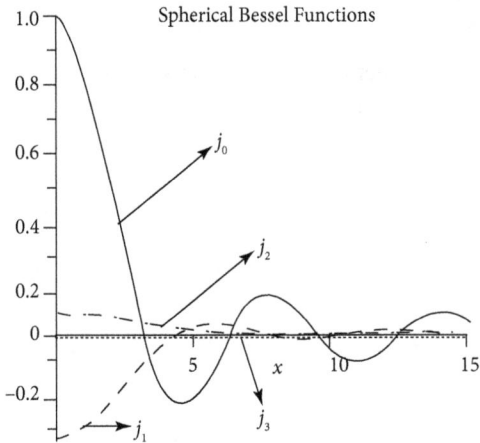

Figure 6.6.3: Plots of the spherical Bessel functions of the first kind.

$plot([n[0],\ n[1],\ n[2],\ n[3]],\ x = 0\ \dots\ 15,\ x = -2\ \dots\ 2,\ title\ = \ 'Spherical\ Bessel\ Functions',\ legend = [n_0, n_1, n_2, n_3], linestyle = [solid, dash, dashdot, dot])$:

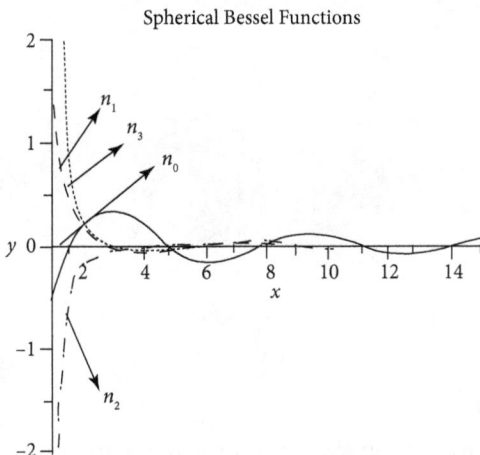

Figure 6.6.4: Plots of the spherical Bessel functions of the second kind.

6.7 MATLAB EXAMPLES

Several special functions common in physics are plotted together in this section. Key MATLAB commands: *besselj, hermiteH, laguerreL, legendreP, plot*

Section 6.3 Legendre Polynomials

The following example plots the first four Legendre polynomials on the same graph. Code is entered into a script file 'legendre.m' and executed at the Command Prompt by typing

```
>> legendre
x=-1:0.05:1;
LagP0=legendreP(0,x);
LagP1=legendreP(1,x);
LagP2=legendreP(2,x);
LagP3=legendreP(3,x);
plot(x,LagP0,'-k',x,LagP1,':k',x,LagP2,'-.k',x,LagP3,'--
    k','LineWidth',2)
legend('P0(x)','P1(x)','P2(x)','P3(x)')
xlabel('x')
```

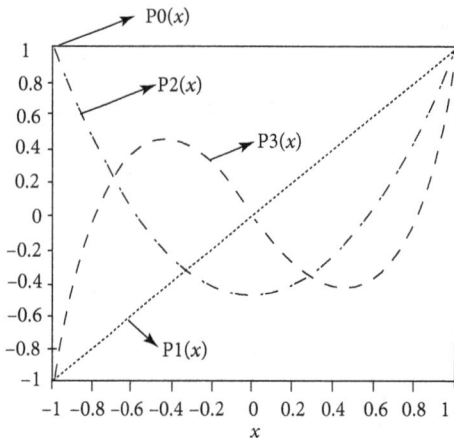

Figure 6.7.1: Plot of Legendre polynomials.

Section 6.4 Laguerre Polynomials

The following example plots the first four Laguerre polynomials on the same graph. Code is entered into a script file 'Laguerre.m' and executed at the Command Prompt by typing

```
>> Laguerre
x=-1.5:.1:5;
L0=laguerreL(0,x);
L1=laguerreL(1,x);
L2=laguerreL(2,x);
L3=laguerreL(3,x);
plot(x,L0,'-k',x,L1,':k',x,L2,'-.k',x,L3,'--k','LineWidth',2)
legend('L0(x)','L1(x)','L2(x)','L3(x)')
xlabel('x')
```

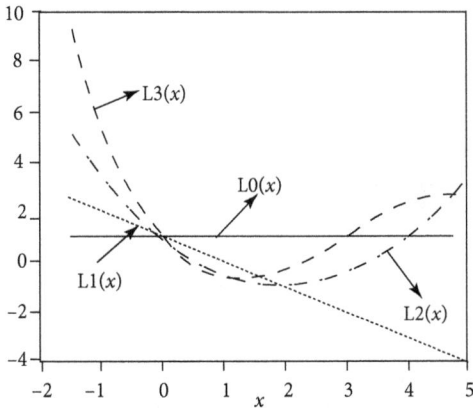

Figure 6.7.2: Plot of Laguerre polynomials.

Section 6.5 Hermite Polynomials

The following example plots the first four Hermite polynomials on the same graph. Code is entered into a script file 'hermite.m' and executed at the Command Prompt by typing

```
>> hermite
x=-1.5:.1:1.5;
H0=hermiteH(0,x);
H1=hermiteH(1,x);
H2=hermiteH(2,x);
H3=hermiteH(3,x);
plot(x,H0,'-k',x,H1,':k',x,H2,'-.k',x,H3,'--k','LineWidth',2)
legend('H0(x)','H1(x)','H2(x)','H3(x)')
title('Hermite polynomials')
xlabel('x')
```

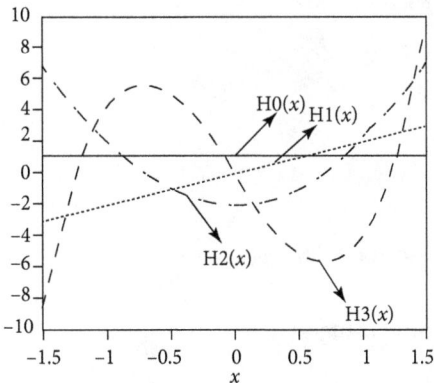

Figure 6.7.3: Plot of Hermite polynomials.

Section 6.6 Bessel Functions

The following example plots the first four Bessel function of the first kind on the same graph. Code is entered at the Command Line.

```
>> x=0:.1:15;
>> plot(x,besselj(0,x),'-k',x,besselj(1,x),':k',x,besselj(2,x),'-
   .k',x,besselj(3,x),'--k','LineWidth',2)
>>  legend('J0(x)','J1(x)','J2(x)','J3(x)')
>> title('Bessel functions')
>> xlabel('x')
```

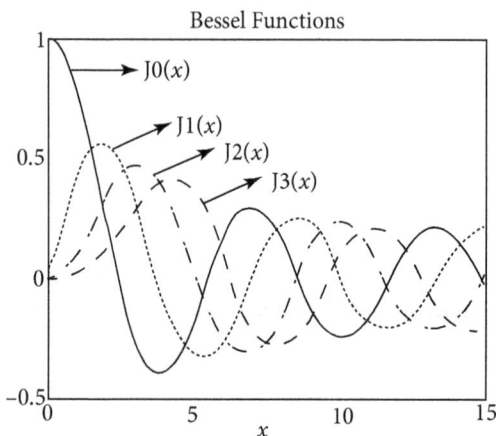

Figure 6.7.4: Plot of Bessel functions.

6.8 EXERCISES

Section 6.1 Dirac Delta Function

1. Evaluate $\displaystyle\int_{-\infty}^{\infty} e^{i(kx-\omega t)}\delta(x-a)dx$

2. Use integration by parts to show that $\displaystyle\int_{-\infty}^{\infty} f(x)\delta'(x-a)dx = -f'(a)$

3. Given a charged spherical shell is described by $\rho(r) = \dfrac{Q}{4\pi a^2}\delta(r-a)$ in spherical coordinates, show that the volume integral $\displaystyle\int_{vol} \rho(r)dv = Q$

4. Given the vector field defined by $\mathbf{F}(r) = \displaystyle\int_{vol} e^{i(\mathbf{k}\cdot\mathbf{r}'-\omega t)}\dfrac{(\mathbf{r}-\mathbf{r}')}{|\mathbf{r}-\mathbf{r}'|^3}dv'$ calculate $\nabla\cdot\mathbf{F}$

5. Given the scalar field defined by the integral $G(\mathbf{r}) = \int_{\text{vol}} \dfrac{e^{i(\mathbf{k}\cdot\mathbf{r}'-\omega t)}}{|\mathbf{r}-\mathbf{r}'|} dv'$ calculate $\nabla^2 G(\mathbf{r})$

6. Show that $\nabla \dfrac{1}{|\mathbf{r}-\mathbf{r}'|} = -\dfrac{\mathbf{r}-\mathbf{r}'}{|\mathbf{r}-\mathbf{r}'|^3}$ in Cartesian coordinates

7. Evaluate $\dfrac{d}{dx}\Theta(x+2)$

8. Plot the functions
   (a) $\Theta(x+1) - \Theta(x-3)$

   (b) $e^{-x^2}\,\Theta(x+1)$

   (c) $\sin(x)\left[\Theta\left(x+\dfrac{\pi}{2}\right) - \Theta\left(x-\dfrac{\pi}{2}\right)\right]$

Section 6.2 Orthogonal Functions

9. Test if $\sin(x)$ and $\cos(2x)$ are orthogonal over the interval $x = [-\pi, \pi]$

10. Test if $e^{i3\phi}$ and $e^{i2\phi}$ are orthogonal over the interval $\phi = [0, 2\pi]$

11. Test if $\sin\left(\dfrac{3\pi x}{L}\right)$ and $\sin\left(\dfrac{2\pi x}{L}\right)$ are orthogonal over the interval $x = [0, L]$

12. Test if e^x and $\cos(x)$ are orthogonal over the interval $x = [0, 2\pi]$

13. Test if x and $\dfrac{1}{\sqrt{1-x^2}}$ are orthogonal over the interval $x = [-1, 1]$

Section 6.3 Legendre Polynomials

14. Calculate the first three Legendre polynomials from the Rodrigues formula

$$P_l(x) = \dfrac{1}{2^l l!}\left(\dfrac{d}{dx}\right)^l (x^2-1)^l$$

15. Calculate the following integrals involving Legendre polynomials

$$\int_0^\pi P_2(\cos\theta)\delta(\cos\theta)\sin\theta\,d\theta$$

$$\int_0^\pi P_2(\cos\theta)P_3(\cos\theta)\sin\theta\,d\theta$$

$$\int_0^\pi P_3(\cos\theta)P_3(\cos\theta)\sin\theta\,d\theta$$

$$\int_0^\pi \left\{ \sum_{\ell=0}^\infty P_\ell(\cos\theta) \right\} P_3(\cos\theta) \sin\theta \, d\theta$$

$$\int_0^\pi \left\{ \sum_{\ell=0}^\infty \left(\frac{R}{r}\right)^\ell P_\ell(\cos\theta) \right\} (1+\cos\theta) \sin\theta \, d\theta$$

$$\int_0^\pi \left\{ \sum_{\ell=0}^\infty \left(\frac{R}{r}\right)^\ell P_\ell(\cos\theta) \right\} \delta(\cos\theta) \sin\theta \, d\theta$$

16. Show that

$$Y_2^0(\theta,\phi) = \frac{1}{4}\sqrt{\frac{5}{\pi}}\left(3\cos^2\theta - 1\right) \text{ and } Y_2^{\pm1}(\theta,\phi) = \frac{1}{4}\sqrt{\frac{5}{\pi}}\left(3\cos^2\theta - 1\right)e^{\pm i\phi}$$

are orthogonal over the surface of a sphere (*hint:* do the ϕ integration first)

17. Show orthogonality between

$$Y_2^0(\theta,\phi) = \frac{1}{4}\sqrt{\frac{5}{\pi}}\left(3\cos^2\theta - 1\right) \text{ and } Y_1^0(\theta,\phi) = \frac{1}{2}\sqrt{\frac{3}{\pi}}\cos\theta$$

over the surface of a sphere.

18. Show that the relation $(\ell+1)P_{\ell+1}(x) - (2\ell+1)xP_\ell(x) + \ell P_{\ell-1}(x) = 0$ is satisfied for $\ell = 2$ by substituting $P_1(x), P_2(x)$ and $P_3(x)$

19. Show that $y = P_2(x)$ is a solution to $(1-x^2)\dfrac{d^2y}{dx^2} - 2x\dfrac{dy}{dx} + 2(2+1)y = 0$

20. Expand $f(x) = \sqrt{|x|}$ as a series of Legendre polynomials over the interval $[-1, 1]$

Section 6.4 Laguerre Polynomials

21. Calculate the following integrals involving Laguerre polynomials

$$\int_0^\infty e^{-x} L_1(x) L_2(x) \, dx$$

$$\int_0^\infty e^{-x} L_1(x)^2 \, dx$$

22. Use Rodrigues' formula $L_n(x) = \dfrac{1}{n!} e^x \dfrac{d^n}{dx^n}\left(x^n e^{-x}\right)$ to compute Laguerre polynomials for $n = 0, 1, 2$

23. Show that $L_n'(x) - nL_{n-1}'(x) + nL_{n-1}(x) = 0$ is satisfied substituting $L_1(x)$ and $L_2(x)$.

24. Show that $y = L_2(x)$ is a solution to $x\dfrac{d^2 L_2}{dx^2} + (1-x)\dfrac{dL_2}{dx} + 2L_2 = 0$.

25. Expand $f(x) = xe^{-x}$ as a series of Laguerre polynomials for $x \geq 0$.

Section 6.5 Hermite Polynomials

26. Compute the following integrals involving Hermite polynomials

$$\int_{-\infty}^{\infty} e^{-x^2} H_1(x) H_2(x)\,dx$$

$$\int_{-\infty}^{\infty} e^{-x^2} H_1(x)^2\,dx$$

27. Use Rodrigues' formula $H_n(x) = (-1)^n e^{x^2}\dfrac{d^n}{dx^n}(e^{-x^2})$ to compute several Hermite polynomials

28. Show that the relation $H_{n+1}(x) - 2xH_n(x) + 2nH_{n-1}(x) = 0$ is satisfied by $H_3(x), H_2(x)$ and $H_1(x)$.

29. Show that $y = H_2(x)$ is a solution to $y'' - 2xy' + 4y = 0$

30. Expand $f(x) = \cos(x)e^{-x^2}$ as a series of Hermite polynomials.

Section 6.6 Bessel Functions

31. Plot the asymptotic form of several Bessel functions. Compare the asymptotic forms with plots of the Bessel functions displayed together

$$J_n(x) \approx \sqrt{\frac{2}{\pi x}}\cos\left(x - \frac{n\pi}{2} - \frac{\pi}{4}\right) \text{ for large } x.$$

32. Create a plot of $J_4(x)$ and display zeros of the function on the same graph

33. Show that $y = J_2(x)$ is a solution to $x^2 y'' + xy' + (x^2 - 4)y = 0$

34. Evaluate the following integrals involving Bessel functions

$$\int_0^{\infty} J_n(\alpha x)\,dx$$

$$\int_0^{\infty} \frac{1}{x} J_1(\alpha x)\,dx$$

$$\int_0^{\infty} \frac{1}{x} J_2(\alpha x)\,dx$$

7 Fourier Series and Transformations

Chapter **7**

Chapter Outline

7.1 Fourier Series
7.2 Fourier Transforms
7.3 Laplace Transforms

7.1 FOURIER SERIES

In this section, Fourier series are introduced, including the Fourier cosine, sine and exponential series. A periodic function $f(x)$ with period $2L$ may be expanded as an infinite sum of sine and cosine functions that form a complete set of orthogonal functions over the interval $[-L, L]$

$$f(x) = \frac{a_0}{2} + \sum_{n=1}^{\infty}\left[a_n \cos\left(\frac{n\pi x}{L}\right) + b_n \sin\left(\frac{n\pi x}{L}\right)\right] \tag{7.1.1}$$

where the coefficients

$$a_n = \frac{1}{L}\int_{-L}^{L} f(x)\cos\left(\frac{n\pi x}{L}\right)dx \tag{7.1.2}$$

331

$$b_n = \frac{1}{L} \int_{-L}^{L} f(x) \sin\left(\frac{n\pi x}{L}\right) dx \qquad (7.1.3)$$

$$a_0 = \frac{1}{L} \int_{-L}^{L} f(x) dx \qquad (7.1.4)$$

The a_0 term represents a "dc offset" that shifts the function up or down. The Fourier series converges to $f(x)$ at points where the function is continuous. The series converges to the average value of $f(x)$ at points of discontinuity. If the interval is instead $[0, 2L]$ then the integral limits are simply from zero to $2L$.

Example 7.1.1

Write integral expressions for the Fourier coefficients of function $f(t)$ with period T.

Solution: The Fourier series is

$$f(t) = \frac{a_0}{2} + \sum_{n=1}^{\infty}\left[a_n \cos\left(\frac{2n\pi t}{T}\right) + b_n \sin\left(\frac{2n\pi t}{T}\right)\right] \qquad (7.1.5)$$

For an interval $[-T/2, T/2]$ the coefficients are

$$a_n = \frac{2}{T} \int_{-T/2}^{T/2} f(t)\cos\left(\frac{2n\pi t}{T}\right) dt \qquad (7.1.6)$$

$$b_n = \frac{2}{T} \int_{-T/2}^{T/2} f(t)\sin\left(\frac{2n\pi t}{T}\right) dt \qquad (7.1.7)$$

$$a_0 = \frac{2}{T} \int_{-T/2}^{T/2} f(t) dt \qquad (7.1.8)$$

These coefficients are often expressed in terms of angular frequency $\omega = 2\pi/T$

$$a_n = \frac{\omega}{\pi} \int_{-\pi/\omega}^{\pi/\omega} f(t)\cos(n\omega t) dt \qquad (7.1.9)$$

$$b_n = \frac{\omega}{\pi} \int_{-\pi/\omega}^{\pi/\omega} f(t)\sin(n\omega t) dt \qquad (7.1.10)$$

$$a_0 = \frac{\omega}{\pi} \int_{-\pi/\omega}^{\pi/\omega} f(t) dt \qquad (7.1.11)$$

7.1.1 Fourier Cosine Series

An even periodic function $f(-x) = f(x)$ may be expanded as an infinite series of cosine functions

$$f(x) = \sum_{n=0}^{\infty} a_n \cos\left(\frac{n\pi x}{L}\right)$$

(7.1.12)

with coefficients

$$a_n = \frac{1}{L} \int_{-L}^{L} f(x)\cos\left(\frac{n\pi x}{L}\right)dx \quad \text{for } n = 1,2,3\ldots$$

(7.1.13)

and

$$a_0 = \frac{1}{2L} \int_{-L}^{L} f(x)dx$$

(7.1.14)

Example 7.1.2

Find the Fourier coefficients a_n corresponding to cosine series of the function

$$f(x) = x^2$$

(7.1.15)

with period 2π over the interval $[-\pi, \pi]$

Solution: $a_n = \dfrac{1}{\pi} \displaystyle\int_{-\pi}^{\pi} x^2 \cos(nx)dx = (-1)^n \dfrac{4}{n^2}$ for $n = 1,2,3\ldots$

(7.1.16)

and

$$a_0 = \frac{1}{2\pi} \int_{-\pi}^{\pi} x^2 dx = \frac{\pi^2}{3}$$

(7.1.17)

7.1.2 Fourier Sine Series

An odd periodic function $f(-x) = f(x)$ may be expanded as a series of sine functions

$$f(x) = \sum_{n=1}^{\infty} b_n \sin\left(\frac{n\pi x}{L}\right)$$

(7.1.18)

over the interval $[-L, L]$ where the coefficients

$$b_n = \frac{1}{L} \int_{-L}^{L} f(x)\sin\left(\frac{n\pi x}{L}\right)dx$$

(7.1.19)

Example 7.1.3

Find the Fourier coefficients b_n corresponding to the sine series of the function

$$f(x) = \begin{cases} -1 & -\pi \leq x < 0 \\ 1 & 0 < x \leq \pi \end{cases} \qquad (7.1.20)$$

with period 2π over the interval $[-\pi, \pi]$

Solution:
$$b_n = -\frac{1}{\pi} \int_{-\pi}^{0} \sin(nx)dx + \frac{1}{\pi} \int_{0}^{\pi} \sin(nx)dx \qquad (7.1.21)$$

$$b_n = \frac{1}{\pi}\frac{1}{n} \cos(nx)\Big|_{-\pi}^{0} - \frac{1}{\pi}\frac{1}{n} \cos(nx)\Big|_{0}^{\pi} \qquad (7.1.22)$$

$$b_n = \frac{2}{n\pi}\left(1 - (-1)^n\right) \qquad (7.1.23)$$

$$b_n = \frac{4}{n\pi} \quad \text{for } n = 1,3,5\ldots \qquad (7.1.24)$$

7.1.3 Fourier Exponential Series

The Fourier series may be expressed more compactly using complex exponentials. A function of time $f(t)$ with period T is expanded as

$$f(t) = \sum_{n=-\infty}^{\infty} a_n \exp\left(i\frac{2\pi n}{T}t\right) \qquad (7.1.25)$$

where the a_n are now complex. Multiplying both sides by $\exp\left(-i\frac{2\pi n'}{T}t\right)$ and integrating over the interval $[-T/2, T/2]$.

$$\int_{-T/2}^{T/2} f(t)\exp\left(i\frac{2\pi n'}{T}t\right)dt = \int_{-T/2}^{T/2} \sum_{n=-\infty}^{\infty} a_n \exp\left(i\frac{2\pi n}{T}t\right)\exp\left(-i\frac{2\pi n'}{T}t\right)dt \qquad (7.1.26)$$

Bringing the sum outside of the integral and combining exponentials on the right

$$\int_{-T/2}^{T/2} f(t)\exp\left(i\frac{2\pi n'}{T}t\right)dt = \sum_{n=-\infty}^{\infty} a_n \int_{-T/2}^{T/2} \exp\left(i\frac{2\pi(n-n')}{T}t\right)dt \qquad (7.1.27)$$

Evaluating the integral on the right

$$\int_{-T/2}^{T/2} \exp\left(i\frac{2\pi(n-n')}{T}t\right)dt = \frac{T}{2i\pi(n-n')}\left[\exp(i\pi(n-n')) - \exp(-i\pi(n-n'))\right]$$

$$(7.1.28)$$

$$\int_{-T/2}^{T/2} \exp\left(i\frac{2\pi(n-n')}{T}t\right)dt = T\frac{\sin\pi(n-n')}{\pi(n-n')} = T\delta_{nn'} \qquad (7.1.29)$$

Thus, the complex a_n are

$$a_n = \frac{1}{T}\int_{-T/2}^{T/2} f(t)\exp\left(-i\frac{2\pi n}{T}t\right)dt \qquad (7.1.30)$$

or in terms of angular frequency $\omega = 2\pi/T$

$$a_n = \frac{\omega}{2\pi}\int_{-\pi/\omega}^{\pi/\omega} f(t)\exp(-in\omega t)dt \qquad (7.1.31)$$

Maple Examples

Even and odd functions are expanded in Fourier cosine and sine series in the Maple worksheet below. The Fourier series expansion is demonstrated for a function that is neither even nor odd. Examples of expansions in terms of Hermite and Legendre polynomials are given.

Key Maple commands: *int, piecewise, plot, sum*

Special functions: HermiteH, LegendreP

Programming: **for** loops, function statements using ' $\rightarrow$ '

restart

with(plots):

Fourier Sine Series (Odd Functions)

$fodd := (x) \rightarrow x^3 \cdot \cos(x)$

$$fodd := (x) \rightarrow x^3 \cdot \cos(x)$$

$$b := n \rightarrow \left(\frac{2}{2\cdot\text{Pi}}\right)\cdot\text{int}\left(fodd(x)\cdot\sin(n\cdot x), x = -\text{Pi}...\text{Pi}\right)$$

$$b := n \mapsto \frac{\int_{-\pi}^{\rho} fodd(x)\sin(nx)dx}{\pi}$$

$B := Array(1 ... 1, 1 ... 3):$

$f\_sinseries := (x, N) \rightarrow sum(b(n)\cdot\sin(n\cdot x), n = 1 ... N):$

for m **from** 1 **to** 3 **do**

$N := m^3:$

$B_{1,m} := plot([fodd(x), f\_sinseries(x, N)], x = -Pi \ldots Pi, legend = [function, sine$
$series(N \ terms)], linestyle = [dash, solid]):$
end do:

$display(B)$

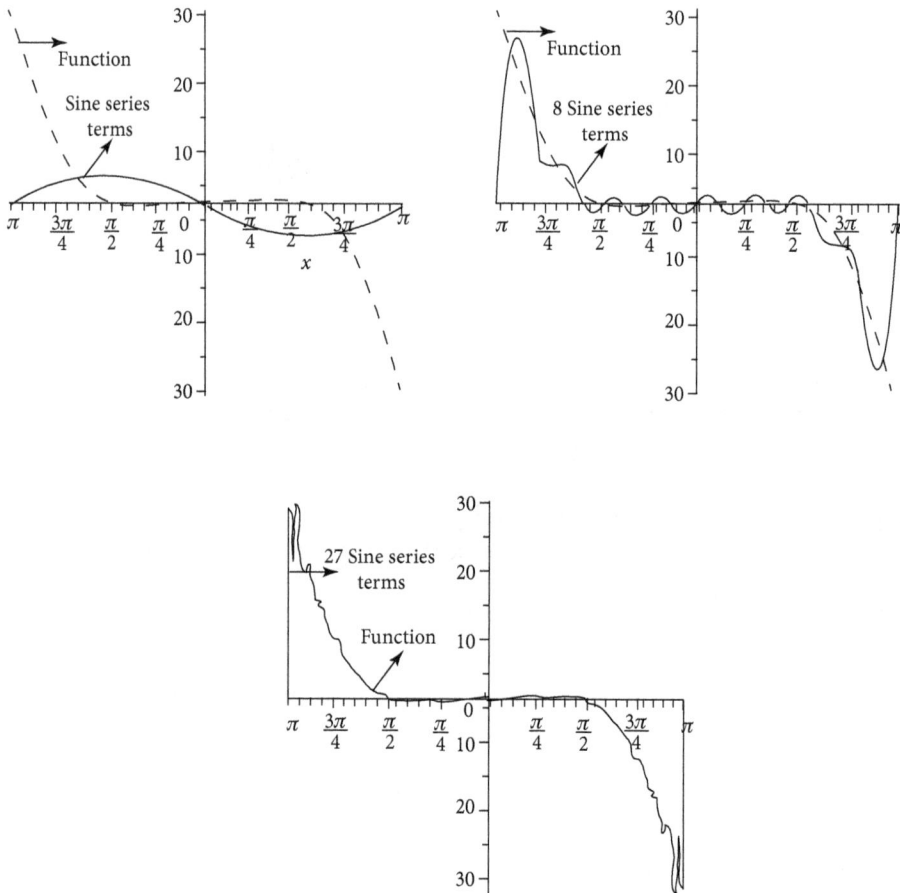

Figure 7.1.1: Fourier sine series plots.

Fourier Cosine Series (Even Functions)

$feven := (x) \rightarrow abs(x) \cdot exp(-x^2)$

$$feven := x \mapsto |x| e^{-x^2}$$

$$a := n \rightarrow \left(\frac{1}{Pi}\right) \cdot int\left(feven(x) \cdot cos(n \cdot x), x = -Pi \ldots Pi\right);$$

$$a := n \mapsto \frac{\int_{-\pi}^{\pi} feven(x)\cos(nx)dx}{\pi}$$

$$f\_\cos series := (x, N) \to \frac{a(0)}{2} + sum\left(a(n) \cdot \cos(n \cdot x), n = 1 \ldots N\right):$$

$C := Array(1 \ldots 1, 1 \ldots 3):$

for m **from** 1 **to** 3 **do**

$N := m^2:$

$C_{1,m} := plot([feven(x), f\_cosseries(x, N)], x = -Pi \ldots Pi, legend = [function,$
  $cosseries(N \ terms)], linestyle = [dash, solid]):$

end do:

$display(C)$

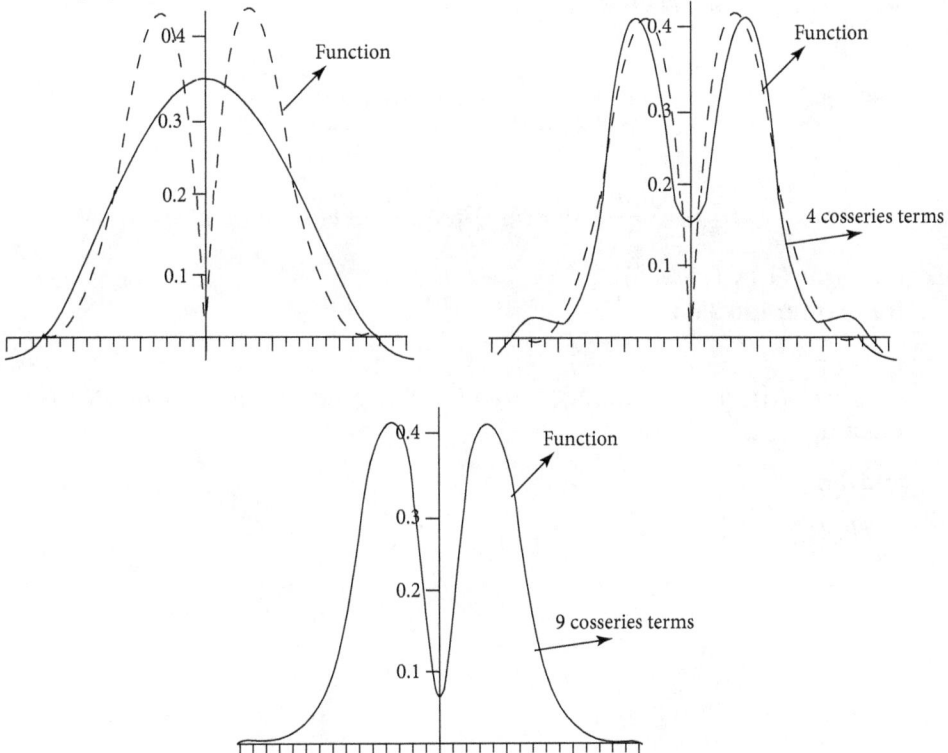

Figure 7.1.2: Fourier cosine series plots.

General Fourier Series (Arbitrary Functions)

$$f := (x) \rightarrow piecewise(x < 0, -3{\cdot}x, x > 0, \ln(x+1)$$

$$f := x \mapsto \begin{cases} -3x & x < 0 \\ \ln(x+1) & 0 < x \end{cases}$$

$$b := n \rightarrow \left(\frac{1}{Pi}\right) {\cdot} int\big(f(x){\cdot}\sin(n{\cdot}x), x = -Pi...Pi\big);$$

$$b := n \mapsto \frac{\int_{-\pi}^{\pi} f(x)\sin(nx)dx}{\pi}$$

$$a := n \rightarrow \left(\frac{1}{Pi}\right) {\cdot} int\big(f(x){\cdot}\cos(n{\cdot}x), x = -Pi...Pi\big);$$

$$a := n \mapsto \frac{\int_{-\pi}^{\pi} f(x)\cos(nx)dx}{\pi}$$

$$f\_series := (x,N) \rightarrow \frac{a(0)}{2} + sum\big(a(n){\cdot}\cos(n{\cdot}x) + b(n){\cdot}\sin(n{\cdot}x), n = 1...N\big):$$

$C := Array(1 ... 1, 1 ... 3):$

for m **from** 1 **to** 3 **do**

$N := m^2:$

$C_{1,m} := plot([f(x), f\_series(x, N)], x = -Pi ... Pi, legend = [function, series(N\ terms)],$
$linestyle = [dash, solid]):$

end do:

$display(C)$

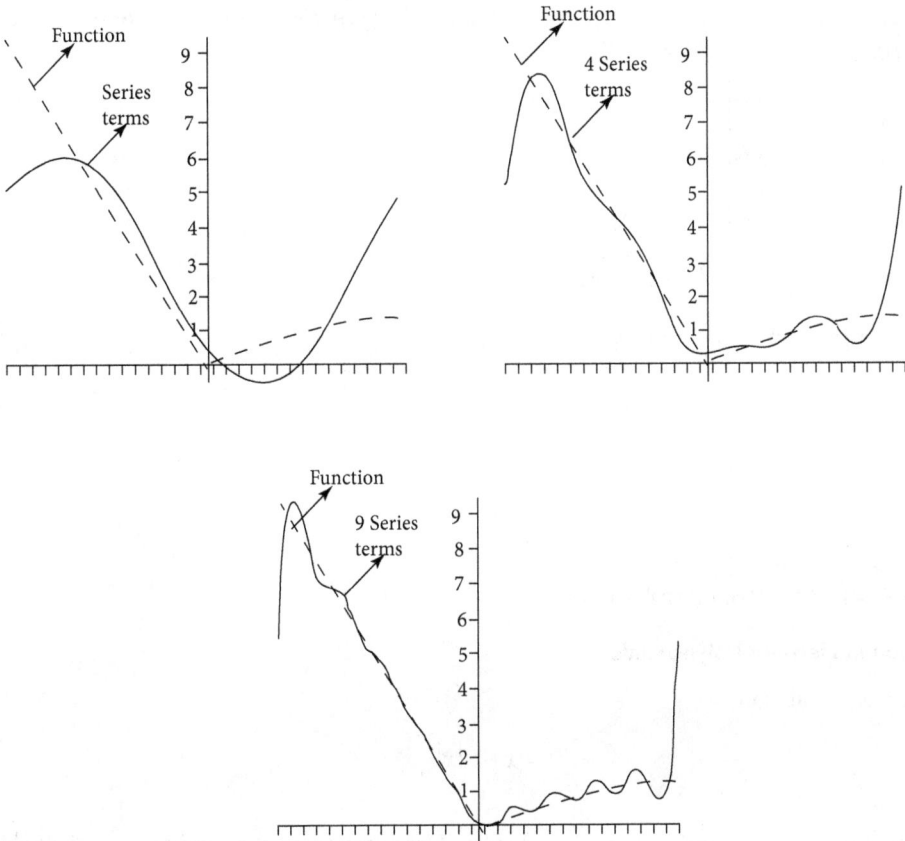

Figure 7.1.3: Fourier series plots.

Expansion in Legendre Polynomials

$$f := (x) \rightarrow \sin(x) \cdot \exp(-x^2)$$

$$f := x \mapsto \sin(x)e^{-x^2}$$

$$a := n \rightarrow \left(\frac{2 \cdot n + 1}{2} \right) \cdot \text{int}\left(f(x) \cdot \text{LegendreP}(n, x), x = -1 \ldots 1 \right);$$

$$a := n \mapsto \left(n + \frac{1}{2} \right) \left(\int_{-1}^{1} f(x) \text{LegendreP}(nx) dx \right)$$

$f\_LegPexp := (x, N) \rightarrow sum(a(n) \cdot \text{LegendreP}(n, x), n = 1 \ldots N):$
$N := 3:$

$plot([f(x), f\_LegPexp(x,N)], x = -1 \ldots 1, legend = [function, Legendre\ Polynomial(N\ terms)], linestyle = [dash, solid])$

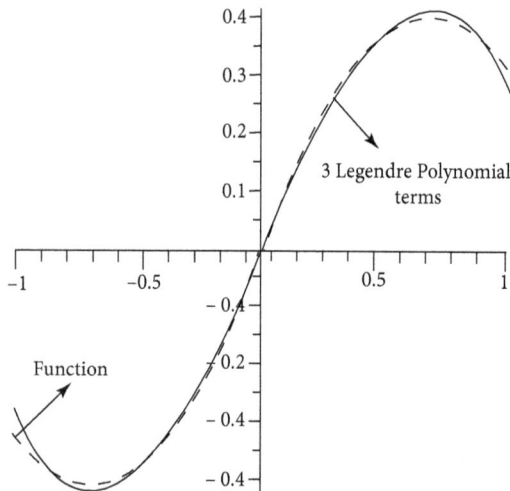

Figure 7.1.4: Expansion in Legendre polynomials.

Expansion in Hermite Polynomials

$f := (x) \rightarrow \text{abs}(x)$

$$f := x \mapsto |x|$$

$$a := n \rightarrow \frac{1}{2^n \cdot n! \cdot \text{sqrt(Pi)}} \cdot \text{int}\left(\exp(-x^2) \cdot f(x) \cdot \text{HermiteH}(n, x), x = -\text{infinity} \ldots \text{infinity}\right);$$

$$a := n \mapsto \frac{\int_{-\infty}^{\infty} e^{-x^2} f(x) \text{HermiteH}(n, x) dx}{2^n n! \sqrt{\pi}}$$

$N := 10:$

$f\_HerHexp := (x, N) \rightarrow sum(a(n) \cdot \text{HermiteH}(n, x), n = 0 \ldots N):$

$plot([f(x), f\_HerHexp(x,N)], x = -1 \ldots 1, legend = [function, Hermite\ Polynomial(N\ terms)], linestyle = [dash, solid])$

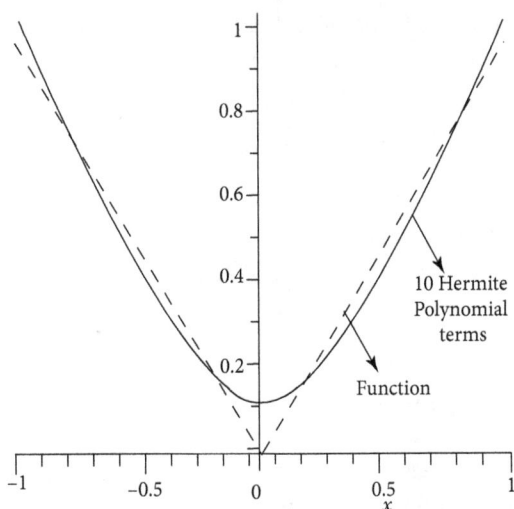

Figure 7.1.5: Fourier expansion in Hermite polynomials.

7.2 FOURIER TRANSFORMS

A periodic function $f(t) = f(t + T)$ with period T is expressed as a Fourier series

$$f(t) = \sum_{n=-\infty}^{\infty} a_n \exp\left(i\frac{2\pi n}{T}t\right) \tag{7.2.1}$$

with complex coefficients

$$a_n = \frac{1}{T} \int_{-T/2}^{T/2} f(t)\exp\left(-i\frac{2\pi n}{T}t\right)dt \tag{7.2.2}$$

that can be written in terms of the angular frequency $\omega = 2\pi/T$. For nonperiodic signals the Fourier coefficients become continuous $a_n \to a(\omega)d\omega$ and we take $T \to \infty$, $2\pi n/T \to \omega$, $1/T \to d\omega/2\pi$ and $\Sigma \to \int$. Thus

$$f(t) = \int_{-\infty}^{\infty} a(\omega)\exp(i\omega t)d\omega \tag{7.2.3}$$

and

$$a(\omega) = \frac{1}{2\pi} \int_{-\infty}^{\infty} f(t)\exp(-i\omega t)dt \tag{7.2.4}$$

The integral for $a(\omega)$ is known as the Fourier transform and the integral for $f(t)$ is the inverse Fourier transform. The factor of $1/2\pi$ is often split so that the integral transforms are symmetric.

Fourier transform:

$$a(\omega) = \frac{1}{\sqrt{2\pi}} \int_{-\infty}^{\infty} f(t)\exp(-i\omega t)dt \qquad (7.2.5)$$

Inverse Fourier transform:

$$f(t) = \frac{1}{\sqrt{2\pi}} \int_{-\infty}^{\infty} a(\omega)\exp(i\omega t)d\omega \qquad (7.2.6)$$

Fourier transforms and inverse transforms enable signals to be observed in time and frequency domains. Signal conditioning can be performed where known interference frequencies are removed from the $a(\omega)$. A filtered time signal $f(t)$ is then obtained from the inverse transform of the modified $a(\omega)$. This procedure is used in digital signal processing, or DSP.

7.2.1 Power Spectrum

Since $a(\omega)$ is a complex function, the power spectrum $S(\omega)$ is often plotted given $f(t)$ where

$$S(\omega) = |a(\omega)|^2 \qquad (7.2.7)$$

The square root of $S(\omega)$ is sometimes called the noise spectrum. A spectrum analyzer is an instrument that displays $S(\omega)$ from a signal input.

7.2.2 Spatial Transforms

Fourier transforms can also be used to determine the wavelengths λ present in a nonperiodic pulse or wave packet described by $f(x)$

$$a(k) = \frac{1}{\sqrt{2\pi}} \int_{-\infty}^{\infty} f(x)\exp(-ikx)dx \qquad (7.2.8)$$

with inverse transform

$$f(x) = \frac{1}{\sqrt{2\pi}} \int_{-\infty}^{\infty} a(k)\exp(ikx)dk \qquad (7.2.9)$$

where $k = 2\pi/\lambda$ is called the spatial frequency. In quantum mechanics, the momentum $p = \hbar k$ where $\hbar$ is Planck's constant divided by 2π. The wavefunction ψ whose magnitude squared gives the probability of locating a particle is described in momentum or coordinate space according to the Fourier transforms

$$\psi(p) = \frac{1}{\sqrt{2\pi}} \int_{-\infty}^{\infty} \psi(x) \exp(-ipx) dx \qquad (7.2.10)$$

and

$$\psi(x) = \frac{1}{\sqrt{2\pi}} \int_{-\infty}^{\infty} \psi(p) \exp(ipx) dp \qquad (7.2.11)$$

Example 7.2.1

Given

$$f(x) = \begin{cases} 1 & -L \le x \le L \\ 0 & |x| > L \end{cases} \qquad (7.2.12)$$

calculate $a(k)$

$$a(k) = \frac{1}{\sqrt{2\pi}} \left[\int_{-\infty}^{-L} 0 + \int_{-L}^{L} e^{-ikx} + \int_{L}^{\infty} 0 \right] dx \qquad (7.2.13)$$

$$a(k) = \frac{1}{\sqrt{2\pi}} \frac{1}{-ik} \left(e^{-ikL} - e^{ikL} \right)$$

$$= \frac{2}{\sqrt{2\pi}} \frac{1}{k} \frac{\left(e^{ikL} - e^{-ikL} \right)}{2i} \qquad (7.2.14)$$

$$= \sqrt{\frac{2}{\pi}} L \frac{\sin(kL)}{kL}$$

Example 7.2.2

Given

$$f(x) = e^{-|x|} \qquad (7.2.15)$$

calculate $a(k)$

$$a(k) = \frac{1}{\sqrt{2\pi}} \int_{-\infty}^{\infty} e^{-|x|} e^{-ikx} dx \qquad (7.2.16)$$

using
$$|x| = \begin{cases} x & x \geq 0 \\ -x & x < 0 \end{cases} \qquad (7.2.17)$$

the integral is broken up as

$$a(k) = \frac{1}{\sqrt{2\pi}} \left[\int_{-\infty}^{0} e^x e^{-ikx} dx + \int_{0}^{\infty} e^{-x} e^{-ikx} dx \right] \qquad (7.2.18)$$

$$a(k) = \frac{1}{\sqrt{2\pi}} \left[\frac{1}{1-ik} e^{(1-ik)x} \right]_{-\infty}^{0} - \frac{1}{\sqrt{2\pi}} \left[\frac{1}{1+ik} e^{-(1+ik)x} \right]_{0}^{\infty} \qquad (7.2.19)$$

$$a(k) = \frac{1}{\sqrt{2\pi}} \left[\frac{1}{1-ik} + \frac{1}{1+ik} \right] \qquad (7.2.20)$$

Plancherel's Formula

Given two functions of time $f(t)$ and $g(t)$ show the integral relation

$$\int_{-\infty}^{\infty} f^*(t)g(t)dt = \int_{-\infty}^{\infty} F^*(\omega)G(\omega)d\omega \qquad (7.2.21)$$

where

$$f(t) = \frac{1}{\sqrt{2\pi}} \int_{-\infty}^{\infty} F(\omega)e^{-i\omega t} d\omega \qquad (7.2.22)$$

and

$$g(t) = \frac{1}{\sqrt{2\pi}} \int_{-\infty}^{\infty} G(\omega)e^{-i\omega t} d\omega \qquad (7.2.23)$$

Writing the integral on the left-hand side in terms of the respective Fourier transforms over frequency

$$\int_{-\infty}^{\infty} f^*(t)g(t)dt = \int_{-\infty}^{\infty} \left[\frac{1}{\sqrt{2\pi}} \int_{-\infty}^{\infty} F(\omega)e^{i\omega t} d\omega \right] \left[\frac{1}{\sqrt{2\pi}} \int_{-\infty}^{\infty} G(\omega')e^{-i\omega' t} d\omega' \right] dt \qquad (7.2.24)$$

ω' is a dummy index. Separating the time components

$$= \int_{-\infty}^{\infty} d\omega' \int_{-\infty}^{\infty} d\omega F(\omega)G(\omega') \left[\frac{1}{2\pi} \int_{-\infty}^{\infty} e^{i(\omega-\omega')t} dt \right] \qquad (7.2.25)$$

gives a delta function

$$= \int_{-\infty}^{\infty} d\omega F^*(\omega) \int_{-\infty}^{\infty} d\omega' G(\omega')\delta(\omega - \omega') \qquad (7.2.26)$$

and we obtain the desired result

$$= \int_{-\infty}^{\infty} F^*(\omega)G(\omega)d\omega \qquad (7.2.27)$$

Maple Examples

Fourier integral transforms are calculated in the Maple worksheet below. Fourier transforms are then used to solve inhomogeneous differential equations.

Key Maple commands: *assume, convert, fourier, int, invfourier, piecewise, simplify, solve*

Maple packages: *with(inttrans)*:

Special functions: Dirac, Heaviside

Programming: **for** loops, function statements using '→'

restart

Fourier Transforms

with(inttrans):
assume(a > 0, omega > 0, k > 0)
fourier(sin(omega·t), t, omega)

$$-I\,\pi\,\text{Dirac}(0)$$

fourier(cos(omega·t), t, omega)

$$\pi\,\text{Dirac}(0)$$

fourier(sin(t), t, omega)

$$-I\,\pi\,\text{Dirac}(\omega\sim - 1)$$

convert(fourier(f(omega·t), t, omega), int)

$$\int_{-\infty}^{\infty} f(\omega \sim t)e^{-I\omega\sim t}\,dt$$

Fourier Transform of Heaviside Functions

$$p := piecewise(x < -1, 0, -1 \le x \le 1, 1, x > 1, 0)$$

$$\begin{cases} 0 & x < -1 \\ 1 & -1 \le x \text{ and } x \le 1 \\ 0 & 1 < x \end{cases}$$

$$f := (x) \rightarrow p;$$

$$x \rightarrow p$$

$$int(f(x) \cdot \exp(-I \cdot k \cdot x), x = -infinity \dots infinity)$$

$$-\frac{I\left(e^{2Ik\sim} - 1\right)e^{-Ik\sim}}{k \sim}$$

$$simplify((convert(\%, trig)))$$

$$\frac{2\sin(k \sim)}{k \sim}$$

$$fourier(\text{Heaviside}(x+1) - \text{Heaviside}(x-1)), x, k)$$

$$\frac{2\sin(k \sim)}{k \sim}$$

Fourier Transform Solution of Differential Equations

$$Deq1 := \frac{d^2}{dt^2} y(t) - y(t) = \sin(at)$$

$$\frac{d^2}{dt^2} y(t) - y(t) = \sin(a \sim t)$$

$$fourier(Deq1, t, s)$$

$$-(s^2 + 1) fourier(y(t), t, s) = I p (-\text{Dirac}(s - a\sim) + \text{Dirac}(s + a\sim))$$

$$solve(\%, fourier(y(t), t, s))$$

$$\frac{I \ (\text{Dirac}(s - a\sim) - \text{Dirac}(s + a\sim))}{}$$

$$invfourier(\%, s, t)$$

$$-\frac{\sin(a \sim t)}{a \sim^2 + 1}$$

$$Deq2 := \frac{d^4}{dt^4} y(t) - y(t) = \text{Dirac}(t-a) \cdot \sin(at)$$

$$\frac{d^4}{dt^4} y(t) - y(t) = \text{Dirac}(t-a \sim)\sin(a \sim t)$$

fourier(Deq2, t, s)

$$(s-1)(s+1)(s^2+1) \, fourier(y(t), t, s) = \sin(a\sim^2)e^{-Ia\sim s}$$

solve(%, fourier(y(t), t, s))

$$\frac{\sin\left(a \sim^2\right)e^{-Ia\sim s}}{(s-1)(s+1)\left(s^2+1\right)}$$

invfourier(%, s, t)

$$\frac{1}{4}\sin\left(a \sim^2\right)\left(-\left(e^{-t+a\sim} + \sin(t-a \sim)\right)\text{Heaviside}(t-a \sim) - \right.$$
$$\left.\left(e^{t-a\sim} - \sin(t-a \sim)\right)\text{Heaviside}(-t+a \sim)\right)$$

7.3 LAPLACE TRANSFORMS

In this section Laplace and inverse Laplace transforms are introduced along with several important properties. These properties enable the solution to initial value problems. Given the function $f(t)$, the Laplace transform is obtained by multiplying by e^{-st} and integrating over time to obtain a new function

$$F(s) = \int_0^\infty e^{-st} f(t)dt \tag{7.3.1}$$

The function $f(t)$ is called the kernel of the transform. This integral transform is expressed as $F(s) = \mathcal{L}[f(t)]$.

Example 7.3.1

Find the Laplace transform of $f(t) = t$

Solution:
$$\mathcal{L}[t] = \int_0^\infty te^{-st}dt = \frac{1}{s^2} \tag{7.3.2}$$

7.3.1 Properties of the Laplace Transform

Important properties of the Laplace transform include
Linearity:

$$\mathcal{L}\left[c_1 f_1(t) + c_2 f_2(t)\right] = c_1 F_1(s) + c_2 F_2(s) \text{ where } c_1 \text{ and } c_2 \text{ are constants.} \qquad (7.3.3)$$

Change of scale:

$$\mathcal{L}\left[f(at)\right] = \frac{1}{a} F\left(\frac{s}{a}\right) \qquad (7.3.4)$$

Shifting properties:

$$\mathcal{L}\left[e^{at} f(t)\right] = F(s-a) \qquad (7.3.5)$$

$$\mathcal{L}\left[f(t-a)\right] = e^{-as} F(s) \qquad (7.3.6)$$

7.3.2 Inverse Laplace Transform

The inverse Laplace transform is defined in terms of a contour integral

$$\mathcal{L}^{-1}\left[F(s)\right] = \frac{1}{2\pi i} \int_{\gamma - i\infty}^{\gamma + i\infty} e^{st} F(s) ds = f(t) \qquad (7.3.7)$$

If $F(s) = \mathcal{L}[f(t)]$ then $\mathcal{L}^{-1}[F(s)] = f(t)$. $\qquad (7.3.8)$

7.3.3 Properties of Inverse Laplace Transforms

The inverse Laplace transform is also a linear operator

$$\mathcal{L}^{-1}\left[c_1 F_1(s) + c_2 F_2(s)\right] = c_1 f_1(t) + c_2 f_2(t) \text{ where } c_1 \text{ and } c_2 \text{ are constants.} \qquad (7.3.9)$$

Other properties of the inverse Laplace transform $\mathcal{L}^{-1}[F(s)] = f(t)$ include the following:

Change of scale:

$$\frac{1}{a} \mathcal{L}^{-1}\left[F\left(\frac{s}{a}\right)\right] = f(at) \qquad (7.3.10)$$

Shifting properties:

$$\mathcal{L}^{-1}\left[F(s-a)\right] = e^{at} f(t) \text{ and } \mathcal{L}^{-1}\left[e^{-as} F(s)\right] = f(t-a) \qquad (7.3.11)$$

7.3.4 Table of Laplace Transforms

Several Laplace transforms and inverse Laplace transforms are shown in Table 7.3.1.

Table 7.3.1: Table of Laplace and inverse Laplace transforms.

Laplace Transform	Inverse Laplace Transform
$\mathcal{L}[f(t)] = F(s)$	$\mathcal{L}^{-1}[F(s)] = f(t)$
$\mathcal{L}(1) = \dfrac{1}{s}$	$\mathcal{L}^{-1}\left(\dfrac{1}{s}\right) = 1$
$\mathcal{L}(t) = \dfrac{1}{s^2}$	$\mathcal{L}^{-1}\left(\dfrac{1}{s^2}\right) = t$
$\mathcal{L}(t^n) = \dfrac{n!}{s^{n+1}}$ (n positive integer)	$\mathcal{L}^{-1}\left(\dfrac{n!}{s^{n+1}}\right) = t^n$
$\mathcal{L}(\sin(\omega t)) = \dfrac{\omega}{s^2 + \omega^2}$	$\mathcal{L}^{-1}\left(\dfrac{\omega}{s^2 + \omega^2}\right) = \sin(\omega t)$
$\mathcal{L}(\cos(\omega t)) = \dfrac{s}{s^2 + \omega^2}$	$\mathcal{L}^{-1}\left(\dfrac{s}{s^2 + \omega^2}\right) = \cos(\omega t)$
$\mathcal{L}(e^{\omega t}) = \dfrac{1}{s - \omega}$	$\mathcal{L}^{-1}\left(\dfrac{1}{s - \omega}\right) = e^{\omega t}$
$\mathcal{L}(t^n e^{\omega t}) = \dfrac{n!}{(s-a)^{n+1}}$ (n positive integer)	$\mathcal{L}^{-1}\left(\dfrac{n!}{(s-a)^{n+1}}\right) = t^n e^{\omega t}$
$\mathcal{L}^{-1}(\delta(t-a)) = e^{-sa}$	$\mathcal{L}^{-1}(e^{-sa}) = \delta(t-a)$
$\mathcal{L}(\Theta(t-a)) = \dfrac{e^{-as}}{s}$	$\mathcal{L}^{-1}\left(\dfrac{e^{-as}}{s}\right) = \Theta(t-a)$

7.3.5 Solving Differential Equations

A very useful property of Laplace transforms is the transformation of derivatives. Transformation of the first derivative of $f(t)$ gives

$$\mathcal{L}\left[\frac{df(t)}{dt}\right] = sF(s) - f(0) \qquad (7.3.12)$$

Transformation of second derivative:

$$\mathcal{L}\left[\frac{d^2 f(t)}{dt^2}\right] = s^2 F(s) - sf(0) - \dot{f}(0) \qquad (7.3.13)$$

The transformation of higher derivatives may also be obtained:

$$\mathcal{L}\left[\frac{d^n f(t)}{dt^n}\right] = s^n F(s) - s^{n-1} f(0) - s^{n-2} \dot{f}(0) - \cdots - f^{(n-1)}(0) \qquad (7.3.14)$$

This property enables the Laplace transform solution to initial value problems.

Example 7.3.2

Solve the differential equation

$$\frac{d^2 y(t)}{dt^2} - \frac{dy(t)}{dt} = t \text{ with initial conditions } y(0) = 0, \ \dot{y}(0) = 1. \qquad (7.3.15)$$

Taking the Laplace transform

$$s^2 Y(s) - sy(0) - \dot{y}(0) - sY(s) + y(0) = \frac{1}{s^2} \qquad (7.3.16)$$

and solving for $Y(s)$

$$Y(s) = \frac{s^2 + 1}{s^4 - s^3} \qquad (7.3.17)$$

The inverse Laplace transform gives

$$y(t) = 2e^t - \frac{1}{2}t^2 - t - 2 \qquad (7.3.18)$$

Maple Examples

Laplace transforms and inverse transforms of functions and derivatives are calculated in the Maple worksheet below. Laplace transforms are used to solve inhomogeneous differential equations with specified initial conditions, a system of differential equations, and an integro-differential equation.

Key Maple commands: *diff, invlaplace, laplace, plot, rhs, solve, subs*

Maple packages: *with(inttrans)*:

Special functions: Dirac, Heaviside

restart

Laplace Transforms

with(*inttrans*):
laplace(cos(*t*), *t*, *s*);

$$\frac{s}{s^2+1}$$

laplace(*t*·cos(*t*), *t*, *s*);

$$\frac{s^2-1}{\left(s^2+1\right)^2}$$

Inverse Laplace Transforms

$$invlaplace\left(\frac{s}{s^2+1}, s, t\right);$$

$$\cos(t)$$

$$invlaplace\left(\frac{s^2-1}{\left(s^2+1\right)^2}, s, t\right);$$

$$t\cos(t)$$

Laplace Transform of Derivatives

laplace(*diff*(*x*(*t*), *t*) *t*, *s*)

$$s\,laplace(x(t), t)\,t, s) - x(0)$$

laplace(*diff*(*x*(*t*), *t*\$2) *t*, *s*)

$$s^2\,laplace(x(t), t)\,t, s) - D(x)(0) - sx(0)$$

Inverse Laplace Transform of Derivatives

invlaplace(*diff*(*x*(*s*), *s*, *s*) *s*, *t*)

$$t^2\,invlaplace(x(s), s, t)$$

invlaplace(*diff*(*x*(*s*), *s*, *s*, *s*, *s*) *s*, *t*)

$$t^4\,invlaplace(x(s), s, t)$$

Laplace Transform Solution of Differential Equations

$$Deq := \frac{d^2}{dt^2}y(t) + \left(\frac{d}{dt}y(t)\right) + y(t) = \sin(t)\cdot\exp(-t);$$

$$\frac{d^2}{dt^2}y(t) + \frac{d}{dt}y(t) + y(t) = \sin(t)e^{-t}$$

laplace(Deq, t, s)

$$s^2 laplace\left(y(t),t,s\right) - D(y)(0) - sy(0) + s\, laplace\left(y(t),t,s\right) - y(0)$$

$$+laplace\left(y(t),t,s\right) = \frac{1}{(s+1)^2 + 1}$$

solve(%, laplace(y(t), t, s))

$$\frac{y(0)s^3 + D(y)(0)s^2 + 3y(0)s^2 + 2D(y)(0)s + 4sy(0) + 2D(y)(0) + 2y(0) + 1}{(s^2 + 2s + 2)(s^2 + s + 1)}$$

subs({y(0) = 1, D(y)(0) = 0}, %)

$$\frac{s^3 + 3s^2 + 4s + 3}{(s^2 + 2s + 2)(s^2 + s + 1)}$$

invlaplace(%, s, t)

$$\frac{2}{3}\sqrt{3}e^{-\frac{1}{2}t}\sin\left(\frac{1}{2}\sqrt{3}t\right) + e^{-t}\cos(t)$$

dsolve({Deq, y(0) = 1, D(y)(0) = 0}, y(t))

$$y(t) = \frac{2}{3}\sqrt{3}e^{-\frac{1}{2}t}\sin\left(\frac{1}{2}\sqrt{3}t\right) + e^{-t}\cos(t)$$

plot(rhs(%), t = 0 ... 20)

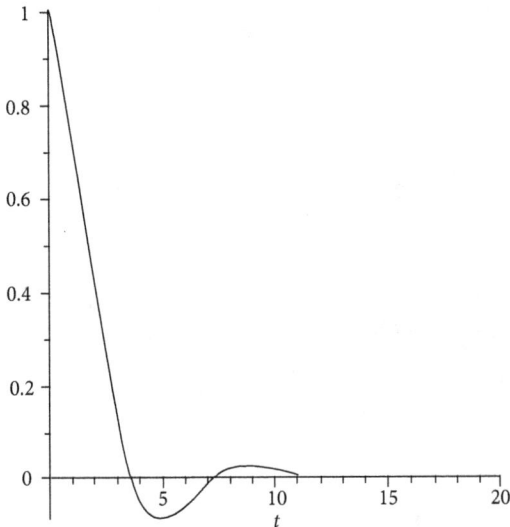

Figure 7.3.1: Laplace transform solution to a differential equation.

Differential Equations Involving Delta Functions

$$Deq := \frac{d^2}{dt^2} y(t) + \left(\frac{d}{dt} y(t) \right) + y(t) = \text{Dirac}(t)$$

$$\frac{d^2}{dt^2} y(t) + \frac{d}{dt} y(t) + y(t) = \text{Dirac}(t)$$

laplace(Deq, t, s)

$$s^2 \, laplace(y(t), t, s) - D(y)(0) - sy(0) + s \, laplace(y(t), t, s) - y(0)$$
$$+ \, laplace(y(t), t, s) = 1$$

solve(%, laplace(y(t), t, s))

$$\frac{sy(0) + D(y)(0) + y(0) + 1}{s^2 + s + 1}$$

subs({y(0) = 1, D(y)(0) = 0}, %)

$$\frac{s+2}{s^2 + s + 1}$$

invlaplace(%, s, t)

$$e^{-\frac{1}{2}t} \left(\cos\left(\frac{1}{2}\sqrt{3}t \right) + \sqrt{3} \sin\left(\frac{1}{2}\sqrt{3}t \right) \right)$$

Differential Equations Involving Step Functions

$$Deq := \frac{d^2}{dt^2}y(t)+3\cdot\left(\frac{d}{dt}y(t)\right)+y(t)=\text{Heaviside}(t)-\text{Heaviside}(t-1)$$

$$\frac{d^2}{dt^2}y(t)+3\left(\frac{d}{dt}y(t)\right)+y(t)=\text{Heaviside}(t)-\text{Heaviside}(t-1)$$

laplace(Deq, t, s)

$$s^2 laplace\left(y(t),t,s\right)-D(y)(0)-sy(0)+3s\,laplace\left(y(t),t,s\right)-3y(0)$$

$$+laplace\left(y(t),t,s\right)=\frac{1-e^{-s}}{s}$$

solve(%, laplace(y(t), t, s))

$$\frac{y(0)s^2+D(y)(0)s+3sy(0)-e^{-s}+1}{s(s^2+3s+1)}$$

subs({y(0) = 1, D(y)(0) = 0}, %)

$$\frac{s^2+1+3s-e^{-s}}{s(s^2+3s+1)}$$

invlaplace(%, s, t)

$$\text{Heaviside}(1-t)+\frac{1}{20}\left(2e^{\frac{1}{2}(\sqrt5-3)(t-1)}+e^{-\frac{1}{2}(\sqrt5+3)(t-1)}(-7+3\sqrt5)\right)$$

$$\text{Heaviside}(t-1)\left(3\sqrt5+5\right)$$

Systems of Differential Equations

$$dEqns:=\left\{\frac{d}{dt}x(t)=2\cdot y(t)+z(t),\frac{d}{dt}y(t)=x(t)-z(t),\frac{d}{dt}z(t)=y(t), x(0)=1,\right.$$

$$\left.y(0)=0, z(0)=0\right\}$$

$$\left\{\frac{d}{dt}x(t)=2y(t)+z(t),\frac{d}{dt}y(t)=x(t)-z(t),\right.$$

$$\left.\frac{d}{dt}z(t)=y(t), x(0)=1, y(0)=0, z(0)=0\right\}$$

dsolve(dEqns, {x(t), y(t), z(t)}, method = laplace)

$$\left\{ x(t) = -\frac{1}{23} \sum_{\_\alpha l = RootOf\left(-Z^3 - \_Z - 1\right)} \left(8\_\alpha l^2 - 12\_\alpha l - 13\right)e^{-\alpha l\, t}, \right.$$

$$y(t) = \frac{1}{23} \sum_{\_\alpha l = RootOf\left(-Z^3 - \_Z - 1\right)} \left(9\_\alpha l^2 - 2\_\alpha l - 6\right)e^{-\alpha l\, t},$$

$$\left. z(t) = -\frac{1}{23} \sum_{\_\alpha l = RootOf\left(-Z^3 - \_Z - 1\right)} \left(6\_\alpha l^2 - 9\_\alpha l - 4\right)e^{-\alpha l\, t} \right\}$$

Integro-Differential Equations

$IGEqns := diff(y(t), t) - 2 \cdot y(t) = Int(y(\text{tau}), \text{tau} = 0 \ldots t)$

$$\frac{d}{dt}y(t) - 2y(t) = \int_0^t y(\tau)d\tau$$

$laplace(IGEqns, t, s)$

$$s\, laplace\left(y(t), t, s\right) - y(0) - 2\, laplace\left(y(t), t, s\right) = \frac{laplace\left(y(t), t, s\right)}{s}$$

$solve(\%, laplace(y(t), t, s))$

$$\frac{y(0)s}{s^2 - 2s - 1}$$

$subs(\{y(0) = 1, D(y)(0) = 0\}, \%)$

$$\frac{s}{s^2 - 2s - 1}$$

$invlaplace(\%, s, t)$

$$\frac{1}{2}e^t \left(\sqrt{2}\sinh\left(t\sqrt{2}\right) + 2\cosh\left(t\sqrt{2}\right)\right)$$

7.4 MATLAB EXAMPLES

The following MATLAB examples demonstrate the calculation of the Fourier series and the fast Fourier transform (FFT). Laplace transforms and Fourier integral transforms are then calculated.

Key MATLAB commands: *abs, fft, fourier,ilaplace, laplace, length, linspace, plot, square, syms, symsum*

Programming: for loops

Section 7.1 Fourier Series

The Fourier sine series (with 16 terms) of a square wave with period $T = 2$ is plotted using the following MATLAB script. A plot of the square wave is superimposed on this figure for comparison.

```
t=0:.01:5;
omega=double(pi);
f=square(omega*t);
g=(4/pi)*symsum(sin((2*n+1)*pi*t)/(2*n+1),n,0,15);

plot(t,g,t,f)

legend('f(t)','Fourier sine series')
xlabel('time')
ylabel('f(t)')
```

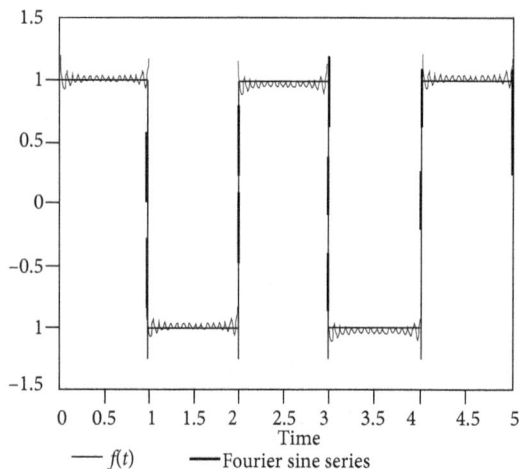

Figure 7.4.1: Fourier sine series of a square wave with period two plotted with 16 terms. The square wave is shown on this figure for comparison.

Section 7.2 Fourier Transforms

Examples using the Symbolic Math Toolbox to calculate Fourier integral transforms are shown below. These calculations are performed at the Command Line.

```
>> fourier(cos(t))
ans =
pi*(dirac(w - 1) + dirac(w + 1))
>> fourier(1/t)
ans =
pi*(2*heaviside(-w) - 1)*i
```

The following script file evaluates the fast Fourier transform (FFT).

```
dt=0.001;
tmax=2.0;
t=0:dt:tmax;
A=2*sin(2*pi*150.0*t)+5*sin(2*pi*333.0*t);
y=fft(A);
plot(abs(y))
xlabel('index')
ylabel('abs(FFT)')
```

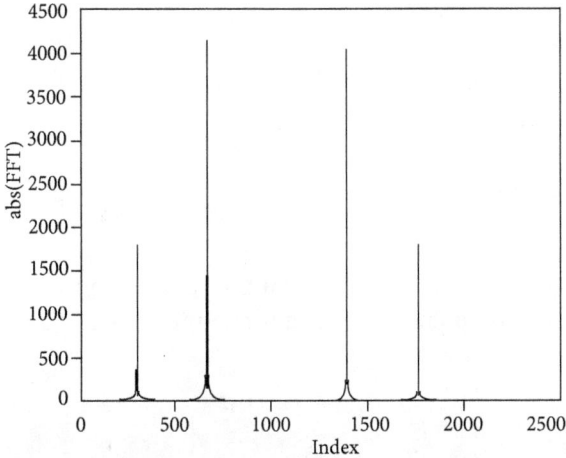

Figure 7.4.2: Double-sided FFT plotted as a function of index. The FFT is symmetric about $N/2$ where the number of points in the time series $N = 2000$.

The following script file evaluates the single-sided fast Fourier transform (FFT) with frequency plotted on the horizontal axis.

```
dt=0.001;
tmax=2.0;
t=0:dt:tmax;
A=2*sin(2*pi*150.0*t)+5*sin(2*pi*333.0*t);
y=fft(A);
fmax=1/(2.0*dt);
f= linspace(0,fmax,length(t)/2);
m=abs(y)/length(f);
m=m(1:length(f));
plot(f,m)
xlabel('frequency')
ylabel('abs(FFT)')
```

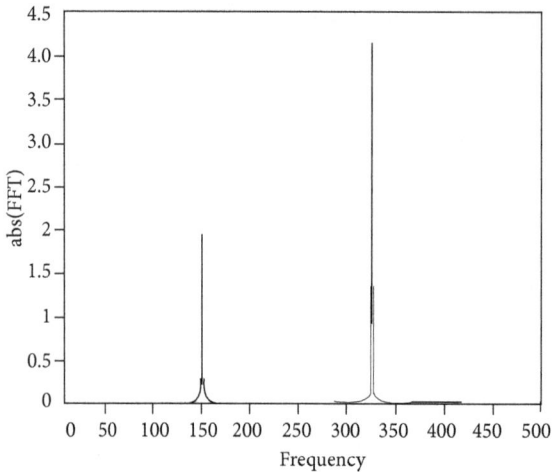

Figure 7.4.3: Single-sided FFT plotted as a function of frequency. The highest frequency, 500 Hz, is equal to half of the sample rate. 1000 points are plotted.

The discrete Fourier transform (DFT) may be calculated if the Fourier transform of a time series is desired at a single frequency or over a limited frequency range as shown below

```
dt=0.001;
tmax=2.0;
t=0:dt:tmax;
A=2*sin(2*pi*150.0*t)+5*sin(2*pi*333.0*t);

nmax=length(t);

fmin = 300;
fmax = 350;
mmax = 1000;

Ftx = zeros(1,mmax);

f = linspace(fmin,fmax,mmax);

for m=1:mmax
    for n=1:nmax
        Ftx(m)=Ftx(m)+A(n)*exp(-1i*2*pi*f(m)*n*dt);
    end
end

Ftx=Ftx/(nmax/2);

plot(f,abs(Ftx))

set(gca,'XTick',fmin:3:fmax)
xlabel('frequency')
ylabel('Fourier transform')
```

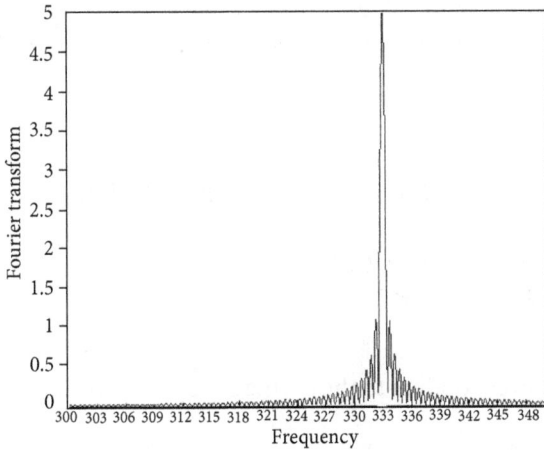

Figure 7.4.4: DFT plotted over a frequency range between 300–350 Hz.

Section 7.3 Laplace Transforms

Laplace integral transformations are evaluated below using MATLAB's Symbolic Math Toolbox.

```
>> clear all
>> syms a s t
>> laplace(t)
ans =
1/s^2
>> laplace(cos(a*t))
ans =
s/(a^2 + s^2)
>> laplace(exp(a*t))
ans =
-1/(a - s)
>> ilaplace(1/s^2)
ans =
t
>> ilaplace(s/(a^2 + s^2))
ans =
cos(a*t)
>> ilaplace(-1/(a - s))
ans =
exp(a*t)
```

7.5 EXERCISES

Section 7.1 Fourier Series

1. Find the Fourier coefficients corresponding to the sine series of the function

$$f(t) = \begin{cases} -1 & -2 < t < 0 \\ 1 & 0 < t < 2 \end{cases}$$

with period $T = 4s$

2. Find the Fourier coefficients corresponding to the cosine series of the function

$$f(x) = \begin{cases} x+1 & -1 < x < 0 \\ -x+1 & 0 < x < 1 \end{cases}$$

with period $L = 2m$

3. Find the Fourier coefficients corresponding to the function

$$f(t) = \begin{cases} -\dfrac{1}{2} & -2 < t < 0 \\ 1 & 0 < t < 2 \end{cases}$$

with period $T = 4s$

Section 7.2 Fourier Transforms

4. Given the following even functions $f(x)$ find the Fourier cosine transform $a(k)$ and plot it

$$f(x) = \frac{1}{x^2 + 9}$$

$$f(x) = e^{-3x}$$

$$f(x) = e^{-3x^2}$$

5. Given the following odd functions $f(x)$ find the Fourier sine transform $a(k)$ and plot it

$$f(x) = \frac{x}{x^2 + 9}$$

$$f(x) = \frac{1}{x}$$

$$f(x) = xe^{-3x^2}$$

6. Given the following function $f(t)$ find the complex Fourier coefficients

$$f(t) = \begin{cases} 2t+1 & -1/2 < t < 0 \\ -t+1 & 0 < t < 1 \end{cases}$$

7. Find the inverse Fourier transform of the following functions to find $f(x)$

$$a(k) = \frac{k}{k^2 + 9}$$

$$a(k) = \frac{1}{\sqrt{k}}$$

$$a(k) = e^{-3k}$$

Section 7.3 Laplace Transforms

8. The Laplace transform of a periodic function is given by

$$\mathcal{L}[f(t)] = \frac{1}{1-e^{-sT}} \int_0^T e^{-st} f(t)dt$$

Calculate the $\mathcal{L}[f(t)]$ where

$$f(t) = \begin{cases} t & 0 \le t < 1/2 \\ 1-t & 1/2 \le t < 1 \end{cases} \quad \text{with} \quad T = 1$$

9. Calculate the following Laplace transforms

(a) $\mathcal{L}[\cos(\omega t)]$

(b) $\mathcal{L}[\cosh(at)]$

(c) $\mathcal{L}[e^{-\lambda t} \cos(\omega t)]$

10. Calculate the following inverse Laplace transforms

(a) $\mathcal{L}^{-1}\left[\dfrac{1}{\sqrt{s-r}+a}\right]$

(b) $\mathcal{L}^{-1}\left[\dfrac{1}{\sqrt{s}(s-r)}\right]$

(c) $\mathcal{L}^{-1}\left[\dfrac{1-e^{-r\sqrt{s}}}{s}\right]$

11. Calculate the following inverse Laplace transforms

(a) $\mathcal{L}^{-1}\left[\dfrac{1}{\sqrt{s^2+r^2}}\right]$

(b) $\mathcal{L}^{-1}\left[\dfrac{s^2}{\left(s^2+r^2\right)^{3/2}}\right]$

(c) $\mathcal{L}^{-1}\left[\dfrac{e^{-r/s}}{s}\right]$

12. Use Laplace transform methods to solve for the charge $q(t)$ in an LC series combination driven by a time-dependent voltage source according to the differential equation

$$L\frac{d^2q}{dt^2}+\frac{q}{C}=V_0 t\sin(\omega t)$$

where $q(0)=\dot{q}(0)=0$

8 PARTIAL DIFFERENTIAL EQUATIONS

Chapter Outline

8.1 TYPES OF PARTIAL DIFFERENTIAL EQUATIONS

A partial differential equation (PDE) involves partial derivatives of an unknown function of several variables. The unknown function may be a scalar or a vector function of position and time such as temperature $T(x, y, z, t)$ or electric field $\mathbf{E}(x, y, z, t)$. The order of the PDE is determined by the order of the highest derivative in the equation. In this section, we review several first and second order PDEs important in physics.

8.1.1 First Order PDEs

Examples of first order PDEs in physics include Maxwell's equations. Maxwell's two divergence equations for electric $\mathbf{E}$ and magnetic $\mathbf{B}$ fields are

$$\nabla \cdot \mathbf{E} = \frac{\rho}{\varepsilon_0} \tag{8.1.1}$$

$$\nabla \cdot \mathbf{B} = 0 \qquad (8.1.2)$$

where ρ is the electric charge density and ε_0 is the permittivity of free space. Maxwell's two curl equations in the absence of sources are

$$\nabla \times \mathbf{E} = -\frac{\partial \mathbf{B}}{\partial t} \qquad (8.1.3)$$

$$\nabla \times \mathbf{B} = \mu_0 \varepsilon_0 \frac{\partial \mathbf{E}}{\partial t} \qquad (8.1.4)$$

These equations show that time-changing magnetic fields give rise to electric fields with curl. Also, time-changing electric fields cause magnetic fields with curl. Another important first order PDE is the continuity equation

$$\nabla \cdot \mathbf{J} = -\frac{\partial \rho}{\partial t} \qquad (8.1.5)$$

where $\mathbf{J}$ is the current density and ρ is the charge density. The continuity equation may also be applied to fluid flow where r is fluid density and $\mathbf{J} = \rho \mathbf{v}$ is the product of fluid density and flow velocity $\mathbf{v}$.

8.1.2 Second Order PDEs

Linear second order partial differential equations of the form

$$A\frac{\partial^2 f}{\partial x^2} + B\frac{\partial^2 f}{\partial x \partial y} + C\frac{\partial^2 f}{\partial y^2} + D\frac{\partial f}{\partial x} + E\frac{\partial f}{\partial y} + Ff = G \qquad (8.1.6)$$

with two independent variables x, y and coefficients A, B, C, D, E, F and G are classified as elliptic if $B^2 - 4AC < 0$, parabolic if $B^2 - 4AC = 0$ or hyperbolic if $B^2 - 4AC > 0$. The PDE is homogeneous if $G = 0$ or inhomogeneous if $G \neq 0$. Common second order PDEs in physics include Laplace's equation, Poisson's equation, the diffusion equation and the wave equation. For diffusion and wave equations with one spatial dimension the variable y would correspond to the time t in the equation above.

8.1.3 Laplace's Equation

The elliptic partial differential equation known as Laplace's equation

$$\frac{\partial^2 \Phi}{\partial x^2} + \frac{\partial^2 \Phi}{\partial y^2} = 0 \qquad (8.1.7)$$

is encountered in electrostatics, magnetostatics, heat flow, and fluid dynamics when there are no sources or sinks. In three dimensions Laplace's equation is

$$\nabla^2\Phi = 0 \qquad (8.1.8)$$

8.1.4 Poisson's Equation

Poisson's equation

$$\frac{\partial^2\Phi}{\partial x^2} + \frac{\partial^2\Phi}{\partial y^2} = -q \qquad (8.1.9)$$

is an inhomogeneous elliptic equation with a source term q proportional to charge density in electrostatic problems or the heat source in steady state heat transfer problems. In three dimensions

$$\nabla^2\Phi = -q \qquad (8.1.10)$$

8.1.5 Diffusion Equation

The diffusion equation

$$\frac{\partial^2\Phi}{\partial x^2} = k\frac{\partial\Phi}{\partial t} \qquad (8.1.11)$$

is a parabolic equation where Φ may represent temperature for transient heat transfer problems or the wavefunction in the Schrödinger equation. In three dimensions, the diffusion equation becomes

$$\nabla^2\Phi = k\frac{\partial\Phi}{\partial t} \qquad (8.1.12)$$

8.1.6 Wave Equation

The hyperbolic PDE

$$\frac{\partial^2\Phi}{\partial x^2} - \frac{1}{c^2}\frac{\partial^2\Phi}{\partial t^2} = 0 \qquad (8.1.13)$$

may describe the propagation of matter waves and electromagnetic waves. In three dimensions using the d'Alembertian operator

$$\Box^2 = \nabla^2 - \frac{1}{c^2}\frac{\partial^2}{\partial t^2} \qquad (8.1.14)$$

the wave equation may be written compactly as

$$\Box^2 \Phi = 0 \tag{8.1.15}$$

8.1.7 Helmholtz Equation

The Helmholtz equation

$$\nabla^2 \Phi + k^2 \Phi = 0 \tag{8.1.16}$$

is a time-independent form of the wave equation useful for the study of vibrations and electromagnetic radiation.

8.1.8 Klein-Gordon Equation

The Klein-Gordon equation

$$(\Box^2 - \alpha^2)\Phi = 0 \tag{8.1.17}$$

is an early relativistic wave equation. In quantum field theory, solutions to the Klein-Gordon equation are scalar fields whose quanta are spin zero particles.

Example 8.1.1

Show that the concentration

$$C(x,t) = \frac{A}{t^{1/2}} \exp\left(-\frac{x^2}{4\mu t}\right) \tag{8.1.18}$$

satisfies the diffusion equation

$$D \frac{\partial^2 C}{\partial x^2} = \frac{\partial C}{\partial t} \tag{8.1.19}$$

and find the constant D in terms of the parameter μ.

Solution: Calculating the x-derivative

$$\frac{\partial C}{\partial x} = \frac{A}{t^{1/2}} \exp\left(-\frac{x^2}{4\mu t}\right)\left(-\frac{x}{2\mu t}\right) \tag{8.1.20}$$

The second derivative is then

$$\frac{\partial^2 C}{\partial x^2} = \frac{A}{t^{1/2}} \exp\left(-\frac{x^2}{4\mu t}\right)\left(-\frac{x}{2\mu t}\right)^2 + \frac{A}{t^{1/2}} \exp\left(-\frac{x^2}{4\mu t}\right)\left(-\frac{1}{2\mu t}\right) \tag{8.1.21}$$

Simplifying we have

$$\frac{\partial^2 C}{\partial x^2} = \frac{A}{t^{1/2}} \exp\left(-\frac{x^2}{4\mu t}\right)\left[\left(\frac{x}{2\mu t}\right)^2 - \frac{1}{2\mu t}\right] \tag{8.1.22}$$

The time derivative of the concentration is then

$$\frac{\partial C}{\partial t} = \frac{A}{t^{1/2}} \exp\left(-\frac{x^2}{4\mu t}\right)\left(\frac{x^2}{4\mu t^2}\right) - \frac{1}{2}\frac{A}{t^{3/2}} \exp\left(-\frac{x^2}{4\mu t}\right) \tag{8.1.23}$$

or

$$\frac{\partial C}{\partial t} = \frac{A}{t^{1/2}} \exp\left(-\frac{x^2}{4\mu t}\right)\left[\left(\frac{x^2}{4\mu t^2}\right) - \frac{1}{2t}\right] \tag{8.1.24}$$

Thus

$$\frac{\partial^2 C}{\partial x^2} = \frac{1}{\mu}\frac{\partial C}{\partial t} \tag{8.1.25}$$

and $D = \mu$.

Maple Examples

The following Maple worksheet demonstrates the solution of first order partial differential equations, the diffusion equation, the wave equation, and the KdV equation. Solutions to PDEs are verified by using the *pdetest* feature where a zero output confirms a given solution. The wave equation is expressed in Jet Notation using the *PDEtools* package.

Key Maple commands: *diff, FromJet, pdetest, pdsolve, ToJet*

Maple packages: *PDEtools*

restart

First Order PDE

PDE := diff(y(x, t), x) = −beta·diff(y(x, t), t)

$$\frac{\partial}{\partial x} y(x,t) = -\beta\left(\frac{\partial}{\partial t} y(x,t)\right)$$

soln := pdsolve(PDE)

$$y(x, t) = \_F1(-\beta x + t)$$

pdetest(soln, PDE)

$$0$$

A zero output of *pdetest* verifies the solution.

Diffusion Equation

PDE := *diff*(*C*(*x, t*), *x, x*) = alpha·*diff*(*C*(*x, t*), *t*)

$$\frac{\partial^2}{\partial x^2} C(x,t) = \alpha \left(\frac{\partial}{\partial t} C(x,t) \right)$$

soln := *pdsolve*(*PDE*)

$$\left(C(x,t) = \_F1(x) \_F2(t) \right)$$

and where $\left[\left\{ \dfrac{d^2}{dx^2} \_F1(x) = \_c_1 \_F1(x), \dfrac{d}{dt} \_F2(t) = \dfrac{\_c_1 \_F2(t)}{\alpha} \right\} \right]$

pdetest(soln, PDE)

$$0$$

Wave Equation

PDE := *diff*(*f*(*x, t*), *t, t*) = *k*²·*diff*(*f*(*x, t*), *x, x*)

$$\frac{\partial^2}{\partial t^2} f(x,t) = k^2 \left(\frac{\partial^2}{\partial x^2} f(x,t) \right)$$

soln := *pdsolve*(*PDE*)

$$f(x, t) = \_F1(kt + x) + \_F2(kt - x)$$

pdetest(soln, PDE)

$$0$$

KdV Equation

PDE := *diff*(*f*(*x, t*), *x, x, x*) + *diff*(*f*(*x, t*), *t*) + 6·*f*(*x, t*)·*diff*(*f*(*x, t*), *x*) = 0

$$\frac{\partial^3}{\partial x^3} f(x,t) + \frac{\partial}{\partial t} f(x,t) + 6 f(x,t) \left(\frac{\partial}{\partial x} f(x,t) \right) = 0$$

soln := *pdsolve*(*PDE*)

$$f(x,t) = -2\_C2^2 \tanh\left(\_C2x + \_C3t + \_C1\right)^2 + \frac{1}{6}\frac{8\_C2^3\_C3}{C2}$$

pdetest(soln, PDE)

$$0$$

Jet Notation

$$PDE := diff(f(x,y), x, x) + diff(f(x,y), y, y) = 0$$

$$\frac{\partial^2}{\partial x^2} f(x,y) + \frac{\partial^2}{\partial y^2} f(x,y) = 0$$

with(PDEtools):
ToJet(PDE,f(x,y))

$$f_{x,x} + f_{y,y} = 0$$

FromJet(%,f(x,y))

$$\frac{\partial^2}{\partial x^2} f(x,y) + \frac{\partial^2}{\partial y^2} f(x,y) = 0$$

8.2 THE HEAT EQUATION

The 3D heat equation is derived in this section. Steady state heat transfer problems are like electrostatic problems where the temperature satisfies Laplace's equation. A Laplace transform solution to the heat equation is then considered.

8.2.1 Transient Heat Flow

In one dimension, the time rate of change of thermal energy Q contained in a rod of cross-sectional area A and temperature gradient $\partial T/\partial x$ is

$$\frac{\partial Q}{\partial t} = \lambda A \frac{\partial T}{\partial x} \tag{8.2.1}$$

where λ is the thermal conductivity. In three dimensions, the change in thermal energy in a volume v bounded by a surface is

$$\frac{\partial Q}{\partial t} = \lambda \int_{surf} \nabla T \cdot d\mathbf{a} \tag{8.2.2}$$

where $\mathbf{F} = -\lambda \nabla T$ is the heat flux vector. We now seek to express the heat equation in terms of one unknown function T. Transforming the surface integral using Gauss's divergence theorem

$$\frac{\partial Q}{\partial t} = \lambda \int_{vol} \nabla \cdot \nabla T \, dv = \lambda \int_{vol} \nabla^2 T \, dv \tag{8.2.3}$$

The thermal energy stored in a mass m at constant temperature T is given by $Q = mCT$. If the temperature or mass density is variable

$$Q = C \int_{vol} T \rho \, dV \tag{8.2.4}$$

where C is the specific heat and ρ is the mass density. Taking the time derivative of Q and equating volume integrals for $\partial Q / \partial t$.

$$c \int_{vol} \frac{\partial T}{\partial t} \rho \, dV = \lambda \int_{vol} \nabla^2 T \, dV \tag{8.2.5}$$

Because the integration region is over an arbitrary volume, we may equate the integrands giving the heat equation

$$\nabla^2 T = \frac{1}{\alpha} \frac{\partial T}{\partial t} \tag{8.2.6}$$

where $\alpha = \lambda / C\rho$. Note that α has units of length^2 / time.

8.2.2 Steady State Heat Flow

For steady state heat flow where $\partial T / \partial t = 0$, the temperature is satisfied by Laplace's equation

$$\nabla^2 T = 0 \tag{8.2.7}$$

where the temperature T is analogous to the electrostatic potential V in electrostatics where

$$\nabla^2 V = 0 \tag{8.2.8}$$

The heat flux vector $\mathbf{F} = -\lambda \nabla T$ is analogous to the electric field $\mathbf{E} = -\nabla V$.

8.2.3 Laplace Transform Solution

Example 8.2.1

A temperature T_0 is applied to the edge of a bar at $t = 0$. Find the temperature $T(x, t)$ of the bar if it is initially at zero temperature.

Solution: The differential equation with initial conditions are

$$\alpha\frac{\partial^2 T}{\partial x^2} = \frac{\partial T}{\partial t} \qquad \begin{bmatrix} T(x,0) \\ T(0,t) \\ T(\infty,t) \end{bmatrix} = \begin{Bmatrix} 0 \\ T_0 \\ 0 \end{Bmatrix} \qquad (8.2.9)$$

Taking the Laplace transform of both the differential equation and the initial conditions gives

$$\alpha\frac{\partial^2 \Theta(s)}{\partial x^2} = s\Theta(s) - T(0) \qquad \begin{bmatrix} \Theta(x,0) \\ \Theta(0,s) \\ \Theta(\infty,s) \end{bmatrix} = \begin{Bmatrix} 0 \\ \dfrac{T_0}{s} \\ 0 \end{Bmatrix} \qquad (8.2.10)$$

where $T(0) = 0$. The solution to this equation is

$$\Theta(s) = A \cdot \exp\left(\sqrt{\frac{s}{\alpha}}x\right) + B \cdot \exp\left(-\sqrt{\frac{s}{\alpha}}x\right) \qquad (8.2.11)$$

Applying the boundary condition $T(x = \infty, t) = 0$ gives $A = 0$ and $B = \dfrac{T_0}{s}$ so that

$$\Theta(s) = \frac{T_0}{s} \cdot \exp\left(-\sqrt{\frac{s}{\alpha}}x\right) \qquad (8.2.12)$$

The inverse Laplace transform gives

$$T(x,t) = T_0 \text{erfc}\left(\frac{1}{2}\sqrt{\frac{1}{\alpha t}}x\right) \qquad (8.2.13)$$

where the complementary error function is given by

$$\text{erfc}(x) = \frac{2}{\sqrt{\pi}}\int_x^\infty e^{-t^2}\, dt \qquad (8.2.14)$$

Maple Examples

The following Maple worksheet demonstrates the separation of variables solution to the heat equation. Solutions to the heat equation are obtained including lateral heat loss and convection. These solutions are displayed using 3D and animated plots.

Key Maple commands: *diff, dsolve, expand, Laplacian, lhs, pds:-animate, pds:-plot3d, pdsolve, rhs, subs*

Maple packages: *with(VectorCalculus)*:

restart

Separation of Variables: Heat Equation 1D

with(VectorCalculus) :
heatEqn := expand(Laplacian(T(x, t), `cartesian`*[x, y, z])) = diff(T(x, t), t)*

$$\frac{\partial^2}{\partial x^2} T(x,t) = \frac{\partial}{\partial t} T(x,t)$$

SubsEqn := expand(subs(T(x, t) = T(x)·phi(t), heatEqn)

$$\left(\frac{d^2}{dx^2} T(x) \right) \phi(t) = T(x) \left(\frac{d}{dt} \phi(t) \right)$$

$$SepEqn := expand \left(\frac{1}{T(x) \cdot phi(t)} \cdot SubsEqn \right)$$

$$\frac{\dfrac{d^2}{dx^2} T(x)}{T(x)} = \frac{\dfrac{d}{dt} \phi(t)}{\phi(t)}$$

dsolve(rhs(SepEqn) = - lambda²)

$$\phi(t) = \_C1 e^{-\lambda^2 t}$$

dsolve(lhs(SepEqn) = − λ²)

$$T(x) = \_C1 \sin(\lambda x) + \_C2 \cos(\lambda x)$$

Heat Equation

$$PDE1 := diff(T(x,t), x, x) = \frac{1}{alpha} \cdot diff(T(x,t), t)$$

$$\frac{\partial^2}{\partial x^2} T(x,t) = \frac{\dfrac{\partial}{\partial t} T(x,t)}{\alpha}$$

ics := T(x, 0) = T₀·sin(x)

$$T(x, 0) = T_0 \sin(x)$$

pdsolve([PDE1, ics])

$$T(x, t) = \sin(x) e^{-\alpha t} T_0$$

Heat Equation: Lateral Heat Loss

$$PDE2 := diff(T(x,t),x,x) = \frac{1}{alpha} \cdot diff(T(x,t),t) + a \cdot T(x,t)$$

$$\frac{\partial^2}{\partial x^2} T(x,t) = \frac{\frac{\partial}{\partial t} T(x,t)}{\alpha} + aT(x,t)$$

$ICS := T(x,0) = \sin(2 \cdot pi \cdot x)$

$$T(x,0) = \sin(2 \pi x)$$

$BCS := T(0,t) = 0, T(1,t) = 0$

$$T(0,t) = 0, T(1,t) = 0$$

$pdsolve([PDE2, ICS, BCS])$

$$T(x,t) = \sum_{-Z1\sim=1}^{\infty} \frac{4(-1)^{1+\_Z1\sim} \pi \_Z1\sim \sin(\pi)\cos(\pi)\sin(\_Z1\sim \pi x)e^{-\alpha t(\pi^2 \_Z1\sim^2 + a)}}{\pi^2 \_Z1\sim^2 - 4\pi^2}$$

Heat Equation: Convection

$PDE3 := diff(T(x,t),x,x) = diff(T(x,t),t) + diff(T(x,t),x)$

$$\frac{\partial^2}{\partial x^2} T(x,t) = \frac{\partial}{\partial t} T(x,t) + \frac{\partial}{\partial x} T(x,t)$$

$ICS := \{T(x,0) = 1, T(0,t) = 0, D[1](T)(1,t) = 0\}$

$$\{T(0,t) = 0, T(x,0) = 1, D_1(T)(1,t) = 0\}$$

$pds := pdsolve(PDE3, ICS, numeric, time = t, range = 0 \ldots 1)$

$$\textbf{module() \ldots end module}$$

$pds:-plot3d(t = 0 \ldots 1, x = 0 \ldots 1)$

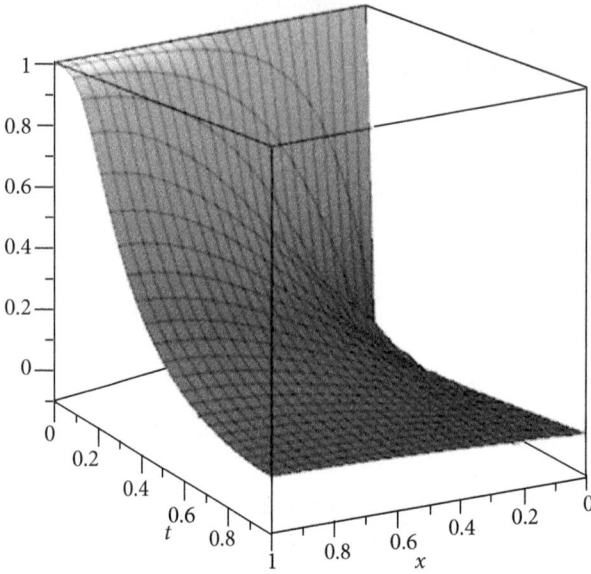

Figure 8.2.1: Solution to a PDE displayed as a surface plot.

Animate the Solution

$$pds\text{:-}animate(t = .2 \ldots 1, frames = 20)$$

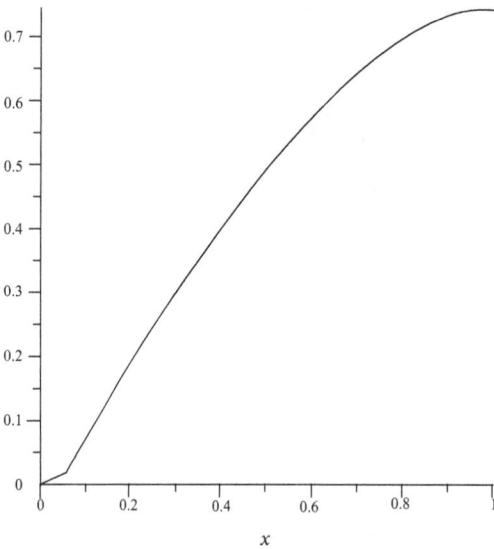

Figure 8.2.2: First frame of an animated solution to the heat equation.

8.3 SEPARATION OF VARIABLES

In this section, we illustrate the technique of separation of variables for solutions of the heat equation, Laplace's equation, the wave equation and the Helmholtz equation.

8.3.1 The Heat Equation

The heat equation in one dimension is

$$\frac{\partial T}{\partial t} = \alpha \frac{\partial^2 T}{\partial x^2} \tag{8.3.1}$$

To separate variables, a product solution is chosen

$$T(x, t) = T(x)\phi(t) \tag{8.3.2}$$

Substitution of the product solution into the heat equation

$$T(x)\frac{\partial \phi(t)}{\partial t} = \alpha \frac{\partial^2 T(x)}{\partial x^2}\phi(t) \tag{8.3.3}$$

and dividing by $T(x)\phi(t)$ gives

$$\alpha \frac{T''(x)}{T(x)} = \frac{\dot{\phi}(t)}{\phi(t)} \tag{8.3.4}$$

Since the left-hand side is only a function of x and the right-hand side is only a function of t each side must be a constant. Choosing our separation constant to be $-k$ we have the differential equations

$$\dot{\phi}(t) = -k\phi(t) \tag{8.3.5}$$

$$T''(x) + \frac{k}{\alpha}T(x) = 0 \tag{8.3.6}$$

with solutions

$$\phi(t) = c_1 \exp(-kt) \tag{8.3.7}$$

$$T(x) = c_2 \cos\sqrt{\frac{k}{\alpha}}x + c_3 \sin\sqrt{\frac{k}{\alpha}}x \tag{8.3.8}$$

where the constants c_1, c_2 and c_3 are determined by boundary conditions.

8.3.2 Laplace's Equation in Cartesian Coordinates

Laplace's equation $\nabla^2 V(x, y) = 0$ in Cartesian coordinates is

$$\frac{\partial^2 V}{\partial x^2} + \frac{\partial^2 V}{\partial y^2} = 0 \qquad (8.3.9)$$

where $-\infty < x < \infty$, $-\infty < y < \infty$. Applying separation of variables, we assume a product solution

$$V(x, y, z) = X(x)Y(y) \qquad (8.3.10)$$

Substituting the product form of V into Laplace's equation and dividing by V gives

$$\frac{1}{X}\frac{\partial^2 X}{\partial x^2} + \frac{1}{Y}\frac{\partial^2 Y}{\partial y^2} = 0 \qquad (8.3.11)$$

The variables x, y may vary independently. Thus, each term above must be equal to a constant $\pm\alpha^2$ if the sum is zero. If we choose

$$\frac{1}{X}\frac{\partial^2 X}{\partial x^2} = -\alpha^2 \quad \frac{1}{Y}\frac{\partial^2 Y}{\partial y^2} = \alpha^2 \qquad (8.3.12)$$

then

$$V(x, y) = \left\{ A\sin(\alpha x) + B\cos(\alpha x) \right\} \left\{ Ce^{-\alpha y} + De^{\alpha y} \right\} \qquad (8.3.13)$$

Expressed in a shorthand notation

$$V(x, y) = \begin{Bmatrix} \sin(\alpha x) \\ \cos(\alpha x) \end{Bmatrix} \begin{Bmatrix} \exp(\alpha y) \\ \exp(-\alpha y) \end{Bmatrix} \qquad (8.3.14)$$

For different values of the constants C and D we may choose hyperbolic functions

$$V(x, y) = \begin{Bmatrix} \sin(\alpha x) \\ \cos(\alpha x) \end{Bmatrix} \begin{Bmatrix} \sinh(\alpha y) \\ \cosh(\alpha y) \end{Bmatrix} \qquad (8.3.15)$$

We may also choose to work with complex exponentials for different values of the constants A and B. When separating variables if we had selected

$$\frac{1}{X}\frac{\partial^2 X}{\partial x^2} = \alpha^2 \quad \frac{1}{Y}\frac{\partial^2 Y}{\partial y^2} = -\alpha^2 \qquad (8.3.16)$$

then the solution to the X equation would have been real exponential functions

$$V(x, y) = \begin{Bmatrix} \exp(\alpha x) \\ \exp(-\alpha x) \end{Bmatrix} \begin{Bmatrix} \sin(\alpha y) \\ \cos(\alpha y) \end{Bmatrix} \qquad (8.3.17)$$

The boundary conditions often motivate the sign choice when separating variables. For example, if $V(x, 0) = 0$ then it would be preferable to work with functions such as $\sin(\alpha y)$ that are zero where $y = 0$.

Example 8.3.1

A square plate is held at a potential of $V = 0$ on the sides $y = 0$ and $y = L$. The potential is $V = V_0$ on sides $x = 0$ and $x = L$. Find the potential inside the square region.

Solution: With the potential of the form

$$V(x, y) = \{A \exp(\alpha x) + B \exp(-\alpha x)\}\{C \sin(\alpha y) + D \cos(\alpha y)\} \qquad (8.3.18)$$

our task is to find the constants. Applying the boundary conditions $V(0, y) = V(L, y) = 0$ we require that $D = 0$ and $\sin(\alpha L) = 0$. Thus $\alpha = n\pi/L$ so that we have different values of the constants for $n = 1, 2, 3, \ldots$. Absorbing the C_n into the A_n and B_n

$$V(x, y) = \sum_{n=0}^{\infty} \left(A_n \exp\left(\frac{n\pi}{L} x\right) + B_n \exp\left(-\frac{n\pi}{L} x\right) \right) \sin\left(\frac{n\pi}{L} y\right) \qquad (8.3.19)$$

Applying the boundary condition $V(0, y) = V_0$ gives

$$V_0 = \sum_{n=0}^{\infty} (A_n + B_n) \sin\left(\frac{n\pi}{L} y\right) \qquad (8.3.20)$$

The condition $V(L, y) = V_0$

$$V_0 = \sum_{n=0}^{\infty} \left(A_n e^{n\pi} + B_n e^{-n\pi} \right) \sin\left(\frac{n\pi}{L} y\right) \qquad (8.3.21)$$

Comparing these expressions

$$A_n e^{n\pi} + B_n e^{-n\pi} = A_n + B_n \qquad (8.3.22)$$

so that we can eliminate the constant

$$B_n = -A_n \frac{(1 - e^{n\pi})}{(1 - e^{-n\pi})} \quad \text{and} \qquad (8.3.23)$$

$$V(x, y) = \sum_{n=0}^{\infty} A_n \left(\exp\left(\frac{n\pi}{L} x\right) - \frac{(1 - e^{n\pi})}{(1 - e^{-n\pi})} \exp\left(-\frac{n\pi}{L} x\right) \right) \sin\left(\frac{n\pi}{L} y\right) \qquad (8.3.24)$$

A_n is found by applying the boundary condition at $V(L, y) = V_0$

$$V_0 = \sum_{n=0}^{\infty} A_n \left(e^{n\pi} + 1 \right) \sin\left(\frac{n\pi}{L} y \right) \tag{8.3.25}$$

Multiplying both sides of this expression by $\sin(n'\pi y/L)$ and integrating

$$V_0 \int_0^L \sin\left(\frac{n'\pi y}{L} \right) dy = \int_0^L \sum_{n=0}^{\infty} A_n \left(e^{n\pi} + 1 \right) \sin\left(\frac{n\pi y}{L} \right) \sin\left(\frac{n'\pi y}{L} \right) dy \tag{8.3.26}$$

Making use of orthogonality and using the identity

$$\sin^2\left(\frac{n\pi y}{L} \right) = \frac{1}{2}\left(1 - \cos\frac{2n\pi y}{L} \right) \tag{8.3.27}$$

and removing the primes since only terms with $n = n'$ are nonzero

$$V_0 \frac{L}{n\pi}\left(1 - \cos(n\pi)\right) = A_n \left(e^{n\pi} + 1 \right)\frac{L}{2} \tag{8.3.28}$$

Now $\cos(n\pi) = (-1)^n$ so that A_n is zero for even n and the sum is over odd n

$$V(x, y) = \frac{4V_0}{\pi} \sum_{n=1,3,5\ldots}^{\infty} \frac{1}{n} \frac{1}{(e^{n\pi}+1)} \left(\exp\left(\frac{n\pi}{L} x \right) - \frac{(1-e^{n\pi})}{(1-e^{-n\pi})} \exp\left(-\frac{n\pi}{L} x \right) \right) \sin\left(\frac{n\pi}{L} y \right) \tag{8.3.29}$$

Note that this solution may also be expressed using hyperbolic functions.

8.3.3 Laplace's Equation in Cylindrical Coordinates

Laplace's equation $\nabla^2 V(r, z) = 0$ in cylindrical coordinates with axial symmetry (no variation in the ϕ-direction) is

$$\frac{1}{r}\frac{\partial}{\partial r}\left(r\frac{\partial V}{\partial r} \right) + \frac{\partial^2 V}{\partial z^2} = 0 \tag{8.3.30}$$

where $r \geq 0$, $-\infty < z < \infty$. Substituting $V(r, z) = R(r)Z(z)$ with separation constant k^2 gives the differential equations

$$\frac{1}{R}\frac{1}{r}\frac{\partial}{\partial r}\left(r\frac{\partial R}{\partial r} \right) = -k^2 \qquad \frac{1}{Z}\frac{\partial^2 Z}{\partial z^2} = k^2 \tag{8.3.31}$$

so that we may construct our solution

$$V(r,z) = \begin{Bmatrix} J_0(kr) \\ N_0(kr) \end{Bmatrix} \begin{Bmatrix} \exp(kz) \\ \exp(-kz) \end{Bmatrix} \tag{8.3.32}$$

where $J_0(kr)$ and $N_0(kr)$ are Bessel functions of zero order. If we had selected

$$\frac{1}{R}\frac{1}{r}\frac{\partial}{\partial r}\left(r\frac{\partial R}{\partial r}\right) = k^2 \qquad \frac{1}{Z}\frac{\partial^2 Z}{\partial z^2} = -k^2 \tag{8.3.33}$$

then

$$V(r,z) = \begin{Bmatrix} I_0(kr) \\ K_0(kr) \end{Bmatrix} \begin{Bmatrix} \sin(kz) \\ \cos(kz) \end{Bmatrix} \tag{8.3.34}$$

where $I_0(kr)$ and $K_0(kr)$ are modified Bessel functions of zero order.

Example 8.3.2

A cylinder of height L and radius R is at a potential of $V = 0$ on the bottom $z = 0$ and top $z = L$ end caps. The potential is $V = V_0$ on the side $r = R$. Find the potential inside the cylindrical region.

Solution: The potential is of the form

$$V(r,z) = \{AI_0(kr) + BK_0(kr)\}\{C\sin(kz) + D\cos(kz)\} \tag{8.3.35}$$

with the constants A, B, C and D to be determined.

We set $B = 0$ since $K_0(0) \to \infty$. Applying the boundary conditions $V(0, r) = V(L, r) = 0$ gives $D = 0$ and $\sin(kL) = 0$. Thus $k_n = n\pi/L$ for $n = 1, 2, 3, \dots$. Absorbing the C_n into the A_n

$$V(r,z) = \sum_{n=0}^{\infty} A_n I_0\left(\frac{n\pi}{L}r\right)\sin\left(\frac{n\pi}{L}z\right) \tag{8.3.36}$$

A_n is found by applying the boundary condition at $V(R, z) = V_0$

$$V_0 = \sum_{n=0}^{\infty} A_n I_0\left(\frac{n\pi}{L}R\right)\sin\left(\frac{n\pi}{L}z\right) \tag{8.3.37}$$

Multiplying both sides of this expression by $\sin(n'\pi z/L)$ and integrating

$$V_0 \int_0^L \sin\left(\frac{n'\pi z}{L}\right)dz = \int_0^L \sum_{n=0}^{\infty} A_n I_0\left(\frac{n\pi}{L}R\right)\sin\left(\frac{n\pi z}{L}\right)\sin\left(\frac{n'\pi z}{L}\right)dz \tag{8.3.38}$$

Removing the primes since only terms with $n = n'$ are nonzero as in the previous example

$$V_0 \frac{L}{n\pi}\left(1 - \cos(n\pi)\right) = A_n I_0\left(\frac{n\pi}{L} R\right)\frac{L}{2} \tag{8.3.39}$$

Now $\cos(n\pi) = (-1)^n$ so that A_n is zero for even n and the sum is over odd n

$$V(r,z) = \frac{4V_0}{\pi} \sum_{n=1,3,5...}^{\infty} \frac{1}{n} \frac{I_0\left(\dfrac{n\pi}{L} r\right)}{I_0\left(\dfrac{n\pi}{L} R\right)} \sin\left(\frac{n\pi}{L} z\right) \tag{8.3.40}$$

8.3.4 Wave Equation

Applying separation of variables for solution of the wave equation

$$\nabla^2 V - \frac{1}{c^2}\frac{\partial^2 V}{\partial t^2} = 0 \tag{8.3.41}$$

we substitute the product

$$V(\mathbf{r},t) \quad V(\mathbf{r}) \quad (t) \tag{8.3.42}$$

into the wave equation. Dividing by the product $V(\mathbf{r})\phi(t)$ then gives

$$\underbrace{\frac{\nabla^2 V(\mathbf{r})}{V(\mathbf{r})}}_{-k^2} - \underbrace{\frac{1}{c^2}\frac{1}{\phi(t)}\frac{\partial^2 \phi(t)}{\partial t^2}}_{-k^2} = 0 . \tag{8.3.43}$$

Choosing the separation constant to be $-k^2$ the differential equation for $\phi(t)$ becomes

$$\frac{\partial^2 \phi(t)}{\partial t^2} = -k^2 c^2 \phi(t) \tag{8.3.44}$$

with solutions $\phi(t) = A\sin(\omega t) + B\cos(\omega t)$ where $c = \omega/k$. The Helmholtz equation gives the spatial part of the wave equation

$$\nabla^2 V(\mathbf{r}) + k^2 V(\mathbf{r}) = 0 \tag{8.3.45}$$

8.3.5 Helmholtz Equation in Cylindrical Coordinates

The Helmholtz equation $(\nabla^2 + k^2)V(r, \phi) = 0$ in cylindrical coordinates with no variation in the z-direction is

$$\frac{1}{r}\frac{\partial}{\partial r}\left(r\frac{\partial V}{\partial r}\right) + \frac{1}{r^2}\frac{\partial^2 V}{\partial \phi^2} + k^2 V = 0 \tag{8.3.46}$$

Substituting $V(r,\phi) = R(r)\Phi(\phi)$ and multiplying by r^2

$$r\frac{\partial}{\partial r}\left(r\frac{\partial R}{\partial r}\right)\Phi + \frac{\partial^2\Phi}{\partial\phi^2}R + k^2r^2R\Phi = 0 \qquad (8.3.47)$$

Dividing by $R\Phi$ we obtain

$$\underbrace{r\frac{\partial}{\partial r}\left(r\frac{\partial R}{\partial r}\right)\frac{1}{R} + k^2r^2}_{n^2} + \underbrace{\frac{\partial^2\Phi}{\partial\phi^2}\frac{1}{\Phi}}_{-n^2} = 0 \qquad (8.3.48)$$

Choosing n^2 as a separation the Φ equation becomes

$$\frac{\partial^2\Phi}{\partial\theta^2} = -n^2\Phi \qquad (8.3.49)$$

so that

$$\Phi(\phi) = C_n\cos(n\phi) + D_n\sin(n\phi) \qquad (8.3.50)$$

The R equation becomes

$$r^2\frac{\partial^2 R}{\partial r^2} + r\frac{\partial R}{\partial r} + (k^2r^2 - n^2)R = 0 \qquad (8.3.51)$$

with solution

$$R(r) = A_n J_n(kr) + B_n N_n(kr) \qquad (8.3.52)$$

so that we may construct our solution

$$V(r,z) = \left\{\begin{array}{c} J_n(kr) \\ N_n(kr) \end{array}\right\}\left\{\begin{array}{c} \sin(n\phi) \\ \cos(n\phi) \end{array}\right\} \qquad (8.3.53)$$

8.3.6 Helmholtz Equation in Spherical Coordinates

In spherical coordinates with axial symmetry, the Helmholtz equation is

$$\frac{1}{r^2}\frac{\partial}{\partial r}\left(r^2\frac{\partial V}{\partial r}\right) + \frac{1}{r^2}\frac{1}{\sin\theta}\frac{\partial}{\partial\theta}\left(\sin\theta\frac{\partial V}{\partial\theta}\right) + k^2V = 0 \qquad (8.3.54)$$

Substituting $V(r,\theta) = R(r)\Theta(\theta)$ with separation constant $\ell(\ell+1)$ we obtain

$$\underbrace{\frac{1}{R}\frac{\partial}{\partial r}\left(r^2\frac{\partial R}{\partial r}\right) + r^2k^2}_{\ell(\ell+1)} + \underbrace{\frac{1}{\sin\theta}\frac{\partial}{\partial\theta}\left(\sin\theta\frac{\partial\Theta}{\partial\theta}\right)}_{-\ell(\ell+1)} = 0 \qquad (8.3.55)$$

The Θ equation is

$$\frac{1}{\sin\theta}\frac{\partial}{\partial\theta}\left(\sin\theta\frac{\partial\Theta}{\partial\theta}\right) = -\ell(\ell+1) \qquad (8.3.56)$$

with the R equation

$$\frac{\partial^2 R}{\partial r^2} + 2r\frac{\partial R}{\partial r} + \left[r^2 k^2 - \ell(\ell+1)\right]R = 0 \qquad (8.3.57)$$

so that our general solution is

$$V(r,\theta) = \begin{Bmatrix} j_\ell(kr) \\ n_\ell(kr) \end{Bmatrix} \begin{Bmatrix} P_\ell(\cos\theta) \\ Q_\ell(\cos\theta) \end{Bmatrix} \qquad (8.3.58)$$

where j_ℓ, n_ℓ are spherical Bessel functions and P_ℓ, Q_ℓ are Legendre functions of the first and second kind, respectively.

Maple Examples

The following Maple worksheet demonstrates the separation of angular and radial components of Laplace's equation in spherical and cylindrical coordinates.

Key Maple commands: *expand, Laplacian, SetCoordinates, subs*

Maple Packages: *with(VectorCalculus)*:

restart

Separation of Variables: Spherical Coordinates

with(VectorCalculus) :
LapEqn := expand(Laplacian(V(r, theta), 'spherical'[r, theta, phi])) = 0

$$\frac{2\left(\frac{\partial}{\partial r}V(r,\theta)\right)}{r} + \frac{\partial^2}{\partial r^2}V(r,\theta) + \frac{\cos(\theta)\left(\frac{\partial}{\partial\theta}V(r,\theta)\right)}{r^2\sin(\theta)} + \frac{\frac{\partial^2}{\partial\theta^2}V(r,\theta)}{r^2} = 0$$

SubsEqn := expand(subs(V(r, theta) = R(r)·Theta(theta), LapEqn))

$$\frac{2\left(\frac{d}{dr}R(r)\right)\Theta(\theta)}{r} + \left(\frac{d^2}{dr^2}R(r)\right)\Theta(\theta) + \frac{\cos(\theta)R(r)\left(\frac{d}{d\theta}\Theta(\theta)\right)}{r^2\sin(\theta)}$$

$$+ \frac{R(r)\left(\frac{d^2}{d\theta^2}\Theta(\theta)\right)}{r^2} = 0$$

$$SepEqn := expand\left(\frac{r^2}{R(r)\cdot Theta(theta)}\cdot SubsEqn\right)$$

$$\frac{2r\left(\dfrac{d}{dr}R(r)\right)}{R(r)} + \frac{r^2\left(\dfrac{d^2}{dr^2}R(r)\right)}{R(r)} + \frac{\cos(\theta)R(r)\left(\dfrac{d}{d\theta}\Theta(\theta)\right)}{\Theta(\theta)\sin(\theta)} + \frac{\dfrac{d^2}{d\theta^2}\Theta(\theta)}{\Theta(\theta)} = 0$$

restart

Separation of Variables: Cylindrical Coordinates

with(VectorCalculus) :
SetCoordinates('cylindrical' [r, phi, z]);

$$cylindrical_{r,\,\phi,\,z}$$

Lap := Laplacian(f(r, z))

$$\frac{\dfrac{\partial}{\partial r}f(r,z)+r\left(\dfrac{\partial^2}{\partial r^2}f(r,z)\right)+r\left(\dfrac{\partial^2}{\partial z^2}f(r,z)\right)}{r}$$

LapEqn := simplify(expand(Lap) = 0)

$$\frac{\dfrac{\partial}{\partial r}f(r,z)}{r}+\frac{\partial^2}{\partial r^2}f(r,z)+\frac{\partial^2}{\partial z^2}f(r,z)=0$$

SubEqn := expand(subs(f(r, z) = R(r)·Z(z), LapEqn))

$$\frac{\left(\dfrac{d}{dr}R(r)\right)Z(z)}{r}+\left(\dfrac{d^2}{dr^2}R(r)\right)Z(z)+R(r)\left(\dfrac{d^2}{dz^2}Z(z)\right)=0$$

$$SepEqn := expand\left(\frac{1}{R(r)\cdot Z(z)}\cdot SubEqn\right)$$

$$\frac{\dfrac{d}{dr}R(r)}{R(r)r}+\frac{\dfrac{d^2}{dr^2}R(r)}{R(r)}+\frac{\dfrac{d^2}{dz^2}Z(z)}{Z(z)}=0$$

8.4 MATLAB EXAMPLES

MATLAB is ideally suited for numerically solving and graphically displaying solutions of partial differential equations. In this section, MATLAB's PDE Toolbox is described and demonstrated in finding the resonant vibrations in a plate. A numerical solution to the transient heat equation is given using the method of finite differences. The solution to a boundary value problem obtained by using separation of variables is then plotted.

Key MATLAB commands: *besseli, mesh, meshc*

Programming: for loops

Section 8.1 Types of Partial Differential Equations

Typing 'pdetool' at the Command Prompt

```
>> pdetool
```

launches the graphical user interface (GUI) of the PDE Toolbox shown in Figure 8.4.1.

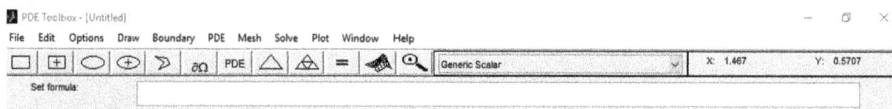

Figure 8.4.1: Toolbar of MATLAB's PDE Toolbox.

The work flow on the toolbar is from left to right in specifying the type of PDE to be solved (Figure 8.4.2), constructing a solution region, assigning boundary conditions, building a finite element mesh, solving the problem and graphically viewing the solution.

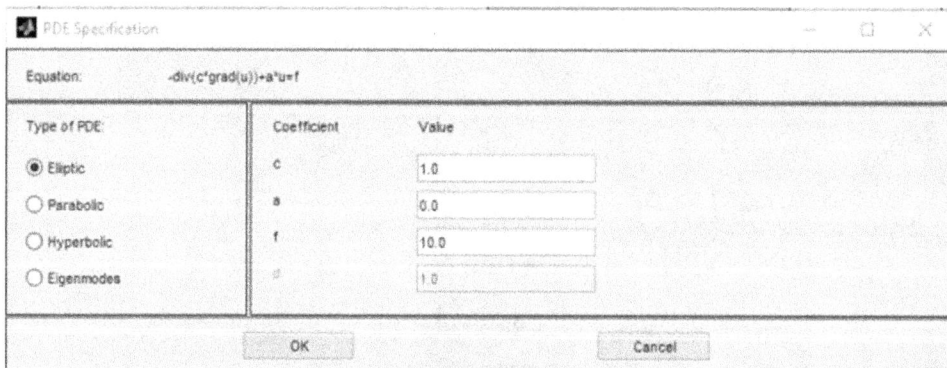

Figure 8.4.2: GUI for selecting type of PDE.

Types of PDEs supported included elliptic, parabolic, hyperbolic, and eigenmode. The elliptic PDE with time-independent u is of the form

$$-\nabla \cdot \left(c\nabla u\right) + au = f$$

The parabolic PDE with time-dependent u is

$$d\frac{\partial u}{\partial t} - \nabla \cdot (c\nabla u) + au = f$$

The hyperbolic PDE with time-dependent u is given by

$$d\frac{\partial^2 u}{\partial t^2} - \nabla \cdot \left(c\nabla u\right) + au = f$$

The eigenvalue equation

$$-\nabla \cdot \left(c\nabla u\right) + au = \lambda du$$

where the eigenvalues λ are calculated by the solver. In the above PDEs, the 2D potential is of form $u(x, y)$ or $u(x, y, t)$. Note that the coefficients c and a may be constant or functions of (x, y) as well. The function f may have time dependence in parabolic and hyperbolic PDEs.

The type of PDE may also be determined by specifying one of the following applications under the Options tab, including:

Generic Scalar
Generic System
Structural Mechanics, Plane Stress
Structural Mechanics, Plane Strain
Electrostatics
Magnetostatics
AC Power Electromagnetics
Conductive Media DC
Heat Transfer
Diffusion

Once the application/type of PDE is chosen, the model is constructed and the boundary conditions are assigned to the edges of the model. Types of boundary conditions include Neumann (derivative of u), Dirichlet (value of u), and Mixed (combination of Neumann and Dirichlet) boundary conditions.

Example 8.4.1

Find the modes of vibration of a membrane in the shape of a polygon with an elliptical hole.

Step I: The problem type is specified as an eigenvalue problem with coefficients $c = 1$, $a = 0$ and $d = 1$.

Step II: The model is drawn as shown in Figure 8.4.3 using the graphical line and ellipse tools. The boundary conditions are specified as Dirichlet with $h = 1$ and $r = 0$ on the straight edges and on the elliptical hole.

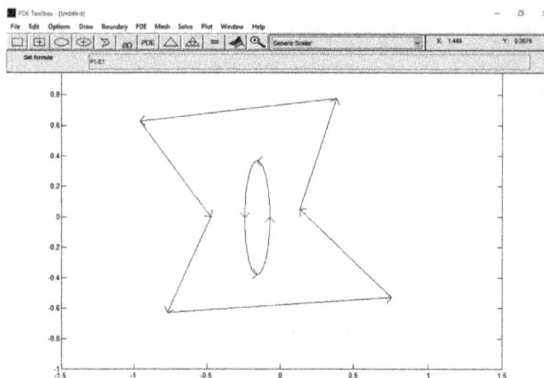

Figure 8.4.3: Drawing the model and specifying boundary conditions.

Step III: The finite element mesh shown in Figure 8.4.4 is generated using the triangular toolbar buttons and refined as desired.

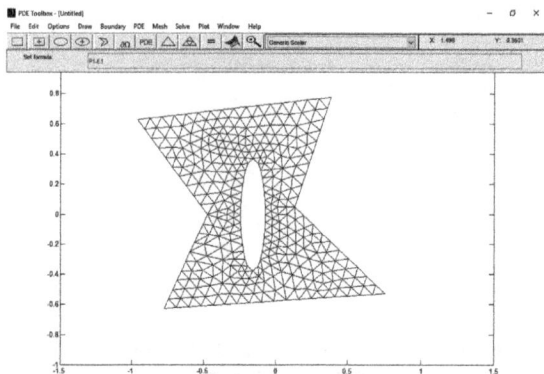

Figure 8.4.4: Finite element mesh.

The refined mesh in Figure 8.4.5 consists of 1234 nodes and 2256 triangular elements.

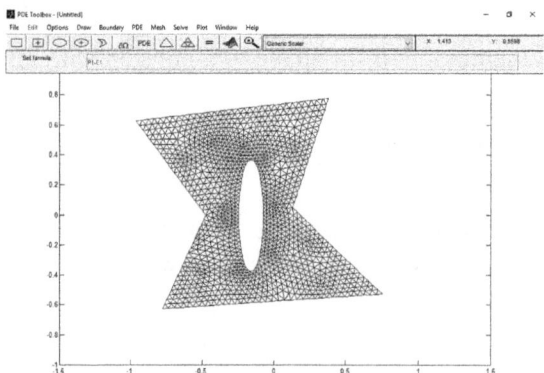

Figure 8.4.5: Refined finite element mesh.

Step IV: The problem is solved by clicking on the '=' button on the toolbar. Solutions corresponding to the first three resonant frequencies (54.34, 66.82, 74.69) are visualized by choosing a contour map of the potential with colormap bone (Figure 8.4.6).

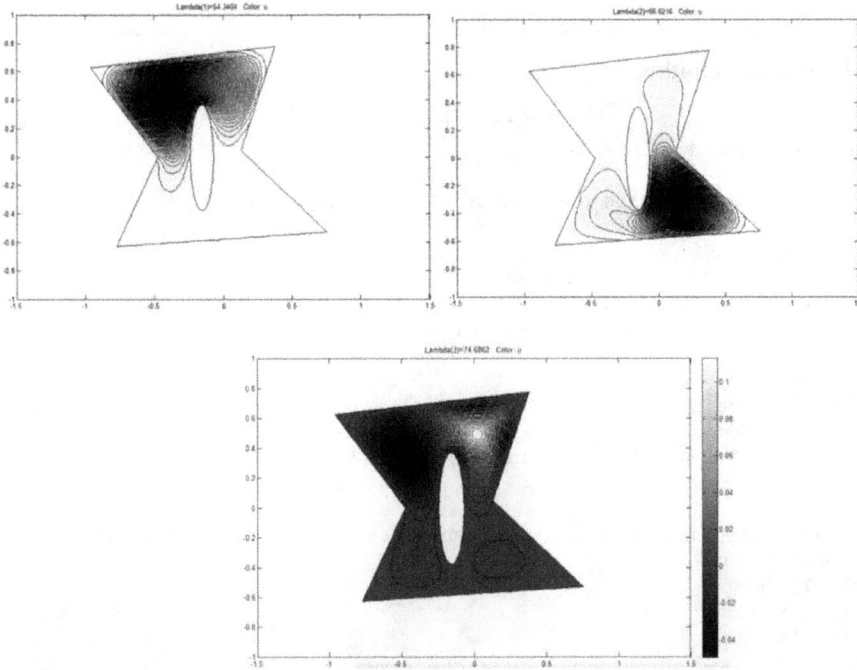

Figure 8.4.6: Shaded contour solutions to the eigenvalue problem.

The fourth resonant frequency (85.08 Hz) is plotted using 3D contour levels (Figure 8.4.7).

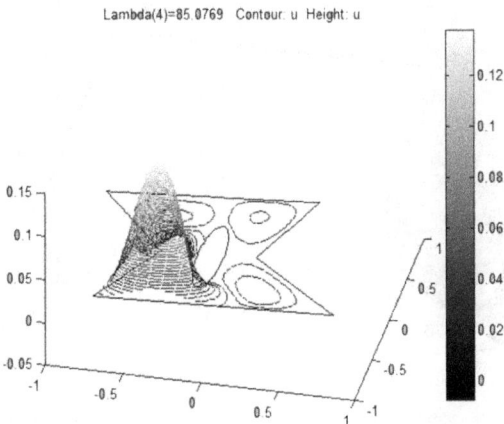

Figure 8.4.7: 3D contour plot of the fourth eigenvalue solution.

Section 8.2 The Heat Equation

The following MATLAB code demonstrates the finite difference method for solving the transient heat equation (also see Figure 8.4.8) for the temperature $T(x, t)$ for a body with initial condition

$$T(x, 0) = 0 \text{ for } 0 < x < L$$

and boundary conditions

$$T(0, t) = T(L, t) = 100$$

```
% parameter r= dt*alpha/dx^2
r=0.2;
% specify the grid size and number of time steps
grid_size = 50;
time_steps= 1000;
% Initialize the temperature matrix
T=zeros(time_steps,grid_size);
% specify boundary temperatures switched on at t=0;
T(:,1)= 100;
T(:,grid_size)=100;
for n = 1:time_steps-1
    for i=2:grid_size-1
        T(n+1,i)= T(n,i)+r*(T(n,i+1)-2*T(n,i)+T(n,i-1));
    end
end
% create a 3D surface plot (with contours) of the time dependent
    temperature
meshc(T)
xlabel('x')
ylabel('t')
zlabel('T')
title('T(x,t)')
colormap bone
```

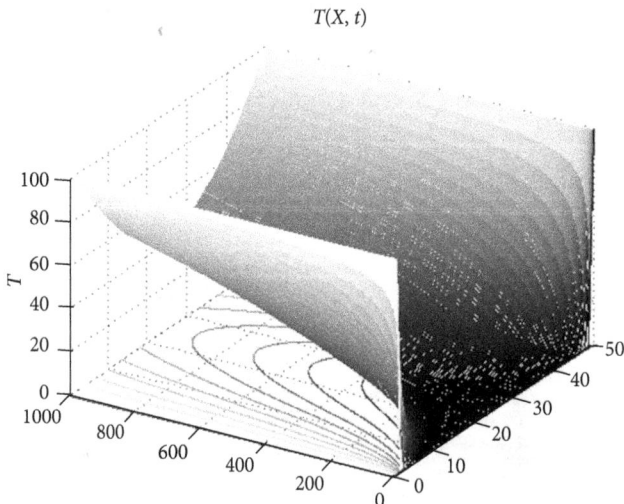

Figure 8.4.8: Solution to the transient heat equation displaced as a surface with contours.

Section 8.3 Separation of Variables

Below the separation of variables solution to Laplace's equation obtained in Section 8.3 for a cylinder with constant potential specified on the cylindrical surface with grounded endcaps is coded in MATLAB (see also Figure 8.4.9).

```
% Analytical solution of Laplace's equation in a cylinder

% specify array dimensions
imax=20;
jmax=20;

% number of term carried out in the infinite sum
nmax=201;

% initialize the array
V=zeros(imax,jmax);

L=1.0;
R=1.0;

% evaluate the potenial over the square region
for i=1:imax
for j=1:jmax

    r=(i-1)/(imax-1);
    z=(j-1)/(jmax-1);
for n=1:2:nmax

    G1=sin(n*pi*z/L)/n;
    G2=besseli(0,n*pi*r/L)/besseli(0,n*pi*R/L);
    V(i,j)=V(i,j)+ G1*G2;
end

end
end

V=V*4/pi;
mesh(V) % plot the results
xlabel('z')
ylabel('r')
zlabel('V')
title('V(r,z)')
colormap bone
```

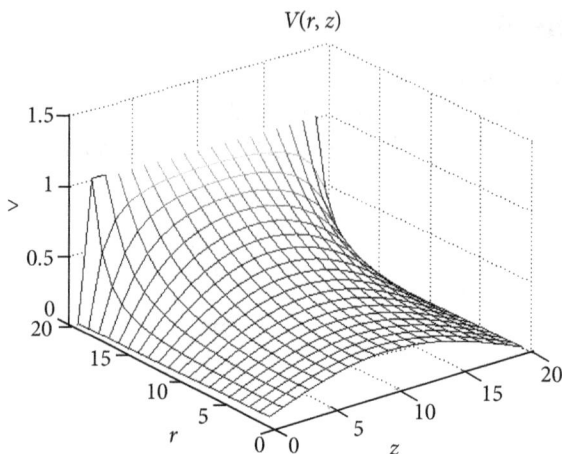

Figure 8.4.9: Solution to Laplace's equation in cylindrical coordinates displayed as a surface plot.

8.5 EXERCISES

Section 8.1 Types of Partial Differential Equations

1. Determine the order of the following partial differential equations. State if the PDE is linear or nonlinear

(a) $\dfrac{\partial^2 \Omega}{\partial u^2} + \dfrac{\partial^2 \Omega}{\partial v^2} + \dfrac{\partial^2 \Omega}{\partial u \partial v} = 6$ with independent variables (u, v) and dependent variable Ω

(b) $\dfrac{\partial^2 y}{\partial x^2} + 2\left(\dfrac{\partial y}{\partial x}\right)^2 + y = 0$ with independent variable x and dependent variable y

(c) $\dfrac{\partial^4 R}{\partial x^4} + 2\dfrac{\partial^2 R}{\partial x \partial y} + \dfrac{\partial R}{\partial y} = 0$ with independent variables (x, y) and dependent variable R

2. State if the following linear PDEs are elliptic, hyperbolic, or parabolic. Also indicate if the PDEs are homogeneous or inhomogeneous

(a) $\dfrac{\partial^2 f}{\partial x^2} + \dfrac{\partial^2 f}{\partial x \partial y} + 3\dfrac{\partial^2 f}{\partial y^2} = 0$ with independent variables (x, y) and dependent variable f

(b) $4\dfrac{\partial^2 R}{\partial r^2} + \dfrac{\partial^2 R}{\partial r \partial t} + \dfrac{\partial^2 R}{\partial t^2} + 2\dfrac{\partial R}{\partial r} = 3R$ with independent variables (r, t) and dependent variable R

(c) $\dfrac{\partial^4 S}{\partial r^4} + \dfrac{\partial^2 S}{\partial r \partial \theta} + 2\dfrac{\partial^2 S}{\partial \theta^2} + 2\dfrac{\partial S}{\partial r} = 3$ with independent variables (r, θ) and dependent variable S

3. Show that $V(x, y) = e^{-2x}\sin 2y$ is a solution to

$$\frac{\partial^2 V}{\partial x^2} + \frac{\partial^2 V}{\partial y^2} = 0$$

$$V\left(x, \frac{\pi}{2}\right) = 0 \quad V(x, 0) = 0$$

4. Show that $\psi(x, t) = e^{-2x}\sin 4t$ is a solution to

$$\frac{\partial^2 \psi}{\partial x^2} = \frac{\partial \psi}{\partial t}$$

$$V(x, 0) = e^{-2x}$$

5. Solve the following differential equations by direct integration

$$\frac{\partial^2 T}{\partial x \partial y} = xy^2$$

$$y\frac{\partial^2 S}{\partial x \partial y} + \frac{\partial S}{\partial y} = 0$$

6. Substitute $\psi(x) = e^{-\lambda t}(c_1 \cos(kx) + c_2 \sin(kx))$ into the PDE

$$a^2 \frac{\partial^2 \psi}{\partial x^2} - \frac{\partial^2 \psi}{\partial t^2} = b\psi$$

to find b in terms of a, λ and k

Section 8.2 The Heat Equation

7. Solve the heat equation

$$\frac{\partial^2}{\partial x^2}T = \frac{1}{\alpha}\frac{\partial T}{\partial t}$$

for a bar of length L with boundary conditions
$T(0, t) = 0, T(L, t) = 0$

and initial conditions

$$T(x,0) = T_0 \sin\left(\frac{\pi x}{L}\right)$$

8. Solve the heat equation

$$\frac{\partial^2}{\partial x^2} T = \frac{1}{\alpha}\frac{\partial T}{\partial t}$$

for a bar of length L with boundary conditions
$T(0, t) = T_0$, $T(L, t) = 0$
and initial conditions

$$T(x,0) = T_0\left(1 - \frac{x}{L}\right)$$

9. Solve the heat equation

$$\frac{\partial^2}{\partial x^2} T = \frac{1}{\alpha}\frac{\partial T}{\partial t}$$

for a bar of length L with initial conditions
$T(x, 0) = 0$
and boundary conditions
$T(0, t) = T_0$, $T(L, t) = 0$

10. Solve the heat equation

$$\frac{\partial^2 T}{\partial r^2} + \frac{1}{r}\frac{\partial T}{\partial r} = \frac{1}{\alpha}\frac{\partial T}{\partial t}$$

for a disk of radius R with initial conditions
$T(r, 0) = 0$
and boundary conditions
$T(R, t) = T_0$

11. Solve the heat equation

$$\frac{\partial^2}{\partial x^2} T = \frac{1}{\alpha}\frac{\partial T}{\partial t}$$

for a semi-infinite bar with initial conditions
$T(x, 0) = 0$
and boundary conditions
$T(0, t) = T_0 \sin(t)$

12. The $x = 0$ plane of a semi-infinite heat conductor is given a brief temperature pulse T_0 at $t = a$ described by a delta function $T_0\delta(t-a)$. Use Laplace transform methods to solve the heat equation

$$\frac{\partial T}{\partial t} = \alpha \frac{\partial^2 T}{\partial x^2}$$

with boundary conditions

$$\begin{bmatrix} T(x,0) \\ T(0,t) \\ T(\infty,t) \end{bmatrix} = \begin{Bmatrix} 0 \\ T_0\delta(t-a) \\ 0 \end{Bmatrix}$$

Section 8.3 Separation of Variables

13. Solve Laplace's equation for the potential V over a square region of length L with boundary conditions $V(0, y) = 0$, $V(x, 0) = 0$, $V(L, y) = V_0$, and $V(x, L) = V_0$.

14. Solve Laplace's equation for the potential V between two cylindrical surfaces of radii a and b with respective boundary values $V(a) = V_0$ and $V(b) = 0$

15. Show that Poisson's equation $\nabla^2 V(r,z) = -\dfrac{1}{\varepsilon_0}\rho(r,z)$ has the form

$$\frac{1}{r}\frac{\partial}{\partial r}\left(r\frac{\partial V}{\partial r}\right) + \frac{\partial^2 V}{\partial z^2} = -\frac{1}{\varepsilon_0}\rho(r,z)$$

in cylindrical coordinates with axial symmetry.

16. Show that Poisson's equation in spherical coordinates $\nabla^2 V(r,\theta) = -\dfrac{1}{\varepsilon_0}\rho(r,\theta)$ has the form

$$\frac{1}{r^2}\frac{\partial}{\partial r}\left(r\frac{\partial V}{\partial r}\right) + \frac{1}{r^2\sin}\frac{\partial}{\partial \theta}\left(\sin\theta\frac{\partial V}{\partial \theta}\right) = -\frac{1}{\varepsilon}\rho(r,\theta)$$

in spherical coordinates with axial symmetry.

9 COMPLEX ANALYSIS

Chapter

Chapter Outline

9.1 Cauchy-Riemann Equations
9.2 Integral Theorems
9.3 Conformal Mapping

9.1 CAUCHY-RIEMANN EQUATIONS

We may express a function of a complex variable $f(z)$, where $z = x + iy$, as a sum of real and imaginary parts

$$f(z) = u(x, y) + iv(x, y) \tag{9.1.1}$$

where $u = \mathrm{Re}(f)$ and $v = \mathrm{Im}(f)$.

The function $f(z)$ is said to be analytic in a region R if its derivative exists everywhere in R. The derivative defined as

$$\frac{d}{dz} f(z) = \lim_{\Delta z \to 0} \frac{f(z + \Delta z) - f(z)}{\Delta z} \tag{9.1.2}$$

exists if the limit is independent of the direction that $\Delta z \to 0$. In the complex plane, there are infinitely many directions that $\Delta z = \Delta x + i\Delta y$ can go to zero. We first consider the direction parallel to the real axis letting $\Delta x \to 0$ while holding y fixed. Separating $f(z)$ into its real and imaginary parts we have

$$\frac{d}{dz} f(z) = \lim_{\Delta x \to 0} \frac{\left[u(x + \Delta x, y) - u(x, y) \right] + i \left[v(x + \Delta x, y) - v(x, y) \right]}{\Delta x} \tag{9.1.3}$$

so that

$$\frac{d}{dz} f(z) = \frac{\partial u}{\partial x} + i \frac{\partial v}{\partial x} \tag{9.1.4}$$

Next, we consider the direction parallel to the imaginary axis, letting $\Delta y \to 0$ while holding x fixed

$$\frac{d}{dz} f(z) = \lim_{\Delta y \to 0} \frac{\left[u(x, y + \Delta y) - u(x, y) \right] + i \left[v(x, y + \Delta y) - v(x, y) \right]}{i \Delta y} \tag{9.1.5}$$

so that

$$\frac{d}{dz} f(z) = -i \frac{\partial u}{\partial y} + \frac{\partial v}{\partial y} \tag{9.1.6}$$

If $f(z)$ is analytic then the derivative obtained each way should be equivalent

$$\frac{\partial u}{\partial x} + i \frac{\partial v}{\partial x} = -i \frac{\partial u}{\partial y} + \frac{\partial v}{\partial y} \tag{9.1.7}$$

Equating real and imaginary parts of this expression gives the Cauchy-Riemann equations

$$\frac{\partial u}{\partial x} = \frac{\partial v}{\partial y} \quad \text{and} \quad \frac{\partial v}{\partial x} = -\frac{\partial u}{\partial y} \tag{9.1.8}$$

9.1 Laplace's Equation

If we differentiate the first Cauchy-Riemann equation with respect to x and the second with respect to y we obtain

$$\frac{\partial^2 u}{\partial x^2} = \frac{\partial^2 v}{\partial x \partial y} \quad \text{and} \quad \frac{\partial^2 v}{\partial y \partial x} = -\frac{\partial^2 u}{\partial y^2} \tag{9.1.9}$$

Differentiating the first equation with respect to y and the second with respect to x

$$\frac{\partial^2 u}{\partial y \partial x} = \frac{\partial^2 v}{\partial y^2} \quad \text{and} \quad \frac{\partial^2 v}{\partial^2 x} = -\frac{\partial^2 u}{\partial x \partial y} \tag{9.1.10}$$

Equating the cross partial derivatives in the two equations above gives

$$\frac{\partial^2 u}{\partial x^2} + \frac{\partial^2 u}{\partial y^2} = 0 \quad \text{and} \quad \frac{\partial^2 v}{\partial x^2} + \frac{\partial^2 v}{\partial y^2} = 0 \tag{9.1.11}$$

This remarkable result shows that both the real and imaginary parts of any analytic function satisfy Laplace's equation.

Maple Examples

The real and imaginary parts of complex variable functions are plotted and shown to satisfy Laplace's equation in the Maple worksheet below.

Key Maple commands: *assume*, Im, *implicitplot*, *Laplacian*, Re, *seq*

Maple packages: *with(VectorCalculus): with(plots):*

restart

Laplacian of Re[F(z)] and Im [F (z)]

$assume(x, real) : assume(y, real) :$
$z := x + I*y$

$$z := x\sim + Iy\sim$$

$F := \exp(z)$

$$F := e^{x\sim + Iy\sim}$$

$u := \text{Re}(F)$

$$u := e^{x\sim} \cos(y\sim)$$

$v := \text{Im}(F)$

$$v := e^{x\sim} \sin(y\sim)$$

$with(VectorCalculus) :$
$Laplacian(u, [x, y])$

$$0$$

$Laplacian(v, [x, y])$

$$0$$

$with(plots) :$

$Re\_contours := seq(u = c, c = 0 \ldots 6) : Im\_contours := seq(v = c, c = -3 \ldots 3) :$
$implicitplot([Re\_contours], x = 0 \ldots 4, y = -2 \ldots 2)$

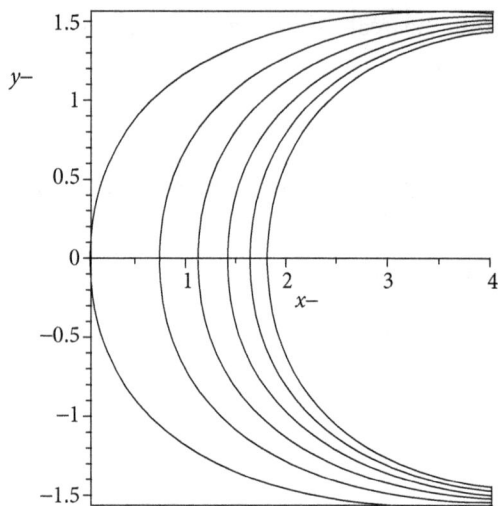

Figure 9.1.1: Contour plot of the real part of $F(z)$.

$$implicitplot([Im\_contours], x = 0 \ldots 4, y = -2 \ldots 2)$$

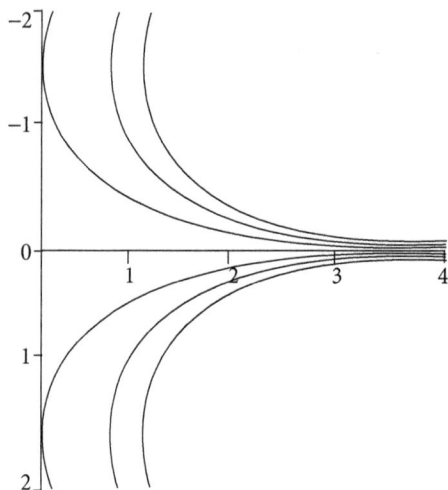

Figure 9.1.2: Contour plot of the imaginary part of $F(z)$.

$G := z^2$

$$G := (x\sim + Iy\sim)^2$$

$u := \text{Re}(G)$

$$u := x\sim^2 - y\sim^2$$

$v := \text{Im}(G)$

$$v := 2x\sim y\sim$$

$Laplacian(u, [x, y])$

$$0$$

$Laplacian(v, [x, y])$

$$0$$

$Re\_contours := seq(u = c, c = -3 \dots 3) : Im\_contours := seq(v = c, c = -3 \dots 3) :$
$p1 := implicitplot([Re\_contours], x = -2 \dots 2, y = -2 \dots 2, linestyle = solid, legend =$
$Real\ Part) :$
$p2 := implicitplot([Re\_contours], x = -2 \dots 2, y = -2 \dots 2, linestyle = dash, legend =$
$Imaginary\ Part) :$

$display(p1, p2)$

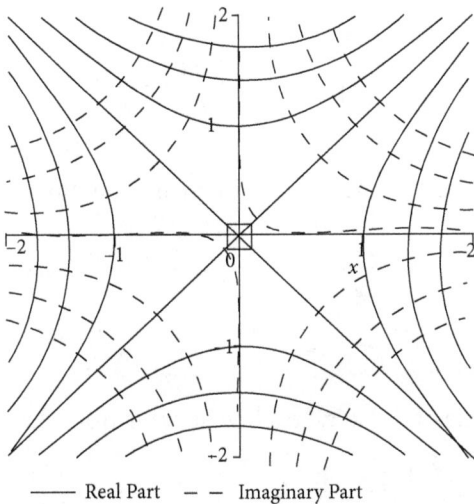

——— Real Part   – – Imaginary Part

Figure 9.1.3: Contour plot of the real and imaginary parts of $G(z)$.

9.2 INTEGRAL THEOREMS

Cauchy's integral theorem and Cauchy's integral formula are first discussed in this section, followed by the Laurent series expansion and types of singularities. Examples include the calculation of residues and integrals using the residue theorem.

9.2.1 Cauchy's Integral Theorem

Cauchy's integral theorem states that the closed line integral of an analytic function is zero

$$\oint f(z)dz = 0 \tag{9.2.1}$$

To show this, we write $f(z) = u(x, y) + iv(x, y)$ and $dz = dx + i\,dy$ so that

$$f(z)dz = (u+iv)(dx+i\,dy) = u\,dx - v\,dy + i(v\,dx + u\,dy) \tag{9.2.2}$$

Separating the integral into real and imaginary parts

$$\oint f(z)dz = \oint (u\,dx - v\,dy) + i\oint (v\,dx + u\,dy) \tag{9.2.3}$$

Green's theorem relates the line integral to the surface integral

$$\oint F(x, y)dx + G(x, y)dy = \int_{\text{surf}} \left(\frac{\partial G}{\partial x} - \frac{\partial F}{\partial y} \right) dx\,dy \tag{9.2.4}$$

Applying Green's theorem, we have

$$\oint (u\,dx - v\,dy) + i\oint (v\,dx + u\,dy) = - \int_{\text{surf}} \left(\frac{\partial v}{\partial x} + \frac{\partial u}{\partial y} \right) dx\,dy + i \int_{\text{surf}} \left(\frac{\partial u}{\partial x} - \frac{\partial v}{\partial y} \right) dx\,dy = 0 \tag{9.2.5}$$

where both surface integrals are zero because of the Cauchy-Riemann equations

$$\frac{\partial u}{\partial x} = \frac{\partial v}{\partial y} \quad \text{and} \quad \frac{\partial v}{\partial x} = -\frac{\partial u}{\partial y} \tag{9.2.6}$$

9.2.2 Cauchy's Integral Formula

Cauchy's integral formula relates the value of $f(z)$ at z_0 to a line integral around z_0

$$f(z_0) = \frac{1}{2\pi i} \oint_{\Gamma} \frac{f(z)}{z - z_0} dz \tag{9.2.7}$$

provided $f(z)$ is analytic inside Γ. To show this consider the contour in Figure 9.2.1

$$\oint_{\Gamma} \frac{f(z)}{z - z_0} dz = \oint_{\Gamma_1} \frac{f(z)}{z - z_0} dz + \oint_{\Gamma_2} \frac{f(z)}{z - z_0} dz \tag{9.2.8}$$

consisting of clockwise and counterclockwise parts.

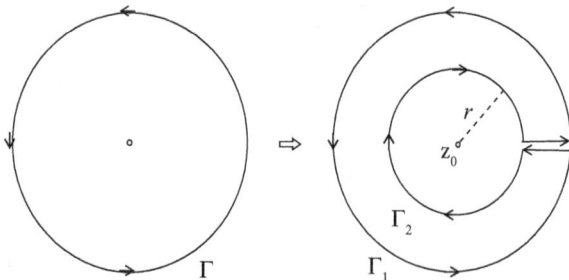

Figure 9.2.1: Contour around z_0.

Since the integrand $f(z)(z - z_0)^{-1}$ is analytic in-between Γ_1 and Γ_2, Cauchy's integral theorem gives

$$\oint_{\Gamma_1} \frac{f(z)}{z - z_0} dz + \oint_{\Gamma_2} \frac{f(z)}{z - z_0} dz = 0 \tag{9.2.9}$$

or

$$\oint_{\Gamma_1} \frac{f(z)}{z - z_0} dz = \oint_{\Gamma_2} \frac{f(z)}{z - z_0} dz \tag{9.2.10}$$

where both integrals are now taken as counterclockwise. Along the path Γ_2, we make the coordinate transformation $z - z_0 = re^{i\theta}$ and $dz = ire^{i\theta} d\theta$

$$\oint_{\Gamma_2} \frac{f(z)}{z - z_0} dz = i \int_0^{2\pi} f(z_0 + re^{i\theta}) d\theta \tag{9.2.11}$$

now

$$\lim_{r \to 0} i \int_0^{2\pi} f(z_0 + re^{i\theta}) d\theta = 2\pi i f(z_0) \tag{9.2.12}$$

Cauchy's integral formula can be extended to derivatives of $f(z)$ at z_0 where the nth derivative

$$f^{(n)}(z_0) = \frac{n!}{2\pi i} \oint_\Gamma \frac{f(z)}{(z - z_0)^{n+1}} dz \tag{9.2.13}$$

9.2.3 Laurent Series Expansion

The Laurent expansion of a complex function $f(z)$ about a point z_0

$$f(z) = \sum_{n=-\infty}^{\infty} a_n (z - z_0)^n \tag{9.2.14}$$

has coefficients corresponding to positive and negative n values

$$f(z) = \cdots + a_{-1}(z - z_0)^{-1} + a_0 + a_1(z - z_0) + a_2(z - z_0)^2 + \cdots \tag{9.2.15}$$

Laurent's theorem gives coefficients

$$a_n = \frac{1}{2\pi i} \oint_\Gamma \frac{f(z)}{(z - z_0)^{n+1}} dz \tag{9.2.16}$$

9.2.4 Types of Singularities

The singularity z_0 of a function $f(z)$ may be classified by the nonzero terms of the Laurent series.

Removable singularity: If all negative coefficients are zero, the singularity z_0 is said to be removable, and the Laurent series is equivalent to a Taylor series.

Essential singularity: The singularity z_0 is essential if there are an infinite number of nonzero coefficients in the Laurent expansion.

Pole: If there are a finite number of nonzero coefficients, the singularity z_0 is classified as a pole.

Example 9.2.1

The function $f(z) = \dfrac{1}{z^n}$ has a pole $z_0 = 0$ of order n

Example 9.2.2

The function $f(z) = \dfrac{1}{z(z-1)^3}$ has a pole $z_0 = 0$ of order one and a pole $z_0 = 1$ of order three

9.2.5 Residues

The coefficient a_{-1} corresponding to $n = -1$ in the Laurent expansion is the residue of $f(z)$ at $z = z_0$. From the integral formula for the coefficients

$$a_n = \frac{1}{2\pi i} \oint_\Gamma \frac{f(z)}{(z - z_0)^{n+1}} dz \tag{9.2.17}$$

we have

$$a_{-1} = \frac{1}{2\pi i} \oint_\Gamma f(z) dz \tag{9.2.18}$$

If z_0 is a pole of order one

$$a_{-1} = \lim_{z \to z_0} (z - z_0) f(z) \tag{9.2.19}$$

If z_0 is a pole of order m

$$a_{-1} = \lim_{z \to z_0} \frac{1}{(m-1)!} \frac{d^{m-1}}{dz^{m-1}} \left[(z - z_0)^m f(z) \right] \tag{9.2.20}$$

Example 9.2.3

Calculate the residues of $f(z) = \dfrac{z^3}{z(z-1)^2}$

Solution:

a_{-1} at $z = 0$: $\displaystyle\lim_{z\to 0}\left[(z-0)f(z)\right] = \lim_{z\to 0}\dfrac{z^3}{(z-1)^2} = 0$ $\qquad\qquad$ (9.2.21)

a_{-1} at $z = 1$: $\displaystyle\lim_{z\to 1}\dfrac{1}{1!}\dfrac{d}{dz}\left[(z-1)^2 f(z)\right] = \lim_{z\to 1}2z^2 = 2$ $\qquad\qquad$ (9.2.22)

9.2.6 Residue Theorem

If Γ is a closed contour and $f(z)$ is analytic inside Γ except at isolated singularities, then

$$\oint_{\Gamma} f(z)dz = 2\pi i \sum a_{-1}$$
$\qquad\qquad$ (9.2.23)

where Σa_{-1} is the sum of residues.

Example 9.2.4

Evaluate the integral

$$\oint_{\Gamma} \frac{z+1}{z-1} dz = 2\pi i \sum a_{-1}$$
$\qquad\qquad$ (9.2.24)

where Γ is a closed contour surrounding $z = 1$

Solution: We evaluate the residue

a_{-1} at $z = 1$: $\displaystyle\lim_{z\to 1}\left[(z-1)f(z)\right] = \lim_{z\to 1}(z+1) = 2$ $\qquad\qquad$ (9.2.25)

$$\oint_{\Gamma} \frac{z+1}{z-1} dz = 4\pi i$$
$\qquad\qquad$ (9.2.26)

Example 9.2.5

Evaluate the integral

$$\oint_{\Gamma} \frac{e^{\pi z}}{z^2+1} dz = 2\pi i \sum a_{-1}$$
$\qquad\qquad$ (9.2.27)

where Γ is the rectangular contour shown in Figure 9.2.2.

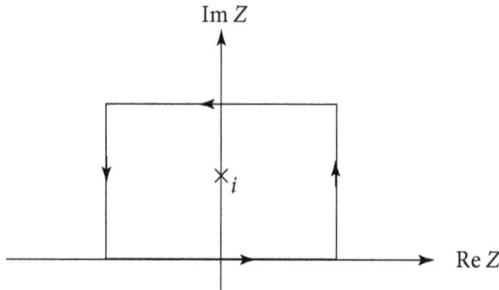

Figure 9.2.2: Rectangular contour in the upper half plane.

Solution: Factoring the denominator

$$\frac{e^{\pi z}}{z^2 + 1} = \frac{e^{\pi z}}{(z+i)(z-i)} \tag{9.2.28}$$

our contour only encloses the pole at $z = i$

$$a_{-1} \text{ at } z = i: \ \lim_{z \to i}\left[(z-i)f(z)\right] = \lim_{z \to i}\frac{e^{\pi z}}{(z+i)} = \frac{e^{i\pi}}{2i} = \frac{-1}{2i} \tag{9.2.29}$$

$$\oint_{\Gamma}\frac{e^{\pi z}}{z^2 + 1}dz = 2\pi i\left(\frac{-1}{2i}\right) = -\pi \tag{9.2.30}$$

Example 9.2.6

Evaluate the integral

$$\oint_{\Gamma}e^{1/z}dz \tag{9.2.31}$$

where Γ is the unit circle surrounding $z = 0$.

Solution: $z = 0$ is an essential singularity. The integrand is expanded as an infinite series

$$e^{1/z} = 1 + \frac{1}{1!}z^{-1} + \frac{1}{2!}z^{-2} + \cdots \tag{9.2.32}$$

where we identify the coefficient of z^{-1} as $a_{-1} = 1$ so that

$$\oint_{\Gamma}e^{1/z}dz = 2\pi i \tag{9.2.33}$$

The residue theorem can be used to evaluate more complicated integrals such as improper integrals and integrals involving trigonometric functions.

9.2.7 Improper Integrals

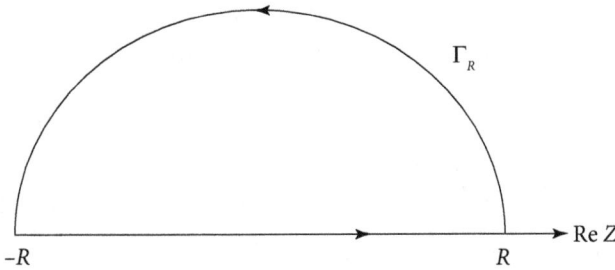

Figure 9.2.3 Semicircular contour closed in the upper half plane

Certain improper integrals of the form $\int\limits_{-\infty}^{\infty} f(x)dx$ can be performed by integrating around infinite semicircular contours in the upper half plane (Figure 9.2.3) where along the x-axis $f(x)dx = f(z)dz$ so that

$$\int\limits_{-\infty}^{\infty} f(x)dx = \oint f(z)dz = \lim_{R\to\infty}\left[\int\limits_{-R}^{R} f(z)dz + \int\limits_{\Gamma_R} f(z)dz\right] = 2\pi i\sum a_{-1} \quad (9.2.34)$$

The estimation lemma gives an upper bound on the contour integral in the upper half plane where the integral over the upper half circle is less than or equal to the length of the contour Γ_R multiplied by the maximum value of $f(z)$ evaluated on Γ_R:

$$\left|\int\limits_{\Gamma_R} f(z)dz\right| \leq \text{length}\left(\Gamma_R\right)\times\max_{z\in\Gamma_R}\left|f(z)\right| \quad (9.2.35)$$

Thus, if $\lim\limits_{R\to\infty} \pi R\times\max\limits_{z\in\Gamma_R}\left|f(z)\right|=0$ then $\lim\limits_{R\to\infty}\int\limits_{\Gamma_R} f(z)dz = 0$ and the indefinite integral of $f(x)$ is given by $2\pi i$ times the sum of residues of $f(z)$ in the upper half plane

$$\int\limits_{-\infty}^{\infty} f(x)dx = 2\pi i \sum_{\text{upper half}} a_{-1} \quad (9.2.36)$$

Example 9.2.7

Evaluate the improper integral

$$\int_{-\infty}^{\infty} \frac{1}{1+x^2} dx \tag{9.2.37}$$

Solution: Taking the contour in the upper half plane (Figure 9.2.4)

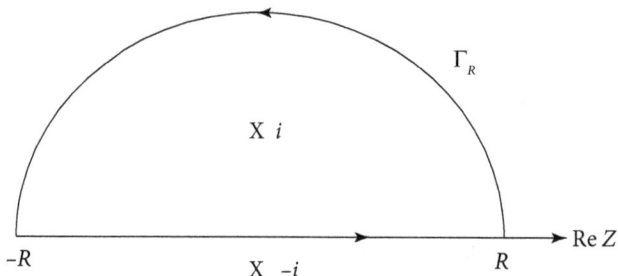

Figure 9.2.4: Semicircular contour enclosing a residue in the upper half plane

$$\oint_{\Gamma} \frac{1}{1+z^2} dz = \int_{-R}^{R} \frac{1}{1+z^2} dz + \int_{\Gamma_R} \frac{1}{1+z^2} dz = 2\pi i \sum a_{-1} \tag{9.2.38}$$

From the estimation lemma, since

$$\lim_{R \to \infty} \text{length}(\Gamma_R) \max \left(\frac{1}{1+z^2} \right)_{z \in \Gamma_R} = 0 \tag{9.2.39}$$

$$\lim_{R \to \infty} \int_{\Gamma_R} \frac{1}{1+z^2} dz = 0 \tag{9.2.40}$$

There is only one residue in the upper half plane

$$a_{-1} = \lim_{z \to i} (z-i) \frac{1}{(z-i)(z+i)} = \frac{1}{2i} \tag{9.2.41}$$

so that

$$\int_{-\infty}^{\infty} \frac{1}{1+x^2} dx = \pi \tag{9.2.42}$$

which agrees with the result obtained by trigonometric substitution.

Example 9.2.8

Evaluate the improper integral

$$\int_{-\infty}^{\infty} \frac{\sin(x)}{x}\,dx \tag{9.2.43}$$

Solution: The integrand has a pole on the x-axis that must be avoided

$$\int_{-\infty}^{\infty} \frac{\sin(x)}{x}\,dx = \operatorname{Im}\lim_{\substack{R\to\infty \\ \varepsilon\to 0}} \left[\int_{-R}^{-\varepsilon} \frac{e^{iz}}{z}\,dz + \int_{\varepsilon}^{R} \frac{e^{iz}}{z}\,dz\right] \tag{9.2.44}$$

Our strategy is to avoid the pole by integrating around the contour shown in Figure 9.2.5

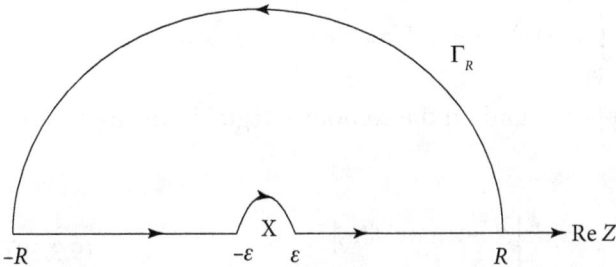

Figure 9.2.5: Closed contour avoiding a pole on the x-axis

and apply Cauchy's integral theorem since our contour encloses no poles

$$\oint_{\Gamma} \frac{e^{iz}}{z}\,dz = \int_{-R}^{-\varepsilon} \frac{e^{iz}}{z}\,dz + \int_{\Gamma_{\varepsilon}} \frac{e^{iz}}{z}\,dz + \int_{\varepsilon}^{R} \frac{e^{iz}}{z}\,dz + \int_{\Gamma_R} \frac{e^{iz}}{z}\,dz = 0 \tag{9.2.45}$$

$$\int_{-\infty}^{\infty} \frac{\sin(x)}{x}\,dx = \operatorname{Im}\lim_{\substack{R\to\infty \\ \varepsilon\to 0}} \left[\int_{-R}^{-\varepsilon} \frac{e^{iz}}{z}\,dz + \int_{\varepsilon}^{R} \frac{e^{iz}}{z}\,dz\right] = -\operatorname{Im}\lim_{\varepsilon\to 0} \int_{\Gamma_{\varepsilon}} \frac{e^{iz}}{z}\,dz \tag{9.2.46}$$

To evaluate the integral over Γ_{ε} we let $z = \varepsilon e^{i\theta}$ and $dz = i\varepsilon e^{i\theta}d\theta$. For small ε, the exponential $e^{iz} \approx 1 + i\varepsilon e^{i\theta}$

$$-\operatorname{Im}\lim_{\varepsilon\to 0}\int_{\Gamma_\varepsilon}\frac{e^{iz}}{z}dz=-\operatorname{Im}\lim_{\varepsilon\to 0}\int_\pi^0\frac{1+i\varepsilon e^{i\theta}}{\varepsilon e^{i\theta}}i\varepsilon e^{i\theta}d\theta=-\operatorname{Im}(-i\pi)\qquad(9.2.47)$$

and

$$\int_{-\infty}^\infty\frac{\sin(x)}{x}dx=\pi\qquad(9.2.48)$$

9.2.8 Fourier Transform Integrals

Improper integrals of the form $\displaystyle\int_{-\infty}^\infty f(x)e^{ikx}dx$ may be evaluated by integrating over infinite semicircular contours taken counterclockwise in the upper half plane for $k>0$

$$\oint f(z)e^{ikz}dz=\lim_{R\to\infty}\left[\int_{-R}^R f(x)e^{ikx}dx+\int_{\Gamma_R}f(z)e^{ikz}dz\right]=2\pi i\sum a_{-1}\qquad(9.2.49)$$

Jordan's lemma gives an upper bound on the contour integral in the upper half plane

$$\left|\int_{\Gamma_R}e^{ikz}f(z)dz\right|\le\frac{\pi}{k}\times\max_{z\in\Gamma_R}\left|f(z)\right|\qquad(9.2.50)$$

Thus, if $\displaystyle\lim_{R\to\infty}\frac{\pi}{k}\times\max_{z\in\Gamma_R}\left|f(z)\right|=0$ then $\displaystyle\lim_{R\to\infty}\int_{\Gamma_R}e^{ikz}f(z)dz=0$ and

$$\int_{-\infty}^\infty f(x)e^{ikx}dx=2\pi i\sum_{\text{upper half}}a_{-1}\qquad(9.2.51)$$

If $k<0$ then the semicircular contour is taken clockwise in the lower half plane

$$\int_{-\infty}^\infty f(x)e^{ikx}dx=-2\pi i\sum_{\text{lower half}}a_{-1}\qquad(9.2.52)$$

Example 9.2.9

Evaluate the improper integral

$$\int_{-\infty}^\infty\frac{\cos 3x}{1+x^2}dx\qquad(9.2.53)$$

Solution: Along the $y = 0$ axis we replace

$$\frac{\cos 3x}{1+x^2} \rightarrow \frac{e^{i3z}}{1+z^2} \tag{9.2.54}$$

and compute the contour integral

$$\oint_\Gamma \frac{e^{i3z}}{1+z^2}dz = \lim_{R\to\infty}\left[\int_{-R}^{R}\frac{e^{i3z}}{1+z^2}dz + \int_{\Gamma_R}\frac{e^{i3z}}{1+z^2}dz\right] = 2\pi i\sum a_{-1} \tag{9.2.55}$$

Γ is taken along the x-axis and closed by semicircular contour Γ_R in the upper half plane. By Jordan's lemma

$$\lim_{R\to\infty}\int_{\Gamma_R}\frac{e^{i3z}}{1+z^2}dz = 0 \tag{9.2.56}$$

and the contour only encloses the pole at $z = i$ with residue

$$a_{-1} = \lim_{z\to i}(z-i)\frac{e^{i3z}}{(z-i)(z+i)} = \frac{e^{-3}}{2i} \tag{9.2.57}$$

so that

$$\oint_\Gamma \frac{e^{iz}}{1+z^2}dz = \int_{-\infty}^{\infty}\frac{\cos 3x}{1+x^2}dx = \frac{\pi}{e^3} \tag{9.2.58}$$

Maple Examples

The following Maple worksheet demonstrates the calculation of residues by series expansion and by using the *residue* command. Improper integrals are calculated using the residue theorem corresponding to infinite semicircular contours closed in the upper and lower half plane.

Key Maple commands: *evalf, expand, factor, residue, series*

Special functions: *Zeta*

restart

Calculation of Residues

series$(\text{Zeta}(z), z = 1, 4)$

$$(z-1) - 1 + \gamma - \gamma(1)(z-1) + \frac{1}{2}\gamma(2)(z-1)^2 - \frac{1}{6}\gamma(3)(z-1)^3 + O\left((z-1)^4\right)$$

residue$(\text{Zeta}(s), s = 1)$

$$series\left(\frac{1}{z^2-1}, z=-1, 4\right)$$

$$-\frac{1}{2}(z+1)^{-1} - \frac{1}{4} - \frac{1}{8}(z+1) - \frac{1}{16}(z+1)^2 + O\left((z+1)^3\right)$$

$$residue\left(\frac{1}{z^2-1}, z=-1\right)$$

$$-\frac{1}{2}$$

$$series\left(\frac{1}{\sin(z)}, z=0, 4\right)$$

$$z^{-1} + \frac{1}{6}z + O(z^3)$$

$$residue\left(\frac{1}{\sin(z)}, z=0\right)$$

$$1$$

$$series\left(\frac{\exp(I \cdot z)}{z^2+1}, z=I, 2\right)$$

$$-\frac{I}{2}e^{-1}(z-I)^{-1} + O()$$

$$residue\left(\frac{\exp(I \cdot z)}{z^2+1}, z=I\right)$$

$$-\frac{I}{2}e^{-1}$$

Improper Integrals (Contours Closed in the Upper Half Plane)

$$\int_{-\infty}^{\infty} \frac{\exp(I \cdot x)}{x^4 + 5 \cdot x^2 + 4}dx$$

$$\frac{\pi e^{-1}}{3} - \frac{\pi e^{-2}}{6}$$

$$factor(x^4 + 5 \cdot x^2 + 4)$$

$$(x^2 + 4)(x^2 + 1)$$

$$2 \cdot pi \cdot I\left(residue\left(\frac{\exp(I \cdot z)}{z^4 + 5 \cdot z^2 + 4}, z = I \right) + residue\left(\frac{\exp(I \cdot z)}{z^4 + 5 \cdot z^2 + 4}, z = 2 \cdot I \right) \right)$$

$$2I\pi\left(-\frac{Ie^{-1}}{6} + \frac{Ie^{-2}}{12} \right)$$

expand(%)

$$\frac{\pi e^{-1}}{3} - \frac{\pi e^{-2}}{6}$$

$$\int_{-\infty}^{\infty} \frac{x^2}{x^6 + 14x^4 + 49x^2 + 36} dx$$

$$\frac{\pi}{60}$$

$$f := (z) \rightarrow \frac{z^2}{z^6 + 14z^4 + 49z^2 + 36}$$

$$f := (z) \rightarrow \frac{z^2}{z^6 + 14z^4 + 49z^2 + 36}$$

factor($z^6 + 14z^4 + 49z^2 + 36$)

$$(z^2 + 9)(z^2 + 4)(z^2 + 1)$$

$2 \cdot pi \cdot I \cdot (residue(f(z), z = I) + residue(f(z), z = 2 \cdot I) + residue(f(z), z = 3 \cdot I))$

$$\frac{\pi}{60}$$

$$\int_{-\infty}^{\infty} \frac{\exp(I \cdot x)}{x^4 + 10 \cdot x^2 + 9} dx$$

$$\frac{\pi e^{-1}}{8} - \frac{\pi e^{-3}}{24}$$

factor($x^4 + 10 \cdot x^2 + 9$)

$$(x^2 + 9)(x^2 + 1)$$

$$2 \cdot pi \cdot I\left(residue\left(\frac{\exp(I \cdot z)}{z^4 + 10 \cdot z^2 + 9}, z = 3 \cdot I \right) + residue\left(\frac{\exp(I \cdot z)}{z^4 + 10 \cdot z^2 + 9}, z = I \right) \right)$$

$$2I\pi\left(\frac{Ie^{-3}}{48} - \frac{Ie^{-1}}{16} \right)$$

expand(%)

$$-\frac{\pi e^{-3}}{24}+\frac{\pi e^{-1}}{8}$$

Improper Integrals (Contours Closed in the Lower Half Plane)

$$\int_{-\infty}^{\infty}\frac{\exp(-3\cdot I\cdot x)}{x^4+5\cdot x^2+4}dx$$

$$-\frac{\pi e^{-6}}{6}+\frac{\pi e^{-3}}{3}$$

$factor(x^4+5\cdot x^2+4)$

$$(x^2+4)(x^2+1)$$

$$-2\cdot pi\cdot I\cdot\left(residue\left(\frac{\exp(-3\cdot I\cdot z)}{z^4+5\cdot z^2+4},z=-I\right)+residue\left(\frac{\exp(-3\cdot I\cdot z)}{z^4+5\cdot z^2+4},z=-2\cdot I\right)\right)$$

$$-2I\pi\left(\frac{Ie^{-3}}{6}-\frac{Ie^{-6}}{12}\right)$$

$expand(\%)$

$$\frac{\pi e^{-3}}{3}-\frac{\pi e^{-6}}{6}$$

$$\int_{-\infty}^{\infty}\frac{1}{x^4+5\cdot x^2+4}dx$$

$$\frac{\pi}{6}$$

$$-2\cdot pi\cdot I\cdot\left(residue\left(\frac{1}{z^4+5\cdot z^2+4},z=-I\right)+residue\left(\frac{1}{z^4+5\cdot z^2+4},z=-2\cdot I\right)\right)$$

$$\frac{\pi}{6}$$

9.3 CONFORMAL MAPPING

Topics in this section include Poisson's integral formulas for a cylinder and half plane, the Schwarz-Christoffel transformation, conformal mappings and mappings on the Riemann sphere.

9.3.1 Poisson's Integral Formulas

Solutions to Laplace's equation may be obtained by integral formulas developed by Poisson. If a function V is specified everywhere along the x-axis, $V(x,0) = F(x')$, then the potential in the upper half plane is

$$V(x,y) = \frac{1}{\pi} \int_{-\infty}^{\infty} \frac{yF(x')dx'}{y^2 + (x-x')^2} \tag{9.3.1}$$

The integral formula to obtain V inside a circle of radius R with boundary condition $V(R,\theta') = G(\theta')$ is

$$V(r,\theta) = \frac{1}{2\pi} \int_{0}^{2\pi} \frac{(R^2 - r^2)G(\theta')d\theta'}{R^2 - 2rR\cos(\theta - \theta') + r^2} \tag{9.3.2}$$

9.3.2 Schwarz-Christoffel Transformation

Polygons with interior angles α_j in the w-plane are mapped to the x-axis of the z-plane by the transformation

$$\frac{dw}{dz} = A(z - x_1)^{\alpha_1/\pi - 1}(z - x_2)^{\alpha_2/\pi - 1} \cdots (z - x_n)^{\alpha_n/\pi - 1} \tag{9.3.3}$$

Integrating this expression gives

$$w = c_1 \int (z - x_1)^{\alpha_1/\pi - 1}(z - x_2)^{\alpha_2/\pi - 1} \cdots (z - x_n)^{\alpha_n/\pi - 1} dz + c_2 \tag{9.3.4}$$

where c_1 and c_2 are constants.

9.3.3 Conformal Mapping

The transformation $w = f(z)$ is conformal over a region R if $f'(z) \neq 0$ on R. If $f(z) = u(x, y) + iv(x, y)$ is analytic over R then u and v satisfy the Cauchy-Riemann equations

$$\frac{\partial u}{\partial x} = \frac{\partial v}{\partial y} \quad \text{and} \quad \frac{\partial v}{\partial x} = -\frac{\partial u}{\partial y} \tag{9.3.5}$$

and therefore satisfy Laplace's equation

$$\frac{\partial^2 u}{\partial x^2} + \frac{\partial^2 u}{\partial y^2} = 0 \quad \text{and} \quad \frac{\partial^2 v}{\partial x^2} + \frac{\partial^2 v}{\partial y^2} = 0 \tag{9.3.6}$$

as discussed in Section 9.1.

Example 9.3.1

The transformation $w = z^2$ maps the first quadrant of the z-plane to the upper half of the w-plane. Find the w components in terms of the original z-coordinates

Solution:

$$u = \mathrm{Re}(w) = x^2 - y^2$$
$$v = \mathrm{Im}(w) = 2xy \qquad (9.3.7)$$

where u and v each satisfy Laplace's equation.

Because contours of constant u are perpendicular to contours of constant v, we may find both electric field lines and equipotentials in electrostatic problems. If a scalar function $V(x, y)$ satisfies Laplace's equation in the x-y plane, then

$$\frac{\partial^2 V}{\partial x^2} + \frac{\partial^2 V}{\partial y^2} = 0 \qquad (9.3.8)$$

A transform that maps a region of the z-plane to a region in the w-plane preserves the form of Laplace's equation written with the new coordinates u and v

$$\frac{\partial^2 V}{\partial u^2} + \frac{\partial^2 V}{\partial v^2} = 0 \qquad (9.3.9)$$

In this way, we may transform an original complicated boundary into a simple boundary. Once $V(u, v)$ is determined, then $V(x, y)$ is found by substituting $V(u(x, y), v(x, y))$. If a complex region can be transformed to a half plane or a unit circle, then Poisson's integral formula can be used to find a solution to Laplace's equation in the transformed region.

6.3.4 Mappings on the Riemann Sphere

Points on the complex plane may be mapped onto a unit sphere with its south pole S coincident with the intersection of the real and imaginary axes. This sphere, known as the Riemann sphere, is illustrated in Figure 9.3.1. A line segment drawn from a point on the complex plane to the north pole N of the sphere maps the point to the intersection of the sphere and the line segment. The regions $|z| < 1$ and $|z| > 1$ are mapped to the southern and northern hemispheres, respectively. The unit circle $|z| = 1$ is mapped to the equator of the sphere. Points at infinity are mapped to the north pole.

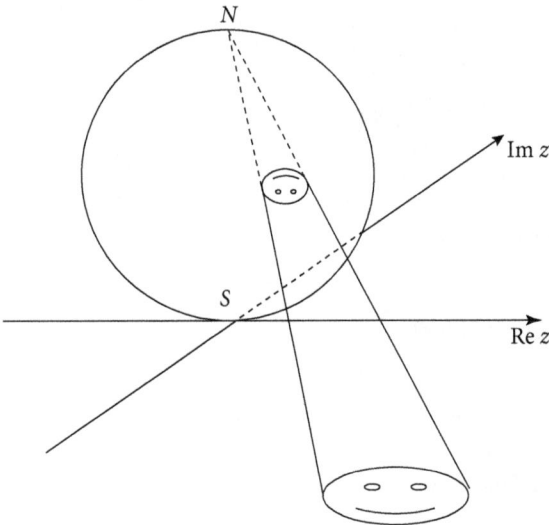

Figure 9.3.1: Mapping on the Riemann sphere.

Maple Examples

The following Maple worksheet demonstrates Poisson's integral formula for calculation of the potential inside the unit circle. Conformal mappings of the first quadrant, upper half plane and all four quadrants as well as mappings on the Riemann sphere are performed.

Key Maple commands: *conformal, conformal3d, evalf*

Maple packages: *with(plots):*

Programming: Functions statements using '→'

restart

Poisson Integral Formula

$G := q \rightarrow \sin(q)$

$$G := q \mapsto \sin(q)$$

$$V := (r, \text{theta}) \rightarrow \left(\frac{1}{2 \cdot \text{Pi}} \text{int} \left(\frac{G(q) \cdot (1 - r^2) \cos(q)}{1 - 2 \cdot r \cdot \cos(\text{theta} - q) + r^2}, q = 0 \ldots 2 \cdot \text{Pi} \right) \right)$$

$$V := (r,\theta) \to \dfrac{\displaystyle\int_0^{2\pi} \dfrac{G(q)\left(-r^2+1\right)\cos(q)}{1-2r\cos\left(\theta-q\right)+r^2}\,dq}{2\pi}$$

evalf$(V(0.99, 0.5))$

$$0.4123628560$$

evalf$(V(0.99, -0.5))$

$$-0.4123628560$$

evalf$(V(0.99, 0))$

$$0$$

Conformal Mappings $w = f(z)$ of the First Quadrant

with(*plots*) :
lower_left := $0 + 0 \cdot I$: *upper_right* := $1 + 1 \cdot I$:
conformal($z^{1.1}$, $z =$ *lower_left* ... *upper_right*, *color* = ["Black", "Gray"])

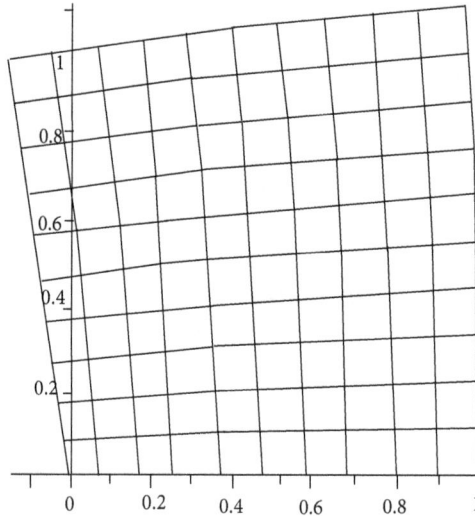

9.3.2(a)

conformal($z^{1.5}$, $z =$ *lower_left* ... *upper_right*, *color* = ["Black", "Grey"])

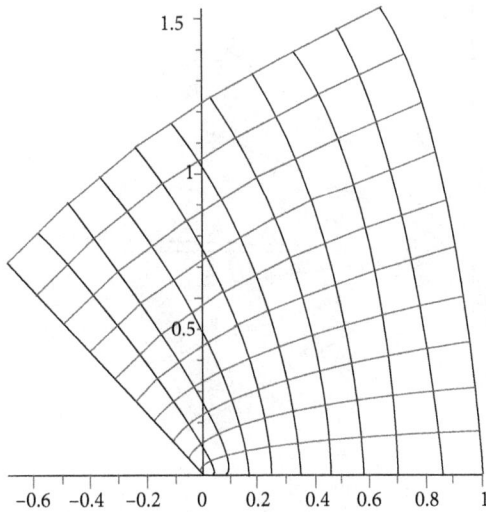

9.3.2(b)

conformal(z^2, z = lower_left ... upper_right, color = [“Black”, “Grey”])

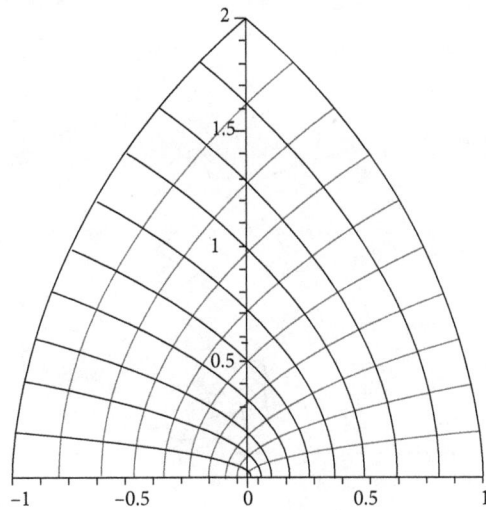

9.3.2(c)

conformal(sin(3·z), z = lower_left ... upper_right, color = [“Black”, “Grey”])

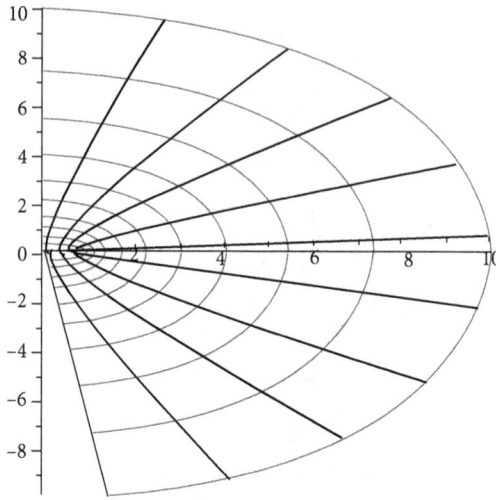

9.3.2(d)

conformal(tanh(*z*), *z* = *lower_left* ... *upper_right, color* = ["Black", "Grey"])

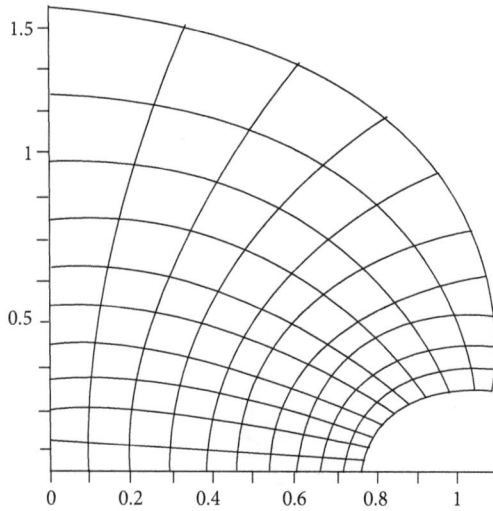

9.3.2(e)

conformal(tan(*z*), *z* = *lower_left* ... *upper_right, color* = ["Black", "Grey"])

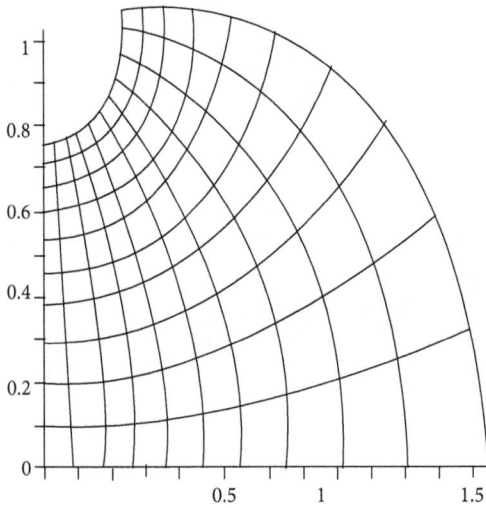

9.3.2(f)

Figure 9.3.2(a–f): Conformal mappings of the first quadrant.

Conformal Mappings *w* = *f(z)* of the Upper Half Plane

lower_left := −1 + 0·*I*: *upper_right* := 1 + 1·*I*:
conformal($z^{1.1}$, *z* = *lower_left* ... *upper_right*, *color* = ["Black", "Grey"])

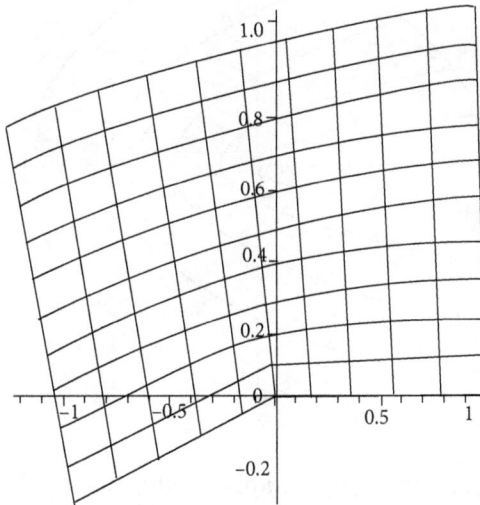

9.3.3(a)

conformal($z^{1.5}$, *z* = *lower_left* ... *upper_right*, *color* = ["Black", "Grey"])

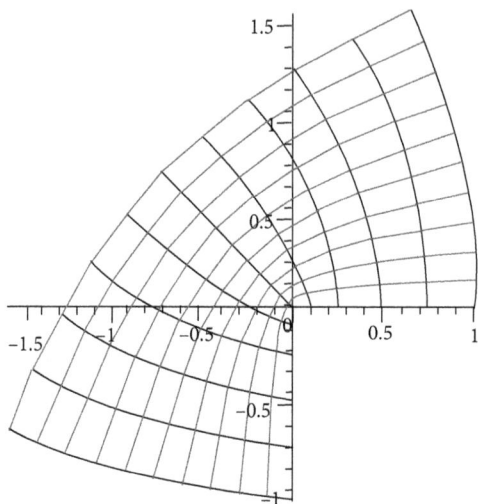

9.3.3(b)

conformal(z^2, $z = lower\_left \ldots upper\_right$, *color* = ["Black", "Grey"])

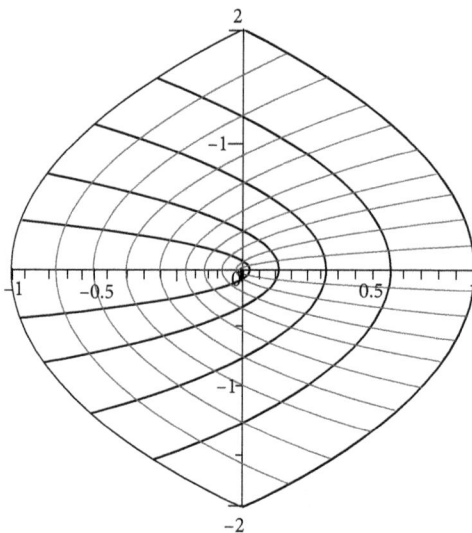

9.3.3(c)

conformal($\sin(3 \cdot z)$, $z = lower\_left \ldots upper\_right$, *color* = ["Black", "Grey"])

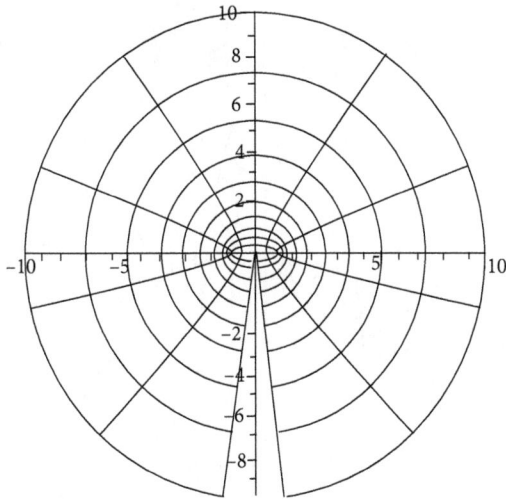

9.3.3(d)

$conformal(\tanh(z), z = lower\_left \ldots upper\_right, color = [\text{"Black"}, \text{"Grey"}])$

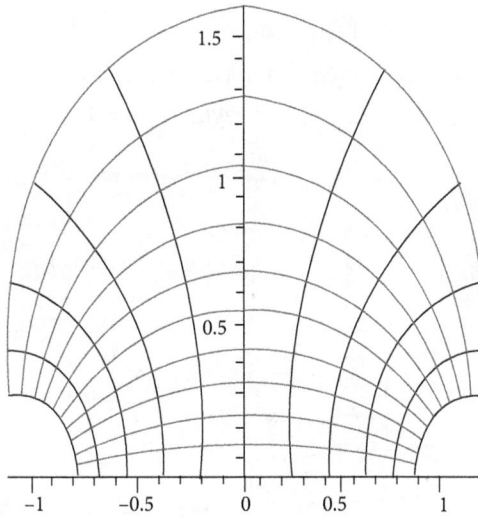

9.3.3(e)

$conformal(\tan(z), z = lower\_left \ldots upper\_right, color = [\text{"Black"}, \text{"Grey"}])$

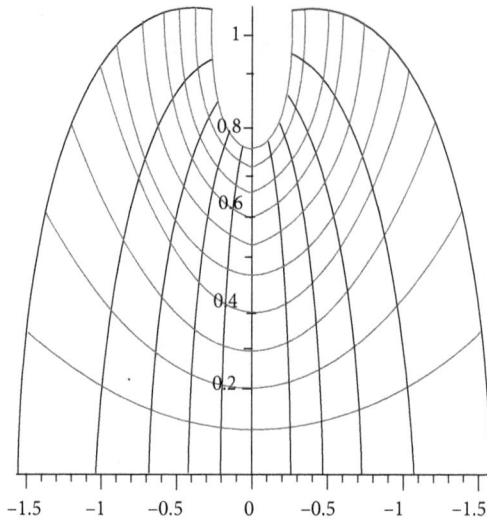

9.3.3(f)

Figure 9.3.3(a–f): Conformal mappings of the upper half plane.

Conformal Mappings $w = f(z)$ of All Quadrants

$lower\_left := -1 - 1 \cdot I$: $upper\_right := 1 + 1 \cdot I$:
$conformal(z^{1.03}, z = lower\_left \dots upper\_right, color = [\text{“Black”, “Grey”}])$

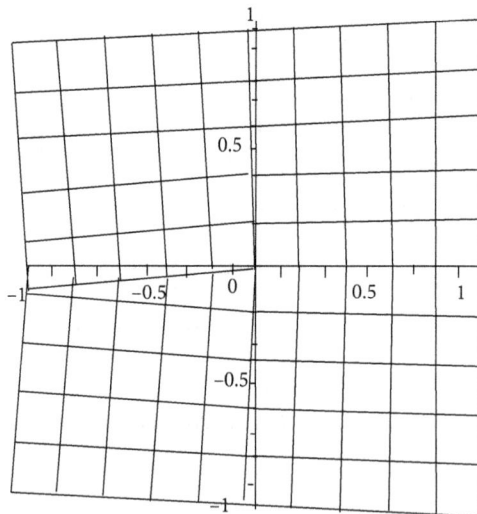

9.3.4(a)

$$conformal\left(\ln\left(\frac{1+z}{1-z}\right), z = lower\_left \dots upper\_right, color = [\text{"Black","Grey"}] \right)$$

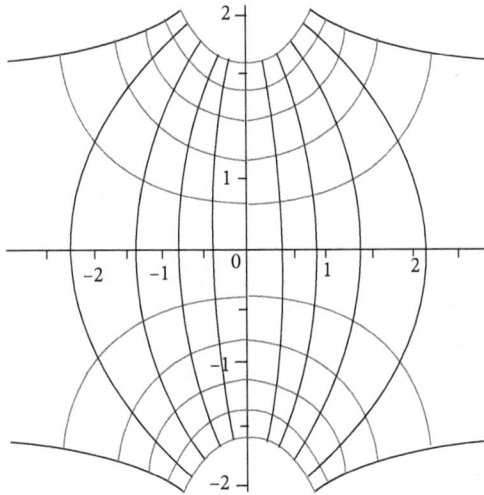

9.3.4(b)

$conformal(\exp(z^2), z = lower\_left \ldots upper\_right, color = [\text{"Black"}, \text{"Grey"}])$

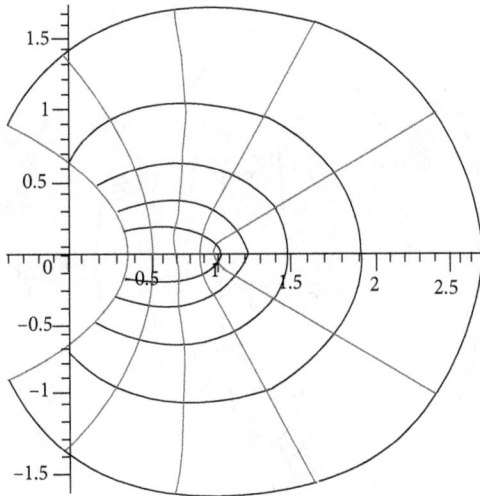

9.3.4(c)

$conformal\left(\dfrac{\sin(z)}{z}, z = lower\_left \ldots upper\_right, color = [\text{"Black"}, \text{"Grey"}]\right)$

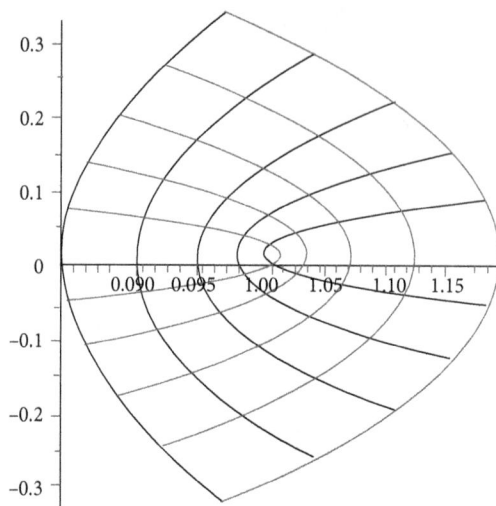

9.3.4(d)

$conformal(\tan(z), z = lower\_left \ldots upper\_right, color = [``Black", ``Grey"])$

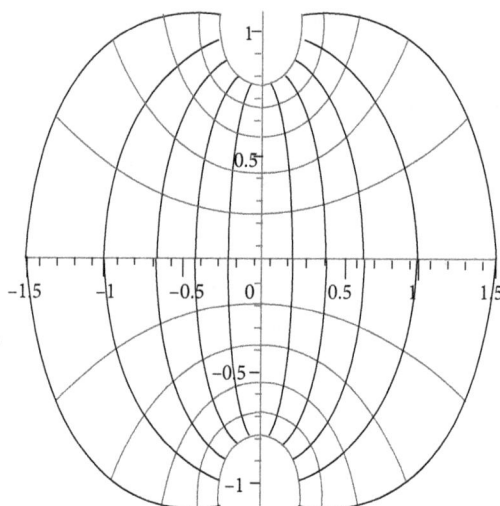

9.3.4(e)

Figure 9.3.4(a–e): Conformal mappings of a square region in all four quadrants.

Conformal Mappings Projected on the Riemann Sphere

lower_left := −1 − 3·*I*: *upper_right* := 3 + 3·*I*:
conformal3d(*z*, *z* = *lower_left* ... *upper_right*, *color* = ["Black", "Grey"], *spherecolor* = "Grey")

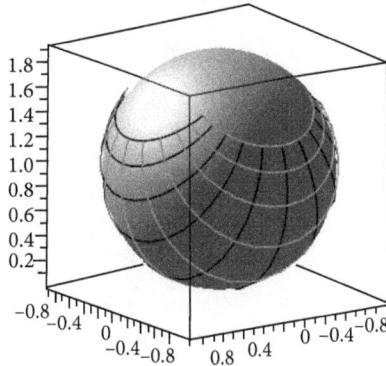

9.3.5(a)

conformal3d(cos(*z*), *z* = *lower_left* ... *upper_right*, *color* = ["Black", "Grey"], *spherecolor* = "Grey")

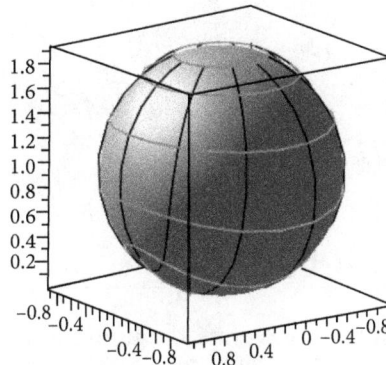

9.3.5(b)

lower_left := −1 − 1·*I*: *upper_right* := 1 + 1·*I*:
for *n* **from** 1 **to** 4 **do**
conformal3d(sin(*n*·*z*), *z* = *lower_left* ... *upper_right*, *color* = ["Black", "Grey"], *spherecolor* = "Grey")

end

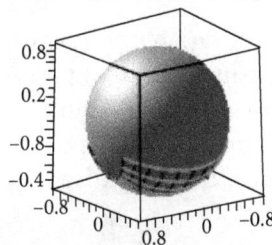

9.3.5(c)

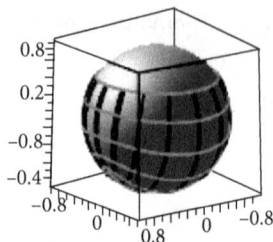

9.3.5(d)

Figure 9.3.5(a–d): Conformal mappings on the Reimann sphere.

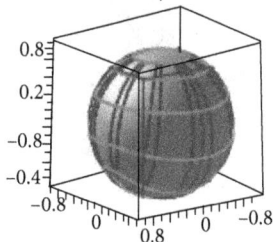

9.3.5(e)

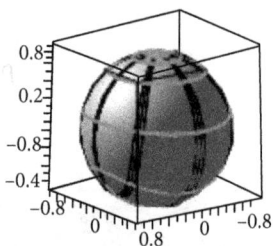

9.3.5(f)

Figure 9.3.5(e–f): Conformal mappings on the Reimann sphere.

9.4 MATLAB EXAMPLES

Examples in this section demonstrate the numerical Laplacian of the real and imaginary parts of complex variable functions satisfying the Cauchy-Riemann equations. The orthogonal streamlines of the real and imaginary components are plotted together.

Key MATLAB commands: *contour, imag, max, meshgrid, real*

Section 9.1 Cauchy-Riemann Equations

Given a function of a complex variable we calculate the discrete Laplacian (`del2`) of the function's real (`u`) and imaginary (`v`) parts. The maximum value of these arrays are calculated as `max(max(del2(v)))` and `max(max(del2(u)))` to show these values are effectively zero within numerical precision. The quantities `max(max(v))` and `max(max(u))` are calculated for comparison.

```
>> [x,y]=meshgrid(-1:0.1:1,-1:0.1:1);
>> z=x+1i*y;
>> z1=exp(z);
>> u=real(z1);
>> v=imag(z1);
>> max(max(del2(u)))
ans =
  1.0248e-005
>> max(max(del2(v)))
ans =

  9.5442e-005

>> max(max(u))

ans =

    2.7183

>> max(max(v))

ans =

    2.2874
```

Section 9.3 Conformal Mapping

Contour plots of complex functions

```
x=-2.0:.05:2.0;
y=-2.0:.05:2.0;
[X,Y]=meshgrid(x,y);
z=X+1i*Y;
z1=sin(z);
u=real(z1);
v=imag(z1);
contour(X,Y,u,'k-')
hold on
contour(X,Y,v,'k:')
hold off
axis equal
legend('Re(sin(z))','Im(sin(z))')
```

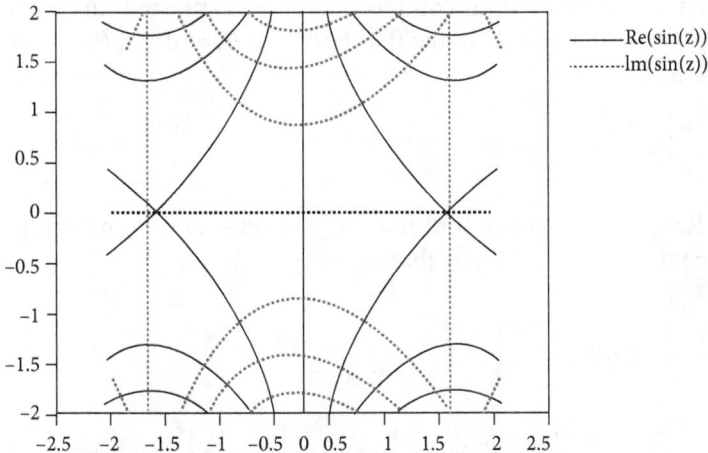

Figure 9.4.1: Contour plot of the real and imaginary parts of sin(z).

9.5 EXERCISES

Section 9.1 Cauchy-Riemann Equations

1. Given $f(z) = e^{z^2}$ create contour plots of Re$f(z) = u(x, y)$ and Im$f(z) = v(x, y)$ displayed together

2. Given $f(z) = e^z$ show that both Re$f(z) = u(x, y)$ and Im$f(z) = v(x, y)$ satisfy Laplace's equation

$$\frac{\partial^2}{\partial x^2}\begin{Bmatrix} u \\ v \end{Bmatrix} + \frac{\partial^2}{\partial y^2}\begin{Bmatrix} u \\ v \end{Bmatrix} = 0$$

3. Show that the Cauchy-Riemann equations in polar coordinates are

$$\frac{\partial u}{\partial r} = \frac{1}{r}\frac{\partial v}{\partial \theta} \text{ and } \frac{\partial v}{\partial r} = -\frac{1}{r}\frac{\partial u}{\partial \theta}$$

where $r = \sqrt{x^2 + y^2}$, $x = r\cos\theta$, $y = r\sin\theta$ and $\theta = \tan^{-1}\left(\dfrac{y}{x}\right)$

Hint: Use the chain rule relations

$$\frac{\partial u}{\partial x} = \frac{\partial u}{\partial r}\frac{\partial r}{\partial x} + \frac{\partial u}{\partial \theta}\frac{\partial \theta}{\partial x}$$

$$\frac{\partial u}{\partial y} = \frac{\partial u}{\partial r}\frac{\partial r}{\partial y} + \frac{\partial u}{\partial \theta}\frac{\partial \theta}{\partial y}$$

with similar expressions for $\dfrac{\partial v}{\partial x}$ and $\dfrac{\partial v}{\partial y}$.

4. Given an analytic function of a complex variable expressed in polar coordinates $f(z) = u(r, \theta) + iv(r, \theta)$ show that both $u(r, \theta)$ and $v(r, \theta)$ satisfy Laplace's equation

$$\frac{\partial^2}{\partial r^2}\begin{Bmatrix} u \\ v \end{Bmatrix} + \frac{1}{r}\frac{\partial}{\partial r}\begin{Bmatrix} u \\ v \end{Bmatrix} + \frac{1}{r}\frac{\partial^2}{\partial \theta^2}\begin{Bmatrix} u \\ v \end{Bmatrix} = 0$$

5. Given $f(z) = z^2$, Re $f(z) = u(x, y)$ and Im$f(z) = v(x, y)$ express $u(x, y)$ and $v(x, y)$ in polar coordinates and verify directly that

$$\frac{\partial^2}{\partial r^2}\begin{Bmatrix} u \\ v \end{Bmatrix} + \frac{1}{r}\frac{\partial}{\partial r}\begin{Bmatrix} u \\ v \end{Bmatrix} + \frac{1}{r}\frac{\partial^2}{\partial \theta^2}\begin{Bmatrix} u \\ v \end{Bmatrix} = 0$$

Section 9.2 Integral Theorems

6. Evaluate the line integral $\int_{\Gamma} z^2 dz$ from $z = 0$ to $z = 1 + i$

7. Show that $\oint_{\Gamma} z^2 dz = 0$ if Γ is

(a) a square with vertices $1 \pm i$ and $-1 \pm i$
(b) the unit circle $|z| = 1$

8. Show that $\oint_{\Gamma} e^z dz = 0$ where Γ is

(a) a square with vertices $1 \pm i$ and $-1 \pm i$
(b) the unit circle $|z| = 1$

9. Use Cauchy's integral formula

$$f(z_0) = \frac{1}{2\pi i} \oint_{\Gamma} \frac{f(z)}{z - z_0} dz$$

to evaluate

$$\frac{1}{2\pi i} \oint_{\Gamma} \frac{\cos(z)}{z - 1} dz \text{ where } \Gamma \text{ is the circle } |z| = 2$$

10. Use Cauchy's integral formula

$$f^{(n)}(z_0) = \frac{n!}{2\pi i} \oint_{\Gamma} \frac{f(z)}{(z - z_0)^{n+1}} dz$$

to evaluate

$$\frac{1}{2\pi i} \oint_{\Gamma} \frac{\cos(z)}{(z - 1)^3} dz = 0 \text{ where } \Gamma \text{ is the circle } |z| = 2$$

11. Find all residues of the following functions

$$\frac{\cosh(z)}{z^2 - 1}$$

$$\frac{z}{z^2 + 1}$$

$$\frac{e^z}{z^2 + z}$$

12. Find residues of the following functions at the indicated values of z

$$\frac{e^z}{z^2 - 1} \text{ at } z = 1$$

$$\frac{\cos(z)}{z^2+1} \text{ at } z=i$$

$$\frac{e^{z^2}}{z^2+z} \text{ at } z=0$$

13. Evaluate the following integrals

$$\oint_\Gamma \frac{\cosh(z)}{z^2-1}dz \text{ where } \Gamma \text{ is the circle } |z|=2$$

$$\oint_\Gamma \frac{z}{z^2+1}dz \text{ where } \Gamma \text{ is the circle } |z|=2$$

14. Evaluate the following improper integrals

$$\int_{-\infty}^{\infty} \frac{x^2 e^{3ix}}{x^4+20x^2+64}dx$$

$$\int_{-\infty}^{\infty} \frac{x^2}{x^4+11x^2+18}dx$$

$$\int_{-\infty}^{\infty} \frac{\sin^2 x}{x^2}dx$$

Section 9.3 Conformal Mapping

15. Create contour plots of the following conformal transformation

$$w = \frac{1}{z} \text{ with } z = \text{the square with vertices } 1\pm i \text{ and } -1\pm i$$

$$w = z^2 \text{ with } z = \text{the square with vertices } 1\pm i \text{ and } -1\pm i$$

$$w = z + \frac{5}{z} \text{ with } z = \text{the square with vertices } 1\pm i \text{ and } -1\pm i$$

16. Create 3D conformal mappings $w = f(z)$ on the Riemann sphere

$$w = \frac{\sin(z)}{z-1} \quad z = \text{the square with vertices } 1\pm i \text{ and } -1\pm i$$

$$w = \frac{\cosh(z)}{z} \quad z = \text{the square with vertices } 1\pm i \text{ and } -1\pm i$$

$$w = z + \frac{5}{z} \quad z = \text{the square with vertices } 1 \pm i \text{ and } -1 \pm i$$

17. Use Poisson's integral formula

$$V(x,y) = \frac{1}{\pi} \int_{-\infty}^{\infty} \frac{yV(x')dx'}{y^2 + (x-x')^2}$$

to plot the potential $V(x, y)$ in the upper half plane where

$$V(x') = \begin{cases} 1 & x' \geq 0 \\ 0 & x' < 0 \end{cases}$$

18. Use Poisson's integral formula

$$V(r,\theta) = \frac{1}{2\pi} \int_0^{2\pi} \frac{(R^2 - r^2)V(\theta')d\theta'}{R^2 - 2rR\cos(\theta - \theta') + r^2}$$

to plot the potential $V(r, \theta)$ inside a circle of radius $R = 1$
$V(\theta') = \cos(3\theta')$

19. Use Maple to create contour plots of the real and imaginary parts of the complex potentials

(a) Uniform flow $V(z) = e^{-3i}z$

(b) Superposition of source and sink $V(z) = \ln\dfrac{z-1}{z+1}$

10 CLASSICAL MECHANICS

Chapter Outline

10.1 VELOCITY-DEPENDENT RESISTIVE FORCES

In this section, we investigate the motion of a body in the presence of a drag force that increases monotonically with the velocity. Resistive motion in a gravitational field is then considered.

10.1.1 Drag Force Proportional to the Velocity

Example 10.1.1

Consider the motion of a body where the only force acting is a resistive drag force proportional to the velocity of the body. Find the velocity as a function of distance.

Solution: Newton's second law gives

$$m\frac{dv}{dt} = -bv \tag{10.1.1}$$

where b is a drag coefficient with units of Ns/m. We first seek the velocity as a function of distance

$$m\frac{dv}{dx}\frac{dx}{dt} = -bv \tag{10.1.2}$$

or

$$m\frac{dv}{dx}v = -bv \tag{10.1.3}$$

Thus, we integrate

$$\frac{dv}{dx} = -\frac{b}{m} \tag{10.1.4}$$

to obtain

$$v(x) = v_0 - \frac{bx}{m} \tag{10.1.5}$$

We can calculate the total distance traveled corresponding to $v(x) = 0$

$$x = \frac{v_0 m}{b} \tag{10.1.6}$$

Example 10.1.2

Find the velocity as a function of time in the example above and then integrate to obtain the distance traveled.

Solution: We may integrate

$$m\frac{dv}{v} = -\frac{b}{m}dt \tag{10.1.7}$$

to find the velocity as a function of time

$$v(t) = v_0 e^{-\frac{b}{m}t} \tag{10.1.8}$$

and then integrate again to obtain the total distance

$$x = \int_0^\infty v(t)\,dt = v_0 \int_0^\infty e^{-\frac{b}{m}t}\,dt = -\frac{v_0 m}{b} e^{-\frac{b}{m}t}\Big|_0^\infty = \frac{v_0 m}{b} \tag{10.1.9}$$

10.1.2 Drag Force on a Falling Body

If the particle is acted on by both gravitational and resistive drag forces, Newton's second law becomes

$$mg - bv = m\frac{dv}{dt} \tag{10.1.10}$$

where the gravitational force is in the opposite direction of the resistive drag force. We can immediately find the terminal velocity where the acceleration is equal to zero

$$v_T = \frac{mg}{b} \tag{10.1.11}$$

Example 10.1.3

Find the velocity and acceleration of a body falling in a resistive medium as a function of time

Solution: To find the velocity as a function of time we separate variables

$$dt = \frac{dv}{g - \frac{b}{m}v} \tag{10.1.12}$$

and integrate

$$\int_0^t dt = \int_0^v \frac{dv}{g - \frac{b}{m}v} \tag{10.1.13}$$

making a u substitution $u = g - \frac{b}{m}v$ so that $du = -\frac{b}{m}dv$ and our integral becomes

$$t = -\frac{m}{b}\int_{g}^{g-\frac{b}{m}v}\frac{du}{u} \qquad (10.1.14)$$

giving

$$-\frac{b}{m}t = \ln u\Big|_{g}^{g-\frac{b}{m}v} = \ln\frac{g-\frac{b}{m}v}{g} \qquad (10.1.15)$$

Exponentiating both sides and using $\ln e^x = x$ gives

$$e^{-\frac{b}{m}t} = 1 - \frac{b}{mg}v \qquad (10.1.16)$$

Solving for the time-dependent velocity

$$v(t) = \frac{mg}{b}\left(1 - e^{-\frac{b}{m}t}\right) \qquad (10.1.17)$$

and we see that at $t = 0$ the velocity $v = 0$ and as $t \to \infty$ we reach the terminal velocity $v \to mg/b$ as calculated previously. Also, we can calculate the acceleration as a function of time

$$a(t) = ge^{-\frac{b}{m}t} \qquad (10.1.18)$$

so we have that $a = g$ at $t = 0$.

Maple Examples

The time-dependent velocity of a body falling with a resistive drag force is calculated in the Maple worksheet below. Projectile motion is considered in the presence of a drag force.

Key Maple commands: *diff, dsolve, expand*

Drag Force Proportional to Velocity

$Eq1 := m \cdot diff(v(t), t) = m \cdot g - beta \cdot v(t)$

$$m\left(\frac{d}{dt}v(t)\right) = mg - \beta v(t)$$

$dsolve(\{Eq1, v(0) = 0\}, v(t));$

$$v(t) = \frac{gm}{\beta} - \frac{e^{\frac{-\beta t}{m}}gm}{\beta}$$

Drag Force Proportional to Velocity Squared

$Eq2 := m \cdot diff(v(t), t) = m \cdot g - C \cdot v(t)^2$

$$m\left(\frac{d}{dt}v(t)\right) = mg - Cv(t)^2$$

$dsolve(\{Eq2, v(0) = 0\}, v(t));$

$$v(t) = \frac{\tanh\left(\dfrac{t\sqrt{Cmg}}{m}\right)\sqrt{Cmg}}{C}$$

Projectile Motion with Drag Proportional to Velocity Squared

$Eq3 := m \cdot diff(y(t), t, t) = -m \cdot g - \text{beta} \cdot diff(y(t), t)$

$$m\left(\frac{d^2}{dt^2}y(t)\right) = -mg - \beta\left(\frac{d}{dt}y(t)\right)$$

$Eq4 := m \cdot diff(x(t), t, t) = -\text{beta} \cdot diff(x(t), t)$

$$m\left(\frac{d^2}{dt^2}x(t)\right) = -\beta\left(\frac{d}{dt}x(t)\right)$$

$dsolve(\{Eq3, Eq4, D(y)(0) = 1, D(x)(0) = 1, y(0) = 0, x(0) = 0\}, \{x(t), y(t)\})$

$$\left\{ x(t) = \frac{m}{\beta} - \frac{me^{\frac{-\beta t}{m}}}{\beta}, \, y(t) = -\frac{\dfrac{me^{\frac{-\beta t}{m}}(gm+\beta)}{\beta} + gmt - \dfrac{m(gm+\beta)}{\beta}}{\beta} \right\}$$

$expand(\%[2])$

$$y(t) = -\frac{m^2 g}{\beta^2 e^{\frac{\beta t}{m}}} - \frac{m}{e^{\frac{\beta t}{m}}\beta} - \frac{gmt}{\beta} + \frac{m^2 g}{\beta} + \frac{m}{\beta}$$

10.2 VARIABLE MASS DYNAMICS

The Tsiolkovsky rocket equation is developed using conservation of linear momentum in this section. Rocket thrust and motion in a gravitational field are calculated.

10.2.1 Rocket Motion

The initial momentum p_i of a rocket with mass $M + \Delta M$ traveling in one dimension with a speed v is

$$p_i = (M + \Delta M)v \qquad (10.2.1)$$

The final momentum p_f of the rocket and the ejected fuel mass ΔM is

$$p_f = M(v + \Delta v)v + \Delta M(v - v_e) \qquad (10.2.2)$$

where v_e is the exhaust speed. Conservation of momentum $p_i = p_f$ gives

$$Mv + \Delta Mv = Mv + M\Delta v + \Delta Mv - \Delta Mv_e \qquad (10.2.3)$$

Simplifying we are left with

$$M\Delta v = v_e \Delta M \qquad (10.2.4)$$

Dividing by Δt gives the rocket thrust as the product of the exhaust speed times the burn rate $\Delta M / \Delta t$.

$$M\frac{\Delta v}{\Delta t} = v_e \frac{\Delta M}{\Delta t} \qquad (10.2.5)$$

We also have a liftoff condition

$$v_e \frac{\Delta M}{\Delta t} > Mg \qquad (10.2.6)$$

where the thrust must exceed the weight of the rocket to leave the launch pad. To calculate the speed of the rocket we first consider the case of zero gravity from equation (10.2.4).

$$\Delta v = v_e \frac{\Delta M}{M} \qquad (10.2.7)$$

for an infinitesimal change in velocity $\Delta v \to dv$ with the rocket losing mass $\Delta M \to -dM$. Integration gives

$$\int_{v_i}^{v_f} dv = -v_e \int_{M_i}^{M_f} \frac{dM}{M} \qquad (10.2.8)$$

and we find

$$v_f = v_i + v_e \ln\left(\frac{M_i}{M_f}\right) \qquad (10.2.9)$$

as obtained by Tsiolkovsky in 1912 showing that the change in velocity is equal to the exhaust speed times the logarithm of the initial to final mass ratio. Considering rocket motion in a gravitational field with $v_i = 0$, $v_f \to v(t)$ and $M_f \to M(t)$.

$$v(t) = v_e \ln\left(\frac{M_i}{M(t)}\right) - gt \tag{10.2.10}$$

Integrating to obtain the height of the rocket as a function of time neglecting air resistance and the variation of g with height we find

$$h(t) = v_e \int_0^t \ln\left(\frac{M_i}{M(t')}\right) dt' - \frac{1}{2}gt^2 \tag{10.2.11}$$

where v_e is assumed constant. We may neglect air resistance during lunar launch where the acceleration of gravity is $g/6$.

Example 10.2.1

Calculate the rocket height as a function of time $h(t)$ for an exponential burn

$$M(t) = M_i e^{-\lambda t}$$

Using the fact that $\ln(e^x) = x$ we integrate the equation

$$h(t) = v_e \int_0^t \ln\left(e^{\lambda t'}\right) dt' - \frac{1}{2}gt^2$$

$$= v_e \lambda \int_0^t t' dt' - \frac{1}{2}gt^2 \tag{10.2.12}$$

$$= \frac{1}{2}(v_e \lambda - g)t^2$$

Thus, we require that $v_e \lambda > g$ for liftoff in this example. Calculating the thrust

$$\text{thrust} = \left| v_e \frac{dM(t)}{dt} \right| = \lambda v_e M_i e^{-\lambda t} \tag{10.2.13}$$

At $t = 0$, thrust $= \lambda v_e M_i$ and we verify our liftoff condition $\lambda v_e M_i > M_i g$ so that $v_e \lambda > g$.

Maple Examples

Rocket motion is modeled with and without resistive drag forces in the Maple worksheet below.

Key Maple commands: *assume, diff, dsolve, simplify*

restart

Rocket Motion: Exponential Burn

$assume(\text{lambda} \geq 0, t \geq 0)$

$M: = (t) \rightarrow Mi \cdot \exp(\text{-lambda} \cdot t)$

$$M: = t \mapsto Mi\, e^{-\lambda t}$$

$$v := (t) \rightarrow ve \cdot \ln\left(\frac{Mi}{M(t)}\right) - g \cdot t$$

$$v := (t) \mapsto ve\, \ln\left(\frac{Mi}{M(t)}\right) - g\, t$$

$Eq1 := diff(h(t), t) = v(t)$

$$Eq1 := \frac{d}{dt \sim} h(t \sim) = ve\lambda \sim t \sim -gt \sim$$

$dsolve(\{Eq1, v(0) = 0, h(0) = 0\}, h(t));$

$$\left\{ h(t \sim) = \frac{1}{2}\lambda \sim t \sim^2 ve - \frac{1}{2} gt \sim^2 \right\}$$

Rocket Motion with Resistive Drag

$withDrag := diff(vel(t), t) = ve \cdot (-diff(M(t), t) - m \cdot g - \text{beta} \cdot vel(t)$

$$withDrag := \frac{d}{dt \sim} vel(t \sim) = veM\lambda \sim e^{-\lambda \sim t \sim} - mg - \beta vel(t \sim)$$

$dsolve(\{withDrag, vel(0) = 0\})$

$$vel(t \sim) = \frac{Mi\lambda \sim ve\, e^{t \sim (\beta - \lambda \sim) - \beta t \sim}}{\beta - \lambda \sim} - \frac{gm}{\beta} + e^{-\beta t \sim}\left(-\frac{veMi\lambda \sim}{\beta - \lambda \sim} + \frac{gm}{\beta} \right)$$

10.3 LAGRANGIAN DYNAMICS

Variational methods find application in many branches of physics, including classical and quantum mechanics, relativity and electromagnetism. Topics in this section include the application of variational techniques to mechanical problems.

10.3.1 Calculus of Variations

The calculus of variations involves finding an unknown function $y(x)$ such that integral

$$J = \int_{x_1}^{x_2} F\left(y, \frac{dy}{dx}, x \right) dx \tag{10.3.1}$$

is an extremum $\delta J = 0$. The integrand $F(y, y', x)$ is a known functional (function of a function) of $y(x)$ and $y'(x)$. The integral J is an extremum if F satisfies the Euler-Lagrange equation

$$\frac{\partial F}{\partial y} - \frac{\partial}{\partial x}\left(\frac{\partial F}{\partial y'}\right) = 0 \qquad (10.3.2)$$

In the case of time-dependent dynamical problems, the functional is the Lagrangian L and

$$S = \int_{t_1}^{t_2} L\left(q, \frac{dq}{dt}, t\right) dt \qquad (10.3.3)$$

is the action integral where $\delta S = 0$ gives the differential equation of motion

$$\frac{\partial L}{\partial q} - \frac{\partial}{\partial t}\left(\frac{\partial L}{\partial \dot{q}}\right) = 0 \qquad (10.3.4)$$

Example 10.3.1

Use the Euler-Lagrange equation to show that the shortest distance between two points is a straight line.

Solution: The line element in two dimensions is

$$ds = \left(dx^2 + dy^2\right)^{1/2} = \left(1 + \left(\frac{dy}{dx}\right)^2\right)^{1/2} dx \qquad (10.3.5)$$

or

$$ds = F(x, y, y') = \left(1 + y'^2\right)^{1/2} \qquad (10.3.6)$$

where $y' = dy/dx$. The Euler-Lagrange equation is

$$\frac{\partial F}{\partial y} - \frac{\partial}{\partial x}\left(\frac{\partial F}{\partial y'}\right) = 0 \qquad (10.3.7)$$

Since $\partial F/\partial y = 0$

$$\frac{\partial}{\partial x}\frac{\partial F}{\partial y'} = \frac{\partial}{\partial x}\left[\frac{y'}{\left(1 + y'^2\right)^{1/2}}\right] = 0 \qquad (10.3.8)$$

and

$$\frac{y'}{\left(1 + y'^2\right)^{1/2}} = \text{const.} \qquad (10.3.9)$$

Squaring both sides and solving gives $y' = c_1$ and

$$y = c_1 x + c_2 \tag{10.3.10}$$

where c_1 and c_2 are constants.

10.3.2 Lagrange's Equations of Motion

The Lagrangian L is defined as

$$L = T - V \tag{10.3.11}$$

where the kinetic energy T and potential energy V are both functions of the generalized coordinates q_i and velocities $\dot{q}_i$ where

$$L = L\left(q_1, q_2, \ldots q_N, \dot{q}_1, \dot{q}_2, \ldots \dot{q}_N, t\right) \tag{10.3.12}$$

for N degrees of freedom. We seek the q_i such that

$$\delta S = \delta \int_{t_1}^{t_2} L \, dt = 0 \tag{10.3.13}$$

with the q_i fixed at the endpoints $\delta q_i(t_1) = \delta q_i(t_2) = 0$.

In one dimension, $L = L(q, \dot{q}, t)$ and the action integral is

$$S = \int_{t_1}^{t_2} L\left(q, \dot{q}, t\right) dt \tag{10.3.14}$$

Calculating the variation of the action

$$\delta S = \int_{t_1}^{t_2} \left[\frac{\partial L}{\partial q} \delta q + \frac{\partial L}{\partial \dot{q}} \delta \dot{q} \right] dt = 0 \tag{10.3.15}$$

with

$$\delta \dot{q} = d\left(\delta q\right) / dt \tag{10.3.16}$$

$$\int_{t_1}^{t_2} \left[\frac{\partial L}{\partial q} \delta q + \frac{\partial L}{\partial \dot{q}} \frac{d}{dt}\left(\delta q\right) \right] dt = 0 \tag{10.3.17}$$

Integrating the second term by parts

$$\int_{t_1}^{t_2} \frac{\partial L}{\partial \dot{q}} \frac{d}{dt}\left(\delta q\right) dt \rightarrow \frac{\partial L}{\partial \dot{q}} \delta q \Big|_{t_1}^{t_2} - \int_{t_1}^{t_2} \frac{d}{dt}\left(\frac{\partial L}{\partial \dot{q}}\right) \delta q \, dt \tag{10.3.18}$$

and since $\delta q(t_1) = \delta q(t_2) = 0$ we have

$$\int_{t_1}^{t_2} \left[\frac{\partial L}{\partial q} \delta q - \frac{d}{dt}\left(\frac{\partial L}{\partial \dot{q}}\right) \delta q \right] dt = \int_{t_1}^{t_2} \left[\frac{\partial L}{\partial q} - \frac{d}{dt}\left(\frac{\partial L}{\partial \dot{q}}\right) \right] \delta q \, dt = 0 \tag{10.3.19}$$

For the integral to be zero for an arbitrary choice of δq the integrand must be zero so

$$\frac{d}{dt}\left(\frac{\partial L}{\partial \dot{q}}\right)-\frac{\partial L}{\partial q}=0 \qquad (10.3.20)$$

This is the Euler-Lagrange equation of motion. In general, we will have one equation for each degree of freedom

$$\frac{d}{dt}\frac{\partial L}{\partial \dot{q}_i}-\frac{\partial L}{\partial q_i}=0 \qquad (10.3.21)$$

Example 10.3.2

Show that $\delta \dot{q}=d(\delta q)/dt$ assuming q depends on a parameter r

Solution:

$$\delta \dot{q}=\delta \frac{dq}{dt}=\frac{\partial}{\partial r}\frac{dq}{dt}\delta r$$

$$=\frac{d}{dt}\frac{\partial q}{\partial r}\delta r \qquad (10.3.22)$$

$$=\frac{d}{dt}\delta q$$

Example 10.3.3

Find the equation of motion of a block on a spring with kinetic energy $T=\frac{1}{2}m\dot{x}^2$ and potential energy $V=\frac{1}{2}kx^2$

Solution: The Lagrangian is

$$L=T-V=\frac{1}{2}m\dot{x}^2-\frac{1}{2}kx^2 \qquad (10.3.23)$$

Lagrange's equation of motion

$$\frac{d}{dt}\frac{\partial L}{\partial \dot{x}}-\frac{\partial L}{\partial x}=0 \qquad (10.3.24)$$

becomes

$$\frac{d}{dt}(m\dot{x})+kx=0 \qquad (10.3.25)$$

or

$$m\ddot{x}=-kx \qquad (10.3.26)$$

Example 10.3.4

Find the equation of motion of a simple pendulum with kinetic energy $T = \frac{1}{2}m\left(\ell\dot\theta\right)^2$ and potential energy $V = mg\ell\left(1-\cos\theta\right)$.

Solution: Constructing the Lagrangian

$$L = \frac{1}{2}m\left(\ell\dot\theta\right)^2 - mg\ell\left(1-\cos\theta\right) \tag{10.3.27}$$

we evaluate

$$\frac{d}{dt}\frac{\partial L}{\partial\dot\theta} - \frac{\partial L}{\partial\theta} = 0 \tag{10.3.28}$$

$$\frac{d}{dt}\left(m\ell^2\dot\theta\right) + mg\ell\sin\theta = 0 \tag{10.3.29}$$

or

$$\ddot\theta = -\frac{g}{\ell}\sin\theta \tag{10.3.30}$$

10.3.3 Lagrange's Equations with Constraints

The Lagrangian L' of a dynamical system subject to the holonomic constraint

$$f\left(q_1,q_2,\ldots q_N,t\right) = 0 \tag{10.3.31}$$

is

$$L' = L + \lambda f\left(q_i\right) \tag{10.3.32}$$

where λ is the Lagrange multiplier to be determined.
Lagrange's equations of motion

$$\frac{d}{dt}\frac{\partial L'}{\partial\dot q_i} - \frac{\partial L'}{\partial q_i} = 0 \tag{10.3.33}$$

may be written as

$$\frac{d}{dt}\frac{\partial L}{\partial\dot q_i} - \frac{\partial L}{\partial q_i} = \lambda\frac{\partial f}{\partial q_i} \tag{10.3.34}$$

Example 10.3.5

Find the equations of motion of a sphere of radius r rolling without slipping down a plane inclined at an angle θ.

Solution: The kinetic energy of the rolling sphere consists of translational and rotational components

$$T = \frac{1}{2}m\dot{x}^2 + \frac{1}{2}I\dot{\phi}^2 \qquad (10.3.35)$$

where I is the moment of inertia of the sphere. The potential energy of the sphere is measured from the bottom of the plane of length L.

$$V = mg(L-x)\sin\theta \qquad (10.3.36)$$

Our constraint equation is

$$x - r\phi = 0 \qquad (10.3.37)$$

with Lagrangian

$$\frac{1}{2}m\dot{x}^2 + \frac{1}{2}I\dot{\phi}^2 - mg(L-x)\sin\theta + \lambda(x-r\phi) = 0 \qquad (10.3.38)$$

The equations of motion become

$$m\ddot{x} + mg\sin\theta - \lambda = 0 \qquad (10.3.39)$$

$$I\ddot{\phi} + \lambda r = 0 \qquad (10.3.40)$$

Maple Examples

Lagrange's equations of motion are found for physical systems in the Maple worksheet below. It is not necessary to substitute dummy variables when differentiating with respect to time derivatives using the Physics package.

Key Maple commands: *diff, Setup, subs*

Maple packages: *with(Physics)*:

restart

Lagrangian: Simple Harmonic Oscillator

$$L := \frac{1}{2}\cdot m\cdot xdot^2 - \frac{1}{2}\cdot k\cdot x^2;$$

$$L := \frac{mxdot^2}{2} - \frac{kx^2}{2}$$

diff(L, xdot)

$$m\,xdot$$

diff(subs(xdot = diff(x(t), t), diff(L, xdot)), t) − subs(x = x(t), diff(L, x)) = 0

$$m\left(\frac{d^2}{dt^2}x(t)\right)+kx(t)=0$$

Lagrangian: Central Potential

restart

$$L:=\frac{1}{2}\cdot m\cdot\left(rdot^2+r^2\cdot phidot^2\right)+\frac{G\cdot m\cdot M}{r};$$

$$L:=\frac{m\left(r^2phidot^2+rdot^2\right)}{2}+\frac{GmM}{r}$$

tsubs := {r = r(t), phi = phi(t), rdot = diff(r(t), t), phidot = diff(phi(t), t)};

$$tsubs:=\left\{\phi=\phi(t),phidot=\frac{d}{dt}\phi(t),r=r(t),rdot=\frac{d}{dt}r(t)\right\}$$

rEqn := diff(subs(tsubs, diff(L, rdot)), t) − subs(tsubs, diff(L, r)) = 0

$$rEqn:=m\left(\frac{d^2}{dt^2}r(t)\right)-m\left(\frac{d}{dt}\phi(t)\right)^2r(t)+\frac{GmM}{r(t)^2}=0$$

Lagrangian: Physics Package

restart
with(Physics) :
Setup(mathematicalnotation=true)

$$[mathematicalnotation=true]$$

$$L:=\frac{1}{2}\cdot m\cdot diff\left(r(t),t\right)^2+\frac{1}{2}\cdot m\cdot r(t)^2\cdot diff\left(phi(t),t\right)^2+\frac{G\cdot m\cdot M}{r(t)}$$

$$L:=\frac{m\left(\frac{d}{dt}r(t)\right)^2}{2}+\frac{mr(t)^2\left(\frac{d}{dt}\phi(t)\right)^2}{2}+\frac{GmM}{r(t)}$$

diff(diff(L, diff(r(t), t)), t) − diff(L, r(t)) = 0

$$m\left(\frac{d^2}{dt^2}r(t)\right)-mr(t)\left(\frac{d}{dt}\phi(t)\right)^2+\frac{GmM}{r(t)^2}=0$$

10.4 HAMILTONIAN MECHANICS

Hamilton's equations of motion are applied to dynamical problems in this section. Hamiltonian mechanics finds utility in several branches of physics and is especially useful in quantum mechanics. We first introduce the concept of generalized momenta and define the Hamiltonian in terms of a Legendre transformation of the Lagrangian function. Hamilton's equations of motion are obtained and expressed using the Poisson bracket formulation. Examples of the simple harmonic oscillator and the simple pendulum are given.

10.4.1 Legendre Transformation

Given the Lagrangian function of generalized coordinates and momenta

$$L = L\left(q_1, q_2, \ldots q_N, \dot{q}_1, \dot{q}_2, \ldots \dot{q}_N\right) \tag{10.4.1}$$

the generalized momenta are

$$p_i = \frac{\partial L}{\partial \dot{q}_i} \tag{10.4.2}$$

The Hamiltonian is constructed as

$$H\left(q_1, q_2, \ldots q_N, \dot{q}_1, \dot{q}_2, \ldots \dot{q}_N\right) = \sum_{i=1}^{N} p_i \dot{q}_i - L\left(q_1, q_2, \ldots q_N, \dot{q}_1, \dot{q}_2, \ldots \dot{q}_N\right) \tag{10.4.3}$$

10.4.2 Hamilton's Equations of Motion

Given a Lagrangian L, the Hamiltonian from the Legendre transformation

$$H = \sum_{i=1}^{N} p_i \dot{q}_i - L \tag{10.4.4}$$

To find Hamilton's equations of motion we evaluate

$$dH = \sum_{i=1}^{N} dp_i \dot{q}_i + \sum_{i=1}^{N} p_i d\dot{q}_i - \sum_{i=1}^{N} \frac{\partial L}{\partial \dot{q}_i} d\dot{q}_i - \sum_{i=1}^{N} \frac{\partial L}{\partial q_i} dq_i \tag{10.4.5}$$

Since the second and third terms sum to zero and $\dot{p}_i = \partial L / \partial q_i$

$$dH = \sum_{i=1}^{N} dp_i \dot{q}_i - \sum_{i=1}^{N} \dot{p}_i dq_i \tag{10.4.6}$$

Given $H = H(p_i, q_i)$ we also have that

$$dH = \sum_{i=1}^{N} \left(\frac{\partial H}{\partial p_i}\right) dp_i + \sum_{i=1}^{N} \left(\frac{\partial H}{\partial q_i}\right) dq_i \tag{10.4.7}$$

Comparing the last two expressions we obtain the equations of motion

$$\dot{p}_i = -\frac{\partial H}{\partial q_i}$$

$$\dot{q}_i = \frac{\partial H}{\partial p_i}$$

(10.4.8)

Hamilton's equations of motion are first order in time while Lagrange's equations are usually second order. The equations of motion are often nonlinear and must be evaluated numerically. Numerical integration schemes can be applied directly to systems of first order equations whereas higher order differential equations must be converted to first order systems before applying integration schemes such as the Euler method or the Runge-Kutta technique.

Example 10.4.1

Find Hamilton's equations of motion of a block of mass m attached to a spring with force constant k.

Solution: The Lagrangian of the block on a spring considered in Section 10.3 is

$$L = T - V = \frac{1}{2}m\dot{x}^2 - \frac{1}{2}kx^2$$

(10.4.9)

The generalized momentum is

$$p_x = \frac{\partial L}{\partial \dot{x}} = m\dot{x}$$

(10.4.10)

The Hamiltonian is obtained from the Legendre transformation

$$H = \dot{x}p_x - L = \frac{1}{2}m\dot{x}^2 + \frac{1}{2}kx^2$$

(10.4.11)

Writing the Hamiltonian in terms of the generalized momentum and position

$$H = \frac{p_x^2}{2m} + \frac{1}{2}kx^2$$

(10.4.12)

Hamilton's equations of motion are thus

$$\dot{p}_x = -\frac{\partial H}{\partial x} = -kx$$

(10.4.13)

$$\dot{x} = \frac{\partial H}{\partial p_x} = \frac{p_x}{m}$$

(10.4.14)

Example 10.4.2

Find Hamilton's equations of motion of the simple pendulum with Lagrangian

$$L = T - V = \frac{1}{2}m\left(\ell\dot{\theta}\right)^2 - mg\ell\left(1 - \cos\theta\right) \qquad (10.4.15)$$

Solution: The generalized momentum is

$$p_\theta = \frac{\partial L}{\partial \dot{\theta}} = m\ell^2\dot{\theta} \qquad (10.4.16)$$

The Hamiltonian

$$H = p_\theta\dot{\theta} - L = \frac{1}{2}m\left(\ell\dot{\theta}\right)^2 + mg\ell\left(1 - \cos\theta\right) \qquad (10.4.17)$$

written in terms of p_θ and θ becomes

$$H = \frac{p_\theta^2}{2m\ell} + mg\ell\left(1 - \cos\theta\right) \qquad (10.4.18)$$

Hamilton's first order equations of motion are

$$\dot{p}_\theta = -\frac{\partial H}{\partial \theta} = -mg\ell\sin\theta \qquad (10.4.19)$$

$$\dot{\theta} = \frac{\partial H}{\partial p_\theta} = \frac{p_\theta}{m\ell} \qquad (10.4.20)$$

Example 10.4.3

Given a Lagrangian such that

$$L = L\left(q,\dot{q}\right) \text{ and } \frac{\partial L}{\partial t} = 0 \qquad (10.4.21)$$

show that the Hamiltonian is a conserved quantity

$$\frac{dH}{dt} = 0 \qquad (10.4.22)$$

Solution: Calculating the total derivative of the Lagrangian

$$\frac{dL}{dt} = \frac{\partial L}{\partial t} + \frac{\partial L}{\partial q}\frac{\partial q}{\partial t} + \frac{\partial L}{\partial \dot{q}}\frac{\partial \dot{q}}{\partial t} \qquad (10.4.23)$$

we have

$$\frac{dL}{dt} = \frac{\partial L}{\partial q}\dot{q} + \frac{\partial L}{\partial \dot{q}}\ddot{q} \qquad (10.4.24)$$

From Lagrange's equation of motion

$$\frac{\partial L}{\partial q} = \frac{d}{dt}\frac{\partial L}{\partial \dot{q}} \tag{10.4.25}$$

we have

$$\frac{dL}{dt} = \frac{d}{dt}\frac{\partial L}{\partial \dot{q}}\dot{q} + \frac{\partial L}{\partial \dot{q}}\ddot{q} = \frac{d}{dt}\left(\frac{\partial L}{\partial \dot{q}}\dot{q}\right) \tag{10.4.26}$$

or

$$\frac{d}{dt}\left(\frac{\partial L}{\partial \dot{q}}\dot{q} - L\right) = 0 \tag{10.4.27}$$

Thus, we have the conserved quantity

$$\frac{dH}{dt} = 0 \tag{10.4.28}$$

where the Hamiltonian is

$$H = \frac{\partial L}{\partial \dot{q}}\dot{q} - L \tag{10.4.29}$$

Also, if

$$L = L\left(q_1, q_2, \ldots q_N, \dot{q}_1, \dot{q}_2, \ldots \dot{q}_N\right) \text{ and } \frac{\partial L}{\partial t} = 0 \tag{10.4.30}$$

the Hamiltonian

$$H = \sum_{i=1}^{N} p_i \dot{q}_i - L \text{ where } p_i = \frac{\partial L}{\partial \dot{q}_i} \tag{10.4.31}$$

is a conserved quantity.

10.4.3 Poisson Brackets

Consider a function $F(q, p)$ of a generalized coordinate q and momentum p. We can calculate the time derivative of F by using the chain rule

$$\frac{dF(q,p)}{dt} = \frac{\partial F}{\partial q}\frac{\partial q}{\partial t} + \frac{\partial F}{\partial p}\frac{\partial p}{\partial t} \tag{10.4.32}$$

or in dot notation

$$\dot{F} = \frac{\partial F}{\partial q}\dot{q} + \frac{\partial F}{\partial p}\dot{p} \tag{10.4.33}$$

Substituting Hamilton's equations

$$\dot{p} = -\frac{\partial H}{\partial q} \tag{10.4.34}$$

$$\dot{q} = \frac{\partial H}{\partial p} \tag{10.4.35}$$

into our time derivative

$$\dot{F} = \frac{\partial F}{\partial q}\frac{\partial H}{\partial p} - \frac{\partial F}{\partial p}\frac{\partial H}{\partial q} = \{F, H\} \tag{10.4.36}$$

where the Poisson bracket between two quantities A and B is defined as

$$\{A, B\} = \frac{\partial A}{\partial q}\frac{\partial B}{\partial p} - \frac{\partial B}{\partial p}\frac{\partial A}{\partial q} \tag{10.4.37}$$

The Poisson bracket of A with the Hamiltonian gives the time derivative

$$\dot{A} = \{A, H\} \tag{10.4.38}$$

Some important Poisson bracket relations include

$$\{p, q\} = -1$$
$$\{q, p\} = 1$$
$$\{p, q^2\} = -2q \tag{10.4.39}$$
$$\{q, p^2\} = 2p$$
$$\{p, p\} = \{q, q\} = 0$$
$$\{p, f(p)\} = \{q, f(q)\} = 0$$

where $f(p)$ and $f(q)$ are arbitrary functions of p and q.

Example 10.4.4

Use Poisson brackets to find Hamilton's equations of motion of the one-dimensional harmonic oscillator with Hamiltonian

$$H = \frac{p_x^2}{2m} + \frac{1}{2}kx^2 \tag{10.4.40}$$

Solution: To obtain the equations of motion we calculate the Poisson brackets

$$\dot{p}_x = \left\{ p_x, \frac{p_x^2}{2m} + \frac{1}{2}kx^2 \right\} = \frac{1}{2}k\{p_x, x^2\} = kx\{p_x, x\} = -kx \tag{10.4.41}$$

$$\dot{x} = \left\{ x, \frac{p_x^2}{2m} + \frac{1}{2}kx^2 \right\} = \frac{1}{2m}\{x, p_x^2\} = \frac{p_x}{m}\{x, p_x\} = \frac{p_x}{m} \qquad (10.4.42)$$

Example 10.4.5

Use Poisson brackets to find Hamilton's equations of motion of the simple pendulum with Hamiltonian

$$H = \frac{p_\theta^2}{2m\ell} + mg\ell(1-\cos\theta) \qquad (10.4.43)$$

Solution: To obtain the equations of motion we calculate the Poisson brackets

$$\dot{p}_\theta = \left\{ p_\theta, \frac{p_\theta^2}{2m\ell} + mg\ell(1-\cos\theta) \right\} = -mg\ell\{p_\theta, \cos\theta\} = -mg\ell\sin\theta \qquad (10.4.44)$$

$$\dot{\theta} = \left\{ \theta, \frac{p_\theta^2}{2m\ell} + mg\ell(1-\cos\theta) \right\} = \frac{1}{2m\ell}\{\theta, p_\theta^2\} = \frac{p_\theta}{m\ell} \qquad (10.4.45)$$

Maple Examples

The Hamiltonian and Hamilton's equations of motion are found for physical systems in the Maple worksheet below. The equations of motion may be found directly using the *DEtools* package. It is not necessary to substitute dummy variables when differentiating with respect to time derivatives using the Physics package.

Key Maple terms: *collect, diff, hamilton_eqs, simplify, solve, subs*

Maple packages: *with(DEtools): with(Physics):*

restart

Hamiltonian from the Lagrangian

$$L := \frac{1}{2} \cdot m \cdot xdot^2 + \frac{1}{2} \cdot m \cdot ydot^2 - \frac{1}{2} \cdot k \cdot x^2 - \frac{1}{2} \cdot k \cdot y^2$$

$$L := \frac{1}{2}mxdot^2 + \frac{1}{2}mydot^2 - \frac{1}{2}kx^2 - \frac{1}{2}ky^2$$

Px = diff(L, xdot)

$$Px = m\,xdot$$

Py = diff(L, ydot)

$$Py = m\,ydot$$

$$H := diff(L, xdot) \cdot xdot + diff(L, ydot) \cdot ydot - L$$

$$H := \frac{1}{2} mxdot^2 + \frac{1}{2} mydot^2 + \frac{1}{2} kx^2 + \frac{1}{2} ky^2$$

$$H := subs\left(\left\{xdot = \frac{Px}{m}, ydot = \frac{Py}{m}\right\}, H\right)$$

$$H := \frac{Px^2}{2m} + \frac{Py^2}{2m} + \frac{kx^2}{2} + \frac{ky^2}{2}$$

Henon-Heiles Hamiltonian

restart
tsubs := $\{x = x(t), y = y(t), Px = Px(t), Py = Py(t)\}$;

$$tsubs := \{Px = Px(t), Py = Py(t), x = x(t), y = y(t)\}$$

$$H := \frac{1}{2}\left(Px^2 + Py^2\right) + \frac{1}{2} \cdot \left(x^2 + y^2\right) + lambda \cdot \left(x^2 \cdot y - \frac{y^3}{3}\right)$$

$$H := \frac{Px^2}{2} + \frac{Py^2}{2} + \frac{x^2}{2} + \frac{y^2}{2} + \lambda\left(x^2 y - \frac{1}{3} y^3\right)$$

Hamilton's Equations of Motion

$$diff(x(t), t) = subs(tsubs, diff(H, Px))$$

$$\frac{d}{dt} x(t) = Px(t)$$

$$diff(y(t), t) = subs(tsubs, diff(H, Py))$$

$$\frac{d}{dt} y(t) = Py(t)$$

$$diff(Px(t), t) = subs(tsubs, -diff(H, x))$$

$$\frac{d}{dt} Px(t) = -2\lambda x(t) y(t) - x(t)$$

$$diff(Py(t), t) = subs(tsubs, -diff(H, y))$$

$$\frac{d}{dt} Py(t) = -y(t) - \lambda\left(x(t)^2 - y(t)^2\right)$$

Hamilton's Equations using the *DEtools* Package

with(DEtools) :

$$H := \frac{1}{2}\left(p1^2 + p2^2\right) + \frac{1}{2}\cdot\left(q1^2 + q2^2\right) + \text{lambda}\cdot\left(q1^2\cdot q2 - \frac{q2^3}{3}\right)$$

$$H := \frac{p1^2}{2} + \frac{p2^2}{2} + \frac{q1^2}{2} + \frac{q2^2}{2} + \lambda\left(q1^2 q2 - \frac{1}{3}q2^3\right)$$

Hamilton_eqs(H)

$$\left[\frac{d}{dt}p1(t) = -2\lambda q1(t)q2(t) - q1(t), \frac{d}{dt}p2(t) = -q2(t) - \lambda\left(q1(t)^2 - q2(t)^2\right),\right.$$

$$\left.\frac{d}{dt}q1(t) = p1(t), \frac{d}{dt}q2(t) = p2(t)\right], \left[p1(t), p2(t), q1(t), q2(t)\right]$$

Hamiltonian of the Double Pendulum from the Lagrangian

restart
with(Physics) :
$x_1 := (t) \rightarrow l\cdot\sin(\theta_1(t))$

$$x_1 := (t) \mapsto l\sin(\theta_1(t))$$

$y_1 := (t) \rightarrow l - l\cdot\cos(\theta_1(t))$

$$y_1 := (t) \mapsto l - l\cos(\theta_1(t))$$

$x_2 := (t) \rightarrow l\cdot\sin(\theta_1(t)) + l\cdot\sin(\theta_2(t))$

$$x_2 := (t) \mapsto l\sin(\theta_1(t)) + l\cdot\sin(\theta_2(t))$$

$y_2 := (t) \rightarrow 2\cdot l - l\cdot\cos(\theta_1(t)) - l\cdot\cos(\theta_2(t))$

$$y_2 := (t) \mapsto 2l - l\cos(\theta_1(t)) - l\cos(\theta_2(t))$$

$$T := \frac{1}{2}\cdot m\cdot\left(\text{diff}\left(x_1(t),t\right)^2 + \text{diff}\left(y_1(t),t\right)^2\right)$$

$$+\frac{1}{2}\cdot m\cdot\left(\text{diff}\left(x_2(t),t\right)^2 + \text{diff}\left(y_2(t),t\right)^2\right):$$

simplify(T, trig)

$$m\left(\left(\frac{d}{dt}\theta_1(t)\right)^2 + \left(\frac{d}{dt}\theta_2(t)\right)\left(\cos(\theta_1(t))\cos(\theta_2(t)) + \sin(\theta_1(t))\sin(\theta_2(t))\right)\right.$$

$$\left.\left(\frac{d}{dt}\theta_1(t)\right) + \frac{\left(\frac{d}{dt}\theta_2(t)\right)^2}{2}\right)t^2$$

$subs(\sin(\theta_1(t))\sin(\theta_2(t)) + \cos(\theta_1(t))\cos(\theta_2(t)) = \cos(\theta_1(t) - \theta_2(t)), \%):$
$T := \%$

$$T := m\left(\left(\frac{d}{dt}\theta_1(t)\right)^2 + \left(\frac{d}{dt}\theta_1(t)\right)\left(\frac{d}{dt}\theta_2(t)\right)\cos\left(\theta_1(t) - \theta_2(t)\right)\right.$$
$$\left. + \frac{\left(\frac{d}{dt}\theta_2(t)\right)^2}{2}\right)t^2$$

$V := simplify(m \cdot g \cdot y_1(t) + m \cdot g \cdot y_2(t)):$
$L := subs(\{m = 1, g = 1, l = 1\}, T - V)$

$$L := \left(\frac{d}{dt}\theta_1(t)\right)^2 + \left(\frac{d}{dt}\theta_1(t)\right)\left(\frac{d}{dt}\theta_2(t)\right)\cos\left(\theta_1(t) - \theta_2(t)\right)$$
$$+ \frac{\left(\frac{d}{dt}\theta_2(t)\right)^2}{2} + 2\cos\left(\theta_1(t)\right) + \cos\left(\theta_2(t)\right) - 3$$

$P_1 := (diff(L, diff(\theta_1(t), t)))$

$$P_1 := 2\frac{d}{dt}\theta_1(t) + \left(\frac{d}{dt}\theta_2(t)\right)\cos\left(\theta_1(t) - \theta_2(t)\right)$$

$P_2 := (diff(L, diff(\theta_2(t), t)))$

$$P_2 := \left(\frac{d}{dt}\theta_1(t)\right)\cos\left(\theta_1(t) - \theta_2(t)\right) + \frac{d}{dt}\theta_2(t)$$

$H := P_1 \cdot diff(\theta_1(t), t) + P_2 \cdot diff(\theta_2(t), t) - L:$
$H := collect(H, \{diff(\theta_1(t), t), diff(\theta_1(t), t)\})$

$$H := \left(\frac{d}{dt}\theta_1(t)\right)^2 + \left(\frac{d}{dt}\theta_1(t)\right)\left(\frac{d}{dt}\theta_2(t)\right)\cos\left(\theta_1(t) - \theta_2(t)\right)$$
$$+ \frac{\left(\frac{d}{dt}\theta_2(t)\right)^2}{2} - 2\cos\left(\theta_1(t)\right) - \cos\left(\theta_2(t)\right) + 3$$

Hamiltonian in Terms of Generalized Momenta

$$solve\left(\{p_1(t) = P_1, p_2(t) = P_2\}, \left\{\frac{d}{dt}\theta_2(t), \frac{d}{dt}\theta_1(t)\right\}\right)$$

$$\begin{cases} \dfrac{d}{dt}\theta_1(t) = \dfrac{p_2(t)\cos\big(\theta_1(t)-\theta_2(t)\big)-p_1(t)}{\cos\big(\theta_1(t)-\theta_2(t)\big)^2-2}, \dfrac{d}{dt}\theta_2(t) \\[4mm] \qquad = -\dfrac{-p_1(t)\cos\big(\theta_1(t)-\theta_2(t)\big)+2p_2(t)}{\cos\big(\theta_1(t)-\theta_2(t)\big)^2-2} \end{cases}$$

$H := simplify(subs(\%, H)):$

Double Pendulum Equations of Motion

$Deq1 := diff(\theta_1(t), t) = diff(H, p_1(t)):$
$Deq2 := diff(\theta_2(t), t) = diff(H, p_2(t)):$
$Deq3 := diff(p_1(t), t) = -simplify(diff(H, theta_1(t))):$
$Deq4 := diff(p_2(t), t) = -simplify(diff(H, theta_2(t))):$
$sol := dsolve(\{Deq1, Deq2, Deq3, Deq4, \theta_1(0) = 0, \theta_2(0) = 0.2, p_1(0) = 0, p_2(0) = 0\},$
$\{\theta_1(t), \theta_2(t), p_1(t), p_2(t)\}, numeric)$
$$sol := \mathbf{proc}(x\_rkf45) \dots \mathbf{end\ proc}$$
$with(plots):$
$withodeplot(sol, [\theta_1(t), p_1(t)], t = 0\dots200, numpoints = 2000)$

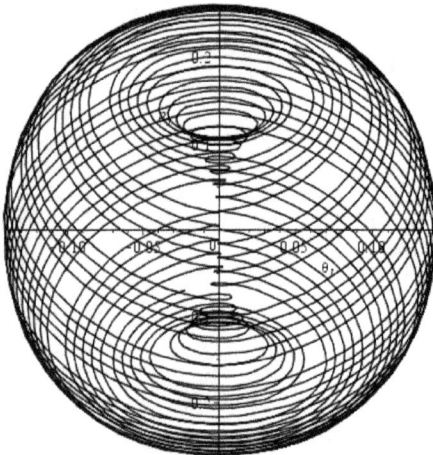

Figure 10.4.1: Phase plot of the double pendulum.

10.5 ORBITAL AND PERIODIC MOTION

Lagrangian formalism is applied to orbital and periodic motion in this section. Integral expressions are developed to calculate the period of motion for different potentials. The theory of small oscillations about equilibria is then discussed.

10.5.1 Kepler Problem

The orbit of a small mass m about a much larger mass $M \gg m$ located at the origin of coordinates may be formulated using Lagrangian mechanics where the kinetic energy in cylindrical coordinates is

$$T = \frac{1}{2}m\left(\dot{r}^2 + r^2\dot{\phi}^2\right) \tag{10.5.1}$$

The potential energy

$$V = -G\frac{Mm}{r} \tag{10.5.2}$$

The Lagrangian is

$$L = T - V = \frac{1}{2}m\left(\dot{r}^2 + r^2\dot{\phi}^2\right) + G\frac{Mm}{r} \tag{10.5.3}$$

Lagrange's equations are

$$\frac{d}{dt}\frac{\partial L}{\partial \dot{r}} - \frac{\partial L}{\partial r} = 0 \tag{10.5.4}$$

$$\frac{d}{dt}\frac{\partial L}{\partial \dot{\phi}} - \frac{\partial L}{\partial \phi} = 0 \tag{10.5.5}$$

The radial equation is

$$\frac{d}{dt}(m\dot{r}) - mr\dot{\phi}^2 + G\frac{Mm}{r^2} = 0 \tag{10.5.6}$$

The angular equation

$$\frac{d}{dt}\left(mr^2\dot{\phi}\right) = 0 \tag{10.5.7}$$

shows the quantity $mr^2\dot{\phi} = \ell$ is a constant so that we may write the radial equation

$$m\ddot{r} - \frac{\ell^2}{mr^3} + G\frac{Mm}{r^2} = 0 \tag{10.5.8}$$

The total energy

$$E = T + V = \frac{1}{2}m\left(\dot{r}^2 + r^2\dot{\phi}^2\right) - G\frac{Mm}{r} \tag{10.5.9}$$

Eliminating $\dot{\phi}$

$$E = \frac{1}{2}m\left(\dot{r}^2 + \frac{\ell^2}{m^2 r^2}\right) - G\frac{Mm}{r} \tag{10.5.10}$$

we may integrate

$$\dot{r} = \sqrt{\frac{2}{m}\left(E + G\frac{Mm}{r}\right) - \frac{\ell^2}{m^2 r^2}} \tag{10.5.11}$$

Separating variables

$$dt = \frac{dr}{\sqrt{\frac{2}{m}\left(E + G\frac{Mm}{r}\right) - \frac{\ell^2}{m^2 r^2}}} \tag{10.5.12}$$

in terms of $\dfrac{d\phi}{dt} = \dfrac{\ell}{mr^2}$ and $dt = \dfrac{mr^2}{\ell}d\phi$

$$d\phi = \frac{\ell}{mr^2}\frac{dr}{\sqrt{\frac{2}{m}\left(E + G\frac{Mm}{r}\right) - \frac{\ell^2}{m^2 r^2}}} = \frac{dr}{\sqrt{\frac{2mE}{\ell^2}r^4 + 2G\frac{Mm^2}{\ell^2}r^3 - r^2}} \tag{10.5.13}$$

This equation may now be integrated.

10.5.2 Periodic Motion

The total energy of a particle of mass moving in a 1D potential $U(x)$ is

$$\frac{1}{2}m\dot{x}^2 + U(x) = E \tag{10.5.14}$$

We seek to find a formula for the period of motion

$$\dot{x}^2 = \frac{2}{m}\left[E - U(x)\right] \tag{10.5.15}$$

Taking the positive square root

$$\frac{dx}{dt} = \sqrt{\frac{2}{m}}\left[E - U(x)\right]^{1/2} \tag{10.5.16}$$

and separating variables

$$\sqrt{\frac{m}{2}}\frac{dx}{\left[E - U(x)\right]^{1/2}} = dt \tag{10.5.17}$$

Half of the period is

$$\sqrt{\frac{m}{2}} \int_{x_1}^{x_2} \frac{dx}{\left[E - U(x) \right]^{1/2}} = \int_0^{T/2} dt = \frac{T}{2} \tag{10.5.18}$$

Example 10.5.1

Find the period of motion of a block of mass m attached to a spring with force constant k

Solution: The force acting on the spring is given by Hooke's law $F = -kx$ with corresponding potential energy function

$$U(x) = \frac{1}{2} kx^2 \tag{10.5.19}$$

If the spring is released from rest at $x = A$ the total mechanical energy is

$$E = \frac{1}{2} kA^2 \tag{10.5.20}$$

The sum of kinetic and potential energy during the motion is equal to the initial potential energy

$$\frac{1}{2} m\dot{x}^2 + \frac{1}{2} kx^2 = \frac{1}{2} kA^2 \tag{10.5.21}$$

Solving for the velocity of the particle

$$\frac{dx}{dt} = \pm \sqrt{\frac{k}{m}} \left[A^2 - x^2 \right]^{1/2} \tag{10.5.22}$$

Notice that the velocity of the particle is zero when the acceleration is maximal $x = \pm A$ while the velocity is maximal where the acceleration is zero at $x = 0$. Separating variables and integrating to obtain the period

$$\frac{T}{2} = \sqrt{\frac{m}{k}} \int_{-A}^{A} \frac{dx}{\left[A^2 - x^2 \right]^{1/2}} \tag{10.5.23}$$

The integral is evaluated using by trigonometric substitution and we obtain

$$\frac{T}{2} = \sqrt{\frac{m}{k}} \pi \tag{10.5.24}$$

Example 10.5.2

Write an integral expression for the period of a particle with total energy E moving in the presence of a potential function that is piecewise defined

$$U(x) = k|x| \tag{10.5.25}$$

Solution: Write

$$T = 2\sqrt{\frac{m}{2}} \int_{x_1}^{x_2} \frac{dx}{\left[E - U(x)\right]^{1/2}} \tag{10.5.26}$$

x_1 and x_2 are the turning points where $E = k|x|$ or $x = \pm E/k$

$$T = 2\sqrt{\frac{m}{2}} \int_{-E/k}^{E/k} \frac{dx}{\left[E - k|x|\right]^{1/2}} \tag{10.5.27}$$

Now

$$|x| = \begin{cases} x & x \geq 0 \\ -x & x < 0 \end{cases} \tag{10.5.28}$$

so that we must break the integral apart

$$T = 2\sqrt{\frac{m}{2}} \left[\int_{-E/k}^{0} \frac{dx}{\left[E + kx\right]^{1/2}} + \int_{0}^{E/k} \frac{dx}{\left[E - kx\right]^{1/2}} \right] \tag{10.5.29}$$

In the first integral, we let $u = E + kx$ so that $du = kdx$. In the second integral $u = E - kx$ and $du = -kdx$.

$$T = 2\sqrt{\frac{m}{2}} \left[\frac{1}{k} \int_{u=0}^{E} u^{-1/2} du - \frac{1}{k} \int_{u=E}^{0} u^{-1/2} du \right] = 8\sqrt{\frac{mE}{2}} \frac{1}{k} \tag{10.5.30}$$

10.5.3 Small Oscillations

Given a physical system described by a Lagrangian function of generalized coordinates q_i and velocities $\dot{q}_i$

$$L = L(q_i, \dot{q}_i) \tag{10.5.31}$$

we consider small oscillations about the equilibria q_{i0} where $\dot{q}_i = 0$. Performing a Taylor expansion of L about the equilibria

$$L = L\left(q_{i0},0\right)+\left(q_i - q_{i0}\right)\frac{\partial L}{\partial q_i}\bigg|_{(q_{i0},0)} +\left(\dot{q}_i - \dot{q}_{i0}\right)\frac{\partial L}{\partial \dot{q}_i}\bigg|_{(q_{i0},0)}$$

$$+\frac{1}{2}\left(q_i - q_{i0}\right)\left(q_j - q_{j0}\right)\frac{\partial^2 L}{\partial q_i \partial q_j}\bigg|_{(q_{i0},0)} +\frac{1}{2}\left(\dot{q}_i - \dot{q}_{i0}\right)\left(\dot{q}_j - \dot{q}_{j0}\right)\frac{\partial^2 L}{\partial \dot{q}_i \partial \dot{q}_j}\bigg|_{(q_{i0},0)} +\cdots$$

$$(10.5.32)$$

The first term is time independent and does not contribute to Lagrange's equations of motion. The second two terms represent the dL and sum to zero. With the substitutions $\eta_i = q_i - q_{i0}$ and $\dot{\eta}_i = \dot{q}_i$ we write the Lagrangian

$$L = \frac{1}{2}T_{ij}\dot{\eta}_i\dot{\eta}_j - \frac{1}{2}V_{ij}\eta_i\eta_j \tag{10.5.33}$$

with kinetic energy matrix

$$T_{ij} = \frac{\partial^2 L}{\partial \dot{q}_i \partial \dot{q}_j}\bigg|_{\dot{q}_i=0} \tag{10.5.34}$$

and potential energy matrix

$$V_{ij} = -\frac{\partial^2 L}{\partial q_i \partial q_j}\bigg|_{\dot{q}_i=0} \tag{10.5.35}$$

Lagrange's equations near the equilibria

$$\frac{d}{dt}\frac{\partial L}{\partial \dot{\eta}_i} - \frac{\partial L}{\partial \eta_i} = 0 \tag{10.5.36}$$

become

$$T_{ij}\ddot{\eta}_j + V_{ij}\eta_j = 0 \tag{10.5.37}$$

To find the characteristic frequencies of motion with harmonic time dependence

$$\eta_j = s_j e^{-i\omega t} \tag{10.5.38}$$

$$-\omega^2 T_{ij}s_j + V_{ij}s_j = 0 \tag{10.5.39}$$

$$\left(V_{ij} - \omega^2 T_{ij}\right)s_j = 0 \tag{10.5.40}$$

We thus have an eigenvalue problem

$$\left(V - \omega^2 T\right)S = 0 \tag{10.5.41}$$

To obtain the characteristic frequencies ω we solve the characteristic equation

$$\left|V - \omega^2 T\right| = 0 \qquad (10.5.42)$$

and then obtain the eigenvectors S.

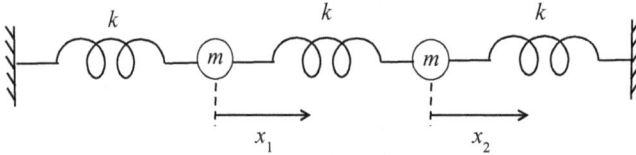

Figure 10.5.1: Two equal masses m connected by three springs with force constant k

Example 10.5.3

As shown in Figure 10.5.1, two equal masses m are connected to three springs with force constant k with Lagrangian

$$L = \frac{1}{2} m \dot{x}_1^2 + \frac{1}{2} m \dot{x}_2^2 - \frac{1}{2} k \left(x_1^2 + (x_1 - x_2)^2 + x_2^2 \right) \qquad (10.5.43)$$

Find the vibrational frequencies for small oscillations.

Solution: Constructing the matrices

$$T_{ij} = \begin{pmatrix} m & 0 \\ 0 & m \end{pmatrix} \quad V_{ij} = \begin{pmatrix} 2k & k \\ k & 2k \end{pmatrix} \qquad (10.5.44)$$

the vibrational frequencies are found from the determinant

$$\left|V - \omega^2 T\right| = \begin{vmatrix} 2k - \omega^2 m & -k \\ -k & 2k - \omega^2 m \end{vmatrix} = 0 \qquad (10.5.45)$$

giving

$$\left(2k - \omega^2 m\right)^2 - k^2 = 0 \qquad (10.5.46)$$

with two frequencies

$$\omega_1 = \sqrt{\frac{k}{m}} \text{ and } \omega_2 = \sqrt{\frac{3k}{m}} \qquad (10.5.47)$$

where ω_1 corresponds to the masses moving in phase (in sync to the left or to the right) and ω_2 out of phase (alternatively toward and away from each other).

10.6 CHAOTIC DYNAMICS

Chaos can be described as erratic but deterministic behavior in low dimensional systems with three or more degrees of freedom. Examples of chaotic systems include orbital motion involving three or more masses, oscillating chemical reactions, turbulent fluid flow, and population dynamics with interaction between species. Chaotic systems are hypersensitive with respect to small changes in initial conditions and are only forecastable for short time durations. An infinitesimal change in starting conditions can result in dramatic differences at later times. This sensitive dependence on initial conditions is known as the butterfly effect.

10.6.1 Strange Attractors

Types of attractors in phase space include stable and unstable fixed points known as elliptic and hyperbolic points, respectively. A classic example is the simple pendulum with an elliptic point at its lowest position and a hyperbolic point where the pendulum is completely inverted. Phase space orbits of dissipative systems tend to spiral into elliptic points or away from hyperbolic points. Trajectories in phase space may also converge to a locus of points known as a limit cycle in two dimensions, or a torus in three dimensions. Chaotic systems are characterized by attractors that do not have integer dimension, known as strange attractors.

10.6.2 Lorenz Model

Perhaps the most famous chaotic system is the Lorenz model consisting of three first order, coupled, nonlinear differential equations

$$\frac{dx}{dt} = -\sigma(x+y)$$
$$\frac{dy}{dt} = -xz + rx - y \qquad (10.6.1)$$
$$\frac{dz}{dt} = xy - bz$$

Typical values of the parameters are $\sigma = 10$, $r = 28$ and $b = 8/3$. This system was first investigated numerically by Edward Lorenz in 1963 to model atmospheric dynamics.

10.6.3 Jerk Systems

Nonlinear differential equations with time-dependent accelerations can exhibit chaos. The first derivative of the acceleration is known as the jerk. For example, the third order system

$$\dddot{x} + \ddot{x} + A\dot{x} + |x| + 1 = 0 \tag{10.6.2}$$

can be written as a system of three first order equations. At least one nonlinear term is required for chaos. Hyper-jerk systems such as

$$\ddddot{x} + \dddot{x} + \ddot{x} + A\dot{x} + |x| + 1 = 0 \tag{10.6.3}$$

that are fourth order or greater may also exhibit chaos.

10.6.4 Time Delay Coordinates

Time delay coordinates can be used to reconstruct phase plots from single time series $x(t)$. A two-dimensional phase plot is obtained by plotting $x(t)$ vs. $x(t - \tau)$ where τ is the time delay. Two time delays may be used to reconstruct a three-dimensional phase with axes $x(t)$, $x(t - \tau)$ and $x(t - 2\tau)$. This technique is useful in the analysis of experimental data where only one sensor reading is available. Reconstructed phase plots show qualitative similarity to the actual phase plot and may be used to obtain information that would not be available by plotting only the time series.

10.6.5 Lyapunov Exponents

Chaotic systems exhibit sensitive dependence on initial conditions. Given nearby initial conditions x_0 and $x_0 + \Delta x_0$, the separation in initial conditions will evolve as

$$\Delta x(t) = \Delta x_0 \exp(\lambda t) \tag{10.6.4}$$

where λ is the Lyapunov exponent. There will be one Lyapunov exponent for each degree of freedom. A block of initial conditions $V_0 = \Delta x_0 \Delta y_0 \Delta z_0$ for a system with three degrees of freedom will evolve as

$$V(t) = V_0 \exp\left(\sum_{i=1}^{3} \lambda_i t\right) \tag{10.6.5}$$

In Section 5.8, we found that the phase volume element

$$V(t) = V_0 \exp\left(\int_0^t \nabla \cdot \mathbf{F} dt'\right) \tag{10.6.6}$$

is obtained from the logarithmic divergence so that if $\nabla \cdot \mathbf{F}$ is independent of time then

$$\nabla \cdot \mathbf{F} = \sum_{i=1}^{3} \lambda_i \tag{10.6.7}$$

Thus, if the sum of Lyapunov exponents is zero, then the system is conservative. If the sum is negative or positive, then the system is dissipative or it blows up, respectively. For a system to exhibit chaos it must have

1. at least one positive Lyapunov exponent
2. at least one nonlinear term
3. at least three degrees of freedom

Time counts as a degree of freedom in nonautonomous systems where it appears explicitly.

Example 10.6.1

Calculate the sum of Lyapunov exponents of the Lorenz system. Is the system conservative, dissipative or does it blow up?

Solution: Evaluating $\nabla \cdot \mathbf{F} = \dfrac{\partial \dot{x}}{\partial x} + \dfrac{\partial \dot{y}}{\partial y} + \dfrac{\partial \dot{z}}{\partial z}$

$$\nabla \cdot \mathbf{F} = \frac{\partial \left[-\sigma \left(x + y \right) \right]}{\partial x} + \frac{\partial \left[-xz + rx - y \right]}{\partial y} + \frac{\partial \left[xy - bz \right]}{\partial z} \tag{10.6.8}$$

$$\nabla \cdot \mathbf{F} = -\sigma - 1 - b \tag{10.6.9}$$

The system is dissipative for positive values of σ and b.

10.6.6 Poincaré Sections

Trajectories of chaotic systems in phase space are often quite complex and difficult to visualize in higher dimensions. Poincaré sections, or strobe plots, are a useful tool where points are plotted where phase space trajectories intersect a specified plane. For example, given a system with three degrees of freedom corresponding to the Cartesian coordinates x, y and z one may plot a point each time the trajectory crosses the x-y plane. The Poincaré section gives a slice of the phase space, analogous to an MRI scan. The strobe plot of periodic or quasi-periodic systems will consist of a few points or a locus of points. For chaotic systems, the resulting plot may be a curve or a fractal.

Maple Examples

A 3D phase plot of the Lorenz model is computed in the Maple worksheet below. Sensitive dependence on initial conditions is demonstrated by comparing time series of the Lorenz system for slightly different initial conditions. Poincaré sections are computed for a Hamiltonian system using the *DEtools* package.

Key Maple terms: *display, dsolve, odeplot, poincare*

Maple packages: *with(plots): with(DEtools):*

restart

Lorenz Model

$$xEqn := \frac{d}{dt}x(t) = 10 \cdot \left(y(t) - x(t)\right)$$

$$xEqn := \frac{d}{dt}x(t) = 10\,y(t) - 10\,x(t)$$

$$yEqn := \frac{d}{dt}y(t) = 28 \cdot x(t) - y(t) - x(t) \cdot z(t)$$

$$yEqn := \frac{d}{dt}y(t) = 28\,x(t) - y(t) - x(t)\,z(t)$$

$$zEqn := \frac{d}{dt}z(t) = x(t) \cdot y(t) - \frac{8}{3} \cdot z(t)$$

$$zEqn := \frac{d}{dt}z(t) = x(t)\,y(t) - \frac{8z(t)}{3}$$

ics := $x(0) = 0.1, y(0) = 0.1, z(0) = 0.1$

$$ics := x(0) = 0.1,\ y(0) = 0.1,\ z(0) = 0.1$$

sol1 := *dsolve*({*xEqn, yEqn, zEqn, ics*}, {$x(t), y(t), z(t)$}, *numeric*)

$$sol1 := \mathbf{proc}(x\_rkf45)\ \ldots\ \mathbf{end\ proc}$$

with(plots) :
odeplot(sol1, [$x(t), y(t), z(t)$], $t = 0 \ldots 30$, *numpoints* = 2000)

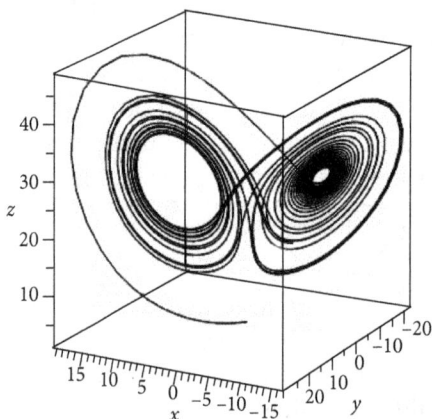

Figure 10.6.1: Lorenz model phase portrait.

Butterfly Effect: Comparing a Small Change in Initial Conditions

$ics2 := x(0) = 0.1001, y(0) = 0.1, z(0) = 0.1$

$$ics2 := x(0) = 0.1001, y(0) = 0.1, z(0) = 0.1$$

$sol2 := dsolve(\{xEqn, yEqn, zEqn, ics2\}, \{x(t), y(t), z(t)\}, numeric)$

$$sol2 := \textbf{proc}(x\_rkf45) \ldots \textbf{end proc}$$

$p1 := odeplot(sol1, [t, x(t)], t = 0\ldots30):$
$p2 := odeplot(sol2, [t, x(t)], t = 0\ldots30, color = blue, linestyle = longdash):$
$display(p1, p2)$

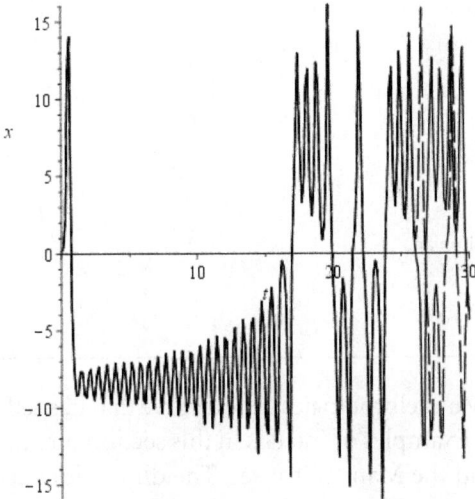

Figure 10.6.2: Divergence in separate time series of the Lorenz model with slightly different initial conditions.

$restart$
$with(DEtools):$

Poincaré Section: Henon-Heiles Hamiltonian

$$H := \frac{\left(p1^2 + p2^2\right)}{2} + \frac{\left(q1^2 + q2^2\right)}{2} + q1^2 \cdot q2 - \frac{q2^3}{3}$$

$$H = \frac{1}{2}p1^2 + \frac{1}{2}p2^2 + \frac{1}{2}q1^2 + \frac{1}{2}q2^2 + q1^2 q2 - \frac{1}{3}q2^3$$

$Pointcare(H, t = 0\ldots20000, \{[0, .2, .2, .2, .33]\}, stepsize = 0.1)$

$H = .11567100, Initial\ conditions:, t = 0, p1 = 0.2, p2 = 0.2, q1 = 0.2, q2 = 0.33$

Number of points found crossing the (p1, q1) plane: 5278
Maximum H deviation: .5667037400 %

Time consumed: 31 seconds

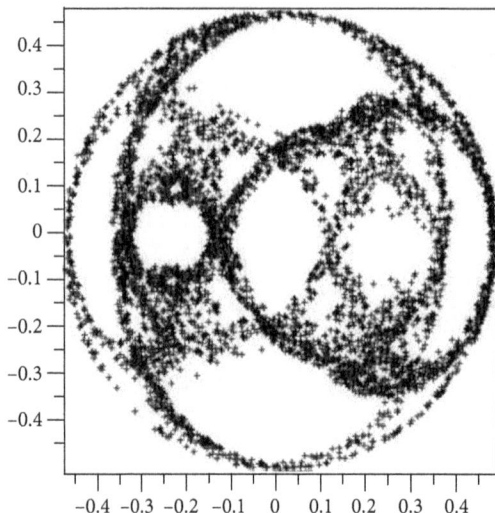

Figure 10.6.3: Poincaré section of the Henon-Heiles Hamiltonian.

10.7 FRACTALS

The term fractal was coined by Benoit Mandelbrot to describe a curve that has self-similar structure over many size scales. Examples of fractals in this section include the Cantor set, the Koch snowflake, and the Mandelbrot set. The dimensionality of fractals and chaotic maps are also discussed.

10.7.1 Cantor Set

The Cantor Set or "Cantor comb" is a fractal formed by first removing the center third of the unit interval. Subsequent thirds are removed from the resulting line segments ad infinitum (Figure 10.7.1). The resulting set consists of infinitely many points with zero length.

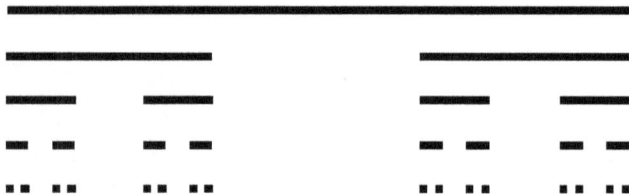

Figure 10.7.1: Iterates of the Cantor comb.

10.7.2 Koch Snowflake

The Koch snowflake (Figure 10.7.2) begins with an equilateral triangle with sides of length L. The center third of the line segments forming the triangle are removed and replaced by two sides of the equilateral triangle with lengths $L/3$. The perimeter has 12 segments each of length $L/3$ after the first iterate and 96 segments of length $L/9$ after the second iterate. The area of the snowflake approaches 8/3 the area of the original triangle with the perimeter length approaching infinity with subsequent iteration.

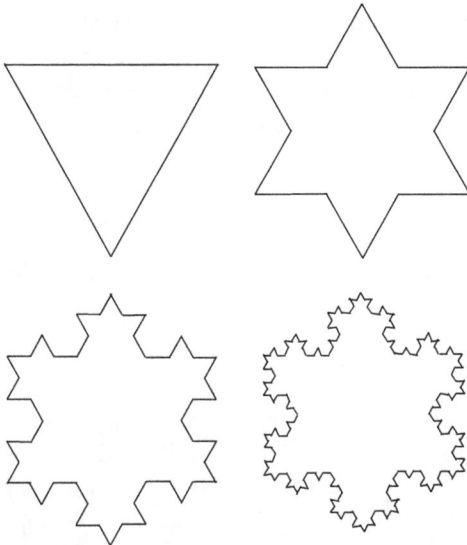

Figure 10.7.2: Iterates of the Koch snowflake. Similar fractals may be drawn using the Base-Motif Fractal Generator in Maple.

10.7.3 Mandelbrot Set

The Mandelbrot set is an escape set consisting of the values of C in the complex plane that do not diverge when repeatedly squared and added to themselves. The repeated iteration is represented by the formula

$$z_{n+1} = z_n^2 + C \qquad (10.7.1)$$

The first three iterations of the above formula are computed as

$$z_1 = C$$
$$z_2 = z_1^2 + C = C^2 + C \qquad (10.7.2)$$
$$z_3 = z_2^2 + C = \left(C^2 + C\right)^2 + C$$

Mandelbrot's formula can be separated into real and imaginary parts, substituting $z_n = x_n + iy_n$, $z_{n+1} = x_{n+1} + iy_{n+1}$ and $C = \mathrm{Re}\,C + i\,\mathrm{Im}\,C$, giving the iterated equations

$$x_{n+1} = x_n^2 + y_n^2 + \mathrm{Re}\,C$$
$$y_{n+1} = 2x_n y_n \quad + \mathrm{Im}\,C \tag{10.7.3}$$

10.7.4 Fractal Dimension

Fractals are geometrical objects with noninteger dimension. There are several ways to define the dimension of a fractal. The simplest definition is the capacity, or "box counting" dimension, that is obtained by finding the number of boxes N required to cover the fractal as a function of the box length ε. For example, the number of boxes required to cover a cube of side L with boxes of side ε would be

$$N(\varepsilon) = L^3 / \varepsilon^3 \tag{10.7.4}$$

In D dimensions we have

$$N(\varepsilon) = L^D (1/\varepsilon)^D \tag{10.7.5}$$

Taking the natural log of both sides

$$\ln N(\varepsilon) = D \ln L + D \ln(1/\varepsilon) \tag{10.7.6}$$

and solving for D

$$D = \frac{\ln N(\varepsilon)}{\ln L + \ln(1/\varepsilon)} \tag{10.7.7}$$

The capacity dimension D_C is defined by taking the limit $\varepsilon \to 0$ as

$$D_C = \lim_{\varepsilon \to 0} \frac{\ln N(\varepsilon)}{\ln(1/\varepsilon)} \tag{10.7.8}$$

Example 10.7.1

Calculate the capacity dimension of the Cantor comb

Solution: For the 0th iterate, we require $N = 1$ boxes with length $\varepsilon = L$ to cover the curve.
first iterate: $N = 2$ $\varepsilon = L/3$
second iterate: $N = 4$ $\varepsilon = L/9$
third iterate: $N = 8$ $\varepsilon = L/27$
nth iterate: $N(\varepsilon) = 2^n$ $\varepsilon = L/3^n$

$$D_C = \lim_{\varepsilon \to 0} \frac{\ln N(\varepsilon)}{\ln(1/\varepsilon)} = \lim_{n \to \infty} \frac{\ln 2^n}{\ln(3^n / L)} = \lim_{n \to \infty} \frac{n \ln 2}{n \ln 3 - \ln L} = \frac{\ln 2}{\ln 3} \quad (10.7.9)$$

Example 10.7.2

Calculate the capacity dimension of the Koch snowflake

Solution: For the 0th iterate we require $N = 3$ boxes with length $\varepsilon = L$ to cover the sides of a triangle.

first iterate: $N = 4 \times 3$ $\varepsilon = L/3$
second iterate: $N = 4^2 \times 3$ $\varepsilon = L/3^2$
third iterate: $N = 4^3 \times 3$ $\varepsilon = L/3^3$
nth iterate: $N(\varepsilon) = 4^n \times 3$ $\varepsilon = L/3^n$

$$D_C = \lim_{\varepsilon \to 0} \frac{\ln N(\varepsilon)}{\ln(1/\varepsilon)} = \lim_{n \to \infty} \frac{\ln(4^n \times 3)}{\ln(3^n / L)} = \lim_{n \to \infty} \frac{n \ln 4 + \ln 3}{n \ln 3 - \ln L} = \frac{\ln 4}{\ln 3} = 2 \frac{\ln 2}{\ln 3}$$

$$(10.7.10)$$

10.7.5 Chaotic Maps

Chaotic trajectories in phase space are the solutions to nonlinear differential equations. Chaotic maps, on the other hand, are generated by iterated finite difference equations. Chaotic maps often resemble Poincaré sections of chaotic flows and can be characterized by a fractal dimension. Whereas three degrees of freedom are necessary for chaos in continuous flows, chaos can occur in one-parameter maps such as

$$x_{n+1} = f(x_n, \mu) \quad (10.7.11)$$

where μ is an adjustable parameter. An example of a chaotic map is the logistic equation that models population dynamics in a limited environment

$$x_{n+1} = \mu x_n (1 - x_n) \quad (10.7.12)$$

Periodic and chaotic oscillations in x_n can result depending on the value of μ. For values of $\mu < 3$, the system reaches a fixed point after several steps where x_n does not change with subsequent iteration. The system goes into a 2-cycle, alternating between two values when μ reaches 3.2. As μ increases, the system progresses through 4-, 8-, 16-, 2^m-cycles, becoming chaotic around $\mu = 3.9$. This type of repeated bifurcation as a system parameter is adjusted is known as the period doubling rout to chaos and is common to other dynamical systems such as the simple driven pendulum with damping.

Maple Examples

Examples of chaotic maps are plotted in the Maple worksheet below, including the Henon, logistic, and standard maps. A bifurcation diagram of the logistic map is plotted. The Mandelbrot set is visualized using 3D surface, 3D contours, and various list plots. The burning ship fractal is shown using the *Fractals:-EscapeTime* package.

Key Maple terms: *Array, BurningShip, Embed, listdensityplot, Matrix, pointplot, seq,* surfdata

Maple packages: *with(plots): with(Fractals:-EscapeTime): with(ImageTools):*

Programming: **for** loops, **if** statements, function statements using '→'

restart

Henon Map

$$with(plots):$$
$$a := 1.4 : b := 0.3$$
$$N := 3000 : x[0] := 0 : y[0] := 0$$
$$\text{for } n \text{ from 0 to } N \text{ do } x[n+1] := b*y[n] + a - x[n]^2; y[n+1] := x[n]; \text{od:}$$
$$pointplot([seq([x[n], y[n]], n = 0 \ldots N)], symbol = point, labels = [``x", ``y"])$$

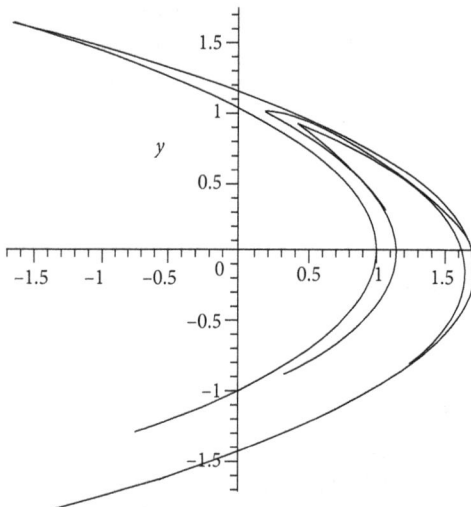

Figure 10.7.3: Henon map fractal.

Bifurcation Diagram of Logistic Map

restart

with(*plots*) :
$f := (x, mu) \rightarrow mu \cdot x \cdot (1 - x)$

$$f := (x, \mu) \mapsto \mu \cdot x \cdot (1 - x)$$

$N := 50 : j := 0 :$
for *i* **from** 0 **to** 2000 **do**
   mu $:= 2.0 + .001 \cdot i;$
   $x[0, i] := 0.5;$
   for *n* **from** 0 **to** *N* **do**
   $x[n + 1, i] := f(x[n, i], \text{mu});$
   if $n > 20$ **then**
   $mm[j] := \text{mu};$
   $xx[j] := x[n + 1, i];$
   $j := j + 1;$
   end if
   od:
   od:
points $:= seq([mm[jn], xx[j]], j = 0 \ldots 60000) :$
pointplot({*points*}, *symbolsize* = 1)

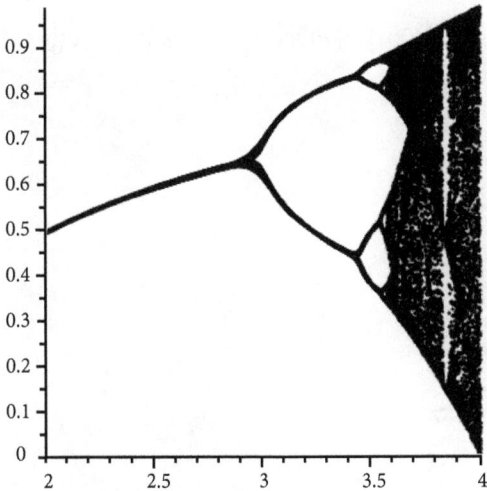

Figure 10.7.4: Bifurcation diagram of the logistic map.

Standard Map

   restart
   with(*plots*) :
   $K := 0.972 :$
   $N := 3000 : z := evalf(2 \cdot \text{Pi}) : m := 0:$

for j **from 0 to 20 do**

$p[0] := 5 - .6 \cdot j$: **theta[0]** $:= 5 - .5 \cdot j$:

for n **from** 0 **to** N **do**

$m := m + 1;$

if theta$[n] > z$ **then**
theta$[n] :=$ theta$[n] - z;$
elif theta$[n] < 0$ **then**
theta$[n] :=$ theta$[n] + z;$
end if

$temp := p[n] + K \cdot \sin(\text{theta}[n]);$
$p[n + 1] := temp;$
theta$[n + 1] :=$ theta$[n] + temp;$
$qq[m] :=$ theta$[n];$
$pp[m] := p[n];$

od:
od:
$pointplot([seq([qq[m], pp[m]], m = 0 \ldots 50000)], symbol = point, labels = [\text{``theta''},$
$\text{``}p\text{''}])$

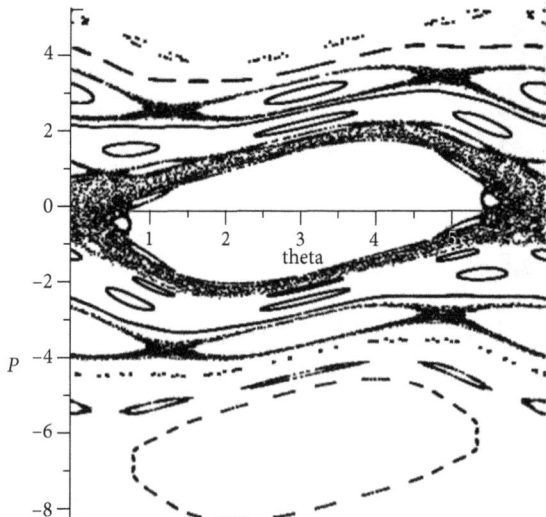

Figure 10.7.5: Plot of the standard map with several initial conditions and K = 0.972.

Mandelbrot Set

> *restart*
> *with(plots)* :

> **for** *i* **from** 0 **to** 200 **do**

> $x0 := -2.0 + \dfrac{2.5}{200.0} \cdot i;$
> $x[0] := 0;$

> **for** *j* **from** 0 **to** 200 **do**

> $y0 := -1.0 + \dfrac{2.0}{200.0} \cdot j;$
> $y[0] := 0;$
> **for** *n* **from** 0 **to** 30 **while** $\mathrm{sqrt}(x[n]^2 + y[n]^2) < 10$ **do**

> $x[n+1] := x[n]^2 - y[n]^2 + x0;$
> $y[n+1] := 2 \cdot x[n] \cdot y[n] + y0;$

> **od:**

> $z[i, j] := n;$

> **od:**
> **od:**

> $M := Matrix(200, 200, z, datatype = float[8])$

$$M := \begin{bmatrix} 200 \; x \; 200 \; Matrix \\ DataType : float_8 \\ Storage : rectangular \\ Order : Fortran\_order \end{bmatrix}$$

> $A := Array(0\ldots200, 0\ldots200, z, datatype = float[8])$

$$A := \begin{bmatrix} 0\ldots200 \; x \; 0\ldots200 \; Array \\ DataType : float_8 \\ Storage : rectangular \\ Order : Fortran\_order \end{bmatrix}$$

> *surfdata(M)*

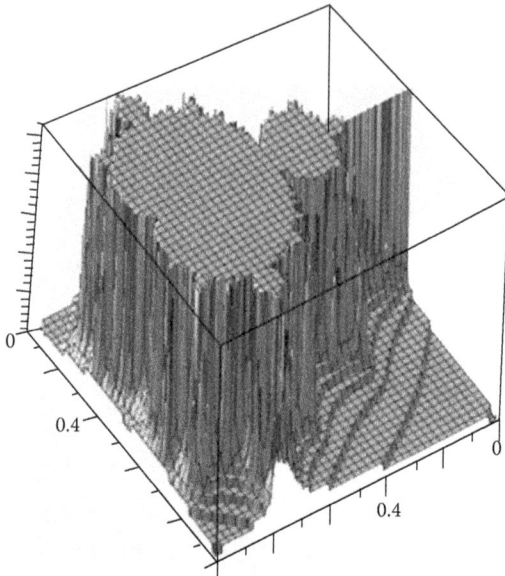

Figure 10.7.6: Landscape plot of the Mandelbrot set.

$$listdensityplot(M, style = point)$$

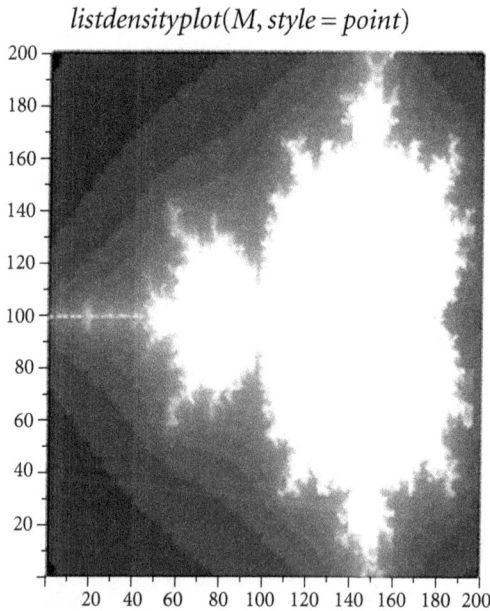

Figure 10.7.7: Listdensity plot of the Mandelbrot set.

Fractals Using EscapeTime

$with(Fractals:-EscapeTime)$

[*BurningShip, Colorize, HSVColorize, Julia, LColorize, Lyapunov, Mandelbrot, Newton*]

$with(ImageTools):$

$size := 500 : lower\_left := -1.8 - 0.2I: upper\_right := -1.5 + 0.1\ I:$

$B := BurningShip(size, lower\_left, upper\_right, output = layer2)$

$$B := \begin{bmatrix} 1...500\ x\ 1...500\ Array \\ DataType : float_8 \\ Storage : rectangular \\ Order : C\_order \end{bmatrix}$$

$Embed(B)$

Figure 10.7.8: Burning ship fractal.

10.8 MATLAB EXAMPLES

In this section, the MATLAB Symbolic Math Toolbox is demonstrated in obtaining Lagrange's equations of motion. A user-defined fourth order Runge-Kutta scheme is used to integrate the Lorenz model equations where a time delay map is constructed. A Julia set fractal is calculated and the Henon map is plotted inside of its basin of attraction.

Key MATLAB commands: *contour, functionalDerivative, imag, linspace,plot, plot3, real, syms*

Programming: for loops, function statements, if statements

Section 10.3 Lagrangian Dynamics

The script 'coupled_oscillators_lag.m' outputs Lagrange's equations of motion for coupled oscillators with kinetic energy

$$T = \frac{1}{2}m\left(\dot{x}_1^2 + \dot{x}_2^2\right)$$ (10.8.1)

and potential energy

$$V = \frac{1}{2}kx_1^2 + \frac{1}{2}kx_2^2 + \frac{1}{2}K\left(x_1 - x_2\right)^2$$ (10.8.2)

```
syms m k K x1(t) x2(t)
T = sym(1)/2*m*diff(x1,t)^2 + sym(1)/2*m*diff(x2,t)^2;
V = sym(1)/2*K*(x1-x2)^2 + sym(1)/2*k*x1^2+sym(1)/2*k*x2^2;
L = T - V

eqn1 = functionalDerivative(L,x1) == 0
eqn2 = functionalDerivative(L,x2) == 0
```

The script is executed at the Command line
```
>> coupled_oscillators_lag

L(t) =

(m*diff(x1(t), t)^2)/2 - (k*x2(t)^2)/2 - (K*(x1(t) - x2(t))^2)/2 -
    (k*x1(t)^2)/2 + (m*diff(x2(t), t)^2)/2


eqn1(t) =

K*x2(t) - K*x1(t) - m*diff(x1(t), t, t) - k*x1(t) == 0

eqn2(t) =

K*x1(t) - m*diff(x2(t), t, t) - K*x2(t) - k*x2(t) == 0
```

Section 10.6 Chaotic Dynamics

MATLAB features built-in ODE solvers such as 'ode45.' User-defined solvers may offer greater flexibility for a given application. For example, the Runge-Kutta algorithm may be used to integrate the first order differential equation with explicit time dependence

$$\frac{dx}{dt} = f(x,t)$$

(10.8.3)

according to

$$x_{n+1} = x_n + \frac{1}{6}(k_1 + 2k_2 + 2k_3 + k_4)$$

$$t_{n+1} = t_n + h$$

(10.8.4)

where

$$k_1 = hf(x_n, t_n)$$

$$k_2 = hf\left(x_n + \frac{k_1}{2}, t_n + \frac{h}{2}\right)$$

$$k_3 = hf\left(x_n + \frac{k_2}{2}, t_n + \frac{h}{2}\right)$$

$$k_4 = hf(x_n + k_3, t_n + h)$$

(10.8.5)

For the Lorenz model in Equation (10.6.1), we have three first order equations so that there will be a set of $k_{1...4}$ values for each differential equation as shown in the MATLAB script below. Also, the t_n do not appear in (10.8.5) unless the differential equations have explicit time dependence.

```
% specify parameters and time step
rho=28;
sigma=10;
beta=8/3;
dt=0.01;
% create function handles
f=@(x,y,z)sigma*(y-x);
g=@(x,y,z)rho*x-y-x*z;
h=@(x,y,z)x*y-beta*z;

% specify number of time steps and create arrays

nmax=5000;

x=zeros(nmax,1);
y=zeros(nmax,1);
z=zeros(nmax,1);
t=zeros(nmax,1);
% arrays for time delay coordinates
x1=zeros(nmax,1);
x2=zeros(nmax,1);
x3=zeros(nmax,1);
% specify initial conditions
x(1)=0.1; y(1)=0.1; z(1)=0.1; t(1)=0.0;
% integrate the Lorenz system using the R-K algorithm
```

```
for n=1:nmax
    k1x = dt*f(x(n),y(n),z(n));
    k2x = dt*f(x(n)+k1x/2,y(n),z(n));
    k3x = dt*f(x(n)+k2x/2,y(n),z(n));
    k4x = dt*f(x(n)+k3x,y(n),z(n));

    k1y = dt*g(x(n),y(n),z(n));
    k2y = dt*g(x(n),y(n)+k1y/2,z(n));
    k3y = dt*g(x(n),y(n)+k2y/2,z(n));
    k4y = dt*g(x(n),y(n)+k3y,z(n));

    k1z = dt*h(x(n),y(n),z(n));
    k2z = dt*h(x(n),y(n),z(n)+k1z/2);
    k3z = dt*h(x(n),y(n),z(n)+k2z/2);
    k4z = dt*h(x(n),y(n),z(n)+k3z);

    x(n+1)=x(n)+ k1x/6 +k2x/3 + k3x/3 + k4x/6;

    y(n+1)=y(n)+k1y/6 + k2y/3 + k3y/3 + k4y/6;

    z(n+1)=z(n)+k1z/6 + k2z/3 + k3z/3 + k4z/6;

    t(n+1)=t(n) + dt;
end

% two time delays of the x-time series
for n=1:nmax-10
    x1(n)=x(n);
    x2(n)=x(n+5);
    x3(n)=x(n+10);
end

% create a 3-D time delay plot
plot3(x1,x2,x3,'k')
xlabel('x(t)')
ylabel('x(t+tau)')
zlabel('x(t+2 tau)')
```

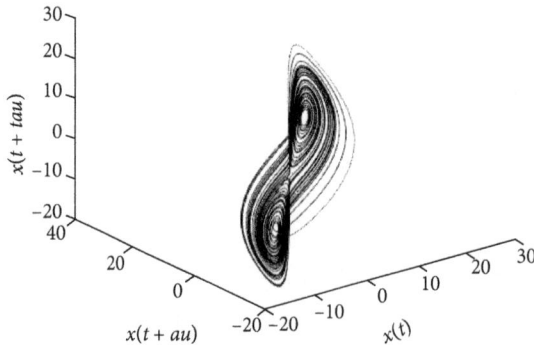

Figure 10.8.1: Lorenz attractor reconstructed from the x-time series using two time delays with five time steps for each time delay.

Section 10.7 Fractals

Julia Sets

Julia set fractals may be computed from Mandelbrot's formula

$$z_{n+1} = z_n^2 + C \tag{10.8.6}$$

The Mandelbrot set are those values of C such that equation (10.8.6) converges upon subsequent iteration. A Julia set consists of the initial values of z_n such that (10.8.6) converges for a given value of the complex number C. Thus, there are infinitely many Julia sets with one Julia set for each point in the Mandelbrot set. Values of C inside the Mandelbrot set form "connected" Julia sets whereas C numbers outside the Mandelbrot set result in disconnected Julia sets that are referred to as "dust." Julia sets may also be found for other iterated formulas of the form

$$z_{n+1} = f(z_n) + C \tag{10.8.7}$$

The MATLAB script 'Julia.m' outputs the fractal shown in Figure 10.8.2, corresponding to iterations of (10.8.6) with $C = -0.75 + 0.02i$

```
nmax  = 100;
x=zeros(nmax,1);
y=zeros(nmax,1);

mmax = 1000;

J = zeros(mmax,mmax);

x0 = linspace(-1.5,1.5,mmax);
y0 = linspace(-1.5,1.5,mmax);

const = -0.75 + 1i*0.02;

for i = 1:mmax
for j = 1:mmax

    x(1)=x0(i);
    y(1)=y0(j);
for n=1:nmax-1
    y(n+1) = 2*x(n)*y(n)+imag(const);
    x(n+1) = x(n)^2 - y(n)^2 +real(const);
end

if sqrt(x(n)^2 + y(n)^2) < 5.0
    J(i,j)= 1;
else
    J(i,j)=0;
end

end
end
```

```
contourf(J)

axis equal
colormap gray
title('Julia Set')
```

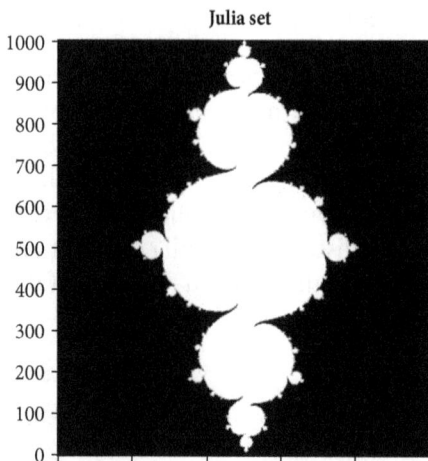

Figure 10.8.2: Julia set fractal with C = -0.75 + 0.02i.

Basin Boundary of the Henon Map

The basin boundary of the Henon map

$$
\begin{aligned}
x_{n+1} &= 1 - ax_n^2 + y_n \\
y_{n+1} &= bx_n
\end{aligned}
$$

(10.8.8)

consists of those initial conditions (x_0, y_0) that converge to the Henon map. In the MATLAB script below, points that converge to the map are given a value of one and those points that diverge are given a value of zero. The Henon map is then displayed in its basin of attraction.

```
nmax = 50;
x=zeros(nmax,1);
y=zeros(nmax,1);

a=1.4; b=0.3;

mmax = 300;

M = zeros(mmax,mmax);

x0 = linspace(-2.0,2.0,mmax);
y0 = linspace(-2.0,2.0,mmax);

for i = 1:mmax
    for j = 1:mmax
```

```
  x(1)=x0(j);
  y(1)=y0(i);

for n=1:nmax-1
y(n+1) = b*x(n);
x(n+1) = 1-a*x(n)^2 + y(n);
end

if sqrt(x(n)^2 + y(n)^2) < 5.0
M(i,j)= 1;
else
    M(i,j)=0;
end

     end
end
contourf(x0,y0,M)
hold on

nmax = 500;
 x(1)=1.2;
 y(1)=0.0;

for n=1:nmax-1
y(n+1) = b*x(n);
x(n+1) = 1-a*x(n)^2 + y(n);
end

plot(x,y,'.k')
axis equal
colormap bone
title('Basin Boundary of the Henon Map')
xlabel('x')
ylabel('y')
```

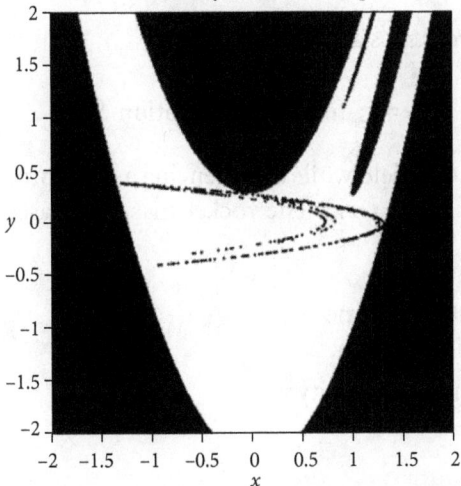

Figure 10.8.3: Henon map plotted inside of its basin of attraction. Initial conditions in the white region converge to the fractal while initial conditions in the black region diverge with subsequent iteration.

10.9 EXERCISES

10.1 Velocity-Dependent Resistive Forces

1. A particle moves with a velocity-dependent drag force in a gravitational field. The differential equation of motion is

$$mg - bv = m\frac{dv}{dt}$$

Calculate the velocity as a function of distance $v(x)$.

2. A particle with initial speed v_0 moves in a medium with resistive drag force proportional to the velocity squared $F_d = Cv^2$. Find the velocity of the particle as a function of position x (excluding gravity).

3. Plot the acceleration $a(t)$ and velocity $v(t)$ in the previous problem.

4. A particle with initial speed v_0 moves in a hypothetical medium with a drag force that depends exponentially on the speed $F_d = K(e^{nv} - 1)$. Find the velocity of the particle as a function of position x.

5. Two particles with mass m and opposite charges q and $-q$ are initially separated by a distance d. The particles are released from rest and allowed to fall together with a resistive drag force proportional to the velocity squared $F_d = Cv^2$. Find the speed of the particles as a function of time (neglect gravitational attraction).

Section 10.2 Variable Mass Dynamics

6. A rocket with initial mass M_i has a time-dependent mass $M(t) = \dfrac{M_i}{1 + t/\tau}$. Find the rocket thrust and velocity of the rocket as a function of time.

7. A rocket burns fuel at a constant rate $\dfrac{dM}{dt} = k$; find the acceleration $\dfrac{dv}{dt}$.

8. A rocket ascends in the earth's gravitation field while experiencing a drag force proportional to the velocity squared $F_{\text{drag}} = \alpha v^2$. The rocket mass decreases exponentially $M(t) = M_i e^{-\lambda t}$
Find differential equations for
(a) the velocity of the rocket as a function of time
(b) the height of the rocket as a function of time

9. Numerically compare the relativistic rocket velocity v_f

$$v_f = v_i + c\tanh\left(\frac{v_e}{c}\ln\frac{M_i}{M_f}\right)$$

to the velocity of a nonrelativistic rocket v_i

$$v_f = v_i + v_e \ln\left(\frac{M_i}{M_f}\right)$$

Compare for the same values of v_e, v_i, M_i, and M_f.

10. A sugar cube falls through an aqueous medium with a resistive drag force proportional to the square of its velocity $F_{drag} = \alpha v^2$. The sugar cube dissolves with a constant rate $\dfrac{dM}{dt} = k$. Find a differential equation describing the velocity of the sugar cube as a function of time.

Section 10.3 Lagrangian Dynamics

11. The kinetic and potential energies of a spherical pendulum (or ice cube sliding inside a frictionless bowl) are expressed in spherical coordinates

$$T = \frac{1}{2}m\ell^2\left(\dot{\theta}^2 + \sin^2\theta\dot{\phi}^2\right)$$

and

$$V = mg\ell(1 - \cos\theta)$$

with Lagrangian

$$L = \frac{1}{2}m\ell^2\left(\dot{\theta}^2 + \sin^2\theta\dot{\phi}^2\right) - mg\ell(1 - \cos\theta)$$

Write down Lagrange's equations of motion. Consider special cases of a conical pendulum (with $\dot{\theta} = 0$) and a simple pendulum (with $\dot{\phi} = 0$)

12. A particle of mass m is projected upward with a speed v_0 from a moon with radius R and mass M. The kinetic energy of the particle is

$$T = \frac{1}{2}m\dot{r}^2$$

with gravitational potential energy

$$V = -G\frac{mM}{r}$$

Find Lagrange's equation of motion of the mass and solve it for $r(0) = R$ and $\dot{r}(0) = v_0$.

13. A particle moves in one dimension with a logarithmic potential energy function with Lagrangian

$$L = \frac{1}{2}m\dot{x}^2 + m\Lambda \ln\left(\frac{x}{x_0}\right)$$

where the constant Λ has units of velocity squared. Find Lagrange's equation of motion and solve it for $x(0) = x_0$ and $\dot{x}(0) = v_0$.

14. For 2D projectile motion in a constant gravitational field the kinetic energy and potential energy functions are

$$T = \frac{1}{2}m\left(\dot{x}^2 + \dot{y}^2\right) \text{ and } V = mgy$$

Show that the solutions to Lagrange's equations give

$$\dot{x} = c_1$$

$$\dot{y} = -gt + c_2$$

where c_1 and c_2 are the initial velocities in the x- and y-directions, respectively.

15. A mass m is attached to a spring with force constant k and equilibrium length r_0 as shown in Figure 10.9.1. The spring pivots about the point O and can only stretch in a radial direction. The Lagrangian of the system is

$$L = \frac{1}{2}m\left(\dot{r}^2 + r^2\dot{\theta}^2\right) - \frac{1}{2}k\left(r - r_0\right)^2 - mg\left(r - r_0\right)\cos\theta$$

Find Lagrange's equations of motion.

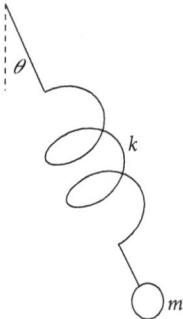

Figure 10.9.1: Mass on a swinging spring.

16. A particle is free to move inside a cone of half angle α as shown in Figure 10.9.2. The cone is coaxial with the z-axis where the distance from the z-axis is given by $r = z \tan \alpha$. Given that the Lagrangian in cylindrical coordinates is

$$L = \frac{1}{2}m\left(\dot{r}^2 + r^2\dot{\phi}^2 + \dot{z}^2\right) - mgz$$

(a) show that

$$L = \frac{1}{2}m\left(\dot{r}^2 \csc^2 \alpha + r^2\dot{\phi}^2\right) - mgr \cot \alpha$$

and write Lagrange's equations of motion for
(b) circular orbits $\dot{r} = 0$ and
(c) zero angular momentum $\dot{\phi} = 0$
Solve the resulting differential equations.

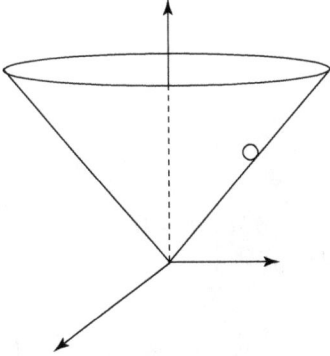

Figure 10.9.2: Mass moving on a conical surface.

17. A mass m is free to slide without friction on the inside of a paraboloid of revolution $z = x^2 + y^2$ as shown in Figure 10.9.3.
   (a) Write the Lagrangian in cylindrical coordinates and find Lagrange's equations of motion
   (b) Write the equations of motion for circular orbits where $\dot{r} = 0$ and zero angular momentum where $\dot{\phi} = 0$
   (c) Solve the resulting differential equations of motion

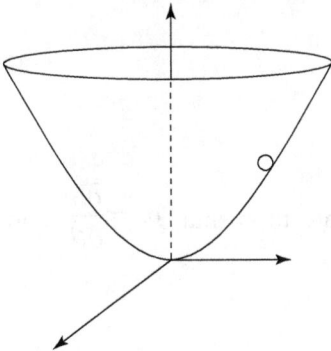

Figure 10.9.3: Mass moving inside a paraboloid of revolution.

18. Find Lagrange's equations of motion for the anharmonic oscillator with Lagrangian

$$L = \frac{1}{2}m\dot{x}^2 - \frac{1}{2}kx^2 - \frac{1}{3}\alpha x^3 - \frac{1}{4}\beta x^4$$

19. A physical system with two degrees of freedom is described by a Lagrangian

$$L = L(q_1, q_2, \dot{q}_1, \dot{q}_2, t)$$

The system is subject to the holonomic constraints

$$f = f(q_1, q_2) = 0$$

such that

$$\delta f = \frac{\partial f}{\partial q_1} \delta q_1 + \frac{\partial f}{\partial q_2} \delta q_2 = 0$$

From the variation of the action integral

$$\delta S = \int_{t_1}^{t_2} \left\{ \left[\frac{\partial L}{\partial q_1} - \frac{d}{dt}\left(\frac{\partial L}{\partial \dot{q}_1} \right) \right] \delta q_1 + \left[\frac{\partial L}{\partial q_2} - \frac{d}{dt}\left(\frac{\partial L}{\partial \dot{q}_2} \right) \right] \delta q_2 \right\} dt = 0$$

show that

$$\frac{\dfrac{\partial L}{\partial q_1} - \dfrac{d}{dt}\left(\dfrac{\partial L}{\partial \dot{q}_1} \right)}{\partial f / \partial q_1} = \frac{\dfrac{\partial L}{\partial q_2} - \dfrac{d}{dt}\left(\dfrac{\partial L}{\partial \dot{q}_2} \right)}{\partial f / \partial q_2} = \text{const.}$$

Set the constant equal to $-\lambda$ and obtain Lagrange's equations of motion.

10.4 Hamiltonian Mechanics

20. A swinging spring has kinetic energy

$$T = \frac{1}{2} m \left(\dot{r}^2 + r^2 \dot{\theta}^2 \right)$$

and potential energy

$$V = \frac{1}{2} k \left(r - \ell_0 \right)^2 + mgr \cos(\theta)$$

Write the Hamiltonian in terms of the conjugate momenta $p_\theta = \dfrac{\partial L}{\partial \dot{\theta}}$ and $p_r = \dfrac{\partial L}{\partial \dot{r}}$ and find Hamilton's equations of motion.

21. A spherical pendulum has kinetic energy

$$T = \frac{1}{2} m \ell^2 \left(\dot{\theta}^2 + \sin^2 \theta \dot{\phi}^2 \right)$$

and potential energy

$$V = mg\ell \left(1 - \cos(\theta) \right)$$

Write the Hamiltonian in terms of the conjugate momenta $p_\theta = \dfrac{\partial L}{\partial \dot{\theta}}$ and $p_\phi = \dfrac{\partial L}{\partial \dot{\phi}}$.

Find Hamilton's equations of motion.

22. Use the Poisson bracket formulation to find Hamilton's equations of motion for projectile motion with Lagrangian

$$L = T - V = \frac{1}{2}m\left(\dot{x}^2 + \dot{y}^2\right) - mgy$$

23. Calculate the following Poisson brackets

$$\{p, \sin q\}$$

$$\left\{q, \frac{1}{p}\right\}$$

$$\{p^2, q^3\}$$

10.5 Orbital and Periodic Motion

24. Write an integral expression for the period of a particle with total energy E moving in the presence of a potential function that is piecewise defined

$$U(x) = ax^2 \qquad x < 0$$
$$U(x) = bx^4 \qquad x \geq 0$$

What are the values of x where the velocity of the particle is zero? You must break the integral apart to calculate the period.

25. Consider the two-body problem consisting of masses m_1 and m_2 orbiting in a plane where each mass is free to move. Find Lagrange's equations of motion of the particles given the kinetic and potential energies

$$T = \frac{1}{2}m_1\left(\dot{x}_1^2 + \dot{y}_1^2\right) + \frac{1}{2}m_2\left(\dot{x}_2^2 + \dot{y}_2^2\right)$$

$$V = -G\frac{m_1 m_2}{\sqrt{\left(x_2 - x_1\right)^2 + \left(y_2 - y_1\right)^2}}$$

26. Two equal masses m are connected to three springs with force constant k, $2k$ and $3k$ with Lagrangian

$$L = \frac{1}{2}m\dot{x}_1^2 + \frac{1}{2}m\dot{x}_2^2 - \frac{1}{2}k\left(x_1^2 + 2\left(x_1 - x_2\right)^2 + 3x_2^2\right).$$

Find the vibrational frequencies for small oscillations.

10.6 Chaotic Dynamics

27. Can a system described by the differential equation $\dddot{y} + \dot{y} + y^2 = 0$ exhibit chaos? Why or why not?

28. Numerically integrate the hyper-jerk system

$$\ddddot{x} + \dddot{x} + \ddot{x} + A\dot{x} + |x| + 1 = 0$$

for $A = 3.9$ and initial conditions $\left(\dddot{x}_0, \ddot{x}_0, \dot{x}_0, x_0\right) = \left(-0.85, 0.26, -0.48, -0.18\right)$

29. Model the three-body problem consisting of masses m_1, m_2 and m_3 orbiting in the $z = 0$ plane where each mass is free to move. $T = \dfrac{1}{2}\displaystyle\sum_{i=1}^{3} m_i \left(\dot{x}_i^2 + \dot{y}_i^2\right)$

$$V = -G\left[\frac{m_1 m_2}{r_{12}} + \frac{m_2 m_3}{r_{23}} + \frac{m_3 m_1}{r_{31}}\right] \text{ where } r_{12} = \sqrt{(x_2 - x_1)^2 + (y_2 - y_1)^2}, \text{ etc.}$$

30. Model the three-body problem above holding m_1 fixed at $(x_1, y_1, z_1) = (0, 0, 0)$. corresponding $m_1 \gg m_2$ and $m_1 \gg m_3$

31. Write Lagrange's equations of motion corresponding to the three-body problem above
(a) holding m_1 fixed at $(x_1, y_1, z_1) = (0, 0, 0)$
(b) allowing all three masses to move
(c) plotting the trajectories of the masses in 3D and creating a Poincaré section plotting points where orbits cross the $z = 0$ plane

10.7 Fractals

32. Compute the first three iterates of the Mandelbrot set for $C = 1 + i$

33. Create plots of the Julia set for values of C inside and outside of the Mandelbrot set.

34. Investigate periodic and chaotic orbits in the "Gingerbread" map

$$x_{n+1} = 1 - y_n + |x_n|$$

$$y_{n+1} = x_n$$

35. The Taylor-Chirikov map

$$p_{n+1} = p_n + K\sin(\theta_n) \quad \mathrm{mod}(2\pi)$$

$$\theta_{n+1} = \theta_n + p_{n+1}$$

is an area-preserving map also known as the standard map. Periodic boundary conditions are applied using $\mathrm{mod}(2\pi)$ conditions so that the map can be projected onto a torus. Create a bifurcation diagram of the map plotting values of θ_n vs. the parameter K with $p_0 = 0$.

Additional Exercises

For the following exercises, the reader is invited to find Lagrange's and Hamilton's equations of motion, plot solutions and find the frequencies of small oscillations where applicable.

36. Two masses m are attached to springs with force constant k and equilibrium length r_0. The springs can only stretch in a radial direction. The top spring pivots about a fixed point O and the bottom spring pivots about the top mass m (Figure 10.9.4).

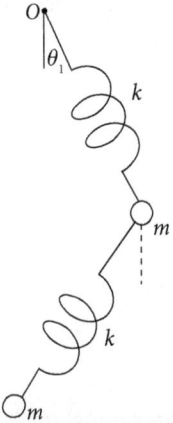

Figure 10.9.4: A flexible double pendulum with two masses and two springs.

37. A triangular wedge of mass M rests on a frictionless surface. A small mass m slides down the wedge without friction. The coordinates x_1 and x_2 locate the position of the wedge along the plane and the small mass along the wedge, respectively (Figure 10.9.5).

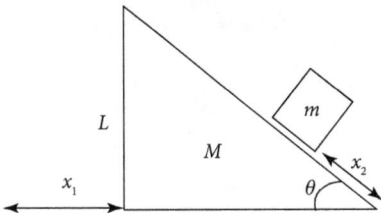

Figure 10.9.5: Block sliding down a wedge on a frictionless surface.

38. A triangular wedge of mass M rests on a frictionless surface. The small mass m slides without friction along the incline and is attached to a spring with force constant k. The spring is attached to the top of the incline. The coordinates x_1 and x_2 locate the position of the wedge along the plane and the small mass along the wedge, respectively (Figure 10.9.6).

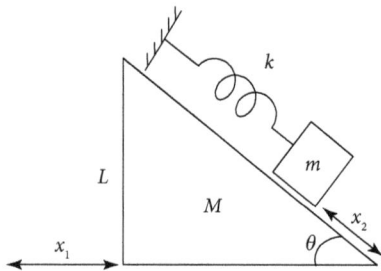

Figure 10.9.6: Block attached to a spring on a wedge sliding on a frictionless surface.

39. The spring and mass configuration in the previous problem is now attached to a fixed spring with force constant k (Figure 10.9.7).

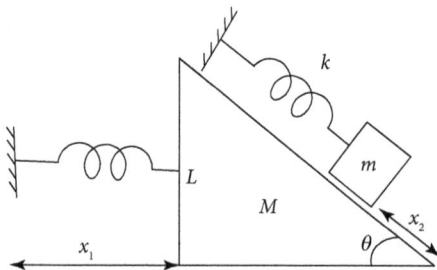

Figure 10.9.7: Block attached to a spring on a wedge sliding on a frictionless surface. The wedge is also attached to a spring.

40. A massless beam of length $2L$ is attached to two springs with force constant k. The beam pivots about an axis O through the center of the beam. The small mass m slides without friction along the massless beam (Figure 10.9.8).

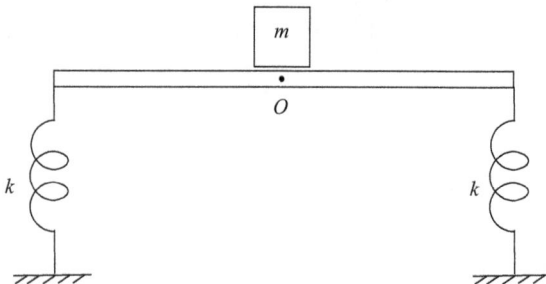

Figure 10.9.8: Block sliding on a beam attached to two springs.

41. A massless beam is attached to two springs with force constant k at opposite ends. A small mass m is attached to one side of the beam by a spring with force constant k and slides without friction along the beam (Figure 10.9.9).

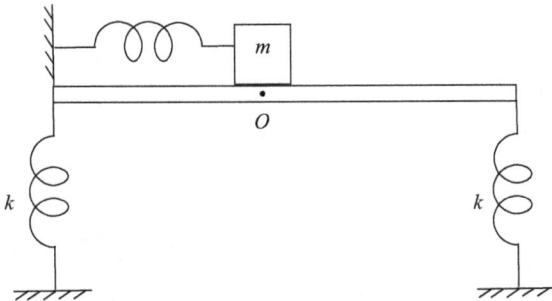

Figure 10.9.9: Block sliding on a beam attached to two springs.

42. A massless beam pivots about a point O and is attached to a spring with force constant k. A small mass m is attached to one side of the beam by a spring with force constant k and slides without friction (Figure 10.9.10).

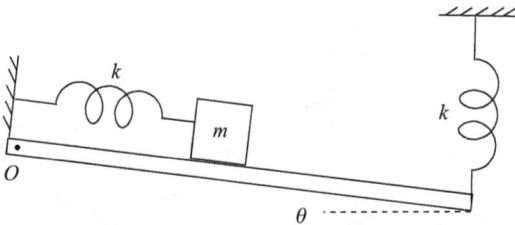

Figure 10.9.10: Mass on a pivoting beam with springs attached.

43. A mass m is attached to a spring with force constant k and slides without friction on a horizontal surface. A simple pendulum consisting of a rigid rod of length L attached to a small mass m swings from a pivot on the sliding mass (Figure 10.9.11).

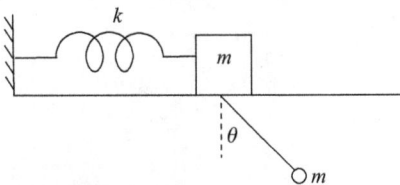

Figure 10.9.11: Pendulum attached to a mass on a spring.

44. A mass M is attached to a spring with force constant k and slides without friction on a horizontal surface. Two simple pendula consisting of massless rigid rods of length L attached to small masses m pivot from two points on the sliding mass. The pendula masses are connected by a spring with force constant k (Figure 10.9.12).

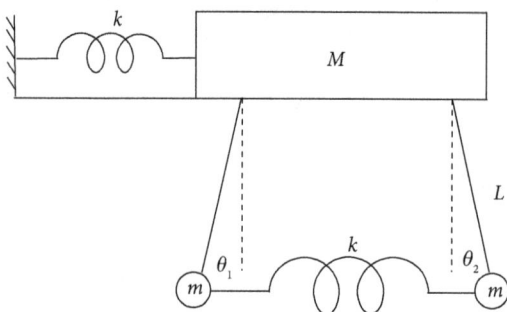

Figure 10.9.12: Coupled pendula attached to block on a spring.

45. Two equal masses M are attached to a spring with force constant k and slide without friction on a horizontal surface. Two simple pendula consisting of massless rigid rods of length L attached to small masses m pivot from points on the sliding masses (Figure 10.9.13).

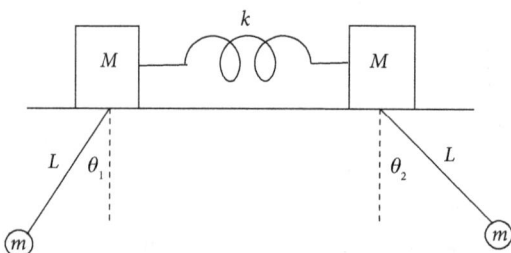

Figure 10.9.13: Pendula attached to two masses connected by a spring.

Chapter **11**

ELECTROMAGNETISM

Chapter Outline

11.1 ELECTROSTATICS IN 1D

One-dimensional electrostatic problems discussed in this section include application of Gauss's law in Cartesian, cylindrical, and spherical coordinates. Examples of Laplace's and Poisson's equation are demonstrated where the potential only depends on one coordinate. Electrostatic energy, capacitance, and induced charge density are discussed.

11.1.1 Integral and Differential Forms of Gauss's Law

Gauss's law states that the electric flux though a closed surface is proportional to the charge enclosed inside the surface. The electric flux through a surface is defined as the normal component of the electric field integrated over the surface. We may use Gauss's law to compute the electric field when the charge density has symmetry in one of the orthogonal coordinate systems such as Cartesian, cylindrical, or spherical coordinates.

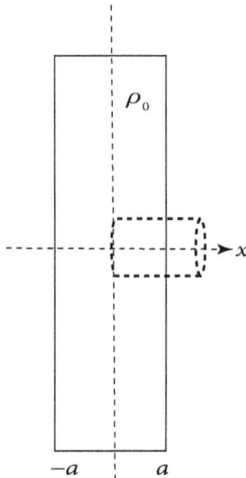

Figure 11.1.1: A charged slab with thickness 2a with a Gaussian pillbox surface for calculating the electric field positioned with one end at $x = 0$.

Example 11.1.1

Given a charged slab as in Figure 11.1.1 with uniform volume charge density

$$\rho(x) = \begin{cases} \rho_0 & -a \leq x \leq a \\ 0 & |x| > a \end{cases} \tag{11.1.1}$$

find the electric field inside the slab using integral and differential forms of Gauss's Law

Solution: From symmetry $\mathbf{E} = 0$ at $x = 0$ so there is no flux through that end cap of the Gaussian surface shown. Also, the electric field is perpendicular to the cylindrical walls where $\mathbf{E} = \hat{\mathbf{n}} = 0$. The total flux EA is through the right endcap so that

$$\int_{\text{surf}} \mathbf{E} \cdot \hat{\mathbf{n}} \, da = \frac{q_{\text{enc}}}{\varepsilon_0} \rightarrow EA = \frac{\rho_0 A a}{\varepsilon_0} \tag{11.1.2}$$

and the electric field

$$\mathbf{E} = \frac{\rho_0 a}{\varepsilon_0} \hat{\mathbf{i}} \text{ for } x \geq a \tag{11.1.3}$$

Also

$$\mathbf{E} = -\frac{\rho_0 a}{\varepsilon_0} \hat{\mathbf{i}} \text{ for } x \leq -a \tag{11.1.4}$$

To find the electric field inside the slab we locate the right-hand side of the Gaussian surface at $x < a$ so that $q_{enc} = \rho_0 A x$ and

$$\mathbf{E} = \frac{\rho_0 x}{\varepsilon_0} \hat{\mathbf{i}} \text{ for } |x| < a \tag{11.1.5}$$

Using the differential form of Gauss's law in one dimension

$$\nabla \cdot \mathbf{E} = \frac{\partial E_x}{\partial x} = \frac{\rho_0}{\varepsilon_0} \tag{11.1.6}$$

Integrating once we obtain

$$E_x = \frac{\rho_0}{\varepsilon_0} x + \text{const.} \tag{11.1.7}$$

where the constant is zero since $E_x = 0$ at $x = 0$.

11.1.2 Laplace's Equation in 1D

Example 11.1.2

Find the potential and electric field in a planar capacitor with boundary conditions $V(0) = 0$ and $V(L) = V_0$

Solution: Laplace's equation

$$\nabla^2 V = \frac{\partial^2 V}{\partial x^2} = 0 \tag{11.1.8}$$

where $V = V(x)$ only has solutions

$$V = c_1 x + c_2 \tag{11.1.9}$$

Applying the boundary conditions to find the constants c_1 and c_2 we have

$$V(0) = 0 \rightarrow c_2 = 0 \tag{11.1.10}$$

and

$$V(L) = c_1 L = V_0 \rightarrow c_1 = \frac{V_0}{L} \tag{11.1.11}$$

and our potential is

$$V(x) = V_0 \left(\frac{x}{L} \right) \tag{11.1.12}$$

with electric field

$$\mathbf{E} = -\frac{\partial V}{\partial x} \hat{\mathbf{i}} = -\frac{V_0}{L} \hat{\mathbf{i}} \tag{11.1.13}$$

Example 11.1.3

Solve Laplace's equation in spherical coordinates to calculate the potential outside of a sphere of radius R and potential $V(R) = V_0$. Calculate the electric field outside the sphere. Check your result using Gauss's law. Show that the capacitance of the sphere is proportional to its radius.

Solution: Laplace's equation without θ or ϕ dependence is

$$\frac{1}{r^2} \frac{\partial}{\partial r} \left(r^2 \frac{\partial V}{\partial r} \right) = 0 \tag{11.1.14}$$

Multiplying by r^2 and integrating once gives

$$r^2 \frac{\partial V}{\partial r} = c_1 \tag{11.1.15}$$

Dividing by r^2 and integrating a second time

$$V(r) = -\frac{c_1}{r} + c_2 \tag{11.1.16}$$

We may determine the constants c_1 and c_2 by applying the boundary conditions $V(R) = V_0$ and $V(\infty) = 0$ so that

$$V(r) = V_0 \left(\frac{R}{r} \right) \tag{11.1.17}$$

Now the electric field is given by $\mathbf{E} = -\nabla V$ or

$$\mathbf{E} = -\frac{\partial V}{\partial r} \hat{\mathbf{r}} = V_0 \frac{R}{r^2} \hat{\mathbf{r}} \tag{11.1.18}$$

We may calculate the surface charge density σ

$$\sigma = -\varepsilon_0 \left. \frac{\partial V}{\partial r} \right|_{r=R} = \varepsilon_0 \frac{V_0}{R} \tag{11.1.19}$$

To compare the electric field obtained from Gauss's law

$$\int_{surf} \mathbf{E} \cdot d\mathbf{a} = - \int_{vol} \cdot dv \tag{11.1.20}$$

or

$$E_r 4\pi r^2 = \frac{1}{\varepsilon_0} \sigma A = \frac{1}{\varepsilon_0} \frac{Q}{4\pi R^2} 4\pi R^2 \tag{11.1.21}$$

and the radial electric field component is

$$E_r = \frac{1}{\varepsilon_0} \frac{Q}{4\pi r^2} \tag{11.1.22}$$

Using the expression for σ calculated form the potential

$$E_r 4\pi r^2 = \frac{1}{\varepsilon_0} \sigma A = \frac{1}{\varepsilon_0} \varepsilon_0 \frac{V_0}{R} 4\pi R^2 \tag{11.1.23}$$

we obtain our previous expression for the electric field

$$E_r = V_0 \frac{R}{r^2} \tag{11.1.24}$$

Now the energy stored in the electric field is

$$W_E = \frac{1}{\varepsilon_0} \int_{vol} |\mathbf{E}|^2 \cdot dv \tag{11.1.25}$$

Performing the volume integral

$$W_E = \frac{1}{\varepsilon_0} \int_0^{2\pi} d\phi \int_0^{\pi} \sin\theta d\theta \int_R^{\infty} \left(V_0 \frac{R}{r^2} \right)^2 r^2 dr \tag{11.1.26}$$

$$W_E = V_0^2 R^2 4\pi \left(-\frac{1}{r} \right)\Big|_R^{\infty} = 2\pi \varepsilon_0 V_0^2 R \tag{11.1.27}$$

The capacitance C is related to the total energy

$$W_E = \frac{1}{2} C V_0^2 \tag{11.1.28}$$

Thus, the capacitance of our sphere is proportional to the radius

$$C = 4\pi\varepsilon_0 R \tag{11.1.29}$$

Example 11.1.4

Find the radial dependence of potential and the electric field outside a cylinder of radius R held at a potential $V(R) = V_0$

Solution: Laplace's equation in cylindrical coordinates with only radial dependence is

$$\nabla^2 V = \frac{1}{r}\frac{\partial}{\partial r}\left(r\frac{\partial V}{\partial r}\right) = 0 \tag{11.1.30}$$

so that

$$r\frac{\partial V}{\partial r} = c_1 \tag{11.1.31}$$

and

$$V(r) = c_1 \ln r + c_2 \tag{11.1.32}$$

We can see from these expressions that the electric field is proportional to $1/r$. Evidently the potential diverges as $r \to \infty$. Note that the cylinder in question is taken to be infinite in length. The potential at a great distance from a cylinder of finite length will in fact go as $1/r$.

Example 11.1.5

Find the potential between two cones of infinite extent (Figure 11.1.2). Use spherical coordinates where the potential only varies in the q direction with $V = V_0$ at $\theta = \alpha$ and $V = -V_0$ at $\theta = \pi - \alpha$

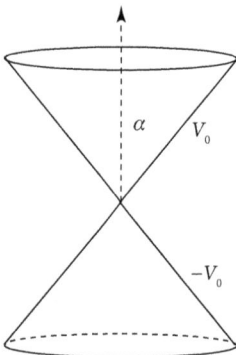

Figure 11.1.2: Two cones coaxial with the z-axis with opposite potentials on the upper and lower cones.

Solution: Laplace's equation in spherical coordinates with $V = V(\theta)$ only is

$$\nabla^2 V = \frac{1}{r^2 \sin^2 \theta} \frac{\partial}{\partial \theta}\left(\sin\theta \frac{\partial V}{\partial \theta}\right) = 0 \tag{11.1.33}$$

Integrating once

$$\sin\theta \frac{\partial V}{\partial \theta} = c_1 \rightarrow \frac{\partial V}{\partial \theta} = \frac{c_1}{\sin\theta} \tag{11.1.34}$$

A second integration gives

$$V(\theta) = c_1 \ln\tan\frac{\theta}{2} + c_2 \tag{11.1.35}$$

Applying boundary conditions $V(\alpha) = V_0$ and $V(\theta_0) = 0$

$$c_1 \ln\tan\frac{\alpha}{2} + c_2 = V_0 \tag{11.1.36}$$

$$c_1 \ln\tan\frac{\theta_0}{2} + c_2 = 0 \tag{11.1.37}$$

Subtracting these equations, we obtain

$$c_1 \ln\tan\frac{\alpha}{2} - c_1 \ln\tan\frac{\theta_0}{2} = V_0 \tag{11.1.38}$$

Solving for c_1

$$c_1 \ln\frac{\tan\dfrac{\alpha}{2}}{\tan\dfrac{\theta_0}{2}} = V_0 \rightarrow c_1 = V_0 \left[\ln\frac{\tan\dfrac{\alpha}{2}}{\tan\dfrac{\theta_0}{2}}\right]^{-1} \tag{11.1.39}$$

The second boundary condition gives c_2

$$c_2 = -c_1 \ln\tan\frac{\theta_0}{2} \tag{11.1.40}$$

and our potential becomes

$$V(\theta) = V_0 \frac{\ln\dfrac{\tan\dfrac{\theta}{2}}{\tan\dfrac{\theta_0}{2}}}{\ln\dfrac{\tan\dfrac{\alpha}{2}}{\tan\dfrac{\theta_0}{2}}} \tag{11.1.41}$$

Whereas the potential is only a function of θ, the electric field will depend on r as well. We may calculate the electric field from the gradient in spherical coordinates

$$\mathbf{E} = -\frac{1}{r}\frac{\partial V}{\partial \theta}\hat{\boldsymbol{\theta}} \qquad (11.1.42)$$

Example 11.1.6

Find the potential and electric field in a wedge with one side at $V(0) = 0$ and $V(\theta_0) = V_0$

Solution: Laplace's equation in cylindrical coordinates where $V = V(\theta)$ only

$$\frac{1}{r^2}\frac{\partial^2 V}{\partial \theta^2} = 0 \qquad (11.1.43)$$

has solutions

$$V = c_1\theta + c_2 \qquad (11.1.44)$$

Applying the boundary condition at $\theta = 0$ and at $\theta = \theta_0$ gives

$$V(0) = 0 \rightarrow c_2 = 0 \qquad (11.1.45)$$

$$V(\theta_0) = c_1\theta_0 = V_0 \rightarrow c_1 = \frac{V_0}{\theta_0} \qquad (11.1.46)$$

so that

$$V(\theta) = V_0\left(\frac{\theta}{\theta_0}\right) \qquad (11.1.47)$$

with electric field

$$\mathbf{E} = -\frac{1}{r}\frac{\partial V}{\partial \theta}\hat{\boldsymbol{\theta}} = -\frac{1}{r}\frac{V_0}{\theta_0}\hat{\boldsymbol{\theta}} \qquad (11.1.48)$$

11.1.3 Poisson's Equation in 1D

Example 11.1.7

Solve Poisson's equation to find the electric field outside of the spherical charge distribution

$$\rho(r) = \frac{Q}{4\pi R^2}\delta(r - R) \qquad (11.1.49)$$

Solution: Poisson's equation

$$\nabla^2 V = -\frac{\rho(r)}{\varepsilon_0} \tag{11.1.50}$$

in spherical coordinates with $V = V(r)$ is

$$\frac{1}{r^2}\frac{\partial}{\partial r}\left(r^2\frac{\partial V}{\partial r}\right) = \frac{1}{\varepsilon_0}\frac{Q}{4\pi R^2}\delta(r-R) \tag{11.1.51}$$

Integrating both sides up to $r > R$

$$\int_0^r \frac{1}{r^2}\frac{\partial}{\partial r}\left(r^2\frac{\partial V}{\partial r}\right)dv = \int_0^r \frac{1}{\varepsilon_0}\frac{Q}{4\pi R^2}\delta(r-R)dv \tag{11.1.52}$$

with $dv = 4\pi r^2 dr$

$$4\pi r^2\frac{\partial V}{\partial r} = -\frac{1}{\varepsilon_0}\frac{Q}{4\pi R^2}4\pi R^2 = -\frac{Q}{\varepsilon_0} \tag{11.1.53}$$

Thus, the electric field for $r \geq R$

$$\mathbf{E} = -\frac{\partial V}{\partial r}\hat{\mathbf{r}} = \frac{Q}{4\pi\varepsilon_0}\frac{\hat{\mathbf{r}}}{r^2} \tag{11.1.54}$$

and $E = 0$ for $r < R$.

Maple Examples

One-dimensional potential problems are demonstrated in the Maple worksheet below. The potential and electric field is calculated between concentric spheres and coaxial cylinders at different potentials. The electric potential and field is calculated at a wedge with one side grounded and the other side at a nonzero potential. The potential and field due to an oblate spheroid at a constant potential is calculated in oblate spheroidal coordinates.

Key Maple terms: *dsolve, expand, factor, Gradient, Laplacian, SetCoordinates, simplify*

Maple packages: *with(VectorCalculus)*:

Electric Potential and Field between Two Spheres

restart
with(VectorCalculus) :
SetCoordinates('spherical'[r, theta, phi])

$$spherical_{r,\theta,\phi}$$

$Lap := Laplacian(V(r))$

$$Lap := \frac{2r\sin(\theta)\left(\dfrac{d}{dr}V(r)\right) + r^2\sin(\theta)\left(\dfrac{d^2}{dr^2}V(r)\right)}{r^2\sin(\theta)}$$

$LapEqn := expand(Lap) = 0$

$$LapEqn := \frac{2\left(\dfrac{d}{dr}V(r)\right)}{r} + \frac{d^2}{dr^2}V(r) = 0$$

$dsolve(\{LapEqn, V(a) = 1, V(b) = 0\})$

$$V(r) = \frac{a}{a-b} - \frac{ab}{(a-b)r}$$

$E = Gradient(rhs(\%))$

$$E = \left(\frac{ab}{(a-b)r^2}\right)\bar{e}_r + (0)\bar{e}_\theta + (0)\bar{e}_\phi$$

Electric Potential in a Wedge

$Lap := Laplacian(V(phi))$

$$Lap := \frac{\dfrac{d^2}{d\phi^2}V(\phi)}{r^2\sin(\theta)^2}$$

$LapEqn := expand(Lap) = 0$

$$LapEqn := \frac{\dfrac{d^2}{d\phi^2}V(\phi)}{r^2\sin(\theta)^2} = 0$$

$dsolve(\{LapEqn, V(alpha) = 1, V(0) = 0\}):$
$simplify(\%)$

$$V(\phi) = \frac{\phi}{\alpha}$$

$E = Gradient(rhs(\%))$

$$E = (0)\bar{e}_r + (0)\bar{e}_\theta + \left(\frac{1}{r\sin(\theta)\alpha}\right)\bar{e}_\phi$$

Electric Potential and Field between Two Cylinders

SetCoordinates('cylindrical'[r, phi, z]);

$$cylindrical_{r,\phi,z}$$

Lap := Laplacian(V(r))

$$Lap := \dfrac{\dfrac{d}{dr}V\,(r)+\left(\dfrac{d^2}{dr^2}V\,(r)\right)r}{r}$$

LapEqn := simplify(expand(Lap) = 0)

$$Lap := \dfrac{\dfrac{d}{dr}V\,(r)}{r}+\dfrac{d^2}{dr^2}V\,(r)=0$$

dsolve({LapEqn, V(a) = 1, V(b) = 0})

$$V(r)=\dfrac{\ln(r)}{\ln(b)-\ln(a)}+\dfrac{\ln(b)}{\ln(b)-\ln(a)}$$

E = Gradient(rhs(%))

$$E=\left(\dfrac{\ln(b)}{(\ln(b)-\ln(a))r}\right)\overline{e}_r+(0)\overline{e}_\phi+(0)\overline{e}_z$$

Electric Potential and Field due to an Oblate Spheroid at Constant Potential

SetCoordinates('oblatespheroidal'[u, v, w]);

$$Oblatespheroidal_{u,v,w}$$

Lap := Laplacian(V(u))

$$Lap := \dfrac{\sinh(u)\sqrt{1-\cos(v)^2}\left(\dfrac{d}{du}V\,(u)\right)+\cosh(u)\sqrt{1-\cos(v)^2}\left(\dfrac{d^2}{du^2}V\,(u)\right)}{\left(\cosh(u)^2-1+\cos(v)^2\right)\cosh(u)\sqrt{1-\cos(v)^2}}$$

LapEqn := simplify(expand(Lap) = 0)

$$Lap := \dfrac{\left(\dfrac{d}{du}V\,(u)\right)\cosh(u)\ \ \sinh(u)\left(\dfrac{d}{du}V\,(u)\right)}{\left(\cosh(u)^2-1+\cos(v)^2\right)\cosh(u)}=0$$

dsolve({LapEqn, V(a) = 1, V(b) = 0})
factor(%)

$$V(u) = -\frac{-\arctan(e^b)+\arctan(e^u)}{\arctan(e^b)-\arctan(e^a)}$$

$E = Gradient(rhs(\%))$

$$E = \left(\frac{e^u}{\sqrt{\cosh(u)^2-1+\cos(v)^2}\left((e^u)^2+1\right)\left(\arctan(e^b)-\arctan(e^a)\right)}\right)\overline{e}_u +(0)\overline{e}_v +(0)\overline{e}_w$$

11.2 LAPLACE'S EQUATION IN CARTESIAN COORDINATES

Electrostatic examples solving Laplace's equation in 3D Cartesian coordinates by separation of variables are first given in this section. The method of images is then discussed.

11.2.1 3D Cartesian Coordinates

Laplace's equation $\nabla^2 V(x, y, z) = 0$ in Cartesian coordinates is written as

$$\frac{\partial^2 V}{\partial x^2}+\frac{\partial^2 V}{\partial y^2}+\frac{\partial^2 V}{\partial z^2}=0 \tag{11.2.1}$$

where $-\infty < x < \infty$, $-\infty < y < \infty$, $-\infty < z < \infty$. Applying separation of variables, we assume a product solution

$$V(x, y, z) = X(x)X(y)Z(z) \tag{11.2.2}$$

Substituting the product form of V into Laplace's equation and dividing by V gives

$$\frac{1}{X}\frac{\partial^2 X}{\partial x^2}+\frac{1}{Y}\frac{\partial^2 Y}{\partial y^2}+\frac{1}{Z}\frac{\partial^2 Z}{\partial z^2}=0 \tag{11.2.3}$$

The variables x, y and z may vary independently. Thus, each term above must be equal to a constant if the sum is zero. One choice of constants is

$$\frac{1}{X}\frac{\partial^2 X}{\partial x^2}=\alpha^2 \quad \frac{1}{Y}\frac{\partial^2 Y}{\partial y^2}=\beta^2 \quad \frac{1}{Z}\frac{\partial^2 Z}{\partial z^2}=-\gamma^2 \tag{11.2.4}$$

where $\alpha^2+\beta^2-\gamma^2=0$. The solutions to these differential equations may be written as linear combinations of sine, cosine and exponential functions.

$$V(x,y,z) = \{A \sin(\alpha x) + B \cos(\alpha x)\}\{C \sin(\beta y) + D \cos(\beta y)\}\{Ee^{\gamma z} + Fe^{-\gamma z}\}$$

$$(11.2.5)$$

with constants A, B, C, D, E, and F determined by the boundary conditions.

Example 11.2.1

A cubical region has boundary condition $V = 0$ on the four sides $y = 0$, $y = L$, $x = 0$ and $x = L$. The sides at $z = -L/2$ and $z = L/2$ are at potentials $V = -V_0$ and $V = V_0$, respectively. Find the potential $V(x, y, z)$ inside the cube.

Solution: Applying the boundary conditions $V(0, y, z) = 0$ and $V(x, 0, z) = 0$ we find that $B = 0$ and $D = 0$. The boundary conditions $V(L, y, z) = 0$ and $V(x, L, z) = 0$ give us $\alpha = n\pi/L$ and $\beta = m\pi/L$ where n and m are integers. Now $\alpha^2 + \beta^2 - \gamma^2 = 0$ so that $\gamma = \dfrac{\pi}{L}\sqrt{n^2 + m^2}$ and our potential is given by the double sum

$$V(x,y,z) = \sum_{n=1}^{\infty}\sum_{m=1}^{\infty} A_{n,m} \sin\left(\frac{n\pi}{L}x\right)\sin\left(\frac{m\pi}{L}y\right)$$

$$\left(E_{n,m}\exp\left(\frac{\pi}{L}\sqrt{n^2+m^2}\,z\right) + F_{n,m}\exp\left(-\frac{\pi}{L}\sqrt{n^2+m^2}\,z\right)\right)$$

$$(11.2.6)$$

Comparing the boundary conditions at $z = L/2$

$$V_0 = \sum_{n=1}^{\infty}\sum_{m=1}^{\infty} A_{n,m} \sin\left(\frac{n\pi}{L}x\right)\sin\left(\frac{m\pi}{L}y\right)$$

$$\left(E_{n,m}\exp\left(\frac{\pi}{2}\sqrt{n^2+m^2}\right) + F_{n,m}\exp\left(-\frac{\pi}{2}\sqrt{n^2+m^2}\right)\right)$$

$$(11.2.7)$$

and at $z = -L/2$

$$-V_0 = \sum_{n=1}^{\infty}\sum_{m=1}^{\infty} A_{n,m} \sin\left(\frac{n\pi}{L}x\right)\sin\left(\frac{m\pi}{L}y\right)$$

$$\left(E_{n,m}\exp\left(-\frac{\pi}{2}\sqrt{n^2+m^2}\right) + F_{n,m}\exp\left(\frac{\pi}{2}\sqrt{n^2+m^2}\right)\right)$$

$$(11.2.8)$$

we find that $F_{n,m} = -E_{n,m}$ and

$$V(x,y,z) = \sum_{n=1}^{\infty}\sum_{m=1}^{\infty} A_{n,m} \sin\left(\frac{n\pi}{L}x\right)\sin\left(\frac{m\pi}{L}y\right)\sinh\left(\frac{\pi}{L}\sqrt{n^2+m^2}\,z\right)$$

$$(11.2.9)$$

where a factor of two and the $E_{n,m}$ have been absorbed into the $A_{n,m}$. To find the $A_{n,m}$ we multiply both sides of the boundary condition at $z = L/2$

$$V_0 = \sum_{n=1}^{\infty} \sum_{m=1}^{\infty} A_{n,m} \sin\left(\frac{n\pi}{L}x\right) \sin\left(\frac{m\pi}{L}y\right) \sinh\left(\frac{\pi}{2}\sqrt{n^2+m^2}\right) \qquad (11.2.10)$$

by $\sin\left(\dfrac{n'\pi}{L}x\right)\sin\left(\dfrac{m'\pi}{L}y\right)$ and integrate over x and y. The left-hand side gives

$$V_0 \int_0^L \int_0^L \sin\left(\frac{n'\pi}{L}x\right)\sin\left(\frac{m'\pi}{L}y\right)dxdy = \frac{L^2}{n'm'\pi^2}(1-\cos n'\pi)(1-\cos m'\pi)$$

$$\qquad (11.2.11)$$

$$= \frac{4L^2}{n'm'\pi^2} \quad (\text{for } n' \text{ and } m' \text{ odd})$$

The right-hand side becomes

$$\int_0^L \int_0^L \sum_{n=1}^{\infty} \sum_{m=1}^{\infty} A_{n,m} \sin\left(\frac{n\pi}{L}x\right)\sin\left(\frac{m\pi}{L}y\right)\sinh\left(\frac{\pi}{2}\sqrt{n^2+m^2}\right)\sin\left(\frac{n'\pi}{L}x\right)\sin\left(\frac{m'\pi}{L}y\right)dxdy$$

$$= \sum_{n=1}^{\infty} \sum_{m=1}^{\infty} A_{n,m}\frac{L^2}{4}\sinh\left(\frac{\pi}{2}\sqrt{n^2+m^2}\right)\delta_{nn'}\delta_{mm'}$$

$$\qquad (11.2.12)$$

Hence, we can drop the primes and find for odd n and m

$$A_{n,m} = \frac{16V_0}{nm\pi^2}\frac{1}{\sinh\left(\dfrac{\pi}{2}\sqrt{n^2+m^2}\right)} \qquad (11.2.13)$$

and

$$V(x,y,z) = \frac{16V_0}{\pi^2}\sum_{n=1,3,5\ldots}^{\infty}\sum_{m=1,3,5\ldots}^{\infty}\frac{1}{nm}\sin\left(\frac{n\pi}{L}x\right)\sin\left(\frac{m\pi}{L}y\right)\frac{\sinh\left(\dfrac{\pi}{L}\sqrt{n^2+m^2}\,z\right)}{\sinh\left(\dfrac{\pi}{2}\sqrt{n^2+m^2}\right)}$$

$$\qquad (11.2.14)$$

11.2.2 Method of Images

The electric field of a charge distribution near a planar conductor can be calculated by calculating the field resulting from the charge distribution and a fictitious image charge distribution on the opposite side of the conductor.

ElectrOMAGNETISM 509

Example 11.2.2

A single line charge with charge per unit length λ is located at height $y' = h$ above a perfectly conducting planar surface at $y = 0$ (Figure 11.2.1). Take the z-axis to be tangent to the conducting plane just under the line charge (perpendicular to the page).

●

Figure 11.2.1: Line charge above a conducting plane.

Solution: The potential of a line charge pointing along the z-axis ($r = 0$) in cylindrical coordinates

$$V(r) = -\frac{\lambda}{2\pi\varepsilon_0}\ln(r) \tag{11.2.15}$$

We may express this potential in Cartesian coordinates

$$V(x,y) = -\frac{\lambda}{2\pi\varepsilon_0}\ln\sqrt{x^2+y^2} \tag{11.2.16}$$

If the line charge is passing through the point (x',y')

$$V(x,y) = -\frac{\lambda}{2\pi\varepsilon_0}\ln\sqrt{(x-x')^2+(y-y')^2} \tag{11.2.17}$$

Use the method of images to find the potential above the conducting plane.

$$V(x,y) = \frac{\lambda}{2\pi\varepsilon_0}\ln\sqrt{x^2+(y-h)^2} - \frac{\lambda}{2\pi\varepsilon_0}\ln\sqrt{x^2+(y+h)^2} \tag{11.2.18}$$

$$V(x,y) = \frac{\lambda}{2\pi\varepsilon_0}\ln\sqrt{\frac{x^2+(y-h)^2}{x^2+(y+h)^2}} \tag{11.2.19}$$

The y-component of the electric field E_y above the conducting plane is

$$E_y = -\frac{\partial V(x,y)}{\partial y} = -\frac{\lambda}{2\pi\varepsilon_0}\frac{(y-h)}{x^2+(y-h)^2} + \frac{\lambda}{2\pi\varepsilon_0}\frac{(y+h)}{x^2+(y+h)^2} \tag{11.2.20}$$

Calculating the induced charge density on the conducting plane as a function of x

$$\sigma = -\varepsilon_0\frac{\partial V(x,y)}{\partial y}\bigg|_{y=0} = \frac{\lambda}{2\pi}\frac{-h}{x^2+h^2} - \frac{\lambda}{2\pi}\frac{h}{x^2+h^2} = -\frac{\lambda}{\pi}\frac{h}{x^2+h^2} \tag{11.2.21}$$

The total charge per unit length induced on the plane is then

$$-\frac{\lambda}{\pi}\int_{-\infty}^{\infty}\frac{h}{x^2+h^2}=-\lambda \qquad (11.2.22)$$

Maple Examples

The electric potential of Example 8.3.1 corresponding to a square region with opposite sides grounded is plotted in the Maple worksheet below. The potential is visualized using a 3D contour plot.

Key Maple terms: *contourplot3d, subs, sum*

Maple packages: *with(plots)*:

Programming: Function statements using '→'

restart

Potential inside a Square

$N := 60 :$

$$V := (x,y) \rightarrow \frac{2 \cdot V0}{Pi} \cdot \left(sum\left(\frac{(1-\cos(n \cdot Pi))}{n \cdot (\exp(n \cdot Pi)+1)} \right. \right.$$

$$\cdot \left(\exp\left(\frac{n \cdot Pi}{L} \cdot x\right) - \frac{(1-\exp(n \cdot Pi))}{(1-\exp(-n \cdot Pi))} \cdot \exp\left(-\frac{n \cdot Pi}{L} \cdot x\right) \right)$$

$$\left. \left. \cdot \sin\left(\frac{n \cdot Pi}{L} \cdot y\right), n = 1 \ldots N \right) \right)$$

$$V := (x,y) \mapsto \frac{2V0\left(\sum_{n=1}^{N} \frac{(1-\cos(n\pi))\left(e^{\frac{n\pi x}{L}} - \frac{(1-e^{n\pi})e^{-\frac{n\pi x}{L}}}{1-e^{-n\pi}} \right) \sin\left(\frac{n\pi y}{L}\right)}{n(e^{n\pi}+1)} \right)}{\pi}$$

$V2 := subs(\{L=1, V0=1\}, V(x,y)) :$
$with(plots) :$
$contourplot3d(V2, x=0\ldots1, y=0\ldots1, color = \text{"Black"}, contours = 30)$

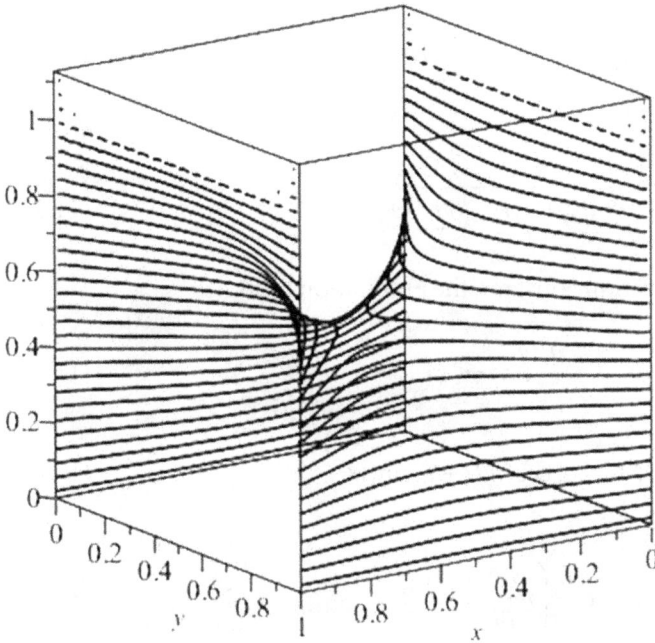

Figure 11.2.2: 3D contour plot of the potential inside a square region with opposite sides grounded and the other two sides at the same potential.

11.3 LAPLACE'S EQUATION IN CYLINDRICAL COORDINATES

Solutions to Laplace's equation in cylindrical coordinates without z-dependence are investigated using separation of variables in this section. Examples of a grounded cylinder in an external field and cylindrical shells with Dirichlet boundary condition are given. Cylindrical solutions to Laplace's equation in three dimensions are discussed. The angular coordinate is represented by θ in the following.

11.3.1 Potentials with Planar Symmetry

If the potential V is only a function of r and θ, Laplace's equation becomes

$$\nabla^2 V\left(r,\theta\right) = \frac{1}{r}\frac{\partial}{\partial r}\left(r\frac{\partial V}{\partial r}\right) + \frac{1}{r^2}\frac{\partial^2 V}{\partial \theta^2} = 0 \qquad (11.3.1)$$

Substituting $V\left(r,\theta\right) = R(r)\Theta(\theta)$ and multiplying by r^2

$$r\frac{\partial}{\partial r}\left(r\frac{\partial R}{\partial r}\right)\Theta + \frac{\partial^2\Theta}{\partial\theta^2}R = 0 \qquad (11.3.2)$$

Dividing by $R\Theta$ we obtain

$$\underbrace{r\frac{\partial}{\partial r}\left(r\frac{\partial R}{\partial r}\right)\frac{1}{R}}_{n^2} + \underbrace{\frac{\partial^2\Theta}{\partial\theta^2}\frac{1}{\Theta}}_{-n^2} = 0 \qquad (11.3.3)$$

Choosing n^2 as the separation constant, the Θ equation becomes

$$\frac{\partial^2\Theta}{\partial\theta^2} = -n^2\Theta \qquad (11.3.4)$$

with solution

$$\Theta(\theta) = c_n\cos(n\theta) + d_n\sin(n\theta) \qquad (11.3.5)$$

The R equation becomes

$$r\frac{\partial}{\partial r}\left(r\frac{\partial R}{\partial r}\right) = n^2R \qquad (11.3.6)$$

or

$$r\frac{\partial R}{\partial r} + r^2\frac{\partial^2 R}{\partial r^2} = n^2R \qquad (11.3.7)$$

It can be verified by direct substitution that r^n and r^{-n} are solutions to this differential equation. Therefore, the general solution to the R equation including the case where $n = 0$ is

$$R(r) = a_n r^n + b_n r^{-n} + a_0\ln(r) + b_0 \qquad (11.3.8)$$

If potential is only a function of r, we have $n = 0$ and

$$V(r) = a_0\ln r + b_0 \qquad (11.3.9)$$

Example 11.3.1

A grounded conducting cylinder of radius R is coaxial with the z-axis. The cylinder is placed in an otherwise uniform electric field $\mathbf{E} = E_0\hat{\mathbf{j}}$ in the y-direction. Solve Laplace's equation for the potential outside of the cylinder. Calculate the induced surface charge density on the cylinder.

Solution: The potential far from the cylinder is

$$V = -E_0 y = -E_0 r\sin\theta \qquad (11.3.10)$$

where $\mathbf{E} = -\nabla V = E_0\hat{\mathbf{j}}$. The general solution for the potential is

$$V(r,\theta) = \sum_{n=1}^{\infty} \left(a_n r^n + b_n r^{-n} \right) \left(c_n \cos n\theta + d_n \sin n\theta \right) \qquad (11.3.11)$$

Applying the boundary condition of zero potential at the surface of the cylinder

$$V(R,\theta) = 0 \Rightarrow a_n R^n + b_n R^{-n} = 0 \qquad (11.3.12)$$

gives us a condition between the a_n and the b_n

$$b_n = -a_n R^{2n} \qquad (11.3.13)$$

For large $r \gg R$

$$\sum_{n=1}^{\infty} a_n r^n \left(c_n \cos n\theta + d_n \sin n\theta \right) = -E_0 r \sin\theta \qquad (11.3.14)$$

Thus $c_n = 0$ and we may absorb the d_n into the a_n. Since $n = 1$ only, we have $a_1 = -E_0$ and the potential

$$V(r,\theta) = -E_0 r \sin\theta + E_0 \frac{R^2}{r} \sin\theta \qquad (11.3.15)$$

Factoring

$$V(r,\theta) = E_0 \left(\frac{R^2}{r} - r \right) \sin\theta \qquad (11.3.16)$$

we see that $V(R, \theta) = 0$. Although the potential is zero at $r = R$, the surface charge density is nonzero

$$\sigma = -\varepsilon_0 \frac{\partial V}{\partial r}\bigg|_{r=R} = 2\varepsilon_0 E_0 \sin\theta \qquad (11.3.17)$$

Example 11.3.2

A cylinder of radius R has a surface potential

$$V(R,\theta) = \begin{cases} V_0 & 0 \le \theta < \pi \\ -V_0 & \pi \le \theta < 2\pi \end{cases} \qquad (11.3.18)$$

Calculate the potential $V(r,\theta)$ for $r \ge R$ and $r < R$.

Solution: Because the potential is odd about $\theta = 0$ we choose the sine series

$$V_<(r,\theta) = \sum_{n=1}^{\infty} a_n r^n \sin(n\theta) \quad (r \le R) \qquad (11.3.19)$$

$$V_>(r,\theta) = \sum_{n=1}^{\infty} b_n r^{-n} \sin(n\theta) \quad (r > R) \tag{11.3.20}$$

We require that the potential is continuous at $r = R$

$$a_n R^n = b_n R^{-n} \Rightarrow b_n = a_n R^{2n} \tag{11.3.21}$$

To solve for the a_n we multiply both sides of

$$V(R,\theta) = \sum_{n=1}^{\infty} a_n R^n \sin(n\theta) \tag{11.3.22}$$

by $\sin(n'\theta)$ and integrate

$$\int_0^{2\pi} V(R,\theta)\sin(n'\theta)\,d\theta = \int_0^{2\pi} \sum_{n=1}^{\infty} a_n R^n \sin(n\theta)\sin(n'\theta)\,d\theta \tag{11.3.23}$$

Left-hand side:

$$\int_0^{2\pi} V(R,\theta)\sin(n'\theta)\,d\theta = V_0 \int_0^{\pi} \sin(n'\theta)\,d\theta - V_0 \int_{\pi}^{2\pi} \sin(n'\theta)\,d\theta$$

$$= -\frac{V_0}{n'}\cos(n'\theta)\Big|_0^{\pi} + \frac{V_0}{n'}\cos(n'\theta)\Big|_{\pi}^{2\pi} \tag{11.3.24}$$

$$= -\frac{V_0}{n'}\Big[(-1)^{n'} - 1\Big] + \frac{V_0}{n'}\Big[1 - (-1)^{n'}\Big]$$

$$= \frac{4V_0}{n'} \quad (n' \text{ odd})$$

Right-hand side:

$$\int_0^{2\pi} \sum_{n=1}^{\infty} a_n R^n \sin(n\theta)\sin(n'\theta)\,d\theta = \sum_{n=1}^{\infty} a_n R^n \int_0^{2\pi} \sin(n\theta)\sin(n'\theta)\,d\theta$$

$$= \sum_{n=1}^{\infty} a_n R^n \frac{\theta}{2}\Big|_0^{2\pi} \delta_{nn'} = a_{n'} R^{n'} \pi \tag{11.3.25}$$

Dropping the primes since $n' = n$

$$a_n = \frac{4V_0}{nR^n \pi} \qquad b_n = \frac{4V_0 R^n}{n\pi} \tag{11.3.26}$$

$$V_<(r,\theta) = \frac{4V_0}{\pi} \sum_{n=1,3,5\ldots}^{\infty} \frac{1}{n}\left(\frac{r}{R}\right)^n \sin(n\theta)$$

and
$$V_>(r,\theta) = \frac{4V_0}{\pi} \sum_{n=1,3,5...}^{\infty} \frac{1}{n} \left(\frac{R}{r}\right)^n \sin(n\theta) \qquad (11.3.27)$$

We can immediately verify $V_<(R,\theta) = V_>(R,\theta)$.

Example 11.3.3

A cylinder of radius R has a surface potential
$$V(R,\theta) = V_0 \cos(3\theta) \qquad (11.3.28)$$

Calculate the potential $V(r,\theta)$ for $r \geq R$ and $r < R$.

Solution: Since the boundary condition is even, we may discard coefficients of $\sin(n\theta)$ so that

$$V_<(r,\theta) = \sum_{n=1}^{\infty} a_n r^n \cos(n\theta) \quad (r \leq R) \qquad (11.3.29)$$

$$V_>(r,\theta) = \sum_{n=1}^{\infty} b_n r^{-n} \cos(n\theta) \quad (r > R) \qquad (11.3.30)$$

We require that the potential is continuous at $r = R$
$$a_n R^n = b_n R^{-n} \Rightarrow b_n = a_n R^{2n} \qquad (11.3.31)$$

Now to solve for the a_n

$$V_0 \cos(3\theta) = \sum_{n=1}^{\infty} a_n R^n \cos(n\theta) \qquad (11.3.32)$$

From inspection, we see that $n = 3$ and $a_3 = V_0/R^3$

$$V_<(r,\theta) = V_0 \left(\frac{r}{R}\right)^3 \cos(3\theta) \quad (r \leq R) \qquad (11.3.33)$$

$$V_>(r,\theta) = V_0 \left(\frac{R}{r}\right)^3 \cos(3\theta) \quad (r > R) \qquad (11.3.34)$$

11.3.2 Potentials in 3D Cylindrical Coordinates

If the potential depends on all three cylindrical coordinates we have general solutions of the form

$$V(r,z,\theta) = \sum_{n=1}^{\infty} \left(a_n J_n(kr) + b_n N_n(kr) \right) \left(c_n \cos n\theta + d_n \sin n\theta \right)$$

$$\left(e_n \sinh kz + f_n \cosh kz \right)$$

(11.3.35)

where the radial dependence is given by Bessel functions. The Neumann Bessel functions $N_n(kr)$ are divergent where $r = 0$ and are discarded in regions including the z-axis.

Maple Examples

The electric potential is calculated inside of a circular region with specified boundary potential and visualized using a contour plot in the Maple worksheet below.

Key Maple terms: *contourplot, int, subs, sum*

Maple packages: *with(plots):*

Programming: Function statements using '→'

restart

Potential inside a Cylinder

$$f := (theta) \rightarrow V0 \cdot \cos(2 \cdot theta) + V0 \cdot \sin(2 \cdot theta)$$

$$f := \theta \mapsto V0 \cos(2\,\theta) + V0 \sin(2\,\theta)$$

$$a := (n) \rightarrow \frac{1}{\pi \cdot R^n} \cdot \int_0^{2 \cdot \pi} f(\theta) \cdot \cos(n \cdot \theta)\,d\theta$$

$$a := n \mapsto \frac{\int_0^{2\pi} f(\theta) \cos(n\theta)\,d\theta}{\pi R^n}$$

$$b := (n) \rightarrow \frac{1}{\pi \cdot R^n} \cdot \int_0^{2 \cdot \pi} f(\theta) \cdot \cos(n \cdot \theta)\,d\theta$$

$$b := n \mapsto \frac{\int_0^{2\pi} f(\theta) \cos(n\theta)\,d\theta}{\pi R^n}$$

$$N := 10 :$$

$$V := sum\left(\left(\frac{r}{R}\right)^n \cdot (a(n) \cdot \cos(n \cdot theta) + b(n) \cdot \sin(n \cdot theta)), n = 0 \ldots N \right)$$

$$V := \frac{2V0r^2 \cos(\theta)^2 - V0r^2 + 2V0r^2 \sin(\theta) \cos(\theta)}{R^4}$$

$V2 := subs(\{R = 1, V0 = 1\}, V)$

$$V2 := 2r^2 \cos(\theta)^2 - r^2 + 2r^2 \sin(\theta) \cos(\theta)$$

$with(plots):$
$contourplot([r, theta, V2], r = 0\ldots1, theta = 0\ldots2\pi, coords = cylindrical, contours =$
$20, color = \text{“Black”})$

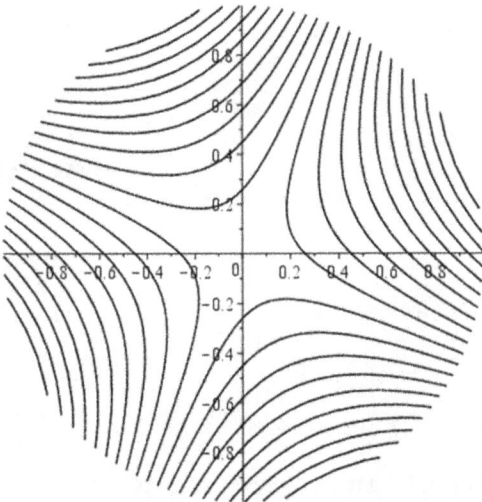

Figure 11.3.1: Equipotentials inside a cylindrical region with specified potential on the cylindrical boundary.

11.4 LAPLACE'S EQUATION IN SPHERICAL COORDINATES

We investigate solutions to Laplace's equation in spherical coordinates without ϕ-dependence using separation of variables in this section. Examples of a hollow ring, nested spheres and a single sphere with Dirichlet boundary conditions are given. Spherical solutions to Laplace's equation with ϕ-dependence are discussed.

11.4.1 Axially Symmetric Potentials

If the electric potential V is only a function of r and θ, Laplace's equation becomes

$$\frac{\partial}{\partial r}\left(r^2 \frac{\partial V}{\partial r} \right) + \frac{1}{\sin\theta} \frac{\partial}{\partial \theta}\left(\sin\theta \frac{\partial V}{\partial \theta} \right) = 0$$

(11.4.1)

Obtaining a separation of variables solution, we substitute $V(r, \theta) = R(r) \, \Theta \, (\theta)$ into Laplace's equation and divide by $R(r) \, \Theta \, (\theta)$

$$\underbrace{\frac{1}{R}\frac{\partial}{\partial r}\left(r^2\frac{\partial R}{\partial r}\right)}_{\ell(\ell+1)}+\underbrace{\frac{1}{\Theta\sin\theta}\frac{\partial}{\partial\theta}\left(\sin\theta\frac{\partial\Theta}{\partial\theta}\right)}_{-\ell(\ell+1)}=0 \qquad (11.4.2)$$

With separation constant $\ell(\ell+1)$, the radial differential equation becomes

$$\frac{\partial}{\partial r}\left(r^2\frac{\partial R}{\partial r}\right)=\ell(\ell+1)R \qquad (11.4.3)$$

with solutions

$$R(r)=A_\ell r^\ell+\frac{B_\ell}{r^{\ell+1}} \qquad (11.4.4)$$

The Θ equation

$$\frac{\partial}{\partial\theta}\left(\sin\theta\frac{\partial\Theta}{\partial\theta}\right)=-\ell(\ell+1)\sin\theta\Theta \qquad (11.4.5)$$

has the general solution

$$\Theta(\theta)=C_\ell P_\ell(\cos\theta)+D_\ell Q_\ell(\cos\theta) \qquad (11.4.6)$$

where $P_\ell(\cos\theta)$ are the Legendre polynomials that can be obtained by Rodrigues' formula.

The $Q_\ell(\cos\theta)$ are the Legendre functions of the second kind. They are usually not considered because they are divergent where $\theta=0$ (on the z-axis). If the z-axis is excluded from a given problem then we must include the $Q_\ell(\cos\theta)$.

Example 11.4.1

Find the potential inside a ring in spherical coordinates held at a potential $V(a,\theta)=V_0$ on its inner surface and grounded on all other surfaces (Figure 11.4.1).

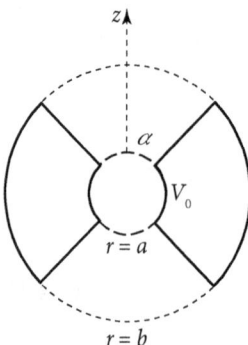

Figure 11.4.1: Ring formed from the intersection of spheres (with radii a and b) and cones (with half angles α and $\pi-\alpha$)

Solution: Because the z-axis is excluded from the solution region the general solution will include the $Q_\ell(\cos\theta)$

$$V(r,\theta) = \sum_{\ell=0}^{\infty} \left(A_\ell r^\ell + B_\ell r^{-(\ell+1)} \right) \left(C_\ell P_\ell(\cos\theta) + D_\ell Q_\ell(\cos\theta) \right) \quad (11.4.7)$$

The boundary conditions are

$$V(b, \theta) = 0, \; V(a, \theta) = V_0, \; V(r, \alpha) = V(r, \pi - \alpha) = 0 \quad (11.4.8)$$

Applying the boundary condition at $r = b$

$$A_\ell b^\ell + B_\ell b^{-(\ell+1)} = 0 \quad (11.4.9)$$

gives a relation between the constants $B_\ell = -A_\ell b^{2\ell+1}$.
With zero potential on the surface of constant α

$$C_\ell P_\ell(\cos\alpha) + D_\ell Q_\ell(\cos\alpha) = 0 \quad (11.4.10)$$

we have

$$D_\ell = -C_\ell \frac{P_\ell(\cos\alpha)}{Q_\ell(\cos\alpha)} \quad (11.4.11)$$

Absorbing the C_ℓ into the A_ℓ, the potential is

$$V(r,\theta) = \sum_{\ell=0}^{\infty} A_\ell \left(r^\ell - b^{2\ell+1} r^{-(\ell+1)} \right) \left[P_\ell(\cos\theta) - \frac{P_\ell(\cos\alpha)}{Q_\ell(\cos\alpha)} Q_\ell(\cos\theta) \right] \quad (11.4.12)$$

Applying the boundary condition at $r = a$

$$V_0 = \sum_{\ell=0}^{\infty} A_\ell \left(a^\ell - b^{2\ell+1} a^{-(\ell+1)} \right) \left[P_\ell(\cos\theta) - \frac{P_\ell(\cos\alpha)}{Q_\ell(\cos\alpha)} Q_\ell(\cos\theta) \right] \quad (11.4.13)$$

As an approximation, we multiply both sides by $P_{\ell'}(\cos\theta)\sin\theta d\theta$ and integrate

$$V_0 \int_0^\pi P_{\ell'}(\cos\theta)\sin\theta d\theta$$

$$= \sum_{\ell=0}^{\infty} A_\ell \left(a^\ell - b^{2\ell+1} a^{-(\ell+1)} \right) \int_0^\pi \left(P_\ell(\cos\theta) - \frac{P_\ell(\cos\alpha)}{Q_\ell(\cos\alpha)} Q_\ell(\cos\theta) \right) P_{\ell'}(\cos\theta)\sin\theta d\theta$$

$$(11.4.14)$$

This gives

$$V_0 \int_{-1}^1 P_{\ell'}(x)dx = \sum_{\ell=0}^{\infty} A_\ell \left(a^\ell - b^{2\ell+1} a^{-(\ell+1)} \right) \frac{2}{2\ell+1} \delta_{\ell\ell'} \quad (11.4.15)$$

and

$$A_\ell = V_0 \frac{2\ell+1}{2\left(a^\ell - b^{2\ell+1}a^{-(\ell+1)}\right)} \int_{-1}^{1} P_\ell(x)dx \qquad (11.4.16)$$

Example 11.4.2

A sphere of radius a has a surface potential

$$V(a, \theta) = V_0 \cos(\theta) \qquad (11.4.17)$$

The sphere is surrounded by a grounded concentric sphere of radius b

$$V(b, \theta) = 0 \qquad (11.4.18)$$

Calculate the potential $V(r, \theta)$ for $r < a$ and $a \leq r \leq b$

Solution: Because the z-axis is included in our solution region, our general solution is

$$V(r,\theta) = \sum_{\ell=0}^{\infty}\left(A_\ell r^\ell + B_\ell r^{-(\ell+1)}\right) P_\ell(\cos\theta) \qquad (11.4.19)$$

Applying the boundary condition at $r = b$ gives

$$A_\ell b^\ell = -B_\ell b^{-(\ell+1)} \qquad (11.4.20)$$

and our potential is

$$V(r,\theta) = \sum_{\ell=0}^{\infty} A_\ell \left(r^\ell - b^{(2\ell+1)} r^{-(\ell+1)}\right) P_\ell(\cos\theta) \qquad (11.4.21)$$

Applying the boundary condition at $r = a$

$$V_0 \cos\theta = \sum_{\ell=0}^{\infty} A_\ell \left(a^\ell - b^{(2\ell+1)} a^{-(\ell+1)}\right) P_\ell(\cos\theta) \qquad (11.4.22)$$

we only have $\ell = 1$ where $P_1(\cos\theta) = \cos\theta$ and

$$A_1 = \frac{V_0}{\left(a - b^3 a^{-2}\right)} \qquad (11.4.23)$$

Writing the potential between the spheres as

$$V(r,\theta) = V_0 \left(\frac{a^2}{a^3 - b^3}\right)\left(\frac{r^3 - b^3}{r^2}\right)\cos\theta \qquad (11.4.24)$$

we can verify that the boundary conditions at $r = a$ and $r = b$ are satisfied.

Example 11.4.3

A sphere of radius a is grounded

$$V(a, \theta) = 0 \qquad (11.4.25)$$

The sphere is surrounded by a larger concentric sphere of radius b with a potential

$$V(b,\theta) = V_0 \left(P_2(\cos\theta) + P_3(\cos\theta) \right) \qquad (11.4.26)$$

Find the potential between the spheres and find the induced surface charge density on the inner sphere.

Solution: Applying the boundary condition at $r = a$

$$V(a,\theta) = \sum_{\ell=0}^{\infty} \left(A_\ell a^\ell + B_\ell a^{-(\ell+1)} \right) P_\ell(\cos\theta) = 0 \qquad (11.4.27)$$

so that $A_\ell a^\ell = -B_\ell a^{-(\ell+1)}$ or $B_\ell = -A_\ell a^{(2\ell+1)}$.

Factoring the A_ℓ and applying the boundary condition at $r = b$

$$V_0 \left(P_2(\cos\theta) + P_3(\cos\theta) \right) = \sum_{\ell=0}^{\infty} A_\ell \left(b^\ell - a^{(2\ell+1)} b^{-(\ell+1)} \right) P_\ell(\cos\theta) \qquad (11.4.28)$$

For $\ell = 2$ we have

$$V_0 P_2(\cos\theta) = A_2 \left(b^2 - a^5 b^{-3} \right) P_2(\cos\theta) \qquad (11.4.29)$$

and

$$A_2 = \frac{V_0}{\left(b^2 - a^5 b^{-3} \right)} = \frac{V_0 b^3}{\left(b^5 - a^5 \right)} \qquad (11.4.30)$$

For $\ell = 3$

$$A_3 = \frac{V_0}{\left(b^3 - a^7 b^{-4} \right)} = \frac{V_0 b^4}{\left(b^7 - a^7 \right)} \qquad (11.4.31)$$

Substituting the constants our potential becomes

$$V(r,\theta) = \frac{V_0 b^3}{(b^5 - a^5)} \left(r^2 - a^5 r^{-3} \right) P_2(\cos\theta) + \frac{V_0 b^4}{(b^7 - a^7)} \left(r^3 - a^7 r^{-4} \right) P_3(\cos\theta) \qquad (11.4.32)$$

and we can verify that the boundary conditions are satisfied at $r = a$ and $r = b$.

Now the surface charge density on the inner sphere

$$\sigma(\theta) = -\varepsilon_0 \left. \frac{\partial V(r,\theta)}{\partial r} \right|_{r=a} \qquad (11.4.33)$$

Calculating the derivatives

$$\sigma(\theta) = -\varepsilon_0 \left[\frac{V_0 b^3}{\left(b^5 - a^5\right)} \left(2r + 3a^5 r^{-4}\right) P_2(\cos\theta) + \frac{V_0 b^4}{\left(b^7 - a^7\right)} \left(3r^2 + 4a^7 r^{-5}\right) P_3(\cos\theta) \right]_{r=a}$$

(11.4.34)

we obtain

$$\sigma(\theta) = -\varepsilon_0 V_0 \left[\frac{5ab^3}{\left(b^5 - a^5\right)} P_2(\cos\theta) + \frac{7a^2 b^4}{\left(b^7 - a^7\right)} P_3(\cos\theta) \right]$$

(11.4.35)

11.4.2 3D Spherical Coordinates

In spherical coordinates without azimuthal symmetry the solution to Laplace's equation is of the form

$$V(r,\theta,\phi) = \sum_{\ell=0}^{\infty} \sum_{m=-\ell}^{\ell} \left(A_{\ell,m} r^{\ell} + B_{\ell,m} r^{-(\ell+1)} \right) Y_{\ell}^{m}(\theta,\phi)$$

(11.4.36)

where the $Y_{\ell}^{m}(\theta,\phi)$ are spherical harmonics.

Maple Examples

The electric potential is calculated inside of a spherical region with specified boundary potential and visualized using a contour plot in the Maple worksheet below.

Key Maple terms: *contourplot3d, int, subs, sum*

Maple packages: *with(plots):*

Programming: Function statements using '→'

Special functions: LegendreP

restart

Potential inside a Sphere

$f := (\text{theta}) \rightarrow V0 \cdot \cos(\text{theta}) \cdot \sin(3 \cdot \text{theta})$
$$f := \theta \mapsto V0 \cos(\theta) \sin(3\,\theta)$$

$$C := (n) \rightarrow \frac{(2 \cdot n + 1)}{2} \cdot \int_0^{\pi} f(\theta) \cdot \text{LegendreP}(n, \cos(\theta)) \cdot \sin(\theta) d\theta$$

$$C := (n) \mapsto \left(n + \frac{1}{2} \right) \left(\int_0^\pi f(\theta) \, \text{LegendreP}(n, \cos(\theta)) \sin(\theta) \, d\theta \right)$$

$N := 7 :$

$$V := sum \left(C(n) \cdot \left(\frac{r}{a} \right)^n \cdot \text{LegendreP}(n, \cos(\text{theta})), n = 0 \ldots N \right)$$

$$V := \frac{3\pi V 0 r \cos(\theta)}{16a} + \frac{21\pi V 0 r^3 \text{LegendreP}(3, \cos(\theta))}{128a^3}$$
$$- \frac{363\pi V 0 r^5 \text{LegendreP}(5, \cos(\theta))}{2048a^5} - \frac{105\pi V 0 r^7 \text{LegendreP}(7, \cos(\theta))}{2048a^7}$$

$V2 := subs(\{a = 1, V0 = 1\}, V) :$
$with(plots) :$
$contourplot3d([r, theta, V2], r = 0 \ldots 1, theta = 0 \ldots 2\pi, coords = cylindrical, contours = 20, color = \text{``White''}, filledregions = true)$

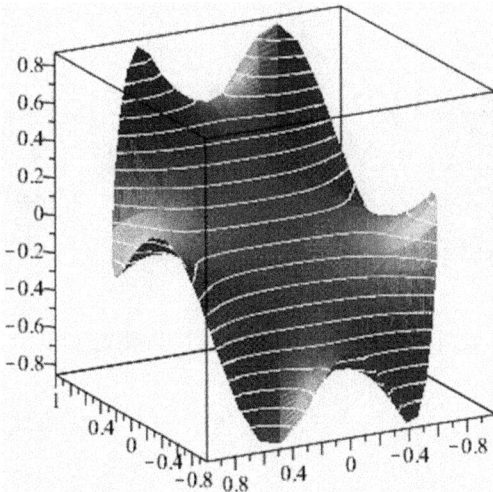

Figure 11.4.2: Equipotentials inside a sphere with specified axially symmetric potential on the surface. The spherical potential with axial symmetry is plotted in cylindrical coordinates.

11.5 MULTIPOLE EXPANSION OF POTENTIAL

We investigate the multipole expansion of potentials with axial symmetry on and off the z-axis below. Examples of a charged needle, disk, and ring are given. The multipole expansion of asymmetric potentials is discussed.

11.5.1 Axially Symmetric Potentials

The multipole expansion technique is an approximation method of calculating the integral from the potential

$$V(\mathbf{r}) = \frac{1}{4\pi\varepsilon_0} \int_{vol} \frac{\rho(\mathbf{r}')}{|\mathbf{r}-\mathbf{r}'|} dv'$$

(11.5.1)

for an arbitrary charge distribution $\rho(\mathbf{r}')$. Since

$$|\mathbf{r}-\mathbf{r}'|^2 = r^2 + r'^2 - 2rr'\cos\alpha = r^2\left(1 + \left(\frac{r'}{r}\right)^2 - 2\left(\frac{r'}{r}\right)\cos\alpha\right)$$

(11.5.2)

where α is the angle between $\mathbf{r}$ and $\mathbf{r}'$, we have

$$\frac{1}{|\mathbf{r}-\mathbf{r}'|} = \frac{1}{r}\left(1 + \left(\frac{r'}{r}\right)^2 - 2\left(\frac{r'}{r}\right)\cos\alpha\right)^{-1/2}$$

(11.5.3)

Expanding for $r > r'$

$$\frac{1}{|\mathbf{r}-\mathbf{r}'|} = \frac{1}{r}\left(1 + \left(\frac{r'}{r}\right)\cos\alpha + \left(\frac{r'}{r}\right)^2\left(\frac{3\cos^2\alpha - 1}{2}\right) + \left(\frac{r'}{r}\right)^3\left(\frac{5\cos^3\alpha - 3\cos\alpha}{2}\right) + \cdots\right)$$

(11.5.4)

we find the coefficients of $(r'/r)^n$ are the Legendre polynomials $P_n(\cos\alpha)$

$$\frac{1}{|\mathbf{r}-\mathbf{r}'|} = \frac{1}{r}\left(\left(\frac{r'}{r}\right)^0 P_0(\cos\alpha) + \left(\frac{r'}{r}\right)P_1(\cos\alpha) + \left(\frac{r'}{r}\right)^2 P_2(\cos\alpha) + \left(\frac{r'}{r}\right)^3 P_3(\cos\alpha) + \cdots\right)$$

(11.5.5)

Thus,

$$\frac{1}{|\mathbf{r}-\mathbf{r}'|} = \frac{1}{r}\sum_{n=0}^{\infty}\left(\frac{r'}{r}\right)^n P_n(\cos\alpha)$$

(11.5.6)

and our integral for potential for $r > r'$

$$V(\mathbf{r}) = \frac{1}{4\pi\varepsilon_0}\int_{vol}\frac{\rho(\mathbf{r}')}{r}\sum_{n=0}^{\infty}\left(\frac{r'}{r}\right)^n P_n(\cos\alpha)dv'$$

(11.5.7)

is evaluated over primed coordinates so that

$$V(\mathbf{r}) = \frac{1}{4\pi\varepsilon_0} \sum_{n=0}^{\infty} \frac{1}{r^{n+1}} \int_{vol} r'^n P_n(\cos\alpha)\rho(\mathbf{r}')dv' \qquad (11.5.8)$$

The first three multipole terms corresponding to $n = 0, 1, 2$ are the monopole term:

$$V(\mathbf{r}) = \frac{1}{4\pi\varepsilon_0} \frac{1}{r} \int_{vol} \rho(\mathbf{r}')dv' \qquad (11.5.9)$$

dipole term:

$$V(\mathbf{r}) = \frac{1}{4\pi\varepsilon_0} \frac{1}{r^2} \int_{vol} r' P_1(\cos\alpha)\rho(\mathbf{r}')dv' \qquad (11.5.10)$$

and the quadrupole term:

$$V(\mathbf{r}) = \frac{1}{4\pi\varepsilon_0} \frac{1}{r^3} \int_{vol} r'^2 P_2(\cos\alpha)\rho(\mathbf{r}')dv' \qquad (11.5.11)$$

Note that an arbitrary, irregular charge distribution $\rho(\mathbf{r}')$ may result in monopole, dipole, quadrupole, and higher multipole terms where we may be only interested in calculating the first few terms. The monopole term is proportional to the total charge. Certain charge distributions may have only even or odd multipole terms.

11.5.2 Off-Axis Trick

For axially symmetric problems, the potential may first be obtained along the z-axis. The potential may then be obtained from points off the axis by making use of the addition theorem of spherical harmonics

$$P_\ell(\cos\gamma) = \frac{4\pi}{2\ell+1} \sum_{m=-\ell}^{\ell} Y_\ell^m(\theta,\phi) Y_\ell^m(\theta',\phi')^* \qquad (11.5.12)$$

or

$$P_\ell(\cos\gamma) = \sum_{m=-\ell}^{\ell} \frac{(\ell-m)!}{(\ell-m)!} P_\ell^m(\cos\theta) P_\ell^m(\cos\theta') e^{im(\phi-\phi')} \qquad (11.5.13)$$

With axial symmetry, we only have $m = 0$, $\cos\gamma \to \cos\alpha$ and

$$P_\ell(\cos\alpha) = P_\ell(\cos\theta) P_\ell(\cos\theta') \qquad (11.5.14)$$

Thus, the expression

$$\frac{1}{|\mathbf{r}-\mathbf{r}'|} = \frac{1}{r}\sum_{n=0}^{\infty}\left(\frac{r'}{r}\right)^n P_n(\cos\alpha) \tag{11.5.15}$$

may be written

$$\frac{1}{|\mathbf{r}-\mathbf{r}'|} = \frac{1}{r}\sum_{n=0}^{\infty}\left(\frac{r'}{r}\right)^n P_n(\cos\theta)P_n(\cos\theta') \tag{11.5.16}$$

We may now calculate the potential along the z-axis where $\theta = 0$

$$V(z) = \frac{1}{4\pi\varepsilon_0}\sum_{n=0}^{\infty}\frac{1}{r^{n+1}}\int_{\text{vol}} r'^n P_n(\cos\theta')\rho(\mathbf{r}')dv' \tag{11.5.17}$$

and then obtain the potential off-axis by multiplying each term in the expansion by $P_n(\cos\theta)$ or

$$V(\mathbf{r}) = \frac{q}{4\pi\varepsilon_0}\sum_{n=0}^{\infty}\left[\frac{1}{r^{n+1}}\int_{\text{vol}} r'^n P_n(\cos\theta')\rho(\mathbf{r}')dv'\right]P_n(\cos\theta) \tag{11.5.18}$$

Example 11.5.1

Calculate the potential due to a charged needle of length $2a$ oriented along the z-axis (Figure 11.5.1)

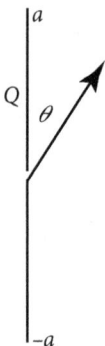

Figure 11.5.1: Needle with charge Q located along the z-axis.

Solution: In the special case where the charge is only oriented along the z-axis $P_n(\cos\alpha) = P_n(\cos\theta)$. Here we have $\mathbf{r}' = z$ with charge density

$$\rho(\mathbf{r}')dv \rightarrow \lambda dz = \frac{Q}{2a}dz \tag{11.5.19}$$

and potential

$$V(\mathbf{r}) = \frac{1}{4\pi\varepsilon_0} \sum_{n=0}^{\infty} \frac{1}{r^{n+1}} \int_{-a}^{a} z^n P_n(\cos\theta) \frac{Q}{2a} dz \tag{11.5.20}$$

Integrating over z

$$V(\mathbf{r}) = \frac{Q}{4\pi\varepsilon_0} \sum_{n=0}^{\infty} \frac{P_n(\cos\theta)}{r^{n+1}} \frac{1}{2a} \frac{a^{n+1} - (-a)^{n+1}}{n+1} \tag{11.5.21}$$

$a^{n+1} - (-a)^{n+1} = 2a^{n+1}$ for even n and we have

$$V(\mathbf{r}) = \frac{Q}{4\pi\varepsilon_0} \frac{1}{r} \sum_{n=0,2,4...}^{\infty} \frac{1}{n+1} \left(\frac{a}{r}\right)^n P_n(\cos\theta) \quad r > a \tag{11.5.22}$$

We can check that the $n = 0$ monopole term gives the potential of a point charge

$$V(r,\theta) = \frac{1}{4\pi\varepsilon_0} \frac{Q}{r} \tag{11.5.23}$$

Example 11.5.2

Calculate the potential due to a uniformly charged disk of radius R coaxial with the z-axis

Solution: In the special case where the charge is only in the $z = 0$ plane $\alpha = \pi/2$ and $P_n(\cos\alpha) = P_n(0)$, we first calculate the potential along the z-axis $\mathbf{r}' = r'\hat{\mathbf{r}}$ and $\mathbf{r} = z\hat{\mathbf{k}}$ so we have that

$$\rho(\mathbf{r}')dv \to \sigma(\mathbf{r}')da' = \frac{Q}{\pi R^2} 2\pi r' dr' \tag{11.5.24}$$

$$V(z) = \frac{1}{4\pi\varepsilon_0} \frac{Q}{\pi R^2} 2\pi \sum_{n=0}^{\infty} \frac{1}{z^{(n+1)}} P_n(0) \int_0^R (r')^{n+1} dr' \tag{11.5.25}$$

$$V(z) = \frac{1}{4\pi\varepsilon_0} \frac{2Q}{z} \sum_{n=0}^{\infty} \frac{1}{n+2} \left(\frac{R}{z}\right)^n P_n(0) \quad z > R \tag{11.5.26}$$

and the potential off the z-axis is then

$$V(r,\theta) = \frac{1}{4\pi\varepsilon_0} \frac{2Q}{r} \sum_{n=0}^{\infty} \frac{1}{n+2} \left(\frac{R}{r}\right)^n P_n(0) P_n(\cos\theta) \quad r > R \tag{11.5.27}$$

We can check that the $n = 0$ monopole term gives the potential of a point charge.

Example 11.5.3

Calculate the potential due to a uniformly charged ring of radius a coaxial with the z-axis

Solution: Expressing the charge density in spherical coordinates

$$\rho(\mathbf{r}') = \frac{q}{2\pi a^2}\delta(r'-a)\delta\left(\theta'-\frac{\pi}{2}\right) \tag{11.5.28}$$

the potential

$$V(\mathbf{r}) = \frac{1}{4\pi\varepsilon_0}\sum_{n=0}^{\infty}\frac{1}{r^{(n+1)}}\int_{\text{vol}}(r')^n P_n(\cos\alpha)\rho(\mathbf{r}')dv' \tag{11.5.29}$$

We first calculate the nonzero multipole terms of the potential along the z-axis $\alpha \to \theta'$. We then calculate the potential everywhere off the axis

$$V(\mathbf{r}) = \frac{1}{4\pi\varepsilon_0}\frac{q}{2\pi a^2}\sum_{n=0}^{\infty}\frac{1}{r^{(n+1)}}\int_0^{2\pi}d\phi\int_0^{\pi}P_n(\cos\theta')\sin\theta'\delta\left(\theta'-\frac{\pi}{2}\right)d\theta\int_0^{\infty}(r')^{n+2}\delta(r'-a)dr' \tag{11.5.30}$$

Performing the integrations

$$V(\mathbf{r}) = \frac{1}{4\pi\varepsilon_0}\frac{q}{r}\sum_{n=0}^{\infty}\left(\frac{a}{r}\right)^n P_n(0) \tag{11.5.31}$$

The potential off the axis is then

$$V(\mathbf{r},\theta) = \frac{1}{4\pi\varepsilon_0}\frac{q}{r}\sum_{n=0,2,4,\ldots}^{\infty}\left(\frac{a}{r}\right)^n P_n(0)P_n(\cos\theta) \tag{11.5.32}$$

Evidently all the odd multipole terms are zero since $P_n(0) = \{1, 0, -1/2, 0, 3/8, 0, \ldots\}$.

11.5.3 Asymmetric Potentials

To compute the multipole expansion of potentials without azimuthal symmetry we use the expansion

$$\frac{1}{|\mathbf{r}-\mathbf{r}'|} = \frac{1}{r}\sum_{\ell=0}^{\infty}\sum_{m=-\ell}^{\ell}\left(\frac{r'}{r}\right)^{\ell}\frac{4\pi}{2\ell+1}Y_{\ell}^m(\theta',\phi')^*Y_{\ell}^m(\theta,\phi) \tag{11.5.33}$$

and our integral for the potential

$$V(\mathbf{r}) = \frac{1}{4\pi\varepsilon_0}\frac{1}{r}\sum_{\ell=0}^{\infty}\frac{1}{r^{\ell}}\frac{4\pi}{2\ell+1}\sum_{m=-\ell}^{\ell}Y_{\ell}^m(\theta,\phi)\int_{\text{vol}}\rho(\mathbf{r}')r'^{\ell}Y_{\ell}^m(\theta',\phi')^*dv' \tag{11.5.34}$$

can be written

$$V(\mathbf{r}) = \frac{1}{4\pi\varepsilon_0} \frac{1}{r} \sum_{\ell=0}^{\infty} \frac{1}{r^\ell} \frac{4\pi}{2\ell+1} \sum_{m=-\ell}^{\ell} q_{\ell m} Y_\ell^m(\theta,\phi) \qquad (11.5.35)$$

where the coefficients

$$q_{\ell m} = \int_{\text{vol}} \rho(r',\theta',\phi') r'^\ell Y_\ell^m(\theta',\phi')^\star dv' \qquad (11.5.36)$$

are the multipole moments.

Maple Examples

The multipole expansion of the potential of a charged ring for points off the z-axis is calculated in the Maple worksheet below.

Key Maple terms: *expand, sum*

Maple packages: *with(orthopoly)*:

Programming: Function statements using '→'

Special functions: LegendreP

restart

Multipole Expansion: Charged Ring

with(orthopoly) :

$$V := (r, theta) \rightarrow \frac{q}{4 \cdot Pi \cdot epsilon \cdot r} \cdot sum\left(\left(\frac{a}{r}\right)^n P(n,0) \cdot P(n, \cos(theta)), n = 0\ldots 4\right);$$

$$V := (r,\theta) \mapsto \frac{q\left(\sum_{n=0}^{4}\left(\frac{a}{r}\right)^n P(n,0) P(n,\cos(\theta))\right)}{4\pi\varepsilon r}$$

V(r, theta)
expand(%)

$$\frac{q}{4\pi\varepsilon r} + \frac{qa^2}{16\pi\varepsilon r^3} - \frac{3qa^2\cos(\theta)^2}{16\pi\varepsilon r^3} + \frac{9qa^4}{256\pi\varepsilon r^5} + \frac{105qa^4\cos(\theta)^4}{256\pi\varepsilon r^5} - \frac{45qa^4\cos(\theta)^2}{128\pi\varepsilon r^5}$$

11.6 ELECTRICITY AND MAGNETISM

Computations in magnetostatics can be more challenging than electrostatics because magnetic sources are vectors while electrostatic sources are scalars. Calculations in magnetics often involve cross products and applications of the right-hand rule. Key integral theorems for electrostatics and magnetostatics are Gauss's divergence theorem and Stokes's theorem, respectively.

11.6.1 Comparison of Electrostatics and Magnetostatics

Table 11.6.1 provides a comparison of electric $\mathbf{E}$ and magnetic $\mathbf{B}$ fields, electric V and magnetic $(\mathbf{A}, \Omega)$ potentials, and electric $(q, \lambda, \sigma, \rho)$ and magnetic $(I, \mathbf{K}, \mathbf{J})$ sources. Analogous expressions for electric and magnetic forces and energies are also presented side by side.

TABLE 11.6.1: Comparison of static electrostatic and magnetostatic equations.

Electrostatics	Magnetostatics
Electric field $\mathbf{E}$ at $\mathbf{r}$ due to a point charge q located at $\mathbf{r}'$ $$\mathbf{E}(\mathbf{r}) = \frac{q}{4\pi\varepsilon_0} \frac{(\mathbf{r}-\mathbf{r}')}{\|\mathbf{r}-\mathbf{r}'\|^3}$$	Magnetic field at $\mathbf{r}$ due to a current element $I d\hat{\ell}$ located at $\mathbf{r}'$ $$\mathbf{B}(\mathbf{r}) = \frac{\mu_0}{4\pi} \frac{I(\mathbf{r}') d\hat{\ell} \times (\mathbf{r}-\mathbf{r}')}{\|\mathbf{r}-\mathbf{r}'\|^3}$$
Electric field due to a line charge λ (with units of C/m) $$\mathbf{E}(\mathbf{r}) = \frac{1}{4\pi\varepsilon_0} \int_\Gamma \frac{\lambda(\mathbf{r}')(\mathbf{r}-\mathbf{r}') d\ell}{\|\mathbf{r}-\mathbf{r}'\|^3}$$	Magnetic field due to a line current (with units of amperes) $$\mathbf{B}(\mathbf{r}) = \frac{\mu_0}{4\pi} \int_\Gamma \frac{I(\mathbf{r}') d\hat{\ell} \times (\mathbf{r}-\mathbf{r}')}{\|\mathbf{r}-\mathbf{r}'\|^3}$$
Electric field due to a surface charge density σ (with units of C/m^2) $$\mathbf{E}(\mathbf{r}) = \frac{1}{4\pi\varepsilon_0} \int_{surf} \frac{\sigma(\mathbf{r}')(\mathbf{r}-\mathbf{r}') da'}{\|\mathbf{r}-\mathbf{r}'\|^3}$$	Magnetic due to a surface current $\mathbf{K}$ (with units of A/m) $$\mathbf{B}(\mathbf{r}) = \frac{\mu_0}{4\pi} \int_{surf} \frac{\mathbf{K}(\mathbf{r}') \times (\mathbf{r}-\mathbf{r}') da'}{\|\mathbf{r}-\mathbf{r}'\|^3}$$
Electric field due to a volume charge density ρ (with units of C/m^3) $$\mathbf{E}(\mathbf{r}) = \frac{1}{4\pi\varepsilon_0} \int_{vol} \frac{\rho(\mathbf{r}')(\mathbf{r}-\mathbf{r}') dv'}{\|\mathbf{r}-\mathbf{r}'\|^3}$$	Magnetic field due to a volume current $\mathbf{J}$ (with units of A/m^2) $$\mathbf{B}(\mathbf{r}) = \frac{\mu_0}{4\pi} \int_{vol} \frac{\mathbf{J}(\mathbf{r}') \times (\mathbf{r}-\mathbf{r}') dv'}{\|\mathbf{r}-\mathbf{r}'\|^3}$$

Electrostatics	Magnetostatics
Integral form of Gauss's law $$\oint_{surf} \mathbf{E} \cdot \hat{\mathbf{n}} da = \frac{1}{\varepsilon_0} Q_{enclosed}$$	Integral form of Ampere's law $$\oint_{\Gamma} \mathbf{B} \cdot d\hat{\ell} = \mu_0 I$$
Differential form of Gauss's law $$\nabla \cdot \mathbf{E} = \frac{\rho}{\varepsilon_0}$$	Differential form of Ampere's law $$\nabla \times \mathbf{B} = \mu_0 \mathbf{J}$$
Electric field from the electric potential $$\mathbf{E}(\mathbf{r}) = -\nabla V(\mathbf{r})$$	Magnetic field from the vector magnetic potential $$\mathbf{B}(\mathbf{r}) = \nabla \times \mathbf{A}(\mathbf{r})$$
Integral computation of electric potential $$V(\mathbf{r}) = \frac{1}{4\pi\varepsilon_0} \int_{vol} \frac{\rho(\mathbf{r}')dv'}{|\mathbf{r} - \mathbf{r}'|}$$	Integral computation of vector magnetic potential $$\mathbf{A}(\mathbf{r}) = \frac{\mu_0}{4\pi} \int_{vol} \frac{\mathbf{J}(\mathbf{r}')dv'}{|\mathbf{r} - \mathbf{r}'|}$$
Multipole expansion of electric potential $$V(\mathbf{r}) = \frac{1}{4\pi\varepsilon_0} \sum_{n=0}^{\infty} \frac{1}{r^{n+1}} \int_{vol} r'^n P_n(\cos\alpha)\rho(\mathbf{r}')dv'$$ where α is the angle between $\mathbf{r}$ and $\mathbf{r}'$	Multipole expansion of vector potential $$\mathbf{A}(\mathbf{r}) = \frac{\mu_0}{4\pi} \sum_{n=0}^{\infty} \frac{1}{r^{n+1}} \int_{vol} r'^n P_n(\cos\alpha)\mathbf{J}(\mathbf{r}')dv'$$ where α is the angle between $\mathbf{r}$ and $\mathbf{r}'$
Poisson's equation for the electric potential $$\nabla^2 V(\mathbf{r}) = -\frac{\rho(\mathbf{r})}{\varepsilon_0}$$	Poisson's equation for the vector magnetic potential $$\nabla^2 \mathbf{A}(\mathbf{r}) = -\mu_0 \mathbf{J}(\mathbf{r})$$
Laplace's equation for the electric potential $\nabla^2 V = 0$ where $\mathbf{E} = -\nabla V$	Laplace's equation for the scalar magnetic potential $\nabla^2 \Omega = 0$ where $\mathbf{H} = -\nabla\Omega$
Force on a point charge in an electric field $$\mathbf{F} = q\mathbf{E}$$	Force on a current element in a magnetic field $$\mathbf{F} = \mathbf{I} \times \mathbf{B}d\ell$$
Force between two point charges $$\mathbf{F} = \frac{q_1 q_2}{4\pi\varepsilon_0} \frac{(\mathbf{r}_1 - \mathbf{r}_2)}{|\mathbf{r}_1 - \mathbf{r}_2|^3}$$	Force between two current loops $$\mathbf{F} = \frac{\mu_0}{4\pi} I_1 I_2 \oint_{\Gamma_1} \oint_{\Gamma_2} \frac{d\hat{\ell}_2 \times d\hat{\ell}_1 \times (\mathbf{r}_1 - \mathbf{r}_2)}{|\mathbf{r}_1 - \mathbf{r}_2|^3}$$
Electric energy in terms of ρ and V $$W_E = \frac{1}{2} \int_{vol} \rho V dv$$	Magnetic energy in terms of $\mathbf{J}$ and $\mathbf{A}$ $$W_B = \frac{1}{2} \int_{vol} \mathbf{J} \cdot \mathbf{A} dv$$

11.6.2 Electrostatic Examples

Example 11.6.1

Calculate the electrostatic energy stored in a collection of point charges and develop an integral expression for the energy stored in a continuous distribution of charge.

Solution: The potential a distance r from a single point charge q is

$$V(r) = \frac{1}{4\pi\varepsilon_0}\frac{q}{r} \qquad (11.6.1)$$

The electrostatic energy of two point charges q_1 and q_2 separated by r_{12} is

$$W_E = \frac{1}{4\pi\varepsilon_0}\frac{q_1 q_2}{r_{12}} \qquad (11.6.2)$$

For a collection of N point charges the electrostatic energy is

$$W_E = \frac{1}{4\pi\varepsilon_0}\frac{1}{2}\sum_{\substack{i,j \\ i\neq j}}^{N}\frac{q_i q_j}{r_{ij}} \qquad (11.6.3)$$

where the factor of $1/2$ accounts for repeated terms such as $q_1 q_2/r_{12}$ and $q_2 q_1/r_{21}$. Factoring the q_i

$$W_E = \frac{1}{2}\sum_i^N q_i \sum_j^N \frac{1}{4\pi\varepsilon_0}\frac{q_j}{r_{ij}} \qquad (11.6.4)$$

we identify the sum over j as $V(\mathbf{r}_i)$ so that

$$W_E = \frac{1}{2}\sum_i^N q_i V(\mathbf{r}_i) \qquad (11.6.5)$$

For a continuous charge distribution, we make the replacements $q_j \rightarrow \rho dv$ and $\Sigma \rightarrow \int$ so that

$$W_E = \int_{\text{vol}} \rho(\mathbf{r}')V(\mathbf{r}')dv' \qquad (11.6.6)$$

Example 11.6.2

From the integral expression for the electrostatic energy stored in a continuous charge distribution show that

$$W_E = \varepsilon_0 \int_{\text{vol}} E^2(\mathbf{r}')dv' \qquad (11.6.7)$$

where the integration volume is over all space.

Solution: Using Maxwell's equation $\nabla \cdot \mathbf{E} = \rho / \varepsilon_0$ the integral is transformed as

$$W_E = \varepsilon_0 \int_{vol} \nabla \cdot \mathbf{E} V dv' \tag{11.6.8}$$

Using the identity for the divergence of a scalar function times a vector function

$$\nabla \cdot (\mathbf{E}V) = V \nabla \cdot \mathbf{E} + \mathbf{E} \cdot \nabla V \tag{11.6.9}$$

we break the integral in two parts

$$W_E = \varepsilon_0 \int_{vol} \nabla \cdot (\mathbf{E}V) dv' - \varepsilon_0 \int_{vol} \mathbf{E} \cdot \nabla V dv' \tag{11.6.10}$$

Applying Gauss's divergence theorem to the first integral and using $\nabla V = -\mathbf{E}$ in the second integral gives

$$W_E = \varepsilon_0 \int_{surf} (\mathbf{E}V) \cdot da' + \varepsilon_0 \int_{vol} \mathbf{E}^2 dv' \tag{11.6.11}$$

For large surface areas $a' \sim r^2$ the surface integral goes to zero as $\sim 1/r$ since $\mathbf{E} \sim 1/r^2$ and $V \sim 1/r$ so we have that

$$W_E = \varepsilon_0 \int_{vol} \mathbf{E}^2 dv' \tag{11.6.12}$$

11.6.3 Magnetostatic Examples

Example 11.6.3

Find the magnetic field along the z-axis of a current loop of radius R carrying a current I. Next find an integral expression for the field along the axis of a washer of inner radius a and outer radius b carrying a uniform current density K.

Solution: The Biot-Savart law for a line current

$$\mathbf{B}(\mathbf{r}) = \frac{\mu_0}{4\pi} \oint \frac{I(\mathbf{r}') d\hat{\ell} \times (\mathbf{r} - \mathbf{r}')}{|\mathbf{r} - \mathbf{r}'|^3} \tag{11.6.13}$$

With the plane of the current loop located at $z = 0$ we have that $Id\hat{\ell} = IRd\phi\hat{\phi}$ and $\mathbf{r} - \mathbf{r}' = z\hat{\mathbf{z}} - R\hat{\mathbf{r}}$

$$Id\hat{\ell} \times (\mathbf{r} - \mathbf{r}') = IRd\phi\hat{\phi} \times (z\hat{\mathbf{z}} - R\hat{\mathbf{r}}) = Izd\phi\hat{\mathbf{r}} + IRd\phi\hat{\mathbf{z}} \tag{11.6.14}$$

The radial component of the magnetic field along the z-axis is zero from symmetry and we only must integrate over ϕ to obtain

$$\mathbf{B}(\mathbf{r}) = \frac{\mu_0}{4\pi} I R^2 \hat{\mathbf{z}} \int_0^{2\pi} \frac{d\phi}{\left(z^2 + R^2\right)^{3/2}} = \frac{\mu_0 I}{2} \frac{R^2}{\left(z^2 + R^2\right)^{3/2}} \hat{\mathbf{z}} \qquad (11.6.15)$$

and

$$B_z = \frac{\mu_0 I}{2} \frac{R^2}{\left(z^2 + R^2\right)^{3/2}} \qquad (11.6.16)$$

For a washer of inner radius a and outer radius b carrying a uniform current density K the magnetic field along the z-axis is

$$B_z = \frac{\mu_0 K}{2} \int_a^b \frac{r^3 dr}{\left(z^2 + r^2\right)^{3/2}} \qquad (11.6.17)$$

Example 11.6.4

Find the vector potential inside and outside of a cylindrical solenoid of radius R carrying a current I with turns per unit length $n = N/L$. The axis of the solenoid coincides with the z-axis.

Solution: The uniform magnetic field inside the solenoid is $\mathbf{B} = \mu_0 n I \hat{\mathbf{z}}$ and $\mathbf{B} = 0$ outside the solenoid. Since $\mathbf{B} = \nabla \times \mathbf{A}$ we can express the magnetic flux inside the solenoid for $r < R$ in terms of A from Stokes's theorem

$$\int \mathbf{B} \cdot \hat{\mathbf{n}} da = \oint \mathbf{A} \cdot d\hat{\ell} \qquad (11.6.18)$$

where $\mathbf{A} = A_\phi \hat{\phi}$ so that

$$\mu_0 n I \left(\pi r^2\right) = A_\phi \left(2\pi r\right) \qquad (11.6.19)$$

Thus, the vector potential inside the solenoid

$$\mathbf{A} = \frac{\mu_0 n I}{2} r \hat{\phi} \qquad (11.6.20)$$

is zero at the center and increases linearly with r. Now to get the vector potential outside the solenoid $r > R$ we have the total flux through a circular contour of radius r as $B(\pi R^2)$. Thus

$$\mu_0 n I \left(\pi R^2\right) = A_\phi \left(2\pi r\right) \qquad (11.6.21)$$

and we see that the vector potential goes as $1/r$ outside the solenoid

$$\mathbf{A} = \frac{\mu_0 n I}{2} \frac{R^2}{r} \hat{\phi} \qquad (11.6.22)$$

Example 11.6.5

From the integral formula for the vector potential of an arbitrary current distribution $\mathbf{J}(\mathbf{r}')$

$$\mathbf{A}(\mathbf{r}) = \frac{\mu_0}{4\pi} \int_{vol} \frac{\mathbf{J}(\mathbf{r}')}{|\mathbf{r}-\mathbf{r}'|} dv' \qquad (11.6.23)$$

calculate the first three multipole moments.

Solution: The integral is only over primed coordinates so that we may factor out the sum

$$\mathbf{A}(\mathbf{r}) = \frac{\mu_0}{4\pi} \sum_{n=0}^{\infty} \frac{1}{r^{n+1}} \int_{vol} r'^n P_n(\cos\alpha) \mathbf{J}(\mathbf{r}') dv' \qquad (11.6.24)$$

The $n = 0$ monopole moment

$$\mathbf{A}(\mathbf{r}) = \frac{\mu_0}{4\pi} \frac{1}{r} \int_{vol} \mathbf{J}(\mathbf{r}') dv' \qquad (11.6.25)$$

will be equal to zero for physical current distributions. For $n = 1$ we have the dipole moment

$$\mathbf{A}(\mathbf{r}) = \frac{\mu_0}{4\pi} \frac{1}{r^2} \int_{vol} r' P_1(\cos\alpha) \mathbf{J}(\mathbf{r}') dv' \qquad (11.6.26)$$

and the quadrupole moment

$$\mathbf{A}(\mathbf{r}) = \frac{\mu_0}{4\pi} \frac{1}{r^3} \int_{vol} r'^2 P_2(\cos\alpha) \mathbf{J}(\mathbf{r}') dv' \qquad (11.6.27)$$

for $n = 2$.

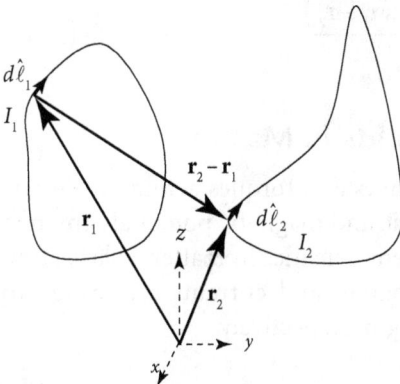

Figure 11.6.1: Two current loops carrying currents I_1 and I_2

Example 11.6.6

Show that the force between two current loops (Figure 11.6.1) is

$$
\mathbf{F} = \frac{\mu_0}{4\pi} I_1 I_2 \oint_{\Gamma_1} \oint_{\Gamma_2} \frac{d\hat{\boldsymbol{\ell}}_2 \times d\hat{\boldsymbol{\ell}}_1 \times (\mathbf{r}_1 - \mathbf{r}_2)}{\left|\mathbf{r}_1 - \mathbf{r}_2\right|^3}
\tag{11.6.28}
$$

Solution: The magnetic field at the point on coil 2 located at $\mathbf{r}_2$ due to a current element at $\mathbf{r}_1$ is

$$
d\mathbf{B}_1(\mathbf{r}_2) = \frac{\mu_0}{4\pi} I_1 \frac{d\hat{\boldsymbol{\ell}}_1 \times (\mathbf{r}_1 - \mathbf{r}_2)}{\left|\mathbf{r}_1 - \mathbf{r}_2\right|^3}
\tag{11.6.29}
$$

The total field at $\mathbf{r}_2$ due to coil 1 is then

$$
\mathbf{B}_1(\mathbf{r}_2) = \frac{\mu_0}{4\pi} I_1 \oint_{\Gamma_1} \frac{d\hat{\boldsymbol{\ell}}_1 \times (\mathbf{r}_1 - \mathbf{r}_2)}{\left|\mathbf{r}_1 - \mathbf{r}_2\right|^3}
\tag{11.6.30}
$$

Now the force on $d\hat{\boldsymbol{\ell}}_2$ due to $\mathbf{B}_1(\mathbf{r}_2)$ is

$$
d\mathbf{F} = I_2 d\hat{\boldsymbol{\ell}}_2 \times \mathbf{B}_1(\mathbf{r}_2)
\tag{11.6.31}
$$

and the total force on coil 2 is

$$
\mathbf{F} = I_2 \int_{\Gamma_2} d\hat{\boldsymbol{\ell}}_2 \times \mathbf{B}_1(\mathbf{r}_2)
\tag{11.6.32}
$$

and we obtain

$$
\mathbf{F} = \frac{\mu_0}{4\pi} I_1 I_2 \oint_{\Gamma_1} \oint_{\Gamma_2} \frac{d\hat{\boldsymbol{\ell}}_2 \times d\hat{\boldsymbol{\ell}}_1 \times (\mathbf{r}_1 - \mathbf{r}_2)}{\left|\mathbf{r}_1 - \mathbf{r}_2\right|^3}
\tag{11.6.33}
$$

11.6.4 Static Electric and Magnetic Fields in Matter

Table 11.6.2 provides a comparison of the forces and torques acting on electric and magnetic dipoles in external fields. Electric and magnetic potentials are then compared for single dipoles and dipole moment densities in matter. Polarization and magnetization resulting from bound charges and currents are related to electric displacement and magnetic field strength, respectively.

TABLE 11.6.2: Comparison of static electrostatic and magnetostatic equations in matter.

Electric Fields in Matter	Magnetic Fields in Matter
Electric dipole moment $\mathbf{p} = q\mathbf{d}$ where $\mathbf{d}$ points from $-q$ to q	Magnetic dipole moment $\mathbf{m} = I\mathbf{a}$ where $\mathbf{a}$ is the area bound by the current $\mathbf{I}$
Torque τ on an electric dipole with moment $\mathbf{p}$ in an electric field $\mathbf{E}$ $\tau = \mathbf{p} \times \mathbf{E}$	Torque τ on a magnetic dipole with moment $\mathbf{m}$ in a magnetic field $\mathbf{B}$ $\tau = \mathbf{m} \times \mathbf{B}$
Force on a dipole in a nonuniform field $\mathbf{F} = (\mathbf{p} \cdot \nabla)\mathbf{E}$	Force on $\mathbf{m}$ in a nonuniform $\mathbf{B}$ field $\mathbf{F} = \nabla(\mathbf{m} \cdot \mathbf{B})$
Energy of an electric dipole in an electric field $W_E = -\mathbf{p} \cdot \mathbf{E}$	Energy of a magnetic dipole in a magnetic field $W_B = -\mathbf{m} \cdot \mathbf{B}$
Potential at $\mathbf{r}$ due to a dipole $\mathbf{p}$ located at $\mathbf{r}'$ $V(\mathbf{r}) = \dfrac{1}{4\pi\varepsilon_0}\dfrac{\mathbf{p} \cdot (\mathbf{r} - \mathbf{r}')}{\|\mathbf{r} - \mathbf{r}'\|^3}$	Vector potential $\mathbf{A}(\mathbf{r})$ due to mat $\mathbf{r}'$ $\mathbf{A}(\mathbf{r}) = \dfrac{\mu_0}{4\pi}\dfrac{\mathbf{m} \times (\mathbf{r} - \mathbf{r}')}{\|\mathbf{r} - \mathbf{r}'\|^3}$
Potential at $\mathbf{r}$ due to an object with dipole moment density $\mathbf{P}(\mathbf{r}')$ where $\mathbf{p} = \mathbf{P}dv'$ $V(\mathbf{r}) = \dfrac{1}{4\pi\varepsilon_0}\displaystyle\int_{\text{vol}}\dfrac{\mathbf{P}(\mathbf{r}') \cdot (\mathbf{r} - \mathbf{r}')}{\|\mathbf{r} - \mathbf{r}'\|^3}dv'$	Vector potential due to an object with dipole moment density $\mathbf{M}(\mathbf{r}')$ where $\mathbf{m} = \mathbf{M}dv'$ $\mathbf{A}(\mathbf{r}) = \dfrac{\mu_0}{4\pi}\displaystyle\int_{\text{vol}}\dfrac{\mathbf{M}(\mathbf{r}') \times (\mathbf{r} - \mathbf{r}')}{\|\mathbf{r} - \mathbf{r}'\|^3}dv'$
The volume integral for the potential is transformed as $V(\mathbf{r}) = \dfrac{1}{4\pi\varepsilon_0}\displaystyle\int_{\text{surf}}\dfrac{\sigma_b(\mathbf{r}')}{\|\mathbf{r} - \mathbf{r}'\|}da' +$ $\dfrac{1}{4\pi\varepsilon_0}\displaystyle\int_{\text{vol}}\dfrac{\rho_b(\mathbf{r}')}{\|\mathbf{r} - \mathbf{r}'\|}dv'$ where $\sigma_b = \mathbf{P} \cdot \hat{\mathbf{n}}$ and $\rho_b = -\nabla \cdot \mathbf{P}$	The volume integral for the vector potential is transformed as $\mathbf{A}(\mathbf{r}) = \dfrac{\mu_0}{4\pi}\displaystyle\int_{\text{surf}}\dfrac{\mathbf{K}_b(\mathbf{r}')}{\|\mathbf{r} - \mathbf{r}'\|}da'$ $+ \dfrac{\mu_0}{4\pi}\displaystyle\int_{\text{vol}}\dfrac{\mathbf{J}_b(\mathbf{r}')}{\|\mathbf{r} - \mathbf{r}'\|}dv'$ where $\mathbf{J}_b = \nabla \times \mathbf{M}$ and $\mathbf{K}_b = \mathbf{M} \times \hat{\mathbf{n}}$
The total charge density is a sum of bound and free charge densities $\rho = \rho_b + \rho_{\text{free}}$	Total current is a sum of bound and free current densities $\mathbf{J} = \mathbf{J}_b + \mathbf{J}_{\text{free}}$
Using the differential form of Gauss's law and the definition of the polarization charge density $\varepsilon_0 \nabla \cdot \mathbf{E} = -\nabla \cdot \mathbf{P} + \rho_{\text{free}}$	Using the differential form of Ampere's law and the definition of the bound current density $\nabla \times \mathbf{B} / \mu_0 = \nabla \times \mathbf{M} + \mathbf{J}_{\text{free}}$

(contd.)

Electric Fields in Matter	Magnetic Fields in Matter
Factoring the divergence terms on the left $$\nabla\cdot\left(\varepsilon_0\mathbf{E}+\mathbf{P}\right)=\rho_{free}$$	Factoring the curl terms on the left $$\nabla\times\left(\mathbf{B}/\mu_0-\mathbf{M}\right)=\mathbf{J}_{free}$$
we identify the electric displacement (with units of C/m²) $$\mathbf{D}=\varepsilon_0\mathbf{E}+\mathbf{P}$$	we identify the magnetic field strength (with units of A/m) $$\mathbf{H}=\mathbf{B}/\mu_0-\mathbf{M}$$
with divergence $$\nabla\cdot\mathbf{D}=\rho_{free}$$	with curl $$\nabla\times\mathbf{H}=\mathbf{J}_{free}$$
Gauss's law for dielectrics: $\mathbf{D}$ is determined by the free charge density. $$\oint_{surf}\mathbf{D}\cdot\hat{\mathbf{n}}da=Q_{free}$$	Ampere's law for magnetic materials: $\mathbf{H}$ is determined by the free current density. $$\oint_{\Gamma}\mathbf{H}\cdot d\hat{\mathbf{l}}=I_{free}$$
For linear dielectrics $\mathbf{P}$ is proportional to $\mathbf{E}$ $$\mathbf{P}=\varepsilon_0\chi_e\mathbf{E}$$	For linear magnetics $\mathbf{M}$ is proportional to $\mathbf{H}$ $$\mathbf{M}=\chi_m\mathbf{H}$$
where χ_e is the electric susceptibility. The electric displacement now becomes $$\mathbf{D}=\varepsilon_0(1+\chi_e)\mathbf{E}$$ with relative permittivity $\varepsilon_r=(1+\chi_e)$ and absolute permittivity $\varepsilon=\varepsilon_0\varepsilon_r$	where χ_m is the magnetic susceptibility. The magnetic flux density is thus $$\mathbf{B}=\mu_0(1+\chi_m)\mathbf{H}$$ with relative permeability $\mu_r=(1+\chi_m)$ and absolute permeability $\mu=\mu_0\mu_r$
The energy stored in an electric field $$W_E=\frac{1}{2}\int_{vol}\mathbf{E}\cdot\mathbf{D}dv$$	The energy stored in a magnetic field $$W_B=\frac{1}{2}\int_{vol}\mathbf{B}\cdot\mathbf{H}dv$$
Surface charge density $$\sigma=-\varepsilon_0\frac{\partial V}{\partial n}$$ where n is the coordinate normal to the surface.	Surface current density $$\mathbf{K}=\mathbf{H}_{tan}$$ in units of A/m².

Boundary conditions at a charged surface
  Potential V is continuous
  Normal component of $\mathbf{D}$ is discontinuous
  Tangential component of $\mathbf{E}$ is continuous
	Boundary conditions at a surface current
	  Vector potential $\mathbf{A}$ is continuous
	  Normal component of $\mathbf{B}$ is continuous
	  Tangential component of $\mathbf{H}$ is discontinuous

11.6.5 Examples: Electrostatic Fields in Matter

Example 11.6.7

A cube of side a located in the first octant of the Cartesian coordinate system has polarization $\mathbf{P}=P_0\left(\dfrac{z}{a}\right)\hat{\mathbf{k}}$. Calculate the total surface charge and volume charge of the polarized cube.

Solution: The bound charge density is $\sigma_b = \mathbf{P} \cdot \hat{\mathbf{n}} = P_0$ on the top of the cube where $\hat{\mathbf{n}} = \hat{\mathbf{k}}$, $\sigma_b = 0$ on the four sides of the cube where $\mathbf{P} \perp \hat{\mathbf{n}}$ and $\sigma_b = 0$ on the bottom where $z = 0$. The charge on top of the cube is

$$Q_{\text{top}} = \int\limits_{\text{surf}} \sigma_b da = \int\limits_0^a \int\limits_0^a P_0 dx dy = P_0 a^2 \qquad (11.6.34)$$

The bound volume charge density is $\rho_b = -\nabla \cdot \mathbf{P} = -P_0/a$. The charge inside the cube is

$$Q_{\text{inside}} = \int\limits_{\text{vol}} \rho_b dv = -\frac{P_0}{a} \int\limits_0^a \int\limits_0^a \int\limits_0^a dx dy dz = -P_0 a^2 \qquad (11.6.35)$$

Thus, the total charge is $Q_{\text{top}} + Q_{\text{inside}} = 0$.

Example 11.6.8

A cube of side a in the first octant has a constant polarization $\mathbf{P} = P_0 \hat{\mathbf{k}}$. Calculate the surface charge and volume charge of the polarized cube

Solution: The bound charge density is $\sigma_b = \mathbf{P} \cdot \hat{\mathbf{n}} = P_0$ on the top of the cube where $\hat{\mathbf{n}} = \hat{\mathbf{k}}$, $\sigma_b = 0$ on the four sides of the cube where $\mathbf{P} \perp \hat{\mathbf{n}}$ and $\sigma_b = -P_0$ on the bottom of the cube where $\hat{\mathbf{n}} = -\hat{\mathbf{k}}$. The total charge on top of the cube is

$$Q_{\text{top}} = \int\limits_{\text{surf}} \sigma_b da = \int\limits_0^a \int\limits_0^a P_0 dx dy = P_0 a^2 \qquad (11.6.36)$$

and on the bottom of the cube

$$Q_{\text{bottom}} = -\int\limits_0^a \int\limits_0^a P_0 dx dy = -P_0 a^2 \qquad (11.6.37)$$

The bound volume charge density is $\rho_b = -\nabla \cdot \mathbf{P} = 0$. The charge inside the cube is

$$Q_{\text{inside}} = \int\limits_{\text{vol}} \rho_b dv = 0 \qquad (11.6.38)$$

Thus, the total charge is $Q_{\text{top}} + Q_{\text{inside}} + Q_{\text{bottom}} = 0$.

Example 11.6.9

A sphere of radius R carries a total free charge Q. The sphere is covered by a dielectric layer of thickness d. Find $\mathbf{D}$, $\mathbf{E}$, and $\mathbf{P}$ inside the dielectric. Calculate the bound charge density on the inner and outer surfaces and in the volume of the dielectric layer.

Solution: Integrating the differential form of Maxwell's equation inside a dielectric

$$\int_{\text{vol}} \nabla \cdot \mathbf{D} dv = \int_{\text{vol}} \rho_{\text{free}} dv \tag{11.6.39}$$

and applying Gauss's divergence theorem

$$\int_{\text{surf}} \mathbf{D} \cdot \hat{\mathbf{n}} da = Q \tag{11.6.40}$$

The electric displacement is obtained by choosing a spherical Gaussian surface inside the dielectric $\mathbf{D} \perp \hat{\mathbf{n}}$ where $\hat{\mathbf{n}} = \hat{\mathbf{r}}$ so that

$$D_r \left(4\pi r^2 \right) = Q \tag{11.6.41}$$

where r is measured from the center of the sphere. Thus, we have

$$\mathbf{D} = \frac{Q}{4\pi r^2} \hat{\mathbf{r}} \tag{11.6.42}$$

Since the electric displacement $\mathbf{D} = \varepsilon_r \varepsilon_0 \mathbf{E}$ we have

$$\mathbf{E} = \frac{Q}{4\pi \varepsilon_0 \varepsilon_r r^2} \hat{\mathbf{r}} \tag{11.6.43}$$

and the polarization $\mathbf{P} = \varepsilon_0 \chi_e \mathbf{E}$

$$\mathbf{P} = \frac{\chi_e Q}{4\pi \varepsilon_r r^2} \hat{\mathbf{r}} \tag{11.6.44}$$

Now the bound charge density on the outer surface of the dielectric is

$$\sigma_{\text{outer}} = \mathbf{P} \cdot \hat{\mathbf{n}} = \frac{\chi_e Q}{4\pi \varepsilon_r \left(R + d \right)^2} \tag{11.6.45}$$

while on the inner surface $\hat{\mathbf{n}} = -\hat{\mathbf{r}}$

$$\sigma_{\text{inner}} = \mathbf{P} \cdot \hat{\mathbf{n}} = -\frac{\chi_e Q}{4\pi \varepsilon_r R^2} \tag{11.6.46}$$

In the volume of the dielectric

$$\rho_b = -\nabla \cdot \mathbf{P} = -\frac{1}{r^2} \frac{\partial}{\partial r} \left(r^2 \frac{\chi_e Q}{4\pi \varepsilon_r r^2} \right) = 0 \tag{11.6.47}$$

Thus, the total charge on the dielectric layer is

$$4\pi \left(R + d \right)^2 \sigma_{\text{outer}} + 4\pi R^2 \sigma_{\text{inner}} = 0 \tag{11.6.48}$$

11.6.6 Examples: Magnetic Fields in Matter

Example 11.6.10

Given the vector potential at $\mathbf{r}$ due to a magnetic dipole located at $\mathbf{r}'$ is

$$A(\mathbf{r}) = \frac{\mu_0}{4\pi} \frac{\mathbf{m}(\mathbf{r}') \times (\mathbf{r} - \mathbf{r}')}{|\mathbf{r} - \mathbf{r}'|^3} \qquad (11.6.49)$$

write an integral expression for the vector potential due to a body with a magnetic dipole per unit volume $\mathbf{M}(\mathbf{r}')$ where $\mathbf{m}(\mathbf{r}') = \mathbf{M}(\mathbf{r}')dv'$. Identify expressions for the bound surface and bound volume currents.

Solution: We have a volume integral for the vector potential due to a continuous magnetization distribution

$$A(\mathbf{r}) = \frac{\mu_0}{4\pi} \int_{vol} \frac{\mathbf{M}(\mathbf{r}') \times (\mathbf{r} - \mathbf{r}')}{|\mathbf{r} - \mathbf{r}'|^3} dv' \qquad (11.6.50)$$

As before $\nabla \dfrac{1}{|\mathbf{r}-\mathbf{r}'|} = -\dfrac{(\mathbf{r}-\mathbf{r}')}{|\mathbf{r}-\mathbf{r}'|^3}$ while $\nabla' \dfrac{1}{|\mathbf{r}-\mathbf{r}'|} = \dfrac{(\mathbf{r}-\mathbf{r}')}{|\mathbf{r}-\mathbf{r}'|^3}$ so

$$A(\mathbf{r}) = \frac{\mu_0}{4\pi} \int_{vol} \mathbf{M}(\mathbf{r}') \times \nabla' \frac{1}{|\mathbf{r}-\mathbf{r}'|} dv' \qquad (11.6.51)$$

Making use of our product rule involving the curl of a scalar times a vector

$$\nabla' \times \left[\mathbf{M}(\mathbf{r}') \frac{1}{|\mathbf{r}-\mathbf{r}'|} \right] = \frac{1}{|\mathbf{r}-\mathbf{r}'|} \nabla' \times \mathbf{M}(\mathbf{r}') - \mathbf{M}(\mathbf{r}') \times \nabla' \frac{1}{|\mathbf{r}-\mathbf{r}'|} \qquad (11.6.52)$$

we write

$$A(\mathbf{r}) = \frac{\mu_0}{4\pi} \left[\int_{vol} \frac{1}{|\mathbf{r}-\mathbf{r}'|} \nabla' \times \mathbf{M}(\mathbf{r}') dv' - \int_{vol} \nabla' \times \frac{\mathbf{M}(\mathbf{r}')}{|\mathbf{r}-\mathbf{r}'|} dv' \right] \qquad (11.6.53)$$

The second integral is transformed into a surface integral

$$A(\mathbf{r}) = \frac{\mu_0}{4\pi} \int_{vol} \frac{\nabla' \times \mathbf{M}(\mathbf{r}')}{|\mathbf{r}-\mathbf{r}'|} dv' + \frac{\mu_0}{4\pi} \oint_{surf} \frac{\mathbf{M}(\mathbf{r}') \times da'}{|\mathbf{r}-\mathbf{r}'|} \qquad (11.6.54)$$

where $da' = \hat{\mathbf{n}} da'$. From these integrals, we identify the bound surface current density $\mathbf{K}_b = \mathbf{M} \times \hat{\mathbf{n}}$ and the bound volume current density $\mathbf{J}_b = \nabla \times \mathbf{M}$.

Example 11.6.11

A cylindrical wire of radius R carries a total free current I. The wire is coated by a layer with relative permeability μ_r and thickness d. Calculate $\mathbf{H}, \mathbf{B}, \mathbf{M}$ and the total energy inside the permeable layer.

Solution: Applying Ampere's law

$$\oint \mathbf{H} \cdot d\hat{\ell} = I \tag{11.6.55}$$

for a circular contour inside the permeable coating $R \le r \le R + d$ gives

$$H(2\pi r) = I \tag{11.6.56}$$

so that

$$\mathbf{H} = \frac{I}{2\pi r} \hat{\phi} \tag{11.6.57}$$

The magnetic field is then

$$\mathbf{B} = \mu_r \mu_0 \mathbf{H} = \frac{\mu_r \mu_0 I}{2\pi r} \hat{\phi} \tag{11.6.58}$$

and the magnetization

$$\mathbf{M} = \frac{\mathbf{B}}{\mu_0} - \mathbf{H} = \left(\mu_r - 1\right) \frac{I}{2\pi r} \hat{\phi} \tag{11.6.59}$$

Notice that $\mathbf{M} = 0$ if $\mu_r = 1$ (air). Now the total magnetic energy is

$$
\begin{aligned}
W_B &= \frac{1}{2} \int_{vol} \mathbf{B} \cdot \mathbf{H} dv \\
&= \frac{1}{2} \int_0^L \int_R^{R+d} \left(\frac{\mu_r \mu_0 I}{2\pi r}\right) \left(\frac{I}{2\pi r}\right) 2\pi r dr dz \\
&= \frac{\mu_0}{4\pi} L I^2 \ln\left(1 + \frac{d}{R}\right)
\end{aligned}
\tag{11.6.60}
$$

Example 11.6.12

A parallel plate conductor carries a current $K\hat{\mathbf{k}}$ on the top plate and $-K\hat{\mathbf{k}}$ on the bottom plate. The space between the plates is filled by a layer with relative permeability μ_r. Calculate $\mathbf{H}, \mathbf{B}$, and $\mathbf{M}$ inside the iron layer.

Solution: Ampere's law

$$\oint \mathbf{H} \cdot d\hat{\ell} = I \tag{11.6.61}$$

for a rectangular contour saddling the top plate gives $HL = I$. Since $K = I/L$ the magnetic field strength is

$$\mathbf{H} = K\hat{\mathbf{i}} \qquad (11.6.62)$$

with magnetic field

$$\mathbf{B} = \mu_r \mu_0 \mathbf{H} = \mu_r \mu_0 K\hat{\mathbf{i}} \qquad (11.6.63)$$

and magnetization

$$\mathbf{M} = \frac{\mathbf{B}}{\mu_0} - \mathbf{H} = (\mu_r - 1)K\hat{\mathbf{i}} \qquad (11.6.64)$$

Maple Examples

The electric potential and field of a line charge and the vector potential and magnetic field of a finite line current is calculated in the Maple worksheet below.

Key Maple terms: *Curl, Gradient, Int, limit, Norm, simplify, value*

Maple packages: *with(Physics[Vectors])*:

restart

with(Physics[Vectors])

> [&x, '+', '.', *ChangeBasis, ChangeCoordinates, Component, Curl, DirectionalDiff, Divergence, Gradient, Identify, Laplacian, Nabla, Norm, Setup, diff*]

Setup(mathematicalnotation = true)

[mathematicalnotation = true]

Finite Line Charge

r_ := x·_i + y·_j + z·_k

$$\vec{r} := x\hat{i} + y\hat{j} + z\hat{k}$$

rp_ := zp·_k

$$\vec{rp} := zp\hat{k}$$

lambda := $\dfrac{Q}{L}$

$$\lambda := \frac{Q}{L}$$

$$V := \frac{1}{4 \cdot Pi \cdot \epsilon_0} \cdot Int\left(\left(\frac{lambda}{Norm(r\_ - rp\_)}\right), zp = -\frac{L}{2} \dots \frac{L}{2}\right)$$

$$V := \frac{\int_{-\frac{L}{2}}^{\frac{L}{2}} \frac{Q}{L\sqrt{x^2+y^2+(z-zp)^2}}\, dzp}{4\pi\epsilon_0}$$

$V := value(V)$

$$V := \frac{1}{4L\pi\epsilon_0}\left(Q\left(-\ln\left(-2z-L+\sqrt{L^2+4zL+4x^2+4y^2+4z^2}\right)\right.\right.$$
$$\left.\left.+\ln\left(-2z+L+\sqrt{L^2-4zL+4x^2+4y^2+4z^2}\right)\right)\right)$$

$V := limit(V, L = 0)$

$$V := \frac{Q}{4\epsilon_0\pi\sqrt{x^2+y^2+z^2}}$$

$E\_ := simplify(-Gradient(V))$

$$\vec{E} := \frac{Q\left(x\hat{i}+y\hat{j}+z\hat{k}\right)}{4\epsilon_0\pi\left(x^2+y^2+z^2\right)^{3/2}}$$

Finite Line Current

$r\_ := x\cdot\_i + y\cdot\_j$

$$\vec{r} := x\hat{i}+y\hat{j}$$

$rp\_ := zp\cdot\_k$

$$\vec{rp} := zp\hat{k}$$

$J\_ := \dfrac{I0}{L}\cdot\_k$

$$\vec{J} := \frac{I0\hat{k}}{L}$$

$$A\_ := \frac{\mu_0}{4\cdot Pi}\cdot Int\left(\left(\frac{J\_}{Norm\left(r\_-rp\_\right)}\right), zp = -\frac{L}{2}\ldots\frac{L}{2}\right)$$

$$\vec{A} := \frac{\mu_0 \left(\int_{-\frac{L}{2}}^{\frac{L}{2}} \frac{I 0 \hat{k}}{L \sqrt{x^2 + y^2 + zp^2}} \, dzp \right)}{4\pi}$$

$A\_ := value(A\_)$

$$\vec{A} := \frac{\mu_0 I 0 \hat{k} \left(-\ln\left(-L + \sqrt{L^2 + 4x^2 + 4y^2} \right) + \ln\left(L + \sqrt{L^2 + 4x^2 + 4y^2} \right) \right)}{4L\pi}$$

$B\_ := simplify(Curl(A\_))$

$$\vec{B} := \frac{2 I 0 \mu_0 \left(\hat{i} y - \hat{j} x \right)}{\sqrt{L^2 + 4x^2 + 4y^2} \, \pi \left(-4x^2 - 4y^2 \right)}$$

11.7 SCALAR ELECTRIC AND MAGNETIC POTENTIALS

In this section, we compare electric and magnetic boundary value problems with dielectrics and permeable materials in external fields. Solutions to Laplace's equation are given for the electric potential V and the scalar magnetic potential Ω for electrostatic and magnetostatic problems, respectively. Additional examples including hollow dielectric and superconducting spheres in external fields are given.

Example 11.7.1

Model dielectric and permeable spheres in external electric and magnetic fields.

Solution: The chart below compares the solutions of a dielectric sphere of radius R in a uniform electric field and a permeable sphere of the same radius in a uniform magnetic field.

Dielectric sphere with $\varepsilon = \varepsilon_r \varepsilon_0$ in an electric field $\mathbf{E} = E_0 \mathbf{k}$	Permeable sphere with $\mu = \mu_r \mu_0$ in a magnetic field $\mathbf{B} = B_0 \mathbf{k}$
Solve Laplace's equation for the electric potential $\nabla^2 V = 0$ where $\mathbf{E} = -\nabla V$	Solve Laplace's equation for the scalar magnetic potential $\nabla^2 \Omega = 0$ where $\mathbf{H} = -\nabla \Omega$

Inside the sphere $r \leq R$ $$V_<(r,\theta) = \sum_{\ell=0}^{\infty} A_\ell r^\ell P_\ell(\cos\theta)$$ Outside the sphere $r > R$ $$V_>(r,\theta) = -E_0 r \cos\theta$$ $$+ \sum_{\ell=0}^{\infty} \frac{B_\ell}{r^{\ell+1}} P_\ell(\cos\theta)$$ Where far from the sphere $V_> = -E_0 z$ so that $\mathbf{E} = E_0 \hat{\mathbf{k}}$	Inside the sphere $r \leq R$ $$\Omega_<(r,\theta) = \sum_{\ell=0}^{\infty} A_\ell r^\ell P_\ell(\cos\theta)$$ Outside the sphere $r > R$ $$\Omega_>(r,\theta) = -\frac{B_0}{\mu_0} r \cos\theta$$ $$+ \sum_{\ell=0}^{\infty} \frac{B_\ell}{r^{\ell+1}} P_\ell(\cos\theta)$$ Where far from the sphere $\Omega_> = -B_0 z / \mu_0$ so that $\mathbf{B} = B_0 \hat{\mathbf{k}}$
Boundary Conditions: Tangential $\mathbf{E}$ is continuous (no free charges) $$V_<(R) = V_>(R)$$ Normal $\mathbf{D}$ is continuous $$\varepsilon \frac{\partial V_<}{\partial r}\Big|_{r=R} = \varepsilon_0 \frac{\partial V_>}{\partial r}\Big|_{r=R}$$	Boundary Conditions: Tangential $\mathbf{H}$ is continuous $$\Omega_<(R) = \Omega_>(R)$$ Normal $\mathbf{B}$ is continuous $$\mu \frac{\partial \Omega_<}{\partial r}\Big|_{r=R} = \mu_0 \frac{\partial \Omega_>}{\partial r}\Big|_{r=R}$$
Only $\ell = 1$ contributes $$A_1 = \frac{-3}{\varepsilon_r + 2} E_0$$ $$B_1 = \frac{\varepsilon_r - 1}{\varepsilon_r + 2} R^3 E_0$$	Only $\ell = 1$ contributes $$A_1 = \frac{-3}{\mu_r + 2} B_0$$ $$B_1 = \frac{\mu_r - 1}{\mu_r + 2} R^3 \frac{B_0}{\mu_0}$$
Inside the sphere $\mathbf{E}$ is uniform $$\mathbf{E} = \frac{3}{\varepsilon_r + 2} E_0 \hat{\mathbf{k}}$$	Inside the sphere $\mathbf{B}$ is uniform $$\mathbf{B} = \frac{3}{\mu_r + 2} B_0 \hat{\mathbf{k}}$$

Example 11.7.2

A hollow dielectric sphere of inner radius R and outer radius $2R$ is placed in a uniform field $\mathbf{E} = E_0 \hat{\mathbf{k}}$. Calculate the potential in the hollow cavity $r < R$ (region I) inside the dielectric $R \leq r \leq 2R$ (region II) and outside the sphere $r > 2R$ (region III)

Solution: The potential in region I: $r < R$

$$V_{\text{I}}(r,\theta) = \sum_{\ell=0}^{\infty} A_\ell r^\ell P_\ell(\cos\theta) \tag{11.7.1}$$

In region II: $R \le r \le 2R$

$$V_{\text{II}}(r,\theta) = \sum_{\ell=0}^{\infty} \left(B_\ell r^\ell + C_\ell r^{-(\ell+1)} \right) P_\ell(\cos\theta) \tag{11.7.2}$$

In region III: $r > 2R$

$$V_{\text{III}}(r,\theta) = \sum_{\ell=0}^{\infty} D_\ell r^{-(\ell+1)} P_\ell(\cos\theta) - E_0 r \cos\theta \tag{11.7.3}$$

The tangential components of **E** and the normal components of **D** are continuous at $r = R$ and $r = 2R$.

Tangential **E** at $r = R$:

$$V_{\text{I}}(R,\theta) = V_{\text{II}}(R,\theta) \tag{11.7.4}$$

$$\sum_{\ell=0}^{\infty} \left(B_\ell R^\ell + C_\ell R^{-(\ell+1)} \right) P_\ell(\cos\theta) = \sum_{\ell=0}^{\infty} A_\ell R^\ell P_\ell(\cos\theta) \tag{11.7.5}$$

$$B_\ell R^\ell + C_\ell R^{-(\ell+1)} = A_\ell R^\ell \tag{11.7.6}$$

Tangential **E** at $r = 2R$:

$$V_{\text{II}}(2R,\theta) = V_{\text{III}}(2R,\theta) \tag{11.7.7}$$

$$\sum_{\ell=0}^{\infty} \left(B_\ell (2R)^\ell + C_\ell (2R)^{-(\ell+1)} \right) P_\ell(\cos\theta)$$
$$= \sum_{\ell=0}^{\infty} D_\ell (2R)^{-(\ell+1)} P_\ell(\cos\theta) - E_0 (2R)\cos\theta \tag{11.7.8}$$

Thus, we have $\ell = 1$ and

$$B_1(2R) + C_1(2R)^{-2} = D_1(2R)^{-2} - E_0(2R) \tag{11.7.9}$$

Normal **D** at $r = R$:

$$-\varepsilon_0 \left.\frac{\partial V_{\text{I}}}{\partial r}\right|_R = -\varepsilon_0 \varepsilon_r \left.\frac{\partial V_{\text{II}}}{\partial r}\right|_R \tag{11.7.10}$$

$$\varepsilon_r \left(B_1 - 2C_1 R^{-3} \right) = A_1 \tag{11.7.11}$$

Normal $\mathbf{D}$ at $r = 2R$:

$$-\varepsilon_0 \varepsilon_r \left. \frac{\partial V_{\mathrm{II}}}{\partial r} \right|_{2R} = -\varepsilon_0 \left. \frac{\partial V_{\mathrm{III}}}{\partial r} \right|_{2R} \tag{11.7.12}$$

$$\varepsilon_r \left(B_1 - 2C_1 (2R)^{-3} \right) = -2D_1 (2R)^{-3} - E_0 \tag{11.7.13}$$

Equations 11.7.6, 11.7.9, 11.7.11, and 11.7.13 are used to find the constants A_1, B_1, C_1, and D_1.

Example 11.7.3

Calculate the magnetic potential Ω and magnetic field B outside a superconducting sphere of radius R with $\mu_r = 0$ in an external field $\mathbf{B} = B_0 \hat{\mathbf{k}}$ where θ is measured with respect to the positive z-axis. Calculate the supercurrent density on the sphere.

Solution: Outside the sphere the scalar magnetic potential is

$$\Omega_> (r, \theta) = -\frac{B_0}{\mu_0} r \cos\theta + \sum_{\ell=0}^{\infty} \frac{B_\ell}{r^{\ell+1}} P_\ell (\cos\theta) \tag{11.7.14}$$

where far from the sphere $\Omega_> = -B_0 z / \mu_0$ so that $\mathbf{B} = B_0 \hat{\mathbf{k}}$. The normal component of $\mathbf{B}$ is continuous at $r = R$

$$\mu_r \mu_0 \left. \frac{\partial \Omega_<}{\partial r} \right|_{r=R} = \mu_0 \left. \frac{\partial \Omega_>}{\partial r} \right|_{r=R} \tag{11.7.15}$$

The relative permeability $\mu_r = 0$ inside a superconductor so that

$$\left. \frac{\partial \Omega_>}{\partial r} \right|_{r=R} = 0 \tag{11.7.16}$$

thus

$$\left[-\frac{B_0}{\mu_0} \cos\theta - \sum_{\ell=0}^{\infty} (\ell+1) \frac{B_\ell}{r^{\ell+2}} P_\ell (\cos\theta) \right]_{r=R} = 0 \tag{11.7.17}$$

This gives

$$-\frac{B_0}{\mu_0} \cos\theta = \sum_{\ell=0}^{\infty} (\ell+1) \frac{B_\ell}{R^{\ell+2}} P_\ell (\cos\theta) \tag{11.7.18}$$

where $P_1(\cos\theta) = \cos\theta$ so that all the B_ℓ are zero except $\ell = 1$ and

$$B_1 = -\frac{B_0}{2\mu_0} R^3 \tag{11.7.19}$$

Outside the sphere

$$\Omega_{>}(r,\theta) = -\frac{B_0}{\mu_0} r \cos\theta - \frac{B_0}{2\mu_0} \frac{R^3}{r^2} \cos\theta \qquad (11.7.20)$$

with magnetic field

$$\mathbf{B} = -\mu_0 \nabla\Omega_{>} = \mu_0 \left[-\frac{\partial\Omega}{\partial r}\hat{\mathbf{r}} - \frac{1}{r}\frac{\partial\Omega}{\partial\theta}\hat{\boldsymbol{\theta}} \right] \qquad (11.7.21)$$

$$\mathbf{B} = B_0 \cos\theta \left(1 - \frac{R^3}{r^3}\right)\hat{\mathbf{r}} - B_0 \sin\theta \left(1 + \frac{R^3}{2r^3}\right)\hat{\boldsymbol{\theta}} \qquad (11.7.22)$$

From this expression, we can see that the normal component of **B** is zero at the surface of the sphere. The magnetic field along the z-axis is obtained by setting $\theta = 0$ where $\hat{\mathbf{r}} = \hat{\mathbf{k}}$

$$B_z = B_0 \left(1 - \frac{R^3}{r^3}\right) \qquad (11.7.23)$$

Now the supercurrent density is given by the tangential component of **H** at $r = R$ in A/m

$$\mathbf{K}_{super} = \mathbf{H}_{tan} = \frac{1}{\mu_0} B_\theta \hat{\boldsymbol{\theta}} \qquad (11.7.24)$$

$$\mathbf{K}_{super} = -\frac{3B_0}{2\mu_0} \sin\theta \hat{\boldsymbol{\theta}} \qquad (11.7.25)$$

Thus, the supercurrent is maximal on the equator of the sphere where $\theta = \pi/2$ and is zero on the poles.

Maple Examples

The scalar potential of a permeable sphere in an otherwise uniform magnetic field is calculated in the Maple worksheet below. Note that the magnetic field is perpendicular to the plotted lines of constant scalar in this example.

Key Maple terms: *implicitplot, piecewise, subs*

Maple packages: *with(plots)*:

restart

Permeable Sphere in an External Field

with(plots) :

$$\Omega_2 := -\frac{B_0}{\mu_0}\cdot r \cdot \cos(\text{theta}) + \frac{B_0}{\mu_0}\cdot\left(\frac{\mu_r-1}{\mu_r+2}\right)\cdot\frac{R^3}{r^2}\cdot\cos(\text{theta})$$

$$\Omega_2 := -\frac{B_0 r \cos(\theta)}{\mu_0} + \frac{B_0(\mu_r-1)R^3\cos(\theta)}{\mu_0(\mu_r+2)r^2}$$

$$\Omega_1 := -3\frac{B_0}{\mu_0\cdot(\mu_r+2)}\cdot r \cdot \cos(\text{theta})$$

$$\Omega_1 := -\frac{3B_0 r \cos(\theta)}{\mu_0(\mu_r+2)}$$

$$\Omega := piecewise\left(r \le R, \Omega_1, r > R, \Omega_2\right)$$

$$\Omega := \begin{cases} -\dfrac{3B_0 r \cos(\theta)}{\mu_0(\mu_r+2)} & r \le R \\ -\dfrac{B_0 r \cos(\theta)}{\mu_0} + \dfrac{B_0(\mu_r-1)R^3\cos(\theta)}{\mu_0(\mu_r+2)r^2} & R < r \end{cases}$$

$$\Omega := subs(\{B_0=1,\mu_0=1,R=1,\mu_r=3\},\Omega)$$

$$\Omega := \begin{cases} -\dfrac{3r\cos(\theta)}{5} & r \le 1 \\ -r\cos(\theta)+\dfrac{2\cos(\theta)}{5r^2} & 1 < r \end{cases}$$

$$implicitplot\left(\left\{seq\left(\Omega=\frac{Contours}{5}, Contours=-5...5\right), r=1\right\},\right.$$
$$\left. r=0...2,\ theta=0...2\cdot Pi,\ coords=\text{polar},\ scaling=\text{constrained}\right)$$

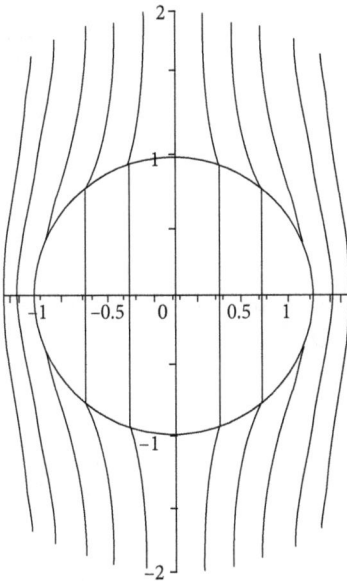

Figure 11.7.1: Scalar magnetic potential of a permeable sphere in a uniform field.

11.8 TIME-DEPENDENT FIELDS

The Ampere-Maxwell equation and its relation to the continuity equation is first discussed in this section. Integral and differential forms of Maxwell's equations are then compared. Examples of the self-inductance of a toroid and the mutual inductance between two current loops are given. Maxwell's wave equation and Maxwell's equations in matter are discussed. Time harmonic fields are then considered. Key equations in electromagnetic theory are compared with the inclusion of hypothetical magnetic monopoles.

11.8.1 The Ampere-Maxwell Equation

The continuity equation relating the electric current density $\mathbf{J}$ and the charge density ρ

$$\nabla \cdot \mathbf{J} = -\frac{\partial \rho}{\partial t} \tag{11.8.1}$$

results from charge conservation. The current flowing out of a region is minus the time rate of change of the total charge within the region. The differential form of Ampere's law relating the magnetic field $\mathbf{B}$ and the current density is

$$\nabla \times \mathbf{B} = \mu_0 \mathbf{J} \tag{11.8.2}$$

Taking the divergence of both sides of this equation

$$\nabla \cdot (\nabla \times \mathbf{B}) = \mu_0 \nabla \cdot \mathbf{J} \tag{11.8.3}$$

and substituting the continuity equation

$$\nabla \cdot (\nabla \times \mathbf{B}) = -\mu_0 \frac{\partial \rho}{\partial t} \tag{11.8.4}$$

Since the divergence of the curl of any vector field is zero we should add to Ampere's law another term whose divergence is $\mu_0 \dfrac{\partial \rho}{\partial t}$.

Considering the differential form of Gauss's law

$$\nabla \cdot \mathbf{E} = \frac{\rho}{\varepsilon_0} \tag{11.8.5}$$

we construct the additional term

$$\mu_0 \frac{\partial \rho}{\partial t} = \mu_0 \varepsilon_0 \frac{\partial}{\partial t} (\nabla \cdot \mathbf{E}) = \nabla \cdot \left(\mu_0 \varepsilon_0 \frac{\partial \mathbf{E}}{\partial t} \right) \tag{11.8.6}$$

The term $\varepsilon_0 \partial \mathbf{E} / \partial t$ is called the displacement current.

11.8.2 Maxwell's Equations

Gauss's law for electric and magnetic fields, Faraday's law and the Ampere-Maxwell equation including the displacement current are collectively referred to as Maxwell's equations. Integral and differential forms of Maxwell's equations are compared in Table 11.8.1. The differential forms of Gauss's law indicate that the charge density ρ is a source of electric field with divergence while the divergence of the magnetic field is zero in the absence of magnetic monopoles. The integral forms of Gauss's law state that the electric flux through any closed surface is proportional to the total charge inside the surface while the total magnetic flux through any closed surface containing no magnetic monopoles is zero. The differential form of Faraday's law shows that a time-changing magnetic field is a source of electric field with curl. The integral form of Faraday's law relates the line integral of the electric field around a closed contour Γ to the time rate of change of magnetic flux Φ_B obtained by integrating the normal component of $\mathbf{B}$ over any open surface bounded by Γ or

$$\Phi_B = \int_{\text{open surf}} \mathbf{B} \cdot \hat{\mathbf{n}} \, da \tag{11.8.7}$$

If the contour Γ coincides with a conducting wire, then there is a voltage V (called the electromotive force) induced in the wire given by

$$V = -\frac{\partial \Phi_B}{\partial t} \tag{11.8.8}$$

The minus sign in this equation indicates that currents induced in the wire will flow in a direction that opposes the change in flux inside the loop per Lenz's law. The differential form of the Ampere-Maxwell equation states that current density and the displacement current are both sources of magnetic field with curl. The integral form of the Ampere-Maxwell equation relates the line integral of **B** around a closed contour Γ to the time rate of change of electric flux Φ_E

$$\Phi_E = \int_{open\ surf} \mathbf{E} \cdot \hat{\mathbf{n}} da \tag{11.8.9}$$

obtained by integrating the normal component of **E** over any open surface bounded by Γ.

TABLE 11.8.1: Comparison of differential and integral forms of Maxwell's equations.

Maxwell's Equation	Differential Form	Integral Form
Gauss's law	$\nabla \cdot \mathbf{E} = \dfrac{\rho}{\varepsilon_0}$	$\oint_{surf} \mathbf{E} \cdot \hat{\mathbf{n}} da = \dfrac{Q}{\varepsilon_0}$
Gauss's law for magnetics	$\nabla \cdot \mathbf{B} = 0$	$\oint_{surf} \mathbf{B} \cdot \hat{\mathbf{n}} da = 0$
Faraday's law	$\nabla \times \mathbf{E} = -\dfrac{\partial \mathbf{B}}{\partial t}$	$\oint_\Gamma \mathbf{E} \cdot d\hat{\ell} = -\dfrac{\partial \Phi_B}{\partial t}$
Ampere-Maxwell equation	$\nabla \times \mathbf{B} = \mu_0 \mathbf{J} + \mu_0 \varepsilon_0 \dfrac{\partial \mathbf{E}}{\partial t}$	$\oint_\Gamma \mathbf{B} \cdot d\hat{\ell} = \mu_0 I + \mu_0 \varepsilon_0 \dfrac{\partial \Phi_E}{\partial t}$

11.8.3 Self-Inductance

The magnetic flux through a current loop is proportional to the current I in the loop

$$\Phi_B = LI \tag{11.8.10}$$

where the proportionality constant L is the self-inductance with units of $Tm^2/A = H$.

Example 11.8.1

The magnetic field inside a toroid with N turns carrying a current I (Figure 11.8.1) is

$$B = \frac{\mu_0 NI}{2\pi r} \tag{11.8.11}$$

where r is measured from the axis of the toroid. Calculate the self-inductance L of the toroid of inner radius a, outer radius b, and height h

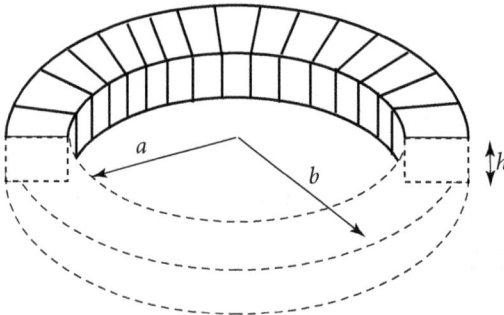

Figure 11.8.1: Toroid with square cross section carrying a current I.

Solution: The magnetic flux through the square cross section of the toroid is

$$\Phi_B = \int B\, da = \int_0^h dz \int_a^b \frac{\mu_0 NI}{2\pi r}\, dr \tag{11.8.12}$$

Integration gives

$$\Phi_B = \frac{\mu_0 NIh}{2\pi} \int_a^b \frac{1}{r}\, dr = \frac{\mu_0 NIh}{2\pi} \ln\!\left(\frac{b}{a}\right) \tag{11.8.13}$$

For a toroid with N turns we have $N\Phi_B = LI$ and the inductance is

$$L = \frac{\mu_0 N^2 h}{2\pi} \ln\!\left(\frac{b}{a}\right) \tag{11.8.14}$$

11.8.4 Mutual Inductance

The magnetic flux through a closed loop Φ_1 is proportional to a current I_2 in a nearby loop

$$\Phi_1 = M_{12} I_2 \tag{11.8.15}$$

where M_{12} is the mutual inductance between the two loops. If instead a current I_1 flows through coil 1 then the flux through coil 2 is $\Phi_2 = M_{21} I_1$.

Example 11.8.2

A current I_1 flows through loop 1 next to a second loop 2. Show that the mutual inductance between loop 1 and loop 2 is given by

$$M_{12} = \frac{\mu_0}{4\pi} \oint \oint \frac{d\vec{\ell}_1 \cdot d\vec{\ell}_2}{|\mathbf{r}_1 - \mathbf{r}_2|} \tag{11.8.16}$$

Solution: To calculate the mutual inductance between the two coils, we calculate the flux through coil 2 due to current flowing in coil 1. From Stokes's theorem we obtain

$$\Phi_2 = \int \mathbf{B}_1 \cdot d\mathbf{a}_2 = \int_{surf} \left(\nabla \times \mathbf{A}_1\right) \cdot d\mathbf{a}_2 = \oint \mathbf{A}_1 \cdot d\hat{\ell}_2 \tag{11.8.17}$$

where the vector potential $\mathbf{A}_1$ at $\mathbf{r}_2$ due to current in coil 1 is

$$\mathbf{A}_1(\mathbf{r}_2) = \frac{\mu_0}{4\pi} I_1 \oint \frac{d\hat{\ell}_1}{|\mathbf{r}_1 - \mathbf{r}_2|} \tag{11.8.18}$$

The flux through coil 2 is then

$$\Phi_2 = \oint \left(\frac{\mu_0}{4\pi} I_1 \oint \frac{d\hat{\ell}_1}{|\mathbf{r}_1 - \mathbf{r}_2|}\right) \cdot d\hat{\ell}_2 \tag{11.8.19}$$

which is proportional to the current flowing in coil 1

$$\Phi_2 = \frac{\mu_0}{4\pi} I_1 \oint \oint \frac{d\hat{\ell}_1 \cdot d\hat{\ell}_2}{|\mathbf{r}_1 - \mathbf{r}_2|} = M_{21} I_1 \tag{11.8.20}$$

where the mutual inductance M_{21} is given by Neumann's formula

$$M_{21} = \frac{\mu_0}{4\pi} \oint \oint \frac{d\hat{\ell}_1 \cdot d\hat{\ell}_2}{|\mathbf{r}_1 - \mathbf{r}_2|} \tag{11.8.21}$$

Interchanging subscripts in this expression we see that $M_{12} = M_{21}$.

11.8.5 Maxwell's Wave Equations

Table 11.8.2 compares Maxwell's equations for electric and magnetic fields in a vacuum and in the absence of sources. Maxwell's wave equations for the electric and magnetic field have the same mathematical form showing that electromagnetic waves propagate with a speed $c = 1/\sqrt{\mu_0 \varepsilon_0} = 2.998 \times 10^8$ m/s in free space.

TABLE 11.8.2: Maxwell's equations in free space.

Electric Fields in Vacuum	Magnetic Fields in Vacuum
Curl and divergence of electric fields in source-free regions $$\nabla \times \mathbf{E} = -\frac{\partial \mathbf{B}}{\partial t}$$ $$\nabla \cdot \mathbf{E} = 0$$	Curl and divergence of magnetic fields in source-free regions $$\nabla \times \mathbf{B} = \mu_0 \varepsilon_0 \frac{\partial \mathbf{E}}{\partial t}$$ $$\nabla \cdot \mathbf{B} = 0$$
Maxwell's wave equation for electric field in source-free regions $$\nabla^2 \mathbf{E} = \frac{1}{c^2} \frac{\partial^2 \mathbf{E}}{\partial t^2}$$	Maxwell's wave equation for magnetic field in source-free regions $$\nabla^2 \mathbf{B} = \frac{1}{c^2} \frac{\partial^2 \mathbf{B}}{\partial t^2}$$

11.8.6 Maxwell's Equations in Matter

The Ampere-Maxwell equation is written

$$\nabla \times \mathbf{B} = \mu_0 \mathbf{J} + \mu_0 \varepsilon_0 \frac{\partial \mathbf{E}}{\partial t} \tag{11.8.22}$$

In matter the total current J may consist of free, bound, and polarization currents

$$\mathbf{J} = \mathbf{J}_{\text{free}} + \mathbf{J}_b + \mathbf{J}_{\text{pol}} \tag{11.8.23}$$

where $\mathbf{J}_{\text{free}}$ is the free current and bound current $\mathbf{J}_b = \nabla \times \mathbf{M}$ is the curl of the magnetization $\mathbf{M}$. The polarization current $\mathbf{J}_{\text{pol}}$ results from the motion of bound charges as a body is becoming polarized or depolarized and is given by the time rate of change of the polarization vector $\mathbf{P}$

$$\mathbf{J}_{\text{pol}} = \frac{\partial \mathbf{P}}{\partial t} \tag{11.8.24}$$

Substituting these forms into the Ampere-Maxwell equation

$$\nabla \times \mathbf{B} = \mu_0 \left(\mathbf{J}_{\text{free}} + \nabla \times \mathbf{M} + \frac{\partial \mathbf{P}}{\partial t} \right) + \mu_0 \varepsilon_0 \frac{\partial \mathbf{E}}{\partial t} \tag{11.8.25}$$

and factoring the curl and the time derivative

$$\nabla \times \left(\frac{\mathbf{B}}{\mu_0} - \mathbf{M} \right) = \mathbf{J}_{\text{free}} + \frac{\partial}{\partial t} \left(\varepsilon_0 \mathbf{E} + \mathbf{P} \right) \tag{11.8.26}$$

we obtain

$$\nabla \times \mathbf{H} = \mathbf{J}_{\text{free}} + \frac{\partial \mathbf{D}}{\partial t} \tag{11.8.27}$$

where the magnetic field strength $\mathbf{H}$ is

$$\mathbf{H} = \left(\mathbf{B} / \mu_0 - \mathbf{M} \right) \qquad (11.8.28)$$

and the electric displacement is

$$\mathbf{D} = \left(\varepsilon_0 \mathbf{E} + \mathbf{P} \right) \qquad (11.8.29)$$

Example 11.8.3

Combine Ohm's law $\mathbf{J} = \sigma \mathbf{E}$, Gauss's law $\nabla \cdot \mathbf{E} = \rho / \varepsilon_0$ and the continuity equation $\nabla \cdot \mathbf{J} = -\dfrac{\partial \rho}{\partial t}$ to show that an initially localized free charge ρ_0 will diffuse exponentially with time in a medium with conductivity σ.

Solution: Taking the divergence of Ohm's law and using the differential form of Gauss's law

$$\nabla \cdot \mathbf{J} = \sigma \nabla \cdot \mathbf{E} = \sigma \frac{\rho}{\varepsilon_0} \qquad (11.8.30)$$

From the continuity equation, we have that

$$\sigma \frac{\rho}{\varepsilon_0} = -\frac{\partial \rho}{\partial t} \qquad (11.8.31)$$

Separating variables

$$\frac{d\rho}{\rho} = -\frac{\sigma}{\varepsilon_0} dt \qquad (11.8.32)$$

and integrating gives

$$\rho(t) = \rho_0 e^{-(\sigma / \varepsilon_0)t} \qquad (11.8.33)$$

11.8.7 Time Harmonic Maxwell's Equations

For electric and magnetic fields with harmonic time dependence

$$\begin{Bmatrix} \mathbf{E}(\mathbf{r},t) \\ \mathbf{B}(\mathbf{r},t) \end{Bmatrix} = \begin{Bmatrix} \mathbf{E}(\mathbf{r}) \\ \mathbf{B}(\mathbf{r}) \end{Bmatrix} \exp(-i\omega t) \qquad (11.8.34)$$

Maxwell's equations in vacuum regions and in the absence of sources become

$$\nabla \cdot \mathbf{B} = 0$$
$$\nabla \cdot \mathbf{E} = 0 \qquad (11.8.35)$$
$$\nabla \times \mathbf{B} = -i\omega \mu_0 \varepsilon_0 \mathbf{E}$$
$$\nabla \times \mathbf{E} = i\omega \mathbf{B}$$

For electromagnetic waves in media with conductivity σ and permittivity ϵ with $\mathbf{J} = \sigma\mathbf{E}$ we have

$$\nabla\times\mathbf{B} = \left(\mu_0\sigma - i\omega\mu_0\varepsilon_0\right)\mathbf{E} \tag{11.8.36}$$

The electric and magnetic fields satisfy the wave equations

$$\nabla^2\left\{\begin{matrix}\mathbf{E}\\\mathbf{B}\end{matrix}\right\} = \gamma^2\left\{\begin{matrix}\mathbf{E}\\\mathbf{B}\end{matrix}\right\} \tag{11.8.37}$$

where the complex propagation constant is

$$\gamma^2 = i\omega\mu\sigma - \omega^2\mu\varepsilon \tag{11.8.38}$$

We write $\gamma = \alpha + i\beta$ and square to solve for α and β

$$\gamma^2 = \alpha^2 - \beta^2 + 2i\alpha\beta = i\omega\mu\sigma - \omega^2\mu\varepsilon \tag{11.8.39}$$

Equating real and imaginary parts

$$\alpha^2 - \beta^2 = -\omega^2\mu\varepsilon \text{ and } 2\alpha\beta = \omega\mu\sigma \tag{11.8.40}$$

Solving these two equations for α and β

$$\begin{pmatrix}\alpha\\\beta\end{pmatrix} = \omega\left[\frac{\mu\varepsilon}{2}\left(\sqrt{1+\left(\frac{\sigma}{\omega\varepsilon}\right)^2}\mp 1\right)\right]^{1/2} \tag{11.8.41}$$

where the upper and lower signs correspond to α and β, respectively. If an electromagnetic wave is polarized such that $\mathbf{E} = (E_x, 0, 0)$ and $\mathbf{B} = (0, B_y, 0)$ we have

$$E_x = E_0 e^{-\alpha z} e^{i(\omega t \pm \beta z)} \qquad B_y = B_0 e^{-\alpha z} e^{i(\omega t \pm \beta z)} \tag{11.8.42}$$

and the wave is attenuated with distance in the direction of $\mathbf{E}\times\mathbf{B}$.

11.8.8 Magnetic Monopoles

Table 11.8.3 compares equations for electric and magnetic fields and forces with electric and magnetic charges (q_e and q_m), charge densities (ρ_e and ρ_m), and current densities ($\mathbf{J}_e$ and $\mathbf{J}_m$). Maxwell's curl and divergence equations become symmetric with the inclusion of magnetic sources ρ_m and $\mathbf{J}_m$. We also have analogous expressions for the continuity equation as well as Coulomb and Lorentz forces if magnetic monopoles exist.

TABLE 11.8.3: Symmetric forms describing electric and magnetic fields and forces in the presence of electric and hypothetical magnetic charges.

Sources of Electric Fields and Forces	Sources of Magnetic Fields and Forces
Differential form of Faraday's law including magnetic currents $$\nabla \times \mathbf{E} = -\mu_0 \mathbf{J}_m - \frac{\partial \mathbf{B}}{\partial t}$$	Differential form of Ampere's law $$\nabla \times \mathbf{B} = \mu_0 \mathbf{J}_e + \mu_0 \varepsilon_0 \frac{\partial \mathbf{E}}{\partial t}$$
Differential form of Coulomb's law for electric fields $$\nabla \cdot \mathbf{E} = \frac{\rho_e}{\varepsilon_0}$$	Differential form of Coulomb's law for magnetic fields $$\nabla \cdot \mathbf{B} = \mu_0 \rho_m$$
Continuity equation for electric charge $$\nabla \cdot \mathbf{J}_e = -\frac{\partial \rho_e}{\partial t}$$	Continuity equation for magnetic charge $$\nabla \cdot \mathbf{J}_m = -\frac{\partial \rho_m}{\partial t}$$
Force between two electric charges $$\mathbf{F}_e = \frac{1}{4\pi\varepsilon_0} \frac{q_{e_1} q_{e_2}}{r^2}$$	Force between two magnetic charges $$\mathbf{F}_m = \frac{\mu_0}{4\pi} \frac{q_{m_1} q_{m_2}}{r^2}$$
Lorentz force on an electric charge moving in electric and magnetic fields $$\mathbf{F}_e = q_e \left(\mathbf{E} + \mathbf{v} \times \mathbf{B} \right)$$	Lorentz force on a magnetic charge moving in electric and magnetic fields $$\mathbf{F}_m = q_m \left(\mathbf{B} - \frac{1}{c^2} \mathbf{v} \times \mathbf{E} \right)$$

Dirac's quantization condition states that given electric and magnetic charges q_e and q_m separated by a distance d the total angular momentum stored in the fields is

$$L = \frac{\mu_0}{4\pi} q_e q_m \tag{11.8.43}$$

where angular momentum is quantized $L = n\hbar$. Thus, electric charge would be quantized

$$q_e = n \left(\frac{4\pi\hbar}{\mu_0 q_m} \right) \quad n = 1, 2, 3\ldots \tag{11.8.44}$$

This is the only known possible explanation for the quantization of electric charge. Even if one monopole exists somewhere in the universe then we require the quantization of electric charge.

11.9 RADIATION

The Poynting vector describing the flow of electromagnetic field energy and momentum is discussed in this section. Wave equations and integral relations are developed for the potentials A and V with time-changing sources. The fields resulting from time harmonic sources are then considered with the simplest example of radiation from a Hertz dipole antenna.

11.9.1 Poynting Vector

The direction of electromagnetic energy propagation is perpendicular to the electric and magnetic field vectors as described by the Poynting vector

$$\mathbf{S} = \frac{1}{\mu_0} \mathbf{E} \times \mathbf{B} \tag{11.9.1}$$

with unit of watts/m². The electromagnetic energy per unit time passing through a surface is given by the power

$$P = \int_{surf} \mathbf{S} \cdot d\mathbf{a} \tag{11.9.2}$$

The divergence of the Poynting vector is

$$\nabla \cdot \mathbf{S} = -\frac{\partial u}{\partial t} \tag{11.9.3}$$

where u is the electromagnetic energy density

$$u = \frac{1}{2} \left(\varepsilon_0 E^2 + \frac{1}{\mu_0} B^2 \right) \tag{11.9.4}$$

The momentum per unit volume **g** is proportional to **S**

$$\mathbf{g} = \mu_0 \varepsilon_0 \mathbf{S} \tag{11.9.5}$$

and the total field momentum

$$\mathbf{P}_{em} = \mu_0 \varepsilon_0 \int_{vol} \mathbf{S} dv \tag{11.9.6}$$

Newton's second law is

$$\frac{d\mathbf{p}_{mech}}{dt} = -\mu_0 \varepsilon_0 \frac{d}{dt} \int_{vol} \mathbf{S} dv + \oint_{surf} \mathbf{T} \cdot d\mathbf{a} \tag{11.9.7}$$

with components of stress energy tensor **T**

$$T_{ij} = \varepsilon_0 \left(E_i E_j - \frac{1}{2}\delta_{ij}E^2 \right) + \frac{1}{\mu_0}\left(B_i B_j - \frac{1}{2}\delta_{ij}B^2 \right)$$ (11.9.8)

If the force **f** per unit volume

$$\mathbf{f} = -\mu_0\varepsilon_0 \frac{d\mathbf{S}}{dt} + \nabla\cdot\mathbf{T}$$ (11.9.9)

is zero, conservation of momentum gives

$$\frac{d\mathbf{g}}{dt} = \nabla\cdot\mathbf{T}$$ (11.9.10)

Example 11.9.1

A potential difference V is applied across a cylindrical resistor of length L and radius a so that a current I flows along the resistor (Figure 11.9.1). Show that the total power

$$\int \mathbf{S}\cdot\hat{n}\, da = IV$$ (11.9.11)

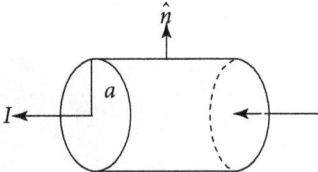

Figure 11.9.1: Cylindrical resistor carrying a current *I*.

Solution: The electric field points along the axis of the resistor in the $\hat{\mathbf{z}}$ -direction

$$\mathbf{E} = \frac{V}{L}\hat{\mathbf{z}}$$ (11.9.12)

while the magnetic field points in the $\hat{\phi}$ -direction

$$\mathbf{B} = \frac{\mu_0 I}{2\pi a}\hat{\phi}$$ (11.9.13)

The Poynting vector is thus in the radial direction

$$\mathbf{S} = \frac{1}{\mu_0}\mathbf{E}\times\mathbf{B} = \frac{1}{\mu_0}\frac{V}{L}\frac{\mu_0 I}{2\pi a}\hat{\mathbf{z}}\times\hat{\phi} = \frac{IV}{2\pi aL}\hat{\mathbf{r}}$$ (11.9.14)

Thus, the total power

$$\int \mathbf{S}\cdot\hat{n}\, da = \frac{IV}{2\pi aL}2\pi aL\hat{\mathbf{r}}\cdot\hat{\mathbf{r}} = IV$$ (11.9.15)

and there is no contribution from $\mathbf{S}\cdot\hat{n}$ on the end caps of the resistor where $\hat{n} = \hat{\mathbf{z}}$.

Example 11.9.2

A plane wave polarized with $\mathbf{E} = (E_x, 0,0)$ and $\mathbf{B} = (0, B_y, 0)$ propagates along the z-axis where

$$B_y = B_0 e^{-az} \exp(i\omega t) \tag{11.9.16}$$

$$E_x = E_0 e^{-az} \exp(i\omega t) \tag{11.9.17}$$

For a plane wave the amplitudes of electric and magnetic fields are related as $B_0 = E_0/c$. Calculate the time-average Poynting vector, momentum density, and energy stored in the field. What is the distance that the wave propagates where the energy density is reduced 50% from its value at $z = 0$?

Solution: The Poynting vector

$$\mathbf{S} = \frac{1}{\mu_0}\mathbf{E}\times\mathbf{B} = \frac{E_0 B_0}{2} e^{-2az} e^{2i\omega t}\hat{\mathbf{k}} \tag{11.9.18}$$

when time averaged $\langle e^{2i\omega t}\rangle = 1/2$ so that

$$\langle\mathbf{S}\rangle = \frac{E_0 B_0}{4\mu_0} e^{-2az}\hat{\mathbf{k}} \tag{11.9.19}$$

The time-averaged momentum density

$$\langle\mathbf{g}\rangle = \mu_0\varepsilon_0\langle\mathbf{S}\rangle = \frac{\varepsilon_0 E_0 B_0}{4} e^{-2az}\hat{\mathbf{k}} \tag{11.9.20}$$

The energy density

$$u = \frac{1}{2}\left(\varepsilon_0 E^2 + \frac{1}{\mu_0}B^2\right) \tag{11.9.21}$$

time averaged with $B_0 = \sqrt{\mu_0\varepsilon_0}E_0$ is

$$\langle u\rangle = \frac{\varepsilon_0}{2}E_0^2 e^{-2az} \tag{11.9.22}$$

To find z such that

$$\frac{\langle u(z)\rangle}{\langle u(0)\rangle} = \frac{1}{2} \tag{11.9.23}$$

we solve

$$e^{-2\alpha z} = \frac{1}{2} \text{ and } z = \frac{\ln 2}{2\alpha} \tag{11.9.24}$$

Example 11.9.3

A single current sheet carries surface charge density $+\sigma$. The sheet is moving with a speed v out of the page in the z-direction. Calculate the Poynting vector.

Solution: Application of Gauss's law with a cylindrical Gaussian surface straddling the charge density gives the electric field

$$\mathbf{E} = \pm\frac{\sigma}{2\varepsilon_0}\hat{\mathbf{j}} \tag{11.9.25}$$

A rectangular Amperian loop gives

$$\mathbf{B} = \mp\frac{\mu_0 K}{2}\hat{\mathbf{i}} \tag{11.9.26}$$

where the top and bottom signs above correspond to $y > 0$ and $y < 0$. On both sides of the sheet the Poynting vector is

$$\mathbf{S} = \frac{1}{\mu_0}\mathbf{E}\times\mathbf{B} = \frac{\sigma K}{4\varepsilon_0}\hat{\mathbf{k}} \tag{11.9.27}$$

Example 11.9.4

A wire with radius a and charge density ρ carries a current $I = \rho v$ in the z-direction where v is the velocity of the charge carriers. Calculate the momentum stored in the fields inside the wire. Note that the fields will have no z-dependence so just do the integral over r and ϕ to get the field momentum per unit length inside the wire.

Solution: To obtain the electric field we consider cylindrical Gaussian surface inside the wire

$$\int \mathbf{E}\cdot\hat{n}\,da = \frac{1}{\varepsilon_0}\int\rho\,dv \tag{11.9.28}$$

$$E2\pi rL = \frac{1}{\varepsilon_0}\rho\pi r^2 L \tag{11.9.29}$$

so that the electric field is radial

$$\mathbf{E} = \frac{\rho r}{2\varepsilon_0}\hat{\mathbf{r}} \tag{11.9.30}$$

Now the magnetic field is obtained using Ampere's law

$$\oint \mathbf{B} \cdot d\hat{\ell} = \int \mathbf{J} \cdot \hat{n} \, da \tag{11.9.31}$$

with a circular contour of radius r and $\hat{n} = \hat{\mathbf{z}}$

$$B \cdot 2\pi r = \mu_0 \rho v \pi r^2 \tag{11.9.32}$$

$$\mathbf{B} = \frac{\mu_0 \rho v r}{2} \hat{\phi} \tag{11.9.33}$$

Now the Poynting vector

$$\mathbf{S} = \frac{1}{\mu_0} \mathbf{E} \times \mathbf{B} = \frac{\rho^2 r^2 v}{4\varepsilon_0} \hat{\mathbf{z}} \tag{11.9.34}$$

The momentum stored in the field is

$$\mathbf{P}_{em} = \mu_0 \varepsilon_0 \int_{vol} \mathbf{S} dv = \mu_0 \varepsilon_0 \int_{vol} S r \, dr \, d\phi \, dz = \frac{\mu_0 \pi L \rho^2 v}{2} \int_0^a r^3 dr \hat{\mathbf{z}} \tag{11.9.35}$$

and the momentum stored per unit length in the wire is

$$\frac{\mathbf{P}_{em}}{L} = \frac{\mu_0 \pi \rho^2 v a^4}{8} \hat{\mathbf{z}} \tag{11.9.36}$$

11.9.2 Inhomogeneous Wave Equations

Wave equations may be developed for the vector and scalar potentials A and V from the Ampere-Maxwell equation

$$\nabla \times \mathbf{B} = \mu_0 \mathbf{J} + \mu_0 \varepsilon_0 \frac{\partial \mathbf{E}}{\partial t} \tag{11.9.37}$$

the differential form of Gauss's law

$$\nabla \cdot \mathbf{E} = \frac{\rho}{\varepsilon_0} \tag{11.9.38}$$

and Faraday's law

$$\nabla \times \mathbf{E} = -\partial \mathbf{B} / \partial t \tag{11.9.39}$$

Substituting the expression for $\mathbf{E}(\mathbf{r}, t)$ and $\mathbf{B}(\mathbf{r}, t)$

$$\mathbf{B}(\mathbf{r},t) = \nabla \times \mathbf{A}(\mathbf{r},t) \tag{11.9.40}$$

$$\mathbf{E}(\mathbf{r},t) = -\frac{\partial}{\partial t} \mathbf{A}(\mathbf{r},t) - \nabla V(\mathbf{r},t) \tag{11.9.41}$$

into Faraday's and Gauss's laws

$$\nabla \times (\nabla \times \mathbf{A}) = \mu_0 \mathbf{J} + \mu_0 \varepsilon_0 \frac{\partial}{\partial t}\left(-\frac{\partial}{\partial t}\mathbf{A} - \nabla V\right) \qquad (11.9.42)$$

$$\nabla \cdot \left(-\frac{\partial}{\partial t}\mathbf{A} - \nabla V\right) = \frac{\rho}{\varepsilon_0} \qquad (11.9.43)$$

and using the identity

$$\nabla \times (\nabla \times \mathbf{A}) = \nabla(\nabla \cdot \mathbf{A}) - \nabla^2 \mathbf{A} \qquad (11.9.44)$$

we obtain

$$\nabla^2 \mathbf{A} - \mu_0 \varepsilon_0 \frac{\partial^2}{\partial t^2}\mathbf{A} = -\mu_0 \mathbf{J} + \nabla\left(\mu_0 \varepsilon_0 \frac{\partial}{\partial t}V + \nabla \cdot \mathbf{A}\right) \qquad (11.9.45)$$

$$\left(-\frac{\partial}{\partial t}\nabla \cdot \mathbf{A} - \nabla^2 V\right) = \frac{\rho}{\varepsilon_0} \qquad (11.9.46)$$

With the Lorentz gauge condition

$$\nabla \cdot \mathbf{A} = -\mu_0 \varepsilon_0 \frac{\partial}{\partial t}V \qquad (11.9.47)$$

we obtain the inhomogeneous wave equations

$$\nabla^2 \mathbf{A} - \mu_0 \varepsilon_0 \frac{\partial^2}{\partial t^2}\mathbf{A} = -\mu_0 \mathbf{J} \qquad (11.9.48)$$

$$\nabla^2 V - \mu_0 \varepsilon_0 \frac{\partial^2}{\partial t^2}V = -\frac{\rho}{\varepsilon_0} \qquad (11.9.49)$$

with source terms on the right. Written in terms of the d'Alembertian operator

$$\Box^2 = \nabla^2 - \mu_0 \varepsilon_0 \frac{\partial^2}{\partial t^2} \qquad (11.9.50)$$

the wave equations become

$$\Box^2 \mathbf{A} = -\mu_0 \mathbf{J} \qquad (11.9.51)$$

$$\Box^2 V = -\frac{\rho}{\varepsilon_0} \qquad (11.9.52)$$

11.9.3 Gauge Transformation

Example 11.9.5

Show that the gauge transformation performed by the replacements

$$\mathbf{A} \rightarrow \mathbf{A} + \nabla \lambda \tag{11.9.53}$$

$$V \rightarrow V - \frac{\partial \lambda}{\partial t} \tag{11.9.54}$$

leave $\mathbf{E}$ and $\mathbf{B}$ unchanged, where λ is any scalar function.

Solution: Substituting the potentials above into $\mathbf{E} = -\nabla V - \partial \mathbf{A} / \partial t$

$$\mathbf{E} = -\nabla \left(V - \frac{\partial \lambda}{\partial t} \right) - \frac{\partial}{\partial t} (\mathbf{A} + \nabla \lambda) = -\nabla V + \nabla \left(\frac{\partial \lambda}{\partial t} \right) - \frac{\partial \mathbf{A}}{\partial t} - \frac{\partial}{\partial t} \nabla \lambda$$

$$= -\nabla V - \frac{\partial \mathbf{A}}{\partial t} \tag{11.9.55}$$

interchanging ∇ and $\partial / \partial t$. Substitution into $\mathbf{B} = \nabla \times \mathbf{A}$ gives

$$\mathbf{B} = \nabla \times (\mathbf{A} + \nabla \lambda) = \nabla \times \mathbf{A} + \nabla \times \nabla \lambda$$

$$= \nabla \times \mathbf{A} \tag{11.9.56}$$

where the curl of the gradient of any scalar function is zero.

11.9.4 Radiation Potential Formulation

Integral relations for vector and scalar potentials for static fields are

$$\mathbf{A}(\mathbf{r}) = \frac{\mu_0}{4\pi} \int \frac{\mathbf{J}(\mathbf{r}')}{|\mathbf{r} - \mathbf{r}'|} dv' \tag{11.9.57}$$

$$V(\mathbf{r}) = \frac{1}{4\pi\varepsilon_0} \int \frac{\rho(\mathbf{r}')}{|\mathbf{r} - \mathbf{r}'|} dv' \tag{11.9.58}$$

The magnetic and electric fields are then obtained from the curl and negative gradient of $\mathbf{A}$ and V

$$\mathbf{B}(\mathbf{r}) = \frac{\mu_0}{4\pi} \int \frac{\mathbf{J}(\mathbf{r}') \times (\mathbf{r} - \mathbf{r}')}{|\mathbf{r} - \mathbf{r}'|^3} dv' \tag{11.9.59}$$

$$\mathbf{E}(\mathbf{r}) = \frac{1}{4\pi\varepsilon_0} \int \frac{\rho(\mathbf{r}')(\mathbf{r} - \mathbf{r}')}{|\mathbf{r} - \mathbf{r}'|^3} dv' \tag{11.9.60}$$

Because of the finite speed of light c, there will be a time delay $|\mathbf{r} - \mathbf{r}'|/c$ between source changes at $\mathbf{r}'$ and field changes at r. The retarded time is thus defined as

$$t_r = t - \frac{|\mathbf{r} - \mathbf{r}'|}{c} \tag{11.9.61}$$

where the potentials are

$$\mathbf{A}(\mathbf{r}, t) = \frac{\mu_0}{4\pi} \int \frac{\mathbf{J}(\mathbf{r}', t_r)}{|\mathbf{r} - \mathbf{r}'|} dv' \tag{11.9.62}$$

$$V(\mathbf{r}, t) = \frac{1}{4\pi\varepsilon_0} \int \frac{\rho(\mathbf{r}', t_r)}{|\mathbf{r} - \mathbf{r}'|} dv' \tag{11.9.63}$$

Integral expressions for $\mathbf{B}(\mathbf{r}, t)$ and $\mathbf{E}(\mathbf{r}, t)$ will not be directly analogous to the static case since the retarded time depends on the position vector $\mathbf{r}$ so that derivatives of the potential will give extra terms. If the sources depend harmonically on time

$$\mathbf{J}(\mathbf{r}', t_r) = \mathbf{J}(\mathbf{r}') e^{-i\omega t_r} \tag{11.9.64}$$

$$\rho(\mathbf{r}', t_r) = \rho(\mathbf{r}') e^{-i\omega t_r} \tag{11.9.65}$$

the potentials will also be time harmonic

$$\mathbf{A}(\mathbf{r}, t) = \mathbf{A}(\mathbf{r}) e^{-i\omega t} \tag{11.9.66}$$

$$V(\mathbf{r}, t) = V(\mathbf{r}) e^{-i\omega t} \tag{11.9.67}$$

We may then separate the spatial part of the potentials

$$\mathbf{A}(\mathbf{r}) = \frac{\mu_0}{4\pi} \int \mathbf{J}(\mathbf{r}') \frac{e^{\frac{i\omega}{c}|\mathbf{r} - \mathbf{r}'|}}{|\mathbf{r} - \mathbf{r}'|} dv' \tag{11.9.68}$$

$$V(\mathbf{r}) = \frac{1}{4\pi\varepsilon_0} \int \rho(\mathbf{r}') \frac{e^{\frac{i\omega}{c}|\mathbf{r} - \mathbf{r}'|}}{|\mathbf{r} - \mathbf{r}'|} dv' \tag{11.9.69}$$

where $\omega/c = 2\pi/\lambda$.

11.9.5 The Hertz Dipole Antenna

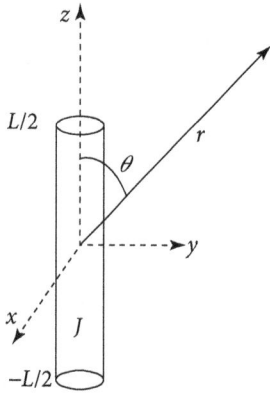

Figure 11.9.2: Hertz dipole carrying a current in the z-direction.

The radiation field of a short dipole antenna was developed by Hertz in 1888. Figure 11.9.2 shows a cylindrical dipole of length L with current density $\mathbf{J} = J_0 \exp(-i\omega t_r)\hat{z}$ in the z-direction. The dipole has a cross section S and a volume LS. To simplify the integrals, we make the approximations that $L \ll \lambda$ and that J_0 is uniform over the volume of the antenna. This approximation is ideal for radio waves on the order of meter wavelengths and greater. For microwaves with $\lambda \approx$ cm the approximation is better for $L \approx$ mm. The integral (11.9.68) over the source coordinates gives the volume of the antenna with $|\mathbf{r} - \mathbf{r}'| \approx r$ and $J_0 S = I$

$$\mathbf{A}(\mathbf{r},t) = \frac{\mu_0}{4\pi} IL \frac{e^{i\frac{\omega}{c}r}}{r} e^{i\omega t}\hat{z} \tag{11.9.70}$$

The term $\exp(i\omega r/c)/r$ is known as the spherical propagation factor. Writing the Cartesian unit vector $\hat{z}$ in spherical coordinates

$$\hat{z} = \cos\theta\hat{r} - \sin\theta\hat{\theta} \tag{11.9.71}$$

we express the vector potential in terms of radial $\hat{r}$ and angular $\hat{\theta}$ components

$$\mathbf{A}(\mathbf{r},t) = \frac{\mu_0 IL}{4\pi} \frac{e^{i\frac{\omega}{c}r}}{r}\left(\cos\theta\hat{r} - \sin\theta\hat{\theta}\right)e^{-i\omega t} \tag{11.9.72}$$

Taking the curl of

$$\nabla \times \mathbf{A} = \left(\frac{1}{r}\frac{\partial}{\partial r}rA_\theta - \frac{1}{r}\frac{\partial}{\partial\theta}A_r\right)\hat{\phi} \tag{11.9.73}$$

gives the magnetic field

$$\mathbf{B}(\mathbf{r},t) = \frac{\mu_0 IL}{4\pi} \frac{e^{i\frac{\omega}{c}r}}{r} \left(\frac{1}{r} - i\frac{\omega}{c}\right) \sin\theta e^{-i\omega t} \hat{\phi} \qquad (11.9.74)$$

that has only a ϕ component. The electric field may be obtained from $\nabla \times \mathbf{E} = -\partial \mathbf{B}/\partial t$ giving

$$\mathbf{E} = \frac{\mu_0}{4\pi} \omega IL \frac{e^{i\frac{\omega}{c}r}}{r} \left[\cos\theta \left(2\left(\frac{c}{r\omega}\right) + 2\left(\frac{c}{r\omega}\right)^2 \right) \hat{r} + \sin\theta \left(-i + \left(\frac{c}{r\omega}\right) + i\left(\frac{c}{r\omega}\right)^2 \right) \hat{\theta} \right] e^{-i\omega t}$$

$$(11.9.75)$$

To obtain the electric and magnetic fields far from the dipole (in the radiation zone) we retain the lowest power of $1/r$ in (11.9.74) and (11.9.75)

$$\mathbf{E} = -i\frac{\mu_0}{4\pi} \omega IL \frac{e^{i\frac{\omega}{c}r}}{r} \sin\theta e^{-i\omega t} \hat{\theta} \qquad (11.9.76)$$

$$\mathbf{B} = -i\frac{\mu_0}{4\pi} \frac{\omega}{c} IL \frac{e^{i\frac{\omega}{c}r}}{r} \sin\theta e^{-i\omega t} \hat{\phi} \qquad (11.9.77)$$

From the time-averaged Poynting vector (in watts/m²)

$$\mathbf{S}_{avg} = \frac{1}{\mu_0} \langle |\mathbf{E} \times \mathbf{B}| \rangle = \frac{1}{2} \mu_0 \frac{\omega^2}{c} \left(\frac{IL}{4\pi}\right)^2 \frac{\sin^2\theta}{r^2} \hat{r} \qquad (11.9.78)$$

we obtain the total power radiated by the dipole (in watts)

$$P = \int \mathbf{S}_{avg} \cdot d\mathbf{a} \qquad (11.9.79)$$

giving

$$P = \frac{\mu_0}{2c} \left(\frac{\omega IL}{4\pi}\right)^2 \int_0^\pi \frac{\sin^2\theta}{r^2} r^2 \sin\theta d\theta \int_0^{2\pi} d\phi = \frac{4\pi\mu_0}{3c} \left(\frac{\omega IL}{4\pi}\right)^2 \qquad (11.9.80)$$

11.10 MATLAB EXAMPLES

In this section, MATLAB's PDE Toolbox is demonstrated solving problems in electromagnetism. Electrostatics, magnetostatics, conductive media DC and AC power electromagnetics problem types are selected under Options → Application. Refer to Section 8.4 and online documentation for a description of the PDE Toolbox.

Section 11.2 Laplace's Equation in Cartesian Coordinates

The GUI for selecting the dielectric constant ε and charge density ρ for the electromagnetics application is shown in Figure 11.10.1.

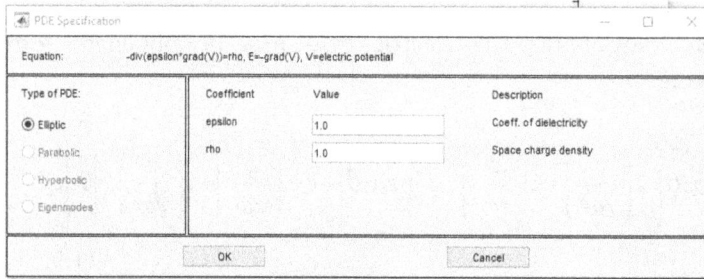

Figure 11.10.1: GUI for selecting the dielectric constant and charge density for electrostatics problems.

Poisson's equations for the scalar electric potential V

$$-\nabla \cdot (\varepsilon \nabla V) = \rho \tag{11.10.1}$$

is solved in this application where $\mathbf{E} = -\Delta V$. Figure 11.10.2 shows a finite element mesh with 1668 nodes and 3168 elements generated over a rectangular solution region with an elliptical hole.

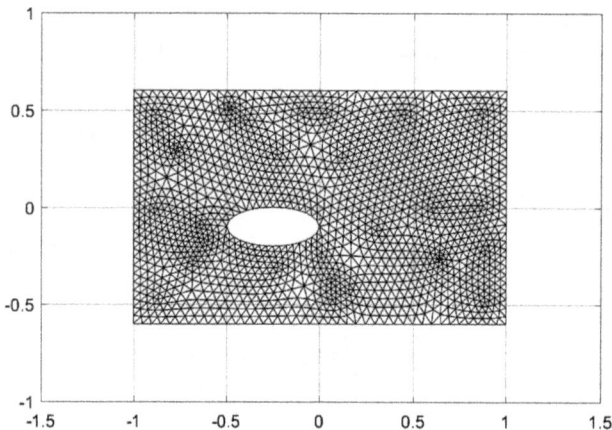

Figure 11.10.2: Finite element mesh generated in the PDE toolbox electrostatic application.

Equipotentials and electric field lines are plotted in Figure 11.10.3 corresponding to the finite element mesh in Figure 11.10.2 with zero potential specified on the rectangular boundary and a positive charge density on the interior elliptical contour.

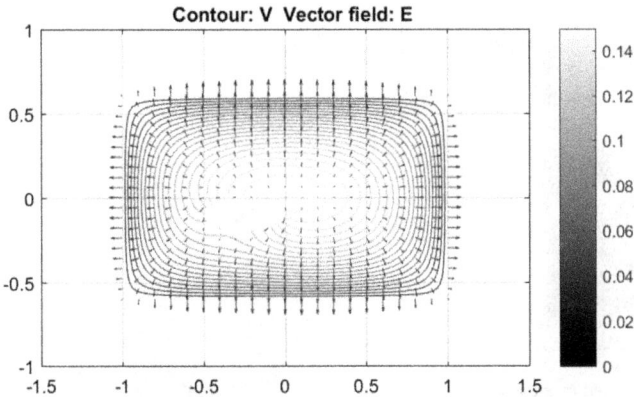

Figure 11.10.3: Equipotentials and electric field vectors are plotted corresponding to a positive charge density inside of a grounded 2D rectangular box.

Section 11.6 Electricity and Magnetism

The GUI for selecting the permeability μ and current density J for the magnetostatics application is shown in Figure 11.10.4.

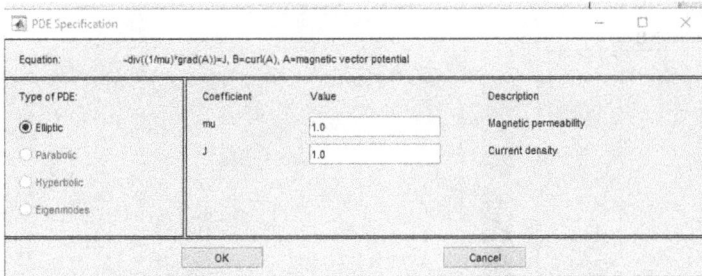

Figure 11.10.4: GUI for selecting the magnetic permeability and current density for the magnetostatics application.

Poisson's equations for the vector potential A

$$-\nabla \cdot \left(\frac{1}{\mu} \nabla A \right) = J \qquad (11.10.2)$$

is solved in this application where $\mathbf{B} = \nabla \times \mathbf{A}$ and A is the component of $\mathbf{A}$ perpendicular to the 2D solution region. Figure 11.10.5 shows a finite element mesh with 2044 nodes and 3872 elements generated over a rectangular solution region with an elliptic superconducting cylinder. The elliptic superconducting region is not meshed since the magnetic field inside of an ideal Type-I superconductor is zero for external fields lower than the critical field of the superconductor.

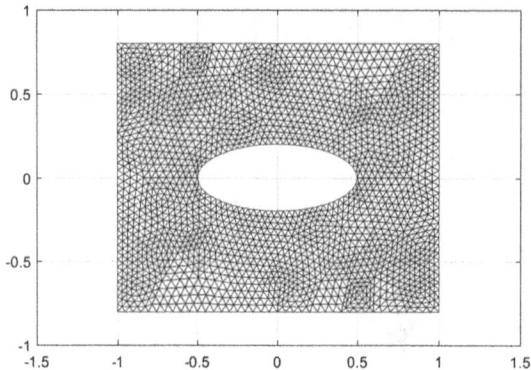

Figure 11.10.5: Finite element mesh generated in the PDE Toolbox magnetostatic application.

Flux lines of constant vector potential and magnetic field vectors are plotted in Figure 11.10.6 corresponding to the finite element mesh in Figure 11.10.5 with opposite values of vector potential specified on the left- and right-hand sides of the model. A zero tangential magnetic field is specified on the top and bottom boundaries of the solution region. A zero-normal field condition is imposed on the elliptical superconducting cylinder.

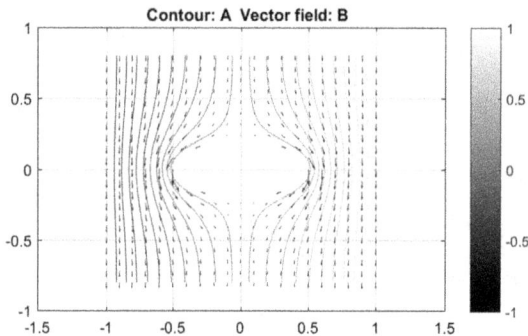

Figure 11.10.6: Flux lines and magnetic field vectors near a superconductor in an external magnetic field.

Section 11.7 Scalar Electric and Magnetic Potentials

The GUI for selecting the material conductivity σ and source currents q for DC current flow problems is shown in Figure 11.10.7.

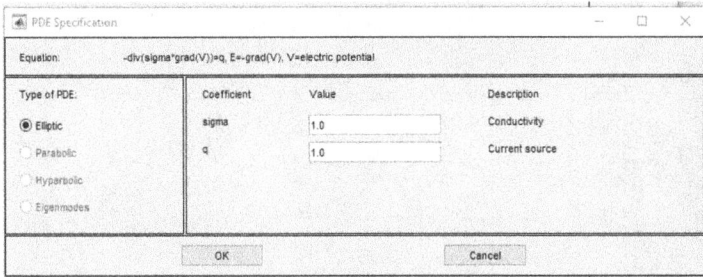

Figure 11.10.7: GUI for selecting the conductivity and current source for the conductive media DC application.

In this application, the scalar potential V is calculated where

$$-\nabla \cdot (\sigma \nabla V) = q \tag{11.10.3}$$

Section 11.8 Time-Dependent Fields

The GUI for selecting the material conductivity σ, relative permeability μ and dielectric constant ε for time harmonic electromagnetic (AC power) problems is shown in Figure 11.10.8

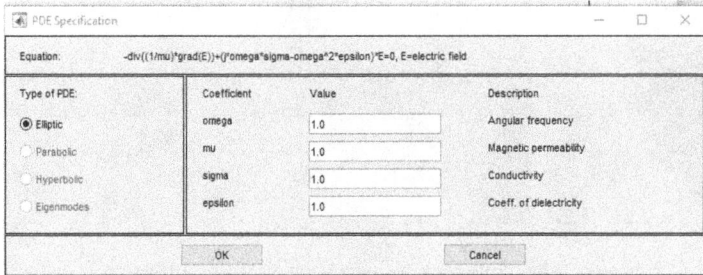

Figure 11.10.8: GUI for selecting the frequency and material properties for the AC power electromagnetics application.

The time harmonic field equation

$$-\nabla \cdot \left(\frac{1}{\mu} \nabla E \right) + \left(i\omega\sigma + \omega^2 \varepsilon \right) E = 0 \tag{11.10.4}$$

is solved in this application where E is the component of the electric field **E** perpendicular to the 2D solution region. Figure 11.10.9 shows a finite element mesh with 2096 nodes and 3904 triangular elements generated over the solution region corresponding to off-axis coaxial cylinders. The interior cylindrical region is not meshed.

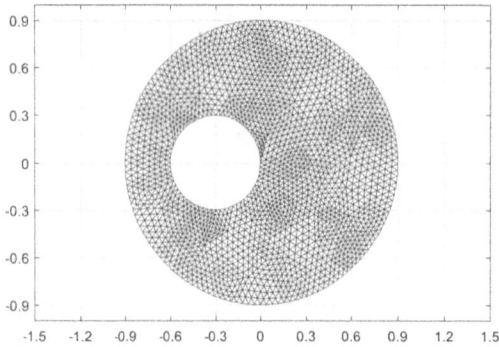

Figure 11.10.9: Finite element mesh generated in the AC power PDE Toolbox application.

The shaded magnitude and lines of constant electric field and magnetic field vectors are plotted in Figure 11.10.10 corresponding to the finite element mesh in Figure 11.10.9, with the outer and inner cylinders at $E = \pm 1.0$ V/m, angular frequency $\omega = 5.0$ rad/s, relative permeability $\mu = 1.0$, conductivity $\sigma = 1.0$ S/m and relative permittivity $\varepsilon = 1.0$.

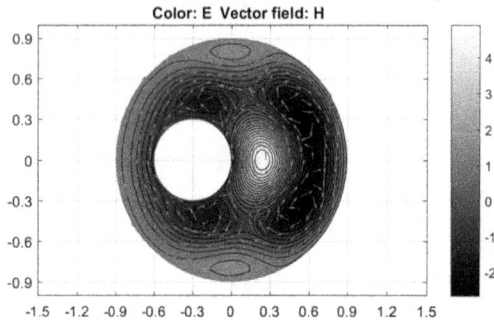

Figure 11.10.10: Electric field (shaded magnitude and constant E contours) and magnetic field vectors corresponding to the finite element AC power model meshed in Figure 11.10.9.

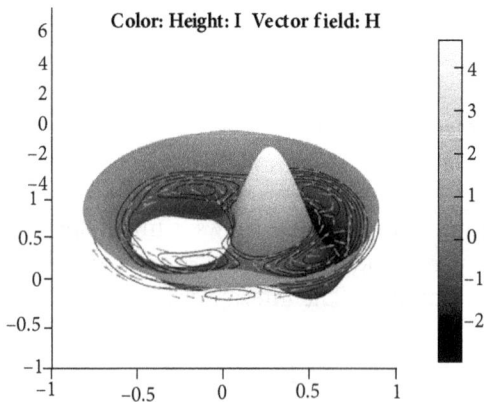

Figure 11.10.11: 3D visualization of electric field (shaded surface and lines of constant E) and magnetic field vectors corresponding to the AC power model meshed in Figure 11.10.9.

11.11 EXERCISES

Section 11.1 Electrostatics in 1D

1. Use the differential and integral forms of Gauss's law to find the electric field inside a hollow spherical shell with uniform charge density

$$\rho(r) = \begin{cases} \rho_0 & a \le r \le b \\ 0 & r < a, \ r > b \end{cases}$$

   Make a plot of the radial electric field as a function of r.

2. Use the differential and integral forms of Gauss's law to find the electric field inside a cylinder with nonuniform charge density

$$\rho(r) = \begin{cases} \rho_0 \dfrac{r}{a} & r \le a \\ 0 & r > a \end{cases}$$

   Make a plot of the radial electric field as a function of r.

3. Use the differential form of Gauss's law to find the electric field inside a sphere of radius a with charge density

$$\rho(r) = \begin{cases} \rho_0 \dfrac{r}{a} & r \le a \\ 0 & r > a \end{cases}$$

   Make a plot of the radial electric field as a function of r.

4. Calculate the potential and electric field due to a thin cone with half angle $\alpha = \pi/8$ over a conducting plane with $\theta_0 = \pi/2$ in spherical coordinates.

5. A coaxial cylinder has an outer conductor grounded $V(b) = 0$ and an inner conductor at $V(a) = V_0$. Show that the potential between the conductors is

$$V(r) = V_0 \frac{\ln\left(\dfrac{r}{b}\right)}{\ln\left(\dfrac{a}{b}\right)}$$

   Find the electric field and the capacitance per unit length.

6. Two concentric spheres have inner and outer radii a and b. The outer sphere is grounded $V(b) = 0$ and the inner sphere has a charge q. Find the potential between the spheres. Find the electric charge density on the inside of the outer sphere. Show that the total charge on the inside of the outer sphere is $-q$.

11.2 Laplace's Equation in Cartesian Coordinates

7. A cube of side a is held at zero potential on four faces ($x = 0, a$ and $y = 0, a$) and at a potential of V_0 on the faces that are parallel to the x-y plane located at $z = -a/2$ and $z = a/2$. Calculate the potential V at points inside the cube.

8. A pipe with a square cross section is closed at one end located at $z = 0$. The pipe is grounded with $V = 0$ on the four sides ($x = 0$, L and $y = 0, L$). The closed end at $z = 0$ is held at a potential $V(x, y, 0) = V_0 \dfrac{xy}{L^2}$. Find the potential $V(x, y, z)$ inside the pipe taking $V(x, y, z \to \infty) = 0$.

9. A point charge q is located a distance d above a conducting plane at zero potential. Use the method of images to find the electric potential above the conducting plane and the induced surface charge density on the plane.

10. A point charge q with a mass m is released a height d above a conducting plane held at zero potential. Calculate the time required for the charge to reach the plane after it is released. Neglect the gravitational force acting on the point charge.

11. Two point charges q and $-q$ are located at distances d and $2d$ above a conducting plane held at zero potential. Find the electric potential above the conducting plane and the induced surface charge density on the plane.

11.3 Laplace's Equation in Cylindrical Coordinates

12. Verify by direct substitution that r^n and r^{-n} are solutions to

$$r\frac{\partial R}{\partial r} + r^2 \frac{\partial^2 R}{\partial r^2} = n^2 R$$

13. A cylinder of radius R has a surface potential

$$V(R, \theta) = \begin{cases} V_0 & 0 \le \theta < \pi/2 \\ 0 & \pi/2 \le \theta < 2\pi \end{cases}$$

Calculate the potential $V(r, \theta)$ for $r \ge R$ and for $r < R$.

14. A cylinder of radius R has a surface potential

$$V(R, \theta) = \begin{cases} V_0 & 0 \le \theta < \pi/2 \\ -V_0 & \pi/2 \le \theta < \pi \\ V_0 & \pi \le \theta < 3\pi/2 \\ -V_0 & 3\pi/2 \le \theta < 2\pi \end{cases}$$

Calculate the potential $V(r, \theta)$ for $r \ge R$ and for $r < R$.

15. A cylinder of radius a has a surface potential
$V(a, \theta) = V_0 \cos(\theta)$
The cylinder is surrounded by a grounded coaxial cylinder of radius b
$V(b, \theta) = 0$
Calculate the potential $V(r, \theta)$ for $r < a$ and for $a \leq r \leq b$.

16. A long hollow cylinder of radius R carries surface charge density $\sigma(R, \theta) = \sigma_0 \cos(2\theta)$. The cylinder is surrounded by a grounded cylinder held at zero potential $V(2R, \theta) = 0$
Calculate the potential $V_<$ inside the inner cylinder $0 < r \leq R$
Calculate the potential $V_>$ between the cylinders $R < r \leq 2R$
Note that the electric field is discontinuous across the surface charge density where the boundary conditions are $V_<(R, \theta) = V_> (R, \theta)$ and

$$\left.\frac{\partial V_<}{\partial r}\right|_{r=R} - \left.\frac{\partial V_>}{\partial r}\right|_{r=R} = \frac{1}{\varepsilon_0}\sigma(R,\theta) .$$

11.4 Laplace's Equation in Spherical Coordinates

17. A thin hollow sphere R is held at a potential

$$V(R,\theta) = \begin{cases} V_0 & 0 \leq \theta < \pi/2 \\ 0 & \pi/2 \leq \theta < \pi \end{cases}$$

Calculate $V(r, \theta)$ both inside and outside the sphere

18. A hollow sphere of radius R carries a potential of $V_0 \cdot P_2(\cos\theta)$. Calculate the potential inside and outside the sphere.

19. A hollow sphere of radius R carries a potential of $V_0(1 + \cos\theta)$. Calculate the potential inside and outside the sphere.

20. A thin hollow sphere of radius $2R$ is held at a potential $V(2R) = V_0 \frac{1}{2}(3\cos^2\theta - 1)$. The sphere is concentric with a grounded (zero potential) sphere of radius R. Calculate the potential between the two spheres $R \leq r \leq 2R$ and outside the sphere $r > 2R$. Calculate the induced charge density on the surface of the grounded sphere.

21. A hollow sphere of radius R carries a potential of $V(R, \theta) = V_0 \cos^2\theta$. The sphere is surrounded by a second sphere of radius $2R$ held at zero potential $V(2R, \theta) = 0$. Calculate the potential in the region between the two spheres.

22. A sphere of radius a is held at a potential
$V(a, \theta) = V_0 \cos(\theta)$
The sphere is surrounded by a grounded concentric sphere of radius b
$V(b, \theta) = 0$

Calculate the potential $V(r, \theta)$ for $r < a$ and for $a \leq r \leq b$.

23. A sphere of radius a is held at a potential

$$V(a, \theta, \phi) = \begin{cases} V_0 & n = 0, 2, 4, 6 \\ 0 & n = 1, 3, 5, 7 \end{cases} \quad \text{where} \quad n\frac{\pi}{4} \leq \phi < (n+1)\frac{\pi}{4}$$

in Figure 11.11.1. Calculate the potential $V(r, \theta, \phi)$ for $r < a$ and $r \geq a$

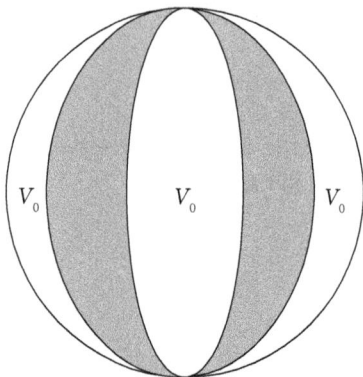

Figure 11.11.1: Sphere of radius a with striped potential.

11.5 Multipole Expansion of Potential

24. A thin hollow sphere of radius R has a surface charge density

$$\sigma(R, \theta) = \begin{cases} \sigma_0 & 0 \leq \theta < \pi/2 \\ -\sigma_0 & \pi/2 \leq \theta < \pi \end{cases}$$

Calculate a multipole expansion of the potential along the z-axis. Obtain an expression for the potential off the z-axis for $r > R$.

25. Calculate the potential everywhere $r > a$ due to a charged needle oriented along the z-axis. The needle carries a charge distribution

$$\sigma(z) = \begin{cases} Q/a & 0 \leq z \leq a \\ -Q/a & -a \leq z < 0 \end{cases}.$$

26. Given a diffuse charge density $\rho(\mathbf{r}') = \dfrac{Q}{r'^6} \exp\left(-\dfrac{\lambda}{r'}\right) P_2(\cos\theta')$

calculate the nonzero multipole terms of the potential along the z-axis. Also, calculate the potential everywhere off the axis.

27. Calculate a multipole expansion of the potential due to point charges q located at $(x, y, z) = (0, 0, \pm a)$.

28. Calculate a multipole expansion of the potential due to point charges q located at $(x, y, z) = (\pm a, 0, 0)$ and $-q$ located at $(x, y, z) = (0, \pm a, 0)$.

29. A thin sphere of radius R carries a surface charge density $\sigma(R,\theta,\phi) = \dfrac{Q}{4\pi R^2} Y_2^1(\theta,\phi)$. Calculate a multipole expansion of the potential outside the sphere.

11.6 Electricity and Magnetism

30. Four equal point charges q are held at the corners of a square by rigid rods of length a. Find the electric potential at the center of the square. Find the energy stored in the charge distribution.

31. A sphere of radius R contains a uniform charge density ρ. Find the electrostatic energy stored in the regions $r \le R$ and $r > R$.

32. The vector potential of an infinite current sheet carrying K in amperes per meter is
$$\mathbf{A} = \mu_0 \frac{K}{2} y\hat{\mathbf{k}} \cdot$$
Compare the magnetic field obtained by computing the curl of $\mathbf{A}$ with the magnetic field calculated from Ampere's law.

33. A thin strip of width L carries a total current I in the z-direction out of the page. Write an integral expression for the vector potential $\mathbf{A}(x,y) = A_z(x,y)\hat{\mathbf{k}}$.

34. A thin strip of width L carries a current $-I$ from $-L/2 < x < 0$ and I from $0 < x < L/2$. Write an integral expression for the vector potential $\mathbf{A}(x,y) = A_z(x,y)\hat{\mathbf{k}}$.

35. Plot the magnetic field and vector potential inside and outside of a long solenoid with radius R as a function of r. The solenoid carries a current per length I/L.

36. Use Ampere's law to calculate the magnetic field B inside $(r < R)$ and outside $(r > R)$ a wire with radius R carrying a total current I.

37. A hollow cylinder of radius R carries a charge Q. Calculate the energy stored per unit length in a surrounding dielectric layer with permittivity ε_1 between $R \le r \le 2R$.

38. A hollow cylinder of radius R carries a charge Q. A surrounding dielectric of thickness $2R$ is divided into four 90-degree quadrants with relative permittivity ε_1, ε_2, ε_3, and ε_4 (Figure 11.11.2). Calculate $\mathbf{D}$, $\mathbf{E}$, and $\mathbf{P}$ in each quadrant.

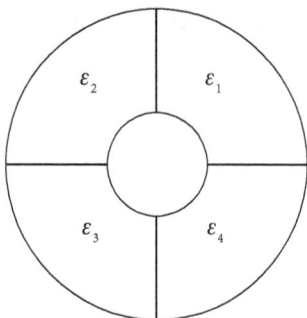

Figure 11.11.2: Charged cylinder surrounded by four dielectric segments.

39. A hollow sphere of radius R carries a charge Q. Calculate the energy stored in a surrounding dielectric shell of thickness R and relative permittivity ε_r.

40. A hollow dielectric sphere with relative permittivity ε_r, inner radius a and outer radius b surrounds a point charge Q located at the center. Calculate **D**, **E**, and **P** inside the dielectric. Calculate the bound charge density on the inner and outer surfaces of the dielectric sphere.

41. The polarization **P** inside a cylinder of radius R and height h is

$$\mathbf{P} = P_0\, 3r^2\hat{\mathbf{r}}$$

The cylinder is coaxial with the z-axis. Calculate the bound volume ρ_b and surface σ_b charge densities. Show that

$$\int_{vol} \rho_b\, dv + \int_{surf} \sigma_b\, da = 0$$

42. A cylindrical wire of radius R carries a total free current I. The wire is coated by a layer with relative permeability μ_r and thickness d. Calculate the bound current density on the inner and outer surfaces of the permeable layer as well as in the volume of the layer.

43. A parallel plate conductor carries a current $K\hat{\mathbf{k}}$ on the top plate and $-K\hat{\mathbf{k}}$ on the bottom plate. The space between the plates is filled by a layer with relative permeability μ_r. Calculate the bound current density on the top and bottom surfaces of the permeable slab as well as in the volume of the slab.

44. Calculate the force between a line current I located a height h above a superconducting plane located at $y = 0$. The line current flows in the z-direction. The magnetic field above the superconductor can be modeled by placing an image current $-I$ below the superconducting plane at $y = -h$.

45. The torque acting on a dipole with moment **m** in a uniform field is $\tau = \mathbf{m} \times \mathbf{B}$. Note that the torque acts to align the dipole with the field direction so that there is no torque when $\mathbf{m} \parallel \mathbf{B}$ and maximum torque when $\mathbf{m} \perp \mathbf{B}$. Calculate the torque acting on a dipole with moment $\mathbf{m} = m_0 \left(\hat{\mathbf{i}} - \hat{\mathbf{k}} \right)$ in a magnetic field $\mathbf{B} = B_0 \left(\hat{\mathbf{j}} - 2\hat{\mathbf{k}} \right)$.

46. The energy of a magnetic dipole $\mathbf{m}$ in a $\mathbf{B}$ field is $W_B = -\mathbf{m} \cdot \mathbf{B}$. The energy is minimized when the dipole points along the direction of the field. Calculate the energy of a dipole $\mathbf{m} = m_0 \left(\hat{\mathbf{i}} - \hat{\mathbf{k}} \right)$ in a field $\mathbf{B} = B_0 \left(\hat{\mathbf{j}} - 2\hat{\mathbf{k}} \right)$.

47. The force acting on a magnetic dipole in a $\mathbf{B}$ field is $\mathbf{F} = \left(\mathbf{m} \cdot \nabla \right) \mathbf{B}$. Note that the force on a dipole $\mathbf{m}$ is zero in a uniform magnetic field. Calculate the force acting on a dipole $\mathbf{m} = m_0 \left(\hat{\mathbf{i}} - \hat{\mathbf{k}} \right)$ in a nonuniform field $\mathbf{B} = B_0 \left(y\hat{\mathbf{j}} - 2z\hat{\mathbf{k}} \right)$.

48. A square slab of thickness d with vertices on the bottom face $[(0, 0, 0), (0, a, 0), (a, a, 0)$ and $(a, 0, 0)]$ and vertices on the top face $[(0,0, d)$, $(0, a, d), (a, a, d)$ and $(a, 0, d)]$ carries a magnetization $\mathbf{M} = M_0 \left(3xy\hat{\mathbf{i}} + 2y\hat{\mathbf{j}} \right)$. Find the bound currents inside the slab, its edges and on the top and bottom faces.

11.7 Scalar Electric and Magnetic Potentials

49. A solid dielectric sphere of radius a is surrounded by a concentric sphere of radius b held at a potential of $V(b, \theta) = V_0 \cos \theta$. Calculate the potential in the dielectric sphere $r < a$ (region I) and between the dielectric sphere and outer spherical shell $a \le r \le b$ (region II).

50. Calculate the electric potential $V(r, \theta)$ inside and outside a dielectric cylinder of permittivity ε_r and an external field $\mathbf{E} = E_0 \hat{\mathbf{i}}$ where θ is measured with respect to the positive x-axis.

51. A hollow dielectric cylinder (inner radius $= a$, outer radius $= b$) with relative dielectric constant ε_r is placed in a uniform electric field $\mathbf{E} = E_0 \hat{\mathbf{i}}$ pointing in the positive x-direction. Calculate the electric potential $V(r, \theta)$ for $0 < r < a$ (region I), $a \le r \le b$ (region II) and $r > b$ (region III).

52. Calculate the magnetic potential Ω and magnetic field B outside a superconducting cylinder of radius R with $\mu_r = 0$ in an external field $\mathbf{B} = B_0 \hat{\mathbf{i}}$ where θ is measured with respect to the positive x-axis. Calculate the supercurrent density on the cylinder.

53. A hollow sphere (inner radius $= a$, outer radius $= b$) with a relative permeability μ_r is placed in a uniform magnetic field $\mathbf{B} = B_0 \hat{\mathbf{k}}$ pointing in the positive z-direction. Calculate the potential $\Omega(r, \theta)$ for $0 < r < a$ (region I), $a \le r \le b$ (region II) and $r > b$ (region III).

54. Calculate the magnetic potential $\Omega(r, \theta)$ inside and outside a cylinder of permeability μ_r and an external field $\mathbf{B} = B_0 \hat{\mathbf{i}}$ where θ is measured with respect to the positive x-axis. The radius of the cylinder is a.

11.8 Time-Dependent Fields

55. Show that $\oint \mathbf{E} \cdot d\hat{\ell} = -\dfrac{\partial \Phi_m}{\partial t}$ is equivalent to $\nabla \times \mathbf{E} = -\dfrac{\partial \mathbf{B}}{\partial t}$

56. Show that $\oint \mathbf{B} \cdot d\hat{\ell} = \mu_0 I + \mu_0 \varepsilon_0 \dfrac{\partial}{\partial t} \iint \mathbf{E} \cdot \hat{n} da$ is equivalent to

$$\nabla \times \mathbf{B} = \mu_0 \mathbf{J} + \mu_0 \varepsilon_0 \dfrac{\partial \mathbf{E}}{\partial t}$$

57. From $\oint \mathbf{B} \cdot d\hat{\ell} = \mu_0 I$ derive $\nabla \times \mathbf{B} = \mu_0 \mathbf{J}$

58. The magnetic field in a region of space is given by $\mathbf{B}(t) = B_0 \sin(kx - \omega t) \hat{\mathbf{k}}$.

Find the electric field component in the y-direction E_y (within a constant) given that the x-component $E_x = 0$.

59. The electric field in a region of space is given by $\mathbf{E}(x, y) = E_0 \dfrac{x}{a} \cos(\omega t) \hat{\mathbf{j}}$. Find the magnetic field (within a constant).

60. Given a static magnetic field $\mathbf{B}(x, y) = B_0 \cos(kx) \exp(-y/y_0) \hat{\mathbf{k}}$. Find the electric current density $\mathbf{J}$.

61. Given that the current in a superconductor is proportional to the vector potential

$$\mathbf{J}_s = -\dfrac{1}{\mu_0 \lambda_L^2} \mathbf{A}$$

derive London's two equations

$$\nabla \times \mathbf{J}_s = -\dfrac{1}{\mu_0 \lambda_L^2} \mathbf{B}$$

$$\dfrac{d\mathbf{J}_s}{dt} = \dfrac{1}{\mu_0 \lambda_L^2} \mathbf{E}$$

by taking the curl and time derivative of $\mathbf{J}_s$ above.

62. A transient current may be produced in a conductor placed in a time-varying magnetic field. Using Ohm's law $\mathbf{J} = \sigma \mathbf{E}$ and neglecting displacement currents inside the metal

$$\nabla \times \mathbf{B} = \mu_0 \sigma \mathbf{E}.$$

Take the curl of this equation and use the relation

$$\nabla \times \mathbf{E} = -\frac{\partial \mathbf{B}}{\partial t}$$

to show that the magnetic field satisfies a diffusion equation

$$\nabla^2 \mathbf{B} = \mu_0 \sigma \frac{\partial \mathbf{B}}{\partial t}.$$

63. Calculate the mutual inductance between a line current coaxial with a toroid of square cross section with inner radius a, outer radius b and height h.

64. Calculate the ratio of forces between two electric and two magnetic monopoles with each pair separated by the same distance r.

11.9 Radiation

65. A parallel plate capacitor carries a surface charge density $+\sigma$ and $-\sigma$ on the upper and lower plates, respectively. The plates are moving with a speed v out of the page (Figure 11.11.3). Calculate the stress energy tensor in between the plates.

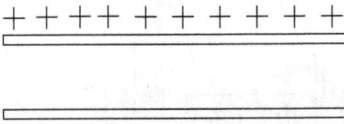

$+ + + + + + + + + + +$

Figure 11.11.3: Oppositely charged parallel plates moving out of the page.

66. Given the expression for the vector potential of a hollow sphere of radius R with total charge Q rotating with angular speed ω is (see D. J. Griffiths, *Introduction to Electrodynamics*, p. 246)

$$\mathbf{A}(r,\theta,\phi) = \frac{\mu_0 R^4 \omega \sigma}{3} \frac{1}{r^2} \sin\theta \hat{\phi}$$

where $4\pi R^2 \sigma = Q$. Calculate the Poynting vector outside of the sphere.

67. The Maxwell stress tensor consists of nine components

$$T_{ij} = \varepsilon_0 \left(E_i E_j - \frac{1}{2}\delta_{ij}E^2 \right) + \frac{1}{\mu_0}\left(B_i B_j - \frac{1}{2}\delta_{ij}B^2 \right)$$

where $T_{11} = T_{xx}$, etc. Write the nonzero terms of T_{ij} for
(a) a single charged sheet with charge Q
(b) a parallel plate capacitor with oppositely charged plates
(c) a long solenoid carrying a current I

68. Find the inhomogeneous wave equations for the potentials $\mathbf{A}$ and V for time harmonic sources.

69. For the Hertz dipole use the AC form of the Lorentz gauge

$$\nabla \cdot \mathbf{A} = i\omega\mu_0\varepsilon_0 V$$

with the divergence in spherical coordinates

$$\nabla \cdot \mathbf{A} = \frac{1}{r^2}\frac{\partial}{\partial r}r^2 A_r + \frac{1}{r\sin\theta}\frac{\partial}{\partial\theta}(\sin\theta A_\theta)$$

to show that the electric potential is

$$V = \frac{1}{4\pi\varepsilon_0}\frac{IL}{\omega}\frac{e^{i\beta r}}{r}\cos\theta\left(\beta + \frac{i}{r}\right)e^{-i\omega t}$$

70. Calculate the electric field of the Hertz dipole using

$$\mathbf{E} = -\frac{\partial \mathbf{A}}{\partial t} - \nabla V$$

where the gradient of V in spherical coordinates

$$\nabla V = \hat{r}\frac{\partial V}{\partial r} + \hat{\theta}\frac{1}{r}\frac{\partial V}{\partial\theta}$$

gives

$$\nabla V = \frac{\mu_0}{4\pi}\omega IL\frac{e^{i\beta r}}{r}\left[\cos\theta\left(i - \frac{2}{\beta r} - \frac{2}{(\beta r)^2}\right)\hat{r} - \sin\theta\left(\frac{1}{\beta r} + \frac{i}{(\beta r)^2}\right)\hat{\theta}\right]e^{-i\omega t}$$

and

$$\frac{\partial \mathbf{A}}{\partial t} = \frac{\mu_0}{4\pi}\omega IL\frac{e^{i\beta r}}{r}\left(-i\cos\theta\hat{r} + i\sin\theta\hat{\theta}\right)e^{-i\omega t}$$

71. The Lagrangian for a nonrelativistic particle of mass m and charge e in electric and magnetic fields is given by

$$L = \frac{1}{2}m\dot{\mathbf{r}}^2 + \frac{e}{c}\mathbf{A}\cdot\dot{\mathbf{r}} - eV$$

Show that L is invariant under the Gauge transformation $\mathbf{A} \rightarrow A + \Delta\Lambda$ and

$V \rightarrow V - \dfrac{\partial\Lambda}{\partial t}$. Find Lagrange's equations of motion.

Chapter **12** **QUANTUM MECHANICS**

Chapter Outline

12.1 Schrödinger Equation
12.2 Bound States I
12.3 Bound States II
12.4 Schrödinger Equation in Higher Dimensions
12.5 Approximation Methods

The time-dependent Schrödinger equation is first discussed in this section followed by the time-dependent Schrödinger equation obtained by separation of variables. Additional topics include quantum mechanical operators and the computation of expectation values. Nonstationary states are discussed followed by the development of the probability current density. Plane wave solutions to the Schrödinger equation are discussed in the context of scattering and quantum mechanical tunneling.

12.1 SCHRÖDINGER EQUATION

12.1.1 Time-Dependent Schrödinger Equation

The time-dependent Schrödinger equation

$$-\frac{\hbar^2}{2m}\nabla^2\Psi\left(\mathbf{r},t\right)+V\left(\mathbf{r}\right)\Psi\left(\mathbf{r},t\right)=i\hbar\frac{\partial}{\partial t}\Psi\left(\mathbf{r},t\right) \qquad (12.1.1)$$

describes the spatial and temporal evolution of the complex wavefunction Ψ in the presence of a potential $V(\mathbf{r})$. The Schrödinger equation is a diffusion equation like Fick's second law and the heat equation. According to the Born interpretation of quantum mechanics, the probability of locating a particle such as an electron in a volume dv is $\Psi^*\Psi dv$ where the complex conjugate Ψ^* is obtained by replacing $i \rightarrow -i$ in the expression for Ψ. Physical properties of the wavefunction include the following:

1. The normalization condition expressed as

$$\left\langle\Psi\big|\Psi\right\rangle=\int_{vol}\Psi*\Psi dv=1 \qquad (12.1.2)$$

    where the probability of locating an electron somewhere in space is unity. Exceptions are plane wave solutions to the Schrödinger that are not normalizable.

2. The wavefunction is a continuous, single valued function of position. Ψ is continuous over regions with varying potential with continuous slope except across delta function potentials where the slope of the wavefunction is discontinuous.

3. Normalizable wavefunctions should be small at large distances with the boundary condition $\Psi \rightarrow 0$ at infinity.

12.1.2 Time-Independent Schrödinger Equation

Writing the wavefunction as a product of spatial and temporal functions

$$\Psi(\mathbf{r},t) = \psi(\mathbf{r})\phi(t) \qquad (12.1.3)$$

we substitute product form into the differential equation

$$-\frac{\hbar^2}{2m}\nabla^2\psi\left(\mathbf{r}\right)\phi\left(t\right)+V\left(\mathbf{r}\right)\psi\left(\mathbf{r}\right)\phi\left(t\right)=i\hbar\psi\left(\mathbf{r}\right)\frac{\partial}{\partial t}\phi\left(t\right) \qquad (12.1.4)$$

and divide by $\psi(\mathbf{r})\phi(t)$

$$-\frac{\dfrac{\hbar^2}{2m}\nabla^2\psi(\mathbf{r})}{\psi(\mathbf{r})}+V(\mathbf{r})=i\hbar\frac{\dfrac{\partial}{\partial t}\phi(t)}{\phi(t)} \qquad (12.1.5)$$

$$\underbrace{\phantom{-\frac{\dfrac{\hbar^2}{2m}\nabla^2\psi(\mathbf{r})}{\psi(\mathbf{r})}+V(\mathbf{r})}}_{\text{function of } \mathbf{r}} \qquad \underbrace{\phantom{i\hbar\frac{\dfrac{\partial}{\partial t}\phi(t)}{\phi(t)}}}_{\text{function of } t}$$

Because the left-hand side is only a function of r and the right-hand side is only a function of time, they must be equal to the same constant E. The ϕ equation

$$\frac{\partial}{\partial t}\phi(t)=\frac{E}{i\hbar}\phi(t) \qquad (12.1.6)$$

gives

$\phi(t)=e^{-i\omega t}$ where $\omega = E/\hbar$. Canceling the complex exponential gives the time-independent Schrödinger equation

$$-\frac{\hbar^2}{2m}\nabla^2\psi(\mathbf{r})+V(\mathbf{r})\psi(\mathbf{r})=E\psi(\mathbf{r}) \qquad (12.1.7)$$

Certain quantum systems are in states where the probability of locating the electron is independent of time. If we can express the wavefunction as

$$\Psi(\mathbf{r},t)=\psi(\mathbf{r})e^{-i\omega t} \qquad (12.1.8)$$

and E is the energy of the electron, then we see that

$$\Psi^*\Psi=\psi(\mathbf{r})^2\, e^{i\omega t}\, e^{-i\omega t}=\psi(\mathbf{r})^2 \qquad (12.1.9)$$

Ψ is then referred to as a *stationary state* satisfying the time-independent Schrödinger equation. Table 12.1.1 compares properties of stationary state and nonstationary state wavefunctions.

TABLE 12.1.1: Stationary vs. nonstationary state wavefunctions in quantum mechanics.

Stationary States	Nonstationary States
Harmonic time dependence $$\Psi(\mathbf{r},t)=\psi(\mathbf{r})e^{-i\omega t}$$	Not separable $$\Psi(\mathbf{r},t)\neq\psi(\mathbf{r})\phi(t)$$
Time-independent probabilities $P(\mathbf{r})dv=\psi(\mathbf{r})^*\,\psi(\mathbf{r})dv$	Time-dependent probabilities $$P(\mathbf{r},t)dv=\Psi(\mathbf{r},t)^*\,\Psi(\mathbf{r},t)dv$$
Time-independent expectation values	Time-dependent expectation values
Sharp (well-defined) energies	Fuzzy (time-dependent) energies

12.1.3 Operators, Expectation Values and Uncertainty

For every observable O of a quantum system there corresponds an operator $\hat{O}$. Table 12.1.2 shows the observables position, momentum, kinetic energy, and total energy with their corresponding operators in 1D and 3D.

TABLE 12.1.2: Quantum mechanical operators corresponding to physical observables in 1D and 3D.

Physical Observable (Name of Operator)	Operator in 1D	Operator in 3D
Position	$\hat{x} = x$	$\hat{\mathbf{r}} = \mathbf{r}$
Momentum	$\hat{p} = \dfrac{\hbar}{i}\dfrac{\partial}{\partial x}$	$\hat{p} = \dfrac{\hbar}{i}\nabla$
Kinetic energy	$\dfrac{\hat{p}^2}{2m} = -\dfrac{\hbar^2}{2m}\dfrac{\partial^2}{\partial x^2}$	$\dfrac{\hat{p}^2}{2m} = -\dfrac{\hbar^2}{2m}\nabla^2$
Total energy (Hamiltonian)	$\hat{H} = -\dfrac{\hbar^2}{2m}\dfrac{\partial^2}{\partial x^2} + V(x)$	$\hat{H} = -\dfrac{\hbar^2}{2m}\nabla^2 + V(\mathbf{r})$
Total energy	$\hat{E} = i\hbar\dfrac{\partial}{\partial t}$	$\hat{E} = i\hbar\dfrac{\partial}{\partial t}$

We may write the time-dependent Schrödinger equation (TDSE) as a scalar operator equation based on conservation of energy where the total energy is given by the sum of kinetic and potential energies

$$\left(\frac{\hat{p}^2}{2m} + V(\mathbf{r})\right)\Psi = \hat{E}\Psi \tag{12.1.10}$$

To calculate the average (or expectation) value of an observable $\langle O \rangle$ we compute

$$\langle O \rangle = \langle \Psi \hat{O} | \Psi \rangle = \int_{\text{vol}} \Psi^* \hat{O}\Psi dv \tag{12.1.11}$$

In one dimension, the expectation values

$$\langle x \rangle = \int_{-\infty}^{\infty} \Psi^* x \Psi dx \qquad \langle x^2 \rangle = \int_{-\infty}^{\infty} \Psi^* x^2 \Psi dx \tag{12.1.12}$$

$$\langle p \rangle = \int_{-\infty}^{\infty} \Psi^* \left(\frac{\hbar}{i}\frac{\partial}{\partial x}\right)\Psi dx \qquad \langle p^2 \rangle = \int_{-\infty}^{\infty} \Psi^* \left(\frac{\hbar}{i}\frac{\partial}{\partial x}\right)\left(\frac{\hbar}{i}\frac{\partial}{\partial x}\right)\Psi dx \tag{12.1.13}$$

with corresponding uncertainties x and p given by their standard deviations

$$\Delta x = \sqrt{\left\langle x^2 \right\rangle - \left\langle x \right\rangle^2} \quad \Delta p = \sqrt{\left\langle p^2 \right\rangle - \left\langle p \right\rangle^2} \tag{12.1.14}$$

The Heisenberg uncertainty principle relates the uncertainties in position and momentum as a product

$$\Delta x \Delta p \geq \frac{\hbar}{2} \tag{12.1.15}$$

It is not possible to measure both position and momentum with infinite precision. A smaller Δx will result in a larger Δp and vice versa. A similar uncertainty product for energy E and time t is given by

$$\Delta E \Delta t \geq \frac{\hbar}{2} \tag{12.1.16}$$

Both Heisenberg's position-momentum and energy-time uncertainty relations may be used to provide estimates of physical quantities.

Example 12.1.1

Use Heisenberg's position-momentum relation to estimate the kinetic energy of a particle of mass m confined to a one-dimensional box of length L.

Solution: Taking the length of the box equal to the uncertainty in position $\Delta x = L$ we estimate the momentum of the particle by Δp.

$$\Delta p \geq \frac{\hbar}{2L} \text{ and the kinetic energy } KE = \frac{p^2}{2m} \approx \frac{\hbar^2}{8mL^2} \tag{12.1.17}$$

In the next section, we find the ground state energy of a particle confined to a 1D box of length L is

$$E_1 = \frac{h^2}{8mL^2} \tag{12.1.18}$$

by solving Schrödinger's equation directly.

12.1.4 Probability Current Density

We will now develop an equation that represents the flow of probability in quantum mechanics. Beginning with the TDSE

$$-\frac{\hbar^2}{2m}\nabla^2\Psi + V\left(\mathbf{r}\right)\Psi = i\hbar\frac{\partial}{\partial t}\Psi \tag{12.1.19}$$

Multiplying on the left by Ψ^*

$$-\frac{\hbar^2}{2m}\Psi * \nabla^2\Psi + V(\mathbf{r})\Psi * \Psi = i\hbar\Psi * \frac{\partial}{\partial t}\Psi \qquad (12.1.20)$$

Taking the complex conjugate of the TDSE and multiply on the left by Ψ

$$-\frac{\hbar^2}{2m}\Psi\nabla^2\Psi * + V(\mathbf{r})\Psi\Psi * = -i\hbar\Psi\frac{\partial}{\partial t}\Psi * \qquad (12.1.21)$$

Subtracting the two equations (12.1.20) and (12.1.21) we obtain

$$-\frac{\hbar^2}{2m}\left(\Psi * \nabla^2\Psi - \Psi\nabla^2\Psi *\right) = i\hbar\left(\Psi * \frac{\partial\Psi}{\partial t} + \Psi\frac{\partial\Psi *}{\partial t}\right) \qquad (12.1.22)$$

The right-hand side is proportional to the time derivative of the probability density

$$\frac{\hbar}{2mi}\left(\Psi^*\nabla^2\Psi - \Psi\nabla^2\Psi^*\right) = -\frac{\partial}{\partial t}\left(\Psi^*\Psi\right) \qquad (12.1.23)$$

This is written as

$$\nabla \cdot \mathbf{J}_{\text{pcd}} = -\frac{\partial}{\partial t}\left(\Psi^*\Psi\right) \qquad (12.1.24)$$

where the probability current density

$$\mathbf{J}_{\text{pcd}} = \frac{\hbar}{2mi}\left(\Psi * \nabla\Psi - \Psi\nabla\Psi^*\right) \qquad (12.1.25)$$

is a vector that gives the direction of probability flow. In 1D

$$\mathbf{J}_{\text{pcd}} = \frac{\hbar}{2mi}\left(\Psi * \frac{\partial}{\partial x}\Psi - \Psi\frac{\partial}{\partial x}\Psi *\right)\hat{\mathbf{i}} \qquad (12.1.26)$$

Example 12.1.2

Calculate the probability current density given the plane wave solution to the TDSE

$$\Psi(x,t) = Ae^{i(kx-\omega t)} \qquad (12.1.27)$$

Solution: For plane wave and stationary state wavefunctions $\mathbf{J}_{\text{pcd}}$ is independent of time. In one dimension

$$\mathbf{J}_{\text{pcd}} = \frac{\hbar}{2mi}\left(\psi * \frac{\partial\psi}{\partial x} - \psi\frac{\partial\psi *}{\partial x}\right)\hat{\mathbf{i}} = \frac{\hbar}{2mi}\left(2ikA * A\right)\hat{\mathbf{i}} = \frac{\hbar k}{m}|A|^2\,\hat{\mathbf{i}} \qquad (12.1.28)$$

corresponding to a particle moving to the right. Note that for a particle moving to the left described by $\psi(x) = Be^{-ikx}$ the probability current density

$$J_{pcd} = -\frac{\hbar k}{m}|B|^2 \,\hat{i} \qquad (12.1.29)$$

Example 12.1.3

Particles with energy E are incident from the left where $V(x) = 0$ for $x < 0$. The particles are confronted with a step in potential energy $V(x) = V_0$ for $x > 0$ where $E > V_0$. Find the reflection coefficient R and the transmission coefficient T.

Solution: For $x < 0$ the wavefunction

$$\psi_1 = A \exp(ik_1 x) + B \exp(-ik_1 x) \text{ where } k_1 = \sqrt{2mE / \hbar^2} \qquad (12.1.30)$$

The incident and reflected probability currents are

$$J_{inc} = \frac{\hbar k_1}{m}|A|^2 \,\hat{i} \text{ and } J_{refl} = -\frac{\hbar k_1}{m}|B|^2 \,\hat{i} \qquad (12.1.31)$$

The reflection coefficient is

$$R = \frac{|J_{refl}|}{|J_{inc}|} = \frac{|B|^2}{|A|^2} \qquad (12.1.32)$$

In the region $x > 0$

$$\psi_2 = C \exp(ik_2 x) \text{ where } k_2 = \sqrt{2m(E - V_0)/\hbar^2} \qquad (12.1.33)$$

The transmitted probability current is

$$J_{trans} = \frac{\hbar k_2}{m}|C|^2 \,\hat{i} \qquad (12.1.34)$$

The transmission coefficient is

$$T = \frac{|J_{trans}|}{|J_{inc}|} = \frac{k_2}{k_1}\frac{|C|^2}{|A|^2} \qquad (12.1.35)$$

The boundary conditions $\psi_1(0) = \psi_2(0)$ give

$$A + B = C \qquad (12.1.36)$$

and $d\psi_1/dx = d\psi_2/dx$ at $x = 0$ gives

$$ik_1(A - B) = ik_2 C \qquad (12.1.37)$$

The reflection coefficient is

$$R = \frac{|B|^2}{|A|^2} = \frac{(k_1 - k_2)^2}{(k_1 + k_2)^2} \qquad (12.1.38)$$

The transmission coefficient is

$$T = \frac{k_2}{k_1} \frac{|C|^2}{|A|^2} = \frac{4k_1 k_2}{(k_1 + k_2)^2}$$

(12.1.39)

and we check that $T + R = 1$.

Maple Examples

Probability current densities as well as reflection and transmission coefficients are computed in the Maple worksheet below for particles incident on a step potential and a barrier.

Key Maple commands: *assign, assume, conjugate, D, diff, evalc, simplify, solve*

Programming: Function statements using '$\rightarrow$'

restart

Transmission and Reflection from a Step Potential

assume$(k_1 > 0, k_2 > 0, m > 0, h > 0)$
assume$(k_1, \text{'real'}, k_2, \text{'real'}, x, \text{'real'}, m, \text{'real'}, h, \text{'real'})$
psiI $:= x \rightarrow A \cdot \exp(I \cdot k_1 \cdot x)$

$$psiI := x \mapsto A e^{I k_1 x}$$

psiR $:= x \rightarrow B \cdot \exp(-I \cdot k_1 \cdot x);$

$$psiR := x \mapsto B e^{-I k_1 x}$$

psiT $:= x \rightarrow C \cdot \exp(I \cdot k_2 \cdot x);$

$$psiT := x \mapsto C e^{I k_2 x}$$

$$pcdR := \left(\frac{hbar}{2 \cdot m \cdot I} \right) \cdot evalc \Big(conjugate \big(psiR(x) \big) \cdot diff \big(psiR(x), x \big) - psiR(x) \Big)$$
$$\cdot diff \big(conjugate \big(psiR(x) \big), x \big)$$

$$pcdR := -\frac{\hbar B^2 k_{1\sim}}{m \sim}$$

$$pcdI := \left(\frac{hbar}{2 \cdot m \cdot I} \right) \cdot evalc \Big(conjugate \big(psiI(x) \big) \cdot diff \big(psiI(x), x \big) - psiI(x) \Big)$$
$$\cdot diff \big(conjugate \big(psiI(x) \big), x \big) \Big)$$

$$pcdI := \frac{\hbar A^2 k_{1\sim}}{m \sim}$$

$$pcdT := \left(\frac{hbar}{2 \cdot m \cdot I}\right) \cdot evalc\left(conjugate\left(psiT(x)\right) \cdot diff\left(psiT(x), x\right) - psiT(x)\right.$$
$$\left. \cdot diff\left(conjugate\left(psiT(x)\right), x\right)\right)$$

$$pcdT := \frac{\hbar C^2 k_{2\sim}}{m \sim}$$

$$T := simplify\left(\frac{pcdT}{pcdI}\right)$$

$$T := \frac{C^2 k_{2\sim}}{A^2 k_{1\sim}}$$

$$R := \frac{-pcdR}{pcdI}$$

$$R := \frac{B^2}{A^2}$$

$$bc1 := psiI(0) + psiR(0) = psiT(0)$$
$$bc1 := A + B = C$$
$$bc2 := D(psiI)(0) + D(psiR)(0) = D(psiT)(0)$$
$$bc2 := IAk_{1\sim} - IBk_{1\sim} = ICk_{2\sim}$$

$$constants := solve(\{bc1, bc2\}, \{A, B\})$$

$$constants := \left\{ A = \frac{C\left(k_{1\sim} + k_{2\sim}\right)}{2k_{1\sim}}, B = \frac{C\left(k_{1\sim} - k_{2\sim}\right)}{2k_{1\sim}} \right\}$$

$$assign(constants)$$
$$simplify(T)$$

$$\frac{4k_{2\sim} k_{1\sim}}{\left(k_{1\sim} + k_{2\sim}\right)^2}$$

$$simplify(R)$$

$$\frac{\left(k_{1\sim} - k_{2\sim}\right)^2}{\left(k_{1\sim} + k_{2\sim}\right)^2}$$

$simplify(T + R)$

$$1$$

Transmission and Reflection from a Barrier

restart
$assume(k_1 > 0, k_2 > 0, k_3 > 0)$
$assume(k_1, 'real', k_2, 'real', L, 'real', x, 'real', k_3, 'real')$
$psiI := x \rightarrow \exp(I \cdot k_1 \cdot x)$

$$psiI := x \mapsto e^{Ik_1 x}$$

$psiR := x \rightarrow R \cdot \exp(-I \cdot k_1 \cdot x);$

$$psiR := x \mapsto Re^{-Ik_1 x}$$

$psiM := x \rightarrow A \cdot \exp(k_2 \cdot x) + B \cdot \exp(k_2 \cdot x);$

$$psiM := x \mapsto Ae^{k_2 x} + Be^{-k_2 x}$$

$psiT := x \rightarrow T \cdot \exp(I \cdot k_1 \cdot x);$

$$psiT := x \mapsto Te^{Ik_1 x}$$

$$pcdR := \left(\frac{hbar}{2 \cdot m \cdot I}\right) \cdot evalc \left(conjugate\left(psiR(x)\right) \cdot diff\left(psiR(x), x\right) - psiR(x)\right.$$
$$\left. \cdot diff\left(conjugate\left(psiR(x)\right), x\right)\right)$$

$$pcdR := -\frac{\hbar R^2 k_{1\sim}}{m}$$

$$pcdI := \left(\frac{hbar}{2 \cdot m \cdot I}\right) \cdot evalc \left(conjugate\left(psiI(x)\right) \cdot diff\left(psiI(x), x\right) - psiI(x)\right.$$
$$\left. \cdot diff\left(conjugate\left(psiI(x)\right), x\right)\right)$$

$$pcdI := \frac{\hbar k_{1\sim}}{m}$$

$$pcdT := \left(\frac{hbar}{2 \cdot m \cdot I}\right) \cdot evalc \left(conjugate\left(psiT(x)\right) \cdot diff\left(psiT(x), x\right) - psiT(x)\right.$$
$$\left. \cdot diff\left(conjugate\left(psiT(x)\right), x\right)\right)$$

$$pcdT := \frac{\hbar T^2 k_{1\sim}}{m}$$

$$Trans := \frac{abs(pcdT)}{abs(pcdI)}$$

$$Trans := |T|^2$$

$$Refl := \frac{abs(pcdR)}{abs(pcdI)}$$

$$Refl := |R|^2$$

$$bc1 := psiI(0) + psiR(0) = psiM(0)$$

$$bc1 := 1 + R = A + B$$

$$bc2 := D(psiI)(0) + D(psiR)(0) = D(psiM)(0)$$

$$bc2 := Ik_{1\sim} - IRk_{1\sim} = Ak_{2\sim} - Bk_{2\sim}$$

$$bc3 := psiM(L) = psiT(L)$$

$$bc3 := Ae^{k_{2\sim}L_\sim} + Be^{-k_{2\sim}L_\sim} = Te^{Ik_{1\sim}L_\sim}$$

$$bc4 := D(psiM)(L) = D(psiT)(L)$$

$$bc4 := Ak_{2\sim}e^{k_{2\sim}L_\sim} - Bk_{2\sim}e^{-k_{2\sim}L_\sim} = ITk_{1\sim}e^{Ik_{1\sim}L_\sim}$$

$constants := solve(\{bc1, bc2, bc3, bc4\}, \{R, A, B, T\}):$
$assign(constants)$
$simplify(expand(T))$

$$\frac{4Ie^{-L_\sim(Ik_{1\sim}-^k{2\sim})}k_{1\sim}k_{2\sim}}{\left(2Ik_{1\sim}k_{2\sim}+k_{1\sim}^2+k_{2\sim}^2\right)e^{2k_{2\sim}L_\sim}+2Ik_{1\sim}k_{2\sim}-k_{1\sim}^2+k_{2\sim}^2}$$

$simplify(expand(R))$

$$\frac{\left(k_{1\sim}^2+k_{2\sim}^2\right)\left(e^{2k_{2\sim}L_\sim}-1\right)}{\left(2Ik_{1\sim}k_{2\sim}+k_{1\sim}^2+k_{2\sim}^2\right)e^{2k_{2\sim}L_\sim}+2Ik_{1\sim}k_{2\sim}-k_{1\sim}^2+k_{2\sim}^2}$$

$simplify(abs(T)^2 + abs(R)^2)$

$$1$$

$simplify(Trans + Refl)$

$$1$$

12.2 BOUND STATES I

In this section, we investigate bound state solutions to the time-independent Schrödinger equation

$$-\frac{\hbar^2}{2m}\frac{\partial^2}{\partial x^2}\psi + V(x)\psi = E\psi \tag{12.2.1}$$

where the potential V is piecewise constant. Examples include a particle in a one-dimensional box, the semi-infinite square well, and a square well with a step.

12.2.1 Particle in a Box

Example 12.2.1

A particle is trapped in a one-dimensional box described by the potential

$$V(x) = \begin{cases} 0 & 0 < x < L \\ \infty & x \le 0 \text{ or } x \ge L \end{cases} \tag{12.2.2}$$

This potential may be used to model the vibrational spectra of atoms as well as light absorption in molecules. Solve the Schrödinger equation and apply boundary conditions at $x = 0$, L to determine the form of the wavefunction within a normalization constant. Find the allowed energy levels. Normalize the wavefunction by requiring unit probability of locating the particle somewhere inside the well.

Solution: Outside the box the potential is infinite where there is no probability of locating the particle. Hence the wavefunction must vanish in regions $x \le 0$ and $x \ge L$. Inside the box where $V(x) = 0$ the Schrödinger equation has the form

$$\frac{\partial^2 \psi}{\partial x^2} = -\frac{2mE}{\hbar^2}\psi \tag{12.2.3}$$

Making the substitution

$$k^2 = \frac{2mE}{\hbar^2} \tag{12.2.4}$$

the general solution to this differential equation is $\psi(x) = A\sin(kx) + B\cos(kx)$. The wavefunction should vanish at the walls $x = 0$ and $x = L$. At $x = 0$, $\psi(0) = A\sin(0) + B\cos(0) = B$ and we have that $B = 0$. At $x = L$, $\psi(L) = A\sin(kL)$ so we require that $kL = n\pi$, $n = 0, 1, 2, 3...$ for $\psi(L) = 0$. For integer values of n, the above condition permits only discrete or quantized energy levels. With $p = \hbar k$ and $E = p^2/2m$ we have that

$$E_n = \left(\frac{\hbar^2 \pi^2}{2mL^2}\right) n^2 \tag{12.2.5}$$

Now that we have the energy levels, we return to the wavefunctions expressed as

$$\psi_n(x) = A \sin\left(\frac{n\pi x}{L}\right) \tag{12.2.6}$$

The constant A is determined by requiring that the wavefunctions are normalized so that the probability of locating the particle somewhere along the x-axis is

$$\int_{-\infty}^{\infty} \psi^* \psi \, dx = 1 \tag{12.2.7}$$

Since the particle is constrained between 0 and L, we only must integrate over this interval and our normalization condition becomes

$$A^2 \int_0^L \sin^2\left(\frac{n\pi x}{L}\right) dx = 1 \tag{12.2.8}$$

Making use of the trigonometric identity

$$\sin^2\left(\frac{n\pi x}{L}\right) = \frac{1}{2}\left(1 - \cos\frac{2n\pi x}{L}\right) \tag{12.2.9}$$

we have

$$\int_0^L \sin^2\left(\frac{n\pi x}{L}\right) dx = \frac{L}{2} \tag{12.2.10}$$

so that

$$A = \sqrt{\frac{2}{L}} \tag{12.2.11}$$

and the normalized wavefunction is now

$$\psi_n(x) = \sqrt{\frac{2}{L}} \sin\left(\frac{n\pi x}{L}\right) \tag{12.2.12}$$

12.2.2 Semi-Infinite Square Well

Example 12.2.2

A particle with energy $E < V_0$ is contained in a semi-infinite square well with potential

$$V(x) = \begin{cases} \infty & x < 0 \\ 0 & 0 \le x \le L \\ V_0 & x > L \end{cases} \tag{12.2.13}$$

Solve the Schrödinger equation and apply boundary conditions at $x = 0$, L and ∞ to determine the form of the wavefunction within an overall constant in each region. Obtain a transcendental equation whose roots determine the allowed energy values.

Solution: The time-independent Schrödinger equation in region I where $V(x) = 0$ for $0 \leq x \leq L$ is written

$$\psi_I'' = -\frac{2mE}{\hbar^2}\psi_I \tag{12.2.14}$$

In region II where $V(x) = V_0$ for $x > L$

$$\psi_{II}'' = \frac{2m(V_0 - E)}{\hbar^2}\psi_{II} \tag{12.2.15}$$

For $x < 0$ we have $\psi = 0$ since $V = \infty$ there. Making the substitution in region I

$$\frac{2mE}{\hbar^2} = k^2 \tag{12.2.16}$$

and in region II we let

$$\frac{2m(V_0 - E)}{\hbar^2} = \alpha^2 \tag{12.2.17}$$

The solutions to $\psi_I'' = -k^2\psi_I$ are

$$\psi_I = A\sin(kx) + B\cos(kx) \tag{12.2.18}$$

while solutions to $\psi_{II}'' = \alpha^2\psi_{II}$ are

$$\psi_{II} = Ce^{\alpha x} + De^{-\alpha x} \tag{12.2.19}$$

where A, B, C and D are constants. Applying the boundary condition at $x = 0$

$$\psi_I(0) = A\sin(k \cdot 0) + B\cos(k \cdot 0) = 0 \tag{12.2.20}$$

gives $B = 0$. The wavefunction must vanish at $x = \infty$ so we have $C = 0$ and

$$\psi_I = A\sin(kx) \tag{12.2.21}$$

$$\psi_{II} = De^{-\alpha x} \tag{12.2.22}$$

At $x = L$ the wavefunction ψ as well as ψ' must be continuous

$$\psi_I(L) = \psi_{II}(L) \tag{12.2.23}$$

$$\psi'_{\mathrm{I}}(L) = \psi'_{\mathrm{II}}(L) \tag{12.2.24}$$

These boundary conditions give

$$A \sin kL = D e^{-\alpha L} \tag{12.2.25}$$

$$A k \cos kL = -D \alpha e^{-\alpha L} \tag{12.2.26}$$

Dividing these two equations results in a transcendental equation whose roots give the energy eigenvalues

$$\tan(kL) = -\frac{k}{\alpha} \tag{12.2.27}$$

In terms of the energies the transcendental equation becomes

$$\tan\left(\sqrt{E}\lambda\right) = -\frac{\sqrt{E}\lambda}{\sqrt{V_0 - E\lambda}} \tag{12.2.28}$$

where $\lambda = \sqrt{2m/\hbar^2}\, L$. Roots of the transcendental equations must be found numerically. The left-hand and right-hand sides of the transcendental equation are plotted as a function of E and the intersection of these curves gives the allowed energy values. Note that we have not found the constants A and D yet. They canceled when we took the ratio of boundary conditions above. To get the constants we use either one of the boundary conditions and the normalization condition

$$\int_{-\infty}^{\infty} \psi^* \psi\, dx = \int_0^L A^2 \sin^2(kx)\, dx + \int_L^{\infty} D^2 e^{-2\alpha x}\, dx = 1 \tag{12.2.29}$$

12.2.3 Square Well with a Step

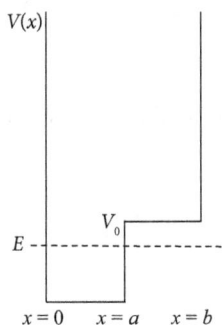

Figure 12.2.1: Square well potential with a step

Example 12.2.3

Consider the potential function shown in Figure 12.2.1 defined by

$$V(x) = \begin{cases} \infty & x < 0 \\ 0 & 0 \leq x < a \\ V_0 & a \leq x < b \\ \infty & b \leq x \end{cases} \qquad (12.2.30)$$

Solve the Schrödinger equation and apply boundary conditions at $x = 0$, a and b to determine the form of the wavefunction within an overall constant in each region for $E < V_0$. Derive a transcendental equation whose roots determine the allowed energy values.

Solution: The boundary condition $\psi_I(0) = 0$ gives

$$\psi_I(x) = A \sin(kx) \qquad (12.2.31)$$

in region I while in region II

$$\psi_{II}(x) = Be^{-\alpha x} + Ce^{\alpha x} \qquad (12.2.32)$$

Continuity of the wavefunction $\psi_I(a) = \psi_{II}(a)$ and its slope $\psi_I'(a) = \psi_{II}'(a)$ give the equations

$$A \sin(ka) = Be^{-\alpha a} + Ce^{\alpha a} \qquad (12.2.33)$$

and

$$Ak \cos(ka) = -B\alpha e^{-\alpha a} + C\alpha e^{\alpha a} \qquad (12.2.34)$$

Because the potential is infinite for $x > b$ we have $\psi_{II}(b) = 0$ or

$$Be^{-\alpha b} + Ce^{\alpha b} = 0 \qquad (12.2.35)$$

giving a relation between the constants

$$B = -Ce^{2\alpha b} \qquad (12.2.36)$$

The transcendental equation is obtained by division of the boundary conditions at $x = a$

$$\frac{\tan ka}{k} = \frac{1}{\alpha} \frac{e^{\alpha a} - e^{-\alpha a}e^{2\alpha b}}{e^{\alpha a} + e^{-\alpha a}e^{2\alpha b}} \qquad (12.2.37)$$

$$\tan ka = \frac{k}{\alpha} \frac{-e^{\alpha(a-b)} + e^{-\alpha(a-b)}}{e^{\alpha(a-b)} + e^{-\alpha(a-b)}} \qquad (12.2.38)$$

$$\tan ka = -\frac{k}{\alpha} \tanh\left[\alpha(b-a)\right] \qquad (12.2.39)$$

Maple Example

The transcendental equation for the semi-infinite square well is obtained in the Maple worksheet below.

Key Maple commands: *assume, diff, lhs, rhs, subs*

Programming: Function statements using ' $\rightarrow$ '

restart

Semi-Infinite Well

$assume(k > 0, \text{alpha} > 0, L > 0)$
$psi1 := (x) \rightarrow A \cdot \sin(k \cdot x)$

$$x \rightarrow A \cdot \sin(k \cdot x)$$

$psi2 := (x) \rightarrow B \cdot \exp(-\text{alpha} \cdot x)$

$$x \rightarrow B e^{-\alpha x}$$

$psiCont := psi1(L) = psi2(L)$

$$A\sin(kL) = Be^{-\alpha L}$$

$dpsiCont := subs(x = L, diff(psi1(x), x) = diff(psi2(x), x))$

$$Ak\cos(kL) = -B\alpha e^{-\alpha L}$$

$$simplify\left(\frac{rhs(psiCont)}{rhs(dpsiCont)} = \frac{lhs(psiCont)}{lhs(dpsiCont)}, trig \right)$$

$$-\frac{1}{\alpha} = \frac{\sin(kL)}{k\cos(kL)}$$

expand(%)

$$-\frac{1}{\alpha} = \frac{\sin(kL)}{k\cos(kL)}$$

12.3 BOUND STATES II

We consider bound state solutions to the time-independent Schrödinger equation (TISE) that are not piecewise constant in this section. Examples include the delta function potential, the quantum bouncer, and the harmonic oscillator. The harmonic oscillator is also described using the operator formalism.

12.3.1 Delta Function Potential

Given the one-dimensional delta function potential located at $x = 0$

$$V(x) = -V_0 \delta(x) \tag{12.3.1}$$

the TISE is

$$-\frac{\hbar^2}{2m} \psi'' - V_0 \delta(x) \psi = E\psi \tag{12.3.2}$$

Example 12.3.1

Find the energy level and normalized wavefunction of the bound state solution with the delta function potential located at the origin.

Solution: Away from the delta function the TISE is of the form

$$-\frac{\hbar^2}{2m} \psi'' = E\psi \tag{12.3.3}$$

For bound state solutions, $E < 0$ and $\psi'' = k^2 \psi$ with

$$k = \sqrt{-\frac{2mE}{\hbar^2}} \tag{12.3.4}$$

we have solutions

$$\psi(x) = C_1 e^{kx} + C_2 e^{-kx} \tag{12.3.5}$$

Requiring that ψ is finite at $\pm\infty$

$$\psi(x) = \begin{cases} C_1 e^{kx} & x \le 0 \\ C_2 e^{-kx} & x \ge 0 \end{cases} \tag{12.3.6}$$

Continuity of ψ at $x = 0$ gives $C_1 = C_2 = C$. To determine C, we apply the normalization condition

$$\int_{-\infty}^{\infty} \psi^2(x) dx = C^2 \left(\int_{-\infty}^{0} e^{2kx} dx + \int_{0}^{\infty} e^{-2kx} dx \right) = 1 \tag{12.3.7}$$

and find $C = \sqrt{k}$. The bound state energy is found by integrating the TISE over the delta function

$$-\frac{\hbar^2}{2m} \int_{-\varepsilon}^{\varepsilon} \psi'' dx - V_0 \int_{-\varepsilon}^{\varepsilon} \psi \delta(x) dx = E \int_{-\varepsilon}^{\varepsilon} \psi dx \tag{12.3.8}$$

This gives

$$-\frac{\hbar^2}{2m} \left[\psi'(x) \right]_{-\varepsilon}^{\varepsilon} - V_0 \psi(0) = 0 \tag{12.3.9}$$

$$\psi'(\varepsilon) - \psi'(-\varepsilon) = -\frac{2m}{\hbar^2} V_0 C \qquad (12.3.10)$$

since $\psi(0) = C$. Taking the limit $\varepsilon \to 0$ we have

$$\lim_{\varepsilon \to 0}\left(-kCe^{-k\varepsilon} - kCe^{k\varepsilon}\right) = -2kC = -\frac{2m}{\hbar^2} V_0 C \qquad (12.3.11)$$

thus

$$k = \frac{mV_0}{\hbar^2} = \sqrt{-\frac{2mE}{\hbar^2}} \qquad (12.3.12)$$

with energy

$$E = -\frac{mV_0^2}{2\hbar^2} \qquad (12.3.13)$$

and normalized wavefunction

$$\psi(x) = \sqrt{\frac{mV_0}{\hbar^2}} \begin{cases} e^{kx} & x \le 0 \\ e^{-kx} & x \ge 0 \end{cases} \qquad (12.3.14)$$

12.3.2 Quantum Bouncer

A quantum bouncing ball with a very small mass m released in a gravitation field g is modeled with the potential energy function

$$V(y) = \begin{cases} mgy & y \ge 0 \\ \infty & y < 0 \end{cases} \qquad (12.3.15)$$

and TISE

$$-\frac{\hbar^2}{2m}\frac{\partial^2}{\partial y^2}\psi + mgy\psi = E\psi \qquad (12.3.16)$$

Example 12.3.2

Find the quantized energy levels and wavefunctions of the quantum bouncer.

Solution: Substituting $\beta = 2m^2 g/\hbar^2$ and $\alpha = 2mE/\hbar^2$ into the TISE

$$\psi'' = (\beta y - \alpha)\psi \qquad (12.3.17)$$

we find the general solution as a linear combination of Airy functions

$$\psi(y) = C_1 \mathrm{Ai}\left(\frac{\beta y - \alpha}{\beta^{2/3}}\right) + C_2 \mathrm{Bi}\left(\frac{\beta y - \alpha}{\beta^{2/3}}\right) \qquad (12.3.18)$$

The constant $C_2 = 0$ since $Bi(x)$ diverges for large x. The quantized energy levels are determined by the boundary condition $\psi(0) = 0$

$$-\alpha/\beta^{2/3} = x_n \tag{12.3.19}$$

where x_n by the roots of $Ai(x) = 0$. The quantized energy levels are

$$E_n = -x_n \left(\frac{mg^2\hbar^2}{2} \right)^{1/3} \tag{12.3.20}$$

The first three energies are given by $x_1 = -2.338$, $x_2 = -4.088$ and $x_3 = -5.521$.

12.3.3 Harmonic Oscillator

Example 12.3.3

Develop a quantum description of a mass subject to a restoring force $F = -kx$ corresponding to a potential $V(x) = kx^2/2$ with one-dimensional Schrödinger equation

$$-\frac{\hbar^2}{2m} \frac{\partial^2 \psi}{\partial x^2} + \frac{1}{2} kx^2 \psi = E\psi \tag{12.3.21}$$

Find the ground state energy and normalized wavefunction.

Solution: The natural frequency of the spring is $\omega = \sqrt{k/m}$ so that $k = m\omega^2$ and

$$\frac{\partial^2 \psi}{\partial x^2} - \frac{m^2\omega^2 x^2}{\hbar^2} \psi = \frac{2mE}{\hbar^2} \psi \tag{12.3.22}$$

We first seek a trial solution of the form

$$\psi(x) = A e^{-\alpha x^2} \tag{12.3.23}$$

Calculating the first derivative

$$\frac{\partial \psi}{\partial x} = -2\alpha x A e^{-\alpha x^2} \tag{12.3.24}$$

and the second derivative

$$\frac{\partial^2 \psi}{\partial x^2} = -2\alpha A e^{-\alpha x^2} + 4\alpha^2 x^2 A e^{-\alpha x^2} \tag{12.3.25}$$

Substitution these derivatives into the Schrödinger equation

$$-2\alpha A e^{-\alpha x^2} + 4\alpha^2 x^2 A e^{-\alpha x^2} + \left(\frac{2mE}{\hbar^2} - \frac{m^2\omega^2 x^2}{\hbar^2} \right) A e^{-\alpha x^2} = 0 \tag{12.3.26}$$

and then canceling the Ae^{-ax^2} and equating constant terms gives

$$2\alpha = \frac{2mE}{\hbar^2} \qquad (12.3.27)$$

Equating terms proportional to x^2

$$4\alpha^2 = \frac{m^2\omega^2}{\hbar^2} \qquad (12.3.28)$$

Thus, we have

$$\alpha = \frac{mE}{\hbar^2} \text{ and } \alpha = \frac{m\omega}{2\hbar} \qquad (12.3.29)$$

so that the energy of the ground state is

$$E = \frac{1}{2}\hbar\omega \qquad (12.3.30)$$

and the ground state wavefunction

$$\psi(x) = Ae^{-\frac{m\omega}{2\hbar}x^2} \qquad (12.3.31)$$

The constant A is determined by normalization

$$\int_{-\infty}^{\infty} \psi^*\psi dx = A^2 \int_{-\infty}^{\infty} e^{-2\alpha x^2} dx = 1 \qquad (12.3.32)$$

This gives

$$A^2\sqrt{\frac{\pi}{2\alpha}} = 1 \text{ or } A = \left(\frac{m\omega}{\pi\hbar}\right)^{1/4} \qquad (12.3.33)$$

12.3.4 Operator Notation

The TISE for the harmonic oscillator

$$-\frac{\hbar^2}{2m}\frac{\partial^2\psi}{\partial x^2} + \frac{1}{2}m\omega^2x^2\psi = E\psi \qquad (12.3.34)$$

can be expressed as $\hat{H}\psi = E\psi$ where the Hamiltonian operator

$$\hat{H} = -\frac{\hbar^2}{2m}\frac{\partial^2}{\partial x^2} + \frac{1}{2}m\omega^2x^2 \qquad (12.3.35)$$

In terms of the quantum mechanical operators

$$\hat{p} = \frac{\hbar}{i}\frac{\partial}{\partial x} \text{ and } \hat{x} = x \qquad (12.3.36)$$

$$\hat{H} = \frac{\hat{p}^2}{2m} + \frac{1}{2}m\omega^2\hat{x}^2 \qquad (12.3.37)$$

Next, we define the operators

$$\hat{a} = \sqrt{\frac{m\omega}{2\hbar}}\left(x + \frac{i\hat{p}}{m\omega}\right) \tag{12.3.38}$$

$$\hat{a}^\dagger = \sqrt{\frac{m\omega}{2\hbar}}\left(x - \frac{i\hat{p}}{m\omega}\right) \tag{12.3.39}$$

where $\hat{a}$ and $\hat{a}^\dagger$ are known as annihilation and creation operators, respectively. Adding and subtracting these operators

$$\hat{a} + \hat{a}^\dagger = 2\sqrt{\frac{m\omega}{2\hbar}}\hat{x}$$

$$\hat{a} - \hat{a}^\dagger = 2\sqrt{\frac{m\omega}{2\hbar}}\frac{i\hat{p}}{m\omega} \tag{12.3.40}$$

and solving for the position and momentum operators

$$\hat{x} = \sqrt{\frac{2\hbar}{m\omega}}\frac{\hat{a} + \hat{a}^\dagger}{2}$$

$$\hat{p} = \sqrt{2\hbar m\omega}\frac{\hat{a} - \hat{a}^\dagger}{2i} \tag{12.3.41}$$

Substituting these into our Hamiltonian

$$\hat{H} = -\frac{1}{4}\hbar\omega\left(\hat{a} - \hat{a}^\dagger\right)^2 + \frac{1}{4}\hbar\omega\left(\hat{a} + \hat{a}^\dagger\right)^2 \tag{12.3.42}$$

$$\hat{H} = \frac{1}{4}\hbar\omega\left[\left(\hat{a} + \hat{a}^\dagger\right)^2 - \left(\hat{a} - \hat{a}^\dagger\right)^2\right] \tag{12.3.43}$$

$$\hat{H} = \frac{1}{4}\hbar\omega\left[\hat{a}^2 + \hat{a}\hat{a}^\dagger + \hat{a}^\dagger\hat{a} + \hat{a}^{\dagger 2} - \hat{a}^2 + \hat{a}\hat{a}^\dagger + \hat{a}^\dagger\hat{a} - \hat{a}^{\dagger 2}\right] \tag{12.3.44}$$

$$\hat{H} = \frac{1}{2}\hbar\omega\left[\hat{a}\hat{a}^\dagger + \hat{a}^\dagger\hat{a}\right] \tag{12.3.45}$$

Consider the commutator between annihilation and creation operators

$$\left[\hat{a}, \hat{a}^\dagger\right] = \frac{m\omega}{2\hbar}\left[\hat{x} + \frac{i\hat{p}}{m\omega}, \hat{x} - \frac{i\hat{p}}{m\omega}\right]$$

$$= \frac{i}{2\hbar}\left(\left[\hat{p}, \hat{x}\right] - \left[\hat{x}, \hat{p}\right]\right) \tag{12.3.46}$$

$$= \frac{i}{2\hbar}\left(-i\hbar - i\hbar\right) = 1$$

Since $\left[\hat{x},\hat{p}\right]=i\hbar$ and $\hat{a}\hat{a}^{\dagger}-\hat{a}^{\dagger}\hat{a}=1$ or $\hat{a}\hat{a}^{\dagger}=1+\hat{a}^{\dagger}\hat{a}$ our Hamiltonian can be written as

$$
\begin{aligned}
\hat{H} &= \frac{1}{2}\hbar\omega\left[1+\hat{a}^{\dagger}\hat{a}+\hat{a}^{\dagger}\hat{a}\right] \\
&= \hbar\omega\left[\hat{a}^{\dagger}\hat{a}+\frac{1}{2}\right]
\end{aligned}
\tag{12.3.47}
$$

12.3.5 Excited States of the Harmonic Oscillator

We can use creation and annihilation operators to deduce wavefunctions and higher energy levels of the harmonic oscillator. If we operate on the ground state $\psi_0(x)$ with $\hat{a}$ we obtain

$$
\begin{aligned}
\hat{a}\psi_0(x) &= \sqrt{\frac{m\omega}{2\hbar}}\left(x+\frac{\hbar}{m\omega}\frac{\partial}{\partial x}\right)\psi_0(x) \\
&= \sqrt{\frac{m\omega}{2\hbar}}\left(\frac{m\omega}{\pi\hbar}\right)^{1/4}\left(x+\frac{\hbar}{m\omega}\frac{\partial}{\partial x}\right)e^{-\frac{m\omega}{2\hbar}x^2} \\
&= \sqrt{\frac{m\omega}{2\hbar}}\left(\frac{m\omega}{\pi\hbar}\right)^{1/4}\left(xe^{-\frac{m\omega}{2\hbar}x^2}-\frac{\hbar}{m\omega}\left(\frac{m\omega}{2\hbar}\right)e^{-\frac{m\omega}{2\hbar}x^2}(2x)\right)=0
\end{aligned}
\tag{12.3.48}
$$

$$
\begin{aligned}
\hat{a}^{\dagger}\psi_0(x) &= \sqrt{\frac{m\omega}{2\hbar}}\left(x-\frac{\hbar}{m\omega}\frac{\partial}{\partial x}\right)\psi_0(x) \\
&= \sqrt{\frac{m\omega}{2\hbar}}\left(\frac{m\omega}{\pi\hbar}\right)^{1/4}\left(x-\frac{\hbar}{m\omega}\frac{\partial}{\partial x}\right)e^{-\frac{m\omega}{2\hbar}x^2} \\
&= \sqrt{\frac{m\omega}{2\hbar}}\left(\frac{m\omega}{\pi\hbar}\right)^{1/4}\left(xe^{-\frac{m\omega}{2\hbar}x^2}+\frac{\hbar}{m\omega}\left(\frac{m\omega}{2\hbar}\right)e^{-\frac{m\omega}{2\hbar}x^2}(2x)\right) \\
&= \sqrt{\frac{m\omega}{2\hbar}}\left(\frac{m\omega}{\pi\hbar}\right)^{1/4}2xe^{-\frac{m\omega}{2\hbar}x^2}=\sqrt{2}\psi_1(x)
\end{aligned}
\tag{12.3.49}
$$

Using the notation $\psi_n(x)=|n\rangle$ we have that $\hat{a}|0\rangle=0$ and $\hat{a}^{\dagger}|0\rangle=\sqrt{2}|1\rangle$. Subsequent operation reveals that

$$
\hat{a}^{\dagger}|n\rangle=\sqrt{n+1}|n+1\rangle
\tag{12.3.50}
$$

$$
\hat{a}|n\rangle=\sqrt{n}|n-1\rangle
\tag{12.3.51}
$$

The creation operator allows us to construct any state $|n\rangle$ from the ground state $|0\rangle$

$$|n\rangle = \frac{\left(\hat{a}^{\dagger}\right)^{n}}{\sqrt{n!}}|0\rangle \qquad (12.3.52)$$

Consider operation on the state $|3\rangle$ first by $\hat{a}$ and then by $\hat{a}^{\dagger}$

$$\hat{a}^{\dagger}\hat{a}|3\rangle = \hat{a}^{\dagger}\sqrt{3}|2\rangle = \sqrt{3}\hat{a}^{\dagger}|2\rangle = \sqrt{3}\sqrt{3}|3\rangle = 3|3\rangle \qquad (12.3.53)$$

In general, $\hat{N}|n\rangle = n|n\rangle$ where the number operator

$$\hat{N} = \hat{a}^{\dagger}\hat{a} \qquad (12.3.54)$$

The Hamiltonian can also be written in terms of the number operator

$$\hat{H} = \hbar\omega\left[\hat{a}^{\dagger}\hat{a} + \frac{1}{2}\right] = \hbar\omega\left(\hat{N} + \frac{1}{2}\right) \qquad (12.3.55)$$

By operating on $|n\rangle$ by the Hamiltonian

$$\hat{H}|n\rangle = \hbar\omega\left(n + \frac{1}{2}\right)|n\rangle \qquad (12.3.56)$$

we find that the energy levels of the harmonic oscillator are evenly spaced

$$E_{n} = \hbar\omega\left(n + \frac{1}{2}\right) \qquad (12.3.57)$$

Maple Examples

The general solution to the quantum bouncer is found in the Maple worksheet below. The wavefunctions corresponding to the first four energy levels of the quantum harmonic oscillator are then plotted. Quantum uncertainties are calculated for several oscillator wavefunctions. A nonstationary superposition state of the quantum harmonic oscillator is animated. Operator notation is demonstrated using the Maple *Physics* package.

Key Maple commands: *animate, Annihilation, assume, Bra, conjugate, Creation, diff, dsolve, eval, expand, int, Ket, plot, simplify, subs*

Maple packages: *with(orthopoly): with(plots): with(Physics):*

Programming: Function statements using '→'

Special functions: HermiteH

restart

Quantum Bouncer

$assume(m > 0, h > 0, g > 0, E > 0);$

$$Deq1 := diff\left(psi(y), y, y\right) = \left(\frac{2m}{h^2}\right) \cdot (m \cdot g \cdot y - E) \cdot psi(y);$$

$$Deq1 := \frac{d^2}{dy^2}\psi(y) = \frac{2m \sim (g \sim m \sim y - E \sim)\psi(y)}{h \sim^2}$$

$dsolve(\{Deq1\}, psi(y));$

$$\left\{ \psi(y) = \_C1 \text{AiryAi}\left(\frac{2^{1/3}\left(g \sim m \sim y - E \sim\right)}{m \sim^{1/3} g \sim^{2/3} h \sim^{2/3}} \right) + \right.$$

$$\left. \_C2 \text{AiryBi}\left(\frac{2^{1/3}\left(g \sim m \sim y - E \sim\right)}{m \sim^{1/3} g \sim^{2/3} h \sim^{2/3}} \right) \right\}$$

$assume(\text{beta} > 0, \text{alpha} > 0);\ Deq2 := diff(psi(y), y, y) = (\text{beta} \cdot y - \text{alpha}) \cdot psi(y);$

$$Deq2 := \frac{d^2}{dy^2}\psi(y) = (\beta \sim y - \alpha \sim)\psi(y)$$

$dsolve(\{Deq2\}, psi(y));$

$$\psi(y) = \_C1 \text{AiryAi}\left(\frac{\beta \sim y - \alpha \sim}{\beta \sim^{2/3}} \right) + \_C2 \text{AiryBi}\left(\frac{\beta \sim y - \alpha \sim}{\beta \sim^{2/3}} \right)$$

Harmonic Oscillator Wavefunctions

$restart$
$with(orthopoly)$

$$[G, H, L, P, T, U]$$

$$w := (x) \to \exp\left(-\frac{x^2}{2}\right)$$

$$w := x \mapsto e^{-\frac{x^2}{2}}$$

$dEqn := diff(y(x), x, x) + (2 \cdot m + 1 - x^2) \cdot y(x) = 0$

$$dEqn := \frac{d^2}{dx^2}y(x) + \left(-x^2 + 2m + 1\right)y(x) = 0$$

$subs(\{y(x) = w(x) \cdot \text{HermiteH}(5, x), m = 5\}, dEqn)$

$$\frac{d^2}{dx^2}\left(e^{-\frac{x^2}{2}}\text{HermiteH}(5,x)\right)+\left(-x^2+11\right)e^{-\frac{x^2}{2}}\text{HermiteH}(5,x)=0$$

$expand(simplify(\%))$

$$0=0$$

Orthogonality of Oscillator Wavefunctions

$$int(w(x)\cdot\text{HermiteH}(3,x)\cdot\text{HermiteH}(2,x),x=\textit{-Infinity}\ldots\textit{Infinity})$$

$$0$$

Plots of Oscillator Wavefunctions and Potential

$$\text{psi}:=(n,x)\rightarrow\frac{1}{\pi^{\frac{1}{4}}\cdot\text{sqrt}\left(2^n\cdot n!\right)}\cdot\exp\left(-\frac{x^2}{2}\right)\cdot\text{HermiteH}(n,x)$$

$$\psi:=(n,x)\rightarrow\frac{e^{-\frac{x^2}{2}}\,\text{HermiteH}(n,x)}{\pi^{1/4}\sqrt{2^n n!}}$$

$$plot\left(\left[seq\left(\text{psi}(n,x)^2+\left(n+\frac{1}{2}\right),n=0\ldots3\right),x^2\right],x=-2\ldots2,legend=\left[\psi_0^2(x),\psi_1^2(x),\psi_2^2(x),\right.\right.$$

$$\left.\left.\psi_3^2(x),V(x)\right],linestyle=\left[dash,dot,dashdot,longdash,solid\right]\right)$$

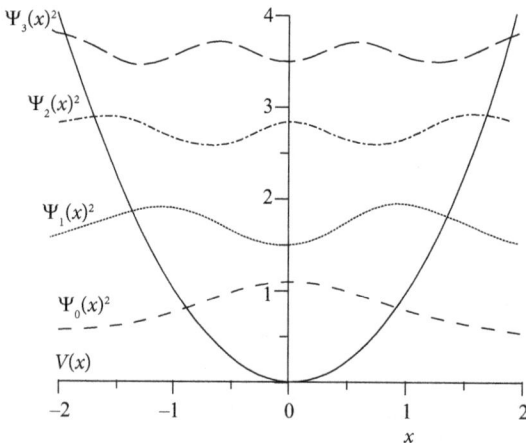

Figure 12.3.1: Harmonic oscillator wavefunctions.

Computation of Uncertainty

$assume(n, \text{'integer'})$

$Unc\_x := (n) \rightarrow sqrt(int(psi(n, x)^2 \cdot x^2, x = -infinity \ldots infinity))$

$$Unc\_x := n \mapsto \sqrt{\int_{-\infty}^{\infty} \psi(n,x)^2 \, x^2 dx}$$

$eval(Unc\_x(3))$

$$\frac{\sqrt{14}}{2}$$

$Unc\_p := (n) \rightarrow sqrt(int(-psi(n, x) \cdot diff(psi(n, x), x, x) \; x = -infinity \ldots infinity))$

$$Unc\_p := n \mapsto \sqrt{\int_{-\infty}^{\infty} -\psi(n,x)\left(\frac{\partial^2}{\partial x^2}\psi(n,x)\right)dx}$$

$eval(Unc\_p(3))$

$$\frac{\sqrt{14}}{2}$$

Minimum Uncertainty of Ground State

$eval(Unc\_x(0) \cdot Unc\_p(0))$

$$\frac{1}{2}$$

Animation of Nonstationary Superposition States

$psiT := (n, x, t) \rightarrow psi(n, x) \cdot exp(-I \cdot n \cdot t)$

$$psiT := (n, x, t) \mapsto \psi(n, x)e^{-Int}$$

$with(plots):$

$animate(plot, [eval(conjugate(psiT(1, x, t) + psiT(2, x, t)) \cdot (psiT(1, x, t) + psiT(2, x, t))), x = -4 \ldots 4], t = 0 \ldots 30, frames = 600)$

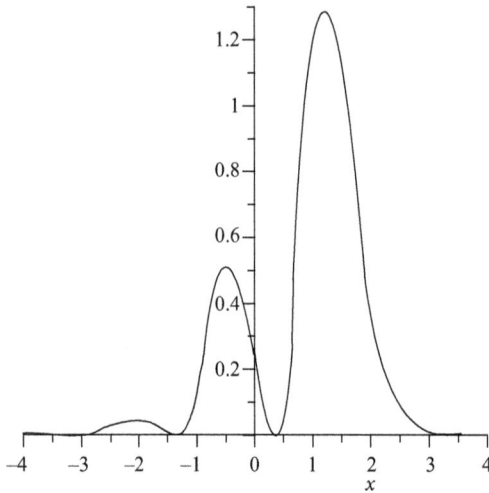

Figure 12.3.2: Animation of a nonstationary superposition state of harmonic oscillator wavefunctions (not normalized).

Operator Notation

restart

with(Physics) :

Setup(mathematicalnotation = true)

$$[mathematicalnotation = true]$$

Ket(v, n)

$$|v_n\rangle$$

Bra(v, m)

$$\langle v_m|$$

Bra(v, m)·Ket(v, n)

$$\delta_{m,n}$$

Annihilation(v)

$$a^-$$

Creation(v)

$$a^+$$

v1 := Annihilation(v)·Ket(v, n)

$$v1 := \sqrt{n}\,|v_{n-1}\rangle$$

Bra(v, m)·v1

$$\sqrt{n}\,\delta_{m,n-1}$$

$Bra(v, m) \cdot Annihilation(v) \cdot Ket(v, n)$

$$\sqrt{m+1}\delta_{n,\, m+1}$$

$Annihilation(v) \cdot Ket(v, 0)$

$$0$$

$Creation(v) \cdot Ket(v, 0)$

$$|v_1\rangle$$

$Creation(v) \cdot \%$

$$\sqrt{2}|v_2\rangle$$

$Creation(v) \cdot \%$

$$\sqrt{6}|v_3\rangle$$

12.4 SCHRÖDINGER EQUATION IN HIGHER DIMENSIONS

In the section, we consider solutions to the time-independent Schrödinger equation in three dimensions. Examples include a particle in a 3D box, Schrödinger's equation in spherical coordinates and the hydrogen radial wavefunctions.

12.4.1 Particle in a 3D Box

We will first consider a particle in a 3D box of side L

$$-\frac{\hbar^2}{2m}\nabla^2\psi(\mathbf{r})+V(\mathbf{r})\psi(\mathbf{r})=E\psi(\mathbf{r}) \qquad (12.4.1)$$

Writing the wavefunction as a product

$$\psi(\mathbf{r})=\psi_1(x)\psi_2(y)\psi_3(z) \qquad (12.4.2)$$

Substituting into the TISE

$$-\frac{\hbar^2}{2m}\frac{\partial^2\psi_1}{\partial x^2}\psi_2\psi_3-\frac{\hbar^2}{2m}\psi_1\frac{\partial^2\psi_2}{\partial y^2}\psi_3-\frac{\hbar^2}{2m}\psi_1\psi_2\frac{\partial^2\psi_3}{\partial z^2}=E\psi_1\psi_2\psi_3 \qquad (12.4.3)$$

Dividing by $\psi_1\psi_2\psi_3$ gives

$$\underbrace{-\frac{\hbar^2}{2m}\frac{1}{\psi_1}\frac{\partial^2\psi_1}{\partial x^2}}_{E_1}\underbrace{-\frac{\hbar^2}{2m}\frac{1}{\psi_2}\frac{\partial^2\psi_2}{\partial y^2}}_{E_2}\underbrace{-\frac{\hbar^2}{2m}\frac{1}{\psi_3}\frac{\partial^2\psi_3}{\partial z^2}}_{E_3}=E \qquad (12.4.4)$$

where the separation constants $E = E_1 + E_2 + E_3$. We now have the three differential equations

$$\frac{\partial^2 \psi_1}{\partial x^2} = -\frac{2mE_1}{\hbar^2}\psi_1 = -k_1{}^2\psi_1 \qquad (12.4.5)$$

$$\frac{\partial^2 \psi_2}{\partial y^2} = -\frac{2mE_2}{\hbar^2}\psi_2 = -k_2{}^2\psi_2 \qquad (12.4.6)$$

$$\frac{\partial^2 \psi_3}{\partial z^2} = -\frac{2mE_3}{\hbar^2}\psi_3 = -k_3{}^2\psi_3 \qquad (12.4.7)$$

Requiring $\psi = 0$ along $x = 0, y = 0$ and $z = 0$ sides

$$\psi_1 = A_1 \sin(k_1 x), \ \psi_2 = A_2 \sin(k_2 y) \text{ and } \psi_3 = A_3 \sin(k_3 z) \qquad (12.4.8)$$

Requiring $\psi = 0$ along $x = L, y = L$ and $z = L$ sides

$$\psi_1 = A_1 \sin(k_1 x), \ \psi_2 = A_2 \sin(k_2 y) \text{ and } \psi_3 = A_3 \sin(k_3 z) \qquad (12.4.9)$$

$$k_1 = \frac{n\pi}{L}, \ k_2 = \frac{m\pi}{L} \text{ and } k_3 = \frac{\ell\pi}{L} \qquad (12.4.10)$$

Combining the constants $A = A_1 A_2 A_3$,

$$\psi = A \sin\left(\frac{n\pi x}{L}\right)\sin\left(\frac{m\pi y}{L}\right)\sin\left(\frac{\ell\pi z}{L}\right) \qquad (12.4.11)$$

Normalization leads to

$$\int_{\text{vol}} \psi^* \psi dv = A^2 \int_0^L \int_0^L \int_0^L \sin^2\left(\frac{n\pi x}{L}\right)\sin^2\left(\frac{m\pi y}{L}\right)\sin^2\left(\frac{\ell\pi z}{L}\right) dx dy dz = 1 \quad (12.4.12)$$

$$A^2 \int_0^L \sin^2\left(\frac{n\pi x}{L}\right)dx \int_0^L \sin^2\left(\frac{m\pi y}{L}\right)dy \int_0^L \sin^2\left(\frac{\ell\pi z}{L}\right)dz = 1 \qquad (12.4.13)$$

with each integral contributing a factor of $L/2$ so we have

$$A = \left(\frac{2}{L}\right)^{3/2} \qquad (12.4.14)$$

The energy levels are

$$E_{n,m,\ell} = \frac{\hbar^2 \pi^2}{2mL^2}\left(n^2 + m^2 + \ell^2\right) \qquad (12.4.15)$$

with the ground state

$$E_{1,1,1} = \frac{3\hbar^2 \pi^2}{2mL^2}$$

(12.4.16)

The next energy level is threefold degenerate where

$$E_{2,1,1} = E_{1,2,1} = E_{1,1,2} = \frac{6\hbar^2 \pi^2}{2mL^2}$$

(12.4.17)

Degeneracy is broken for a box with unequal sides L_1, L_2 and L_3 where

$$E_{n,m,\ell} = \frac{\hbar^2 \pi^2}{2m}\left[\left(\frac{n}{L_1}\right)^2 + \left(\frac{m}{L_2}\right)^2 + \left(\frac{\ell}{L_3}\right)^2\right]$$

(12.4.18)

12.4.2 Schrödinger Equation in Spherical Coordinates

The time-independent Schrödinger equation in 3D is

$$-\frac{\hbar^2}{2m}\nabla^2 \psi + V(r)\psi = E\psi$$

(12.4.19)

In spherical coordinates, the Laplacian is

$$\nabla^2 = \frac{1}{r^2}\left(\frac{\partial}{\partial r}r^2\frac{\partial}{\partial r}\right) + \frac{1}{r^2}\left[\frac{1}{\sin\theta}\frac{\partial}{\partial \theta}\left(\sin\theta\frac{\partial}{\partial \theta}\right) + \frac{1}{\sin^2\theta}\frac{\partial^2}{\partial \phi^2}\right]$$

(12.4.20)

The solution to Schrödinger's equation is obtained by applying separation of variables in spherical coordinates. Substituting

$$\psi(r,\theta,\phi) = R(r)\Theta(\theta)\Phi(\phi)$$

(12.4.21)

into

$$\nabla^2 \psi - \frac{2m}{\hbar^2}V(r)\psi = -\frac{2mE}{\hbar^2}\psi$$

(12.4.22)

and dividing by $R\Theta\Phi$ gives

$$\frac{1}{r^2 R(r)}\left(\frac{\partial}{\partial r}r^2\frac{\partial R(r)}{\partial r}\right) + \frac{1}{r^2 \Theta(\theta)}\left[\frac{1}{\sin\theta}\frac{\partial}{\partial \theta}\left(\sin\theta\frac{\partial\Theta(\theta)}{\partial \theta}\right) + \frac{1}{\sin^2\theta\Phi(\phi)}\frac{\partial^2\Phi(\phi)}{\partial \phi^2}\right]$$
$$= \frac{2m}{\hbar^2}(V(r) - E)\psi$$

(12.4.23)

Separating angular and radial terms

$$
\underbrace{\frac{1}{\Theta(\theta)}\left[\frac{1}{\sin\theta}\frac{\partial}{\partial\theta}\left(\sin\theta\frac{\partial\Theta(\theta)}{\partial\theta}\right)+\frac{1}{\sin^2\theta\Phi(\phi)}\frac{\partial^2\Phi(\phi)}{\partial^2\phi}\right]}_{C}
$$
$$
=\underbrace{-\frac{1}{R(r)}\left(\frac{\partial}{\partial r}r^2\frac{\partial R(r)}{\partial r}\right)+\frac{2m}{\hbar^2}r^2\left(V(r)-E\right)}_{C}
$$

(12.4.24)

The angular equation is further separated

$$
\underbrace{\frac{1}{\Theta(\theta)}\sin\theta\frac{\partial}{\partial\theta}\left(\sin\theta\frac{\partial\Theta(\theta)}{\partial\theta}\right)-C\sin^2\theta}_{D}=\underbrace{-\frac{1}{\Phi(\phi)}\frac{\partial^2\Phi(\phi)}{\partial^2\phi}}_{D}
$$

(12.4.25)

The Φ equation

$$
\frac{\partial^2\Phi(\phi)}{\partial^2\phi}=-D\Phi(\phi)
$$

(12.4.26)

has solutions

$$
\Phi(\phi)=\exp\left(\pm i\sqrt{D}\phi\right)
$$

(12.4.27)

The Θ equation is

$$
\frac{1}{\Theta(\theta)}\sin\theta\frac{\partial}{\partial\theta}\left(\sin\theta\frac{\partial\Theta(\theta)}{\partial\theta}\right)-C\sin^2\theta=D
$$

(12.4.28)

Choosing the separation constants $C = -\ell(\ell+1)$ and $D = m_\ell^2$ solutions for $\Phi(\Theta)$ are associated Legendre polynomials $\Theta(\theta)=P_\ell^{m_\ell}(\cos\theta)$ with allowed values $m_\ell = -\ell, -\ell+1, \ldots 0, \ldots, \ell-1, \ell$. With $\Phi(\phi) = \exp(im_\ell\phi)$ the product $\Theta\Phi$ are the spherical harmonics

$$
Y_\ell^m(\theta,\phi)=\sqrt{\frac{2\ell+1}{4\pi}\frac{(\ell-m)!}{(\ell+m)!}}P_\ell^{m_\ell}(\cos\theta)e^{im_\ell\phi}
$$

(12.4.29)

where the normalization constant is such that

$$
\int_0^{2\pi}\int_0^{\pi}Y_\ell^m(\theta,\phi)\,^*Y_{\ell'}^{m'}(\theta,\phi)\sin(\theta)\,d\theta d\phi=\delta_{\ell\ell'}\delta_{mm'}
$$

(12.4.30)

The angular momentum operators

$$\hat{L}^2 = -\hbar^2 \left\{ \frac{1}{\sin\theta} \left(\sin\theta \frac{\partial}{\partial\theta} \right) + \frac{1}{\sin^2\theta} \frac{\partial^2}{\partial^2\phi} \right\} \qquad (12.4.31)$$

and

$$\hat{L}_z = i\hbar \frac{\partial}{\partial\phi} \qquad (12.4.32)$$

acting on $Y_\ell^m(\theta,\phi)$ give

$$\hat{L}^2 Y_\ell^m = \hbar^2 \ell(\ell+1) Y_\ell^m \qquad (12.4.33)$$

so that the magnitude of the angular momentum vector $L = \hbar\sqrt{\ell(\ell+1)}$ is constant. The operator

$$\hat{L}_z Y_\ell^{m_\ell} = m_\ell \hbar Y_\ell^{m_\ell} \qquad (12.4.34)$$

gives the z-component of angular moment $L_z = \hbar m_\ell$. The angle θ that the angular momentum vector makes with respect to the z-axis is also quantized where

$$\cos\theta = \frac{L_z}{L} = \frac{m_\ell}{\sqrt{\ell(\ell+1)}} \qquad (12.4.35)$$

This is known as space quantization.

Example 12.4.1

Find the allowed orientations of the angular momentum vector for $\ell = 3$

Solution: With the allowed values of m_ℓ = -3, -2, -1, 0, 1, 2, 3 and $\sqrt{\ell(\ell+1)} = 2\sqrt{3}$ we have

$$\theta = \cos^{-1}\left\{ \frac{-3}{2\sqrt{3}}, \frac{-1}{\sqrt{3}}, \frac{-1}{2\sqrt{3}}, 0, \frac{1}{2\sqrt{3}}, \frac{1}{\sqrt{3}}, \frac{3}{2\sqrt{3}} \right\} \qquad (12.4.36)$$

or

$$\theta = \left\{ 150°, 125°, 107°, 90.0°, 73.2°, 54.7°, 30.0° \right\} \qquad (12.4.37)$$

12.4.3 Radial Equation

If the potential $V(r)$ is only a function of the radial coordinate, then $R(r)$ is given by solutions to

$$\left(\frac{\partial}{\partial r} r^2 \frac{\partial R(r)}{\partial r} \right) + \frac{2m}{\hbar^2} (E - V(r)) r^2 R(r) - \ell(\ell+1) R(r) = 0 \qquad (12.4.38)$$

For a free particle $V(r) = 0$ and the radial equation

$$-\frac{\hbar^2}{2m}\frac{1}{r^2}\frac{\partial}{\partial r}\left(r^2\frac{\partial R(r)}{\partial r}\right)+\frac{\hbar^2\ell(\ell+1)}{2mr^2}R(r)=ER(r)$$

(12.4.39)

with solutions given by the spherical Bessel functions

$$R(r) = A_\ell j_\ell(kr)+B_\ell n_\ell(kr)$$

(12.4.40)

where $k = \sqrt{2mE/\hbar^2}$. The functions $n_l(kr)$ are divergent at $r = 0$ and are discarded in regions including the origin.

Example 12.4.2

Find the wavefunctions $\psi(r,\theta,\phi)$ and allowed energies of a particle contained in a spherical well with

$$V(r)=\begin{cases}0 & r\le a \\ \infty & r>a\end{cases}$$

(12.4.41)

Solution: Since the solution region includes the origin we have

$$R(r) = A_\ell j_\ell(kr)$$

(12.4.42)

Applying the boundary condition $\psi(r,\theta,\phi) = 0$ gives

$$j_\ell(ka)=0$$

(12.4.43)

so that $ka = \lambda_{n\ell}$ where $\lambda_{n\ell}$ are the roots of $j_\ell(\lambda_{n\ell}) = 0$. Thus, our quantized energy levels are

$$E_{n\ell} = \frac{\hbar^2}{2m}\left(\frac{\lambda_{n\ell}}{a}\right)^2$$

(12.4.44)

$$\psi(r,\theta,\phi) = A_{n\ell}j_\ell\left(\lambda_{n\ell}\frac{r}{a}\right)Y_\ell^m(\theta,\phi)$$

(12.4.45)

The constants $A_{n\ell}$ are obtained from normalization.

12.4.4 Hydrogen Radial Wavefunctions

The radial equation part of the wave equation for hydrogen with potential

$$V(r)=-\frac{1}{4\pi\varepsilon_0}\frac{e^2}{r}$$

(12.4.46)

is

$$-\frac{\hbar^2}{2m}\frac{1}{r^2}\frac{\partial}{\partial r}\left(r^2\frac{\partial R(r)}{\partial r}\right)+\frac{\hbar^2 \ell(\ell+1)}{2mr^2}R(r)-\frac{1}{4\pi\varepsilon_0}\frac{e^2}{r}R(r)=ER(r) \quad (12.4.47)$$

Setting $u(r) = rR(r)$ we have

$$-\frac{\hbar^2}{2m}\frac{\partial^2}{\partial r^2}u(r)+\frac{\hbar^2 \ell(\ell+1)}{2mr^2}u(r)-\frac{1}{4\pi\varepsilon_0}\frac{e^2}{r}u(r)=Eu(r) \quad (12.4.48)$$

Arranging terms

$$\frac{\partial^2}{\partial r^2}u(r)+\left[\frac{-\ell(\ell+1)}{r^2}+\frac{2me^2}{4\pi\varepsilon_0\hbar^2}\frac{1}{r}\right]u(r)=-\frac{2mE}{\hbar^2}u(r) \quad (12.4.49)$$

with the Bohr radius $a_0 = \dfrac{4\pi\varepsilon_0\hbar^2}{me^2}$ and $k = \sqrt{-\dfrac{2mE}{\hbar^2}}$ for bound state solutions E < 0 so $k > 0$.

Note that k has dimension of inverse length so that we may define a dimensionless radius $\rho = kr$

$$k^2\frac{\partial^2}{\partial\rho^2}u(\rho)+\left[\frac{-\ell(\ell+1)}{r^2}+\frac{2}{a_0}\frac{1}{r}\right]u(\rho)=k^2 u(\rho) \quad (12.4.50)$$

Dividing by k^2

$$\frac{\partial^2}{\partial\rho^2}u(\rho)+\left[\frac{-\ell(\ell+1)}{\rho^2}+\frac{\rho_0}{\rho}\right]u(\rho)=u(\rho) \quad (12.4.51)$$

where $\rho_0 = 2/ka_0$. For large values of ρ, the quantity in square brackets is small and

$$\frac{\partial^2}{\partial\rho^2}u(\rho)\approx u(\rho) \quad (12.4.52)$$

with solution $u(\rho)\approx c_1 e^{-\rho}+c_2 e^{\rho}$ \quad (12.4.53)

Since e^{ρ} diverges for large values of ρ, we require that $c_2 = 0$. For small values of ρ

$$\frac{\partial^2}{\partial\rho^2}u(\rho)\approx\frac{\ell(\ell+1)}{\rho^2}u(\rho) \quad (12.4.54)$$

with solutions

$$u(\rho)\approx c_3\rho^{-\ell}+c_4\rho^{\ell+1} \quad (12.4.55)$$

Since $\rho^{-\ell}$ diverges for small ρ, we require that $c_3 = 0$.

Thus, we take as a general solution

$$u(\rho) = \rho^{\ell+1} e^{-\rho} f(\rho) \tag{12.4.56}$$

where $f(\rho)$ is a polynomial in ρ.

To find the form of $f(\rho)$ we first evaluate the derivatives

$$\frac{\partial u}{\partial \rho} = (\ell+1)\rho^\ell e^{-\rho} f - \rho^{\ell+1} e^{-\rho} f + \rho^{\ell+1} e^{-\rho} \frac{\partial f}{\partial \rho} \tag{12.4.57}$$

and

$$\frac{\partial^2 u}{\partial \rho^2} = \ell(\ell+1)\rho^{\ell-1} e^{-\rho} f - (\ell+1)\rho^\ell e^{-\rho} f + (\ell+1)\rho^\ell e^{-\rho} \frac{\partial f}{\partial \rho}$$

$$- (\ell+1)\rho^\ell e^{-\rho} f + \rho^{\ell+1} e^{-\rho} f - \rho^{\ell+1} e^{-\rho} \frac{\partial f}{\partial \rho} \tag{12.4.58}$$

$$+ (\ell+1)\rho^\ell e^{-\rho} \frac{\partial f}{\partial \rho} - \rho^{\ell+1} e^{-\rho} \frac{\partial f}{\partial \rho} + \rho^{\ell+1} e^{-\rho} \frac{\partial^2 f}{\partial \rho^2}$$

factoring

$$\frac{\partial^2 u}{\partial \rho^2} = \left[\left(\ell(\ell+1)\rho^{\ell-1} - 2(\ell+1)\rho^\ell + \rho^{\ell+1} \right) f + 2\left((\ell+1)\rho^\ell - \rho^{\ell+1} \right) \frac{\partial f}{\partial \rho} \right.$$

$$\left. + \rho^{\ell+1} \frac{\partial^2 f}{\partial \rho^2} \right] e^{-\rho} \tag{12.4.59}$$

and substitution into the differential equation gives

$$\frac{\partial^2 f}{\partial \rho^2} + \left(\frac{2(\ell+1)}{\rho} - 2 \right) \frac{\partial f}{\partial \rho} + \left(\frac{\rho_0}{\rho} - \frac{2(\ell+1)}{\rho} \right) f = 0 \tag{12.4.60}$$

This will be in the form of Laguerre's differential equation letting $x = 2\rho$

$$x \frac{\partial^2 f}{\partial x^2} + \left(2(\ell+1) - x \right) \frac{\partial f}{\partial x} + \left(\frac{\rho_0}{2} - (\ell+1) \right) f = 0 \tag{12.4.61}$$

The associated Laguerre polynomials $L_{n'}^m(x)$ are solutions to

$$x \frac{\partial^2}{\partial x^2} L_{n'}^m(x) + (m+1-x) \frac{\partial}{\partial x} L_{n'}^m(x) + n' L_{n'}^m(x) = 0 \tag{12.4.62}$$

so that we have $m = 2\ell + 1$ and $n' = \dfrac{\rho_0}{2} - (\ell+1)$.

Since n' is an integer, $\rho_0/2$ is also an integer n. Writing $\rho_0 = 2n$ with $\rho_0 = 2/ka_0$ we have

$$k = \sqrt{-\frac{2mE}{\hbar^2}} = \frac{1}{na_0} \qquad (12.4.63)$$

This gives our quantized energy levels

$$E = \frac{E_1}{n^2} \text{ with } E_1 = -\frac{\hbar^2}{2ma_0^2} = -13.6 \text{ eV} \qquad (12.4.64)$$

Our normalized radial wavefunctions with $\rho_0 = kr = r/na_0$ are

$$R_{n,\ell}(r) = \sqrt{\left(\frac{2}{na_0}\right)^3 \frac{(n-\ell-1)!}{2n\left[(n+\ell)!\right]^3}} (2r/na_0)^{\ell+1} e^{-r/na_0} L_{n-\ell-1}^{2\ell+1}(2r/na_0) \quad (12.4.65)$$

Including angular dependence our wavefunctions are

$$\psi(r,\theta,\phi) = R_{n,\ell} Y_\ell^m(\theta,\phi) \qquad (12.4.66)$$

Allowed values of the three quantum numbers n(principal), ℓ(orbital), and m (magnetic) required to specify the wavefunction in three dimensions are shown in Table 12.4.1.

TABLE 12.4.1: Allowed values of the principal, orbital, and magnetic quantum numbers.

Quantum Number	Name	Allowed Values
n	Principal	$n = 1, 2, 3, \ldots$
ℓ	Orbital	$\ell = 0, 1, 2, \ldots, n-1$
m_ℓ	Magnetic	$m_\ell = -\ell, -\ell+1, \ldots, \ell$

Maple Examples

Hydrogen atom wavefunctions are demonstrated in the Maple worksheet below. Calculations include the evaluation of the normalized radial wavefunctions, expectation value of position, and density plots of the wavefunction.

Key Maple commands: *assume, conjugate, densityplot, integrate*

Maple packages: *with(plots): with(orthopoly):*

Programming: Function statements using '$\rightarrow$'

Special functions: SphericalY, L (Laguerre polynomial), altL (alternate Laguerre polynomial)

restart

Hydrogen Atom Radial Wavefunctions

$with(orthopoly):$

$altL := (n, a, x) \rightarrow (-1)^a \cdot n! \cdot orthopoly[L](n - a, a, x):$

$$R := (n,l,r) \rightarrow -\left(\frac{2}{n \cdot a_0}\right)^{\frac{3}{2}} \cdot sqrt\left(\frac{(n-l-1)!}{2 \cdot n \cdot ((n+1)!)^3}\right) \cdot \exp\left(-\frac{r}{n \cdot a_0}\right) \cdot \left(\frac{2 \cdot r}{n \cdot a_0}\right)^l$$

$$\cdot altL\left(n+l, 2l+1, \frac{2 \cdot r}{n \cdot a_0}\right)$$

$$R := (n,l,r) \mapsto 2\sqrt{2}\left(\frac{1}{na_0}\right)^{3/2}\sqrt{\frac{(n-l-1)!}{2n(n+l)!^3}}\,e^{-\frac{r}{na_0}}\left(\frac{2r}{na_0}\right)altL\left(n+l, 2l+1, \frac{2r}{na_0}\right)$$

$R(1, 0, r)$

$$2\left(\frac{1}{a_0}\right)^{3/2}e^{-\frac{r}{a_0}}$$

$R(2, 0, r)$

$$-\frac{\sqrt{2}\left(\dfrac{1}{a_0}\right)^{3/2}e^{-\frac{r}{2a_0}}\left(-4+\dfrac{2r}{a_0}\right)}{8}$$

$R(2, 1, r)$

$$-\frac{\left(\dfrac{1}{a_0}\right)^{3/2}\sqrt{6}e^{-\frac{r}{2a_0}}r}{12a_0}$$

Normalization of Radial Wavefunctions

$assume(a_0 > 0):$

$integrate(R(19, 15, r)^2 \cdot r^2, r = 0 \ldots infinity)$

$$1$$

$integrate(R(5, 0, r)^2 \cdot r^2, r = 0 \ldots infinity)$

$$1$$

Expectation Value <r>

$integrate(R(3, 0, r)^2 \cdot r^3, r = 0 \ldots infinity)$

$$\frac{27a_{0-}}{2}$$

Hydrogen Probability Density

$R1 := (n, l, r) \rightarrow subs(a_0 = 1, R(n, l, r))$

$$R1 := (n, l, r) \mapsto subs(a_0 = 1, R(n, l, r))$$

$\psi := (n, l, m, r, \text{theta}, \text{phi}) \rightarrow R1(n, l, r) \cdot \text{SphericalY}(l, m, \text{theta}, \text{phi})$

$$\psi := (n, l, m, r, \theta, \phi) \mapsto R1(n, l, r) \cdot \text{SphericalY}(l, m, \theta, \phi)$$

$with(plots):$

$densityplot(r^2 \cdot \psi(4, 1, 1, r, \text{theta}, 0) \cdot conjugate(\psi(4, 1, 1, r, \text{theta}, 0)), r = 0 \dots 10,$
$\text{theta} = 0 \dots 2 \cdot \text{Pi}, colorstyle = SHADING, coords = \text{polar})$

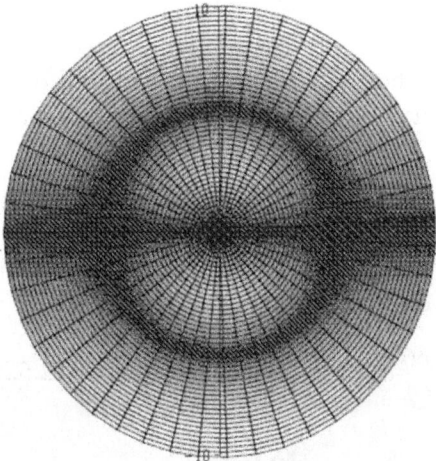

Figure 12.4.1: Shaded plot of hydrogen probability density state (4,1,1).

$densityplot(r^2 \cdot \psi(4, 2, 1, r, \text{theta}, 0) \cdot conjugate(\psi(4, 2, 1, r, \text{theta}, 0)), r = 0 \dots 40,$
$\text{theta} = 0 \dots 2 \cdot \text{Pi}, coords = \text{polar})$

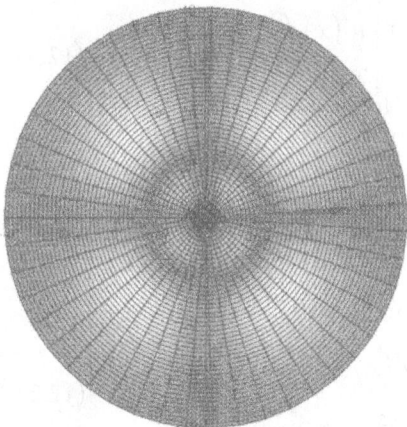

Figure 12.4.2: Shaded plot of hydrogen probability density state (4,2,1).

12.5 APPROXIMATION METHODS

Approximation methods in quantum mechanics are investigated in this section. The WKB (Wentzel, Kramers, and Brillouin) approximation is first discussed. Time-independent perturbation theory is then considered for estimating energies and wavefunctions of systems that can be modeled as modified bound states with known energies and wavefunctions. Methods of degenerate perturbation theory are then developed for problems where the unperturbed systems have multiple states with the same energy.

12.5.1 WKB Approximation

The WKB approximation is a general method of approximating second order linear differential equations such as the time-independent Schrödinger equation

$$-\frac{\hbar^2}{2m}\frac{\partial^2}{\partial x^2}\psi(x)+V(x)\psi(x)=E\psi(x) \tag{12.5.1}$$

where the wavefunction $\psi(x)$ is slowly varying compared to the de Broglie wavelength. The WKB wavefunction is of the form

$$\psi(x)=u(x)e^{iS(x)/\hbar} \tag{12.5.2}$$

Taking the first derivative of ψ

$$\frac{\partial\psi}{\partial x}=u'(x)e^{iS(x)/\hbar}+u(x)e^{iS(x)/\hbar}iS'(x)/\hbar \tag{12.5.3}$$

and taking the second derivative gives

$$\frac{\partial^2\psi}{\partial x^2}=u''(x)e^{iS(x)/\hbar}+u'(x)e^{iS(x)/\hbar}iS'(x)/\hbar+u'(x)e^{iS(x)/\hbar}iS'(x)/\hbar$$
$$+u(x)e^{iS(x)/\hbar}\left(iS'(x)/\hbar\right)^2+u(x)e^{iS(x)/\hbar}iS''(x)/\hbar \tag{12.5.4}$$

Also, we have that

$$\frac{\partial^2\psi}{\partial x^2}=-\frac{2m}{\hbar^2}\left(E-V(x)\right)u(x)e^{iS(x)/\hbar} \tag{12.5.5}$$

Setting expressions for ψ'' equal and canceling the exponential factors

$$-\frac{2m}{\hbar^2}\left(E-V(x)\right)u(x)=u''(x)+2u'(x)iS'(x)/\hbar+$$
$$+u(x)\left(iS'(x)/\hbar\right)^2+u(x)iS''(x)/\hbar \tag{12.5.6}$$

Equating the real and imaginary parts of this expression gives

$$-\frac{2m}{\hbar^2}\left(E-V\left(x\right)\right)u\left(x\right)=u''\left(x\right)-u\left(x\right)\left(S'\left(x\right)/\hbar\right)^2 \tag{12.5.7}$$

and

$$0=2u'\left(x\right)S'\left(x\right)/\hbar+u\left(x\right)S''\left(x\right)/\hbar \tag{12.5.8}$$

For $u(x)$ slowly varying we make the approximation that $u''\left(x\right)\ll u\left(x\right)$ so that

$$S'\left(x\right)=\pm\sqrt{2m\left(E-V\left(x\right)\right)}=\pm p\left(x\right) \tag{12.5.9}$$

Integration gives

$$S\left(x\right)=\pm\int p\left(x\right)dx+\text{const.} \tag{12.5.10}$$

$$0=2u'\left(x\right)p\left(x\right)+u\left(x\right)p'\left(x\right) \tag{12.5.11}$$

Next, we solve

$$\frac{1}{u\left(x\right)}\left(u\left(x\right)^2 p\left(x\right)\right)'=0 \tag{12.5.12}$$

giving

$$u\left(x\right)^2 p\left(x\right)=\text{const.} \tag{12.5.13}$$

or

$$u\left(x\right)=\pm\frac{\text{const.}}{\sqrt{p\left(x\right)}} \tag{12.5.14}$$

Our WKB wavefunction is then obtained

$$\psi\left(x\right)=\frac{c_1}{\sqrt{p\left(x\right)}}e^{\frac{i}{\hbar}\int p(x)dx}+\frac{c_2}{\sqrt{p\left(x\right)}}e^{-\frac{i}{\hbar}\int p(x)dx} \tag{12.5.15}$$

where c_1 and c_2 are constants. The WKM method can be used to estimate the transmission probability through a barrier with slowly varying potential. The probability T that a particle incident from the left with energy E will surmount a barrier where $V(x) > E$ from x_1 to x_2 is

$$T=\exp\left[-\frac{2}{\hbar}\int_{x_1}^{x_2}\left|p\left(x'\right)\right|dx'\right] \tag{12.5.16}$$

Example 12.5.1

A particle with energy $E = V_0/2$ is incident from the left on a step potential

$$V(x) = \begin{cases} V_0 & |x| \le a \\ 0 & |x| > a \end{cases} \qquad (12.5.17)$$

Find the probability that the particle will tunnel through the barrier.

Solution: Using the WKB approximation

$$T = \exp\left[-\frac{2}{\hbar} \int_{x_1}^{x_2} \sqrt{2m(V(x') - E)}\,dx' \right]$$

$$= \exp\left[-\frac{2}{\hbar} \int_{-a}^{a} \sqrt{2m(V_0 - V_0/2)}\,dx' \right] = \exp\left(-\frac{4\sqrt{mV_0}\,a}{\hbar} \right) \qquad (12.5.18)$$

12.5.2 Time-Independent Perturbation Theory

The TISE for the unperturbed system is written as

$$\hat{H}_0 \psi_n^0 = E_n^0 \psi_n^0 \qquad (12.5.19)$$

The Hamiltonian for the system including the perturbation is

$$\hat{H} = \hat{H}_0 + \hat{H}' \qquad (12.5.20)$$

where $\hat{H}_0$ is the Hamiltonian of the unperturbed system and $\hat{H}'$ is the Hamiltonian of the perturbation. Now we rewrite

$$\hat{H} = \hat{H}_0 + \varepsilon \hat{H}' \qquad (12.5.21)$$

where ε is a small dimensionless parameter to be expanded about. The corrected energies and wavefunctions are thus

$$E_n = E_n^0 + \varepsilon E_n^1 + \varepsilon^2 E_n^2 + \cdots \qquad (12.5.22)$$

$$\psi_n = \psi_n^0 + \varepsilon \psi_n^1 + \varepsilon^2 \psi_n^2 + \cdots \qquad (12.5.23)$$

The TISE $\hat{H}\psi_n = E_n \psi_n$ is now

$$\left(\hat{H}_0 + \varepsilon \hat{H}' \right)\left(\psi_n^0 + \varepsilon \psi_n^1 + \varepsilon^2 \psi_n^2 + \cdots \right) = \left(E_n^0 + \varepsilon E_n^1 + \varepsilon^2 E_n^2 + \cdots \right)$$
$$\left(\psi_n^0 + \varepsilon \psi_n^1 + \varepsilon^2 \psi_n^2 + \cdots \right) \qquad (12.5.24)$$

Equating terms multiplying equal powers of ε on the left and right sides we obtain an infinite set of equations

$$\left.\begin{array}{c} \hat{H}_0 \psi_n^0 = E_n^0 \psi_n^0 \\ \varepsilon \hat{H}_0 \psi_n^1 + \varepsilon \hat{H}' \psi_n^0 = \varepsilon E_n^1 \psi_n^0 + \varepsilon E_n^0 \psi_n^1 \\ \varepsilon^2 \hat{H}_0 \psi_n^2 + \varepsilon^2 \hat{H}' \psi_n^1 = \varepsilon^2 E_n^0 \psi_n^2 + \varepsilon^2 E_n^1 \psi_n^1 + \varepsilon^2 E_n^2 \psi_n^0 \\ \vdots \end{array}\right\}$$ (12.5.25)

The first equation gives the TISE for the unperturbed system. The second and third equations give the first and second order corrections, and so on. Approximations to the perturbed wavefunctions and their respective energies can then be found. We consider the first order correction

$$E_n^1 \psi_n^0 = \hat{H}' \psi_n^0 + \left(\hat{H}_0 - E_n^0\right)\psi_n^1$$ (12.5.26)

Since ψ_n^1 can be expanded in terms of any orthogonal functions, we choose to expand them as a linear combination of the unperturbed wavefunctions ψ_n^0

$$\psi_n^1 = \sum_{m=0}^{\infty} a_{nm} \psi_m^0$$ (12.5.27)

$$E_n^1 \psi_n^0 = \hat{H}' \psi_n^0 + \left(\hat{H}_0 - E_n^0\right)\sum_{m=0}^{\infty} a_{nm} \psi_m^0$$ (12.5.28)

Since $\hat{H}_0 \psi_m^0 = E_m^0 \psi_m^0$ the last term

$$\left(\hat{H}_0 - E_n^0\right)\sum_{m=0}^{\infty} a_{nm} \psi_m^0 = \sum_{m=0}^{\infty} a_{nm}\left(\hat{H}_0 - E_n^0\right)\psi_m^0 = \sum_{m=0}^{\infty} a_{nm}\left(E_m^0 - E_n^0\right)\psi_m^0$$ (12.5.29)

Multiplying both sides of our first order correction by $\psi_n^0{}^*$ and integrating

$$E_n^1 \left\langle \psi_n^0 \middle| \psi_n^0 \right\rangle = \left\langle \psi_n^0 \middle| \hat{H}' \psi_n^0 \right\rangle + \sum_{m=0}^{\infty} a_{nm}\left(E_m^0 - E_n^0\right)\left\langle \psi_n^0 \middle| \psi_m^0 \right\rangle$$ (12.5.30)

The sum vanishes for $n \neq m$ because of orthogonality and for $n = m$ where $\left(E_m^0 - E_n^0\right) = 0$. Thus, we have the first order correction to the energy

$$E_n^1 = \left\langle \psi_n^0 \middle| \hat{H}' \psi_n^0 \right\rangle$$ (12.5.31)

To determine the a_{nm} we multiply the first order correction by $\psi_m^0{}^*$ and integrate to obtain

$$E_n^1 \left\langle \psi_m^0 \middle| \psi_n^0 \right\rangle = \left\langle \psi_m^0 \middle| \hat{H}' \psi_n^0 \right\rangle + \sum_{m=0}^{\infty} a_{nm}\left(E_m^0 - E_n^0\right)\left\langle \psi_m^0 \middle| \psi_m^0 \right\rangle$$ (12.5.32)

The term on the left vanishes for $m \neq n$ because of orthogonality and

$$a_{nm} = \frac{\left\langle \psi_m^0 \left| \hat{H}' \psi_n^0 \right\rangle \right.}{E_n^0 - E_m^0} \tag{12.5.33}$$

and the first order correction to our wavefunction is

$$\psi_n^1 = \sum_{m=0}^{\infty} \frac{\left\langle \psi_m^0 \left| \hat{H}' \psi_n^0 \right\rangle \right.}{E_n^0 - E_m^0} \psi_m^0 \tag{12.5.34}$$

Example 12.5.2

Consider a particle in a 1D potential well (Figure 12.5.1)

$$V(x) = \begin{cases} \infty & x < 0 \\ W_0 & 0 \leq x < a \\ 0 & a \leq x \leq L \\ \infty & x > L \end{cases} \tag{12.5.35}$$

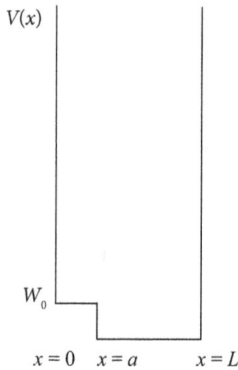

Figure 12.5.1: Square well with a small step

Treat the small step of height W_0 as a perturbation of the infinite square potential. Calculate the perturbed energy levels and wavefunctions.

Solution: The first order correction to the energy levels is

$$E_n^1 = \left\langle \psi_n^0 \left| \hat{H}' \psi_n^0 \right\rangle \right. \tag{12.5.36}$$

$$\psi_n^0(x) = \sqrt{\frac{2}{L}} \sin\left(\frac{n\pi x}{L}\right) \text{ with } \hat{H}' = W_0 \text{ for } 0 \leq x \leq a \tag{12.5.37}$$

$$E_n^1 = W_0 \frac{2}{L} \int_0^a \sin^2\left(\frac{n\pi x}{L}\right) dx = W_0 \frac{2}{L} \int_0^a \frac{1}{2}\left(1 - \cos\frac{2n\pi x}{L}\right) dx$$

$$= \frac{W_0}{L}\left(a - \frac{L}{2n\pi}\sin\frac{2n\pi a}{L}\right) \tag{12.5.38}$$

The first order correction to the wavefunctions is

$$\psi_n^1 = \sum_{m=0}^{\infty} \frac{\left\langle \psi_m^0 \middle| \hat{H}' \psi_n^0 \right\rangle}{E_n^0 - E_m^0} \psi_m^0 \quad \text{with} \quad E_n = \left(\frac{\hbar^2 \pi^2}{2mL^2}\right) n^2 \tag{12.5.39}$$

$$\psi_n^1 = \sum_{m=0}^{\infty} \frac{\left\langle \psi_m^0 \middle| \hat{H}' \psi_n^0 \right\rangle}{E_n^0 - E_m^0} \psi_m^0 = \frac{2mL^2}{\hbar^2 \pi^2} \sum_{m=0}^{\infty} \frac{W_0 \frac{2}{L} \int_0^a \sin\left(\frac{m\pi x}{L}\right) \sin\left(\frac{n\pi x}{L}\right) dx}{n^2 - m^2} \sqrt{\frac{2}{L}} \sin\left(\frac{m\pi x}{L}\right)$$

$$\tag{12.5.40}$$

12.5.3 Degenerate Perturbation Theory

Different quantum states of a system with the same energy are degenerate where g is the number of states with the same energy. For example, the degeneracy of the first excited state of a cubical square well with side L is $g = 3$ where $E_{2,1,1} = E_{1,2,1} = E_{1,1,2}$. Our first order correction to the wavefunction is not defined for degenerate wavefunctions where $E_n^0 = E_m^0$. To apply perturbation theory to a system with degeneracy, we expand the wavefunction as a linear combination of degenerate states ϕ_{ni}.

$$\psi_n = \sum_{i=1}^{g} c_{ni} \phi_{ni} \tag{12.5.41}$$

where g is the degeneracy of the nth level and $\hat{H}\psi_n = E_n \psi_n$ is now

$$\left(\hat{H}_0 + \hat{H}'\right) \sum_{i=1}^{g} c_{ni} \phi_{ni} = E_n \sum_{i=1}^{g} c_{ni} \phi_{ni} \tag{12.5.42}$$

Since $\hat{H}_0 \psi_n = E_n^0 \psi_n$

$$\sum_{i=1}^{g} c_{ni} \left(E_n^0 \phi_{ni} + \hat{H}' \phi_{ni}\right) = E_n \sum_{i=1}^{g} c_{ni} \phi_{ni} \tag{12.5.43}$$

Multiplying on the left by $\phi_{nj}^{\;*}$ and integrating

$$\sum_{i=1}^{g} c_{ni}\left(E_n^0\left\langle\phi_{nj}\middle|\phi_{ni}\right\rangle+\left\langle\phi_{nj}\middle|\hat{H}'\phi_{ni}\right\rangle\right)=E_n\sum_{i=1}^{g}c_{ni}\left\langle\phi_{nj}\middle|\phi_{ni}\right\rangle \tag{12.5.44}$$

since $\left\langle\phi_{nj}\middle|\phi_{ni}\right\rangle=\delta_{ij}$

$$c_{ni}E_n^0+\sum_{i=1}^{g}c_{ni}\left\langle\phi_{nj}\,\hat{H}'\middle|\phi_{ni}\right\rangle=c_{ni}E_n \tag{12.5.45}$$

and

$$\sum_{i=1}^{g}c_{ni}\left\langle\phi_{nj}\,\hat{H}'\middle|\phi_{ni}\right\rangle=c_{ni}\left(E_n-E_n^0\right) \tag{12.5.46}$$

This equation can be expressed in matrix form with $\Delta E_n = E_n - E_n^0$

$$\begin{pmatrix}\left\langle\phi_{n1}\hat{H}'\middle|\phi_{n1}\right\rangle-\Delta E_n & \left\langle\phi_{n2}\hat{H}'\middle|\phi_{n1}\right\rangle & \cdots & \left\langle\phi_{ng}\hat{H}'\middle|\phi_{n1}\right\rangle \\ \left\langle\phi_{n1}\hat{H}'\middle|\phi_{n2}\right\rangle & \left\langle\phi_{n2}\hat{H}'\middle|\phi_{n2}\right\rangle-\Delta E_n & \cdots & \left\langle\phi_{ng}\hat{H}'\middle|\phi_{n2}\right\rangle \\ \vdots & \vdots & \ddots & \vdots \\ \left\langle\phi_{n1}\hat{H}'\middle|\phi_{ng}\right\rangle & \left\langle\phi_{n2}\hat{H}'\middle|\phi_{ng}\right\rangle & \cdots & \left\langle\phi_{ng}\hat{H}'\middle|\phi_{ng}\right\rangle-\Delta E_n\end{pmatrix}\begin{pmatrix}c_{n1}\\c_{n2}\\\vdots\\c_{ng}\end{pmatrix}=0 \tag{12.5.47}$$

12.5.4 Stark Effect

The shifting of atomic and molecular energy levels due to application of an electric field E is known as the Stark effect. The shift in energy levels results in a splitting of spectral lines. Here we consider the linear Stark effect where shifts in energy levels are proportional to the magnitude of the electric field. Second order effects are proportional to the square of the electric field.

A hydrogen atom in the $n = 2$ state in an electric field oriented in the z-direction $\mathbf{E} = E\hat{\mathbf{k}}$ with perturbation $\hat{H}' = eEz$

$$eE\begin{pmatrix}\left\langle\phi_{200}z\middle|\phi_{200}\right\rangle & \left\langle\phi_{200}z\middle|\phi_{210}\right\rangle & \left\langle\phi_{200}z\middle|\phi_{211}\right\rangle & \left\langle\phi_{200}z\middle|\phi_{21-1}\right\rangle \\ \left\langle\phi_{210}z\middle|\phi_{200}\right\rangle & \left\langle\phi_{210}z\middle|\phi_{210}\right\rangle & \left\langle\phi_{210}z\middle|\phi_{211}\right\rangle & \left\langle\phi_{210}z\middle|\phi_{21-1}\right\rangle \\ \left\langle\phi_{211}z\middle|\phi_{200}\right\rangle & \left\langle\phi_{211}z\middle|\phi_{210}\right\rangle & \left\langle\phi_{211}z\middle|\phi_{211}\right\rangle & \left\langle\phi_{211}z\middle|\phi_{21-1}\right\rangle \\ \left\langle\phi_{21-1}z\middle|\phi_{200}\right\rangle & \left\langle\phi_{21-1}z\middle|\phi_{210}\right\rangle & \left\langle\phi_{21-1}z\middle|\phi_{211}\right\rangle & \left\langle\phi_{21-1}z\middle|\phi_{21-1}\right\rangle\end{pmatrix}\begin{pmatrix}c_{21}\\c_{22}\\c_{23}\\c_{24}\end{pmatrix}=\Delta E_2\begin{pmatrix}c_{21}\\c_{22}\\c_{23}\\c_{24}\end{pmatrix} \tag{12.5.48}$$

With the nonzero matric elements

$$\langle \phi_{200} \, z \, | \, \phi_{210} \rangle = \langle \phi_{210} \, z \, | \, \phi_{200} \rangle = 3Eea_0 \qquad (12.5.49)$$

our matrix equation becomes

$$\begin{pmatrix} -\Delta E_2 & 3Eea_0 & 0 & 0 \\ 3Eea_0 & -\Delta E_2 & 0 & 0 \\ 0 & 0 & -\Delta E_2 & 0 \\ 0 & 0 & 0 & -\Delta E_2 \end{pmatrix} \begin{pmatrix} c_{21} \\ c_{22} \\ c_{23} \\ c_{24} \end{pmatrix} = 0 \qquad (12.5.50)$$

Solving for ΔE_2

$$\begin{vmatrix} -\Delta E_2 & 3Eea_0 & 0 & 0 \\ 3Eea_0 & -\Delta E_2 & 0 & 0 \\ 0 & 0 & -\Delta E_2 & 0 \\ 0 & 0 & 0 & -\Delta E_2 \end{vmatrix} = 0 \qquad (12.5.51)$$

we find $\Delta E_2 = 3Eea_0, -3Eea_0, 0, 0$.

Substituting $\Delta E_2 = +3Eea_0$ into our matrix equation we find $c_{21} = -c_{22}$ and

$$\psi_2^+ = \frac{1}{\sqrt{2}} \left(\phi_{200} - \phi_{210} \right) \qquad (12.5.52)$$

for $\Delta E_2 = -3Eea_0$ we have $c_{21} = c_{22}$ and

$$\psi_2^- = \frac{1}{\sqrt{2}} \left(\phi_{200} + \phi_{210} \right) \qquad (12.5.53)$$

The states ϕ_{211} and $\phi_{21\text{-}1}$ are unperturbed by the electric field.

Maple Examples

Approximation methods in quantum mechanics are demonstrated in the Maple worksheet below. Calculations include the WKB probability of tunneling through barriers and the perturbed energy levels for nondegenerate and degenerate wavefunctions.

Key Maple commands: *Array, assume, Eigenvectors, plot*

Maple packages: *with(LinearAlgebra): with(Student[LinearAlgebra]):*

Programming: Function statements using '→'

restart

WKB Tunneling through a Ramp Potential

$$V := (x) \rightarrow \begin{cases} (1-\text{abs}(x)) & -1 < x < 1 \\ 0 & otherwise \end{cases}$$

$$V := x \mapsto \begin{cases} 1-|x| & -1 < x < 1 \\ 0 & otherwise \end{cases}$$

$$E := \frac{1}{2} : x_1 := \frac{-1}{2} : x_2 := \frac{1}{2} : V_0 := 1 :$$

$plot([V(x), E], x = -2 \dots 2, legend = [\text{``}V(x)\text{''}, \text{``}E\text{''}], linestyle = [solid, dash])$

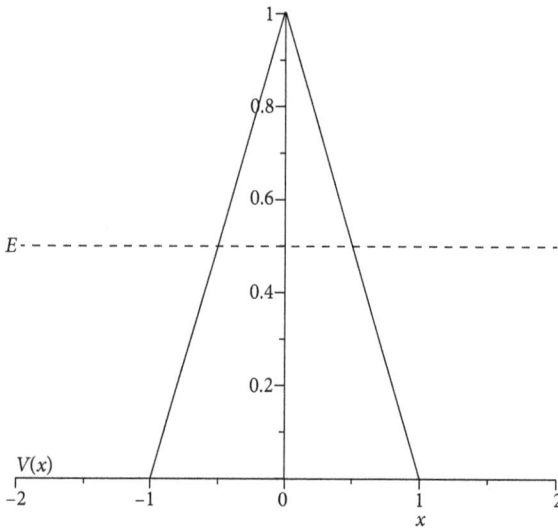

Figure 12.5.2: Ramped potential

$$T = \exp\left(-\frac{2 \cdot \sqrt{2 \cdot m}}{\hbar} \cdot \int_{x_1}^{x_2} \sqrt{(V(x) - E)} \, dx \right)$$

$$T = e^{-\frac{4\sqrt{m}}{3\hbar}}$$

WKB Tunneling through a Hump Potential

$$V2 := (x) \rightarrow \begin{cases} \cos(x) & -\frac{\text{pi}}{2} < x < \frac{\text{pi}}{2} \\ 0 & otherwise \end{cases}$$

$$V2 := x \mapsto \begin{cases} \cos(x) & -\dfrac{\pi}{2} < x < \dfrac{\pi}{2} \\ 0 & \textit{otherwise} \end{cases}$$

$$plot([V2(x), E], x = -2 \ldots 2, legend = [\text{``}V(x)\text{''}, \text{``}E\text{''}], linestyle = [\text{``}solid\text{''}, \text{``}dash\text{''}])$$

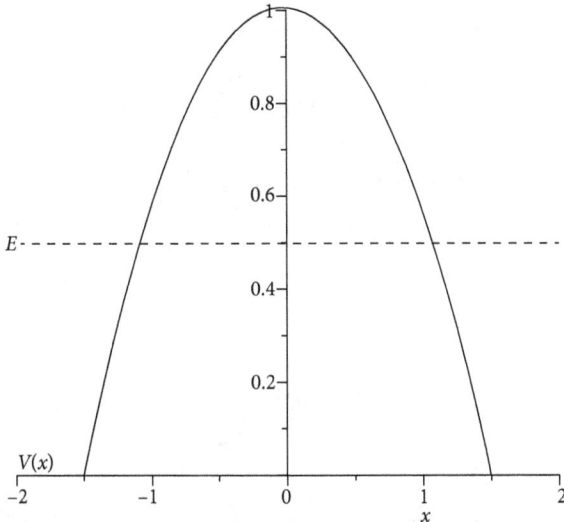

Figure 12.5.3: Humped potential

$$solve(V2(x) = E)$$

$$\frac{\pi}{3}$$

$$x_1 := -\frac{Pi}{3} : x_2 := \frac{Pi}{3}$$

$$x_2 := \frac{\pi}{3}$$

$$T = \exp\left(-\frac{2 \cdot \sqrt{2 \cdot m}}{\hbar} \cdot \int_{x_1}^{x_2} \sqrt{(V2(x) - E)}\, dx\right)$$

$$T = e^{\dfrac{2\sqrt{2}\sqrt{m}\left(-3\sqrt{2}\,\text{Elliptick}\left(\frac{1}{2}\right) + 4\sqrt{2}\,\text{Elliptick}\left(\frac{1}{2}\right)\right)}{\hbar}}$$

Nondegenerate Perturbation: Delta Function inside a 3D Rectangular Well

restart

$$\text{assume}(n :: \mathbb{Z}, m :: \mathbb{Z}, l :: \mathbb{Z}, a > 0, b > 0, c > 0)$$

$$psi := (x, y, z) \to \sqrt{\frac{2}{a}} \cdot \sin\left(\frac{n \cdot \text{Pi} \cdot x}{a}\right) \cdot \sqrt{\frac{2}{b}} \cdot \sin\left(\frac{m \cdot \text{Pi} \cdot y}{b}\right) \cdot \sqrt{\frac{2}{c}} \cdot \sin\left(\frac{l \cdot \text{Pi} \cdot z}{c}\right)$$

$$\Psi := (x, y, z) \mapsto \sqrt{\frac{2}{a}} \sin\left(\frac{n\pi x}{a}\right) \sqrt{\frac{2}{b}} \sin\left(\frac{m\pi y}{b}\right) \sqrt{\frac{2}{c}} \sin\left(\frac{l\pi x}{c}\right)$$

$$Hp := (x, y, z) \to V_0 \cdot \text{Dirac}\left(x - \frac{a}{2}\right) \cdot \text{Dirac}\left(y - \frac{b}{2}\right) \cdot \text{Dirac}\left(z - \frac{c}{2}\right)$$

$$Hp := (x, y, z) \mapsto V_0 \text{Dirac}\left(x - \frac{a}{2}\right) \text{Dirac}\left(y - \frac{b}{2}\right) \text{Dirac}\left(z - \frac{c}{2}\right)$$

$$\Delta E = simplify\left(\int_0^a \int_0^b \int_0^c Hp(x, y, z) \cdot psi(x, y, z)^2 \, dz \, dy \, dx \right)$$

$$\Delta E = \frac{8 V_0 \sin\left(\frac{l \sim \pi}{2}\right)^2 \sin\left(\frac{m \sim \pi}{2}\right)^2 \sin\left(\frac{n \sim \pi}{2}\right)^2}{a \sim b \sim c \sim}$$

Degenerate Perturbation: Delta Function inside a 2D Box

restart
with(LinearAlgebra) :
assume$(L > 0, V_0 > 0)$

$$\Psi_a := (x, y) \to \frac{2}{L} \cdot \sin\left(\frac{\text{Pi} \cdot x}{L}\right) \cdot \sin\left(\frac{2\text{Pi} \cdot y}{L}\right)$$

$$\Psi_a := (x, y) \mapsto \frac{2 \sin\left(\frac{\pi x}{L}\right) \sin\left(\frac{2\pi y}{L}\right)}{L}$$

$$\Psi_b := (x, y) \to \frac{2}{L} \cdot \sin\left(\frac{2 \cdot \text{Pi} \cdot x}{L}\right) \cdot \sin\left(\frac{\text{Pi} \cdot y}{L}\right)$$

$$\Psi_b := (x, y) \mapsto \frac{2 \sin\left(\frac{2\pi x}{L}\right) \sin\left(\frac{\pi y}{L}\right)}{L}$$

$$Hp := (x, y) \mapsto V_0 \cdot \text{Dirac}\left(x - \frac{L}{3}\right) \text{Dirac}\left(y - \frac{L}{4}\right)$$

QUANTUM MECHANICS **635**

$$Hp := (x, y) \mapsto V_0 \text{Dirac}\left(x - \frac{L}{3}\right)\text{Dirac}\left(y - \frac{L}{4}\right)$$

$A := Array(1 \ldots 2, 1 \ldots 2) :$

$$A_{1,1} := \int_0^L \int_0^L \Psi_a(x,y) \cdot Hp(x,y) \cdot \Psi_a(x,y) \mathrm{dx\ dy}$$

$$A_{1,1} := \frac{3V_{0\sim}}{L\sim^2}$$

$$A_{1,2} := \int_0^L \int_0^L \Psi_a(x,y) \cdot Hp(x,y) \cdot \Psi_b(x,y) \mathrm{dx\ dy}$$

$$A_{1,2} := \frac{3V_{0\sim}\sqrt{2}}{2L\sim^2}$$

$$A_{2,1} := \int_0^L \int_0^L \Psi_b(x,y) \cdot Hp(x,y) \cdot \Psi_a(x,y) \mathrm{dx\ dy}$$

$$A_{2,1} := \frac{3V_{0\sim}\sqrt{2}}{2L\sim^2}$$

$$A_{2,2} := \int_0^L \int_0^L \Psi_b(x,y) \cdot Hp(x,y) \Psi_b(x,y) \mathrm{dx\ dy}$$

$$A_{2,2} := \frac{3V_{0\sim}}{2L\sim^2}$$

$with(Student[LinearAlgebra]) :$
$Eigenvectors(A)$

$$\begin{bmatrix} \dfrac{9V_{0\sim}}{2L_\sim^2} \\ 0 \end{bmatrix}, \begin{bmatrix} \sqrt{2} & -\dfrac{\sqrt{2}}{2} \\ 1 & 1 \end{bmatrix}$$

12.6 MATLAB EXAMPLES

Section 12.4 Schrödinger Equation in Higher Dimensions

The PDE Toolbox may be used to find a numerical solution to the 2D time-independent Schrödinger equation

$$\nabla^2 \psi = -\frac{2mE}{\hbar^2}\psi \tag{12.8.1}$$

The eigenvalue equation to be solved is

$$-\nabla \cdot \left(c\nabla u\right) + au = \lambda du \tag{12.8.2}$$

where $u = \psi$, $c = 1$, $a = 0$, $d = 1$, $\lambda = \dfrac{2mE}{\hbar^2}$

Figure 12.6.1 shows a finite element mesh generated over an elliptical region with a ratio of semi-minor to semi-major axes of 0.6. The Dirichlet boundary condition $u = 0$ is applied to the elliptical boundary to model a particle of mass m confined to the interior.

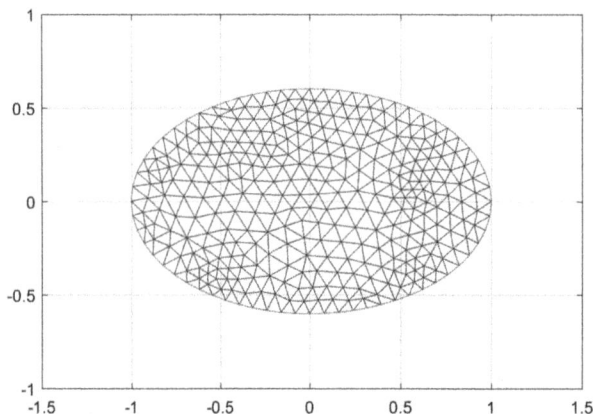

Figure 12.6.1: Finite element mesh for the solution of TISE describing a particle confined to a 2D elliptical well.

Contour and shaded surface plots of the second and fourth excited state wavefunctions are shown in Figure 12.6.2 calculated with the finite element mesh in Figure 12.6.1. The resulting eigenvalues are $\lambda = 10.9, 21.2, 34.5, 36.0, 50.1, 55.7, 70.0, 73.6, 80.6, 94.3, 95.8$ with energy levels $E = \hbar^2\lambda/2m$.

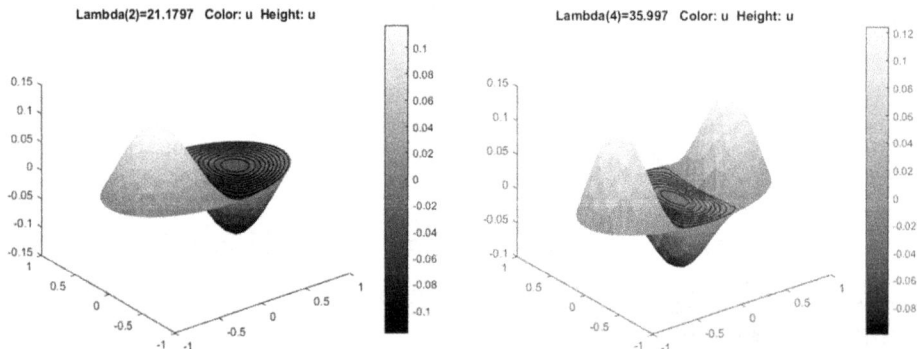

Figure 12.6.2 Wavefunctions corresponding to the second and fourth energy levels inside a 2D elliptical well.

12.7 EXERCISES

Section 12.1 Schrödinger Equation

1. An electron is represented by the time-independent wavefunction

$$\psi(x) = Ae^{-\alpha|x|}$$

where α has units of inverse length.
(a) Find the normalization constant A
(b) Determine the probability of finding the electron in the range $x_1 = 0$ to $x_2 = \alpha^{-1}$
(c) Find the probability that the electron is outside the interval above

2. A trial wave function for a particle is

$$\psi(x) = Ax(x - L)$$

for $L > 0$ and $0 \leq x \leq L$
(a) Find the normalization constant A
(b) Calculate the quantum uncertainty in position Δx

3. A particle is described by the wavefunction

$$\psi(x) = \begin{cases} Ae^{-x/L} & x \geq 0 \\ 0 & x < 0 \end{cases}$$

(a) Find the normalization constant A
(b) Determine the probability of finding the particle in the range $x = 0$ to $x = L$.
(c) Calculate the quantum uncertainty in position Δx

4. Given a trial time-independent wavefunction

$$\psi(x) = \begin{cases} Ax/L & 0 \leq x \leq L \\ 0 & \text{elsewhere} \end{cases}$$

(a) Find the normalization constant A
(b) Calculate the quantum uncertainty in position Δx

5. An electron is represented by the time-independent wavefunction

$$\psi(x) = \begin{cases} Ax^2 e^{-\beta x} & x \geq 0 \\ 0 & x < 0 \end{cases}$$

(a) Find the normalization constant A
(b) Find the quantum uncertainty in position Δx
(c) Estimate the quantum uncertainty in momentum $\Delta p \approx \hbar / \Delta x$ based on your answer in (b)

6. An electron is represented by the time-independent wavefunction
$$\psi(x) = Axe^{-\alpha x^2}$$
(a) Find the normalization constant A
(b) Find the quantum uncertainties in position and momentum. Write the uncertainty product $\Delta x \Delta p$

7. Calculate $\Psi^*\Psi$ where $\Psi(x,t) = \psi_1(x)e^{i\omega_1 t} + \psi_2(x)e^{i\omega_2 t}$.

8. Given the nonstationary wavefunction of a particle
$$\Psi(x,t) = \frac{1}{\sqrt{2}}\psi_1(x)e^{-i\frac{E_1}{\hbar}t} + \frac{1}{\sqrt{2}}\psi_2(x)e^{-i\frac{E_2}{\hbar}t}$$
calculate $\langle E \rangle = \int_0^L \Psi^*\left(i\hbar\frac{\partial}{\partial t}\right)\Psi dx$ where ψ_1 and ψ_2 are orthogonal and normalized

9. Given that the wavefunction
$$\Psi(x,t) = Ae^{i(kx-\omega t)}$$
satisfies the time-dependent Schrödinger equation for $V(x) = 0$
(a) Find a relation between k and ω
(b) Calculate $\Psi^*\Psi$ and $\hat{p}\Psi$

10. An electron of mass m is confined to a 1D box. Use Heisenberg's uncertainty principle to calculate the length of the box L such that the uncertainty in velocity is one-tenth the speed of light $\Delta v = 0.1c$.

11. Spherical C-60 Bucky balls ($m = 1.2 \times 10^{-24}$ kg) are fired through a slit of two-nanometer width ($d = 2 \times 10^{-9}$ m) with a speed of 10^6 m/s in the x-direction as shown in Figure 12.7.1.

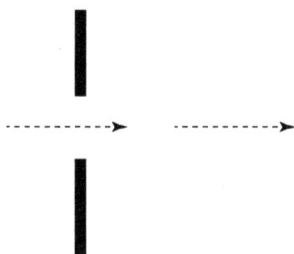

Figure 12.7.1: Bucky balls fired through a slit.

(a) Calculate the y-momentum uncertainty of the Bucky balls in the slit (perpendicular to the incident direction)
(b) Use the y-momentum uncertainty obtained in (a) to estimate the spread in position of Bucky balls impacting a screen located 1.0 m to the right of the slit.

12. Show that $\nabla \cdot (\Psi^* \nabla \Psi - \Psi \nabla \Psi^*) = \Psi^* \nabla^2 \Psi - \Psi \nabla^2 \Psi^*$

13. Particles with energy $E = V_0$ are incident from the left where $V(x) = 0$ into a region where $V(x) = -V_0$. Find the particles' reflection and transmission probabilities.

14. Particles with energy $E = 2V_0$ are incident from the left on a potential

$$V(x) = \begin{cases} 0 & x < 0 \\ -V_0 & 0 \le x \le a \\ 0 & x > a \end{cases}$$

Find the particles' reflection and transmission probabilities.

15. Write a program to animate the diffusive Gaussian wave packet with probability density

$$|\psi(x,t)|^2 = \sqrt{\frac{2}{\pi^2 a^4 + 4\hbar^2 \pi^2 t^2 / m^2}} \exp\left(-\frac{2(x - v_g t)^2}{a^2 + 4\hbar^2 t^2 / m^2 a^2} \right)$$

with group velocity $v_g = \hbar k / m$. Calculate the time-dependent width $\Delta x(t)$.

Section 12.2 Bound States I

16. A particle in a 1D square well is represented by the time-independent wavefunction

$$\psi_n(x) = \sqrt{\frac{2}{L}} \sin\left(\frac{n\pi x}{L} \right) \qquad (0 \le x \le L)$$

(a) Calculate the average values $\langle x \rangle$ and $\langle x^2 \rangle$
(b) Calculate the average values $\langle p \rangle$ and $\langle p^2 \rangle$
(c) Calculate the uncertainty product $\Delta x \Delta p$ as a function of n

17. Determine if the time-dependent wavefunctions

$$\Psi_n(x,t) = \sqrt{\frac{2}{L}} \sin\left(\frac{n\pi x}{L} \right) \exp\left(\frac{-iE_n t}{\hbar} \right) \text{ are an eigenfunction of}$$

(a) $\hat{p}^2 = \left(\frac{\hbar}{i} \frac{\partial}{\partial x} \right)\left(\frac{\hbar}{i} \frac{\partial}{\partial x} \right)$. If so what are the corresponding eigenvalue?

(b) $\hat{p} = \left(\frac{\hbar}{i} \frac{\partial}{\partial x} \right)$. If so what are the corresponding eigenvalue?

(c) $\hat{E} = \left(i\hbar \frac{\partial}{\partial t} \right)$. If so what are the corresponding eigenvalue?

18. Consider the potential function $V(x) =$
$$\begin{cases} \infty & x < -b \\ V_0 & -b \le x < -a \\ 0 & -a \le x \le a \\ V_0 & a < x \le b \\ \infty & x > b \end{cases}$$

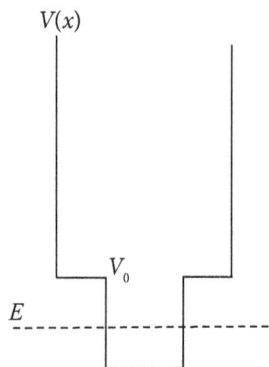

Figure 12.7.2: Piecewise constant potential for 18.

(a) Write the form of the wavefunction in each region for $E < V_0$
(b) Apply boundary conditions on the wavefunction at $x = \pm a$ and $x = \pm b$
(c) Obtain a transcendental equation whose roots satisfy the energy spectrum
Partial Solution:
$\psi_I(x) = Ae^{\alpha x} + Be^{-\alpha x}$
$\psi_{II}(x) = C\cos kx + D\sin kx$ (consider even and odd wavefunctions separately)
$\psi_{III}(x) = Ee^{\alpha x} + Fe^{-\alpha x}$
$\psi_I(-b) = Ae^{-ab} + Be^{ab} = 0$
$A = -Be^{2ab}$
$\psi_{III}(b) = Ee^{ab} + Fe^{-ab} = 0$
$F = -Ee^{2ab}$

19. Repeat the previous exercise for $E > V_0$

20. Show that the energy levels of the infinite 1D square well are obtained from the transcendental equation
$$\tan(kL) = -\frac{k}{\alpha}$$
describing energy levels of the semi-infinite square well in the limit $V_0 \to \infty$
where
$$\alpha = \sqrt{\frac{2m(V_0 - E)}{\hbar^2}}$$

21. Consider the potential function $V(x) = \begin{cases} \infty & x < 0 \\ W_0 & 0 \leq x < a \\ 0 & a \leq x \leq d \\ V_0 & d < x < \infty \end{cases}$

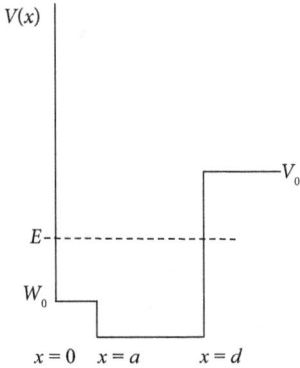

Figure 12.7.3: Piecewise constant potential for 21.

(a) Write the form of the wavefunction in each region for $W_0 < E < V_0$

(b) Apply boundary conditions on the wavefunction at $x = 0$, $x = a$, $x = d$ and infinity.

(c) Obtain a transcendental equation whose roots satisfy the energy spectrum

22. Covalent bonds can be modeled by a double well potential. Consider the double well function symmetric about $x = 0$ shown in Figure 12.7.4 with potential

$$V(x) = \begin{cases} \infty & x < -d \\ 0 & -d \leq x < -a \\ V_0 & -a \leq x \leq a \\ 0 & a < x \leq d \\ \infty & x > d \end{cases}$$

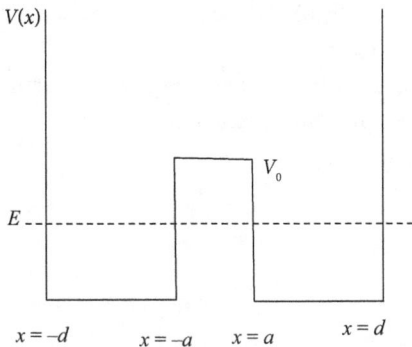

Figure 12.7.4: Piecewise constant potential for 22.

(a) Write the form of the wavefunction in each region for $E < V_0$. Consider even and odd wavefunctions separately.

(b) Apply boundary conditions on the even and odd wavefunctions at $x = 0$, $x = \pm a$ and $x = \pm d$.

(c) Obtain transcendental equations whose roots satisfy the energy spectrum for both even and odd wavefunctions.

Section 12.3 Bound States II

23. Consider the potential function defined by

$$V(x) = \begin{cases} \infty & x < 0 \\ V_0 \dfrac{x}{a} & 0 \le x < a \\ \infty & a \le x \end{cases}$$

where $E < V_0$ as shown in Figure 12.7.5.

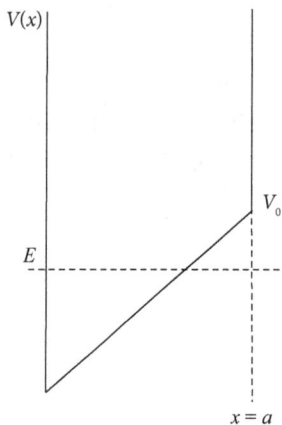

$V(x)$

V_0

E

$x = a$

Figure 12.7.5: Piecewise defined potential for 23.

(a) Write down the form of the wavefunction for $0 \le x < a$

(b) Apply boundary conditions at $x = 0$ and $x = a$

(c) Obtain a transcendental equation whose roots specify the energy spectrum.

(d) Write an integral expression for the normalization constant.

24. Consider the potential function defined by

$$V(x) = \begin{cases} \infty & x < 0 \\ 0 & 0 \le x < a \\ c_1 x + c_2 & a \le x \le d \\ \infty & x > d \end{cases}$$

where $c_1 = \dfrac{V_0 - W_0}{d - a}$, $c_2 = \dfrac{-V_0 a + W_0 d}{d - a}$ and $E < V_0$ as in Figure 12.7.6.

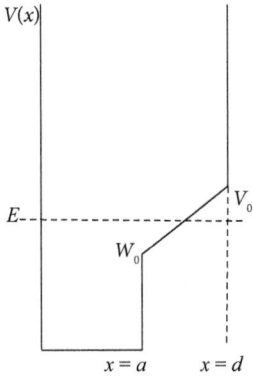

Figure 12.7.6: Piecewise defined potential for 24.

    (a) Write down the form of the wavefunction for $0 \le x \le a$ and $a \le x \le d$

    (b) Apply boundary conditions at $x = 0$, $x = a$ and $x = d$.

    (c) Obtain a transcendental equation whose roots specify the energy spectrum.

    (d) Write an integral expression for the normalization constant.

25. The potential of an infinite square well with a delta function centered in the well is

$$V(x) = \begin{cases} \infty & x < -L/2 \\ -V_0 \delta(x) & -L/2 \le x \le L/2 \\ \infty & x > L/2 \end{cases}$$

    (a) Write the form of the wavefunctions in the well.

    (b) Obtain a transcendental equation whose roots determine the energy spectra.

26. The potential of a double delta function is given by

$$V(x) = -V_0 \left[\delta\left(x - \frac{L}{2}\right) + \delta\left(x + \frac{L}{2}\right) \right]$$

    (a) Write the form of even and odd wavefunctions in the well

    (b) Obtain transcendental equations whose roots determine the energy spectra of the even and odd wavefunctions

27. Consider the potential function defined by

$$V(x) = \begin{cases} \infty & x < 0 \\ V_0 \dfrac{x}{a} & 0 \leq x < a \\ V_0 & a \leq x \end{cases}$$

where $E < V_0$ as in Figure 12.7.7.

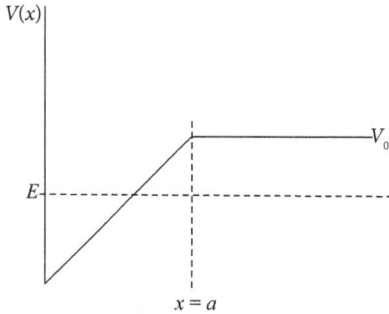

Figure 12.7.7: Piecewise defined potential for 27.

(a) Write down the form of the wavefunction for $0 \leq x \leq a$ and $a \leq x$
(b) Apply boundary conditions at $x = 0$, $x = a$ and infinity.
(c) Obtain a transcendental equation whose roots specify the energy spectrum.
(d) Write an integral expression for the normalization constant.

28. Show that the wavefunctions of the ground state $\psi_0(x)$ and the first excited state $\psi_1(x)$ of the simple harmonic oscillator are orthogonal where

$$\psi_0(x) = A_0 \exp\left(-m\omega x^2 / 2\hbar\right)$$

$$\psi_1(x) = A_1 \sqrt{\frac{m\omega}{\hbar}} \, x \, \exp\left(-m\omega x^2 / 2\hbar\right)$$

29. Given the ground state of the harmonic oscillator is $\psi_0(x) = A_0 \exp(-m\omega x^2 / 2\hbar)$
(a) Calculate the average values $\langle x \rangle$ and $\langle x^2 \rangle$
(b) Calculate the average values $\langle p \rangle$ and $\langle p^2 \rangle$
(c) Calculate the uncertainty product $\Delta x \Delta p$

30. Show that the ground state of the harmonic oscillator
is in an eigenstate of $\hat{p}^2 = \left(\dfrac{\hbar}{i}\dfrac{\partial}{\partial x}\right)\left(\dfrac{\hbar}{i}\dfrac{\partial}{\partial x}\right)$

31. Show that $\left[\hat{N}, \hat{a}\right] = \hat{a}$

$$\left[\hat{N}, \hat{a}^\dagger\right] = \hat{a}^\dagger$$

Section 12.4 Schrödinger Equation in Higher Dimensions

32. Consider a particle moving in a two-dimensional space defined by $V = 0$ for $0 \le x \le L$ and $0 \le y \le L$ and $V = \infty$ elsewhere
    (a) Write down the wavefunctions for the particle in this well
    (b) Calculate the probability that the particle is in the corner region $0 \le x \le L/4$ and $0 \le y \le L/4$

33. Consider a particle constrained to a 2D box with $V = 0$ for $0 \le x \le a$ and $0 \le y \le b$ and $V = \infty$ elsewhere
    (a) What are the normalized wavefunctions inside the box?
    (b) What are the energy levels?

34. Consider a particle moving in a 3D space defined by $V = 0$ for $0 \le x \le L$. What percent of the time is the particle located within a distance $L/10$ from the sides of the box?

35. For hydrogen in the $l = 3$ state calculate the magnitude of L and the allowed values of L_z and θ.

36. Show that $\dfrac{1}{r^2}\dfrac{\partial}{\partial r}\left(r^2 \dfrac{\partial R(r)}{\partial r}\right) = \dfrac{1}{r}\dfrac{\partial^2 u(r)}{\partial r^2}$ where $u(r) = rR(r)$

37. For a spherically symmetric state of a hydrogen atom, the Schrödinger equation in spherical coordinates is

$$-\frac{\hbar^2}{2m}\left(\frac{d^2\psi}{dr^2}+\frac{2}{r}\frac{d\psi}{dr}\right)-\frac{ke^2}{r}\psi = E\psi$$

    (a) Show that the 1s wavefunction for an electron in hydrogen,

$$\psi_{1s}(r) = \frac{1}{\sqrt{\pi a_0^3}}\exp(-r/a_0)$$

    satisfies the Schrödinger equation
    (b) Show that the 1s wavefunction is normalized
    (c) Calculate the probability that the electron is at a distance greater than $5a_0$ from the nucleus
    (d) Given the charge density of the ground state of hydrogen $\rho(r) = q\psi_{1s}^2(r)$ use Gauss's law to show that the electric field

$$\mathbf{E}(r) = \frac{1}{4\pi\varepsilon_0 r^2}\left[1-e^{-\frac{2r}{a_0}}\left(1+\frac{2r}{a_0}+\frac{2r^2}{a_0^2}\right)\right]\hat{\mathbf{r}}$$

38. The simplest probability distribution for hydrogen corresponding to the 1s state $(n = 1, \ell = 0)$ is $P_{1s}(r) = 4\pi r^2 \Psi^*\Psi$ or

$$P_{1s}(r) = \frac{4r^2}{a_0^{\,3}} \exp\left(-\frac{2r}{a_0}\right)$$

Calculate the uncertainty Δr in the state where

$$\langle r \rangle = \int_0^\infty P_{1s} r\, dr \qquad \langle r^2 \rangle = \int_0^\infty P_{1s} r^2\, dr \qquad \Delta r = \sqrt{\langle r^2 \rangle - \langle r \rangle^2}$$

Note that the integrals required above are not Gaussian but have the form

$$\int_0^\infty r^n \exp(-\beta r)\, dr = \frac{n!}{\beta^{n+1}}$$

Calculate the probability that the electron is within one Bohr radius of the nucleus.

39. For the 1s state of hydrogen, evaluate the most probable value of r by setting $dP_{1s}/dr = 0$ and solving for r.

40. The radial part of the wavefunction for the hydrogen atom in the 2p state is given by

$$R_{2p} = \frac{1}{\sqrt{24a_0}} r e^{-r/2a_0}$$

where a_0 is the Bohr radius. Calculate the uncertainty Δr

41. The wavefunction for the hydrogen atom in the 2s state is given by

$$\psi_{2s} = \frac{1}{4\sqrt{2\pi}a_0^{3/2}}\left(2 - \frac{r}{a_0}\right) e^{-r/2a_0}$$

where a_0 is the Bohr radius. Find the most probable radial position of the electron.

42. Find the wavefunctions $\psi(r, \theta, \phi)$ and allowed energies of a particle contained in a finite spherical well with potential

$$V(r) = \begin{cases} 0 & r \le a \\ V_0 & r > a \end{cases}$$

43. Find the wavefunctions $\psi(r, \theta, \phi)$ and allowed energies of a particle contained in a spherical shell potential

$$V(r) = \begin{cases} \infty & r < a \\ 0 & a \le r \le b \\ \infty & r > b \end{cases}$$

Section 12.5 Approximation Methods

44. A particle with energy $E = V_0/2$ is incident from the left on a ramped potential

$$V(x) = \begin{cases} V_0 x/a & 0 \le x \le a \\ 0 & \text{elsewhere} \end{cases}$$

Find the WKB probability that the particle will tunnel through the barrier.

45. A particle with energy $E = V_0/2$ is incident from the left on a hump potential

$$V(x) = \begin{cases} V_0 \left[1 - (x/a)^2 \right] & |x| \le a \\ 0 & |x| > a \end{cases}$$

Find the WKB probability that the particle will tunnel through the barrier.

46. Obtain general expressions for the second and third order corrections to the energy levels and wavefunctions for the nondegenerate time-independent perturbation theory.

47. Consider a particle in a 1D potential well as in Figure 12.7.8. Treat the delta potential of height W_0 as a perturbation of the infinite square potential. Find the first order perturbation of the energy levels and wave functions.

$$V(x) = \begin{cases} \infty & x < 0 \\ W_0 \delta(x-d) & 0 \le x \le L \\ \infty & x > L \end{cases}$$

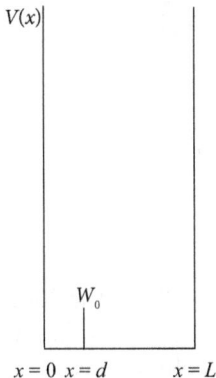

$x = 0 \; x = d \qquad x = L$

Figure 12.7.8: Square well with a delta function.

48. Consider a particle in a 1D potential well as in Figure 12.7.9.

$$V(x) = \begin{cases} \infty & x < 0 \\ W_0 (1 - x/L) & 0 \le x \le L \\ \infty & x > L \end{cases}$$

Treat the tilted floor with slope $-W_0/L$ as a perturbation of the infinite square potential. Find the first order perturbation of the energy levels and wavefunctions.

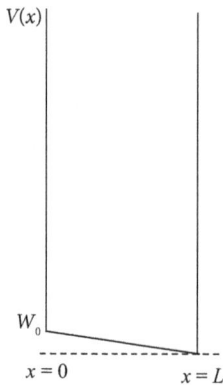

Figure 12.7.9: Square well with a ramped floor.

49. Consider a particle in a 1D potential well as in Figure 12.7.10. Treat the delta potential of height W_0 as a perturbation of the semi-infinite square potential. Find the first order perturbation of the energy levels and wavefunctions.

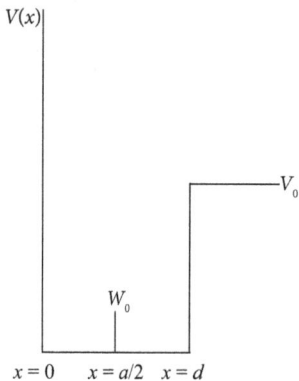

Figure 12.7.10: Semi-infinite well with a delta function.

50. Consider a particle in a 1D potential well as in Figure 12.7.11. Treat the delta potential of height W_0 as a perturbation of the simple harmonic oscillator. Find the first order perturbation of the energy levels and wavefunctions.

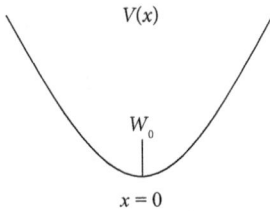

Figure 12.7.11: Harmonic oscillator with a delta function.

51. Use degenerate perturbation theory to find the perturbed energy levels and wavefunctions of a 2D harmonic oscillator with a delta function perturbation of height W_0 at $(x, y) = (0, 0)$

52. A three-dimensional square well with unperturbed wave functions

$$\psi(x, y, z) = \left(\frac{2}{L}\right)^{3/2} \sin\left(\frac{n\pi x}{L}\right)\sin\left(\frac{m\pi y}{L}\right)\sin\left(\frac{\ell \pi y}{L}\right)$$

and energy levels $E(n, m, \ell) = \dfrac{\hbar^2 \pi^2}{2mL^2}\left(n^2 + m^2 + \ell^2\right)$ is given a corner perturbation $H' = W_0 \left(0 \le x \le a, \, 0 \le y \le a, \, 0 \le z \le a\right)$ as shown in Figure 12.7.12. Use degenerate perturbation theory to find perturbed energy levels and wavefunctions.

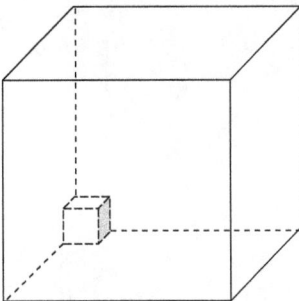

Figure 12.7.12: 3D square well with a corner perturbation.

53. Show that the nondegenerate perturbation of the hydrogen atom in the $n = 1$ state due to an electric field with perturbing Hamiltonian $\hat{H}' = eEz$ is zero.

Chapter 13 STATISTICAL MECHANICS

Chapter Outline

13.1 MICROCANONICAL ENSEMBLE

The microcanonical ensemble is introduced in this section. The Shannon entropy formula is then computed from the Boltzmann equation and the number of microstates. The entropy of an array of spins is calculated.

13.1.1 Number of Microstates and the Entropy

The microcanonical ensemble describes systems with fixed total energy E, volume V and number of particles N. Consider a system with N particles and M possible states. The total number of possible arrangements of the particles, or the number of microstates, is given by

$$\Omega = \frac{N!}{n_1! n_2! \cdots n_M!}$$

(13.1.1)

where the total number of particles is given by

$$N = \sum_{i=1}^{M} n_i \tag{13.1.2}$$

The number of particles in each state is n_i. The entropy S is then given by the Boltzmann formula $S = k_B \ln \Omega$

$$S = k_B \ln \frac{N!}{n_1! n_2! \cdots n_M!} \tag{13.1.3}$$

Expanding the log of the ratio as a difference of logs

$$S = k_B \left[\ln N! - \ln \left(n_1! n_2! \cdots n_M! \right) \right] \tag{13.1.4}$$

Applying Stirling's approximation to ln $N!$ and expanding the log of the product in the second term gives

$$S = k_B \left(N \ln N - N - \sum_{i=1}^{M} \ln n_i! \right) \tag{13.1.5}$$

The last term is then expanded using Stirling's approximation

$$S = k_B \left(N \ln N - N - \sum_{i=1}^{M} \left(n_i \ln n_i - n_i \right) \right) \tag{13.1.6}$$

Using $\Sigma n_i = N$ the factor of N cancels

$$S = k_B \left(N \ln N - \sum_{i=1}^{M} n_i \ln n_i \right) \tag{13.1.7}$$

Factoring N from this expression

$$S = N k_B \left(\ln N - \sum_{i=1}^{M} \frac{n_i}{N} \ln n_i \right) \tag{13.1.8}$$

and combining terms

$$S = N k_B \left(-\sum_{i=1}^{M} \frac{n_i}{N} \ln \frac{n_i}{N} \right) \tag{13.1.9}$$

We now write the occupation probability $P_i = n_i/N$

$$S = -N k_B \sum_{i=1}^{M} P_i \ln P_i \tag{13.1.10}$$

This formula is known as the Shannon entropy or the information entropy. The entropy is a positive quantity since $P_i \leq 1$ and $\ln P_i \leq 0$.

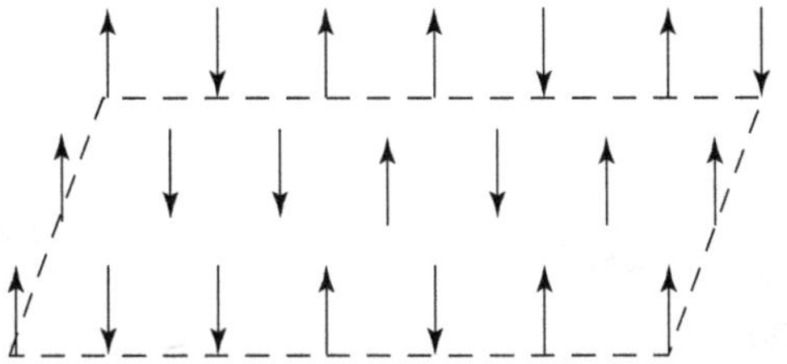

Figure 13.1.1: Array of spins with two possible orientations.

Example 13.1.1

Find the maximum and minimum entropy of an array of N spins where each spin can be up or down (Figure 13.1.1).

Solution: $N = n_u + n_d$ where n_u and n_d are the respective number of spins pointing up and down.

The number of possible spin orientations in the array is

$$\Omega = \frac{N!}{n_u! n_d!} \qquad (13.1.11)$$

For example, the number of possible ways to arrange $n_u = N/2$ and $n_d = N/2$ spins with $N = 100$ is

$$\Omega = \frac{100!}{50!50!} \approx 10^{29} \qquad (13.1.12)$$

corresponding to the state of maximum entropy $S = 66.8\ k_B$. Minimum entropy states with all the spins up ($n_u = N$, $n_d = 0$) or all the spins down ($n_u = 0$, $n_d = N$) have $S = k_B \ln 1 = 0$ where

$$\Omega = \frac{N!}{N!0!} = \frac{N!}{0!N!} = 1 \qquad (13.1.13)$$

since there is only one way in which all the dipoles can either point up or down.

Maple Examples

The Shannon entropy of the logistic map is calculated as a function of the parameter μ in the Maple worksheet below.

Key Maple commands: *add, Array, plot*

Maple packages: *with(plots)*:

Programming: **for** loops, **if** statements

restart

Shannon Entropy of the Logistic Map

with(*plots*) :
$f := (x, \text{mu}) \rightarrow \text{mu} \cdot x \cdot (1-x)$

$$f := (x, \mu) \mapsto \mu x (1-x)$$

$N := 1000 : bins := 40 : length\_mu := 300 :$
$muu := Array(1 \ldots length\_mu):$
$p := Array(1 \ldots bins)$

$$p := \begin{vmatrix} 1..40\,Array \\ Data\ Type : anything \\ Storage : rectangular \\ Order : Fortran\_order \end{vmatrix}$$

$S := Array(1 \ldots length\_mu):$

for *j* **from** 1 **to** *length_mu* **do**
$x[0] := 0.5;$
$muu[j] := 2.49 + 0.005 \cdot j;$
for *n* **from** 0 **to** *N* **do**
$x[n+1] := f(x[n], muu[j]);$
od:

for *n* **from** 20 **to** *N* **do**
for *nn* **from** 1 **to** *bins* **do**
if $\dfrac{nn-1}{bins} \le x[n] < \dfrac{nn}{bins}$ **then**
$p[nn] := p[nn] + 1;$
end if
od:
od:

for *n* **from** 20 **to** *N* **do**
for *nn* **from** 1 **to** *bins* **do**
if $\dfrac{nn-1}{bins} \le x[n] < \dfrac{nn}{bins}$ **then**
$p[nn] := p[nn] + 1;$

end if
od:
od:
$total := add(p[nn], nn = 1 \dots bins);$

$S[j] := 0;$
for nn **from** 1 **to** $bins$ **do**
if $p[nn] > 0$ **then**

$$S[j] := S[j] - evalf\left(\frac{p[nn]}{total} \cdot \log\left(\frac{p[nn]}{total}\right)\right);$$

$p[nn] := 0;$
end if
od:
od:
$plot(muu, S, labels = ['\text{mu}", "\text{S}"], title = "\text{Entropy of the Logistic Map}")$

Entropy of the Logistic Map

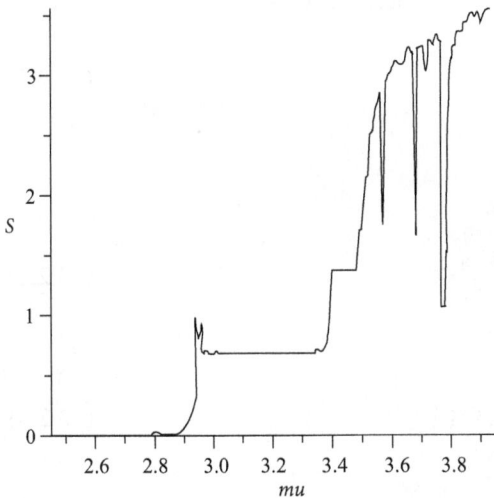

Figure 13.1.2: Shannon entropy of the logistic map as a function of the parameter μ.
The downward spikes in entropy occur in periodic windows after the onset of chaos above about μ = 3.57.

13.2 CANONICAL ENSEMBLE

The partition function for the canonical ensemble is demonstrated as a sum of Boltzmann factors in this section. Thermodynamic quantities including the free energy, entropy, and specific heat are all obtained from the partition function.

Examples are given including a particle in a box, the simple harmonic oscillator and the rigid rotator. Separable partition functions are discussed for composite systems with the example of a stretched rubber band.

13.2.1 Boltzmann Factor and Partition Function

The canonical ensemble applies to physical systems with a fixed number of particles N and volume V in thermodynamic equilibrium at a fixed temperature T. The probability P_i that a given state will have energy E_i is given by the Boltzmann factor

$$P_i \sim \exp\left(-\frac{E_i}{k_B T}\right) \tag{13.2.1}$$

The probability is normalized by the partition function Z

$$P_i = \frac{\exp(-\beta E_i)}{Z} \tag{13.2.2}$$

where $\beta = 1/k_B T$ and Z is the sum of all Boltzmann factors

$$Z = \sum_i e^{-\beta E_i} \tag{13.2.3}$$

13.2.2 Average Energy

The average energy of states in a system

$$\langle E \rangle = \sum_i P_i E_i \tag{13.2.4}$$

becomes

$$\langle E \rangle = \frac{1}{Z} \sum_i E_i \exp(-\beta E_i) \tag{13.2.5}$$

This can be written compactly as

$$\langle E \rangle = -\frac{\partial \ln Z}{\partial \beta} \tag{13.2.6}$$

Using $\partial / \partial \beta = -k_B T^2 \partial / \partial T$, we can express the average energy as

$$\langle E \rangle = k_B T^2 \frac{\partial \ln Z}{\partial T} \tag{13.2.7}$$

The total energy is the number of particles times the average energy $E = N\langle E \rangle$.

13.2.3 Free Energy and Entropy

The free energy F is less than the total energy E

$$F = E - TS \qquad (13.2.8)$$

where S is the entropy of the system. F represents the energy available to do useful work that decreases as the entropy S and temperature T increase. The entropy may be obtained by differentiation with respect to temperature

$$S = -\left(\frac{\partial F}{\partial T}\right)_{N,V} \qquad (13.2.9)$$

for fixed volume and number of particles denoted by the subscripts N, V. We may write the free energy in terms of the partition function

$$F = -Nk_B T \ln Z \qquad (13.2.10)$$

and the entropy by differentiating F with respect to T

$$S = Nk_B \left(\ln Z + \beta \langle E \rangle\right) \qquad (13.2.11)$$

From this expression, we see that the entropy is zero at $T = 0\ K$ in accord with the third law of thermodynamics.

13.2.4 Specific Heat

The temperature of a system will increase as energy is added. The specific heat C is numerically equivalent to the energy required to increase the system temperature by 1 K. Written as

$$C = \frac{\partial \langle E \rangle}{\partial T} \qquad (13.2.12)$$

we see that the units of C are J/K. An equivalent formula is obtained by differentiating the average energy with respect to β

$$C = -\frac{1}{k_B T^2}\frac{\partial}{\partial \beta}\langle E \rangle \qquad (13.2.13)$$

13.2.5 Rigid Rotator

The energy levels of a rigid rotator such as a diatomic molecule is

$$E_\ell = \frac{L^2}{2I} = \frac{\hbar^2}{2I}\ell(\ell+1) \qquad (13.2.14)$$

where L is the angular momentum and I is the moment of inertia. The degeneracy of the ℓth state is $g_\ell = 2\ell + 1$, thus the partition function

$$Z = \sum_{\ell=0}^{\infty} g_\ell \exp\left(-\beta E_\ell\right) \qquad (13.2.15)$$

becomes

$$Z = \sum_{\ell=0}^{\infty} (2\ell+1) \exp\left(-\frac{\beta \hbar^2}{2I} \ell(\ell+1)\right) \qquad (13.2.16)$$

At low temperatures, we may approximate Z by taking the first two terms where most of the molecules are in the ground state or the first excited state

$$Z = 1 + 3\exp\left(-\frac{\beta \hbar^2}{I}\right) \qquad (13.2.17)$$

The average energy is

$$\langle E \rangle = -\frac{\partial \ln Z}{\partial \beta} = \frac{3\hbar^2 / I}{3 + \exp\left(\beta \hbar^2 / I\right)} \qquad (13.2.18)$$

with specific heat

$$C = -\frac{1}{k_B T^2} \frac{\partial}{\partial \beta} \left[\frac{3\hbar^2 / I}{3 + \exp\left(\beta \hbar^2 / I\right)}\right] = \frac{1}{k_B T^2} \frac{3\left(\hbar^2 / I\right)^2}{\left[3 + \exp\left(\beta \hbar^2 / I\right)\right]^2} \qquad (13.2.19)$$

The free energy is

$$F = -N k_B T \ln Z = -N k_B T \ln\left(1 + 3\exp\left(-\frac{\beta \hbar^2}{I}\right)\right) \qquad (13.2.20)$$

with the entropy

$$S = -\frac{\partial F}{\partial T} = \frac{\partial}{\partial T} N k_B T \ln Z = N k_B \left[\ln\left(1 + 3e^{-\beta \hbar^2 / I}\right) + \frac{3\beta \hbar^2 / I}{\left(1 + 3e^{-\beta \hbar^2 / I}\right)}\right] \qquad (13.2.21)$$

13.2.6 Harmonic Oscillator

A harmonic oscillator potential $U(x) = kx^2/2$ corresponding to a linear restoring force $F(x) = -kx$ can be used to model the vibrational energy levels of molecules. Solutions to the Schrödinger equation with harmonic oscillator potential give evenly spaced energy

$$E_n = \left(n + \frac{1}{2}\right)\hbar\omega \qquad n = 0,\ 1,\ 2,\dots \qquad (13.2.22)$$

The resulting partition function

$$Z = \sum_{n=0}^{\infty} \exp\left[-\left(n+\frac{1}{2}\right)\beta\hbar\omega\right]$$ (13.2.23)

can be written factoring the zero-point energy term

$$Z = \exp\left(-\frac{\beta\hbar\omega}{2}\right)\sum_{n=0}^{\infty}\exp(-n\beta\hbar\omega)$$ (13.2.24)

This is a geometric series

$$\sum_{n=0}^{\infty} r^n = \frac{1}{1-r}$$ (13.2.25)

with $r = \exp(-\beta\hbar\omega)$ so that

$$Z = \frac{\exp\left(-\frac{\beta\hbar\omega}{2}\right)}{1-\exp(-\beta\hbar\omega)}$$ (13.2.26)

The average energy, specific heat, free energy, and entropy may now be calculated.

13.2.7 Composite Systems

Consider a system of N noninteracting particles each with possible states $1i$, $2i$, ..., Ni and with possible energies E_{1i} of particle 1, E_{2i} of particle 2, etc. Possible energy levels of the composite system are $E_{1i} + E_{2i} + ... + E_{Ni}$ so that the partition function

$$Z_{total} = \sum_{1i,2i,\cdots,Ni} e^{-\beta(E_{1i}+E_{2i}+\cdots+E_{Ni})} = \sum_{1i,2i,\cdots,Ni} e^{-\beta E_{1i}}e^{-\beta E_{2i}}\cdots e^{-\beta E_{Ni}}$$ (13.2.27)

may be factored into N sums

$$Z_{total} = \underbrace{\sum_{1i} e^{-\beta E_{1i}}}_{Z_1} \underbrace{\sum_{2i} e^{-\beta E_{2i}}}_{Z_2} \cdots \underbrace{\sum_{Ni} e^{-\beta E_{Ni}}}_{Z_N} = \prod_{k=1}^{N} Z_k$$ (13.2.28)

As an example, the partition function of a diatomic gas can be separated into translational, rotational and vibrational components.

13.2.8 Stretching a Rubber Band

As an example of a one-dimensional composite system we consider a polymer chain consisting of N segments each of length L. The fully stretched length of the chain is NL. When stretched small distances x, the restoring force is f corresponding to a potential energy function

$$U(x) = -f \cdot x \qquad (13.2.29)$$

The total length of the chain is

$$x = L \sum_{i=1}^{N} \sigma_i \qquad (13.2.30)$$

where $\sigma_i = 1$ corresponds to the segment pointing to the right and $\sigma_i = -1$ corresponds to the segment pointing to the left. Given the partition function

$$Z = \exp\left(\beta f L \sum_{i=1}^{N} \sigma_i \right) \qquad (13.2.31)$$

it may be shown that

$$Z = \left(2\cosh \beta f L \right)^{N} \qquad (13.2.32)$$

to calculate the average energy, entropy, free energy and specific heat of the chain.

Maple Examples

The partition function, average energy and specific heat of a particle in a square well is computed in the Maple worksheet below. The partition function, average energy, free energy and entropy of a simple harmonic oscillator and a two-state system are also calculated.

Key Maple commands: *convert, diff, expand, factor, plot, semilogplot, subs, sum*

Maple packages: *with(plots):*

restart

Particle in a Square Well

$E := (n) \to E0 \cdot n^2$

$$E := (n) \mapsto E0\, n^2$$

$Z := Sum(\exp(-beta \cdot E(n)), n = 1 \dots infinity)$

$$Z := \sum_{n=1}^{\infty} e^{-\beta E0 n^2}$$

$AvgE := -diff(\log(Z), beta)$

$$AvgE := -\frac{\displaystyle\sum_{n=1}^{\infty} \left(-E0 n^2 e^{-\beta E0 n^2} \right)}{\displaystyle\sum_{n=1}^{\infty} e^{-\beta E0 n^2}}$$

$$C := -expand\left(\frac{diff(AvgE, beta)}{kB \cdot T^2}\right)$$

$$C := \frac{E0^2\left(\displaystyle\sum_{n=1}^{\infty} \frac{n^4}{e^{\beta E0 n^2}}\right)}{kBT^2\left(\displaystyle\sum_{n=1}^{\infty} \frac{1}{e^{\beta E0 n^2}}\right)} - \frac{E0^2\left(\displaystyle\sum_{n=1}^{\infty} \frac{n^2}{e^{\beta E0 n^2}}\right)^2}{kBT^2\left(\displaystyle\sum_{n=1}^{\infty} \frac{1}{e^{\beta E0 n^2}}\right)^2}$$

$$subs\left(\left\{E0 = 1, beta = \frac{1}{T}\right\}, AvgE\right)$$

$$-\frac{\displaystyle\sum_{n=1}^{\infty}\left(-n^2 e^{-\frac{n^2}{T}}\right)}{\displaystyle\sum_{n=1}^{\infty} e^{-\frac{n^2}{T}}}$$

$$C1 := (T) \rightarrow subs\left(\left\{E0 = 1, beta = \frac{1}{T}, kB = 1\right\}, C\right)$$

$$C1 := T \mapsto subs\left(\left\{E0 = 1, \beta = \frac{1}{T}, kB = 1\right\}, C\right)$$

$with(plots):$
$semilogplot(C1(T), T = 0.1 \dots 100)$

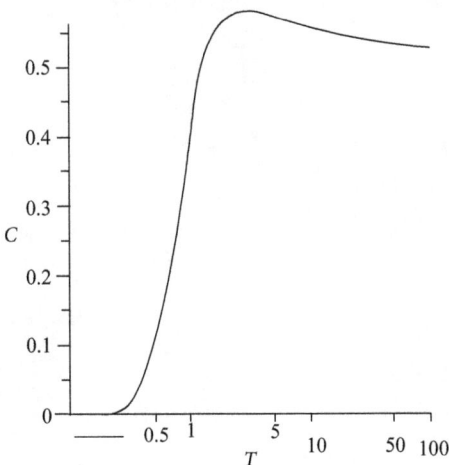

Figure 13.2.1: Specific heat vs. temperature of a particle in a square well.

$assume(beta > 0, h > 0, \omega > 0)$

Harmonic Oscillator

$$E := (n) \to \left(n + \frac{1}{2}\right) \hbar \cdot \omega$$

$$E := (n) \mapsto \left(n + \frac{1}{2}\right) \hbar \omega$$

$$Z := sum(\exp(-beta \cdot E(n)), n = 0 \ldots infinity)$$

$$Z := \frac{e^{\frac{\beta \sim \hbar \sim \omega \sim}{2}}}{e^{\beta \sim \hbar \sim \omega \sim} - 1}$$

$$E_{avg} := -factor(diff(\log(Z), beta))$$

$$E_{avg} := \frac{\hbar \sim \omega \sim \left(e^{\beta \sim \hbar \sim \omega} + 1\right)}{2\left(e^{\beta \sim \hbar \sim \omega \sim} - 1\right)}$$

$$F := -k_B \cdot T \cdot expand(\ln(Z))$$

$$F := -k_B T \left(-\ln\left(e^{\beta \sim \hbar \sim \omega} - 1\right) + \frac{\beta \sim \hbar \sim \omega \sim}{2}\right)$$

$$F := subs\left(beta = \frac{1}{k_B \cdot T}, F\right)$$

$$F := -k_B T \left(-\ln\left(e^{\frac{\hbar \sim \omega \sim}{k_B T}} - 1\right) + \frac{\hbar \sim \omega \sim}{2 k_B T}\right)$$

$$S := expand(diff(F, T))$$

$$S := k_B \ln\left(e^{\frac{\hbar \sim \omega \sim}{k_B T}} - 1\right) - \frac{\hbar \sim \omega \sim e^{\frac{\hbar \sim \omega \sim}{k_B T}}}{T\left(e^{\frac{\hbar \sim \omega \sim}{k_B T}} - 1\right)}$$

restart

Two-State System

$$Z := \exp(-beta \cdot epsilon) + \exp(beta \cdot epsilon)$$

$$Z := e^{-\beta \epsilon} + e^{\beta \epsilon}$$

$$E_{avg} := -factor(diff(\ln(Z), beta))$$

$$E_{avg} := \frac{\epsilon \left(e^{\beta \epsilon} - e^{\beta \epsilon}\right)}{e^{\beta \epsilon} + e^{\beta \epsilon}}$$

$E_{avg} := convert(E_{avg}, trig)$

$$E_{avg} := \frac{\epsilon \sinh(\beta \epsilon)}{\cosh(\beta \epsilon)}$$

$F := -N \cdot k_B \cdot T \cdot expand(\ln(Z))$

$$F := -N k_B T \ln(e^{-\beta \epsilon} + e^{\beta \epsilon})$$

$$F := subs\left(beta = \frac{1}{k_B \cdot T}, F \right)$$

$$F := -N k_B T \ln\left(e^{-\frac{\epsilon}{k_B T}} + e^{\frac{\epsilon}{k_B T}} \right)$$

$S := -diff(F, T)$

$$S := N k_B \ln\left(e^{-\frac{\epsilon}{k_B T}} + e^{\frac{\epsilon}{k_B T}} \right) + \frac{N k_B T \left(\dfrac{\epsilon\, e^{-\frac{\epsilon}{k_B T}}}{k_B T^2} - \dfrac{\epsilon\, e^{\frac{\epsilon}{k_B T}}}{k_B T^2} \right)}{e^{-\frac{\epsilon}{k_B T}} + e^{\frac{\epsilon}{k_B T}}}$$

$S := expand(convert(S, trig))$

$$S := N k_B \ln(2) + N k_B \ln\left(\cosh\left(\frac{\epsilon}{k_B T} \right) \right) - \frac{N \epsilon \sinh\left(\dfrac{\epsilon}{k_B T} \right)}{T \cosh\left(\dfrac{\epsilon}{k_B T} \right)}$$

13.3 CONTINUOUS ENERGY DISTRIBUTIONS

In this section, thermodynamic quantities are calculated for systems with continuous energy distributions, partition function and average energy. Examples of a particle in a box, the Maxwell-Boltzmann speed distribution of an ideal gas and a relativistic gas are given.

13.3.1 Partition Function and Average Energy

The partition function corresponding to a system with a continuous energy distribution is obtained by integrating the Boltzmann factor over all internal degrees of freedom. If the energy is only a function of the parameter r then

$$Z = \int_{all\ r} e^{-\beta E(r)} dr \tag{13.3.1}$$

The probability that the system is between a and b is then

$$P\left(a \leq r \leq b\right) = \frac{1}{Z}\int_a^b e^{-\beta E(r)}dr \qquad (13.3.2)$$

The average energy $\langle E \rangle$ of a system with continuous energy spectrum in thermodynamic equilibrium is similarly obtained

$$\langle E \rangle = \frac{1}{Z}\int_{\text{all }r} E\left(r\right)e^{-\beta E(r)}dr \qquad (13.3.3)$$

13.3.2 Particle in a Box

Systems with discrete energy levels can be modeled as having a continuous energy distribution if the energy levels are closely spaced. The energy levels of a particle contained in a 1D box of length L are

$$E_n = \frac{\hbar^2 \pi^2}{2mL^2}n^2 = E_1 n^2 \qquad (13.3.4)$$

so that our partition function becomes

$$Z = \sum_{n=1}^{\infty} \exp\left(-\beta E_1 n^2\right) \qquad (13.3.5)$$

For large L the energy levels are more closely spaced so that we may approximate the infinite sum by the Gaussian integral

$$Z = \int_0^{\infty} \exp\left(-\beta E_1 n^2\right)dn \qquad (13.3.6)$$

with solution

$$Z = \frac{\sqrt{\pi}}{2}\sqrt{\frac{2mL^2 k_B T}{\hbar^2 \pi^2}} = \sqrt{\frac{2\pi m k_B T}{h^2}}L \qquad (13.3.7)$$

The square root has units of inverse length since Z is dimensionless.

13.3.3 Maxwell-Boltzmann Distribution

In a gas at temperature T, the distribution of molecular speeds is proportional to a Boltzmann factor where the probability that the ith molecule has energy E_i is proportional to

$$\exp\left(-\frac{E_i}{k_B T}\right) = \exp\left(-\frac{mv^2}{2k_B T}\right) \qquad (13.3.8)$$

where the energy $E_i = mv^2/2$. Gas molecules obeying Maxwell-Boltzmann statistics have a normalized distribution of speeds described by

$$n(v)\,dv = 4\pi \left(\frac{m}{2\pi k_B T}\right)^{3/2} v^2 \exp\left(-\frac{mv^2}{2k_B T}\right) dv \tag{13.3.9}$$

To calculate the average speed, we evaluate the integral

$$\langle v \rangle = \int_0^\infty v n(v)\,dv = \sqrt{\frac{8 k_B T}{\pi m}} \tag{13.3.10}$$

To calculate the average of v^2 we evaluate

$$\langle v^2 \rangle = \int_0^\infty v^2 n(v)\,dv = \frac{3 k_B T}{m} \tag{13.3.11}$$

The root mean square speed is then given by

$$v_{\text{rms}} = \sqrt{\langle v^2 \rangle} = \sqrt{\frac{3 k_B T}{m}} \tag{13.3.12}$$

13.3.4 Relativistic Gas

There is no maximum speed limit of molecules in the Maxwell-Boltzmann distribution where the tail region of the $n(v)$ plot extends to infinity. However, the maximum speed of molecules should not be greater than the speed of light c in accordance with special relativity. Using the relativistic form of the kinetic energy $(\gamma - 1)mc^2$ for a hypothetical gas

$$n(v)\,dv = A\left(4\pi v^2\right)\exp\left(-\frac{(\gamma-1)mc^2}{k_B T}\right) dv \tag{13.3.13}$$

with $\gamma = 1/\sqrt{1 - v^2/c^2}$ we have

$$n(v)\,dv = A\left(4\pi v^2\right)\exp\left(\frac{mc^2}{k_B T}\left(1 - \frac{1}{\sqrt{1 - \dfrac{v^2}{c^2}}}\right)\right) dv \tag{13.3.14}$$

The normalization constant A is determined by numerically evaluating

$$\int_0^c n(v)\,dv = 1 \tag{13.3.15}$$

In terms of $u = v/c$ and $T_R = mc^2/k_B$

$$n(u)du = \frac{u^2 \exp\left(\dfrac{T_R}{T}\left(1 - \dfrac{1}{\sqrt{1-u^2}}\right)\right)du}{\displaystyle\int_0^1 u'^2 \exp\left(\dfrac{T_R}{T}\left(1 - \dfrac{1}{\sqrt{1-u'^2}}\right)\right)du'} \tag{13.3.16}$$

the average speed

$$\langle v \rangle = c \int_0^1 n(u)u\,du \tag{13.3.17}$$

and average speed squared

$$\langle v^2 \rangle = c^2 \int_0^1 n(u)u^2\,du \tag{13.3.18}$$

Note that this analysis neglects collisional effects and particle pair creation and annihilation at high energies.

Maple Examples

The normalized Maxwell-Boltzmann (M-B) velocity distribution of a gas at three temperatures is plotted in the Maple worksheet below. The average velocity and average squared velocity of the Maxwell-Boltzmann distribution is then computed. The velocity distribution of a relativistic gas is then computed and plotted at three temperatures. The average velocity of the relativistic gas is then computed.

Key Maple commands: *assume, evalf, plot, simplify*

restart

Maxwell-Boltzmann Distribution

assume$(m > 0, k_B > 0, T > 0)$

$$n := (m, T, v) \rightarrow 4 \cdot Pi \cdot \left(\frac{m}{2 \cdot Pi \cdot k_B \cdot T}\right)^{3/2} \cdot v^2 \cdot \exp\left(-\frac{m \cdot v}{2 \cdot k_B \cdot T}\right)$$

$$n := (m, T, v) \mapsto \pi\sqrt{2}\left(\frac{m}{\pi k_B T}\right)^{3/2} v^2\, e^{-\frac{mv^2}{2k_B T}}$$

Plot of Maxwell-Boltzmann Distribution

$k_B := 1$

$$k_B := 1$$

$plot([n(1, 10, v), n(1, 100, v), n(1, 1000, v)], v = 0 \ldots 120, linestyle = [solid,$
$\quad dash, dashdot], legend = ['kT = 10', 'kT = 100', 'kT = 1000'])$

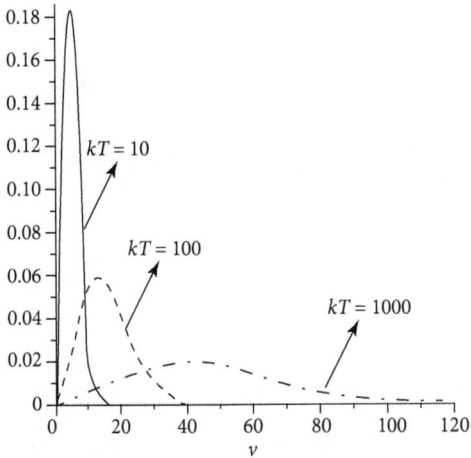

Figure 13.3.1: Plot of the M-B velocity distribution at three temperatures.

Show That the M-B Distribution Is Normalized

$$\int_0^\infty n(m, T, v) dv$$

$$\frac{\pi^{3/2} \left(\dfrac{m \sim}{\pi T \sim} \right)^{3/2}}{\left(\dfrac{m \sim}{T \sim} \right)^{3/2}}$$

$simplify(\%)$

$$1$$

Calculate <v^2>

$$\int_0^\infty v^2 \cdot n(m, T, v) dv$$

$$\frac{3\pi^{3/2}\left(\dfrac{m\sim}{\pi T\sim}\right)^{3/2}}{\left(\dfrac{m\sim}{T\sim}\right)^{5/2}}$$

simplify(%)

$$\frac{3T\sim}{m\sim}$$

Calculate <v>

$$\int\limits_{0}^{\infty} v \cdot n(m,T,v)\,dv$$

$$\frac{2\pi\sqrt{2}\left(\dfrac{m\sim}{\pi T\sim}\right)^{3/2} T\sim^{2}}{m\sim^{2}}$$

simplify(%)

$$\frac{2\sqrt{2}\sqrt{T\sim}}{\sqrt{\pi}\sqrt{m\sim}}$$

Relativistic Gas

restart

$m := 1.67{\cdot}10^{-27}; c := 3.0{\cdot}10^{8}; k := 1.38{\cdot}10^{-23};$

$$m := 1.670000000\ 10^{-27}$$

$$c := 3.000000000\ 10^{8}$$

$$k := 1.380000000\ 10^{-23}$$

$$\text{alpha} := \frac{m \cdot c^2}{k}$$

$$\alpha := 1.089130435\ 10^{13}$$

$$nrel := (T,u) \rightarrow \frac{u^2 \cdot \exp\left(\dfrac{1}{T}\left(1 - \dfrac{1}{\mathrm{sqrt}(1-u^2)}\right)\right)}{evalf\left(\displaystyle\int_0^1 w^2 \cdot \exp\left(\dfrac{1}{T}\left(1 - \dfrac{1}{\mathrm{sqrt}(1-w^2)}\right)\right)dw\right)}$$

$$nrel := (T, u) \mapsto \frac{u^2 \, e^{\frac{1 - \frac{1}{\sqrt{-u^2 + 1}}}{T}}}{evalf\left(\int_0^1 w^2 \, e^{\frac{1 - \frac{1}{\sqrt{-w^2 + 1}}}{T}} \, dw\right)}$$

$$plot\left([nrel(0.01, u), nrel(.1, u), nrel(1, u)], u = 0 \cdots 1, linestyle = [solid, dash, dashdot],\right.$$

$$\left. legend = \left['T = 0.01 \cdot \frac{mc^2}{k}', 'T = 0.1 \cdot \frac{mc^2}{k}', 'T = \frac{mc^2}{k}',\right]\right.$$

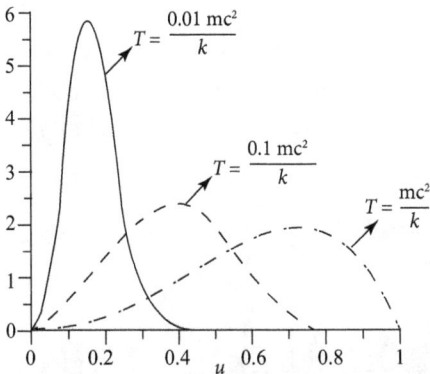

Figure 13.3.2: Plot of the velocity distribution of a relativistic gas at three temperatures.

Normalization of Relativistic Velocity Distribution

$$evalf\left(\int_0^1 nrel(.001, v) \, dv\right)$$

1.000000000

Average Velocity of Relativistic Gas

$$evalf\left(\int_0^1 v \cdot nrel(.1, v) \, dv\right)$$

0.3973286862

13.4 GRAND CANONICAL ENSEMBLE

The partition function for the grand canonical ensemble is evaluated as a sum of Gibbs factors in this section. Calculated thermodynamic quantities include average values of energy and particle number. Bose-Einstein and Fermi-Dirac statistics are discussed with examples including black-body radiation and the Debye theory of specific heat.

13.4.1 Gibbs Factor

The grand canonical ensemble describes open systems with fixed temperature T, chemical potential μ and volume V where the total number of particles is variable. For multiple species, we express the Gibbs factor like the Boltzmann factor

$$P_{i,j} \sim \exp\left(-\frac{E_{i,j} - \mu_j N_j}{k_B T}\right) \tag{13.4.1}$$

where $E_{i,j}$ is the ith energy level, μ_j is the chemical potential and N_j is the number of particles of the jth species. The Gibbs factor is proportional to the probability $P_{i,j}$ of finding a particle of a given species with energy $E_{i,j}$. The probability is normalized by the grand canonical partition function

$$Z_G = \sum_{i,j} \exp\left(-\beta E_{i,j} + \beta \mu_j N_j\right) \tag{13.4.2}$$

with $\beta^{-1} = k_B T$ so that

$$P_{i,j} = \frac{\exp\left(-\beta E_{i,j} + \beta \mu_j N_j\right)}{Z_G} \tag{13.4.3}$$

and

$$\sum_{i,j} P_{i,j} = 1 \tag{13.4.4}$$

13.4.2 Average Energy and Particle Number

The average energy is calculated in the grand canonical ensemble as

$$\langle E \rangle = \sum_{i,j} P_{i,j} E_{i,j} \tag{13.4.5}$$

or

$$\langle E \rangle = \frac{1}{Z_G} \sum_{i,j} E_{i,j} \exp\left(-\beta E_{i,j} + \beta \mu_j N_j\right) \tag{13.4.6}$$

This can be expressed as a derivative

$$\langle E \rangle = -\frac{\partial \ln Z_G}{\partial \beta} = k_B T^2 \frac{\partial \ln Z_G}{\partial T} \tag{13.4.7}$$

We may also compute the average particle number N

$$\langle N \rangle = \frac{1}{Z_G} \sum_{i,j} N_{i,j} \exp\left(-\beta E_{i,j} + \beta \mu_j N_j\right) \tag{13.4.8}$$

13.4.3 Single Species

For a single species with chemical potential μ the partition function is a single sum

$$Z_G = \sum_i \exp\left(-\beta E_i\right) \exp\left(\beta \mu N\right) \tag{13.4.9}$$

defining the parameter $\gamma = \beta \mu$. We can calculate the average particle number

$$\langle N \rangle = \frac{\partial \ln Z_G}{\partial \gamma} \tag{13.4.10}$$

13.4.4 Grand Potential

From the grand potential defined as

$$\Omega = -k_B T \ln Z_G \tag{13.4.11}$$

we may compute the entropy S at constant μ and V

$$S = -\left(\frac{\partial \Omega}{\partial T}\right)_{\mu,V} \tag{13.4.12}$$

and the pressure P at constant μ and T

$$P = -\left(\frac{\partial \Omega}{\partial V}\right)_{\mu,T} \tag{13.4.13}$$

13.4.5 Comparison of Canonical and Grand Canonical Ensembles

Table 13.4.1 compares the canonical and grand canonical ensembles. In both ensembles, once the partition function is found then thermodynamic quantities such as average energy and entropy may be calculated.

TABLE 13.4.1: Comparison of thermodynamic quantities evaluated in the canonical and grand canonical ensembles.

Canonical Ensemble	Grand Canonical Ensemble
Boltzmann factor $$P_i \sim \exp\left(-\beta E_i\right)$$	Gibbs factor $$P_{ij} \sim \exp\left(-\beta E_{i,j} + \beta \mu_j N_j\right)$$

(contd.)

Canonical Ensemble	Grand Canonical Ensemble
Partition function $$Z = \sum_i e^{-\beta E_i}$$	Grand canonical partition function $$Z_G = \sum_{i,j} \exp\left(-\beta E_{i,j} + \beta \mu_j N_j\right)$$
Normalized probability $$P_i = \frac{\exp\left(-\beta E_i\right)}{Z}$$	Normalized probability $$P_{i,j} = \frac{\exp\left(-\beta E_{i,j} + \beta \mu_j N_j\right)}{Z_G}$$
Average energy $$\langle E \rangle = -\frac{\partial \ln Z}{\partial \beta} = k_B T^2 \frac{\partial \ln Z}{\partial T}$$	Average energy $$\langle E \rangle = -\frac{\partial \ln Z_G}{\partial \beta} = k_B T^2 \frac{\partial \ln Z_G}{\partial T}$$
Helmholtz free energy $$F = -N k_B T \ln Z$$	Grand potential $$\Omega = -k_B T \ln Z_G$$

13.4.6 Bose-Einstein Statistics

Bose-Einstein statistics are applicable to integer spin particles such as photons and phonons. Photons are spin one particles while phonons are quanta of vibration without spin (spin zero). Such particles do not obey the Pauli exclusion principle so there is no limit to the number of particles that may be in each state n_s. We write the partition function summing over all possible configurations R

$$Z_G = \sum_R \exp\left(-\beta E_R + \beta \mu N\right) \tag{13.4.14}$$

with

$$E_R = n_1 E_1 + n_2 E_2 + \cdots = \sum_s n_s E_s \tag{13.4.15}$$

The variable number of particles N can be written

$$N = n_1 + n_2 + \cdots = \sum_s n_s \tag{13.4.16}$$

allowing us to factor the partition function

$$Z_G = \sum_{n_1, n_2, \cdots} e^{-\beta(n_1 E_1 + n_2 E_2 + \cdots)} e^{\beta \mu(n_1 + n_2 + \cdots)} \tag{13.4.17}$$

The factor

$$Z_G = \left(\sum_{n_1=0}^{\infty} e^{-\beta n_1 E_1 + \beta \mu n_1}\right)\left(\sum_{n_2=0}^{\infty} e^{-\beta n_2 E_2 + \beta \mu n_2}\right)\cdots \tag{13.4.18}$$

is written as an infinite product

$$Z_G = \prod_{k=0}^{\infty}\left(\sum_{n_k=0}^{\infty} e^{-\beta n_k E_k + \beta \mu n_k}\right) \qquad (13.4.19)$$

Evaluating the sum

$$Z_G = \prod_{k=0}^{\infty}\left(\frac{1}{1-e^{-\beta E_k + \beta\mu}}\right) \qquad (13.4.20)$$

and taking the logarithm

$$\ln Z_G = \sum_{k=0}^{\infty} \ln\left(\frac{1}{1-e^{-\beta E_k + \beta\mu}}\right) = -\sum_{k=0}^{\infty}\ln\left(1-e^{-\beta E_k + \beta\mu}\right) \qquad (13.4.21)$$

The average number of particles with energy E_k is shown as

$$\langle n_k \rangle = -\frac{1}{\beta}\frac{\partial}{\partial E_k}\ln Z_G = \frac{e^{-\beta E_k + \beta\mu}}{1-e^{-\beta E_k + \beta\mu}} = \frac{1}{e^{\beta(E_k-\mu)}-1} \qquad (13.4.22)$$

This formula expressed as a continuous function of energy $f_{BE}(E)=\langle n_k\rangle$ is known as the Bose-Einstein distribution where

$$f_{BE}(E) = \frac{1}{e^{\beta(E-\mu)}-1} \qquad (13.4.23)$$

13.4.7 Black-Body Radiation

By treating emitters of radiation as discrete (as opposed to continuous), Planck obtained a very good fit to measured black-body spectra. For a cubical cavity of side L with volume $V = L^3$, the allowed wavelengths are $\lambda = 2L/n$ corresponding to the energy levels

$$E = hf = \frac{hc}{\lambda} = \frac{hc}{2L}n \qquad (13.4.24)$$

where $n^2 = n_x^2 + n_y^2 + n_z^2$. The number of standing electromagnetic waves in a cubical cavity of side L between n and $n + dn$ expressed as the first octant of a spherical shell

$$g(n)dn = 2\frac{1}{8}4\pi n^2 dn = \pi n^2 dn \qquad (13.4.25)$$

with a factor of 2 for two photon polarizations and $n = 2Lf/c$

$$g(f)df = \left(\frac{2L}{c}\right)^3 f^2 df \qquad (13.4.26)$$

The energy density $u(f, T)df$ is obtained by multiplying the average energy $\langle E \rangle$ by $g(f)df$ divided by the volume L^3

$$u(f,T)df = \frac{8\pi f^2}{c^3}\langle E \rangle df \qquad (13.4.27)$$

The average energy of photons may be obtained using the Bose-Einstein distribution with zero chemical potential

$$u(f,T)df = \frac{8\pi f^2}{c^3}\left(\frac{hf}{\exp(hf/k_BT)-1}\right)df \qquad (13.4.28)$$

The radiated flux, or power per area P/A, is obtained from integrating $u(f, T)$ times $c/4$

$$\frac{P}{A} = \frac{c}{4}\int_0^\infty u(f,T)df \qquad (13.4.29)$$

This integral is evaluated in Section 3.5 giving

$$\frac{P}{A} = \sigma T^4 \qquad (13.4.30)$$

where σ is the Stephan-Boltzmann constant. To find the frequency corresponding to the maximum power we evaluate

$$\frac{d}{df}u(f,T) = 0 \qquad (13.4.31)$$

This gives a transcendental equation that must be numerically solved. Expressed in terms of the maximum wavelength $\lambda_{max} = c/f_{max}$ Wien's law is thus obtained

$$\lambda_{max}T = 2.9\times10^{-3}\,\text{mK} \qquad (13.4.32)$$

13.4.8 Debye Theory of Specific Heat

The Debye theory of specific heat treats lattice vibrations as spin zero quanta that obey Bose-Einstein statistics. Whereas the maximum mode number is infinite for photons, in a solid the maximum mode number will be limited by the separation between atoms. Vibrational energy levels are given by

$$E = hf = \frac{hv_s}{\lambda} \qquad (13.4.33)$$

where v_s is the speed of sound in the solid. For a cubical solid of side L with volume $V = L^3$, the allowed wavelengths are $\lambda = 2L/n$ corresponding to the energy levels

$$E = \frac{h v_s}{2L} n \tag{13.4.34}$$

The maximum energy level

$$E_{max} = \frac{h v_s}{\lambda_{min}} \tag{13.4.35}$$

with the minimum phonon wavelength taken as twice the atomic separation $\lambda_{min} = 2L/n_{max}$. The maximum mode number for phonons in a solid with N atoms is $n_{max} = \sqrt[3]{N}$. The total energy of lattice vibrations is thus

$$U = \sum_{n_x=1}^{n_{max}} \sum_{n_y=1}^{n_{max}} \sum_{n_z=1}^{n_{max}} E(n) \langle N(E(n)) \rangle \tag{13.4.36}$$

where $n^2 = n_x^2 + n_y^2 + n_z^2$. The average number of phonons with energy $E(n)$

$$\langle N(E(n)) \rangle = \frac{3}{e^{E(n)/k_B T} - 1} \tag{13.4.37}$$

with the factor of three for vibrations in x-, y- and z- directions. Making an integral approximation to the sum by treating the mode numbers as continuous

$$U = 3\sum_{n_x=1}^{n_{max}} \sum_{n_y=1}^{n_{max}} \sum_{n_z=1}^{n_{max}} \frac{E(n)}{e^{E(n)/k_B T} - 1} \approx 3 \int_0^{n_{max}} \int_0^{n_{max}} \int_0^{n_{max}} \frac{E(n)}{e^{E(n)/k_B T} - 1} dn_x dn_y dn_z \tag{13.4.38}$$

This integral is converted to spherical coordinates integrating over the first octant of a sphere with positive n_x, n_y and n_z

$$U = 3 \int_0^{\pi/2} \int_0^{\pi/2} \int_0^R \frac{E(n)}{e^{E(n)/k_B T} - 1} n^2 dn \sin\theta d\theta d\phi \tag{13.4.39}$$

Integration over angular coordinates gives $\pi/2$ and

$$U = \frac{3\pi}{2} \int_0^R \frac{h v_s n}{2L} \frac{n^2}{e^{h v_s n/2L k_B T} - 1} dn \tag{13.4.40}$$

The integral is written in dimensionless form making the substitution

$$x = \frac{h v_s n}{2L k_B T} \quad \text{so that} \quad n^3 dn = \left(\frac{2L k_B T}{h v_s}\right)^4 x^3 dx \tag{13.4.41}$$

and

$$U = \frac{3\pi}{2}\left(\frac{2Lk_B}{h\nu_s}\right)^3 k_B T^4 \int_0^{x_{max}} \frac{x^3}{e^x - 1} dx \qquad (13.4.42)$$

We define the Debye temperature T_D such that

$$x_{max} = \frac{h\nu_s R}{2Lk_B T} = \frac{T_D}{T} \text{ and } T_D = \frac{h\nu_s}{2Lk_B}\sqrt[3]{\frac{6N}{\pi}} \qquad (13.4.43)$$

giving

$$U = 9Nk_B \frac{T^4}{T_D^3}\int_0^{\frac{T_D}{T}} \frac{x^3}{e^x - 1} dx \qquad (13.4.44)$$

The specific heat C is then obtained

$$C = \frac{dU}{dT} = \frac{3\pi}{2}\left(\frac{h\nu_s}{2L}\right)^2 \frac{1}{k_B T^2}\int_0^R \frac{n^4 e^{h\nu_s n/2Lk_B T}}{\left(e^{h\nu_s n/2Lk_B T} - 1\right)^2} dn \qquad (13.4.45)$$

Expressed in dimensionless form

$$C = 9Nk_B \frac{T^3}{T_D^3}\int_0^{\frac{T_D}{T}} \frac{x^4 e^x}{\left(e^x - 1\right)^2} dx \qquad (13.4.46)$$

13.4.9 Fermi-Dirac Statistics

Fermi-Dirac statistics apply to half integer spin particles such as electrons that obey the Pauli exclusion principle. A given state can be vacant $n_i = 0$ or occupied $n_i = 1$. The factored partition function

$$Z_G = \left(\sum_{n_1=0}^1 e^{-\beta n_1 E_1 + \beta \mu n_1}\right)\left(\sum_{n_2=0}^1 e^{-\beta n_2 E_2 + \beta \mu n_2}\right)\cdots \qquad (13.4.47)$$

is written as an infinite product

$$Z_G = \prod_{k=0}^\infty \left(\sum_{n_k=0}^1 e^{-\beta n_k E_k + \beta \mu n_k}\right) \qquad (13.4.48)$$

Carrying out the summation

$$Z_G = \prod_{k=0}^{\infty}\left(1+e^{-\beta E_k + \beta\mu}\right) \tag{13.4.49}$$

and taking the logarithm

$$\ln Z_G = \sum_{k=0}^{\infty}\ln\left(1+e^{-\beta E_k + \beta\mu}\right) \tag{13.4.50}$$

we may calculate the average number of particles with energy E_k

$$\langle n_k\rangle = -\frac{1}{\beta}\frac{\partial}{\partial E_k}\ln Z_G = \frac{e^{-\beta E_k + \beta\mu}}{1+e^{-\beta E_k + \beta\mu}} \tag{13.4.51}$$

Dividing by the exponential gives

$$\langle n_k\rangle = \frac{1}{e^{\beta E_k - \beta\mu}+1} \tag{13.4.52}$$

This formula expressed as a continuous function of energy $f_{FD}(E)=\langle n_k\rangle$ is known as the Fermi-Dirac distribution where the chemical potential is identified at the Fermi energy $\mu = E_F$. When $E = E_F$ we have $f_{FD}(E) = 1/2$ for all temperatures. At absolute zero

$$f_{FD}(E) = \frac{1}{e^{\beta(E-E_F)}+1} \rightarrow \begin{cases} 1 & E < E_F \\ 0 & E > E_F \end{cases} \tag{13.4.53}$$

The number of electronic standing waves in a cubical cavity of side L between n and $n + dn$ expressed as the first octant of a spherical shell is

$$g(n)dn = 2\frac{1}{8}4\pi n^2 dn = \pi n^2 dn \tag{13.4.54}$$

with two spin states of the electron. Setting the wavelength of the nth mode $\lambda = 2L/n$ equal to the de Broglie wavelength $\lambda = h/p$ the nonrelativistic energy is

$$E = \frac{p^2}{2m} = \frac{1}{2m}\left(\frac{nh}{2L}\right)^2 \tag{13.4.55}$$

Solving for n

$$n = \frac{2L}{h}\sqrt{2mE} \tag{13.4.56}$$

we have

$$dn = \frac{\sqrt{2mL}}{h} E^{-1/2} dE \cdot$$

(13.4.57)

The electronic density of states is

$$g(E)dE = \pi n^2 dn = \frac{8\sqrt{2}\pi L^3 m^{3/2}}{h^3} E^{1/2} dE$$

(13.4.58)

The number of states per unit volume with energy between E and $E + dE$ is

$$n(E)dE = g(E) f_{FD}(E) dE = \underbrace{\frac{8\sqrt{2}\pi m^{3/2}}{h^3} \sqrt{E}}_{g(E)} \underbrace{\frac{1}{e^{\beta(E-E_F)}+1}}_{f_{FD}} dE$$

(13.4.59)

Integrating we obtain

$$\frac{N}{V} = \frac{8\sqrt{2}\pi m^{3/2}}{h^3} \int_0^\infty \frac{\sqrt{E}\,dE}{e^{(E-E_F)/k_B T}+1}$$

(13.4.60)

Below the Fermi energy $f_{FD} = 1$ and the integral simplifies to

$$\frac{N}{V} = \frac{8\sqrt{2}\pi m^{3/2}}{h^3} \int_0^{E_F} \sqrt{E}\,dE = \frac{8\sqrt{2}\pi m^{3/2}}{h^3} \left(\frac{2}{3} E_F^{3/2}\right)$$

(13.4.61)

We may thus evaluate the Fermi energy from this expression

$$E_F = \frac{h^2}{2m_e} \left(\frac{3N}{8\pi V}\right)^{2/3}$$

(13.4.62)

The Fermi temperature is defined as

$$T_F = \frac{E_F}{k_B}$$

(13.4.63)

Only a small fraction of electrons within $k_B T$ of E_F may be excited thermally with electronic contribution to the specific heat $C_{el} \approx 3RT/T_F$ which is about 1% of the specific heat at room temperature.

Maple Examples

The Fermi-Dirac distribution is plotted at three temperatures as a function of E/E_F in the Maple worksheet below. The Debye specific heat and internal energy is then numerically computed and plotted as a function of T/T_D.

Key Maple commands: *evalf, plot*

restart

Fermi-Dirac Distribution

$$f := (E, r) \rightarrow \frac{1}{\exp(r \cdot (E-1)) + 1}$$

$$f := (E, r) \mapsto \frac{1}{e^{r(E-1)} + 1}$$

plot([f(E, 100), f(E, 10), f(E, 5)], E = 0 ... 3, labels = ["E/E_F", "f(E)"], linestyle = [solid, dash, dashdot], legend = [beta E_F = 100, beta E_F = 10, beta E_F = 5])

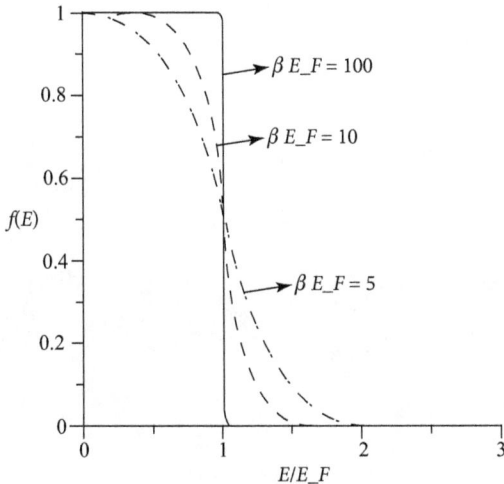

Figure 13.4.1: Plot of the Fermi-Dirac distribution.

Debye Specific Heat

$$C := (T) \rightarrow 9 \cdot T^3 \cdot evalf\left(\int_0^{\frac{1}{T}} \frac{\exp(x) \cdot x^4}{(\exp(x) - 1)^2} \, dx \right)$$

$$C := (T) \mapsto 9 T^3 evalf\left(\int_0^{\frac{1}{T}} \frac{e^x \, x^4}{(e^x - 1)^2} \, dx \right)$$

$C(1)$

$$2.855196201 - 3.491860243 \ 10^{-7}I$$

$plot(\text{Re}(C(T)), T = .001 \ldots 1, labels = [\text{"T/T\_D"}, \text{"C/(N k\_B)"}])$

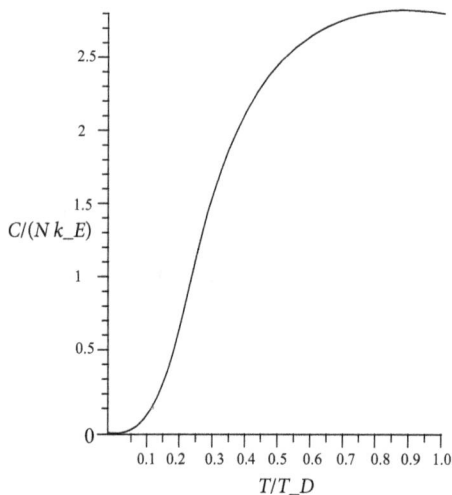

Figure 13.4.2: Specific heat in the Debye model.

Debye Internal Energy

$$U := (T) \rightarrow 9 \cdot T^4 \cdot evalf \left(\int_0^{\frac{1}{T}} \frac{x^3}{\exp(x)-1} dx \right)$$

$$U := (T) \mapsto 9T^4 evalf \left(\int_0^{\frac{1}{T}} \frac{x^3}{e^x - 1} dx \right)$$

$plot(\text{Re}(U(T)), T = .001 \ldots 1, labels = [\text{"T/T\_D"}, \text{"U/(N k\_B T\_D)"}])$

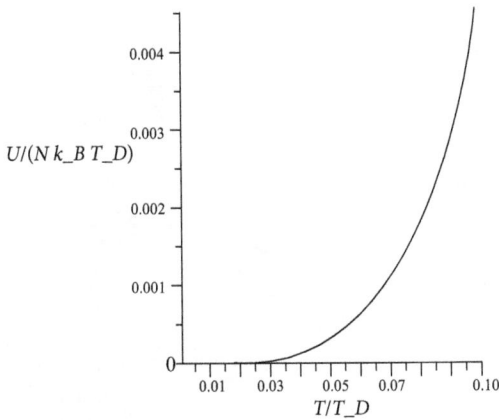

Figure 13.4.3: Internal energy in the Debye model.

13.5 MATLAB EXAMPLES

A Shannon entropy calculation of the tent map and plots of the Maxwell-Boltzmann distribution are demonstrated in this section.

Key MATLAB commands: *hist, sum, line, plot*

Section 13.1 Microcanonical Ensemble

Figure 13.5.1 shows a cobweb plot showing 21 iterates of the tent map for $\mu = 1.6$.

$$x_{n+1} = \begin{cases} \mu x_n & 0 \le x_n \le 1/2 \\ \mu(1-x_n) & 1/2 < x_n \le 1 \end{cases} \tag{13.6.1}$$

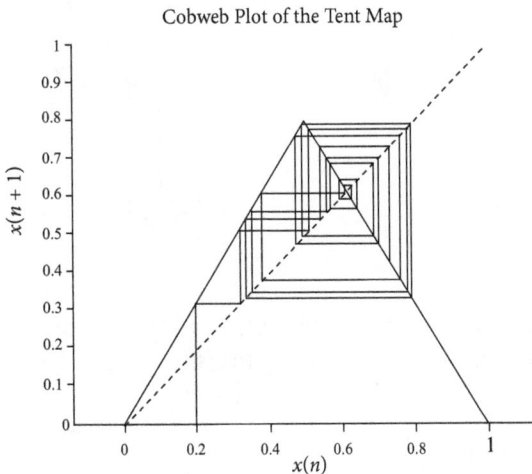

Figure 13.5.1: Cobweb plot showing 21 iterates of the tent map $\mu = 1.6$.

```
% parameters and initial conditions
mu=1.6;
x0=0.2;
Nmax=21;

x=zeros(Nmax,1);
t=zeros(Nmax,1);
y=zeros(Nmax,1);

x(1)=x0;

% iterate the tent map
for i=1:Nmax
    if x(i)<0.5 && x(i)>0
        x(i+1)=mu*x(i);
    elseif x(i)<1.0 && x(i)>=0.5
        x(i+1)=mu*(1-x(i));
    end
end

for i=1:Nmax
    t(i)=(i-1)/(Nmax-1);
     if t(i)<0.5 && t(i)>0
        y(i)=mu*t(i);
    elseif t(i)<1.0 && t(i)>=0.5
        y(i)=mu*(1-t(i));
    end
end

hold on
% plot the function and the diagonal line

plot(t,t,t,y)
% create the cobweb plot
line([x(1) x(1)], [0 x(2)])

for i=1:Nmax-1
    line([x(i) x(i+1)], [x(i+1) x(i+1)])
    line([x(i+1) x(i+1)], [x(i+1) x(i+2)])
end

hold off
axis equal
title("Cobweb plot of the tent map")
xlabel("x(n)")
ylabel("x(n+1)")
colormap bone
```

Once the MATLAB file is executed, the entropy calculation is performed at the Command line. A histogram of the x_n time series with 20 bins is first created and normalized by the length of the time series.

```
>> y=hist(x,20)/length(x)
y =
```

```
  Columns 1 through 16
0.0020         0          0          0     0.0459     0.0539     0.0339     0.0499
0.0479     0.0479     0.0679     0.0739     0.0778     0.0579     0.0599     0.0719
  Columns 17 through 20
0.0818     0.0838     0.0679     0.0758
```

Bins that have a value of '0' are assigned a value of '1' using the command 'y(y==0)=1' to properly calculate the Shannon entropy.

```
>> y(y == 0) = 1
y =
  Columns 1 through 16
0.0020     1.0000     1.0000     1.0000     0.0459     0.0539     0.0339     0.0499
0.0479     0.0479     0.0679     0.0739     0.0778     0.0579     0.0599     0.0719
  Columns 17 through 20
0.0818     0.0838     0.0679     0.0758
>> S=-sum(y.*log(y))
S =
    2.7541
```

Section 13.3 Continuous Energy Distribution

The following script plots the Maxwell-Boltzmann velocity distribution of molecular oxygen and carbon dioxide at the same temperature.

```
% molecular masses of O2 and CO2

m_O2=5.31*10^-26;
m_CO2=7.32*10^-26;

k_B=1.38e-23; % Boltzmann constant

T=400; % temperature in K

%normalize the O2 and CO2 M-B distributions

norm_O2=4*pi*(m_O2/(2*pi*k_B*T))^(3/2);

norm_CO2=4*pi*(m_CO2/(2*pi*k_B*T))^(3/2);

v=1:10:2000;      % velocity range in m/s

%compute the O2 and CO2 distributions

n_O2=norm_O2*(v.^2).*exp(-m_O2*v.^2/(2*k_B*T));

n_CO2=norm_CO2*(v.^2).*exp(-m_CO2*v.^2/(2*k_B*T));

%compare O2 and CO2 MB distributions at the same temperature

hold on

plot(v,n_O2,'-.k');
```

```
plot(v,n_CO2,'k');

hold off

xlabel('velocity (m/s)')
ylabel('n(v)')
legend('O2','CO2')
title('M-B Velocity Distribution')
```

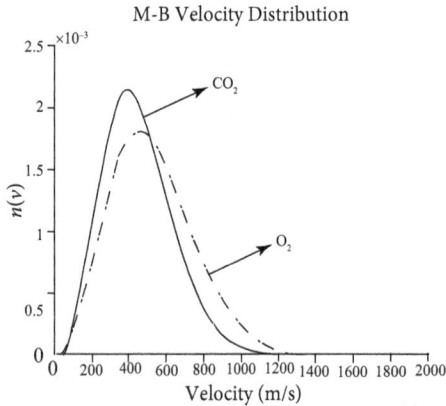

Figure 13.5.2: Maxwell-Boltzmann velocity distribution of two gases.

13.6 EXERCISES

Section 13.1 Microcanonical Ensemble

1. A collection of $N = 100$ particles can exist in three spin states (up, down and zero). Find the maximum and minimum entropy of this system.

2. Use the Shannon entropy formula to plot the entropy of the tent map in Section 13.5 as a function of the parameter μ.

3. Mixing dissimilar substances with particle numbers N_1 and N_2 where $N = N_1 + N_2$ will result in a change in entropy

$$\Delta S = k_B \ln \frac{N!}{N_1! N_2!}$$

Use Sterling's formula to show that

$$\Delta S = -k_B \left(N_1 \ln \frac{N_1}{N} + N_2 \ln \frac{N_2}{N} \right).$$

Section 13.2 Canonical Ensemble

4. Calculate the average energy in a two-state system with $E_1 = 3.0J$ and $E_2 = -3.0J$ with $P_1 = 0.65$ and $P_2 = 0.35$. Note that $\langle E \rangle \neq (E_1 + E_2)/2$ except when $P_1 = P_2 = 1/2$.

5. A two-state system has energy levels $E_1 = 3.0J$ and $E_2 = -3.0J$. At what temperature are one-fourth of the particles in state 1?

6. Show that $\dfrac{\partial}{\partial T} = -\dfrac{1}{k_B T^2}\dfrac{\partial}{\partial \beta}$

7. Use the Shannon entropy formula $S = -Nk_B \sum_i P_i \ln P_i$ with $P_i = \exp(-\beta E_i)/Z$ in the canonical ensemble to show that

$$S = \frac{N}{T}\langle E_i \rangle + Nk_B \ln Z$$

8. Suppose that a system of N particles can be in three allowed states S_1, S_2 and S_3 with energies $E_1 = -\varepsilon$, $E_2 = 0$ and $E_3 = \varepsilon$. Derive expressions for P_1, P_2 and P_3 at temperature T. Calculate the average energy from the formula

$$\langle E \rangle = \sum_i E_i P_i$$

Calculate the entropy using the Shannon formula.

9. A population of N magnetic dipoles with two possible orientations of magnetic moment $\mathbf{\mu}$ are placed in a magnetic field $B\hat{\mathbf{k}}$ at a temperature T. The interaction energies are $(-\mu_z B, +\mu_z B)$. Calculate the partition function Z, the average energy $<E>$, the free energy F, the specific heat C and the entropy S with the dipoles in thermodynamic equilibrium.

Section 13.3 Continuous Energy Distributions

10. For a particle in a 3D box with side L modeled as having a continuous energy distribution, show that the partition function

$$Z = \left(L/L_Q\right)^3$$

where the quantum length is defined as $L_Q = h/\sqrt{2\pi m k_B T}$.

11. Show that the average energy of a system with a continuous energy distribution is

$$\langle E \rangle = \frac{\displaystyle\int_0^\infty P(E)E\,dE}{\displaystyle\int_0^\infty P(E)\,dE} = k_B T$$

where the Boltzmann factor $P(E) = \exp(-E/k_B T)$.

12. The partition function for the rigid rotator is

$$Z = \sum_{\ell=0}^{\infty} (2\ell+1)\exp\left(-\frac{\beta\hbar^2}{2I}\ell(\ell+1)\right)$$

Make an integral approximate to Z where the energy levels are more closely spaced at high temperatures. Treat $\ell(\ell + 1)$ as a continuous variable with $2\ell + 1 = d(\ell(\ell + 1))$. Calculate the average energy, specific heat, free energy, and entropy of the rigid rotator in the high-temperature limit.

13. By setting $dn(v)/dv = 0$ show that the most probable speed of molecules in the Maxwell-Boltzmann distribution is $v_p = \sqrt{2k_B T/m}$

14. Calculate the uncertainty in speed $\Delta v = \sqrt{\langle v^2\rangle - \langle v\rangle^2}$ of the Maxwell-Boltzmann distribution. Plot the uncertainty as a function of the temperature T.

15. Calculate the average energy at temperature T of a classical harmonic oscillator with energy as a function of speed and position

$$E = \frac{1}{2}mv^2 + \frac{1}{2}kx^2$$

with partition function

$$Z = A\int_{-\infty}^{\infty}\int_{-\infty}^{\infty} e^{-\beta E}\,dx\,dp$$

where A is a normalization factor.

16. Calculate the average energy of a relativistic harmonic oscillator with

$$E = \sqrt{p^2c^2 + m^2c^4} + \frac{1}{2}kx^2$$

at temperature T.

17. Find the root mean square speed v_{rms} of a relativistic gas where the kinetic energy $(\gamma-1)mc^2 = \frac{3}{2}k_B T$ and $\gamma = \dfrac{1}{\sqrt{1-\dfrac{v_{\mathrm{rms}}^2}{c^2}}}$.

18. Find the most probable speed of a relativistic gas $v_p = cu_p$ by setting $dn(u)/du = 0$.

19. Write the partition function for a relativistic electron in a box. Write an integral form of the partition function treating the energy levels as continuous. Find expressions for the free energy, entropy, and specific heat.

Section 13.4 Grand Canonical Ensemble

20. Given that the average energy of photons in the Bose-Einstein distribution with $\mu = 0$ is

$$\langle E \rangle = \frac{\displaystyle\sum_{n=0}^{\infty} E_n e^{-\beta E_n}}{\displaystyle\sum_{n=0}^{\infty} e^{-\beta E_n}}$$

where $E_n = nhf$ show that

$$\langle E \rangle = \left(\frac{hf}{\exp(hf / k_B T) - 1} \right)$$

21. Model the average energy of phonons with frequency f in a solid at temperature T using the Bose-Einstein formula

$$\langle E \rangle = \left(\frac{hf}{\exp(hf / k_B T) - 1} \right)$$

to calculate Einstein's expression for the specific heat.

$$C = \frac{dU}{dT}$$

where $U = 3N \langle E \rangle$. Evaluate the high-temperature limit of $\langle E \rangle$ to find C at high temperatures.

22. Show that the high-frequency approximation to Planck's black-body formula where $\exp(hf / k_B T) \gg 1$ gives

$$u(f, T) \approx \frac{8\pi h f^3}{c^3} \exp\left(-\frac{hf}{k_B T} \right)$$

while at low frequencies where $hf \ll k_B T$ so $\exp(hf / k_B T) \approx 1 + hf / k_B T$

$$u(f, T) \approx \frac{8\pi f^2}{c^3} k_B T$$

This formula is known as the Rayleigh-Jeans law and is commonly used in radio astronomy. The high-frequency (short-wavelength) discrepancy of the Rayleigh-Jeans law from observed black-body spectra was known as the "ultraviolet catastrophe."

23. Find the low-temperature limit of the specific heat C in the Debye theory where $T_D/T \to \infty$. Find C at high temperatures using $e^x \approx 1+x$ where $x \ll 1$. The high-temperature limit of C is known as the Dulong-Petit law.

24. Show that the average energy at absolute zero in the Fermi-Dirac distribution is

$$\langle E \rangle = \frac{3E_F}{5}$$

where

$$\langle E \rangle = \frac{\dfrac{8\sqrt{2}\pi m^{3/2}}{h^3} \displaystyle\int_0^\infty \dfrac{E\sqrt{E}\,dE}{e^{(E-E_F)/k_B T}+1}}{N/V}$$

Chapter # 14   SPECIAL RELATIVITY

Chapter Outline

14.1 KINEMATICS

Topics in this section include the postulates of special relativity, relativistic time dilation, length contraction, the relativistic Doppler effect, Galilean and Lorentz transformations, relativistic velocity addition, and 4-vector notation.

14.1.1 Postulates of Special Relativity

Maxwell's wave equations

$$\nabla^2 \mathbf{E} - \mu_0 \varepsilon_0 \frac{\partial^2}{\partial t^2} \mathbf{E} = 0$$

$$\nabla^2 \mathbf{B} - \mu_0 \varepsilon_0 \frac{\partial^2}{\partial t^2} \mathbf{B} = 0$$

$$(14.1.1)$$

predict that electromagnetic waves propagate with the speed of light $c = 1/\sqrt{\mu_0 \varepsilon_0}$. The wave equations do not describe how the measured light speed should depend on the velocities of sources or observers in relative motion. In the late 1800s, it was assumed that light traveled through a medium called the ether that pervaded all of space. Experiments by Michelson and Morley subsequently failed to detect variations in the measured speed of light due to the earth's motion through the supposed ether. In 1905 Einstein postulated that light propagates with the same speed in all inertial frames of reference independent of the motion of the source or the observer. Furthermore, there is no preferred frame of reference, and the laws of physics are the same in all inertial frames. Because of the constancy of the speed of light, observers in relative motion disagree on the order of events as well as time intervals and lengths measured in their respective frames.

14.1.2 Time Dilatation

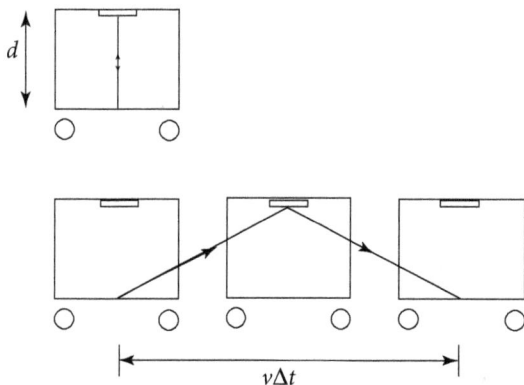

Figure 14.1.1: (top) A light ray is projected from the bottom of a train car and reflected off a mirror located at the top of the car. (bottom): The light ray travels a greater distance in the earth frame.

To illustrate the effect of time dilation, consider the geometry of Figure 14.1.1. In this figure, a light ray is reflected off a mirror in a moving train car with a height of d. On the train, the light travels a distance $2d$ in a time $\Delta t_p = 2d/c$. The train travels at a speed v and distance $v\Delta t$ in the earth frame. The distance that light travels in the earth frame is $c\Delta t$. Per Einstein's second postulate, light travels the same speed in both reference frames. Thus, it takes a longer time $\Delta t > \Delta t_p$ for light to travel a greater distance in the earth frame. We can calculate Δt from the right triangle shown in Figure 14.1.2.

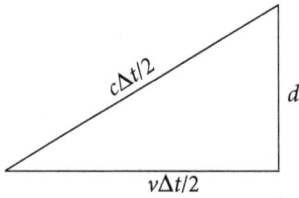

Figure 14.1.2: Right triangle for computing the time dilation formula.

The Pythagorean theorem gives

$$\left(\frac{c\Delta t}{2}\right)^2 = \left(\frac{v\Delta t}{2}\right)^2 + d^2 \tag{14.1.2}$$

Solving for Δt we have

$$\Delta t^2 = \frac{4d^2}{c^2 - v^2} = \frac{4d^2}{c^2}\frac{1}{1 - v^2/c^2} \tag{14.1.3}$$

so that

$$\Delta t = \gamma \Delta t_p \quad \text{where} \quad \gamma = \frac{1}{\sqrt{1 - v^2/c^2}} \tag{14.1.4}$$

14.1.3 Length Contraction

An observer at rest with respect to two stars measures proper distance L_p and a time interval Δt required for a rocket to travel between the two stars. An astronaut in the spaceship traveling at a speed v measures the proper time Δt_p and a distance between the two stars $L = v\Delta t_p$. Since $\Delta t = \gamma \Delta t_p$, the proper length $L_p = v\Delta t = \gamma v\Delta t_p = \gamma L$. Thus, the distance between the two stars in the rocket frame is shorter than the proper distance by a factor of γ

$$L = \frac{1}{\gamma}L_p = \sqrt{1 - \frac{v^2}{c^2}}L_p \tag{14.1.5}$$

14.1.4 Relativistic Doppler Effect

Consider an observer moving with speed v toward a source of light with frequency f_{source} and wavelength $\lambda_{source} = c/f_{source}$. The observed wavelength λ_{obs} is shortened by $v\Delta t$

$$\lambda_{obs} = \lambda_{source} - v\Delta t \tag{14.1.6}$$

and with $f_{source} = 1/\Delta t$ we have

$$\lambda_{obs} = c\Delta t - v\Delta t \tag{14.1.7}$$

Now Δt in the observer frame is related to Δt_p in the source frame by the time dilation formula $\Delta t = \gamma \Delta t_p$ so that

$$\lambda_{obs} = (c - v)\gamma \Delta t_p \tag{14.1.8}$$

Since $\lambda_{obs} = c/f_{obs}$ we have

$$\frac{c}{f_{obs}} = (c-v)\gamma \frac{1}{f_{source}} \tag{14.1.9}$$

Rearranging gives

$$f_{obs} = f_{source} \frac{\sqrt{1 - \dfrac{v^2}{c^2}}}{\left(1 - \dfrac{v}{c}\right)} \tag{14.1.10}$$

or

$$f_{obs} = f_{source} \sqrt{\frac{1 + \dfrac{v}{c}}{1 - \dfrac{v}{c}}} \tag{14.1.11}$$

14.1.5 Galilean Transformation

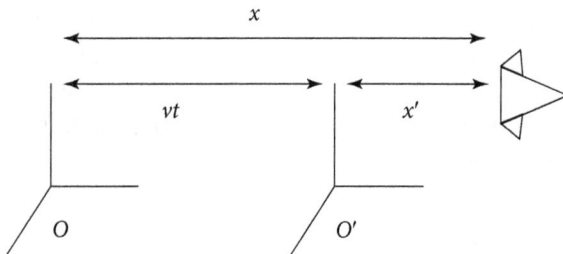

Figure 14.1.3: Frames O and O' are coincident at $t = 0$. A rocket ship moving to the right is located by x and x' in the unprimed and primed frames, respectively.

Consider the two reference frames in relative motion in Figure 14.1.3. The frame O' moves at speed v with respect to O. The two frames are coincident at $t = 0$ and are separated by a distance vt at time t. The coordinates x' and x locate a rocket ship in the O' and O frames, respectively. The Galilean transformation relates the primed and unprimed coordinates

$$x' = x - vt \tag{14.1.12}$$

We may write the inverse transformation

$$x = x' + vt \qquad (14.1.13)$$

by solving algebraically for x or by switching the primes and changing the sign of v.

The velocity of the rocket in the primed frame $u'_x = dx'/dt$ and in the unprimed frame $u_x = dx/dt$ are related by the Galilean velocity transformation

$$u'_x = u_x - v \qquad (14.1.14)$$

with the inverse transformation

$$u_x = u'_x + v \qquad (14.1.15)$$

14.1.6 Lorentz Transformations

To obtain the relativistic transformation we guess a form

$$x' = \gamma(x - vt) \qquad (14.1.16)$$

We seek the functional form of γ where $\gamma \rightarrow 1$ for $v \ll c$. The inverse transformation should also be of the form

$$x = \gamma(x' + vt') \qquad (14.1.17)$$

Substituting x' into x gives

$$x = \gamma^2(x - vt) + \gamma vt' \qquad (14.1.18)$$

and solving for t' gives

$$t' = \frac{1}{\gamma v}(1 - \gamma^2)x + \gamma t \qquad (14.1.19)$$

Now we can express the velocity of the rocket in the primed frame in terms of the velocity in the unprimed frame

$$u'_x = \frac{dx'}{dt'} = \frac{\gamma(dx - vdt)}{\frac{1}{\gamma v}(1 - \gamma^2)dx + \gamma dt} \qquad (14.1.20)$$

and dividing by γdt

$$u'_x = \frac{\frac{dx}{dt} - v}{\frac{1}{v}\left(\frac{1}{\gamma^2} - 1\right)\frac{dx}{dt} + 1} = \frac{u_x - v}{\frac{1}{v}\left(\frac{1}{\gamma^2} - 1\right)u_x + 1} \qquad (14.1.21)$$

Now, instead of a rocket, we consider a light ray moving to the right with the same speed in both frames. Setting $u'_x = u_x = c$ gives

$$c = \frac{c - v}{\frac{1}{v}\left(\frac{1}{\gamma^2} - 1\right)c + 1}$$

(14.1.22)

Solving for γ we obtain

$$\gamma = \frac{1}{\sqrt{1 - \frac{v^2}{c^2}}}$$

(14.1.23)

We can now write the equation for ct' more compactly as

$$ct' = \gamma\left(ct - \frac{v}{c}x\right)$$

(14.1.24)

The Lorentz transformation equations are then

$$\begin{pmatrix} x' \\ y' \\ z' \\ ct' \end{pmatrix} = \begin{pmatrix} \gamma(x - vt) \\ y \\ z \\ \gamma\left(ct - \frac{v}{c}x\right) \end{pmatrix}$$

(14.1.25)

The inverse Lorentz transformation is obtained by switching the primes and changing the sign of v so

$$\begin{pmatrix} x \\ y \\ z \\ ct \end{pmatrix} = \begin{pmatrix} \gamma(x' + vt') \\ y' \\ z' \\ \gamma\left(ct' + \frac{v}{c}x'\right) \end{pmatrix}$$

(14.1.26)

Example 14.1.1

Obtain the length contraction formula from the Lorentz transformation equations

$$x'_1 = \gamma(x_1 - vt_1)$$
$$x'_2 = \gamma(x_2 - vt_2)$$

(14.1.27)

Solution: If measurements of x_1 and x_2 are made simultaneously in the unprimed frame, then $t_1 = t_2$ and

$$x_2' - x_1' = \gamma(x_2 - x_1) \tag{14.1.28}$$

and $L = L_p/\gamma$ with $L_p = x_2' - x_1'$ and $L = x_2 - x_1$.

Example 14.1.2

Show the time dilation formula follows from the Lorentz transformation equations

$$ct_2' = \gamma\left(ct_2 - \frac{v}{c}x_2\right)$$
$$ct_1' = \gamma\left(ct_1 - \frac{v}{c}x_1\right) \tag{14.1.29}$$

Solution: If measurements of t_1 and t_2 are made at the same location in the unprimed frame $x_1 = x_2$ and

$$t_2' - t_1' = \gamma(t_2 - t_1) \tag{14.1.30}$$

and $\Delta t_p = \gamma\Delta t$ with $\Delta t_p = t_2' - t_1'$ and $\Delta t = t_2 - t_1$.

14.1.7 Relativistic Addition of Velocities

Formulas for the relativistic addition of velocities are obtained from the Lorentz transformation equation. If v is along the x-direction

$$u_x' = \frac{dx'}{dt'} = \frac{u_x - v}{1 - \dfrac{u_x v}{c^2}} \tag{14.1.31}$$

Evaluating $u_y' = dy'/dt'$ and $u_z' = dz'/dt'$ we obtain

$$u_y' = \frac{u_y}{\gamma\left(1 - \dfrac{u_x v}{c^2}\right)} \tag{14.1.32}$$

$$u_z' = \frac{u_z}{\gamma\left(1 - \dfrac{u_x v}{c^2}\right)} \tag{14.1.33}$$

To obtain the velocities u_x, u_y, and u_z in terms of u_x', u_y', and u_z' we simply interchange the primes and change the sign of v.

Example 14.1.3

Consider a frame O' moving with a speed $v = c/2$ with respect to O. A rocket travels with a speed $u'_x = c/2$ with respect to O'. What is the speed of the rocket in the O frame?

Solution

$$u_x = \frac{u'_x + v}{1 + \frac{u_x v}{c^2}} \tag{14.1.34}$$

$$u_x = \frac{c/2 + c/2}{1 + \frac{(c/2)(c/2)}{c^2}} = \frac{c}{1 + \frac{1}{4}} = \frac{4}{5}c \tag{14.1.35}$$

This differs from the Galilean velocity addition where $u_x = u'_x + v = c$.

Example 14.1.4

Consider an observer on earth (unprimed frame) with $u_x = 0$. A mass m is projected upward with a speed u_y. According to an observer (primed frame) moving with a speed v the velocity u'_y is given by

$$u'_y = \frac{u_y}{\gamma\left(1 - \frac{0 \cdot v}{c^2}\right)} = \frac{u_y}{\gamma} \tag{14.1.36}$$

Suppose the mass in question is replaced by a ray of light. The speed of light should be the same in both frames. The speed of light in the unprimed frame is $u_y = c$. Note that the light ray also has an x-velocity component in the primed frame $u' = -v$ The speed of light in the primed frame is

$$|\mathbf{u'}| = \sqrt{u'^2_y + u'^2_x} = \sqrt{\left(\frac{c}{\gamma}\right)^2 + (-v)^2} = c \tag{14.1.37}$$

so that both observers measure the same speed of light.

14.1.8 Velocity Addition Approximation

A simple relativistic velocity addition shortcut may be applied in the special case where relativistic speeds v/c are represented as a decimal followed by a series of nines. The shortcut is useful for board work examples in introductory physics courses and for quickly comparing the results of relativistic to Galilean velocity addition.

Consider two bodies traveling in one dimension attached to inertial frames O and O' with speeds v_{OE} and $v_{O'E}$ relative to a third "earth" frame E. The Galilean velocity addition formula is $v_{oo'} = v_{OE} + v_{EO'}$. The relativistic formula is given by

$$v_{oo'} = \frac{v_{OE} + v_{EO'}}{1 + \dfrac{v_{OE} v_{EO'}}{c^2}} \tag{14.1.38}$$

where c is the speed of light. For two bodies approaching at $v_{OE} = 0.999\ c$ and $v_{EO'} = 0.99\ c$ the velocity addition formula gives $v_{oo'} = 0.999995\ c$ compared to $v_{oo'} = 1.989\ c$ in the Galilean case. Thus, to perform the velocity addition to very good approximation we simply "add the nines." Similarly, one can find that any combination of nines will sum in a straightforward way as shown in Table 14.1.1.

TABLE 14.1.1: Velocity $v_{oo'}$ (third column) obtained from the addition of velocities v_{OE} (first column) and $v_{Eo'}$ (second column).

v_{OE}/c	$v_{EO'}/c$	$v_{oo'}/c$
0.999	0.999	0.9999995
0.9999	−0.99	0.98
0.99999	0.99	0.99999995
0.99999	−0.99	0.998

Another shortcut can be used to quickly obtain the relativistic factor $\gamma = \left(1 - v^2/c^2\right)^{-1/2}$. For velocities represented as an odd number of nines $v/c = 0.9, 0.999, 0.99999 \ldots$ we have $\gamma = 2.29, 22.4, 224, \ldots$. For an even number of nines $v/c = 0.99, 0.9999, 0.999999 \ldots$ we have $\gamma = 7.1, 70.7, 707 \ldots$. Hence for every two nines added to the velocity, the relativistic factor γ increases very nearly an order of magnitude. Thus, examples using relativistic velocity addition can be extended to quickly find time dilations and length contractions in different reference frames.

14.1.9 4-Vector Notation

The Lorentz transformation may be expressed in matrix form

$$\begin{pmatrix} x^{0'} \\ x^{1'} \\ x^{2'} \\ x^{3'} \end{pmatrix} = \begin{pmatrix} \gamma & -\gamma v/c & 0 & 0 \\ -\gamma v/c & \gamma & 0 & 0 \\ 0 & 0 & 1 & 0 \\ 0 & 0 & 0 & 1 \end{pmatrix} \begin{pmatrix} x^0 \\ x^1 \\ x^2 \\ x^3 \end{pmatrix} \tag{14.1.39}$$

where the 4-vectors are

$$\begin{pmatrix} x^0 \\ x^1 \\ x^2 \\ x^3 \end{pmatrix} = \begin{pmatrix} ct \\ x \\ y \\ z \end{pmatrix} \text{ and } \begin{pmatrix} x^{0'} \\ x^{1'} \\ x^{2'} \\ x^{3'} \end{pmatrix} = \begin{pmatrix} ct' \\ x' \\ y' \\ z' \end{pmatrix} \tag{14.1.40}$$

The inverse transformation is

$$\begin{pmatrix} x^0 \\ x^1 \\ x^2 \\ x^3 \end{pmatrix} = \begin{pmatrix} \gamma & \gamma v/c & 0 & 0 \\ \gamma v/c & \gamma & 0 & 0 \\ 0 & 0 & 1 & 0 \\ 0 & 0 & 0 & 1 \end{pmatrix} \begin{pmatrix} x^{0'} \\ x^{1'} \\ x^{2'} \\ x^{3'} \end{pmatrix} \tag{14.1.41}$$

The inverse transformation matrix is such that

$$\begin{pmatrix} \gamma & \gamma v/c & 0 & 0 \\ \gamma v/c & \gamma & 0 & 0 \\ 0 & 0 & 1 & 0 \\ 0 & 0 & 0 & 1 \end{pmatrix} \begin{pmatrix} \gamma & -\gamma v/c & 0 & 0 \\ -\gamma v/c & \gamma & 0 & 0 \\ 0 & 0 & 1 & 0 \\ 0 & 0 & 0 & 1 \end{pmatrix} = \begin{pmatrix} 1 & 0 & 0 & 0 \\ 0 & 1 & 0 & 0 \\ 0 & 0 & 1 & 0 \\ 0 & 0 & 0 & 1 \end{pmatrix} \tag{14.1.42}$$

The Lorentz transformation equations can be written compactly as

$$x^{\mu'} = \Lambda^{\mu}_{\nu} x^{\nu} \tag{14.1.43}$$

This corresponds to four equations for each value μ. Since the ν are repeated they are summed over according to the Einstein summation convention

$$\begin{aligned} x^{0'} &= \Lambda^0_0 x^0 + \Lambda^0_1 x^1 + \Lambda^0_2 x^2 + \Lambda^0_3 x^3 \\ x^{1'} &= \Lambda^1_0 x^0 + \Lambda^1_1 x^1 + \Lambda^1_2 x^2 + \Lambda^1_3 x^3 \\ x^{2'} &= \Lambda^2_0 x^0 + \Lambda^2_1 x^1 + \Lambda^2_2 x^2 + \Lambda^2_3 x^3 \\ x^{3'} &= \Lambda^3_0 x^0 + \Lambda^3_1 x^1 + \Lambda^3_2 x^2 + \Lambda^3_3 x^3 \end{aligned} \tag{14.1.44}$$

Maple Examples

Operations involving Lorentz transformation matrices are demonstrated in the Maple worksheet below. The dot product of a 4-vector is shown to be invariant under successive Lorentz transformations, or boosts, along the x-axis. An equivalent transformation matrix is found corresponding to successive boosts.

Key Maple commands: $\cdot$ (matrix multiplication), *simplify*, *subs*

restart

Lorentz Transformation Matrix

$$\gamma_1 := \frac{1}{\text{sqrt}\left(1 - \frac{v_1^2}{c^2}\right)}:$$

$$\lambda_1 := \left\langle \left\langle \gamma_1 \left| -\frac{\gamma_1 \cdot v_1}{c} \right| 0 \middle| 0 \right\rangle, \left\langle -\frac{\gamma_1 \cdot v_1}{c} \middle| \gamma_1 \middle| 0 \middle| 0 \right\rangle, \langle 0|0|1|0 \rangle, \langle 0|0|0|1 \rangle \right\rangle$$

$$\lambda_1 := \begin{bmatrix} \dfrac{1}{\sqrt{1 - \dfrac{v_1^2}{c^2}}} & -\dfrac{v_1}{\sqrt{1 - \dfrac{v_1^2}{c^2}}c} & 0 & 0 \\[2em] -\dfrac{v_1}{\sqrt{1 - \dfrac{v_1^2}{c^2}}} & \dfrac{1}{\sqrt{1 - \dfrac{v_1^2}{c^2}}} & 0 & 0 \\[2em] 0 & 0 & 1 & 0 \\[0.5em] 0 & 0 & 0 & 1 \end{bmatrix}$$

Inverse Lorentz Transformation Matrix

$$\lambda_{1\_inv} := \left\langle \left\langle \gamma_1 \left| \frac{\gamma_1 \cdot v_1}{c} \right| 0 \middle| 0 \right\rangle, \left\langle \frac{\gamma_1 \cdot v_1}{c} \middle| \gamma_1 \middle| 0 \middle| 0 \right\rangle, \langle 0|0|1|0 \rangle, \langle 0|0|0|1 \rangle \right\rangle$$

$$\lambda_{1\_inv} := \begin{bmatrix} \dfrac{1}{\sqrt{1 - \dfrac{v_1^2}{c^2}}} & \dfrac{v_1}{\sqrt{1 - \dfrac{v_1^2}{c^2}}c} & 0 & 0 \\[2em] \dfrac{v_1}{\sqrt{1 - \dfrac{v_1^2}{c^2}}} & \dfrac{1}{\sqrt{1 - \dfrac{v_1^2}{c^2}}} & 0 & 0 \\[2em] 0 & 0 & 1 & 0 \\[0.5em] 0 & 0 & 0 & 1 \end{bmatrix}$$

$simplify(\lambda_1 \cdot \lambda_{1\_inv})$

$$\begin{vmatrix} 1 & 0 & 0 & 0 \\ 0 & 1 & 0 & 0 \\ 0 & 0 & 1 & 0 \\ 0 & 0 & 0 & 1 \end{vmatrix}$$

Transformation of a 4-Vector

$A := \langle A_0, A_1, A_2, A_3 \rangle$

$$A := \begin{bmatrix} A_0 \\ A_1 \\ A_2 \\ A_3 \end{bmatrix}$$

$A_{p1} := \lambda_1 \cdot A$

$$A_{p1} := \begin{bmatrix} \dfrac{A_0}{\sqrt{1-\dfrac{v_1^2}{c^2}}} - \dfrac{v_1 A_1}{\sqrt{1-\dfrac{v_1^2}{c^2}}c} \\[30pt] -\dfrac{v_1 A_0}{\sqrt{1-\dfrac{v_1^2}{c^2}}c} + \dfrac{A_1}{\sqrt{1-\dfrac{v_1^2}{c^2}}} \\[30pt] A_2 \\ A_3 \end{bmatrix}$$

Invariance of a Dot Product

$simplify(A_{p1}[1] \cdot A_{p1}[1] - A_{p1}[2] \cdot A_{p1}[2] - A_{p1}[3] \cdot A_{p1}[3] - A_{p1}[4] \cdot A_{p1}[4])$

$$A_0^2 - A_1^2 - A_2^2 - A_3^2$$

Successive Lorentz Transformations (Boosts along the x-Direction)

$$\gamma_2 := \frac{1}{\text{sqrt}\left(1-\dfrac{v_2^2}{c^2}\right)}$$

$$\gamma_2 := \frac{1}{\sqrt{1-\dfrac{v_2^2}{c^2}}}$$

$$\lambda_2 := \left\langle \left\langle \gamma_2 \left| -\frac{\gamma_2 \cdot v_2}{c} \right| 0 \middle| 0 \right\rangle, \left\langle -\frac{\gamma_2 \cdot v_2}{c} \middle| \gamma_2 \middle| 0 \middle| 0 \right\rangle, \langle 0|0|1|0\rangle, \langle 0|0|0|1\rangle \right\rangle$$

$$\lambda_2 := \begin{bmatrix} \dfrac{1}{\sqrt{1-\dfrac{v_2^2}{c^2}}} & -\dfrac{v_2}{\sqrt{1-\dfrac{v_1^2}{c^2}}c} & 0 & 0 \\[20pt] -\dfrac{v_2}{\sqrt{1-\dfrac{v_2^2}{c^2}}c} & \dfrac{1}{\sqrt{1-\dfrac{v_2^2}{c^2}}} & 0 & 0 \\[20pt] 0 & 0 & 1 & 0 \\[6pt] 0 & 0 & 0 & 1 \end{bmatrix}$$

$A_{p2} := simplify(\lambda_2 \cdot A_{p1})$

$$A_{p2} := \begin{bmatrix} \dfrac{A_0 c^2 - A_1(v_1 + v_2)c + A_0 v_1 v_1}{\sqrt{\dfrac{c^2 - v_1^2}{c^2}}\sqrt{\dfrac{c^2 - v_2^2}{c^2}}c^2} \\[24pt] \dfrac{A_1 c^2 - A_0(v_1 + v_2)c + A_1 v_1 v_2}{\sqrt{\dfrac{c^2 - v_1^2}{c^2}}\sqrt{\dfrac{c^2 - v_2^2}{c^2}}c^2} \\[24pt] A_2 \\[6pt] A_3 \end{bmatrix}$$

Invariance of a Dot Product under Two Boosts

$simplify(A_{p2}[1]\cdot A_{p2}[1] - A_{p2}[2]\cdot A_{p2}[2] - A_{p2}[3]\cdot A_{p2}[3] - A_{p2}[4]\cdot A_{p2}[4])$

$$A_0^2 - A_1^2 - A_2^2 - A_3^2$$

Equivalent Transformation Matrix for Two Boosts

$simplify(\lambda_1 \cdot \lambda_2)$

$$\begin{bmatrix} \dfrac{c^2+v_1+v_2}{\sqrt{\dfrac{c^2-v_2^2}{c^2}}\sqrt{\dfrac{c^2-v_1^2}{c^2}}c^2} & -\dfrac{v_1+v_2}{\sqrt{\dfrac{c^2-v_2^2}{c^2}}\sqrt{\dfrac{c^2-v_1^2}{c^2}}c} & 0 & 0 \\[2em] -\dfrac{v_1+v_2}{\sqrt{\dfrac{c^2-v_2^2}{c^2}}\sqrt{\dfrac{c^2-v_1^2}{c^2}}c} & \dfrac{c^2+v_1 v_2}{\sqrt{\dfrac{c^2-v_2^2}{c^2}}\sqrt{\dfrac{c^2-v_1^2}{c^2}}c^2} & 0 & 0 \\[2em] 0 & 0 & 1 & 0 \\[1em] 0 & 0 & 0 & 1 \end{bmatrix}$$

$subs(v_1 = 0.9 \cdot c, v_2 = 0.9 \cdot c, \%)$

$$\begin{bmatrix} 9.526315784 & -9.473684204 & 0 & 0 \\ -9.473684204 & 9.526315784 & 0 & 0 \\ 0 & 0 & 1 & 0 \\ 0 & 0 & 0 & 1 \end{bmatrix}$$

14.2 ENERGY AND MOMENTUM

In this section, relativistic forms of momentum and Newton's second law are discussed. The mass energy relation is derived from the work kinetic energy theorem. The classical form of kinetic energy is obtained at low velocities. The energy-momentum relation is then shown to be a Lorentz invariant. Examples of particle decay and completely inelastic collisions are given.

14.2.1 Newton's Second Law

The relativistic form of momentum

$$\mathbf{p} = \gamma m \mathbf{v} \tag{14.2.1}$$

insures that momentum is conserved in all inertial reference frames in the absence of external forces. The form of Newton's second law

$$\mathbf{F} = \frac{d\mathbf{p}}{dt} \tag{14.2.2}$$

is also correct for relativistic velocities. We may then obtain the relation between force and acceleration

$$\mathbf{F} = \gamma^3 m \frac{d\mathbf{v}}{dt} = \left(1 - \frac{v^2}{c^2}\right)^{-3/2} m \frac{d\mathbf{v}}{dt} \qquad (14.2.3)$$

and we have a velocity-dependent acceleration

$$\frac{d\mathbf{v}}{dt} = \frac{\mathbf{F}}{m} \left(1 - \frac{v^2}{c^2}\right)^{3/2} \qquad (14.2.4)$$

For a constant force per mass ratio the speed of light will be approached asymptotically with $dv/dt \rightarrow 0$ as $v \rightarrow c$. The speed of light is analogous to the terminal speed of a particle moving through a resistive medium under a constant applied force.

14.2.2 Mass Energy and Kinetic Energy

The work done on a mass m by an applied force of magnitude F

$$W = \int_{x_i}^{x_f} F dx = m \int_{x_i}^{x_f} \left(1 - \frac{v^2}{c^2}\right)^{-3/2} \frac{dv}{dt} dx = m \int_{v_i}^{v_f} \left(1 - \frac{v^2}{c^2}\right)^{-3/2} dv \qquad (14.2.5)$$

is equal to the change in kinetic energy of the particle. If the particle starts from rest $v_i = 0$ and $W = \Delta KE = KE$. Taking $v_f = v$ the u-substitution integral gives the kinetic energy

$$KE = (\gamma - 1) mc^2 \quad \text{or} \quad \gamma mc^2 = mc^2 + KE \qquad (14.2.6)$$

where we identify γmc^2 = total energy and mc^2 = rest mass energy of the particle.

Example 14.2.1

Find the speed of a particle whose rest mass energy is equal to its kinetic energy.

Solution: Equating rest mass and kinetic energies

$$mc^2 = (\gamma - 1) mc^2 \qquad (14.2.7)$$

gives $\gamma = 2$ corresponding to $v = \sqrt{1 - \frac{1}{\gamma^2}} c = \frac{\sqrt{3}}{2} c$.

Table 14.2.1 shows a comparison of physical quantities expressed in Newtonian and relativistic forms, including time intervals, lengths, linear momentum, Newton's second law, force and acceleration, kinetic energy, and rest mass energy.

TABLE 14.2.1: Comparison of physical quantities expressed in Newtonian and relativistic forms.

Physical Quantity	Newtonian Form	Relativistic Form
Time interval measured in different inertial frames	$\Delta t = \Delta t_p$	$\Delta t = \gamma \Delta t_p$
Length measured in different inertial frames	$L = L_p$	$L = \dfrac{L_p}{\gamma}$
Linear momentum	$\mathbf{p} = m\mathbf{v}$	$\mathbf{p} = \gamma m \mathbf{v}$
Newton's second law	$\mathbf{F} = \dfrac{d\mathbf{p}}{dt}$	$\mathbf{F} = \dfrac{d\mathbf{p}}{dt}$
Relation between force and acceleration	$\mathbf{F} = m\mathbf{a}$	$\mathbf{F} = \gamma^3 m\mathbf{a}$
Kinetic energy	$KE = \dfrac{1}{2}mv^2$	$KE = (\gamma - 1)mc^2$
Rest mass energy	-----	$E_r = mc^2$

14.2.3 Low Velocity Approximation

For low velocities $v \ll c$ and we can use the binomial theorem to approximate

$$\gamma = \frac{1}{\sqrt{1 - \left(\dfrac{v}{c}\right)^2}} \approx 1 + \frac{1}{2}\left(\frac{v}{c}\right)^2 \tag{14.2.8}$$

neglecting powers of velocity higher than v^2. Multiplying both sides by mc^2

$$\gamma mc^2 \approx mc^2 + \frac{1}{2}mv^2 \tag{14.2.9}$$

and we see that the total energy is sum of rest mass and kinetic energies.

14.2.4 Energy Momentum Relation

From the total energy $E = \gamma mc^2$ and linear momentum $p = \gamma mv$ we find that

$$E^2 - p^2c^2 = m^2c^4 \tag{14.2.10}$$

is the same in all inertial reference frames. Such a quantity is called a Lorentz invariant. For massless particles, we have $E = pc$ and for particles at rest $E = mc^2$. The energy momentum relation may be represented by the right triangle shown in Figure 14.2.1 with hypotenuse E and sides pc and mc^2.

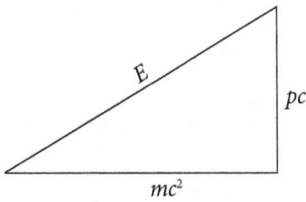

Figure 14.2.1: Energy momentum triangle.

Example 14.2.2

Demonstrate the mass energy relation using 4-vector notation

Solution: The 4-momenta $p^\mu = \gamma(mc, \mathbf{mv})$ and $p_\mu = \gamma(mc, -\mathbf{mv})$

$$p^\mu p_\mu = \gamma^2 m^2 \left(c^2, -v_x^2, -v_y^2, -v_z^2\right)$$

$$= \frac{E^2}{c^2} - p^2$$

$$= m^2 c^4 \tag{14.2.11}$$

14.2.5 Completely Inelastic Collisions

Example 14.2.3

Consider two "ideal" blobs of clay with equal masses m moving toward each other with a speed v relative to the laboratory frame. Find the final mass of the blobs after they collide.

Solution: The momentum of the masses is conserved before and after the collision where

$$p_i = \gamma_1 m v_1 + \gamma_2 m v_2 \tag{14.2.12}$$

Now $p_i = 0$ since $v_1 = -v_2$ and $\gamma_1 = \gamma_2 = \gamma$. Momentum conservation $p_i = p_f$ gives $p_f = 0$.

The initial energy before the collision is

$$E_i = \gamma m c^2 + \gamma m c^2 \tag{14.2.13}$$

After the collision, the combined masses are at rest in the laboratory frame

$$E_f = Mc^2 \tag{14.2.14}$$

Conservation of energy $E_i = E_f$ gives the final mass $M = 2\gamma m$.

Consider a completely inelastic collision of two equal ideal masses with differing speeds. We must now solve the following system of equations for momentum conservation

$$\gamma_1 m v_1 + \gamma_2 m v_2 = \gamma_f M v_f \tag{14.2.15}$$

and energy conservation

$$\gamma_1 m c^2 + \gamma_2 m c^2 = \gamma_f M c^2 \tag{14.2.16}$$

to determine the final combined mass M and velocity v_f.

14.2.6 Particle Decay

Example 14.2.4

A hypothetical particle of mass M traveling at a speed $v = c/3$ spontaneously decays into two particles each with mass $M/3$. Find the final speed of the particles and the angle that each particle makes with respect to the x-axis.

Solution: The initial and final energies of the particles are shown as

$$E_i = \gamma M c^2 \tag{14.2.17}$$

$$E_f = \gamma' \frac{M}{3} c^2 + \gamma' \frac{M}{3} c^2 \tag{14.2.18}$$

Conservation of energy $E_i = E_f$ gives $\gamma \quad -\gamma'$ where

$$\gamma = \sqrt{\frac{1}{1 - \dfrac{(c/3)^2}{c^2}}} = \sqrt{\frac{1}{1 - \dfrac{1}{9}}} = \frac{3\sqrt{2}}{4} \tag{14.2.19}$$

so $\gamma' = \dfrac{3}{2}\gamma = \dfrac{9\sqrt{2}}{8}$ and

$$v' = \sqrt{1 - \frac{1}{\gamma'^2}}\, c = \sqrt{1 - \frac{32}{81}}\, c = \frac{7}{9} c \tag{14.2.20}$$

The initial and final x-component momenta of the particles are

$$p_{ix} = \gamma M \frac{c}{3} \tag{14.2.21}$$

$$p_{xf} = \gamma' \frac{M}{3} v' \cos\theta + \gamma' \frac{M}{3} v' \cos\theta \tag{14.2.22}$$

Setting $p_{ix} = p_{xf}$ we obtain

$$\gamma M \frac{c}{3} = \frac{3}{2}\gamma \frac{M}{3}\frac{7}{9} c \cos\theta + \frac{3}{2}\gamma \frac{M}{3}\frac{7}{9} c \cos\theta \tag{14.2.23}$$

giving $\cos\theta = \dfrac{3}{7}$.

14.2.7 Energy Units

The S.I. units of energy and mass are joules and kilograms, respectively. In these units, the rest mass energy of the electron is

$$m_e c^2 = \left(9.11 \cdot 10^{-31}\ \text{kg}\right)\left(2.998 \cdot 10^8\ \text{m/s}\right)^2 = 8.19 \cdot 10^{-14}\ \text{J}$$

$$(14.2.24)$$

To avoid large negative exponents when dealing with elementary particles it is more convenient to express their rest mass energy in millions of electron volts (MeV) where $1\ \text{eV} = 1.602 \cdot 10^{-19}\ \text{J}$ is the potential energy of an electron across a potential difference of 1 V. $1\ \text{MeV} = 1.602 \cdot 10^{-13}\ \text{J}$ and $m_e c^2 = 0.511\ \text{MeV}$. Mass is then expressed in units of MeV/c^2.

Maple Examples

A one-dimensional completely inelastic collision is modeled in the Maple worksheet below. A mass m traveling with a speed v_1 impacts an identical mass m initially at rest. The final speed and combined mass is calculated. The energy-momentum relation is verified.

Key Maple terms: *evalf, simplify, solve, subs*

restart

Completely Inelastic Collision

Initial Momentum of *m* before the Collision

$$\gamma_1 := \frac{1}{sqrt\left(1 - \dfrac{v_1^2}{c^2}\right)}$$

$$p_i := m \cdot \gamma_1 \cdot v_1$$

$$p_i := \frac{m v_1}{\sqrt{1 - \dfrac{v_1^2}{c^2}}}$$

Final Momentum of Combined Masses after Collision

$$\gamma_f := \frac{1}{sqrt\left(1 - \dfrac{v_f^2}{c^2}\right)} :$$

$$p_f := M \cdot \gamma_f \cdot v_f$$

$$p_f := \frac{M v_f}{\sqrt{1 - \dfrac{v_f^2}{c^2}}}$$

Conservation of Momentum and Energy Equations

$$E_i := m \cdot \gamma_1 \cdot c^2 + m \cdot c^2 : E_f := m \cdot \gamma_f \cdot c^2 :$$
$$pEqn := p_i = p_f :$$
$$EEqn := E_i = E_f :$$
$$pEqn := subs\left(v_1 = \frac{c}{2}, pEqn \right)$$

$$pEqn := \frac{m\sqrt{3}\sqrt{4}c}{6} = \frac{M v_f}{\sqrt{1 - \dfrac{v_f^2}{c^2}}}$$

$$EEqn := subs\left(v_1 = \frac{c}{2}, EEqn \right)$$

$$EEqn := \frac{c^2 m\sqrt{3}\sqrt{4}c}{3} + mc^2 = \frac{Mc^2}{\sqrt{1 - \dfrac{v_f^2}{c^2}}}$$

Final Velocity v_f and Combined Mass M after Collision

$$solve(\{pEqn, EEqn\}, \{v_f, M\})$$

$$\left\{ M = \frac{2m\sqrt{3}\sqrt{4\sqrt{3}-6}}{3} + m\sqrt{4\sqrt{3}-6}, v_f = -\sqrt{3}c + 2c \right\}$$

$$evalf(\%)$$

$$\{M = 2.075909701\, m, v_f = 0.267949192c\}$$

Energy-Momentum Relation

$$E := \gamma_1 \cdot m \cdot c^2$$

$$E := \frac{c^2 m}{\sqrt{1 - \dfrac{v_1^2}{c^2}}}$$

$$p := \gamma_1 \cdot m \cdot v_1$$

$$p := \frac{m v_1}{\sqrt{1 - \frac{v_1^2}{c^2}}}$$

$$simplify(E^2 - p^2 \cdot c^2)$$

$$c^4 m^2$$

14.3 ELECTROMAGNETICS IN RELATIVITY

In this section, we investigate how electric and magnetic fields transform in different reference frames in relative motion. Simple examples of charged plates and solenoids in relative motion are examined. Maxwell's equations and the Lorentz force acting on charged particles moving in electric and magnetic forces are then written using relativistic notation.

14.3.1 Relativistic Transformation of Fields

Because of Lorentz contractions, a charged body will have a greater charge density in reference frames moving with respect to the body.

Example 14.3.1

Find the electric field in a frame at rest with respect to a large current sheet located in the $y = 0$ plane that carries a surface charge density $+\sigma$ moving with a velocity $\mathbf{v}$ out of the page in the z-direction (Figure 14.3.1).

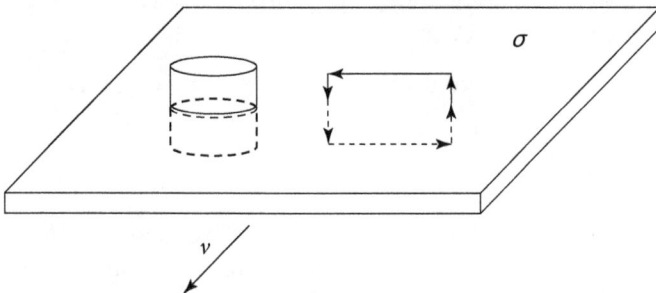

Figure 14.3.1: Charged sheet moving out of the page. The Gaussian pillbox and Amperian loop are used to calculate the electric and magnetic fields, respectively.

Solution: In the unprimed frame in which the charged sheet is at rest

$$\mathbf{E} = \frac{\sigma}{2\varepsilon_0} \hat{\mathbf{j}} \tag{14.3.1}$$

In a primed frame in which the charged sheet is moving the charge density is increased by a factor of γ because the sheet is contracted along the direction of motion so that the electric field in the primed frame is

$$\mathbf{E} = \frac{\gamma\sigma}{2\varepsilon_0}\hat{\mathbf{j}} \qquad (14.3.2)$$

The electric field is increased in a direction perpendicular to the direction of motion. From Ampere's law, the magnetic field is

$$\mathbf{B} = \mu_0 \frac{K}{2}\hat{\mathbf{i}} = \mu_0 \frac{\gamma\sigma v}{2}\hat{\mathbf{i}} \qquad (14.3.3)$$

Example 14.3.2

Find the electric field inside a parallel plate capacitor with upper and lower plates having charge densities $\pm\sigma$. Let the capacitor move in a direction perpendicular to the area of the plates $\mathbf{v} = v_y\hat{\mathbf{j}}$.

Solution: The electric field between the plates is

$$\mathbf{E} = \frac{\sigma}{\varepsilon_0}\hat{\mathbf{j}} \qquad (14.3.4)$$

The distance between the plates will be smaller because of Lorentz contraction. The electric field inside the capacitor will be unchanged, however. In this simple example, the electric field along the direction of motion is unchanged.

Example 14.3.3

Find the magnetic field inside a solenoid carrying a current I coaxial with the x-axis that moves with a velocity $\mathbf{v} = v_x\hat{\mathbf{i}}$.

Solution: In a frame at rest with respect to the solenoid the magnetic field is

$$\mathbf{B} = \mu_0 \frac{N}{L}I\hat{\mathbf{i}} \qquad (14.3.5)$$

In the laboratory frame the solenoid is length-contracted so that $L' = L/\gamma$. Current in the solenoid is reduced by a factor of γ due to time dilation $I' = \Delta q/\Delta t'$ where $\Delta t' = \gamma\Delta t$. Thus, the magnetic field along the direction of motion is unchanged.

In general, the electric and magnetic fields are related in the primed and unprimed frames

$$\mathbf{E}'_\| = \mathbf{E}_\|$$

$$\mathbf{E}'_\perp = \gamma\left(\mathbf{E}_\perp + \mathbf{v}\times\mathbf{B}\right)$$

$$\mathbf{B}'_\| = \mathbf{B}_\| \tag{14.3.6}$$

$$\mathbf{B}'_\perp = \gamma\left(\mathbf{B}_\perp - \frac{1}{c^2}\mathbf{v}\times\mathbf{E}\right)$$

where $\|$ and $\perp$ refer to field components parallel and perpendicular to $\mathbf{v}$. To transform from the primed frame to the unprimed frame simply interchange the primes everywhere and replace $\mathbf{v}$ with $-\mathbf{v}$.

Example 14.3.4

Find the magnetic field resulting from the moving charged sheet in Example 14.3.1 using equation (14.3.5)

Solution: From equation (14.3.5) the magnetic field is

$$\mathbf{B}' = -\frac{\gamma}{c^2}\mathbf{v}\times\mathbf{E} = -\frac{\gamma v}{c^2}\frac{\sigma}{2\varepsilon_0}\hat{\mathbf{k}}\times\hat{\mathbf{j}} \tag{14.3.7}$$

or

$$\mathbf{B}' = \mu_0\frac{\gamma v\sigma}{2}\hat{\mathbf{i}} \tag{14.3.8}$$

in agreement with the result obtained from Ampere's law.

14.3.2 Covariant Formulation of Maxwell's Equations

Although electric and magnetic fields depend on the inertial frame, Maxwell's equations are correct in all inertial frames. The electromagnetic field tensor

$$F^{\mu\nu} = \partial^\mu A^\nu - \partial^\nu A^\mu \tag{14.3.9}$$

is obtained from the vector 4-potential

$$A^\mu = \left(V/c, A_x, A_y, A_z\right) \tag{14.3.10}$$

In terms of the field components

$$F^{\mu\nu} = \begin{pmatrix} 0 & -E_x/c & -E_y/c & -E_z/c \\ E_x/c & 0 & -B_z & B_y \\ E_y/c & B_z & 0 & -B_x \\ E_z/c & -B_y & B_x & 0 \end{pmatrix} \tag{14.3.11}$$

For example,

$$F^{01} = \partial^0 A^1 - \partial^1 A^0 = \frac{1}{c}\frac{\partial A_x}{\partial t} + \frac{1}{c}\frac{\partial V}{\partial x} = -\frac{1}{c}E_x \qquad (14.3.12)$$

$$F^{12} = \frac{\partial A^2}{\partial x_1} - \frac{\partial A^1}{\partial x_2} = \underbrace{-\frac{\partial A_y}{\partial x} + \frac{\partial A_x}{\partial y}}_{(\nabla \times \mathbf{A})_z} = -B_z \qquad (14.3.13)$$

The diagonal elements $F^{00} = F^{11} = F^{22} = F^{33} = 0$.
The electromagnetic 4-current is

$$J^\mu = \left(c\rho, J_x, J_y, J_z\right) \qquad (14.3.14)$$

The inhomogeneous Maxwell equations (with sources) are given by the formula

$$\frac{\partial F^{\mu\nu}}{\partial x^\mu} = \mu_0 J^\nu \qquad (14.3.15)$$

Example 14.3.5

Write out explicitly

$$\frac{\partial F^{\mu 0}}{\partial x^\mu} = \mu_0 J^0 \qquad (14.3.16)$$

Solution: Summing over the repeated index

$$\frac{\partial F^{00}}{\partial x^0} + \frac{\partial F^{10}}{\partial x^1} + \frac{\partial F^{20}}{\partial x^2} + \frac{\partial F^{30}}{\partial x^3} = \mu_0 J^0 \qquad (14.3.17)$$

we obtain

$$0 + \frac{1}{c}\frac{\partial E_x}{\partial x} + \frac{1}{c}\frac{\partial E_y}{\partial y} + \frac{1}{c}\frac{\partial E_z}{\partial z} = \mu_0 c\rho \qquad (14.3.18)$$

which is equivalent to

$$\nabla \cdot \mathbf{E} = \frac{\rho}{\varepsilon_0} \qquad (14.3.19)$$

where the constants $\mu_0 c^2 = 1/\varepsilon_0$. The remaining three equations

$$\frac{\partial F^{\mu i}}{\partial x^\mu} = \mu_0 J^i \quad i = 1,2,3 \qquad (14.3.20)$$

give the Ampere-Maxwell relation

$$\nabla \times \mathbf{B} = \mu_0 \mathbf{J} + \mu_0 \varepsilon_0 \frac{\partial \mathbf{E}}{\partial t} \tag{14.3.21}$$

14.3.3 Homogeneous Maxwell Equations

The homogeneous Maxwell equations (without sources) are obtained from

$$\frac{\partial G^{\mu\nu}}{\partial x^{\mu}} = 0 \tag{14.3.22}$$

where the dual field tensor is

$$G^{\mu\nu} = \begin{pmatrix} 0 & -B_x & -B_y & -B_z \\ B_x & 0 & E_z/c & -E_y/c \\ B_y & -E_z/c & 0 & E_x/c \\ B_z & E_y/c & -E_x/c & 0 \end{pmatrix} \tag{14.3.23}$$

Example 14.3.6

Write out explicitly $\dfrac{\partial G^{\mu 3}}{\partial x^{\mu}} = 0$

Solution:

$$\frac{\partial G^{03}}{\partial x^0} + \frac{\partial G^{13}}{\partial x^1} + \frac{\partial G^{23}}{\partial x^2} + \frac{\partial G^{33}}{\partial x^3} = 0 \tag{14.3.24}$$

$$\frac{1}{c}\frac{\partial B_y}{\partial t} - \frac{1}{c}\frac{\partial E_z}{\partial x} + 0 + \frac{1}{c}\frac{\partial E_x}{\partial z} = 0 \tag{14.3.25}$$

$$\underbrace{\frac{\partial E_x}{\partial z} - \frac{\partial E_z}{\partial x}}_{(\nabla \times \mathbf{E})_y} = -\frac{\partial B_y}{\partial t} \tag{14.3.26}$$

Combined with $\dfrac{\partial G^{1\nu}}{\partial x^{\nu}} = 0$ and $\dfrac{\partial G^{2\nu}}{\partial x^{\nu}} = 0$ we obtain the differential form of Faraday's law

$$\nabla \times \mathbf{E} = -\frac{\partial \mathbf{B}}{\partial t} \tag{14.3.27}$$

Example 14.3.7

Show that $\dfrac{\partial A^{\mu}}{\partial x^{\mu}} = 0$ gives the Lorentz gauge condition

Solution: Summing over repeated indices

$$\frac{\partial A^0}{\partial x^0} + \frac{\partial A^1}{\partial x^1} + \frac{\partial A^2}{\partial x^2} + \frac{\partial A^3}{\partial x^3} = 0 \tag{14.3.28}$$

$$\frac{1}{c}\frac{\partial}{\partial t}\left(\frac{V}{c}\right) + \underbrace{\frac{\partial A_x}{\partial x} + \frac{\partial A_y}{\partial y} + \frac{\partial A_z}{\partial z}}_{\nabla \cdot \mathbf{A}} = 0 \tag{14.3.29}$$

and we have the Lorentz gauge condition

$$\nabla \cdot \mathbf{A} = -\frac{1}{c^2}\frac{\partial V}{\partial t} \tag{14.3.30}$$

Example 14.3.8

Show that $\dfrac{\partial J^\mu}{\partial x^\mu} = 0$ gives the continuity equation

$$\frac{\partial J^0}{\partial x^0} + \frac{\partial J^1}{\partial x^1} + \frac{\partial J^2}{\partial x^2} + \frac{\partial J^3}{\partial x^3} = 0 \tag{14.3.31}$$

$$\frac{1}{c}\frac{\partial}{\partial t}(\rho c) + \underbrace{\frac{\partial J_x}{\partial x} + \frac{\partial J_y}{\partial y} + \frac{\partial J_z}{\partial z}}_{\nabla \cdot \mathbf{J}} = 0 \tag{14.3.32}$$

$$\nabla \cdot \mathbf{J} = -\frac{\partial \rho}{\partial t} \tag{14.3.33}$$

14.3.4 Lorentz Force Equation

A covariant form of the Lorentz force equation

$$\frac{dp^\mu}{d\tau} = qF^{\mu\nu}U_\nu \tag{14.3.34}$$

where the 4-momentum $p^\mu = (\gamma mc, p_x, p_y, p_z)$, τ is the proper time and 4-velocity $U_\nu = \gamma(c, -v_x, -v_y, -v_z)$. Thus, we have four equations

$$\frac{d}{d\tau}\begin{pmatrix} \gamma mc \\ \gamma m v_x \\ \gamma m v_y \\ \gamma m v_z \end{pmatrix} = \gamma \begin{pmatrix} 0 & -E_x/c & -E_y/c & -E_z/c \\ E_x/c & 0 & -B_z & B_y \\ E_y/c & B_z & 0 & -B_x \\ E_z/c & -B_y & B_x & 0 \end{pmatrix}\begin{pmatrix} c \\ -v_x \\ -v_y \\ -v_z \end{pmatrix} \tag{14.3.35}$$

The first equation

$$\frac{d}{d\tau}\gamma mc = \frac{\gamma}{c}\left(E_x v_x + E_y v_y + E_z v_z\right) \tag{14.3.36}$$

$$\frac{d}{d\tau}\left(\gamma mc^2\right) = \gamma\left(\mathbf{v}\cdot\mathbf{E}\right) \tag{14.3.37}$$

$$\frac{dE}{dt} = \mathbf{v}\cdot\mathbf{E} \tag{14.3.38}$$

where $d/d\tau = \gamma d/dt$ and the total energy $E = \gamma mc^2$
The second equation is

$$\frac{d}{d\tau}\left(\gamma m v_x\right) = \gamma q\left(E_x + B_z v_y - B_y v_z\right) = \gamma q\left(\mathbf{E} + \mathbf{v}\times\mathbf{B}\right)_x \tag{14.3.39}$$

Similarly, the third and fourth equations give

$$\frac{d}{d\tau}\left(\gamma m v_y\right) = \gamma q\left(\mathbf{E} + \mathbf{v}\times\mathbf{B}\right)_y \tag{14.3.40}$$

$$\frac{d}{d\tau}\left(\gamma m v_z\right) = \gamma q\left(\mathbf{E} + \mathbf{v}\times\mathbf{B}\right)_z \tag{14.3.41}$$

Combining the last three equations

$$\frac{d\mathbf{p}}{dt} = q\left(\mathbf{E} + \mathbf{v}\times\mathbf{B}\right) \tag{14.3.42}$$

where $\mathbf{p} = \gamma m\mathbf{v}$.

The Lorentz force equations with metric signature $(-, +, +, +)$ are obtained from

$$\frac{d}{d\tau}\begin{pmatrix} \gamma mc \\ \gamma m v_x \\ \gamma m v_y \\ \gamma m v_z \end{pmatrix} = \gamma\begin{pmatrix} 0 & E_x/c & E_y/c & E_z/c \\ -E_x/c & 0 & -B_z & B_y \\ -E_y/c & B_z & 0 & -B_x \\ -E_z/c & -B_y & B_x & 0 \end{pmatrix}\begin{pmatrix} -c \\ v_x \\ v_y \\ v_z \end{pmatrix} \tag{14.3.43}$$

where the signs of the $F^{\mu\nu}$ electric field components and 4-velocity components are opposite from the signs of the $(+, -, -, -)$ signature. Representations of the electromagnetic field tensor can vary among textbooks depending on the metric signature and system of units used. For example, the $(-, +, +, +)$ is used in David Griffith's textbook *Introduction to Electrodynamics*, while the $(+, -, -, -)$ signature is used in J. D. Jackson's textbook *Classical Electrodynamics*.

Maple Example

Transformation of the electric and magnetic field components for a Lorentz boost Λ along the x- axis given by $\Lambda^T F^{\mu\nu} \Lambda$ is shown in the Maple worksheet below.

Key Maple commands: $\cdot$ (matrix multiplication), *assume*, *simplify*, *Transpose*

Maple packages: *with(LinearAlgebra)*:

restart

Relativistic Transformation of Fields

with(LinearAlgebra) :

$$F := \left\langle \left\langle 0 \left| \frac{-E_x}{c} \right| \frac{-E_y}{c} \right| \frac{-E_z}{c} \right\rangle, \left\langle \frac{E_x}{c} \left| 0 \right| -B_z \left| B_y \right\rangle, \left\langle \frac{E_y}{c} \right| B_z \left| 0 \right| -B_x \right\rangle, \left\langle \frac{E_z}{c} \right| -B_y \left| B_x \right| 0 \right\rangle \right\rangle$$

$$F := \begin{bmatrix} 0 & -\dfrac{E_x}{c} & -\dfrac{E_y}{c} & -\dfrac{E_z}{c} \\[2mm] \dfrac{E_x}{c} & 0 & -B_z & B_y \\[2mm] \dfrac{E_y}{c} & B_z & 0 & -B_x \\[2mm] \dfrac{E_z}{c} & -B_y & B_x & 0 \end{bmatrix}$$

$$\gamma_1 := \frac{1}{sqrt\left(1 - \dfrac{v^2}{c^2}\right)} :$$

$$\Lambda := \left\langle \left\langle \gamma_1 \left| -\frac{\gamma_1 \cdot v}{c} \right| 0 \right| 0 \right\rangle, \left\langle -\frac{\gamma_1 \cdot v}{c} \left| \gamma_1 \right| 0 \right| 0 \right\rangle, \langle 0 | 0 | 1 | 0 \rangle, \langle 0 | 0 | 0 | 1 \rangle \right\rangle$$

$$\Lambda := \begin{bmatrix} \dfrac{1}{\sqrt{1 - \dfrac{v^2}{c^2}}} & -\dfrac{v}{\sqrt{1 - \dfrac{v^2}{c^2}}\,c} & 0 & 0 \\[4mm] -\dfrac{v}{\sqrt{1 - \dfrac{v^2}{c^2}}\,c} & \dfrac{1}{\sqrt{1 - \dfrac{v^2}{c^2}}} & 0 & 0 \\[4mm] 0 & 0 & 1 & 0 \\[2mm] 0 & 0 & 0 & 1 \end{bmatrix}$$

$Ft := simplify(Transpose(\Lambda).F.\Lambda)$

$$Ft := \begin{bmatrix} 0 & -\dfrac{E_x}{c} & \dfrac{vB_z - E_y}{\sqrt{\dfrac{c^2-v^2}{c^2}}\,c} & -\dfrac{vB_y + E_z}{\sqrt{\dfrac{c^2-v^2}{c^2}}\,c} \\[3em] \dfrac{E_x}{c} & 0 & \dfrac{-B_z c^2 + vE_y}{\sqrt{\dfrac{c^2-v^2}{c^2}}\,c^2} & \dfrac{B_y c^2 + vE_z}{\sqrt{\dfrac{c^2-v^2}{c^2}}\,c^2} \\[3em] \dfrac{-vB_z + E_y}{\sqrt{\dfrac{c^2-v^2}{c^2}}\,c} & \dfrac{B_z c^2 - vE_y}{\sqrt{\dfrac{c^2-v^2}{c^2}}\,c^2} & 0 & -B_x \\[3em] \dfrac{vB_y + E_z}{\sqrt{\dfrac{c^2-v^2}{c^2}}\,c} & -\dfrac{B_y c^2 + vE_z}{\sqrt{\dfrac{c^2-v^2}{c^2}}\,c^2} & B_x & 0 \end{bmatrix}$$

$assume(c > 0)$
$simplify(Ft, sqrt)$

$$Ft := \begin{bmatrix} 0 & -\dfrac{E_x}{c \sim} & \dfrac{vB_z - E_y}{\sqrt{c\sim^2 - v^2}} & -\dfrac{vB_y + E_z}{\sqrt{c\sim^2 - v^2}} \\[2.5em] \dfrac{E_x}{c \sim} & 0 & \dfrac{-B_z c\sim^2 + vE_y}{c \sim \sqrt{c\sim^2 - v^2}} & \dfrac{B_y c\sim^2 + vE_z}{c \sim \sqrt{c\sim^2 - v^2}} \\[2.5em] \dfrac{-vB_z + E_y}{\sqrt{c\sim^2 - v^2}} & \dfrac{B_z c\sim^2 - vE_y}{c \sim \sqrt{c\sim^2 - v^2}} & 0 & -B_x \\[2.5em] \dfrac{vB_y + E_z}{\sqrt{c\sim^2 - v^2}} & -\dfrac{B_y c\sim^2 + vE_z}{c \sim \sqrt{c\sim^2 - v^2}} & B_x & 0 \end{bmatrix}$$

14.4 RELATIVISTIC LAGRANGIAN FORMULATION

The Lagrangian formulation is applied to relativistic problems below, including the free particle, simple harmonic oscillator, and a particle in electric and magnetic fields. Lagrange's equations of motion are then obtained. The relativistic Hamiltonian and Hamilton's equations of motion are then demonstrated.

14.4.1 Lagrangian of a Free Particle

The relativistic Lagrangian of a free particle moving in one dimension is

$$L = -mc^2 \sqrt{1 - \frac{\dot{x}^2}{c^2}} \qquad (14.4.1)$$

From the generalized momentum

$$p_x = \frac{\partial L}{\partial \dot{x}} = -\frac{1}{2} \frac{mc^2}{\sqrt{1 - \frac{\dot{x}^2}{c^2}}} \left(-2 \frac{\dot{x}}{c^2} \right) = \frac{m\dot{x}}{\sqrt{1 - \frac{\dot{x}^2}{c^2}}} \qquad (14.4.2)$$

the Hamiltonian is obtained using a Legendre transformation

$$H = p\dot{x} - L = \frac{m\dot{x}^2}{\sqrt{1 - \frac{\dot{x}^2}{c^2}}} + mc^2 \sqrt{1 - \frac{\dot{x}^2}{c^2}} \qquad (14.4.3)$$

Obtaining a common denominator

$$H = \frac{mc^2}{\sqrt{1 - \frac{\dot{x}^2}{c^2}}} \qquad (14.4.4)$$

14.4.2 Relativistic 1D Harmonic Oscillator

The relativistic Lagrangian for the 1D simple harmonic oscillator is

$$L = -mc^2 \sqrt{1 - \frac{\dot{x}^2}{c^2}} - \frac{1}{2} kx^2 \qquad (14.4.5)$$

with corresponding Hamiltonian

$$H = \frac{mc^2}{\sqrt{1 - \frac{\dot{x}^2}{c^2}}} + \frac{1}{2} kx^2 \qquad (14.4.6)$$

14.4.3 Charged Particle in Electric and Magnetic Fields

The relativistic Lagrangian of a free particle of mass m and charge q moving in electric and magnetic fields with scalar potential V and magnetic vector potential **A** is

$$L = -mc^2 \sqrt{1 - \frac{v^2}{c^2}} + q\mathbf{v} \cdot \mathbf{A} - qV \qquad (14.4.7)$$

where $v^2 = \dot{x}^2 + \dot{y}^2 + \dot{z}^2$ and $\mathbf{v} \cdot \mathbf{A} = \dot{x}A_x + \dot{y}A_y + \dot{z}A_z$. Lagrange's equations in Cartesian coordinates are

$$\frac{d}{dt}\frac{\partial L}{\partial \dot{x}} - \frac{\partial L}{\partial x} = 0, \ \frac{d}{dt}\frac{\partial L}{\partial \dot{y}} - \frac{\partial L}{\partial y} = 0 \text{ and } \frac{d}{dt}\frac{\partial L}{\partial \dot{z}} - \frac{\partial L}{\partial z} = 0 \qquad (14.4.8)$$

corresponding to

$$\frac{d}{dt}\left[\frac{m\dot{x}}{\sqrt{1-\dfrac{v^2}{c^2}}} + qA_x\right] - q\frac{\partial}{\partial x}(\mathbf{v} \cdot \mathbf{A}) + q\frac{\partial V}{\partial x} = 0 \qquad (14.4.9)$$

$$\frac{d}{dt}\left[\frac{m\dot{y}}{\sqrt{1-\dfrac{v^2}{c^2}}} + qA_y\right] - q\frac{\partial}{\partial y}(\mathbf{v} \cdot \mathbf{A}) + q\frac{\partial V}{\partial y} = 0 \qquad (14.4.10)$$

$$\frac{d}{dt}\left[\frac{m\dot{z}}{\sqrt{1-\dfrac{v^2}{c^2}}} + qA_z\right] - q\frac{\partial}{\partial z}(\mathbf{v} \cdot \mathbf{A}) + q\frac{\partial V}{\partial z} = 0 \qquad (14.4.11)$$

Combining these equations

$$\frac{d}{dt}\left[\frac{m\mathbf{v}}{\sqrt{1-\dfrac{v^2}{c^2}}} + q\mathbf{A}\right] - q\nabla(\mathbf{v} \cdot \mathbf{A}) + q\nabla V = 0 \qquad (14.4.12)$$

and using

$$\mathbf{v} \times \mathbf{B} = \nabla(\mathbf{v} \cdot \mathbf{A}) - (\mathbf{v} \cdot \nabla)\mathbf{A} \qquad (14.4.13)$$

with

$$\frac{d\mathbf{A}}{dt} = \frac{\partial \mathbf{A}}{\partial t} + (\mathbf{v} \cdot \nabla)\mathbf{A} \qquad (14.4.14)$$

we obtain the Lorentz force equation

$$\frac{d}{dt}(\gamma m\mathbf{v}) = q\left(-\frac{\partial \mathbf{A}}{\partial t} - \nabla V + \mathbf{v} \times \mathbf{B}\right) \qquad (14.4.15)$$

or

$$\frac{d}{dt}(\gamma m \mathbf{v}) = q(\mathbf{E} + \mathbf{v} \times \mathbf{B}) \qquad (14.4.16)$$

Maple Examples

Lagrangian and Hamiltonian formalisms are demonstrated for the free particle and simple harmonic oscillator in the Maple worksheet below.

Key Maple commands: *diff, dsolve, hamilton_eqs, simplify, solve, subs*

Maple packages: *with(Physics): with(DETools):*

restart

with(Physics) :

Setup(mathematicalnotation = true)

$$[mathematicalnotation = true]$$

Lagrangian of a Free Particle

$$L := -m \cdot c^2 \cdot sqrt\left(1 - \frac{diff(x(t),t)^2}{c^2}\right) - \frac{1}{2} \cdot k \cdot x(t)^2$$

$$L := -mc^2 \sqrt{1 - \frac{\left(\frac{d}{dt}x(t)\right)^2}{c^2}} - \frac{kx(t)^2}{2}$$

LagEqn := diff(diff(L, diff(x(t), t)), t) = 0

$$LagEqn := \frac{m\left(\frac{d}{dt}x(t)\right)^2\left(\frac{d^2}{dt^2}x(t)\right)}{\left(1 - \frac{\left(\frac{d}{dt}x(t)\right)^2}{c^2}\right)^{3/2} c^2} + \frac{m\left(\frac{d^2}{dt^2}x(t)\right)}{\sqrt{1 - \frac{\left(\frac{d}{dt}x(t)\right)^2}{c^2}}} = 0$$

dsolve(LagEqn)

$$x(t) = \_C1\, t + \_C2$$

$H := diff(L, diff(x(t), t)) \cdot diff(x(t), t) - L$

$$H := \frac{m \left(\dfrac{d}{dt} x(t)\right)^2}{\sqrt{1 - \dfrac{\left(\dfrac{d}{dt} x(t)\right)^2}{c^2}}} + mc^2 \sqrt{1 - \frac{\left(\dfrac{d}{dt} x(t)\right)^2}{c^2}}$$

$simplify(\%)$

$$\frac{mc^2}{\sqrt{\dfrac{-\left(\dfrac{d}{dt} x(t)\right)^2 + c^2}{c^2}}}$$

$solve(p(t) := diff(L, diff(\cdot(t), t)) \cdot diff(x(t), t))$

$$\frac{p(t)c}{\sqrt{c^2 m^2 + p(t)^2}}, -\frac{p(t)c}{\sqrt{c^2 m^2 + p(t)^2}}$$

$H := simplify(subs(diff(x(t), t) = \%[1], H))$

$$H := \frac{mc^2}{\sqrt{\dfrac{c^2 m^2}{c^2 m^2 + p(t)^2}}}$$

$diff(x(t), t) = simplify(diff(H, p(t)))$

$$\frac{d}{dt} x(t) = \frac{mc^2 p(t)}{\left(c^2 m^2 + p(t)^2\right) \sqrt{\dfrac{c^2 m^2}{c^2 m^2 + p(t)^2}}}$$

Lagrangian and Hamiltonian of a Simple Harmonic Oscillator

$$L := -m \cdot c^2 \cdot sqrt\left(1 - \frac{diff(x(t), t)^2}{c^2}\right) - \frac{1}{2} \cdot k \cdot x(t)^2$$

$$L := -mc^2 \sqrt{1 - \frac{\left(\dfrac{d}{dt} x(t)\right)^2}{c^2}} - \frac{kx(t)^2}{2}$$

$diff(diff(L, diff(x(t), t)), t) - diff(L, x(t)) = 0$

$$\frac{m\left(\dfrac{d}{dt}x(t)\right)^2\left(\dfrac{d^2}{dt^2}x(t)\right)}{\left(1-\dfrac{\left(\dfrac{d}{dt}x(t)\right)^2}{c^2}\right)^{3/2}c^2}+\frac{m\left(\dfrac{d^2}{dt^2}x(t)\right)}{\sqrt{1-\dfrac{\left(\dfrac{d}{dt}x(t)\right)^2}{c^2}}}+kx(t)=0$$

$H := diff(L, diff(x(t), t)) \cdot diff(x(t), t) - L$

$$H := \frac{m\left(\dfrac{d}{dt}x(t)\right)^2}{\sqrt{1-\dfrac{\left(\dfrac{d}{dt}x(t)\right)^2}{c^2}}}+mc^2\sqrt{1-\dfrac{\left(\dfrac{d}{dt}x(t)\right)^2}{c^2}}+\frac{kx(t)^2}{2}$$

$solve(p(t) := diff(L, diff(x(t), t)), diff(x(t), t))$

$$\frac{p(t)c}{\sqrt{c^2m^2+p(t)^2}}, -\frac{p(t)c}{\sqrt{c^2m^2+p(t)^2}}$$

$H := simplify(subs(diff(x(t), t) = \%[1], H))$

$$H := \frac{kx(t)^2\sqrt{\dfrac{c^2m^2}{c^2m^2+p(t)^2}}+2mc^2}{2\sqrt{\dfrac{c^2m^2}{c^2m^2+p(t)^2}}}$$

$diff(x(t), t) = simplify(diff(H, p(t)))$

$$\frac{d}{dt}x(t)=\frac{mc^2p(t)}{\left(c^2m^2+p(t)^2\right)\sqrt{\dfrac{c^2m^2}{c^2m^2+p(t)^2}}}$$

$diff(p(t), t) = -simplify(diff(H, x(t)))$

$$\frac{d}{dt}p(t) = -kx(t)$$

Hamilton's Equations Using the *DETools* Package

$HH := subs(\{x(t) = q1, p(t) = p1\}, H))$

$$HH := \frac{kq1^2 \sqrt{\dfrac{c^2m^2}{c^2m^2 + pl^2}} + 2mc^2}{2\sqrt{\dfrac{c^2m^2}{c^2m^2 + pl^2}}}$$

$with(DETools):$
$hamilton\_eqs(HH)$

$$\left[\frac{d}{dt}p1(t) = -kq1(t), \frac{d}{dt}q1(t) := -\frac{kq1(t)^2\, p1(t)}{2\left(c^2m^2 + p1(t)^2\right)} \right.$$

$$\left. + \frac{\left(kq1(t)^2 \sqrt{\dfrac{c^2m^2}{c^2m^2 + p1(t)^2}} + 2mc^2\right)c^2m^2\, p1(t)}{2\left(\dfrac{c^2m^2}{c^2m^2 + p1(t)^2}\right)^{3/2}\left(c^2m^2 + p1(t)^2\right)^2}, \left[p1(t), q1(t) \right] \right]$$

$simplify(\%[1][2])$

$$\frac{d}{dt}q1(t) = \frac{mc^2\, p1(t)}{\sqrt{\dfrac{c^2m^2}{c^2m^2 + p1(t)^2}}\left(c^2m^2 + p1(t)^2\right)}$$

14.5 MATLAB EXAMPLES

Section 14.4 Relativistic Lagrangian Formulation

Lagrange's relativistic equation of motion describing a point charge in an electric field is obtained with the MATLAB script 'relativistic_lagrangian.m.' The equation of motion is then solved with the point charge initially at rest.

Key MATLAB commands: *assume, dsolve, functionalDerivative, syms*

```
syms m q E c x(t)
T =  -m*c^2*sqrt(1-diff(x,t)^2/c^2);
V = q*E*x;
```

```
L = T - V
eqn = functionalDerivative(L,x) == 0
assume(m>0);
assume(q>0);
assume(E>0);
assume(c>0);
Dx=diff(x,t);
conds=[x(0)==0,Dx(0)==0];
dsolve(eqn,conds)
The above script is exectuted at the Command line.
>> relativistic_lagrangian
L(t) =
- E*q*x(t) - c^2*m*(1 - diff(x(t), t)^2/c^2)^(1/2)
eqn(t) =
-(m*diff(x(t), t, t) + E*q*(1 - diff(x(t), t)^2/c^2)^(3/2))/(1 -
    diff(x(t), t)^2/c^2)^(3/2) == 0
ans =
(c*(c^2*m^2)^(1/2))/(E*q) - (c*(E^2*q^2*t^2 + c^2*m^2)^(1/2))/(E*q)
>> pretty(ans)
        2   2              2   2  2    2  2
c sqrt(c  m )     c sqrt(E  q  t  + c  m )
-------------  -  ------------------------
     E q                   E q
```

14.6 EXERCISES

Section 14.1 Kinematics

1. At what speed does a meter stick move with respect to an observer who measures its length to be 1.0 cm?

2. A 1-km-long train approaches a tunnel that is 100 m long. How fast must the train travel relative to the tunnel for it to fit completely inside for a brief instant in a reference frame at rest with respect to the tunnel?

3. According to an observer O in the lab frame particle A moves at $c/2$ and particle B moves at $-c/2$. What is the speed of A with respect to B?

4. An observer O in the lab frame measures particle A with a velocity of $0.999c$ and particle B with a velocity of $-0.99c$. What is the speed of A with respect to B?

Section 14.2 Energy and Momentum

5. A proton with mass $m = 938$ MeV/c^2 moves at $0.99c$. Calculate its kinetic energy and total energy.

6. What is the speed of a particle with relativistic momentum $p = \gamma mv$ equal to three times its classical momentum $p = mv$? Leave your answer in terms of the speed of light c.

7. A rocket's kinetic energy is three times its rest mass energy in a frame at rest with respect to the background stars. For every second that ticks by in the rocket frame how many seconds pass by in the frame at rest with respect to the background stars?

8. Calculate the speed v a particle of mass m in a frame where (a) its total energy is equal to three times its rest mass energy and (b) its kinetic energy is equal to its rest mass energy

9. Given $p = \gamma mv$ show that $\dfrac{dp}{dt} = \gamma^3 m \dfrac{dv}{dt}$

10. Given $E = \gamma mc^2$ and $p = \gamma mv$ show that $E^2 = p^2 c^2 + m^2 c^4$

11. A hypothetical particle of mass M initially at rest in the lab decays into particles of mass m moving with a speed $c/2$ relative to the laboratory frame. What was the initial mass M in terms of the mass m of the decay products?

12. Two hypothetical blobs of clay both of mass M are moving toward each other at speeds $c/2$ and $c/3$ relative to the laboratory frame. What is the final mass of the combined blobs after they stick together? What is the final speed of the combined blobs?

13. A particle of mass M, initially at rest in the laboratory frame, decays into two masses m_1 and m_2 moving in opposite directions at speeds v_1 and v_2.
    (a) Write down relativistic equations for the conservation of energy and momentum.
    (b) For the special case where $v_1 = v_2 = v$ write down an expression for m_1 and m_2 in terms of M and v.

14. An X-ray photon with wavelength λ and frequency f scatters off an electron at an angle θ. The scattered X-ray wavelength and frequencies are λ' and f'. The electron is scattered at an angle ϕ with an energy E_e and momentum p_e. Apply energy conservation $hf + m_e c^2 = hf' + E_e$
with momentum conservation in the x-direction

$$\frac{h}{\lambda} = p_e \cos\phi + \frac{h}{\lambda'}\cos\theta$$

and momentum conservation in the y-direction

$$0 = p_e \sin\phi - \frac{h}{\lambda'}\sin\theta$$

using the relation

$$E_e^2 = p_e^2 c^2 + m_e^2 c^4$$

to show Compton's scattering formula

$$\lambda' = \lambda + \lambda_c \left(1 - \cos\theta\right)$$

where $\lambda_c = \dfrac{h}{m_e c} = 0.00243$ nm .

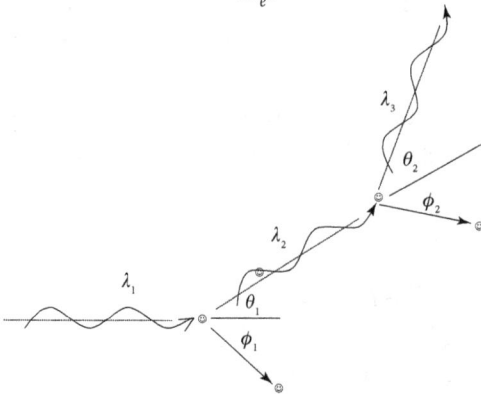

Figure 14.6.1: A photon scatters off two electrons.

15. A photon Compton scatters off two electrons as shown in Figure 14.6.1. Calculate (a) the energy and (b) the wavelength λ_3 of the photon after scattering off the second electron in terms of λ_1 of θ_1 and θ_2.

16. A photon Compton scatters off three electrons with each scattering angle equal to 60 degrees. What fraction of the initial energy of the photon remains after the third scattering event?

17. An X-ray Compton scatters off a stationary electron. By how much does the wavelength of the X-ray increase if the scattering angle is 180 degrees. Find the momentum and the kinetic energy of the electron.

18. A particle of mass M traveling along the x-axis at $v_i = c/3$ decays into two particles of mass $m = M/3$. Each particle moves away with speed v_f at an angle θ above and below the x-axis.
   (a) Write down the relativistic equations for conservation of total energy and momentum.
   (b) Obtain values for v_f and the angle θ.

19. An electron and a positron (each with $mc^2 = 0.511$ MeV) combine and annihilate with negligible initial kinetic energy. What is the frequency and wavelength of the emitted photons?

20. A positron traveling along the x-axis with a speed of $0.999c$ strikes a stationary electron producing two very high-energy photons. Calculate the energy of each photon and the angle θ above and below the x-axis that the photons travel after the collision.

Section 14.3 Electromagnetics in Relativity

21. An electron is accelerated across a potential difference of 10^8 V. What fraction of the speed of light v/c does the electron attain?

22. Over what potential difference must an electron be accelerated so that it attains a speed of $0.999c$?

23. Two identical masses M each have a charge Q and are separated by a distance d. Apply conservation of energy with the relativistic form of the kinetic energy to calculate the final speed of the masses long after they are released (neglect their gravitational attraction).

24. A particle of mass m and charge Q travels with a relativistic velocity $\mathbf{v}$ perpendicular to the direction of magnetic field $\mathbf{B}$. What is the radius of curvature of the particle's path?

25. A particle with charge Q and mass m is placed in an electric field $\mathbf{E}$. Write a relativistic expression for the particle's acceleration. What is the particle's speed as a function of time?

26. A point charge q travels with a velocity $\mathbf{v} = v_0\hat{\mathbf{i}}$ in the x-direction. Write the components of the electric field in Cartesian coordinates. Create a vector plot of the electric field in the x-y plane.

27. A point charge q of mass m is released from rest at $x = 0$ in an electric field $\mathbf{E} = E_0\hat{\mathbf{i}}$. Find the velocity of the charge as a function of position in the lab frame.

28. Write out explicitly $F^{32} = \dfrac{\partial A^2}{\partial x_3} - \dfrac{\partial A^3}{\partial x_2}$

29. Write $\dfrac{\partial G^{3\nu}}{\partial x^\nu} = 0$ out explicitly to obtain $\nabla \cdot \mathbf{B} = 0$

30. Show that the inhomogeneous wave equations $\Box^2 \mathbf{A} = -\mu_0\mathbf{J}$ and $\Box^2 V = -\dfrac{\rho}{\varepsilon_0}$ can be written as $\Box^2 A^\mu = -\mu_0 J^\mu$.

31. Show that

$$\frac{\partial F_{\mu\nu}}{\partial x^\lambda} + \frac{\partial F_{\nu\lambda}}{\partial x^\mu} + \frac{\partial F_{\lambda\mu}}{\partial x^\nu} = 0$$

32. Given the electromagnetic displacement tensor

$$D^{\mu\nu} = \begin{pmatrix} 0 & D_x c & D_y c & D_z c \\ -D_x/c & 0 & H_z & -H_y \\ -D_y/c & -H_z & 0 & H_x \\ -D_z/c & H_y & -H_x & 0 \end{pmatrix}$$

and the magnetization-polarization tensor

$$M^{\mu\nu} = \begin{pmatrix} 0 & P_x c & P_y c & P_z c \\ -P_x/c & 0 & M_z & -M_y \\ -P_y/c & -M_z & 0 & M_x \\ -P_z/c & M_y & -M_x & 0 \end{pmatrix}$$

show that $D^{\mu\nu} = \dfrac{1}{\mu_0} F^{\mu\nu} - M^{\mu\nu}$

33. Given the 4-current J^ν is a sum of free and bound currents

$$J^\nu = J_{\text{free}}^\nu + J_b^\nu \text{ where } J_{\text{free}}^\nu = \left(c\rho_{\text{free}}, \mathbf{J}_{\text{free}}\right) \text{ and } J_b^\nu = \left(c\rho_b, \mathbf{J}_b\right)$$

show that $\partial_\mu D^{\mu\nu} = J_{\text{free}}^\nu$ gives Maxwell's equations in matter

$$\nabla \cdot \mathbf{D} = \rho_{\text{free}}$$

$$\nabla \times \mathbf{H} = \mathbf{J}_{\text{free}} + \frac{\partial \mathbf{D}}{\partial t}$$

Section 14.4 Relativistic Lagrangian Formulation

34. Show the relativistic Lagrangian of a free particle $L = -mc^2 \sqrt{1 - \dfrac{v^2}{c^2}}$ becomes
$$L \approx -mc^2 + \frac{1}{2}mv^2$$

for $v \ll c$

35. Write an expression for the relativistic Lagrangian and Hamiltonian of a simple harmonic oscillator in two dimensions and find the equations of motion.

36. Given the Lagrangian for the electromagnetic field

$$L = -\frac{1}{4} F_{\mu\nu} F^{\mu\nu} - J^\mu A_\mu$$

show that

$$\frac{\partial}{\partial x^\mu} \frac{\partial L}{\partial A_{\mu,\nu}} - \frac{\partial L}{\partial A_\mu} = 0$$

gives Maxwell's equations in the form

$$\partial_\mu F^{\mu\nu} = J^\mu$$

where $A_{\mu,\nu} = \partial A_\mu/\partial x^\nu$

15 GENERAL RELATIVITY

Chapter

Chapter Outline

15.1 THE EQUIVALENCE PRINCIPLE

The results of special relativity are extended in the general theory that includes acceleration and gravity. The laws of physics are the same in all inertial and accelerated reference frames. There is no way to distinguish an accelerating frame from a frame at rest in a gravitational field as stated by the equivalence principle. Because of the equivalence between gravitational and inertial mass, a small mass m will experience the same acceleration as a large mass M released in a gravitational field where $ma = mg \rightarrow a = g$ and $Ma = Mg \rightarrow a = g$.

A famous thought experiment illustrates an important consequence of the equivalence principle. Observer 1 is in an upward accelerated elevator in the absence of gravity. Observer 2 is in an identical elevator at rest in a gravitational field. Observer 1 releases a mass that appears to fall toward the floor. Observer 2 releases an identical mass that falls in the gravitational field. The measurements of Observers 1 and 2 are indistinguishable. Observer 1 now projects a laser horizontally across the elevator. The beam strikes slightly lower on the opposite wall since the elevator has traveled upward during the travel time of the laser beam. Observer 2 repeats the experiment and obtains the same results. It is concluded that the light ray is deviated by gravity.

15.1.1 Classical Approximation to Gravitational Redshift

Consider a photon of initial energy $E_i = hf$ projected upward from the surface of the earth. When the photon reaches a height H it has an energy $E_f = m_{eff} gH + hf'$ where the effective mass

$$m_{eff} = \frac{hf}{c^2} \tag{15.1.1}$$

Applying conservation of energy $E_i = E_f$

$$hf = \frac{hf}{c^2} gH + hf' \tag{15.1.2}$$

so that the frequency at height H has decreased

$$f' = f\left(1 - \frac{gH}{c^2}\right) \tag{15.1.3}$$

From the relation $\lambda f = c$

$$\frac{c}{\lambda'} = \frac{c}{\lambda}\left(1 - \frac{gH}{c^2}\right) \tag{15.1.4}$$

we find that the wavelength is redshifted $\lambda' > \lambda$ at height H

$$\lambda' = \frac{\lambda}{\left(1 - \frac{gH}{c^2}\right)} \tag{15.1.5}$$

15.1.2 Photon Emitted from a Spherical Star

Now consider a photon of frequency f emitted from the photosphere of a spherical star of radius R and mass M. The gravitational potential energy a distance r from the center of the star is

$$U_g(r) = -m_{eff}\frac{GM}{r} \tag{15.1.6}$$

where $r \geq R$. At the surface of the star

$$U_g(R) = -m_{eff}\frac{GM}{R} \tag{15.1.7}$$

and our energy conservation equation is

$$hf - \left(\frac{hf}{c^2}\right)\frac{GM}{R} = hf' - \left(\frac{hf}{c^2}\right)\frac{GM}{r} \tag{15.1.8}$$

Far from the star $r \gg R$

$$f' = f\left(1 - \frac{GM}{Rc^2}\right) \tag{15.1.9}$$

This result differs from the correct expression obtained from general relativity

$$f' = f\left(1 - \frac{2GM}{Rc^2}\right) \tag{15.1.10}$$

15.1.3 Gravitational Time Dilation

We can infer an expression for the time dilation in the gravitational field at the surface of a spherical body of radius R and mass M. Taking $\Delta t' = 1/f'$ and $\Delta t = 1/f$

$$\Delta t' = \frac{\Delta t}{\left(1 - \frac{2GM}{Rc^2}\right)} \tag{15.1.11}$$

for every second that ticks by on the surface of the gravitational mass $\left(1 - \frac{2GM}{Rc^2}\right)^{-1} > 1$ seconds tick by on an observer's watch far away from the mass. Hence time slows down near a massive body. Also, when $R = \frac{2GM}{c^2}$ we have $\left(1 - \frac{2GM}{Rc^2}\right)^{-1} \rightarrow \infty$. Evidently time stands still at the event horizon of a black hole relative to a distant observer. Note that gravitational time dilation is not symmetric as in the case of bodies in relative motion. In special relativity, observers in relative motion would see each other's clocks ticking slowly.

15.1.4 Comparison of Time Dilation Factors

Example 15.1.1

At what speed would gravitational time dilation be comparable to the time dilation experienced by observers in relative motion?

Solution: Equating the time dilation factors

$$\frac{1}{\sqrt{1-\left(\dfrac{v}{c}\right)^2}} = \frac{1}{1-\dfrac{2GM}{Rc^2}} \tag{15.1.12}$$

and solving for v

$$v = \sqrt{1-\left(1-\dfrac{2GM}{Rc^2}\right)^2}\, c \tag{15.1.13}$$

For sufficiently weak fields where $\dfrac{2GM}{Rc^2} \ll 1$ the binomial approximation gives

$$v \approx \sqrt{2}\, v_{\text{escape}} \text{ where } v_{\text{escape}} = \sqrt{\dfrac{2GM}{R}} \tag{15.1.14}$$

Signals from global positioning satellites must be corrected to account for both special and general relativistic time dilations.

15.2 TENSOR CALCULUS

General relativity offers a description of gravity as a curvature of spacetime. As John A. Wheeler succinctly summarizes: "Matter tells space how to curve and space tells matter how to move." The mathematical description of gravity in general relativity requires the use of tensors. Topics in this section include tensor notation, line element and spacetime interval, the metric tensor in Cartesian and spherical coordinates, the raising and lowering of indices, dot and cross product, the Levi-Civita tensor, transformation properties of tensors, the quotient rule and covariant derivatives. The metric tensor is used in the computation of covariant derivatives involving Christoffel symbols.

15.2.1 Tensor Notation

A tensor can be represented as an array of elements. The dimensionality of the array gives the rank (or order) of the tensor that can be determined by the number of indices. Scalars such as λ, β, etc., have no indices and are rank zero tensors. Vectors such as x^μ, A_i, etc., are tensors of rank one. Greek indices ($\mu, \nu, \alpha, \beta, \sigma$, etc.) usually have four values—e.g., $x^\mu = (x^0, x^1, x^2, x^3)$. Latin indices ($i, j, k, n, m$, etc.) usually correspond to three values—e.g., $A_i = (A_1, A_2, A_3)$.

Rank two tensors with two indices can be formed by the product of two rank one tensors. For example, the product of the contravariant vectors A^μ and B^ν gives the $N \times N$ matrix $C^{\mu\nu} = A^\mu B^\nu$ for $\mu = 0, 1, 2, 3$ and $\nu = 0, 1, 2, 3$; we have the 16 components

$$C^{\mu\nu} = \begin{pmatrix} A^0 B^0 & A^0 B^1 & A^0 B^2 & A^0 B^3 \\ A^1 B^0 & A^1 B^1 & A^1 B^2 & A^1 B^3 \\ A^2 B^0 & A^2 B^1 & A^2 B^2 & A^2 B^3 \\ A^3 B^0 & A^3 B^1 & A^3 B^2 & A^3 B^3 \end{pmatrix} \tag{15.2.1}$$

The trace of $C^{\mu\nu}$ is found by setting two indices equal $\mu = \nu$

$$\mathrm{Tr} C^{\mu\nu} = C^{\mu\mu} = A^0 B^0 + A^1 B^1 + A^2 B^2 + A^3 B^3 \tag{15.2.2}$$

where we sum over repeated indices.

15.2.2 Line Element and Spacetime Interval

The line element in Cartesian coordinates

$$d\ell^2 = dx^2 + dy^2 + dz^2 \tag{15.2.3}$$

with spacetime interval

$$ds^2 = c^2 dt^2 - dx^2 - dy^2 - dz^2 \tag{15.2.4}$$

is also expressed as

$$ds^2 = \left(dx^0\right)^2 - \left(dx^1\right)^2 - \left(dx^2\right)^2 - \left(dx^3\right)^2 \tag{15.2.5}$$

where $x^0 = ct, x^1 = x, x^2 = y, x^3 = z$.
The spacetime interval may be written in tensor notation

$$ds^2 = g_{\mu\nu} dx^\mu dx^\nu \tag{15.2.6}$$

where $g_{\mu\nu}$ is the metric tensor. In flat Minkowski spacetime, $g_{\mu\nu} = \mathrm{diag}(1, -1, -1, -1)$ or

$$g_{\mu\nu} = \begin{pmatrix} 1 & 0 & 0 & 0 \\ 0 & -1 & 0 & 0 \\ 0 & 0 & -1 & 0 \\ 0 & 0 & 0 & -1 \end{pmatrix} \tag{15.2.7}$$

with signature $(+, -, -, -)$. The signature $(-, +, +, +)$ is also common. The flat space metric is often written as $\eta_{\mu\nu}$. It is also common to refer to the spacetime interval as a "line element."

15.2.3 Raising and Lowering Indices

A covariant vector has a lower index such as $A_\mu = (A_0, A_1, A_2, A_3)$
A contravariant vector is written with an upper index $B^\mu = (B^0, B^1, B^2, B^3)$
The index of a contravariant vector may be lowered using the metric tensor

$$A_\mu = g_{\mu\nu} A^\nu \tag{15.2.8}$$

To raise the index of a covariant vector

$$A^\nu = g^{\mu\nu} A_\mu \tag{15.2.9}$$

where the inverse of $g_{\mu\nu}$ is written with superscripts $g^{\mu\nu}$. It has the property that

$$g_{\mu\sigma} g^{\sigma\nu} = \delta^\nu_\mu \tag{15.2.10}$$

where δ^ν_μ is the Kronecker delta function. For an arbitrary vector $A^\mu = (A^0, A^1, A^2, A^3)$

$$A_0 = A^0 \quad A_1 = -A^1 \quad A_2 = -A^2 \quad A_3 = -A^3 \tag{15.2.11}$$

15.2.4 Metric Tensor in Spherical Coordinates

The line element in spherical coordinates is
$$d\ell^2 = dr^2 + r^2 d\theta^2 + r^2 \sin^2\theta d\phi^2 \tag{15.2.12}$$
with spacetime interval
$$ds^2 = c^2 dt^2 - dr^2 - r^2 d\theta^2 - r^2 \sin^2\theta d\phi^2 \tag{15.2.13}$$
In spherical coordinates, the metric tensor is

$$g_{\mu\nu} = \begin{pmatrix} 1 & 0 & 0 & 0 \\ 0 & -1 & 0 & 0 \\ 0 & 0 & -r^2 & 0 \\ 0 & 0 & 0 & -r^2\sin^2\theta \end{pmatrix} \tag{15.2.14}$$

with inverse

$$g^{\mu\nu} = \begin{pmatrix} 1 & 0 & 0 & 0 \\ 0 & -1 & 0 & 0 \\ 0 & 0 & -\dfrac{1}{r^2} & 0 \\ 0 & 0 & 0 & -\dfrac{1}{r^2\sin^2\theta} \end{pmatrix} \tag{15.2.15}$$

so that $g_{\mu\sigma} g^{\sigma\nu} = \delta_{\mu}^{\nu}$

15.2.5 Dot Product

The metric tensor may be used to form the dot product between two vectors A^{μ} and B^{μ}

$$\mathbf{A}\cdot\mathbf{B} = g_{\mu\nu} A^{\nu} B^{\mu} = A_{\mu} B^{\mu} = A_0 B^0 - A_1 B^1 - A_2 B^2 - A_3 B^3 \qquad (15.2.16)$$

15.2.6 Cross Product

The cross product of vectors $\mathbf{C} = (C_1, C_2, C_3)$ and $\mathbf{D} = (D_1, D_2, D_3)$ may be expressed in tensor notion using the Levi-Civita tensor

$$(\mathbf{C}\times\mathbf{D})_i = \varepsilon_{ijk} C_j D_k \qquad (15.2.17)$$

In three dimensions, the Levi-Civita tensor can be shown as

$$\varepsilon_{ijk} = \begin{cases} 1 & \text{even permutations} \\ -1 & \text{odd permutations} \\ 0 & \text{repeated indices} \end{cases} \qquad (15.2.18)$$

For even permutations $\varepsilon_{123} = \varepsilon_{231} = \varepsilon_{312} = 1$, for odd permutations $\varepsilon_{321} = \varepsilon_{213} = \varepsilon_{132} = -1$ and for repeated indices $\varepsilon_{311} = \varepsilon_{223} = $ etc. $= 0$. The Levi-Civita tensor can be generalized to N dimensions. The four-dimensional Levi-Civita tensor $\varepsilon_{\mu\nu\sigma\gamma}$ may be used to form a tensor such as

$$S_{\mu\nu} = \varepsilon_{\mu\nu\sigma\gamma} A^{\sigma} B^{\gamma} \qquad (15.2.19)$$

For example, the component

$$S_{12} = A^3 B^4 - A^4 B^3 \qquad (15.2.20)$$

15.2.7 Transformation Properties of Tensors

Components of a contravariant vector A^{ν} expressed in a primed coordinate system are

$$A^{\mu'} = \frac{\partial x^{\mu'}}{\partial x^{\nu}} A^{\nu} \qquad (15.2.21)$$

The components of a covariant vector A_{ν} transform as

$$A_{\mu'} = \frac{\partial x^{\nu}}{\partial x^{\mu'}} A_{\nu} \qquad (15.2.22)$$

Contravariant, covariant and mixed tensors of rank two transform as

$$T^{\mu'\nu'} = \frac{\partial x^{\mu'}}{\partial x^{\mu}} \frac{\partial x^{\nu'}}{\partial x^{\nu}} T^{\mu\nu} \quad \text{(contravariant)} \tag{15.2.23}$$

$$T_{\mu'\nu'} = \frac{\partial x^{\mu}}{\partial x^{\mu'}} \frac{\partial x^{\nu}}{\partial x^{\nu'}} T_{\mu\nu} \quad \text{(covariant)} \tag{15.2.24}$$

$$T^{\mu'}_{\nu'} = \frac{\partial x^{\mu'}}{\partial x^{\mu}} \frac{\partial x^{\nu}}{\partial x^{\nu'}} T^{\mu}_{\nu} \quad \text{(mixed)} \tag{15.2.25}$$

Mixed tensors of arbitrary rank transform as

$$T^{\mu'\nu'\sigma'\cdots}_{\alpha'\beta'\gamma'\cdots} = \frac{\partial x^{\mu'}}{\partial x^{\mu}} \frac{\partial x^{\nu'}}{\partial x^{\nu}} \frac{\partial x^{\sigma'}}{\partial x^{\sigma}} \cdots \frac{\partial x^{\alpha}}{\partial x^{\alpha'}} \frac{\partial x^{\beta}}{\partial x^{\beta'}} \frac{\partial x^{\gamma}}{\partial x^{\gamma'}} \cdots T^{\mu\nu\sigma\cdots}_{\alpha\beta\gamma\cdots} \tag{15.2.26}$$

Objects that do not transform in accordance with the above rules are nontensors.

15.2.8 Quotient Rule for Tensors

Given that the contraction $B^{\lambda}A_{\lambda\mu\nu}$ is a tensor for any arbitrary vector B^{λ}, then $A_{\lambda\mu\nu}$ is a tensor. The quotient rule for tensors holds if $A_{\lambda\mu\nu}$ is replaced by a quantity with any number of upper or lower indices. If the contraction of an entity $A^{\mu\nu\sigma\cdots}_{\alpha\beta\gamma\cdots}$ with an arbitrary tensor produces another tensor, then $A^{\mu\nu\sigma\cdots}_{\alpha\beta\gamma\cdots}$ is a tensor.

Example 15.2.1

Given that the dot product between two vectors is invariant under a coordinate transformation

$$\mathbf{A} \cdot \mathbf{B} = g_{\mu\nu} A^{\nu} B^{\mu} = g_{\mu'\nu'} A^{\nu'} B^{\mu'} \tag{15.2.27}$$

show that $g_{\mu\nu}$ is a tensor

Solution: We write

$$g_{\mu\nu} A^{\nu} B^{\mu} = g_{\mu'\nu'} \frac{\partial x^{\nu'}}{\partial x^{\nu}} A^{\nu} \frac{\partial x^{\mu'}}{\partial x^{\mu}} B^{\mu} = g_{\mu'\nu'} \frac{\partial x^{\nu'}}{\partial x^{\nu}} \frac{\partial x^{\mu'}}{\partial x^{\mu}} A^{\nu} B^{\mu} \tag{15.2.28}$$

This must hold for any arbitrary A^{μ} and B^{μ} so that

$$g_{\mu\nu} = g_{\mu'\nu'} \frac{\partial x^{\nu'}}{\partial x^{\nu}} \frac{\partial x^{\mu'}}{\partial x^{\mu}} \tag{15.2.29}$$

Thus $g_{\mu\nu}$ is a tensor.

15.2.9 Covariant Derivatives

With metric signature $(+, -, -, -)$, the derivative operator with respect to covariant components is

$$\partial^\mu = \frac{\partial}{\partial x_\mu} = \left(\frac{\partial}{\partial x^0}, -\frac{\partial}{\partial x^1}, -\frac{\partial}{\partial x^2}, -\frac{\partial}{\partial x^3} \right) = \left(\frac{\partial}{\partial x^0}, -\nabla \right) \qquad (15.2.30)$$

The derivative operator with respect to contravariant components is

$$\partial_\mu = \frac{\partial}{\partial x^\mu} = \left(\frac{\partial}{\partial x^0}, \frac{\partial}{\partial x^1}, \frac{\partial}{\partial x^2}, \frac{\partial}{\partial x^3} \right) = \left(\frac{\partial}{\partial x^0}, \nabla \right) \qquad (15.2.31)$$

The gradient of a scalar function S with respect to a contravariant component gives a covariant vector

$$\frac{\partial S}{\partial x^\mu} = \partial_\mu S \qquad (15.2.32)$$

This is written even more compactly using a comma (,) shorthand notation $\partial_\mu S = S_{,\mu}$.

The transformation rule for a covariant vector is

$$A_{\mu'} = \frac{\partial x^\nu}{\partial x^{\mu'}} A_\nu \qquad (15.2.33)$$

If we take the derivative of $A_{\mu'}$ with respect to $x^{\lambda'}$

$$\frac{\partial A_{\mu'}}{\partial x^{\lambda'}} = \frac{\partial x^\nu}{\partial x^{\mu'}} \frac{\partial x^\rho}{\partial x^{\lambda'}} \frac{\partial A_\mu}{\partial x^\rho} + \frac{\partial^2 x^\nu}{\partial x^{\lambda'} \partial x^{\mu'}} A_\nu \qquad (15.2.34)$$

or using the comma notation

$$A_{\mu',\lambda'} = x^\nu_{,\mu'} x^\rho_{,\lambda'} A_{\mu,\rho} + x^\nu_{,\lambda',\mu'} A_\nu \qquad (15.2.35)$$

Because of the second term, the derivative $A_{\mu',\lambda'}$ does not transform as a tensor. We define the covariant derivative D_ν of a covariant vector

$$D_\nu A_\mu = A_{\mu,\nu} - \Gamma^\sigma_{\mu\nu} A_\sigma \qquad (15.2.36)$$

where the Christoffel symbol of the second kind

$$\Gamma^\sigma_{\mu\nu} = g^{\sigma r} \Gamma_{r\mu\nu} = \frac{1}{2} g^{\sigma r} \left(\frac{\partial g_{\mu r}}{\partial x^\nu} + \frac{\partial g_{\nu r}}{\partial x^\mu} - \frac{\partial g_{\mu\nu}}{\partial x^r} \right) \qquad (15.2.37)$$

is obtained from the Christoffel symbol of the first kind

$$\Gamma_{\sigma\mu\nu} = \frac{1}{2}\left(\frac{\partial g_{\mu\sigma}}{\partial x^{\nu}} + \frac{\partial g_{\nu\sigma}}{\partial x^{\mu}} - \frac{\partial g_{\mu\nu}}{\partial x^{\sigma}}\right)$$

(15.2.38)

by contraction with the metric tensor. The Christoffel symbols are nontensors. Examination of the above reveals $\Gamma_{\sigma\mu\nu} = \Gamma_{\sigma\nu\mu}$ and $\Gamma^{\sigma}_{\mu\nu} = \Gamma^{\sigma}_{\nu\mu}$. A colon (:) shorthand notation denotes a covariant derivative $D_{\nu}A_{\mu} = A_{\mu:\nu}$. The semicolon (;) is also a common notation for the covariant derivative. The covariant derivative of a contravariant vector is

$$A^{\mu}_{:\nu} = A^{\mu}_{,\nu} + \Gamma^{\mu}_{\sigma\nu}A^{\sigma}$$

(15.2.39)

The covariant derivative of a second rank tensor with covariant indices is

$$T_{\mu\nu:\sigma} = T_{\mu\nu,\sigma} - \Gamma^{\alpha}_{\nu\sigma}T_{\mu\alpha} - \Gamma^{\alpha}_{\mu\sigma}T_{\alpha\nu}$$

(15.2.40)

For a mixed tensor with upper and lower indices

$$T^{\mu}_{\nu:\sigma} = T^{\mu}_{\nu,\sigma} - \Gamma^{\alpha}_{\nu\sigma}T^{\mu}_{\alpha} + \Gamma^{\mu}_{\alpha\sigma}T^{\alpha}_{\nu}$$

(15.2.41)

In general

$$T^{\mu\cdots}_{\nu\cdots:\sigma} = T^{\mu\cdots}_{\nu\cdots,\sigma} - \Gamma \text{ term (for each lower index)} + \Gamma \text{ term (for each upper index)}$$

(15.2.42)

Maple Examples

Tensors are defined in the Maple worksheet below. Examples include the raising and lowering of indices, dot product, the Levi-Civita tensor, cross product, covariant derivatives and Christoffel symbols. The Maple command '$g\_[lineelement]$' outputs the spacetime interval for a given metric tensor. Note that the default metric tensor

$$g_{\mu\nu} = \begin{pmatrix} -1 & 0 & 0 & 0 \\ 0 & -1 & 0 & 0 \\ 0 & 0 & -1 & 0 \\ 0 & 0 & 0 & 1 \end{pmatrix}$$

in Maple corresponds to the spacetime interval

$$ds^2 = dt^2 - dx^2 - dy^2 - dz^2$$

with $c = 1$ and the indices μ, ν run from one to four and $g_{00} \rightarrow g_{44} = 1$. This convention is consistent with the signature used in this textbook where $g_{\mu\nu} = \text{diag}(1, -1, -1, -1)$. For a given metric tensor, Maple will simply output g_{44} if g_{00} is input at the Command line.

Key Maple commands:
convert, D_[mu], d_[mu], Define, g_[ ] , g_[lineelement] ,Geodesics, LeviCivita, *Simplify*, SumOverRepeatedIndices
Maple packages: *with(Physics)*:

restart

with(Physics) :

setup(mathematicalnotation = true)

$$setup[mathematicalnotation = true]$$

Metric Tensor for Minkowski Space

g_[]

$$g_{\mu\nu} = \begin{pmatrix} -1 & 0 & 0 & 0 \\ 0 & -1 & 0 & 0 \\ 0 & 0 & -1 & 0 \\ 0 & 0 & 0 & 1 \end{pmatrix}$$

g_[alpha, beta]·g_[~beta, ~nu]

$$g_{\alpha\beta}\, g^{\beta,\nu}$$

Simplify(%)

$$g_{\alpha}^{\;\nu}$$

Spacetime Interval from Metric Tensor

g_[lineelement]

Systems of spacetime Coordinates are: $\{X = (x1, x2, x3, x4)\}$

$$-\partial(x1)^2 - \partial(x2)^2 - \partial(x3)^2 + \partial(x4)^2$$

Define Tensors

Define(A, B, N, S)

Defined objects with tensor properties

$$\{A, B, N, S, \gamma_\mu, \sigma_\mu, X_\mu, \partial_\mu, g_{\mu,\nu}, \delta_{\mu,\nu}, \in_{\alpha,\beta,\mu,\nu}\}$$

Lowering the Index of a Contravariant Vector

g_[mu, nu]·A_[~nu]

$$g_{\mu,\nu}\, A^\nu$$

Simplify(%)

$$A_\mu$$

Raising the Index of a Covariant Vector

g_[~mu, ~nu]·A[nu]

$$g^{\mu,\nu} A_\nu$$

Simplify(%)

$$A^\mu$$

Second Rank Mixed Tensor

B[mu]·A[~nu]

$$B_\mu A^\nu$$

Dot Product

g_[mu, nu]·B[~nu]·A[~mu]

$$g_{\mu,\nu} B^\nu A^\mu$$

SumOverRepeatedIndices(%)

$$-A^1 B^1 - A^2 B^2 - A^3 B^3 + A^4 B^4$$

4-Index Levi-Civita Symbols

LeviCivita(a, b, c, d)

$$\epsilon_{a, b, c, d}$$

LeviCivita(~a, ~b, ~c, ~d)

$$\epsilon^{a, b, c, d}$$

LeviCivita[1, 2, 3, 4]

$$1$$

LeviCivita[~1, ~2, ~3, ~4]

$$-1$$

LeviCivita[2, 1, 3, 4]

$$-1$$

LeviCivita[1, 1, 3, 4]

$$0$$

LeviCivita[1, 2, 3, 3]

$$0$$

3-Index Levi-Civita Symbols

$LeviCivita[i, j, k]$

$$\epsilon_{i, j, k}$$

$LeviCivita[1, 2, 3]$

$$1$$

$LeviCivita[3, 2, 1]$

$$-1$$

$LeviCivita[1, 2, 2]$

$$0$$

Cross Product

$LeviCivita[1, 2, sigma, gamma] \cdot A[\sim sigma] \cdot B[\sim gamma]$

$$-\epsilon_{1,2,\gamma,\sigma}\, A^\sigma\, B^\gamma$$

$SumOverRepeatedIndices(\%)$

$$A^3 B^4 - A^4 B^3$$

Tolman Metric

$g\_[Tolman]$

The Tolman metric in spherical coordinates

Default differentiation variables for $d\_$, $D\_$ and dAlembertian are: $\{X = (r, \theta, \phi, t)\}$

Systems of spacetime Coordinates are: $\{X = (r, \theta, \phi, t)\}$

$$g_{\mu\nu} = \begin{pmatrix} -\dfrac{\left(\dfrac{\partial}{\partial r}R(t,r)\right)^2}{1+2E(r)} & 0 & 0 & 0 \\ 0 & -R(t,r)^2 & 0 & 0 \\ 0 & 0 & -R(t,r)^2 \sin(\theta)^2 & 0 \\ 0 & 0 & 0 & 1 \end{pmatrix}$$

Spacetime Interval from Metric Tensor

$g\_[lineelement]$

$$\frac{-\left(\dfrac{\partial}{\partial r}R(t,r)\right)^2 \partial(r)^2 - (1+2E(r))R(t,r)^2\left(\sin(\theta)^2\,\partial(\phi)^2 + \partial(\theta)^2\right)}{1+2E(r)} + \partial(t)^2$$

Derivative of a Scalar

$d\_[mu](S(X))$

$$\partial_\mu(S(X))$$

Derivative of a Vector

$d\_(A[mu](X))$

$$\partial_\nu(A_\mu(X))\partial(X^\nu)$$

Covariant Derivative of a Scalar

$D\_[mu](S(X))$

$$\partial_\mu(S(X))$$

Covariant Derivative of a Covariant Vector

$D\_[mu](A[sigma](X))$

$$\mathcal{D}_\mu(A_\sigma(X))$$

$convert(\%, d\_)$

$$\partial_\mu\left(A_\sigma(X)\right) - \Gamma^\nu_{\mu,\sigma}A_\nu(X)$$

Covariant Derivative of a Contravariant Vector

$D\_[mu](A[\sim sigma](X))$

$$\mathcal{D}_\mu(A^\sigma(X))$$

$convert(\%, d\_)$

$$\partial_\mu\left(A^\sigma(X)\right) + \Gamma^\sigma_{\mu,\nu}A^\nu(X)$$

Covariant Derivative of a Mixed Tensor

$D\_[mu](N[gamma, \sim sigma](X))$

$$\mathcal{D}_\mu\left(N^\sigma_\gamma(X)\right)$$

$convert(\%, d\_)$

$$\partial_\mu\left(N^\sigma_\gamma(X)\right) - \Gamma^\nu_{\gamma,\mu}N^\sigma_\nu(X) + \Gamma^\sigma_{\mu,\nu}N^\nu_\gamma(X)$$

15.3 EINSTEIN'S EQUATIONS

The geodesic equations of motion of a body in curved spacetime are first developed in this section. The geodesic equations of motion are obtained from the Schwarzschild metric. Next, the Ricci tensor, Ricci scalar and the Einstein tensor are obtained from the Riemann curvature tensor. Einstein's field equations are obtained from the Einstein tensor and the stress energy tensor.

15.3.1 Geodesic Equations of Motion

Example 15.3.1

Find the geodesic equations of motion by requiring that the variation of the action integral is zero or

$$\delta S = \delta \int ds = \delta \int \frac{ds}{d\tau} d\tau = \delta \int \sqrt{g_{\mu\nu} \frac{dx^\mu}{d\tau} \frac{dx^\nu}{d\tau}} d\tau = 0 \qquad (15.3.1)$$

where the Lagrangian

$$L = \sqrt{g_{\mu\nu} \frac{dx^\mu}{d\tau} \frac{dx^\nu}{d\tau}} \qquad (15.3.2)$$

Solution: Making the substitutions $\dot{x}^\mu = \dfrac{dx^\mu}{d\tau}$, $\dot{x}^\nu = \dfrac{dx^\nu}{d\tau}$ and $\eta = g_{\mu\nu}\dot{x}^\mu\dot{x}^\nu$ into L we calculate

$$\frac{d}{d\tau}\frac{\partial L}{\partial \dot{x}^\mu} - \frac{\partial L}{\partial x^\mu} = 0 \qquad (15.3.3)$$

Evaluating the derivatives

$$\frac{\partial L}{\partial x^\mu} = \frac{1}{2}\eta^{-1/2} g_{\rho\nu,\mu}\dot{x}^\rho \dot{x}^\nu \qquad (15.3.4)$$

$$\frac{\partial L}{\partial \dot{x}^\mu} = \frac{1}{2}\eta^{-1/2}\frac{\partial \eta}{\partial \dot{x}^\mu} = \eta^{-1/2} g_{\mu\nu}\dot{x}^\nu \qquad (15.3.5)$$

$$\frac{d}{d\tau}\frac{\partial L}{\partial \dot{x}^\mu} = \frac{d}{d\tau}\left(\eta^{-1/2} g_{\mu\nu}\dot{x}^\nu\right) = \eta^{-1/2} g_{\mu\nu}\ddot{x}^\nu + \eta^{-1/2} g_{\mu\nu,\rho}\dot{x}^\rho \dot{x}^\nu - \frac{1}{2}\eta^{-1}\dot{\eta} g_{\mu\nu}\dot{x}^\nu \quad (15.3.6)$$

where application of the chain rule above $\dfrac{\partial g_{\mu\nu}}{\partial \tau} = \dfrac{\partial g_{\mu\nu}}{\partial x^\rho}\dfrac{\partial x^\rho}{\partial \tau} = g_{\mu\nu,\rho}\dot{x}^\rho$.

Writing $g_{\mu\nu,\rho}$ as the average $(g_{\mu\nu,\rho} + g_{\mu\rho,\nu})/2$ the Euler-Lagrange equation becomes

$$\eta^{-1/2} g_{\mu\nu}\ddot{x}^\mu + \eta^{-1/2}\frac{1}{2}\left(g_{\mu\nu,\rho} + g_{\mu\rho,\nu}\right)\dot{x}^\rho \dot{x}^\nu - \frac{1}{2}\eta^{-1/2} g_{\rho\nu,\mu}\dot{x}^\rho \dot{x}^\nu = \frac{1}{2}\eta^{-1}\dot{\eta} g_{\mu\nu}\dot{x}^\nu$$

$$(15.3.7)$$

Now $\dot{\eta}$ is proportional to $\dot{L} = 0$. After multiplying by $\sqrt{\eta}g^{\sigma\mu}$ we have

$$g^{\sigma\mu}g_{\mu\nu}\ddot{x}^{\mu} + \frac{1}{2}g^{\sigma\mu}\left(g_{\mu\nu,\rho} + g_{\mu\rho,\nu} - g_{\rho\nu,\mu}\right)\dot{x}^{\rho}\dot{x}^{\nu} = 0, \tag{15.3.8}$$

Since $g^{\sigma\mu}g_{\mu\nu} = \delta^{\sigma}_{\nu}$ and $\dot{x}^{\rho}\dot{x}^{\nu}$ is contracted with a Christoffel symbol of the second kind, the geodesic equations are

$$\frac{d^2 x^{\sigma}}{d\tau^2} + \Gamma^{\sigma}_{\rho\nu}\frac{dx^{\rho}}{d\tau}\frac{dx^{\nu}}{d\tau} = 0 \tag{15.3.9}$$

These can be written as a first order system

$$\frac{dv^{\sigma}}{d\tau} + \Gamma^{\sigma}_{\rho\nu}v^{\rho}v^{\nu} = 0 \tag{15.3.10}$$

For a particle moving in both gravitational and electromagnetic fields, the geodesic equations of motion are

$$\frac{d^2 x^{\sigma}}{d\tau^2} + \Gamma^{\sigma}_{\rho\nu}\frac{dx^{\rho}}{d\tau}\frac{dx^{\nu}}{d\tau} = \frac{q}{m}F^{\sigma\nu}\frac{dx^{\rho}}{d\tau}g_{\rho\nu} \tag{15.3.11}$$

15.3.2 Alternative Lagrangian

The geodesic equations of motions may also be obtained if we require

$$\delta\int g_{\mu\nu}\frac{dx^{\mu}}{d\tau}\frac{dx^{\nu}}{d\tau}d\tau = 0 \tag{15.3.12}$$

where our alternative Lagrangian $L' = g_{\mu\nu}\dot{x}^{\mu}\dot{x}^{\nu}$

Example 15.3.2

Find the geodesic equation directly from the Schwarzschild metric in spherical coordinates with spacetime interval

$$ds^2 = \left(1 - \frac{2GM}{c^2 r}\right)c^2 dt^2 - \left(1 - \frac{2GM}{c^2 r}\right)^{-1}dr^2 - r^2 d\theta^2 - r^2\sin^2\theta d\phi^2 \tag{15.3.13}$$

Solution: We obtain the geodesic equation from the alternate Lagrangian

$$L' = \left(1 - \frac{r_s}{r}\right)c^2\dot{t}^2 - \left(1 - \frac{r_s}{r}\right)^{-1}\dot{r}^2 - r^2\dot{\theta}^2 - r^2\sin^2\theta\dot{\phi}^2 \tag{15.3.14}$$

where $r_s = 2GM/c^2$. The Euler-Lagrange equation of motion for the t-equation is

$$\frac{d}{d\tau}\frac{dL'}{d\dot{t}} - \frac{dL'}{dt} = 0$$

$$-2\frac{d}{d\tau}\left[\left(1-\frac{r_s}{r}\right)c^2\dot{t}\right] = 0$$

(15.3.15)

Thus, we have a constant of the motion proportional to the energy E

$$\left(1-\frac{r_s}{r}\right)\dot{t} = \frac{E}{mc^2}$$

(15.3.16)

The r-equation is

$$\frac{d}{d\tau}\frac{dL'}{d\dot{r}} - \frac{dL'}{dr} = 0$$

$$-2\frac{d}{d\tau}\left[\left(1-\frac{r_s}{r}\right)^{-1}\dot{r}\right] - \left[\frac{r_s}{r^2}c^2\dot{t}^2 + \left(1-\frac{r_s}{r}\right)^{-2}\frac{r_s}{r^2}\dot{r}^2 - 2r\dot{\theta}^2 - 2r\sin^2\theta\dot{\phi}^2\right] = 0 \quad (15.3.17)$$

$$2\left(1-\frac{r_s}{r}\right)^{-1}\ddot{r} + \frac{r_s}{r^2}c^2\dot{t}^2 - \left(1-\frac{r_s}{r}\right)^{-2}\frac{r_s}{r^2}\dot{r}^2 - 2r\dot{\theta}^2 - 2r\sin^2\theta\dot{\phi}^2 = 0$$

The θ-equation is

$$\frac{d}{d\tau}\frac{dL'}{d\dot{\theta}} - \frac{dL'}{d\theta} = 0$$

$$r^2\ddot{\theta} + 2r\dot{r}\dot{\theta} - \sin\theta\cos\theta r^2\dot{\phi}^2 = 0$$

(15.3.18)

The ϕ-equation is

$$\frac{d}{d\tau}\frac{dL'}{d\dot{\phi}} - \frac{dL'}{d\phi} = 0$$

$$\frac{d}{d\tau}\left(2r^2\sin^2\theta\dot{\phi}\right) = 0$$

(15.3.19)

with a constant of the motion proportional to the angular momentum ℓ

$$\ell = mr^2\sin^2\theta\dot{\phi}$$

(15.3.20)

15.3.3 Riemann Curvature Tensor

To form the Reimann curvature tensor, we subtract the second covariant derivatives

$$A_{\mu:\nu:\alpha} - A_{\mu:\alpha:\nu} = \left(A_{\mu,\nu} - \Gamma^\sigma_{\mu\nu}A_\sigma\right)_{:\alpha} - \left(A_{\mu,\alpha} - \Gamma^\sigma_{\mu\alpha}A_\sigma\right)_{:\nu}$$

(15.3.21)

Evaluating the derivatives

$$
\begin{aligned}
A_{\mu:\nu:\alpha} &= \left(A_{\mu,\nu} - \Gamma^{\sigma}_{\mu\nu}A_{\sigma}\right)_{,\nu} - \Gamma^{\beta}_{\mu\alpha}\left(A_{\beta,\nu} - \Gamma^{\sigma}_{\beta\nu}A_{\sigma}\right) - \Gamma^{\beta}_{\nu\alpha}\left(A_{\mu,\beta} - \Gamma^{\sigma}_{\mu\beta}A_{\sigma}\right) \\
&= A_{\mu,\nu,\alpha} - \Gamma^{\sigma}_{\mu\nu}A_{\sigma,\alpha} - \Gamma^{\beta}_{\mu\alpha}A_{\beta,\nu} - \Gamma^{\beta}_{\nu\alpha}A_{\mu,\beta} + \left(\Gamma^{\beta}_{\mu\alpha}\Gamma^{\sigma}_{\beta\nu} + \Gamma^{\beta}_{\nu\alpha}\Gamma^{\sigma}_{\mu\beta} - \Gamma^{\sigma}_{\mu\nu,\alpha}\right)A_{\sigma}
\end{aligned}
$$

$$(15.3.22)$$

while

$$
\begin{aligned}
A_{\mu:\alpha:\nu} &= \left(A_{\mu,\alpha} - \Gamma^{\sigma}_{\mu\alpha}A_{\sigma}\right)_{,\nu} - \Gamma^{\beta}_{\mu\nu}\left(A_{\beta,\alpha} - \Gamma^{\sigma}_{\beta\alpha}A_{\sigma}\right) - \Gamma^{\beta}_{\alpha\nu}\left(A_{\mu,\beta} - \Gamma^{\sigma}_{\mu\beta}A_{\sigma}\right) \\
&= A_{\mu,\alpha,\nu} - \Gamma^{\sigma}_{\mu\alpha}A_{\sigma,\nu} - \Gamma^{\beta}_{\mu\nu}A_{\beta,\alpha} - \Gamma^{\beta}_{\alpha\nu}A_{\mu,\beta} + \left(\Gamma^{\beta}_{\mu\nu}\Gamma^{\sigma}_{\beta\alpha} + \Gamma^{\beta}_{\alpha\nu}\Gamma^{\sigma}_{\mu\beta} - \Gamma^{\sigma}_{\mu\alpha,\nu}\right)A_{\sigma}
\end{aligned}
$$

$$(15.3.23)$$

Subtracting expression (15.3.23) from (15.3.22)

$$
\begin{aligned}
A_{\mu:\nu:\alpha} - A_{\mu:\alpha:\nu} &= A_{\mu,\nu,\alpha} - \Gamma^{\sigma}_{\mu\nu}A_{\sigma,\alpha} - \Gamma^{\beta}_{\mu\alpha}A_{\beta,\nu} - \Gamma^{\beta}_{\mu\alpha}A_{\mu,\beta} \\
&+ \left(\Gamma^{\beta}_{\mu\alpha}\Gamma^{\sigma}_{\beta\nu} + \Gamma^{\beta}_{\nu\alpha}\Gamma^{\sigma}_{\mu\beta} - \Gamma^{\sigma}_{\mu\nu,\alpha}\right)A_{\sigma} - \left(A_{\mu,\alpha,\nu} - \Gamma^{\sigma}_{\mu\alpha}A_{\sigma,\nu} - \Gamma^{\beta}_{\mu\nu}A_{\beta,\alpha} - \Gamma^{\beta}_{\alpha\nu}A_{\mu,\beta}\right. \\
&+ \left.\left(\Gamma^{\beta}_{\mu\nu}\Gamma^{\sigma}_{\beta\alpha} + \Gamma^{\beta}_{\alpha\nu}\Gamma^{\sigma}_{\mu\beta} - \Gamma^{\sigma}_{\mu\alpha,\nu}\right)A_{\sigma}\right)
\end{aligned}
$$

$$(15.3.24)$$

and canceling terms

$$
A_{\mu:\nu:\alpha} - A_{\mu:\alpha:\nu} = \left(\Gamma^{\beta}_{\mu\alpha}\Gamma^{\sigma}_{\beta\nu} - \Gamma^{\beta}_{\mu\nu}\Gamma^{\sigma}_{\beta\alpha} - \Gamma^{\sigma}_{\mu\nu,\alpha} + \Gamma^{\sigma}_{\mu\alpha,\nu}\right)A_{\sigma} \tag{15.3.25}
$$

$$
A_{\mu:\nu:\alpha} - A_{\mu:\alpha:\nu} = R^{\sigma}_{\mu\nu\alpha}A_{\sigma} \tag{15.3.26}
$$

where the Reimann tensor is

$$
R^{\sigma}_{\mu\nu\alpha} = \Gamma^{\beta}_{\mu\alpha}\Gamma^{\sigma}_{\beta\nu} - \Gamma^{\beta}_{\mu\nu}\Gamma^{\sigma}_{\beta\alpha} - \Gamma^{\sigma}_{\mu\nu,\alpha} + \Gamma^{\sigma}_{\mu\alpha,\nu} \tag{15.3.27}
$$

The Reimann tensor has 256 components with 20 independent components. All the components are zero in flat spacetime and may be nonzero in curved spacetime.

15.3.4 Ricci Tensor

The 16-component Ricci tensor $R_{\mu\nu}$ is obtained by contracting the first and last indices of the Reimann tensor.

$$
R_{\mu\nu} = R^{\sigma}_{\mu\nu\sigma} = R^{0}_{\mu\nu0} + R^{1}_{\mu\nu1} + R^{2}_{\mu\nu2} + R^{3}_{\mu\nu3} \tag{15.3.28}
$$

The Ricci tensor is symmetric as $R_{\mu\nu} = R_{\nu\mu}$.

15.3.5 Ricci Scalar

The Ricci scalar R is formed by contraction of the Ricci tensor

$$R = g^{\mu\nu} R_{\mu\nu} \qquad (15.3.29)$$

Both the Ricci tensor and Ricci scalar are zero in flat spacetime and may vanish in curved spacetime.

15.3.6 Einstein Tensor

The Einstein tensor $G_{\mu\nu}$ is composed of the Ricci tensor, metric tensor and the Ricci scalar

$$G_{\mu\nu} = R_{\mu\nu} - \frac{1}{2} g_{\mu\nu} R \qquad (15.3.30)$$

and has the property $D_\mu G_{\mu\nu} = G_{\mu\nu;\mu} = 0$.

15.3.7 Einstein's Field Equations

Written with covariant components, the field equations

$$R_{\mu\nu} - \frac{1}{2} g_{\mu\nu} R = \frac{8\pi G}{c^4} T_{\mu\nu} \qquad (15.3.31)$$

where the stress energy tensor $T_{\mu\nu}$ satisfies the continuity equation

$$D_\mu T_{\mu\nu} = T_{\mu\nu;\mu} = 0 \qquad (15.3.32)$$

15.3.8 Friedman Cosmology

In the Friedmann cosmology, the stress tensor is taken as that of a perfect fluid with energy density ρ and pressure p

$$T_{\mu\nu} = diag\left(\rho c^2, p, p, p\right) \qquad (15.3.33)$$

The spacetime interval is

$$ds^2 = c^2 dt^2 - a^2\left(t\right)\left[\frac{1}{1-kr^2} dr^2 - r^2 d\theta^2 - r^2 \sin^2\theta d\phi^2\right] \qquad (15.3.34)$$

where a is the scale factor related to the Hubble constant

$$H = \frac{\dot{a}}{a} \qquad (15.3.35)$$

Values of $k = -1, 0, 1$ correspond to open, flat and closed universes, respectively.

15.3.9 Killing Vectors

Vectors V_μ that satisfy Killing's equation

$$D_\mu V_\nu - D_\nu V_\mu = 0 \qquad (15.3.36)$$

are known as Killing vectors, pointing in directions that leave the metric tensor unchanged. The Killing vectors may also be written in contravariant form.

Maple Examples

Examples of Einstein's equations are given in the Maple worksheet below. Computed quantities include the geodesic equations of motion, Riemann tensor, Ricci tensor, and Ricci scalar for several metric tensors. Einstein's field equations are obtained from the Einstein tensor and the stress energy tensor.

Key Maple commands: *Christoffel, Define, dsolve, Einstein, g_[ ] , Geodesics, KillingVectors, Ricci, Riemann, Setup, SumOverRepeatedIndices*

Maple packages: *with(Physics)*:

Programming: for loops

restart

with(Physics) :

Setup(mathematicalnotation = true)

Schwarzschild Metric

$g\_[sc]$

Systems of spacetime Coordinates are: $\{X = (r, \theta, \phi, t)\}$
Default differentiation variables for d_, D_ and dAlembertian are: $\{X = (r, \theta, \phi, t)\}$
The Schwarzschild metric in coordinates $[r, \theta, \phi, t]$
Parameters: $[m]$

$$g_{\mu,\nu} = \begin{pmatrix} \dfrac{r}{-r+2m} & 0 & 0 & 0 \\ 0 & -r^2 & 0 & 0 \\ 0 & 0 & -r^2 \sin(\theta)^2 & 0 \\ 0 & 0 & 0 & \dfrac{r-2m}{r} \end{pmatrix}$$

Schwarzschild Metric: Christoffel Symbols

Christoffel[1, nu, alpha, *matrix*]

$$\Gamma_{1,\nu,\alpha} = \begin{pmatrix} \dfrac{m}{(-r+2m)^2} & 0 & 0 & 0 \\ 0 & r & 0 & 0 \\ 0 & 0 & r\sin(\theta)^2 & 0 \\ 0 & 0 & 0 & -\dfrac{m}{r^2} \end{pmatrix}$$

Christoffel[2, nu, alpha, *matrix*]

$$\Gamma_{2,\nu,\alpha} = \begin{pmatrix} 0 & -r & 0 & 0 \\ -r & 0 & 0 & 0 \\ 0 & 0 & r\sin(\theta)\cos(\theta) & 0 \\ 0 & 0 & 0 & 0 \end{pmatrix}$$

Christoffel[3, nu, alpha, *matrix*]

$$\Gamma_{3,\nu,\alpha} = \begin{pmatrix} 0 & 0 & -r\sin(\theta)^2 & 0 \\ 0 & 0 & -r^2\sin(\theta)\cos(\theta) & 0 \\ -r\sin(\theta)^2 & -r^2\sin(\theta)\cos(\theta) & 0 & 0 \\ 0 & 0 & 0 & 0 \end{pmatrix}$$

Christoffel[4, nu, alpha, *matrix*]

$$\Gamma_{4,\nu,\alpha} = \begin{pmatrix} 0 & 0 & 0 & \dfrac{m}{r^2} \\ 0 & 0 & 0 & 0 \\ 0 & 0 & 0 & 0 \\ \dfrac{m}{r^2} & 0 & 0 & 0 \end{pmatrix}$$

Geodesic Equations

Geodesics(*tensornotation*)

$$\frac{d^2}{d\tau^2}X^\mu(\tau) + \Gamma^\mu_{\alpha,\nu}\left(\frac{d}{d\tau}X^\nu(\tau)\right)\left(\frac{d}{d\tau}X^\alpha(\tau)\right)$$

Schwarzschild Geodesic Equations for Planar Orbits

$$subs\left(\left\{\frac{d}{d\tau}\theta(\tau)=0,\theta(\tau)=\frac{Pi}{2}\right\},Geodesics()[1]\right)$$

$$\frac{d^2}{d\tau^2}\phi(\tau)=-\frac{2\left(\frac{d}{d\tau}\phi(\tau)\right)\left(\frac{d}{d\tau}r(\tau)\right)}{r(\tau)}$$

$$simplify\left(subs\left(\left\{\frac{d}{d\tau}\theta(\tau)=0,\theta(\tau)=\frac{Pi}{2}\right\},Geodesics()[3]\right)\right)$$

$$\frac{d^2}{d\tau^2}r(\tau)=\frac{1}{(-r(\tau)+2m)r(\tau)^3}\left(-4r(\tau)^3\left(-\frac{r(\tau)}{2}+m\right)^2\left(\frac{d}{d\tau}\phi(\tau)\right)^2\right.$$

$$\left.+4m\left(-\frac{r(\tau)}{2}+m\right)^2\left(\frac{d}{d\tau}t(\tau)\right)^2-m\left(\frac{d}{d\tau}r(\tau)\right)^2r(\tau)^2\right)$$

$$subs\left(\left\{\frac{d}{d\tau}\theta(\tau)=0,\theta(\tau)=\frac{Pi}{2}\right\},Geodesics()[4]\right)$$

$$\frac{d^2}{d\tau^2}t(\tau)=-\frac{2m\left(\frac{d}{d\tau}r(\tau)\right)\left(\frac{d}{d\tau}t(\tau)\right)}{r(\tau)(r(\tau)-2m)}$$

Metric Search by Author

$Setup(metric = Robertson)$

[12, 9, 1] = ["Authors" = ["Robertson(1929, 1935, 1936)", "Walker(1936)"],
"PrimaryDescription" = "Generic", "SecondaryDescription" = ["FRW"], "Comments"
= ["k = 1. Use the orthonormal tetrad to calculate Killing vectors and then transform
to coordinate frame",
"The parameter _s is the dtdt component of the trace-free Ricci tensor. The side
condtion _s <> 0 fixes the Segre type"]]

[12, 9, 2] = ["Authors" = ["Robertson(1929, 1935, 1936)", "Walker(1936)"],
"PrimaryDescription" = "Generic", "SecondaryDescription" = ["FRW"], "Comments"
= ["k = 0", "The parameter _s is the dtdt component of the trace-free Ricci tensor. The
side condtion _s <> 0 fixes the Segre type"]]

[12, 9, 3] = ["Authors" = ["Robertson(1929, 1935, 1936)", "Walker(1936)"],
  "PrimaryDescription" = "Generic", "SecondaryDescription" = ["FRW"], "Comments"
  = ["k = -1. Use [i] the orthonormal tetrad and [ii] the coordinate ordering [t, phi,
  theta, r] to calculate Killing vectors. Then transform to coordinate frame", "The
  parameter _s is the dtdt component of the trace-free Ricci tensor. The side condtion _s
  <> 0 fixes the Segre type"]]

[12, 9, 4] = ["Authors" = ["Robertson(1929, 1935, 1936)", "Walker(1936)"],
  "PrimaryDescription" = "Generic", "SecondaryDescription" = ["FRWL, Friedmann-
  Lemaître-Robertson-Walker"], "Comments" = ["Reduced-circumference polar
  coordinates of the FLRW metric"]]

[12, 9, 5] = ["Authors" = ["Robertson(1929, 1935, 1936)", "Walker(1936)"],
  "PrimaryDescription" = "Generic" "SecondaryDescription" = ["FRWL, Friedmann-
  Lemaître-Robertson-Walker"], "Comments" = ["Hyperspherical coordinates, Case 1,
  _k > 0"]]

[12, 9, 6] = ["Authors" = ["Robertson(1929, 1935, 1936)", "Walker(1936)"],
  "PrimaryDescription"= "Generic", "SecondaryDescription" = ["FRWL, Friedmann-
  Lemaître-Robertson-Walker"], "Comments" = ["Hyperspherical coordinates,
  Case 2, _k = 0"]]

[12, 9, 7] = ["Authors" = ["Robertson(1929, 1935, 1936)", "Walker(1936)"],
  "PrimaryDescription" = "Generic", "SecondaryDescription" = ["FRWL, Friedmann-
  Lemaître-Robertson-Walker"], "Comments" = ["Hyperspherical coordinates, Case 3,
  _k < 0"]]

[37, 13, 1] = ["Authors" = ["Robertson-Walker"], "PrimaryDescription" = "Generic",
  "SecondaryDescription" = ["Robertson-Walker"]]

*Warning, found more than one match for the keyword 'Robertson', as seen
above. Please refine your 'keyword' or re-enter the metric 'g_[...]' with the
list of three numbers identifying the metric, for example as in g_[12, 9, 1] or
Setup(metric = [12, 9, 1])*

Selecting a Metric from the Maple Database

Setup(metric = [12, 9, 1])
  Systems of spacetime Coordinates are: $\{X = (t, r, \theta, \phi)\}$
Default differentiation variables for d_, D_ and dAlembertian are: $\{X = (t, r, \theta, \phi)\}$

The Robertson(1929, 1935, 1936), Walker(1936)metric in coordinates $[t, r, \theta, \phi]$
Parameters: $[c, k, a(t)]$
Comments : Hyperspherical coordinates, Case 3, _k < 0
Resetting the signature of spacetime from "... +" *to* '- + + +' *in order to match the signature in the database of metrics:*

$$
\left[metric = \left\{ (1,1) = -c^2, (2,2) = a(t)^2, (3,3) = \frac{a(t)^2 \sinh\left(r\sqrt{|k|}\right)^2}{|k|}, \right.\right.
$$

$$
\left.\left. (4,4) = -\frac{a(t)^2 \left(\cosh\left(r\sqrt{|k|}\right)^2 \cos(\theta)^2 - \cosh\left(r\sqrt{|k|}\right)^2 - \cos(\theta)^2 + 1\right)}{|k|} \right\}\right]
$$

Display the Metric in Matrix Form

$$g\_[]$$

$$
g_{\mu\nu} = \left[\left[-c^2, 0, 0, 0\right], \left[0, a(t)^2, 0, 0\right], \left[0, 0, \frac{a(t)^2 \sinh\left(r\sqrt{|k|}\right)^2}{|k|}, 0\right], \right.
$$

$$
\left. \left[0, 0, 0, -\frac{a(t)^2 \left(\cosh\left(r\sqrt{|k|}\right)^2 \cos(\theta)^2 - \cosh\left(r\sqrt{|k|}\right)^2 - \cos(\theta)^2 + 1\right)}{|k|}\right] \right]
$$

Kerr Metric: Rotating Black Holes

Setup(metric = [5, 29, 1])
Systems of spacetime Coordinates are: $\{X = (t, r, \theta, \phi)\}$
Default differentiation variables for d_, D_ and dAlembertian are: $\{X = (t, r, \theta, \phi)\}$
The Kerr(1963) metric in coordinates $[t, r, \theta, \phi]$
Parameters: $[a, m]$
Comments : Boyer-Lindquist coordinates

$$
\left[metric = \left\{ (1,1) = \frac{2mr - r^2 - a^2 \cos(\theta)^2}{r^2 + a^2 \cos(\theta)^2}, (1,4) = \frac{2mra \sin(\theta)^2}{r^2 + a^2 \cos(\theta)^2}, (2,2) \right.\right.
$$

$$
= \frac{r^2 + a^2 \cos(\theta)^2}{a^2 - 2mr + r^2}, (3,3) = r^2 + a^2 \cos(\theta)^2, (4,4)
$$

$$
\left.\left. = \frac{\left(a^2 \left(a^2 - 2mr + r^2\right)\cos(\theta)^2 + \left(2mr + r^2\right)a^2 + r^4\right)\sin(\theta)^2}{r^2 + a^2 \cos(\theta)^2} \right\}\right]
$$

$g\_[]$

$$g_{\mu\nu} = \left[\left[\frac{2mr - r^2 - a^2\cos(\theta)^2}{r^2 + a^2\cos(\theta)^2}, 0, 0, -\frac{2mra\sin(\theta)^2}{r^2 + a^2\cos(\theta)^2}\right], \left[0, \frac{r^2 + a^2\cos(\theta)^2}{a^2 - 2mr + r^2}, 0, 0\right], \left[0, 0, r^2 + a^2\cos(\theta)^2, 0\right],\right.$$

$$\left.\left[-\frac{2mra\sin(\theta)^2}{r^2 + a^2\cos(\theta)^2}, 0, 0, \frac{\left(a^2\left(a^2 - 2mr + r^2\right)\cos(\theta)^2 + \left(2mr + r^2\right)a^2 + r^4\right)\sin(\theta)^2}{r^2 + a^2\cos(\theta)^2}\right]\right]$$

Kerr Metric: Inner and Outer Surfaces of Infinite Redshift

$surf\_redshift := solve(g\_[1, 1] = 0, r)$

Warning, solve may be ignoring assumptions on the input variables.

$$surf\_redshift := m + \sqrt{m^2 - a^2\cos(\theta)^2}, m - \sqrt{m^2 - a^2\cos(\theta)^2}$$

$sr1 := subs(\{m = 1, a = 0.99\}, surf\_redshift[1])$

$$sr1 := 1 + \sqrt{1 - 0.9801\cos(\theta)^2}$$

$sr2 := subs(\{m = 1, a = 0.99\}, surf\_redshift[2])$

$$sr2 := 1 - \sqrt{1 - 0.9801\cos(\theta)^2}$$

Kerr Metric: Middle Two Event Horizons

$surf\_horrizon := solve(g\_[\sim2, \sim2] = 0, r)$

Warning, solve may be ignoring assumptions on the input variables.

$$surf\_horrizon := m + \sqrt{-a^2 + m^2}, m - \sqrt{a^2 + m^2}$$

$sh1 := subs(\{m = 1, a = 0.99\}, surf\_horrizon [1])$

$$sh1 := 1.141067360$$

$sh2 := subs(\{m = 1, a = 0.99\}, surf\_horrizon [2])$

$$sh2 := 0.8589326402$$

$$plot3d\left([sr1, sr2, sh1, sh2], \text{theta} = 0\ldots\text{Pi}, \text{phi} = -\frac{1.4\text{Pi}}{2}\ldots\frac{1.4\text{Pi}}{2}, coords = spherical,\right.$$

$$\left.scaling = constrained, color = gray\right)$$

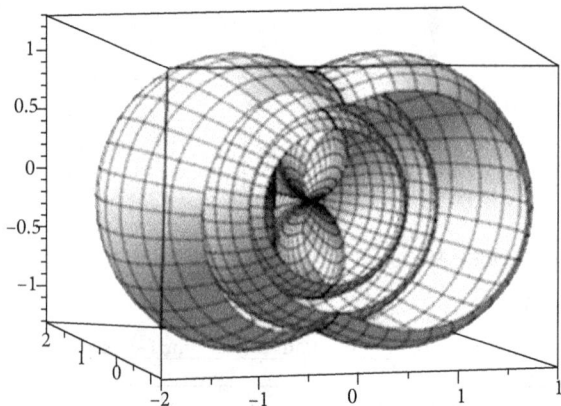

Figure 15.3.1: Two event horizons (middle two surfaces) and two surfaces of infinite redshift (inner and outer surfaces) of a rotating black hole described by the Kerr metric.

Godel Metric

$Setup(metric = Goedel)$

Systems of spacetime Coordinates are: $\{X = (t, x, y, z)\}$

Default differentiation variables for $d\_$, $D\_$ and dAlembertian are: $\{X = (t, x, y, z)\}$

The Goedel(1949) metric in coordinates $[t, x, y, z]$

parameters: $[]$

$$\left[metric = \left\{ (1,1) = -a^2, (1,4) = -a^2 e^x, (2,2) = a^2, (3,3) = a^2, (4,4) = -\frac{a^2 e^{2x}}{2} \right\} \right]$$

$g\_[]$

$$g_{\mu,\nu} = \begin{bmatrix} -a^2 & 0 & 0 & -a^2 e^x \\ 0 & a^2 & 0 & 0 \\ 0 & 0 & a^2 & 0 \\ -a^2 e^x & 0 & 0 & -\dfrac{a^2 e^{2x}}{2} \end{bmatrix}$$

Godel Metric: Ricci Tensor

$Ricci[\text{mu, nu}, matrix]$

$$R_{\mu,\nu} = \begin{bmatrix} 1 & 0 & 0 & e^x \\ 0 & 0 & 0 & 0 \\ 0 & 0 & 0 & 0 \\ e^x & 0 & 0 & e^{2x} \end{bmatrix}$$

Godel Metric: Einstein Tensor

$Einstein[\text{mu}, \text{nu}, matrix]$

$$G_{\mu,\nu} = \begin{bmatrix} \dfrac{1}{2} & 0 & 0 & \dfrac{e^x}{2} \\[2ex] 0 & \dfrac{1}{2} & 0 & 0 \\[2ex] 0 & 0 & \dfrac{1}{2} & 0 \\[2ex] \dfrac{e^x}{2} & 0 & 0 & \dfrac{3e^{2x}}{4} \end{bmatrix}$$

Godel Metric: Killing Vectors

$Define(V)$

$\qquad Defined\ objects\ with\ tensor\ properties$

$$\{V, D_\mu, \gamma_\mu, \sigma_\mu, R_{\mu,\nu}, R_{\mu,\nu,\alpha,\beta}, C_{\mu,\nu,\alpha,\beta}, X_\mu, \partial_\mu, g_{\mu,\nu}, \Gamma_{\mu,\nu,\alpha}, G_{\mu,\nu}, \delta_{\mu,\nu}, \in_{\alpha,\beta,\mu,\nu}\}$$

$KillingVectors(V)$

$$\left[V^\mu = [1,0,0,0], V^\mu = \left[e^{-x}, -\frac{z}{2}, 0, \frac{z^2}{2} - \frac{e^{-2x}}{2} \right], V^\mu = [0,0,1,0], V^\mu = [0,1,0,-z], V^\mu = [0,0,0,1], \right]$$

$Setup(coordinates = spherical)$

Spherical Coordinates: Robertson-Walker

$$dsRW := c^2 dt^2 - a(t)^2 \cdot r^2 \cdot dtheta^2 - a(t)^2 \cdot r^2 \cdot \sin(theta)^2\, dphi^2 - \frac{a(t)^2}{1 - K \cdot r^2} \cdot dr^2$$

$$dsRW := c^2 dt^2 - a(t)^2 r^2 d\theta^2 - a(t)^2 r^2 \sin(\theta)^2\, d\phi^2 - \frac{a(t)^2\, dr^2}{-Kr^2 + 1}$$

$Setup(metric = dsRW)$

$$\left[metric = \left\{ (1,1) = c^2, (2,2) = \frac{a(t)^2}{Kr^2 - 1}, (3,3) = -a(t)^2 r^2, (4,4) = -a(t)^2 r^2 \sin(\theta)^2 \right\} \right]$$

g_[]

$$g_{\mu,\nu} = \begin{bmatrix} c^2 & 0 & 0 & 0 \\ 0 & \dfrac{a(t)^2}{Kr^2-1} & 0 & 0 \\ 0 & 0 & -a(t)^2 r^2 & 0 \\ 0 & 0 & 0 & -a(t)^2 r^2 \sin(\theta)^2 \end{bmatrix}$$

Ricci Tensor from the Riemann Tensor

$SumOverRepeatedIndices(Riemann[\sim mu, 1, mu, 1])$

$$-\frac{3\left(\dfrac{d^2}{dt^2}a(t)\right)}{a(t)}$$

$SumOverRepeatedIndices(Riemann[\sim mu, 2, mu, 2])$

$$-\frac{2Kc^2 + \left(\dfrac{d^2}{dt^2}a(t)\right)a(t) + 2\left(\dfrac{d}{dt}a(t)\right)^2}{c^2(Kr^2-1)}$$

$SumOverRepeatedIndices(Riemann[\sim mu, 3, mu, 3])$

$$\frac{\left(2Kc^2 + \left(\dfrac{d^2}{dt^2}a(t)\right)a(t) + 2\left(\dfrac{d}{dt}a(t)\right)^2\right)r^2}{c^2}$$

$SumOverRepeatedIndices(Riemann[\sim mu, 4, mu, 4])$

$$\frac{\left(2Kc^2 + \left(\dfrac{d^2}{dt^2}a(t)\right)a(t) + 2\left(\dfrac{d}{dt}a(t)\right)^2\right)\sin(\theta)^2 r^2}{c^2}$$

$simplify(Ricci[\text{mu}, \text{nu}, matrix])$

$$R_{\mu\nu} = \left[\left[-\frac{3\left(\dfrac{d^2}{dt^2}a(t)\right)}{a(t)}, 0, 0, 0\right],\right.$$

$$\left[0, \frac{-2Kc^2 - \left(\dfrac{d^2}{dt^2}a(t)\right)a(t) - 2\left(\dfrac{d}{dt}a(t)\right)^2}{c^2(Kr^2-1)}, 0, 0\right],$$

$$\left[0, 0, \frac{\left(2Kc^2 + \left(\dfrac{d^2}{dt^2}a(t)\right)a(t) + 2\left(\dfrac{d}{dt}a(t)\right)^2\right)r^2}{c^2}, 0\right],$$

$$\left.\left[0, 0, 0, \frac{\left(2Kc^2 + \left(\dfrac{d^2}{dt^2}a(t)\right)a(t) + 2\left(\dfrac{d}{dt}a(t)\right)^2\right)\sin(\theta)^2 r^2}{c^2}\right]\right]$$

Stress Energy Tensor

$T := array(symmetric, sparse, 1 \ldots 4, 1 \ldots 4):$

$T[1,1] := p : T[2,2] := p : T[3,3] := p : T[4,4] := \text{rho}\cdot c^2 :$

Einstein's Field Equations

for mu **from 1 to 4 do** $expand(Einstein[\text{mu}, \text{mu}]) = \dfrac{8\cdot pi \cdot G \cdot T[\text{mu}, \text{mu}]}{c^4}$ **end**

$$\frac{3Kc^2}{a(t)^2} + \frac{3\left(\dfrac{d^2}{dt^2}a(t)\right)^2}{a(t)^2} = \frac{8\pi Gp}{c^4}$$

$$\frac{K}{Kr^2-1} + \frac{\left(\dfrac{d}{dt}a(t)\right)^2}{c^2(Kr^2-1)} + \frac{2a(t)\left(\dfrac{d^2}{dt^2}a(t)\right)}{c^2(Kr^2-1)} = \frac{8\pi Gp}{c^4}$$

$$-Kr^2 - \frac{\left(\dfrac{d}{dt}a(t)\right)^2 r^2}{c^2} - \frac{2a(t)r^2\left(\dfrac{d^2}{dt^2}a(t)\right)}{c^2} = \frac{8\pi Gp}{c^4}$$

$$-Kr^2\sin(\theta)^2 - \frac{\left(\dfrac{d}{dt}a(t)\right)^2 r^2 \sin(\theta)^2}{c^2} - \frac{2a(t)r^2\sin(\theta)^2\left(\dfrac{d^2}{dt^2}a(t)\right)}{c^2} = \frac{8\pi Gp}{c^2}$$

Wave Metric

$Setup(metric = [25, 61, 1])$

Systems of spacetime Coordinates are: $\{X = (t, x, y, z)\}$

Default differentiation variables for d_, D_ and dAlembertian are: $\{X = (t, x, y, z)\}$

The Stephani metric in coordinates $[t, x, y, z]$

Parameters: $[\Psi 1(t, z), \gamma 1(t, z), A(t, z)]$

Comments: We assume here that gamma1 = gamma1(t, z) and Psi1 = Psi1(t, z)

$[metric = \{(1, 1) = -e^{2\gamma 1(t,z) - 2\Psi 1(t,z)}, (2, 2) = e^{2\Psi 1(t,z)}, (3, 3) = t^2 e^{-2\Psi 1(t,z)}, (4, 4)$
$= e^{2\gamma 1(t,z) - 2\Psi 1(t,z)}\}]$

$g\_[]$

$$g_{\alpha,\nu} = \begin{bmatrix} -e^{2\gamma 1(t,z)-2\Psi 1(t,z)} & 0 & 0 & 0 \\ 0 & e^{2\Psi 1(t,z)} & 0 & 0 \\ 0 & 0 & t^2 e^{-2\Psi 1(t,z)} & 0 \\ 0 & 0 & 0 & e^{2\gamma 1(t,z)-2\Psi 1(t,z)} \end{bmatrix}$$

$Define(V)$

Defined as tensors

$$\left\{ D_\mu, \gamma_\mu, \sigma_\mu, R_{\mu,\nu}, R_{\mu,\nu,\alpha,\beta}, V^\mu, C_{\mu,\nu,\alpha,\beta}, X_\nu, \partial_\mu, g_{\mu,\nu}, \Gamma_{\mu,\nu,\alpha}, G_{\mu,\nu}, \delta_{\mu,\nu}, \in_{\alpha,\beta,\mu,\nu} \right\}$$

$KillingVectors(V)$

$$[V^n = [0, 1, 0, 0], V^n = [0, 0, 1, 0]]$$

Einstein Tensor Component

$Einstein[1, 1]$

$$\frac{-\left(\frac{\partial}{\partial t}\Psi 1(t,z)\right)^2 t - \left(\frac{\partial}{\partial z}\Psi 1(t,z)\right)^2 t + \frac{\partial}{\partial t}\gamma 1(t,z)}{t}$$

Reduced Dimensions

$Setup(dimension = 2)$

The dimension and signature of the tensor space are set to : $[2, - +]$

Detected 't', the time variable, in position 1. Changing the signature of the spacetime metric accordingly, to: + -

Systems of spacetime Coordinates are: $\{X = (t, x)\}$

$[dimension = 2]$

Coordinates(X)

> *Systems of spacetime Coordinates are:* $\{X = (x1, x2)\}$
> $\{X\}$

$xy\_metric := dx1^2 + \sin(x1)^2 \cdot dx2^2$

> $xy\_metric := dx1^2 + \sin(x1)^2\, dx2^2$

Setup(metric = xy_metric)

> $[metric = \{(1, 1) = 1, (2, 2) = \sin(x1)^2\}]$

Geodesics( )

$$\left[\frac{d^2}{d\tau^2} x1(\tau) = \sin(x1(\tau))\cos(x1(\tau))\left(\frac{d}{d\tau}x2(\tau)\right)^2, \right.$$

$$\left. \frac{d^2}{d\tau^2} x2(\tau) = -\frac{2\cos(x1(\tau))\left(\frac{d}{d\tau}x2(\tau)\right)\left(\frac{d}{d\tau}x1(\tau)\right)}{\sin(x1(\tau))} \right]$$

Christoffel[nonzero]

> $[\Gamma_{\alpha,\beta,\nu} = \{(1, 2, 2) = -\sin(x1)\cos(x1), (2, 1, 2) = \sin(x1)\cos(x1),$
> $(2, 2, 1) = \sin(x1)\cos(x1)\}$

Higher Dimensions

Setup(dimension = 5)

> *The dimension and signature of the tensor space are set to :* $[5, + - - -]$
> *Systems of spacetime Coordinates are:* $\{X = (x1, x2, x3, x4, x5)\}$
> $[dimension = 5]$

$5metric := dx1^2 - dx2^2 - dx3^2 - dx4^2 - dx5^2$

> $5metric := dx1^2 - dx2^2 - dx3^2 - dx4^2 - dx5^2$

Setup(metric = 5 metric)

> $[metric = \{(1, 1) = 1, (2, 2) = -1, (3, 3) = -1, (4, 4) = -1, (5, 5) = -1\}]$

g_[]

$$g_{\alpha,\nu} = \begin{bmatrix} 1 & 0 & 0 & 0 & 0 \\ 0 & -1 & 0 & 0 & 0 \\ 0 & 0 & -1 & 0 & 0 \\ 0 & 0 & 0 & -1 & 0 \\ 0 & 0 & 0 & 0 & -1 \end{bmatrix}$$

Geodesics( )

$$\left[\frac{d^2}{d\tau^2} x5(\tau) = 0, \frac{d^2}{d\tau^2} x4(\tau) = 0, \frac{d^2}{d\tau^2} x3(\tau) = 0, \frac{d^2}{d\tau^2} x2(\tau) = 0, \frac{d^2}{d\tau^2} x1(\tau) = 0 \right]$$

dsolve(%)

$$\{x1(\tau) = \_C9\tau + \_C10, \ x2(\tau) = \_C7\tau + \_C8, x3(\tau) = \_C5\tau + \_C6, x4(\tau) = \_ \\ C3\tau + \_C4, x5(\tau) = \_C1\tau + \_C2)\}$$

15.4 MATLAB EXAMPLES

A numerical solution to the geodesic equation of motion describing the orbit around a Schwarzschild black hole is demonstrated in this section.

Key MATLAB commands: *function, global, ode45, odeset, polarplot, RelTol*

Section 15.3 Einstein's Equations

The alternate Lagrangian describing Schwarzschild geodesics with $\theta = \pi/2$ and in the equatorial plane is

$$L' = \left(1 - \frac{r_s}{r}\right)c^2\dot{t}^2 - \left(1 - \frac{r_s}{r}\right)^{-1}\dot{r}^2 - r^2\dot{\phi}^2 \tag{15.5.1}$$

From the *t*-equation

$$\frac{d}{d\tau}\frac{dL'}{d\dot{t}} - \frac{dL'}{dt} = 0 \tag{15.5.2}$$

The constant of the motion proportional to the energy is

$$\left(1 - \frac{r_s}{r}\right)\dot{t} = \frac{E}{mc^2} \tag{15.5.3}$$

From the ϕ-equation

$$\frac{d}{d\tau}\frac{dL'}{d\dot{\phi}} - \frac{dL'}{d\phi} = 0$$

$$\frac{d}{d\tau}\left(2r^2\dot{\phi}\right) = 0 \tag{15.5.4}$$

We have a constant of the motion proportional to the angular momentum $\ell = mr^2\dot{\phi}$. The *r*-equation is

$$\frac{d}{d\tau}\frac{dL'}{d\dot{r}} - \frac{dL'}{dr} = 0$$

$$2\left(1 - \frac{r_s}{r}\right)^{-1}\ddot{r} + \frac{r_s}{r^2}c^2\dot{t}^2 - \left(1 - \frac{r_s}{r}\right)^{-2}\frac{r_s}{r^2}\dot{r}^2 - 2r\dot{\phi}^2 = 0 \tag{15.5.5}$$

Substituting energy and angular momentum constants of motion to eliminate $\dot{t}$ and $\dot{\phi}$ in the r-equation

$$\ddot{r}+\frac{r_s}{2r^2}\left(\frac{E}{mc}\right)^2\left(1-\frac{r_s}{r}\right)^{-1}-\left(1-\frac{r_s}{r}\right)^{-1}\frac{r_s}{2r^2}\dot{r}^2-\frac{2\ell^2}{m^2r^3}\left(1-\frac{r_s}{r}\right)=0 \qquad (15.5.6)$$

To find a numerical solution describing the Schwarzschild orbits, we obtain the first order system

$$\dot{r}=v$$

$$\dot{v}=-\frac{r_s}{2r^2}\left(\frac{E}{mc}\right)^2\left(1-\frac{r_s}{r}\right)^{-1}+\left(1-\frac{r_s}{r}\right)^{-1}\frac{r_s}{2r^2}v^2+\frac{2\ell^2}{m^2r^3}\left(1-\frac{r_s}{r}\right) \qquad (15.5.7)$$

$$\dot{\phi}=\frac{\ell}{mr^2}$$

Taking $L=\ell/m, \alpha=E/mc$ and $r_s=1$

$$\dot{r}=v$$

$$\dot{v}=\frac{v^2-\alpha^2}{2r(r-1)}+\frac{2L^2}{r^3}\left(1-\frac{1}{r}\right) \qquad (15.5.8)$$

$$\dot{\phi}=\frac{L}{r^2}$$

The derivative vector is specified in the MATLAB function 'schwarzschild.m' where $(r,\dot{r},\phi)\rightarrow\left[y(1),y(2),y(3)\right]$

```
function yp=schwarzschild(t,y)

global alpha L

yp= [y(2);(y(2)^2-alpha^2)/(2*y(1)*(y(1)-1))+2*L^2*(1-1/y(1))/y(1)^3;
     L/y(1)^2];
```

The MATLAB script 'runsc.m' calls the function 'schwarzschild.m' and numerically integrates the geodesic equations of motion over a time range $t = [0, 1000]$ with initial conditions $(r_0,\dot{r}_0,\phi_0)\rightarrow\left[10,-0.5,0\right]$ using a Runge-Kutta algorithm. Increased numerical accuracy is achieved with smaller values of the relative tolerance 'RelTol.' A polar plot of $(\phi,r)\rightarrow[y(:,3),y(:,1)]$ illustrates the orbital precession in Figure 15.4.1.

```
global alpha L
alpha = 3;
L=5;
opts=odeset('RelTol', 1e-5);
[t,y]=ode45(@schwarzschild,[0, 1000],[10;-0.5;0],opts);
```

```
polarplot(y(:,3),y(:,1))
title('Schwarzschild orbit')
```

The above script is executed at the Command Line

```
>runsc
```

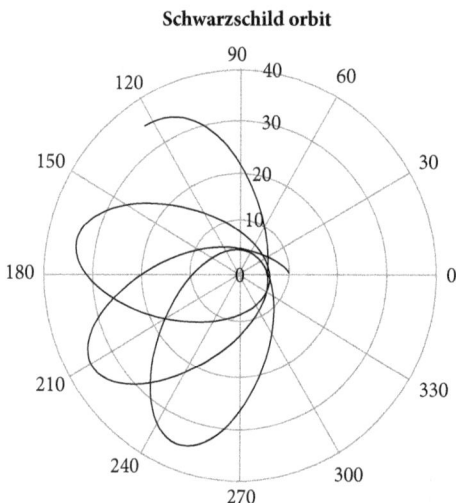

Figure 15.4.1: MATLAB polar plot of the Schwarzschild orbit demonstrating precession of the semi-major axis.

15.5 EXERCISES

Section 15.1 The Equivalence Principle

1. A photon is projected down in the earth's gravitational field from a height H above the surface with initial wavelength λ. Using the classical approximation, show that the photon wavelength is blueshifted $\lambda' < \lambda$ when it reaches the surface with a wavelength

$$\lambda' = \frac{\lambda}{\left(1 + \dfrac{gH}{c^2}\right)}$$

2. Calculate the redshift $(\lambda' - \lambda)/\lambda$ of a photon with initial wavelength λ emitted from the surface of a spherical star of radius R and mass M using the classical approximation
   (a) at a distance of $r = 2R$ from the center of the star
   (b) very far from the star $r \gg R$

Section 15.2 Tensor Calculus

3. Verify the relations involving the Levi-Civita tensor

$$\varepsilon_{ijk}\varepsilon_{ij\ell} = 2\delta_{k\ell}$$
$$\varepsilon_{ijk}\varepsilon_{ijk} = 6$$

4. Show that $\partial^{\mu}A_{\mu} = \partial_{\mu}A^{\mu}$

5. Given that

$$F^{\mu\nu} = \begin{pmatrix} 0 & -E_x/c & -E_y/c & -E_z/c \\ E_x/c & 0 & -B_z & B_y \\ E_y/c & B_z & 0 & -B_x \\ E_z/c & -B_y & B_x & 0 \end{pmatrix}$$

show that

$$F_{\mu\nu} = g_{\mu\alpha}F^{\alpha\beta}g_{\beta\nu} = \begin{pmatrix} 0 & E_x/c & E_y/c & E_z/c \\ -E_x/c & 0 & -B_z & B_y \\ -E_y/c & B_z & 0 & -B_x \\ -E_z/c & -B_y & B_x & 0 \end{pmatrix}$$

6. Verify the duality transformation

$$G^{\mu\nu} = \frac{1}{2}\varepsilon^{\mu\nu\alpha\beta}F_{\alpha\beta}$$

where $\varepsilon^{\mu\nu\alpha\beta}$ is the Levi-Civita tensor with four indices and

$$G^{\mu\nu} = \begin{pmatrix} 0 & -B_x & -B_y & -B_z \\ B_x & 0 & E_z/c & -E_y/c \\ B_y & -E_z/c & 0 & E_x/c \\ B_z & E_y/c & -E_x/c & 0 \end{pmatrix}$$

7. Evaluate the quantities $F^{\mu\nu}F_{\mu\nu}$ and $G^{\mu\nu}F_{\mu\nu}$.

8. Take the covariant derivative of the product $D_{\sigma}(A^{\mu}B_{\mu})$ to show that the covariant derivative of a contravariant vector is $A^{\mu}_{\;:\sigma} = A^{\mu}_{\;,\sigma} + \Gamma^{\mu}_{\nu\sigma}A^{\nu}$

9. Show that $\Gamma_{\sigma\mu\nu} + \Gamma_{\mu\sigma\nu} = \dfrac{\partial g_{\sigma\mu}}{\partial x^{\nu}}$

 (a) using a symmetry property of $\Gamma_{\sigma\mu\nu} = \dfrac{1}{2}\left(\dfrac{\partial g_{\mu\sigma}}{\partial x^{\nu}} + \dfrac{\partial g_{\nu\sigma}}{\partial x^{\mu}} - \dfrac{\partial g_{\mu\nu}}{\partial x^{\sigma}}\right)$

 (b) using the relation $g_{\sigma\mu:\nu} = 0$

Section 15.3 Einstein's Equations

10. Show that the geodesic equations of motion

$$\frac{d^2x^\sigma}{d\tau^2} + \Gamma^\sigma_{\rho\nu} \frac{dx^\rho}{d\tau} \frac{dx^\nu}{d\tau} = 0$$

can be obtained from an alternate Lagrangian of the form

$$L' = g_{\mu\nu} \frac{dx^\mu}{d\tau} \frac{dx^\nu}{d\tau}$$

11. Find the nonzero Christoffel symbols corresponding to the 2D line element in polar coordinates
$ds^2 = dr^2 + r^2\, d\theta^2$.

Write the corresponding geodesic equations of motion directly from Lagrange's equations where the alternate Lagrangian
$L' = \dot{r}^2 + r^2\dot\theta^2$

12. Find the Christoffel symbols $\Gamma^1_{\alpha,\beta}$, $\Gamma^2_{\alpha,\beta}$, $\Gamma^3_{\alpha,\beta}$ and $\Gamma^4_{\alpha,\beta}$ corresponding to the Schwarzschild metric tensor.

13. We may parameterize the geodesic equations using the angular coordinate instead of the proper time τ. From our line element $ds = cd\tau$ with $\theta = \pi/2$ for equatorial orbits

$$c^2 d\tau^2 = \left(1 - \frac{r_s}{r}\right)c^2 dt^2 - \left(1 - \frac{r_s}{r}\right)^{-1} dr^2 - r^2 d\phi^2$$

we have

$$c^2 = \left(1 - \frac{r_s}{r}\right)c^2\dot{t}^2 - \left(1 - \frac{r_s}{r}\right)^{-1}\dot{r}^2 - r^2\dot\phi^2$$

Substitute $\left(1 - \frac{r_s}{r}\right)\dot{t} = \frac{E}{mc^2}$ and $\ell = mr^2\dot\phi$ above to show that

$$\dot{r}^2 = \frac{E^2}{m^2c^2} - \left(c^2 + \frac{\ell^2}{m^2r^2}\right)\left(1 - \frac{r_s}{r}\right)$$

and divide $\dot{r}^2$ by the expression for $\dot\phi^2$ to obtainwe obtain

$$\left(\frac{dr}{d\phi}\right)^2 = \frac{E^2}{c^2\ell^2}r^4 - \left(\frac{m^2c^2}{\ell^2}r^4 + r^2\right)\left(1 - \frac{r_s}{r}\right)$$

14. The spacetime interval corresponding to the Kerr metric describing rotating black holes can be expressed as

$$ds^2 = -\frac{\Delta}{\rho^2}\left(dt - a\sin^2\theta\, d\phi\right)^2 + \frac{\sin^2\theta}{\rho^2}\left(\left(r^2 + a^2\right)d\phi - a\, dt\right)^2 + \frac{\rho^2}{\Delta}dr^2 + \rho^2 d\theta^2$$

where $\Delta = r^2 - 2Mr + a^2$ and $\rho^2 = r^2 + a^2\cos^2\theta$.

Find the null geodesic equations with $ds^2 = 0$ for equatorial trajectories $\theta = \pi/2$.

15. Calculate the Ricci tensor, Ricci scalar and Einstein tensor for the Kerr metric.

16. Calculate the scalar quantity

$$R_{\beta\mu\nu\alpha}R^{\beta\mu\nu\alpha}$$

for the Schwarzschild and Kerr metrics.

17. Given the spacetime interval in spherical coordinates

$$ds^2 = e^{2\nu(r)}\,dt^2 - e^{2\lambda(r)}\,dr^2 - r^2\,d\theta^2 - r^2\sin^2\theta\,d\phi^2.$$

calculate the corresponding Ricci tensor, Ricci scalar and Einstein tensor.

18. Show that $D_\mu(g^{\mu\nu}R) = g^{\mu\nu}\partial_\mu R$

19. The evolution of the scale factor in the Friedman model of a flat universe is

$$\dot{a}^2 = \frac{8\pi G\rho}{3c^2}a^2$$

For a matter-dominated universe where the density $\rho \sim a^{-3}$ show that

$$a(t) = \text{const} \times t^{2/3}$$

with the age of the universe

$$t = \frac{1}{\sqrt{6\pi G\rho_m}} = \frac{2}{3}H^{-1}$$

where $H = \dot{a}/a$.

Chapter

16 RELATIVISTIC QUANTUM MECHANICS

Chapter Outline
16.1 Early Models
16.2 Dirac Equation
16.3 Solutions to the Dirac Equation

16.1 EARLY MODELS

In this section, early relativistic models of the wave nature of matter are developed. The de Broglie wavelengths are combined with the relativistic energy-momentum relationship to obtain phase and group velocities of matter waves. Energy levels of the 1D square well are obtained. The Klein-Gordon equation, also based on the energy-momentum relation, is introduced. The Klein-Gordon probability current and Lagrangian density are explored.

16.1.1 de Broglie waves

For massless particles $E = pc$ and $E = hf = hc/\lambda$. Thus, the wavelength of a photon is inversely proportional to its momentum

$$\lambda = \frac{h}{p} \tag{16.1.1}$$

Louis de Broglie hypothesized that the wave particle duality of light extended to massive particles such as electrons. For a massive particle $p = \gamma mv$ so the de Broglie wavelength of a particle such as an electron is given by

$$\lambda = \frac{h}{mv}\sqrt{1 - \frac{v^2}{c^2}} \tag{16.1.2}$$

The wave nature of particles was later confirmed by the Davisson-Germer experiment. To model the wave nature of massive particles, we substitute de Broglie's relation $p = \hbar k$ for particles with wave vector $k = 2\pi/\lambda$ and energy $E = \hbar\omega$ into the mass energy relation $E^2 = p^2 c^2 + m^2 c^4$

$$\hbar\omega = \sqrt{\left(c\hbar k\right)^2 + m^2 c^4} \tag{16.1.3}$$

The wave speed, or phase velocity, is

$$v_p = \frac{\omega}{k} = c\sqrt{1 + \left(\frac{mc}{\hbar k}\right)^2} \tag{16.1.4}$$

That is evidently greater than the speed of light. The group velocity

$$v_g = \frac{d\omega}{dk} = \frac{d}{dk}\sqrt{c^2 k^2 + \frac{m^2 c^4}{\hbar^2}} \tag{16.1.5}$$

or

$$v_g = \frac{c}{\sqrt{1 + \left(\frac{mc}{\hbar k}\right)^2}} \tag{16.1.6}$$

is interpreted as the actual speed of the particle that is less than the speed of light. The product of the group and phase velocities $v_g v_p = c^2$.

Example 16.1.1

Use the relativistic form of the momentum $p = \gamma mv$ and the energy-momentum relation to find the phase and group velocities of a matter wave

Solution: The phase velocity is

$$v_p = \frac{\omega}{k} = \frac{E}{p} = \sqrt{c^2 + \frac{m^2 c^4}{p^2}} \tag{16.1.7}$$

Substituting $p = \gamma mv$

$$v_p = \frac{\omega}{k} = \sqrt{c^2 + \frac{m^2 c^4}{m^2 v^2}\left(1 - \frac{v^2}{c^2}\right)} = \sqrt{c^2 + \frac{c^4}{v^2} - c^2} = \frac{c^2}{v} \tag{16.1.8}$$

The group velocity is then

$$v_g = \frac{\partial \omega}{\partial k} = \frac{\partial E}{\partial p} = \frac{\partial}{\partial p}\sqrt{p^2 c^2 + m^2 c^4} \tag{16.1.9}$$

$$v_g = \frac{c^2}{\sqrt{c^2 + \frac{m^2 c^4}{m^2 v^2}\left(1 - \frac{v^2}{c^2}\right)}} = \frac{c^2}{\sqrt{c^2 + \frac{c^4}{v^2}\left(1 - \frac{v^2}{c^2}\right)}} = v \tag{16.1.10}$$

As before we find that $v_g v_p = c^2$.

Example 16.1.2

Use the relativistic energy-momentum relation to develop an expression for the energy levels of a particle with mass m inside a 1D square well using the de Broglie relation $p = \hbar k$ with wave vector $k = 2\pi/\lambda = n\pi/L$. Use the binomial expansion theorem to approximate the energy levels for large L.

Solution: The energy-momentum relation

$$E = \sqrt{\hbar^2 k^2 c^2 + m^2 c^4} \tag{16.1.11}$$

gives the quantized energy levels

$$E_n = \sqrt{\frac{\hbar^2 \pi^2 c^2}{L^2} n^2 + m^2 c^4} \tag{16.1.12}$$

Factoring mc^2 from the square root

$$E_n = mc^2 \sqrt{1 + \frac{\hbar^2 \pi^2}{m^2 c^2 L^2} n^2} \tag{16.1.13}$$

For values of $L \gg \dfrac{mc}{n\hbar\pi}$, the binomial theorem gives

$$E_n \approx mc^2 \left(1 + \frac{1}{2}\frac{\hbar^2 \pi^2}{m^2 c^2 L^2} n^2\right) \tag{16.1.14}$$

or

$$E_n \approx mc^2 + \frac{\hbar^2 \pi^2}{2mL^2} n^2 \tag{16.1.15}$$

This expression gives the energy levels obtained from the Schrödinger equation plus the rest mass energy of the electron.

16.1.2 Klein-Gordon Equation

The Klein-Gordon (KG) equation was initially proposed as a relativistic equation of the electron. It was discarded because of its prediction of negative probability densities. Schrödinger went on to develop the nonrelativistic equation that bears his name. It turns out the KG equation is useful for describing massive particles with spin zero. To derive the KG equation for free particles in the absence of a position-dependent potential we begin with the energy-momentum relation

$$E^2 = p^2 c^2 + m^2 c^4 \tag{16.1.16}$$

Now we make the quantum replacements

$$\hat{E} = i\hbar \frac{\partial}{\partial t} \quad \text{and} \quad \hat{p} = \frac{\hbar}{i}\nabla \tag{16.1.17}$$

Substituting into the energy-momentum relation gives

$$-\hbar^2 \frac{\partial^2}{\partial t^2} = -\hbar^2 c^2 \nabla^2 + m^2 c^4 \tag{16.1.18}$$

Multiplying both sides by ψ gives

$$\left(\nabla^2 - \frac{1}{c^2}\frac{\partial^2}{\partial t^2}\right)\Psi = \frac{m^2 c^2}{\hbar^2}\Psi \tag{16.1.19}$$

This may be written more compactly using the D'Alembertian operator

$$\Box^2 \equiv \nabla^2 - \frac{1}{c^2}\frac{\partial^2}{\partial t^2} \tag{16.1.20}$$

so that

$$\left(\Box^2 - \alpha^2\right)\psi = 0 \tag{16.1.21}$$

where

$$\alpha^2 = \frac{m^2 c^2}{\hbar^2} \tag{16.1.22}$$

16.1.3 Probability Current Density

We obtain the probability current density by multiplying the Klein-Gordon equation by Ψ^*

$$\Psi * \nabla^2 \Psi - \frac{1}{c^2} \Psi * \frac{\partial^2 \Psi}{\partial t^2} = \alpha^2 \Psi * \Psi \tag{16.1.23}$$

and subtracting the complex conjugate of the Klein-Gordon equation multiplied by Ψ

$$\Psi \nabla^2 \Psi * - \frac{1}{c^2} \Psi \frac{\partial^2 \Psi *}{\partial t^2} = \alpha^2 \Psi \Psi * \tag{16.1.24}$$

Since $\Psi * \Psi = \Psi \Psi *$ we obtain

$$\Psi * \nabla^2 \Psi - \Psi * \nabla^2 \Psi = \frac{1}{c^2} \left(\Psi * \frac{\partial^2 \Psi}{\partial t^2} - \Psi \frac{\partial^2 \Psi *}{\partial t^2} \right) \tag{16.1.25}$$

Substituting the probability current density

$$\mathbf{J}_{pcd} = \frac{\hbar}{2mi} \left(\Psi * \nabla \Psi - \Psi \nabla \Psi * \right) \tag{16.1.26}$$

we obtain

$$\frac{2mi}{\hbar} \nabla \cdot \mathbf{J}_{pcd} = -\frac{1}{c^2} \frac{\partial}{\partial t} \left(\Psi \frac{\partial \Psi *}{\partial t} - \Psi * \frac{\partial \Psi}{\partial t} \right) \tag{16.1.27}$$

Comparing the continuity equation

$$\nabla \cdot \mathbf{J}_{pcd} = -\frac{\partial \rho}{\partial t} \tag{16.1.28}$$

we find the probability density is

$$\rho = \frac{1}{c^2} \frac{\hbar}{2mi} \left(\Psi \frac{\partial \Psi *}{\partial t} - \Psi * \frac{\partial \Psi}{\partial t} \right) \tag{16.1.29}$$

Here, ρ can either be positive or negative and is not positive definite.

16.1.4 Lagrangian Formulation of the Klein-Gordon Equation

Example 16.1.3

Show that the Klein-Gordon equation is obtained by varying the Lagrangian

$$L = \frac{1}{2} \partial_\mu \phi \partial^\mu \phi - m^2 \frac{1}{2} \phi^2 \tag{16.1.30}$$

with respect to ϕ

Solution: Lagrange's equation of motion is

$$\frac{\partial L}{\partial \phi} - \frac{\partial}{\partial x^\mu} \frac{\partial L}{\partial (\partial_\mu \phi)} = 0 \tag{16.1.31}$$

giving the Klein-Gordon equation in natural units

$$\frac{1}{2}\partial_\mu\partial^\mu\phi+m^2\phi=0 \tag{16.1.32}$$

Although the Klein-Gordon equation does not provide an adequate relativistic wave equation, it does find utility in the context of quantum field theory where ϕ is interpreted as a scalar field.

Maple Examples

The following Maple worksheet illustrates the general solution to linear and nonlinear Klein-Gordon equations.

Key Maple commands: *diff, pdsolve*

restart

Klein-Gordon Equation

assume(c > 0)

$$PDE1:= diff\left(psi(x,t),t,t\right)-\frac{1}{c^2}\cdot diff\left(psi(x,t),x,x\right)=\frac{m^2\cdot c^2}{\hbar^2}\cdot psi(x,t)$$

$$PDE1:=\frac{\partial^2}{\partial t^2}\psi(x,t)-\frac{\dfrac{\partial^2}{\partial x^2}\psi(x,t)}{c\sim^2}=\frac{m^2c\sim^2\psi(x,t)}{\hbar^2}$$

pdsolve(PDE1)

$$\left(\psi(x,t)=\_F1(\_\xi1)\_F2(\_\xi2)\right)$$

$$\&\ where\ \left[\left\{\frac{d}{d\_\xi1}\_F1(\_\xi1)=\_c_1\_F1(\_\xi1),\frac{d}{d\_\xi2}\_F2(\_\xi2)\right.\right.$$

$$=\frac{m^2c\sim^3\_F2(\_\xi2)}{2\_c_1\hbar^2}\right\},\&\ and\ \left(\left\{\_\xi1=-c\sim x+t,\_\xi2=\frac{x}{2}+\frac{t}{2c\sim}\right\}\right)\right]$$

Nonlinear Klein-Gordon Equation

PDE2 := diff(psi(x, t), t, t) − alpha·*diff*(psi(x, t), x, x) + gamma²·psi(x, t) = beta·psi(x, t)²

$$PDE2:=\frac{\partial^2}{\partial t^2}\psi(x,t)-\alpha\left(\frac{\partial^2}{\partial x^2}\psi(x,t)\right)+\gamma^2\psi(x,t)=\beta\psi(x,t)^2$$

pdsolve(PDE2)

$$\psi(x,t) = -\frac{3\gamma^2 \tanh\left(-\frac{\sqrt{4 - C2^2\alpha - \gamma^2}\,t}{2} + \_C2x + \_C1\right)}{2\beta} + \frac{3\gamma^2}{2\beta}$$

16.2 DIRAC EQUATION

In the previous section, we obtained the Klein-Gordon equation with operator substitution into the energy-momentum relation. Because time derivatives in the Klein-Gordon equation are second order, it is necessary to specify both the initial value of the wavefunction $\Psi(x,0)$ and its time derivative $\dot{\Psi}(x,0)$. Also, the calculated probability density was not positive definite. Below, the Dirac equation is introduced that is first order in space and time derivatives. Thus only $\Psi(x,0)$ is required to obtain a general solution. One of the crowning achievements of the relativistic theory of the electron is the prediction of antimatter.

16.2.1 Derivation of a First Order Equation

Seeking a relativistic equation that is first order in time that is consistent with the Klein-Gordon equation we try

$$\sqrt{\hat{p}^2 c^2 + m^2 c^4}\,\psi = \hat{E}\psi = i\hbar\frac{\partial\Psi}{\partial t} \tag{16.2.1}$$

where $\hat{p}^2 = \hat{p}_x^{\,2} + \hat{p}_y^{\,2} + \hat{p}_y^{\,2}$. The square root on the left-hand side

$$\sqrt{\hat{p}_x^{\,2}c^2 + \hat{p}_y^{\,2}c^2 + \hat{p}_y^{\,2}c^2 + m^2c^4}\,\Psi = \left(c\alpha_x\hat{p}_x + c\alpha_y\hat{p}_y + c\alpha_z\hat{p}_z + \beta mc^2\right)\Psi \tag{16.2.2}$$

with coefficients $\alpha_x, \alpha_y, \alpha_z$ and β on the right-hand side should satisfy

$$\hat{p}_x^{\,2}c^2 + \hat{p}_y^{\,2}c^2 + \hat{p}_y^{\,2}c^2 + m^2c^4 = \left(c\alpha_x\hat{p}_x + c\alpha_y\hat{p}_y + c\alpha_z\hat{p}_z + \beta mc^2\right)^2 \tag{16.2.3}$$

Expanding the right-hand side
$$\hat{p}_x^{\,2}c^2 + \hat{p}_y^{\,2}c^2 + \hat{p}_y^{\,2}c^2 + m^2c^4 = c^2\alpha_x^{\,2}\hat{p}_x^{\,2} + c^2\alpha_y^{\,2}\hat{p}_y^{\,2} \tag{16.2.4}$$
$$+ c^2\alpha_z^{\,2}\hat{p}_z^{\,2} + \beta^2 m^2c^4 + \text{cross terms}$$

For this equation to be satisfied, all the cross terms should cancel and $\alpha_x^{\,2} = \alpha_y^{\,2} = \alpha_z^{\,2} = \beta^2 = 1$. These conditions can be satisfied if α_i ($\alpha_{x,y,z} = \alpha_{1,2,3}$) and β are matrices and we find that

$$\alpha_i\alpha_j + \alpha_j\alpha_i = 2\delta_{ij}$$
$$\alpha_i\beta + \beta\alpha_i = 0$$
$$\alpha_i^2 = 1$$
$$\beta^2 = 1$$

(16.2.5)

There is no unique set of matrices that satisfies the conditions above. A common set of 4×4 matrices known as the Dirac-Pauli matrices are

$$\alpha_1 = \begin{pmatrix} 0 & 0 & 0 & 1 \\ 0 & 0 & 1 & 0 \\ 0 & 1 & 0 & 0 \\ 1 & 0 & 0 & 0 \end{pmatrix} \quad \alpha_2 = \begin{pmatrix} 0 & 0 & 0 & -i \\ 0 & 0 & i & 0 \\ 0 & -i & 0 & 0 \\ i & 0 & 0 & 0 \end{pmatrix}$$

$$\alpha_3 = \begin{pmatrix} 0 & 0 & 1 & 0 \\ 0 & 0 & 0 & -1 \\ 1 & 0 & 0 & 0 \\ 0 & -1 & 0 & 0 \end{pmatrix} \quad \beta = \begin{pmatrix} 1 & 0 & 0 & 0 \\ 0 & 1 & 0 & 0 \\ 0 & 0 & -1 & 0 \\ 0 & 0 & 0 & -1 \end{pmatrix}$$

(16.2.6)

In terms of the Pauli spin matrices

$$\sigma_1 = \begin{pmatrix} 0 & 1 \\ 1 & 0 \end{pmatrix} \quad \sigma_2 = \begin{pmatrix} 0 & -i \\ i & 0 \end{pmatrix} \quad \sigma_3 = \begin{pmatrix} 1 & 0 \\ 0 & -1 \end{pmatrix}$$

(16.2.7)

The α_i and β are written more compactly as

$$\alpha_1 = \begin{pmatrix} 0 & \sigma_1 \\ \sigma_1 & 0 \end{pmatrix} \quad \alpha_2 = \begin{pmatrix} 0 & \sigma_2 \\ \sigma_2 & 0 \end{pmatrix} \quad \alpha_3 = \begin{pmatrix} 0 & \sigma_3 \\ \sigma_3 & 0 \end{pmatrix} \quad \beta = \begin{pmatrix} 1 & 0 \\ 0 & -1 \end{pmatrix}$$

(16.2.8)

where each element above is a 2×2 matrix. Writing out the individual terms on the left-hand side of the Dirac equation

$$c\alpha_1\hat{p}_x\Psi = \begin{pmatrix} 0 & 0 & 0 & cp_x \\ 0 & 0 & cp_x & 0 \\ 0 & cp_x & 0 & 0 \\ cp_x & 0 & 0 & 0 \end{pmatrix}\Psi \quad c\alpha_2\hat{p}_y\Psi = \begin{pmatrix} 0 & 0 & 0 & -icp_y \\ 0 & 0 & icp_y & 0 \\ 0 & -icp_y & 0 & 0 \\ icp_y & 0 & 0 & 0 \end{pmatrix}\Psi$$

$$c\alpha_3\hat{p}_z\Psi = \begin{pmatrix} 0 & 0 & cp_z & 0 \\ 0 & 0 & 0 & -cp_z \\ cp_z & 0 & 0 & 0 \\ 0 & -cp_z & 0 & 0 \end{pmatrix}\Psi \quad \beta mc^2\Psi = \begin{pmatrix} mc^2 & 0 & 0 & 0 \\ 0 & mc^2 & 0 & 0 \\ 0 & 0 & -mc^2 & 0 \\ 0 & 0 & 0 & -mc^2 \end{pmatrix}\Psi$$

(16.2.9)

where Ψ is a four-component object known as a spinor

$$\Psi = \begin{pmatrix} \psi_1 \\ \psi_2 \\ \psi_3 \\ \psi_4 \end{pmatrix} \qquad (16.2.10)$$

Combining the terms above we write the Dirac equation explicitly as

$$\begin{pmatrix} mc^2 & 0 & c\hat{p}_z & c\hat{p}_x - ic\hat{p}_y \\ 0 & mc^2 & c\hat{p}_x + ic\hat{p}_y & -c\hat{p}_z \\ c\hat{p}_z & c\hat{p}_x - ic\hat{p}_y & -mc^2 & 0 \\ c\hat{p}_x + ic\hat{p}_y & -c\hat{p}_z & 0 & -mc^2 \end{pmatrix} \begin{pmatrix} \psi_1 \\ \psi_2 \\ \psi_3 \\ \psi_4 \end{pmatrix} = \hat{E} \begin{pmatrix} \psi_1 \\ \psi_2 \\ \psi_3 \\ \psi_4 \end{pmatrix} \qquad (16.2.11)$$

16.2.2 Probability Current

To obtain an expression for the probability current we write our first order Dirac equation

$$\left(c\alpha_x \hat{p}_x + c\alpha_y \hat{p}_y + c\alpha_z \hat{p}_z + mc^2 \beta \right) \Psi = i\hbar \frac{\partial \Psi}{\partial t} \qquad (16.2.12)$$

compactly using index notation

$$c\frac{\hbar}{i}\alpha_k \frac{\partial \Psi}{\partial x^k} + mc^2 \beta \Psi = i\hbar \frac{\partial \Psi}{\partial t} \qquad (16.2.13)$$

where the repeated index k is summed from 1 to 3. Multiplying (16.2.13) on the left by $\Psi^\dagger$

$$c\frac{\hbar}{i}\Psi^\dagger \alpha_k \frac{\partial \Psi}{\partial x^k} + mc^2 \Psi^\dagger \beta \Psi = i\hbar \Psi^\dagger \frac{\partial \Psi}{\partial t} \qquad (16.2.14)$$

Forming the Hermitian conjugate of (16.2.13) and right multiplying by Ψ

$$-c\frac{\hbar}{i}\alpha_k \frac{\partial \Psi^\dagger}{\partial x^k} \Psi + mc^2 \beta \Psi^\dagger \Psi = -i\hbar \frac{\partial \Psi^\dagger}{\partial t} \Psi \qquad (16.2.15)$$

where $\alpha_k = \alpha_k^\dagger$ and $\beta = \beta^\dagger$ are Hermitian matrices. Subtracting (16.2.14) and (16.2.15) we obtain

$$c\frac{\hbar}{i}\frac{\partial}{\partial x^k}\left(\Psi^\dagger \alpha_k \Psi \right) = i\hbar \frac{\partial}{\partial t}\left(\Psi^\dagger \Psi \right) \qquad (16.2.16)$$

that is in the form of the continuity equation $\nabla \cdot \mathbf{J} = -\partial \rho / \partial t$ where the components of **J** are

$$J^k = c \Psi^\dagger \alpha_k \Psi \tag{16.2.17}$$

and our now positive definite probability density

$$\rho = \Psi^\dagger \Psi \tag{16.2.18}$$

16.2.3 Gamma Matrices

Writing out the $\hat{p}_x$, $\hat{p}_y$ and $\hat{p}_z$ operators and dividing by c

$$-i\hbar \left(\alpha_x \frac{\partial}{\partial x} + \alpha_y \frac{\partial}{\partial y} + \alpha_z \frac{\partial}{\partial z} \right) \Psi + \beta m c \Psi = i\hbar \frac{1}{c} \frac{\partial}{\partial t} \Psi \tag{16.2.19}$$

with the gamma matrices defined as $\gamma^0 = \beta$ and $\gamma^i = \beta \alpha_i$

$$\gamma^0 = \begin{pmatrix} 1 & 0 & 0 & 0 \\ 0 & 1 & 0 & 0 \\ 0 & 0 & -1 & 0 \\ 0 & 0 & 0 & -1 \end{pmatrix} \quad \gamma^1 = \begin{pmatrix} 0 & 0 & 0 & 1 \\ 0 & 0 & 1 & 0 \\ 0 & -1 & 0 & 0 \\ -1 & 0 & 0 & 0 \end{pmatrix}$$

$$\gamma^2 = \begin{pmatrix} 0 & 0 & 0 & -i \\ 0 & 0 & i & 0 \\ 0 & i & 0 & 0 \\ -i & 0 & 0 & 0 \end{pmatrix} \quad \gamma^3 = \begin{pmatrix} 0 & 0 & 1 & 0 \\ 0 & 0 & 0 & -1 \\ -1 & 0 & 0 & 0 \\ 0 & 1 & 0 & 0 \end{pmatrix} \tag{16.2.20}$$

The Dirac equation becomes

$$\left(i\hbar \gamma^\mu \partial_\mu - mc \right) \Psi = 0 \tag{16.2.21}$$

where repeated indices are summed over

$$\gamma^\mu \partial_\mu = \gamma^0 \frac{\partial}{\partial x^0} + \gamma^1 \frac{\partial}{\partial x^1} + \gamma^2 \frac{\partial}{\partial x^2} + \gamma^3 \frac{\partial}{\partial x^3} \tag{16.2.22}$$

This notation can be condensed using the "Feynman slash" notation where quantities contracted with the gamma matrices are abbreviated $\gamma^\mu a_\mu = \not{a}$. In natural units where $\hbar = c = 1$ the Dirac equation appears in its most compact form $\left(i\not{\partial} - m \right) \Psi = 0$.

16.2.4 Positive and Negative Energies

For particles at rest

$$\left(i\hbar\gamma^0\partial_0 - mc\right)\Psi = 0 \tag{16.2.23}$$

where $c\partial_0 = \partial/\partial t$ and

$$i\hbar \begin{pmatrix} 1 & 0 & 0 & 0 \\ 0 & 1 & 0 & 0 \\ 0 & 0 & -1 & 0 \\ 0 & 0 & 0 & -1 \end{pmatrix} \frac{\partial}{\partial t} \begin{pmatrix} \psi_1 \\ \psi_2 \\ \psi_3 \\ \psi_4 \end{pmatrix} = mc^2 \begin{pmatrix} \psi_1 \\ \psi_2 \\ \psi_3 \\ \psi_4 \end{pmatrix} \tag{16.2.24}$$

With solutions proportional to $\exp(-iEt/\hbar)$ we have

$$\psi_1 = \exp\left(-i\frac{mc^2}{\hbar}t\right) \begin{pmatrix} 1 \\ 0 \\ 0 \\ 0 \end{pmatrix} \quad \text{and} \quad \psi_2 = \exp\left(-i\frac{mc^2}{\hbar}t\right) \begin{pmatrix} 0 \\ 1 \\ 0 \\ 0 \end{pmatrix} \tag{16.2.25}$$

with positive energy $E = +mc^2$ while

$$\psi_3 = \exp\left(i\frac{mc^2}{\hbar}t\right) \begin{pmatrix} 0 \\ 0 \\ 1 \\ 0 \end{pmatrix} \quad \text{and} \quad \psi_4 = \exp\left(i\frac{mc^2}{\hbar}t\right) \begin{pmatrix} 0 \\ 0 \\ 0 \\ 1 \end{pmatrix} \tag{16.2.26}$$

are negative energy solutions with $E = -mc^2$. The Pauli exclusion principle states that no two electrons can occupy the same quantum state. Dirac thus postulated that electrons with positive energy couldn't simply cascade down to negative energies if all the negative energy states were already filled. The supposed filled negative energy states became know as the Dirac Sea. Excitations of negative energy states could produce positive energy electrons since the positive energy states are not filled, however. A vacancy (with positive energy) in the negative energy sea would be left behind, corresponding to the anti-electron. The anti-electron was later discovered and Dirac was awarded the Nobel Prize in physics for his relativistic formulation of the electron. The anti-proton was later discovered. All particles in nature have associated anti particles, many of which are routinely produced in particle accelerators.

16.2.5 Lagrangian Formulation of the Dirac Equation

Example 16.2.1

Show that the Dirac equation is obtained by varying the Lagrangian

$$L = i\hbar\overline{\Psi}\gamma^{\mu}\partial_{\mu}\Psi - mc^2\overline{\Psi}\Psi \tag{16.2.27}$$

with respect to $\overline{\Psi}$

Solution: Lagrange's equation of motion is

$$\frac{\partial L}{\partial\overline{\Psi}} - \frac{\partial}{\partial x^{\mu}}\frac{\partial L}{\partial\left(\partial_{\mu}\overline{\Psi}\right)} = 0 \tag{16.2.28}$$

The second term is zero since the Lagrangian does not depend explicitly on $\partial_{\mu}\overline{\Psi}$. The Dirac equation is then obtained from

$$\frac{\partial L}{\partial\overline{\Psi}} = i\hbar\gamma^{\mu}\partial_{\mu}\Psi - mc^2\Psi = 0 \tag{16.2.29}$$

The conjugate of the Dirac equation is similarly obtained by varying L with respect to Ψ

$$\frac{\partial L}{\partial\Psi} - \frac{\partial}{\partial x^{\mu}}\frac{\partial L}{\partial\left(\partial_{\mu}\Psi\right)} = 0 \tag{16.2.30}$$

Maple Examples

The following Maple worksheet demonstrates the standard representation of gamma matrices. Trace and commutation operations are carried out with the gamma matrices and the related Pauli spin matrices.

Key Maple commands: *AntiCommutator, Commutator, Dagger, Dgamma, dimension, mathematicalnotation, Psigma, Setup, Trace, SumOverRepeatedIndices*

Maple packages: *with(Physics)*:

restart

Gamma Matrices

with(Physics) :
Setup(mathematicalnotation = true)
                    [*mathematicalnotation = true*]
Setup(Dgammarepresentation = standard)
                Setting lowercaselatin letters to represent spinor indices

Defined Dirac gamma matrices (Dgamma) in standard representation, $\gamma_1, \gamma_2, \gamma_3, \gamma_4$

$$[Dgammarepresentation = standard]$$

Dgamma[0]

$$\gamma_4$$

Dgamma[0][ ]

$$(\gamma_4)_{a,b} = \begin{bmatrix} 1 & 0 & 0 & 0 \\ 0 & 1 & 0 & 0 \\ 0 & 0 & -1 & 0 \\ 0 & 0 & 0 & -1 \end{bmatrix}$$

Dgamma[4][ ]

$$(\gamma_4)_{a,b} = \begin{bmatrix} 1 & 0 & 0 & 0 \\ 0 & 1 & 0 & 0 \\ 0 & 0 & -1 & 0 \\ 0 & 0 & 0 & -1 \end{bmatrix}$$

Dgamma[1][ ]

$$(\gamma_1)_{a,b} = \begin{bmatrix} 0 & 0 & 0 & -1 \\ 0 & 0 & -1 & 0 \\ 0 & 1 & 0 & 0 \\ 1 & 0 & 0 & 0 \end{bmatrix}$$

Dgamma[2][ ]

$$(\gamma_2)_{a,b} = \begin{bmatrix} 0 & 0 & 0 & I \\ 0 & 0 & -I & 0 \\ 0 & -I & 0 & 0 \\ I & 0 & 0 & 0 \end{bmatrix}$$

Dgamma[3][ ]

$$(\gamma_3)_{a,b} = \begin{bmatrix} 0 & 0 & -1 & 0 \\ 0 & 0 & 0 & 1 \\ 1 & 0 & 0 & 0 \\ 0 & -1 & 0 & 0 \end{bmatrix}$$

Trace Relations

$map(Trace, [Dgamma[0]^2, Dgamma[1]^2, Dgamma[2]^2, Dgamma[3]^2])$

$$[4, -4, -4, -4]$$

$Trace(Dgamma[mu] \cdot Dgamma[nu])$

$$4g_{\mu, \nu}$$

$Dgamma[1][\ ] \cdot Dgamma[2][\ ]$

$$\left(\gamma_1 \right)_{a,b} \left(\gamma_2 \right)_{\sim a, \sim b} = \begin{bmatrix} -I & 0 & 0 & 0 \\ 0 & I & 0 & 0 \\ 0 & 0 & -I & 0 \\ 0 & 0 & 0 & I \end{bmatrix}$$

$Dgamma[2][\ ] \cdot Dgamma[1][\ ]$

$$\left(\gamma_1 \right)_{a,b} \left(\gamma_2 \right)_{\sim a, \sim b} = \begin{bmatrix} I & 0 & 0 & 0 \\ 0 & -I & 0 & 0 \\ 0 & 0 & I & 0 \\ 0 & 0 & 0 & -I \end{bmatrix}$$

Gamma 5 Matrix

$I \cdot Dgamma[1][\ ] \cdot Dgamma[2][\ ] \cdot Dgamma[3][\ ] \cdot Dgamma[4][\ ]$

$$I \left(\gamma_1 \right)_{a,b} \left(\gamma_2 \right)_{a,b} \left(\gamma_3 \right)_{a,b} \left(\gamma_4 \right)_{a,b} = \begin{bmatrix} 0 & 0 & 1 & 0 \\ 0 & 0 & 0 & 1 \\ 1 & 0 & 0 & 0 \\ 0 & 1 & 0 & 0 \end{bmatrix}$$

$Dgamma[5][\ ]$

$$\left(\gamma_5 \right)_{a,b} = \begin{bmatrix} 0 & 0 & 1 & 0 \\ 0 & 0 & 0 & 1 \\ 1 & 0 & 0 & 0 \\ 0 & 1 & 0 & 0 \end{bmatrix}$$

Dgamma[5]·*Dgamma*[5][ ]

$$\begin{bmatrix} 1 & 0 & 0 & 0 \\ 0 & 1 & 0 & 0 \\ 0 & 0 & 1 & 0 \\ 0 & 0 & 0 & 1 \end{bmatrix}$$

Trace(*Dgamma*[1]·*Dgamma*[2]·*Dgamma*[3]·*Dgamma*[4])

$$0$$

Dgamma[mu]·*Dgamma*[mu]

$$\gamma_\mu \gamma^\mu$$

SumOverRepeatedIndices(%)

$$\gamma_1 \gamma^1 + \gamma_2 \gamma^2 + \gamma_3 \gamma^3 + \gamma_4 \gamma^4$$

Dagger(*Dgamma*[1])

$$-\gamma_1$$

Dagger(*Dgamma*[mu]·*Dgamma*[nu]·*Dgamma*[sigma])

$$\gamma_\sigma{}^\dagger, \gamma_\nu{}^\dagger \gamma_\mu{}^\dagger$$

AntiCommutator(*Dgamma*[1], *Dgamma*[2])

$$[\gamma_1, \gamma_2]_+$$

Commutator(*Dgamma*[1], *Dgamma*[2])

$$[\gamma_1, \gamma_2]_-$$

Gamma Matrices (Reduced Dimension)

restart
with(*Physics*) :

Setup(*mathematicalnotation* = *true*)

$$[mathematicalnotation = true]$$

Setup(*dimension* = [3, '-'], *Dgammarepresentation* = *standard*)
  The dimension and signature of the tensor space are set to: [3, - - +]
    Setting lowercaselatin letters to represent spinor indices
  Defined Dirac gamma matrices (Dgamma) in standard representation, $\gamma_1, \gamma_2, \gamma_3$

$$[Dgammarepresentation = standard, dimension = 3, signature = - - +]$$

Dgamma[1][ ]

$$\left(\gamma_1\right)_{a,b} = \begin{bmatrix} 0 & -I \\ -I & 0 \end{bmatrix}$$

Dgamma[2][ ]

$$\left(\gamma_2\right)_{a,b} = \begin{bmatrix} 0 & -1 \\ 1 & 0 \end{bmatrix}$$

Dgamma[3][ ]

$$\left(\gamma_3\right)_{a,b} = \begin{bmatrix} 1 & 0 \\ 0 & -1 \end{bmatrix}$$

Pauli Spin Matrices

for *l* from 1 to 3 do *Psigma[l]* = *eval(Psigma[l])* end

$$\sigma_1 = \begin{bmatrix} 0 & 1 \\ 1 & 0 \end{bmatrix}$$

$$\sigma_2 = \begin{bmatrix} 0 & -I \\ I & 0 \end{bmatrix}$$

$$\sigma_3 = \begin{bmatrix} 1 & 0 \\ 0 & -1 \end{bmatrix}$$

Commutator(Psigma[1], *Psigma*[2])

$$\left[\sigma_1, \sigma_2\right]_-$$

AntiCommutator(Psigma[1], *Psigma*[2])

$$\left[\sigma_1, \sigma_2\right]_+$$

$Trace(Psigma[1])$

$$0$$

16.3 SOLUTIONS TO THE DIRAC EQUATION

Plane wave solutions to the Dirac equation are first explored in this section. The Dirac equation is then written in a two-component form to separate positive and negative energy solutions. The two-component form of the Dirac equation is then developed for an electron in an electromagnetic field.

16.3.1 Plane Wave Solutions

For plane solutions, we substitute

$$\Psi = \begin{pmatrix} \psi_1 \\ \psi_2 \\ \psi_3 \\ \psi_4 \end{pmatrix} = \begin{pmatrix} u_1 \\ u_2 \\ u_3 \\ u_4 \end{pmatrix} \exp\left(i\frac{\mathbf{p} \cdot \mathbf{r}}{\hbar} - i\frac{Et}{\hbar} \right) \tag{16.3.1}$$

into the Dirac equation

$$\begin{pmatrix} mc^2 & 0 & c\hat{p}_z & c\hat{p}_x - ic\hat{p}_y \\ 0 & mc^2 & c\hat{p}_x + ic\hat{p}_y & -c\hat{p}_z \\ c\hat{p}_z & c\hat{p}_x - ic\hat{p}_y & -mc^2 & 0 \\ c\hat{p}_x + ic\hat{p}_y & -c\hat{p}_z & 0 & -mc^2 \end{pmatrix} \begin{pmatrix} \psi_1 \\ \psi_2 \\ \psi_3 \\ \psi_4 \end{pmatrix} = \hat{E} \begin{pmatrix} \psi_1 \\ \psi_2 \\ \psi_3 \\ \psi_4 \end{pmatrix} \tag{16.3.2}$$

with $\hat{E} = i\hbar \partial / \partial t$ and $\hat{p} = -i\hbar \nabla$ to obtain

$$\begin{pmatrix} mc^2 & 0 & cp_z & cp_x - icp_y \\ 0 & mc^2 & cp_x + icp_y & -cp_z \\ cp_z & cp_x - icp_y & -mc^2 & 0 \\ cp_x + icp_y & -cp_z & 0 & -mc^2 \end{pmatrix} \begin{pmatrix} \psi_1 \\ \psi_2 \\ \psi_3 \\ \psi_4 \end{pmatrix} = \begin{pmatrix} E & 0 & 0 & 0 \\ 0 & E & 0 & 0 \\ 0 & 0 & E & 0 \\ 0 & 0 & 0 & E \end{pmatrix} \begin{pmatrix} \psi_1 \\ \psi_2 \\ \psi_3 \\ \psi_4 \end{pmatrix}$$

$$\tag{16.3.3}$$

where $\hat{p}_x \Psi = p_x \Psi$, etc. Cancelling the exponential on both sides we obtain an eigenvalue equation for the energies and amplitudes

$$\begin{pmatrix} mc^2 - E & 0 & cp_z & cp_x - icp_y \\ 0 & mc^2 - E & cp_x + icp_y & -cp_z \\ cp_z & cp_x - icp_y & -mc^2 - E & 0 \\ cp_x + icp_y & -cp_z & 0 & -mc^2 - E \end{pmatrix} \begin{pmatrix} u_1 \\ u_2 \\ u_3 \\ u_4 \end{pmatrix} = 0 \qquad (16.3.4)$$

16.3.2 Nonplane Wave Solutions

For nonplane solutions with harmonic time dependence, we take

$$\Psi = \begin{pmatrix} \psi_1(\mathbf{r}) \\ \psi_2(\mathbf{r}) \\ \psi_3(\mathbf{r}) \\ \psi_4(\mathbf{r}) \end{pmatrix} \exp\left(-i\frac{Et}{\hbar}\right) \qquad (16.3.5)$$

If $\Psi = \Psi(x, t)$ then the Dirac equation becomes

$$\begin{pmatrix} mc^2 & 0 & 0 & -i\hbar c\, \partial/\partial x \\ 0 & mc^2 & -i\hbar c\, \partial/\partial x & 0 \\ 0 & -i\hbar c\, \partial/\partial x & -mc^2 & 0 \\ -i\hbar c\, \partial/\partial x & 0 & 0 & -mc^2 \end{pmatrix} \begin{pmatrix} \psi_1(x) \\ \psi_2(x) \\ \psi_3(x) \\ \psi_4(x) \end{pmatrix}$$

$$= \begin{pmatrix} E & 0 & 0 & 0 \\ 0 & E & 0 & 0 \\ 0 & 0 & E & 0 \\ 0 & 0 & 0 & E \end{pmatrix} \begin{pmatrix} \psi_1 \\ \psi_2 \\ \psi_3 \\ \psi_4 \end{pmatrix} \qquad (16.3.6)$$

The components ψ_1 and ψ_4 are related by

$$\left(mc^2 - E\right)\psi_1 = i\hbar c\frac{\partial \psi_4}{\partial x}$$
$$\left(-mc^2 - E\right)\psi_4 = i\hbar c\frac{\partial \psi_1}{\partial x} \qquad (16.3.7)$$

and the components ψ_2 and ψ_3 are related by

$$\left(mc^2 - E\right)\psi_2 = i\hbar c\frac{\partial \psi_3}{\partial x}$$
$$\left(-mc^2 - E\right)\psi_3 = i\hbar c\frac{\partial \psi_2}{\partial x} \qquad (16.3.8)$$

16.3.3 Nonrelativistic Limit

To develop the nonrelativistic limit of the Dirac equation we begin with the Hamiltonian form

$$\hat{H} = c\alpha_1\hat{p}_x + c\alpha_2\hat{p}_y + c\alpha_3\hat{p}_z + \beta mc^2 = c\alpha \cdot \mathbf{p} + \beta mc^2 \tag{16.3.9}$$

and express the α and β matrices as

$$\alpha_i = \begin{pmatrix} 0 & \sigma_i \\ \sigma_i & 0 \end{pmatrix} \quad \beta = \begin{pmatrix} 1 & 0 \\ 0 & -1 \end{pmatrix} \tag{16.3.10}$$

with each element representing a 2×2 matrix and $\sigma = (\sigma_1, \sigma_2, \sigma_3)$ as the Pauli spin matrices. Writing the four-component Ψ in terms of "upper" $\tilde{\varphi}$ and "lower" $\tilde{\chi}$ components

$$\Psi = \begin{pmatrix} \tilde{\varphi} \\ \tilde{\chi} \end{pmatrix} \tag{16.3.11}$$

where $\tilde{\varphi}$ and $\tilde{\chi}$ are each two component spinors corresponding to positive and negative energy solutions, respectively. The Dirac equation is now expressed in terms of these upper and lower components

$$i\hbar \frac{\partial}{\partial t}\begin{pmatrix} \tilde{\varphi} \\ \tilde{\chi} \end{pmatrix} = c\frac{\hbar}{i}\begin{pmatrix} 0 & \sigma \\ \sigma & 0 \end{pmatrix} \cdot \nabla \begin{pmatrix} \tilde{\varphi} \\ \tilde{\chi} \end{pmatrix} + \begin{pmatrix} 1 & 0 \\ 0 & -1 \end{pmatrix} mc^2 \begin{pmatrix} \tilde{\varphi} \\ \tilde{\chi} \end{pmatrix} \tag{16.3.12}$$

The top equation is

$$i\hbar \frac{\partial\tilde{\varphi}}{\partial t} = \left(c\frac{\hbar}{i}\sigma \cdot \nabla \right)\tilde{\chi} + mc^2\tilde{\varphi} \tag{16.3.13}$$

and the bottom equation is

$$i\hbar \frac{\partial\tilde{\chi}}{\partial t} = \left(c\frac{\hbar}{i}\sigma \cdot \nabla \right)\tilde{\varphi} - mc^2\tilde{\chi} \tag{16.3.14}$$

In the nonrelativistic limit, the time dependence of Ψ is dominated by the negative exponential

$$\begin{pmatrix} \tilde{\varphi} \\ \tilde{\chi} \end{pmatrix} = e^{-i\frac{mc^2}{\hbar}t}\begin{pmatrix} \varphi \\ \chi \end{pmatrix} \tag{16.3.15}$$

where φ and χ are more slowly varying in time. Since

$$i\hbar \frac{\partial}{\partial t}\begin{pmatrix} \tilde{\varphi} \\ \tilde{\chi} \end{pmatrix} = \left[mc^2\begin{pmatrix} \varphi \\ \chi \end{pmatrix} + i\hbar\frac{\partial}{\partial t}\begin{pmatrix} \varphi \\ \chi \end{pmatrix} \right]e^{-i\frac{mc^2}{\hbar}t} \tag{16.3.16}$$

our system is now

$$ i\hbar \frac{\partial}{\partial t} \begin{pmatrix} \varphi \\ \chi \end{pmatrix} = c \frac{\hbar}{i} \begin{pmatrix} 0 & \sigma \\ \sigma & 0 \end{pmatrix} \cdot \nabla \begin{pmatrix} \varphi \\ \chi \end{pmatrix} - 2mc^2 \begin{pmatrix} 0 \\ \chi \end{pmatrix} \tag{16.3.17} $$

For $mc^2\chi$ large compared to $\hbar\dot\chi$, the bottom equation becomes

$$ 2mc^2\chi \approx \left(c\frac{\hbar}{i}\sigma \cdot \nabla \right)\varphi \tag{16.3.18} $$

The small component χ is on the order of v/c smaller than the large component φ.

The bottom equation can now be used to eliminate χ from the top equation where the time dependence of φ is given by

$$ i\hbar \frac{\partial\varphi}{\partial t} \approx \frac{1}{2m}\left(\frac{\hbar}{i}\sigma \cdot \nabla \right)^2 \varphi \tag{16.3.19} $$

16.3.4 Dirac Equation in an Electromagnetic Field

In an electromagnetic field with vector potential A and scalar potential V

$$ i\hbar \frac{\partial}{\partial t}\Psi = \left[c\alpha \cdot \left(\frac{\hbar}{i}\nabla - \frac{e}{c}\mathbf{A} \right) + \beta mc^2 + eV \right]\Psi \tag{16.3.20} $$

The Dirac equation is expressed in terms of the positive and negative energy solutions

$$ i\hbar \frac{\partial}{\partial t}\begin{pmatrix} \tilde\varphi \\ \tilde\chi \end{pmatrix} = c\begin{pmatrix} 0 & \sigma \\ \sigma & 0 \end{pmatrix}\cdot\left(\frac{\hbar}{i}\nabla - \frac{e}{c}\mathbf{A} \right)\begin{pmatrix} \tilde\varphi \\ \tilde\chi \end{pmatrix} + \begin{pmatrix} 1 & 0 \\ 0 & -1 \end{pmatrix}mc^2\begin{pmatrix} \tilde\varphi \\ \tilde\chi \end{pmatrix} + eV\begin{pmatrix} \tilde\varphi \\ \tilde\chi \end{pmatrix} \tag{16.3.21} $$

The top equation is

$$ i\hbar \frac{\partial\tilde\varphi}{\partial t} = c\sigma \cdot \left(\frac{\hbar}{i}\nabla - \frac{e}{c}\mathbf{A} \right)\tilde\chi + \left(eV + mc^2 \right)\tilde\varphi \tag{16.3.22} $$

and the bottom equation is

$$ i\hbar \frac{\partial\tilde\chi}{\partial t} = c\sigma \cdot \left(\frac{\hbar}{i}\nabla - \frac{e}{c}\mathbf{A} \right)\tilde\varphi + \left(eV - mc^2 \right)\tilde\chi \tag{16.3.23} $$

Factoring the negative exponential time dependence in the nonrelativistic limit with $mc^2\chi$ large compared to $\hbar\dot\chi$ and eV, the bottom equation is approximately

$$ \chi \approx \frac{c\sigma}{2mc^2} \cdot \left(\frac{\hbar}{i}\nabla - \frac{e}{c}\mathbf{A} \right)\varphi \tag{16.3.24} $$

Substitution into the top equation gives

$$ i\hbar\frac{\partial\varphi}{\partial t}=\frac{1}{2m}\left[\sigma\cdot\left(\frac{\hbar}{i}\nabla-\frac{e}{c}\mathbf{A}\right)\right]^2\varphi+eV\varphi \qquad (16.3.25) $$

Using the identity

$$ \left(\mathbf{a}\cdot\sigma\right)\left(\mathbf{b}\cdot\sigma\right)=\left(\mathbf{a}\cdot\mathbf{b}\right)+i\sigma\cdot\left(\mathbf{a}\times\mathbf{b}\right) \qquad (16.3.26) $$

we expand

$$ \left[\sigma\cdot\left(\frac{\hbar}{i}\nabla-\frac{e}{c}\mathbf{A}\right)\right]^2=\left(\frac{\hbar}{i}\nabla-\frac{e}{c}\mathbf{A}\right)^2+i\sigma\cdot\left[\left(\frac{\hbar}{i}\nabla-\frac{e}{c}\mathbf{A}\right)\times\left(\frac{\hbar}{i}\nabla-\frac{e}{c}\mathbf{A}\right)\right] \qquad (16.3.27) $$

The cross product becomes

$$ \left(\frac{\hbar}{i}\nabla-\frac{e}{c}\mathbf{A}\right)\times\left(\frac{\hbar}{i}\nabla-\frac{e}{c}\mathbf{A}\right)=-\frac{\hbar e}{ic}\nabla\times\mathbf{A}-\frac{\hbar e}{ic}\mathbf{A}\times\nabla \qquad (16.3.28) $$

and we obtain the Pauli equation

$$ i\hbar\frac{\partial\varphi}{\partial t}\approx\left[\frac{1}{2m}\left(\frac{\hbar}{i}\nabla-\frac{e}{c}\mathbf{A}\right)^2-\frac{e\hbar}{2mc}\sigma\cdot\mathbf{B}+eV\right]\varphi \qquad (16.3.29) $$

describing a nonrelativistic electron in an electromagnetic field where φ is the two-component Pauli spinor.

Maple Example

The following Maple worksheet illustrates a plane wave solution to the Dirac equation.

Key Maple commands: *algsubs, Eigenvalues, Eigenvectors, Matrix*

Maple packages: *with(LinearAlgebra)*:

restart

Plane Wave Solutions

with(LinearAlgebra) :
A:= Matrix([[m·c², 0, c·p_z, c·p_x − I·c·p_y], [0, m·c², c·p_x + I·c·p_y, −c·p_z], [c·p_z, c·p_x − I·c·p_y, −m·c², 0], [c·p_x + I·c·p_y, −c·p_z 0, −m·c²]])

$$A := \begin{bmatrix} mc^2 & 0 & cp_z & cp_x - Icp_y \\ 0 & mc^2 & cp_x + Icp_y & -cp_z \\ cp_z & cp_x - Icp_y & -mc^2 & 0 \\ cp_x + Icp_y & -cp_z & 0 & -mc^2 \end{bmatrix}$$

Eigenvectors(A)[2]

$$\left[\left[\frac{cp_z}{-mc^2 + \sqrt{c^2m^2 + p_x^2 + p_y^2 + p_z^2}\,c}, -\frac{c\left(Ip_y - p_x\right)}{-mc^2 + \sqrt{c^2m^2 + p_x^2 + p_y^2 + p_z^2}\,c},\right.\right.$$

$$\left.-\frac{c\left(Ip_y - p_x\right)}{-mc^2 - \sqrt{c^2m^2 + p_x^2 + p_y^2 + p_z^2}\,c}, \frac{cp_z}{-mc^2 - \sqrt{c^2m^2 + p_x^2 + p_y^2 + p_z^2}\,c}\right],$$

$$\left[\frac{c\left(Ip_y + p_x\right)}{-mc^2 + \sqrt{c^2m^2 + p_x^2 + p_y^2 + p_z^2}\,c}, -\frac{cp_z}{-mc^2 + \sqrt{c^2m^2 + p_x^2 + p_y^2 + p_z^2}\,c},\right.$$

$$\left.-\frac{cp_z}{-mc^2 - \sqrt{c^2m^2 + p_x^2 + p_y^2 + p_z^2}\,c}, \frac{c\left(Ip_y + p_x\right)}{-mc^2 - \sqrt{c^2m^2 + p_x^2 + p_y^2 + p_z^2}\,c}\right],$$

$$\left.\left[1,0,0,1\right],\left[1,0,0,1\right]\right]$$

algsubs$\left(p_x^2 + p_t^2 + p_z^2 = p^2, \%\right)$

$$\left[\left[\frac{p_z}{-cm + \sqrt{c^2m^2 + p^2}}, -\frac{c\left(Ip_y - p_x\right)}{-mc^2 + \sqrt{c^2m^2 + p^2}\,c}, -\frac{c\left(Ip_y - p_x\right)}{-mc^2 - \sqrt{c^2m^2 + p^2}\,c},\right.\right.$$

$$\left.-\frac{p_z}{cm + \sqrt{c^2m^2 + p^2}}\right],$$

$$\left[\frac{c\left(Ip_y + p_x\right)}{-mc^2 + \sqrt{c^2m^2 + p^2}\,c}, -\frac{p_z}{-cm + \sqrt{c^2m^2 + p^2}}, \frac{p_z}{cm + \sqrt{c^2m^2 + p^2}}, \frac{c\left(Ip_y + p_x\right)}{-mc^2 - \sqrt{c^2m^2 + p^2}\,c}\right],$$

$$\left.\left[1,0,0,1\right],\left[0,1,1,0\right]\right]$$

Eigenvalues(*A*)

$$\begin{bmatrix} \sqrt{c^2m^2 + p_x^2 + p_y^2 + p_z^2}\,c \\ -\sqrt{c^2m^2 + p_x^2 + p_y^2 + p_z^2}\,c \\ \sqrt{c^2m^2 + p_x^2 + p_y^2 + p_z^2}\,c \\ -\sqrt{c^2m^2 + p_x^2 + p_y^2 + p_z^2}\,c \end{bmatrix}$$

$algsubs\left(p_x^2 + p_t^2 + p_z^2 = p^2,\%\right)$

$$\begin{bmatrix} \sqrt{c^2m^2 + p^2}\,c \\ -\sqrt{c^2m^2 + p^2}\,c \\ \sqrt{c^2m^2 + p^2}\,c \\ -\sqrt{c^2m^2 + p^2}\,c \end{bmatrix}$$

16.4 MATLAB EXAMPLES

A leapfrog finite difference time domain (FDTD) numerical approximation to the 1D Dirac equation is demonstrated in this section. The routine below is for demonstration purposes in applying a leapfrog method to a system of first order equations. Numerical solution of the Dirac equation for various applications is an active area of research. The reader is encouraged to consult additional references for numerical codes in pursuing research problems. Comparison of the results of numerical simulation with analytical solutions and periodically checking normalization is recommended.

Key MATLAB commands: *meshc, subplot*

Programming: for loops, function statements

Section 16.3 Solutions to the Dirac Equation

Nonplane wave solutions to the 1D Dirac equation are obtained from

$$\begin{pmatrix} mc^2 & 0 & 0 & c\hat{p}_x \\ 0 & mc^2 & c\hat{p}_x & 0 \\ 0 & c\hat{p}_x & -mc^2 & 0 \\ c\hat{p}_x & 0 & 0 & -mc^2 \end{pmatrix} \begin{pmatrix} \psi_1(x,t) \\ \psi_2(x,t) \\ \psi_3(x,t) \\ \psi_4(x,t) \end{pmatrix} = \hat{E} \begin{pmatrix} \psi_1(x,t) \\ \psi_2(x,t) \\ \psi_3(x,t) \\ \psi_4(x,t) \end{pmatrix} \qquad (16.5.1)$$

with $\hat{E} = i\hbar\partial/\partial t$ and $\hat{p} = -i\hbar\partial/\partial x$. The two coupled sets of equations are

$$mc^2\psi_1 - i\hbar c\frac{\partial\psi_4}{\partial x} = i\hbar\frac{\partial\psi_1}{\partial t}$$
$$-mc^2\psi_4 - i\hbar c\frac{\partial\psi_1}{\partial x} = i\hbar\frac{\partial\psi_4}{\partial t}$$

(16.5.2)

and

$$mc^2\psi_2 - i\hbar c\frac{\partial\psi_3}{\partial x} = i\hbar\frac{\partial\psi_2}{\partial t}$$
$$-mc^2\psi_3 - i\hbar c\frac{\partial\psi_2}{\partial x} = i\hbar\frac{\partial\psi_3}{\partial t}$$

(16.5.3)

We will solve

$$\psi_1 - i\frac{\hbar}{mc}\frac{\partial\psi_4}{\partial x} = i\frac{\hbar}{mc^2}\frac{\partial\psi_1}{\partial t}$$
$$-\psi_4 - i\frac{\hbar}{mc}\frac{\partial\psi_1}{\partial x} = i\frac{\hbar}{mc^2}\frac{\partial\psi_4}{\partial t}$$

(16.5.4)

rescaling x and t

$$\frac{\partial\psi_1}{\partial t} = -i\psi_1 - \frac{\partial\psi_4}{\partial x}$$
$$\frac{\partial\psi_4}{\partial t} = i\psi_4 - \frac{\partial\psi_1}{\partial x}$$

(16.5.6)

Using the MATLAB script below, a FDTD method solves for $\psi_1(x, t)$ and $\psi_4(x, t)$ with initial conditions $\psi_1(x, 0)$ and $\psi_4(x, 0)$.
A forward difference scheme is used for time derivatives

$$\dot{\psi} \approx \frac{\psi(x_n, t_{m+1}) - \psi(x_n, t_m)}{\Delta t}$$

(16.5.7)

Using a central difference scheme

$$\psi' \approx \frac{\psi(x_{n+1}, t_{m+1/2}) - \psi(x_{n-1}, t_{m+1/2})}{2\Delta x}$$

(16.5.8)

to calculate spatial derivatives, our finite difference equations are

$$\psi_1(x_n, t_{m+1}) = \psi_1(x_n, t_m) - \left[i\psi_1(x_n, t_m) + \frac{\psi_4(x_{n+1}, t_{m+1/2}) - \psi_4(x_{n-1}, t_{m+1/2})}{2\Delta x}\right]\Delta t$$

$$\psi_4(x_n, t_{m+1}) = \psi_4(x_n, t_m) + \left[i\psi_4(x_n, t_m) - \frac{\psi_1(x_{n+1}, t_{m+1/2}) - \psi_1(x_{n-1}, t_{m+1/2})}{2\Delta x}\right]\Delta t$$

(16.5.9)

In one dimension, the Courant stability condition is $\Delta t \leq \Delta x / c$. The notation $t_{m+1/2}$ indicates that the spatial derivatives are calculated in-between time steps t_m and t_{m+1}. The real and imaginary parts of ψ_1 and ψ_4 are plotted in Figure 16.4.1 as a function of x and t.

```
Nmax =200; % number of spatial steps
Mmax=1500;  % number of time steps

% initialize arrays
psi1=zeros(Nmax,Mmax);
psi4=zeros(Nmax,Mmax);
dpsi1=zeros(Nmax,Mmax);
dpsi4=zeros(Nmax,Mmax);

% time and space steps
dt=0.005;
dx=0.05;
% initialize psi1
f=@(x)exp(-x^2)*(2/pi)^(1/4);

for n=1:Nmax
x=-5.0+n*dx;
psi1(n,1)=f(x);
end

% perform the finite difference equations
for m=1:Mmax-1

for n=2:Nmax-1
dpsi1(n,m)=(psi1(n+1,m)-psi1(n-1,m))/2*dx;
end

for n=2:Nmax-1
psi4(n,m+1)=psi4(n,m)+(1i*psi1(n,m)-dpsi1(n,m))*dt;
end

for n=2:Nmax-1
dpsi4(n,m)=(psi4(n+1,m+1)-psi4(n-1,m+1))/2*dx;
end

for n=2:Nmax-1
psi1(n,m+1)=psi1(n,m)-(1i*psi1(n,m)+dpsi4(n,m))*dt;
end

end

% plot the real and imaginary parts of psi1 and psi4
figure
colormap(bone)

subplot(2,2,1)
meshc(real(psi1))

title('Re(psi1)')
```

```
xlabel('time step')
ylabel('x')


subplot(2,2,2)
meshc(imag(psi1))
title('Im(psi1)')
xlabel('time step')
ylabel('x')

subplot(2,2,3)
meshc(real(psi4))
title('Re(psi4)')
xlabel('time step')
ylabel('x')

subplot(2,2,4)
meshc(imag(psi4))
title('Im(psi4)')
xlabel('time step')
ylabel('x')
```

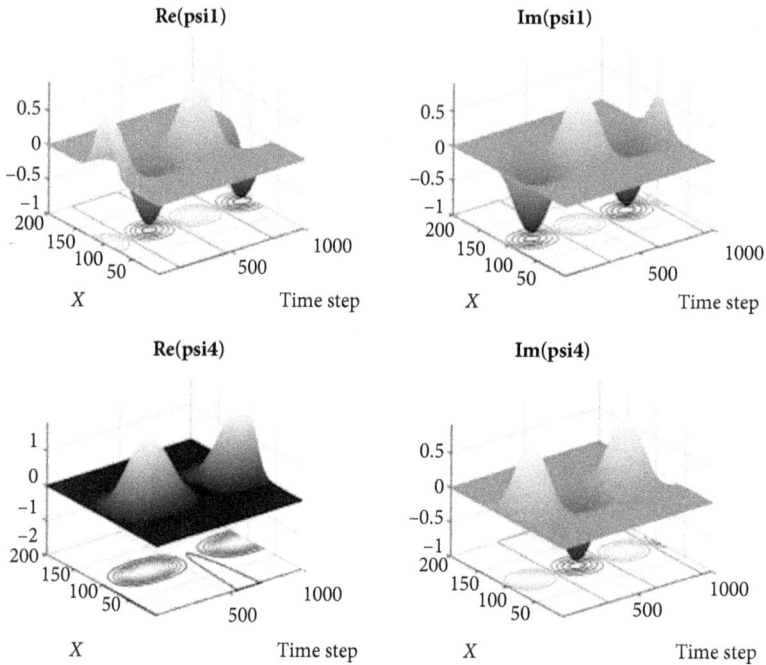

Figure 16.4.1: Spatial and temporal evolution of the Dirac spinor components $\psi_1(x, t)$ and $\psi_4(x, t)$.

(d) Find $\langle x \rangle = \int_0^L \Psi^\dagger x \Psi \, dx = \int_0^L \begin{pmatrix} \psi_1{}^* & 0 & 0 & \psi_4{}^* \end{pmatrix} \begin{pmatrix} x & 0 & 0 & 0 \\ 0 & x & 0 & 0 \\ 0 & 0 & x & 0 \\ 0 & 0 & 0 & x \end{pmatrix} \begin{pmatrix} \psi_1 \\ 0 \\ 0 \\ \psi_4 \end{pmatrix} dx$

(e) Find $\langle p \rangle = \int_0^L \Psi^\dagger \hat{p} \Psi \, dx = \int_0^L \begin{pmatrix} \psi_1{}^* & 0 & 0 & \psi_4{}^* \end{pmatrix}$

$$\begin{pmatrix} -i\hbar\dfrac{\partial}{\partial x} & 0 & 0 & 0 \\ 0 & -i\hbar\dfrac{\partial}{\partial x} & 0 & 0 \\ 0 & 0 & -i\hbar\dfrac{\partial}{\partial x} & 0 \\ 0 & 0 & 0 & -i\hbar\dfrac{\partial}{\partial x} \end{pmatrix} \begin{pmatrix} \psi_1 \\ 0 \\ 0 \\ \psi_4 \end{pmatrix} dx$$

15. For a particle in the 1D square well above

take $\psi_3(x) = A_3 \sin\left(\dfrac{n\pi x}{L}\right)$ with $\psi_1 = \psi_4 = 0$ and $n = 1, 2, 3, \ldots$

(a) Use the relations $\begin{aligned} \left(mc^2 - E\right)\psi_2 &= i\hbar c \,\dfrac{\partial \psi_3}{\partial x} \\ \left(-mc^2 - E\right)\psi_3 &= i\hbar c \,\dfrac{\partial \psi_2}{\partial x} \end{aligned}$ to find $\psi_2(x)$ and the energy

levels E

(b) Find $A_3 * A_3$ from normalization
(c) Find $\langle x \rangle$
(d) Find $\langle p \rangle$
Note that

$$\Psi_a = \begin{pmatrix} \psi_1 \\ 0 \\ 0 \\ \psi_4 \end{pmatrix} \exp\left(-i\frac{E_a t}{\hbar}\right) \quad \text{and} \quad \Psi_b = \begin{pmatrix} 0 \\ \psi_2 \\ \psi_3 \\ 0 \end{pmatrix} \exp\left(-i\frac{E_b t}{\hbar}\right)$$

are interpreted as different spin states where the electron may exist in a superposition of these orthogonal states. In this simple model, ψ_4 and ψ_3 are proportional to $\sin(n\pi x/L)$ and are properly zero at $x = 0, L$. However, ψ_1 and ψ_2 are proportional to $\cos(n\pi x/L)$ and suffer a discontinuity in magnitude at $x = 0, L$. If the difference in energy levels is negligible compared to the rest mass energy, then ψ_1 and ψ_2 are small compared to ψ_4 and ψ_3. This analysis may also be extended to model an electron confined to a finite potential well where the components of Ψ are nonzero at $x = 0, L$.

16. Given the Dirac spinor $\Psi = \Psi(x, y, t)$ with harmonic time dependence $\exp(-iEt/\hbar)$ find differential equations relating the components
    (a) ψ_1 and ψ_4
    (b) ψ_2 and ψ_3
    Create a code to numerically evaluate the spinor components with arbitrary time dependence.

17. Create an animation or a surface plot of the time-dependent solution to the Dirac equation

$$\psi_1(t,x) = A \frac{\Gamma}{\sqrt{\pi}} \int_{-\infty}^{\infty} dp \exp\left(-\Gamma^2 p^2 + ipx\right)\left[\cos(Et) - i\frac{mc^2}{E}\sin(Et)\right]$$

$$\psi_4(t,x) = -iA \frac{\Gamma}{\sqrt{\pi}} \int_{-\infty}^{\infty} dp \exp\left(-\Gamma^2 p^2 + ipx\right)\frac{cp}{E}\sin(Et)$$

where $E = \sqrt{p^2 c^2 + m^2 c^4}$, Γ is the width of the wave packet and A is a normalization factor. The integrals above must be evaluated numerically.

A concise introduction to the Dirac equation including the square well potential is available in *Introduction to Quantum Mechanics* by Chalmers W. Sherwin. See the bibliography for more information.

BIBLIOGRAPHY

MATHEMATICAL METHODS IN PHYSICS

Boas, Mary L. *Mathematical Methods in the Physical Sciences*. 3rd Ed. New York: Wiley, 2006.

Arfken, George B., Hans J. Weber, and Frank E. Harris. *Mathematical Methods for Physicists: a Comprehensive Guide*. 7th Ed. Academic press, 2011.

Spiegel, Murray R. *Advanced Mathematics*. McGraw-Hill, Incorporated, 1991.

Riley, K. F., Hobson, M. P., Bence, S. J., *Mathematical Methods for Physics and Engineering: A Comprehensive Guide*. 2nd Ed. Cambridge University Press, 2006.

Steiner, Robert V. and Schmidt, Philip P. *Schaum's Outline: Mathematics for Physics Students*, McGraw-Hill, 2011.

Harper, Charlie, Introduction to Mathematical Physics, Prentice-Hall, Englewood Cliffs, N. J. (1976)

MAPLE REFERENCES

Wang, Frank Y. *Physics with Maple: The Computer Algebra Resource for Mathematical Methods in Physics*. John Wiley & Sons, 2008.

Enns, Richard H., and George C. McGuire. *Nonlinear Physics with Maple for Scientists and Engineers*. Springer Science & Business Media, 2012.

Robertson, John S. *Engineering Mathematics with Maple*. McGraw-Hill, Inc., 1995.

Harris, F. E., *Mathematics for Physical Science and Engineering: Symbolic Computing Applications in Maple and Mathematica*, 1st Ed. Elsevier, 2014.

MATLAB REFERENCES

Hahn, B. and Valentine, D. *Essential MATLAB for Engineers and Scientists*, 6th Ed., Academic Press /Elsevier, 2017.

Higham, D. J., Higham, N. J., MATLAB *Guide*, 3rd Ed., SIAM, 2016.

CHAPTER 1 FUNDAMENTALS

Fleisch, Daniel, and Julia Kregenow. *A Student's Guide to the Mathematics of Astronomy*. Cambridge University Press, 2013.

CHAPTER 2 VECTORS AND MATRIX METHODS

Fleisch, Daniel A. *A Student's Guide to Vectors and Tensors*. Cambridge University Press, 2011.

CHAPTER 3 CALCULUS

Das, Biman. *Mathematics for Physics with Calculus*. Pearson Prentice Hall, 2005.

CHAPTER 4 VECTOR CALCULUS

Schey, Harry Moritz. *Div, Grad, Curl, and All That. An informal Text on Vector Calculus*. 4th Ed. Norton, 2004.

CHAPTER 5 ORDINARY DIFFERENTIAL EQUATIONS

Farlow, Stanley J. *An Introduction to Differential Equations and their Applications*. Dover Books on Mathematics, 1994.

CHAPTER 6 SPECIAL FUNCTIONS

Spiegel, Murray R. *Mathematical Handbook of Formulas and Tables.* (1968).

CHAPTER 7 FOURIER SERIES AND TRANSFORMATIONS

Spiegel, Murray R. *Schaum's Outline of Fourier Analysis with Applications to Boundary Value Problems.* 1st Ed.

CHAPTER 8 PARTIAL DIFFERENTIAL EQUATIONS

Farlow, Stanley J. *Partial Differential Equations for Scientists and Engineers.* Dover Books on Mathematics, 1993.

CHAPTER 9 COMPLEX ANALYSIS

Brown, James Ward, Ruel Vance Churchill, and Martin Lapidus. *Complex Variables and Applications.* Vol. 7. New York: McGraw-Hill, 1996.

CHAPTER 10 CLASSICAL MECHANICS

Goldstein, Herbert, Charles P. Poole, and John L. Safko. *Classical Mechanics: Pearson New International Edition.* Pearson Higher Ed, 2011.

Fowles, Grant R., and George L. Cassiday. *Analytical mechanics.* Saunders college, 1999.

Baker, Gregory L., and Jerry P. Gollub. *Chaotic dynamics: an introduction.* Cambridge University Press, 1996.

Marion, Jerry B., and Stephen T. Thornton. *Classical dynamics of particles and systems.* Cengage Learning, 2003.

CHAPTER 11 ELECTROMAGNETICS

Griffiths, David Jeffrey. *Introduction to Electrodynamics.* 4th Ed. Upper Saddle River, NJ: Prentice Hall, 2012.

Jackson, John David. *Classical Electrodynamics.* Wiley, 1999.

Fleisch, Daniel. *A Student's Guide to Maxwell's equations.* Cambridge University Press, 2008.

CHAPTER 12 QUANTUM MECHANICS

Dirac, Paul Adrien Maurice. *The Principles of Quantum Mechanics.* No. 27. Oxford university press, 1981.

Griffiths, David Jeffrey. *Introduction to Quantum Mechanics.* Upper Saddle River, NJ: Prentice Hall, 2004.

Susskind, Leonard, and Art Friedman. *Quantum Mechanics: the Theoretical Minimum.* Vol. 2. Basic Books, 2015.

Liboff, Richard L. *Introductory Quantum Mechanics.* Addison-Wesley, 2003.

Feynman, Richard P., Robert B. Leighton, and Matthew Sands. *The Feynman Lectures on Physics,* Vol. III. Addison-Wesley, 1965.

Davies, P. C. W., *Quantum Mechanics,* Student Physics Series, Routledge & Kegan Paul, London, 1984.

CHAPTER 13 STATISTICAL MECHANICS

Schroeder, Daniel V. *An Introduction to Thermal Physics.* Vol. 60. New York: Addison Wesley, 2000.

Glazer, Mike, and Justin Wark. *Statistical Mechanics.* Oxford University Press, 2002.

CHAPTER 14 SPECIAL RELATIVITY

Katti, Ashok N. *Mathematical Theory of Special and General Relativity.* Create space Independent Publishing, 2013.

CHAPTER 15 GENERAL RELATIVITY

Dirac, Paul Adrien Maurice. *General Theory of Relativity*. Princeton University Press, 1996.

Weinberg, Steven. *Gravitation and Cosmology: Principles and Applications of the General Theory of Relativity*. Vol. 1. New York: Wiley, 1972.

Misner, Charles W., Kip S. Thorne, and John Archibald Wheeler. *Gravitation*. Macmillan, 1973.

Peebles, P. J. E., *Principles of Physical Cosmology*, Princeton University Press, Princeton N. J, 1993.

CHAPTER 16 RELATIVISTIC QUANTUM MECHANICS

Lancaster, Tom, and Stephen J. Blundell. *Quantum Field Theory for the Gifted Amateur*. OUP Oxford, 2014.

Sherwin, Chalmers W., *Introduction to Quantum Mechanics*, Holt, Rinehart and Winston, New York, 1959. Available in paperback by Andesite Press, 2015.

APPENDIX

Legendre's Differential Equation: $\left(1-x^2\right)y'' - 2xy' + \ell\left(\ell+1\right)y = 0$,

Legendre Polynomials: $y = A_\ell P_\ell(x) + B_\ell Q_\ell(x)$, $x = \cos\theta$. The $Q_\ell(0) \to \infty$ and are often discarded.

Rodrigues' Formula:

$$P_\ell(x) = \frac{1}{2^\ell\,\ell!}\frac{d^\ell}{dx^\ell}\left(x^2-1\right)^\ell$$

$$P_{\text{odd }\ell}\left(0\right) = 0,\ P_{\text{even }\ell}\left(0\right) = \left(-1\right)^{\ell/2}\frac{(\ell-1)!!}{\ell!!}$$

$$P_\ell\left(1\right) = 1,\ P_\ell\left(-x\right) = \left(-1\right)^\ell P_\ell\left(x\right)$$

Generating Function:

$$\frac{1}{\sqrt{1-2rx+r^2}} = \sum_{\ell=0}^{\infty} P_\ell(x)r^\ell$$

Orthogonality Relation:

$$\int_{-1}^{1} P_\ell(x)P_{\ell'}(x)dx = \frac{2}{2\ell+1}\delta_{\ell\ell'}$$

Associated Legendre's Differential Equation:

$$\left(1-x^2\right)y'' - 2xy' + \left[\ell\left(\ell+1\right) - \frac{m^2}{1-x^2}\right]y = 0$$

Associated Legendre Polynomials: $y = A_{\ell m}P_\ell^m(x) + B_{\ell m}Q_\ell^m(x)$, $x = \cos\theta$,

Spherical Harmonics: $Y_\ell^m(\theta,\phi) = \sqrt{\frac{2\ell+1}{4\pi}\frac{(\ell-m)!}{(\ell+m)!}}P_\ell^m(\cos\theta)e^{im\phi}$

Rodrigues' Formula:

$$P_\ell^m(x) = \frac{\left(1-x^2\right)^{m/2}}{2^\ell \ell!} \frac{d^{\ell+m}}{dx^{\ell+m}}\left(x^2-1\right)^\ell$$

$$Q_\ell^m(x) = \left(1-x^2\right)^{m/2} \frac{d^m}{dx^m} Q_\ell(x)$$

Generating Function:

$$\frac{(2m)!\left(1-x^2\right)^{m/2} r^m}{2^m m!\left(1-2rx+r^2\right)^{m+1/2}} = \sum_{\ell=m}^{\infty} P_\ell^m(x)r^\ell$$

Orthogonality Relation:

$$\int_{-1}^{1} P_\ell^m(x)P_{\ell'}^{m'}(x)dx = \frac{2}{2\ell+1}\frac{(\ell+m)!}{(\ell-m)!}\delta_{\ell\ell'}\delta_{mm'}$$

Laguerre's Differential Equation: $xy'' + \left(1-x\right)y' + ny = 0$

Laguerre Polynomials: $y = L_n(x)$

Rodrigues' Formula:

$$L_n(x) = e^x \frac{d^n}{dx^n}\left(x^n e^{-x}\right)$$

Generating Function:

$$\frac{e^{-rx/(1-r)}}{1-r} = \sum_{n=0}^{\infty} \frac{L_n(x)r^n}{n!}$$

Orthogonality Relation:

$$\int_{0}^{\infty} e^{-x} L_n(x)L_{n'}(x)dx = \delta_{nn'}(n!)^2$$

Associated Laguerre's Differential Equation:

$$xy'' + \left(m+1-x\right)y' + \left(n-m\right)y = 0$$

Associated Laguerre Polynomials: $y = L_n^m(x)$

Rodrigues' Formula:

$$L_n^m(x) = \frac{1}{n!} e^x x^{-m} \frac{d^n}{dx^n} \left(x^{n+m} e^{-x} \right)$$

Generating Function:

$$\frac{(-1)^m r^m}{(1-r)^{m+1}} e^{-rx/(1-r)} = \sum_{n=m}^{\infty} \frac{L_n^m(x) r^n}{n!}$$

Orthogonality Relation:

$$\int_0^{\infty} e^{-x} x^m L_n^m(x) L_{n'}^m(x) dx = \delta_{nn'} \frac{(n!)^3}{(n-m)!}$$

Hermite's Differential Equation: $y'' - 2xy' + 2ny = 0$

Hermite Polynomials: $y = H_n(x)$

Rodrigues' Formula:

$$H_n(x) = (-1)^n e^{x^2} \frac{d^n}{dx^n} \left(e^{-x^2} \right)$$

$$H_{\text{odd } n}(0) = 0$$

$$H_{\text{even } n}(0) = (-1)^{n/2} (2)^{n/2} (n-1)!!$$

$$H_n(-x) = (-1)^n H_n(x)$$

Generating Function:

$$e^{2rx - r^2} = \sum_{n=0}^{\infty} \frac{H_n(x) r^n}{n!}$$

Orthogonality Relation:

$$\int_{-\infty}^{\infty} e^{-x^2} H_n(x) H_{n'}(x) dx = 2^n n! \sqrt{\pi} \delta_{nn'}$$

Index

Symbols

3D Cartesian Coordinates, 506
3D conformal mappings, 430
3D delta function, 289
3D Spherical Coordinates, 522
4-velocity, 714

A

abs
  Maple, 24
  MATLAB, 355
action integral, 442
add
  Maple, 145, 653
addition theorem of spherical
        harmonics, 525
Airy's Differential Equation, 245
Algebra, 1, 42
algsubs
  Maple, 787
Ampere-Maxwell Equation, 551, 556
Amperes' law for magnetic materials, 538
angular momentum matrices, 92
angular momentum operators, 617
angular momentum quantization, 44
animate
  Maple, 608
Annihilation
  Maple, 608
annihilation and creation operators, 606
anticommutator, 61
AntiCommutator

Maple, 778
Array
  Maple, 299, 309, 313, 653
assign
  Maple, 592
associated Laguerre polynomials, 620
associated Legendre functions of the
        second kind, 297
Associated Legendre Polynomials, 297
associated Legendre polynomials of the first
        kind, 297
assume
  Maple, 8, 31, 119, 345, 397, 592, 601, 608,
        666, 716
  MATLAB, 723
Assume
  Maple, 183
atmospheric pressure, 280
Autonomous Systems, 264
Average Energy and Particle Number,
        670–671
Axially Symmetric Potentials, 517, 524

B

Benoit Mandelbrot, 468
Bernoulli's equation, 208
Bernoullis equation, 43
besseli
  MATLAB, 384
BesselI
  Maple, 318
besselj
  MATLAB, 323

www.ingramcontent.com/pod-product-compliance
Lightning Source LLC
Chambersburg PA
CBHW061922190326
41458CB00009B/2629

9781683920984